T0210539

Lecture Notes in Computer Science   9574

Commenced Publication in 1973
Founding and Former Series Editors:
Gerhard Goos, Juris Hartmanis, and Jan van Leeuwen

Roman Wyrzykowski · Ewa Deelman
Jack Dongarra · Konrad Karczewski
Jacek Kitowski · Kazimierz Wiatr (Eds.)

Parallel Processing and Applied Mathematics

11th International Conference, PPAM 2015
Krakow, Poland, September 6–9, 2015
Revised Selected Papers, Part II

 Springer

Editors

Roman Wyrzykowski
Czestochowa University of Technology
Czestochowa
Poland

Ewa Deelman
Department of Computer Science
University of Southern California
Marina Del Rey, CA
USA

Jack Dongarra
Electrical Engineering and Computer
  Science
University of Tennessee
Knoxville, TN
USA

Konrad Karczewski
Institute of Computer and Information
  Science
Czestochowa University of Technology
Czestochowa
Poland

Jacek Kitowski
Department of Computer Science
AGH University of Science and Technology
Krakow
Poland

Kazimierz Wiatr
AGH University of Science and Technology
Krakow
Poland

ISSN 0302-9743   ISSN 1611-3349   (electronic)
Lecture Notes in Computer Science
ISBN 978-3-319-32151-6   ISBN 978-3-319-32152-3   (eBook)
DOI 10.1007/978-3-319-32152-3

Library of Congress Control Number: 2016934929

LNCS Sublibrary: SL1 – Theoretical Computer Science and General Issues

Printed on acid-free paper

This Springer imprint is published by Springer Nature
The registered company is Springer International Publishing AG Switzerland

Preface

This volume comprises the proceedings of the 11th International Conference on Parallel Processing and Applied Mathematics – PPAM 2015, which was held in Krakow, Poland, September 6–9, 2015. It was organized by the Department of Computer and Information Science of the Częstochowa University of Technology together with with the AGH University of Science and Technology, under the patronage of the Committee of Informatics of the Polish Academy of Sciences, in cooperation with the ICT COST Action IC1305 "Network for Sustainable Ultrascale Computing (NESUS)." The main organizer was Roman Wyrzykowski.

PPAM is a biennial conference. Ten previous events have been held in different places in Poland since 1994. The proceedings of the last six conferences have been published by Springer in the *Lecture Notes in Computer Science* series (Nałęczów, 2001, vol. 2328; Częstochowa, 2003, vol. 3019; Poznań, 2005, vol. 3911; Gdańsk, 2007, vol. 4967; Wrocław, 2009, vols. 6067 and 6068; Toruń, 2011, vols. 7203 and 7204; Warsaw, 2013, vols. 8384 and 8385).

The PPAM conferences have become an international forum for exchanging ideas between researchers involved in parallel and distributed computing, including theory and applications, as well as applied and computational mathematics. The focus of PPAM 2015 was on models, algorithms, and software tools that facilitate efficient and convenient utilization of modern parallel and distributed computing architectures, as well as on large-scale applications, including big data problems.

This meeting gathered more than 190 participants from 33 countries. A strict refereeing process resulted in the acceptance of 111 contributed presentations, while approximately 43 % of the submissions were rejected. Regular tracks of the conference covered important fields of parallel/distributed/cloud computing and applied mathematics such as:

- Numerical algorithms and parallel scientific computing
- Parallel non-numerical algorithms
- Tools and environments for parallel/distributed/cloud computing
- Applications of parallel computing
- Applied mathematics, neural networks, evolutionary computing, and metaheuristics

The plenary and invited talks were presented by:

- David A. Bader from the Georgia Institute of Technology (USA)
- Costas Bekas from IBM Research — Zurich (Switzerland)
- Pete Beckman from the Argonne National Laboratory (USA)
- Christopher Carothers from the Rensselaer Polytechnic Institute (USA)
- Barbara Chapman from the University of Houston (USA)
- Willem Deconinck from the European Centre for Medium-Range Weather Forecast (UK)
- Geoffrey C. Fox from Indiana University (USA)

- Dieter Kranzlmueller from the Ludwig-Maximilians-Universität München (Germany)
- Vladik Kreinovich from the University of Texas at El Paso (USA)
- Alexey Lastovetsky from the University College Dublin (Ireland)
- Carlos Osuna from ETH Zurich (Switzerland)
- Srinivasan Parthasarathy from the Ohio State University (USA)
- Enrique S. Quintana-Orti from the Universidad Jaime I (Spain)
- Thomas Rauber from the University of Bayreuth (Germany)
- Daniel Reed from the University of Iowa (USA)
- Rizos Sakellariou from the University of Manchester (UK)
- Boleslaw K. Szymanski from the Rensselaer Polytechnic Institute (USA)
- Manuel Ujaldon from Nvidia
- Jeffrey Vetter from the Oak Ridge National Laboratory and Georgia Institute of Technology (USA)
- Richard W. Vuduc from the Georgia Institute of Technology (USA)
- Torsten Wilde from the Leibnitz Supercomputing Centre (LRZ) (Germany)

Important and integral parts of the PPAM 2015 conference were the workshops:

- Minisympsium on GPU Computing organized by José R. Herrero from the Universitat Politecnica de Catalunya (Spain), Enrique S. Quintana-Ortí from the Universidad Jaime I (Spain), and Robert Strzodka from Heidelberg University (Germany).
- The Third Workshop on Models, Algorithms and Methodologies for Hierarchical Parallelism in New HPC Systems organized by Giulliano Laccetti and Marco Lapegna from the University of Naples Federico II (Italy), and Raffaele Montella from the University of Naples Parthenope (Italy).
- Workshop on Power and Energy Aspects of Computation organized by Jee Choi from the IBM T.J. Watson Research Center (USA), Piotr Luszczek from the University of Tennessee (USA), Leonel Sousa from the Technical University of Lisbon (Portugal), and Richard W. Vuduc from the Georgia Institute of Technology (USA).
- Workshop on Scheduling for Parallel Computing — SPC 2015 organized by Maciej Drozdowski from the Poznań University of Technology (Poland).
- The 6th Workshop on Language-Based Parallel Programming Models — WLPP 2015 organized by Ami Marowka from the Bar-Ilan University (Israel).
- The 5th Workshop on Performance Evaluation of Parallel Applications on Large-Scale Systems organized by Jan Kwiatkowski from the Wrocław University of Technology (Poland).
- Workshop on Parallel Computational Biology — PBC 2015 organized by Bertil Schmidt from the University of Mainz (Germany) and Jarosław Żola from the University at Buffalo (USA).
- Workshop on Applications of Parallel Computations in Industry and Engineering organized by Raimondas Čiegis from the Vilnius Gediminas Technical University (Lithuania) and Julius Žilinskas from the Vilnius University (Lithuania).

– Minisymposium on HPC Applications in Physical Sciences organized by Grzegorz Kamieniarz and Wojciech Florek from the A. Mickiewicz University in Poznań (Poland).
– The Second Workshop on Applied High-Performance Numerical Algorithms in PDEs organized by Piotr Krzyżanowski and Leszek Marcinkowski from Warsaw University (Poland) and Talal Rahman from Bergen University College (Norway).
– Minisymposium on High-Performance Computing Interval Methods organized by Bartłomiej J. Kubica from the Warsaw University of Technology (Poland).
– Workshop on Complex Collective Systems organized by Paweł Topa and Jarosław Wąs from the AGH University of Science and Technology (Poland).
– Special Session on Efficient Algorithms for Problems with Matrix and Tensor Decompositions organized by Marian Vajtersic from the University of Salzburg (Austria) and Gabriel Oksa from the Slovak Academy of Sciences.
– Special Session on Algorithms, Methodologies, and Frameworks for HPC in Geosciences and Weather Prediction organized by Zbigniew Piotrowski from the Institute of Meteorology and Water Management (Poland) and Krzysztof Rojek from the Częstochowa University of Technology (Poland).

The PPAM 2015 meeting began with four tutorials:

– Scientific Computing with GPUs, by Dominik Göddeke from the University of Stuttgart (Germany), Robert Strzodka from Heidelberg University (Germany), and Manuel Ujaldon from the University of Malaga (Spain) and Nvidia.
– Advanced Scientific Visualization with VisNow, by Krzysztof Nowiński, Bartosz Borucki, Kerstin Kantiem, and Szymon Jaranowski from the University of Warsaw (Poland).
– Parallel Computing in Java, by Piotr Bała from the Warsaw University of Technology (Poland) and Marek Nowicki, Łukasz Górski, Magdalena Ryczkowska from the Nicolaus Copernicus University (Poland).
– Introduction to Programming with Intel Xeon Phi, by Krzysztof Rojek and Łukasz Szustak from the Częstochowa University of Technology (Poland).

An integral part of the GPU Tutorial was the CUDA quiz with participants challenged to maximize the performance on a common GPU model. The winner was Miłosz Ciżnicki from the Poznan Supercomputing and Networking Center. The winner received the prize of a Tesla K40 GPU generously donated by Nvidia for the conference given its role of PPAM sponsor. The second and third prizes were granted, respectively, to Michał Antkowiak and Łukasz Kucharski, both from the A. Mickiewicz University in Poznań.

Nvidia also donated another prize, GeForce GTX480 GPU, for the authors of the best paper presented at the Minisymposium on GPU Computing. This prize was awarded to Jan Gmys, Mohand Mezmaz, Nouredine Melab, and Daniel Tuyttens from the University of Mons, who presented the paper "IVM-Based Work Stealing for Parallel Branch-and-Bound on GPU."

Special Session on Algorithms, Methodologies, and Frameworks for HPC in Geosciences and Weather Prediction: Contemporary and future applications of numerical weather prediction, climate research, and studies in geosciences demand multidisciplinary advancements in computing methodologies, including the use of multi-/manycore processors and accelerators, scalable and energy-efficient frameworks, and big data strategies, as well as new or improved numerical algorithms. This includes, for example, development of scalable, high-resolution methods for integration of fluid PDEs and efficient iterative solvers, highly optimized ports to modern hardware (CPU, GPU, Xeon Phi), code development and portability strategies, and libraries for handling geophysical datasets.

The special session served as a multidisciplinary forum for the discussion of state-of-the-art research and development toward the next-generation geophysical fluid solvers and weather/climate prediction applications.

The special session featured a number of invited and contributed talks, covering recent advances in numerical algorithms, accelerator methodologies, energy-efficent computing, and large dataset managements, including:

- Algorithms and tools for the extreme-scale numerical weather prediction (invited plenary talk by Willem Deconinck et al.)
- Adaptation of COSMO Consortium weather and climate numerical models to hybrid architecures (invited plenary talk by Carlos Osuna et al.)
- Highly efficient port of the GCR solver using high-level stencil framework on multi- and many-core architectures (by M. Ciżnicki et al.)
- Autotuned scheduler for time/energy optimization for a fully three-dimensional MPDATA advection scheme on the hybrid CPU-GPU clusters (by K. Rojek et al.)
- Parallel alternating direction implicit preconditioners for all-scale atmospheric models (by Z. Piotrowski et al.)

The organizers are indebted to the PPAM 2015 sponsors, whose support was vital to the success of the conference. The main sponsor was Intel Corporation and the other sponsors were: Nvidia, Action S.A., and Gambit. We thank all the members of the international Program Committee and additional reviewers for their diligent work in reviewing the submitted papers. Finally, we thank all of the local organizers from the Częstochowa University of Technology and the AGH University of Science and Technology, who helped us run the event very smoothly. We are especially indebted to Grażyna Kołakowska, Urszula Kroczewska, Łukasz Kuczyński, Adam Tomaś, and Marcin Woźniak from the Częstochowa University of Technology; and to Krzysztof Zieliński, Kazimierz Wiatr, and Jacek Kitowski from the AGH University of Science and Technology.

We hope that this volume will be useful to you. We would like everyone who reads it to feel invited to the next conference, PPAM 2017, which will be held during September 10–13, 2017, in Lublin, the largest Polish city east of the Vistula River.

January 2016

Roman Wyrzykowski
Jack Dongarra
Ewa Deelman
Konrad Karczewski
Jacek Kitowski
Kazimierz Wiatr

Organization

Program Committee

Jan Węglarz	Poznań University of Technology, Poland, (Honorary Chair)
Roman Wyrzykowski	Częstochowa University of Technology, Poland, (Program Chair)
Ewa Deelman	University of Southern California, USA, (Program Co-chair)
Francisco Almeida	Universidad de La Laguna, Spain
Pedro Alonso	Universidad Politecnica de Valencia, Spain
Peter Arbenz	ETH, Zurich, Switzerland
Cevdet Aykanat	Bilkent University, Ankara, Turkey
Piotr Bała	Warsaw University, Poland
David A. Bader	Georgia Institute of Technology, USA
Michael Bader	TU München, Germany
Olivier Beaumont	Inria Bordeaux, France
Włodzimierz Bielecki	West Pomeranian University of Technology, Poland
Paolo Bientinesi	RWTH Aachen, Germany
Radim Blaheta	Institute of Geonics, Czech Academy of Sciences, Czech Republic
Jacek Błażewicz	Poznań University of Technology, Poland
Pascal Bouvry	University of Luxembourg
Jerzy Brzeziński	Poznań University of Technology, Poland
Marian Bubak	AGH Kraków, Poland, and University of Amsterdam, The Netherlands
Tadeusz Burczyński	Polish Academy of Sciences, Warsaw
Christopher Carothers	Rensselaer Polytechnic Institute, USA
Jesus Carretero	Universidad Carlos III de Madrid, Spain
Raimondas Čiegis	Vilnius Gediminas Technical University, Lithuania
Andrea Clematis	IMATI-CNR, Italy
Zbigniew Czech	Silesia University of Technology, Poland
Jack Dongarra	University of Tennessee and ORNL, USA, and University of Manchester, UK
Maciej Drozdowski	Poznań University of Technology, Poland
Mariusz Flasiński	Jagiellonian University, Poland
Tomas Fryza	Brno University of Technology, Czech Republic
Jose Daniel Garcia	Universidad Carlos III de Madrid, Spain
Pawel Gepner	Intel Corporation, USA
Domingo Gimenez	University of Murcia, Spain

Mathieu Giraud	LIFL and Inria, France
Jacek Gondzio	University of Edinburgh, Scotland, UK
Andrzej Gościński	Deakin University, Australia
Laura Grigori	Inria, France
Adam Grzech	Wroclaw University of Technology, Poland
Inge Gutheil	Forschungszentrum Juelich, Germany
Georg Hager	University of Erlangen-Nuremberg, Germany
José R. Herrero	Universitat Politecnica de Catalunya, Barcelona, Spain
Ladislav Hluchy	Slovak Academy of Sciences, Bratislava, Slovakia
Sasha Hunold	Vienna University of Technology, Austria
Florin Isaila	Universidad Carlos III de Madrid, Spain
Ondrej Jakl	Institute of Geonics, Czech Academy of Sciences, Czech Republic
Emmanuel Jeannot	Inria, France
Bo Kagstrom	Umea University, Sweden
Christos Kartsaklis	Oak Ridge National Laboratory, USA
Eleni Karatza	Aristotle University of Thessaloniki, Greece
Ayse Kiper	Middle East Technical University, Turkey
Jacek Kitowski	Institute of Computer Science, AGH, Poland
Joanna Kołodziej	Cracow University of Technology, Poland
Jozef Korbicz	University of Zielona Góra, Poland
Stanislaw Kozielski	Silesia University of Technology, Poland
Dieter Kranzlmueller	Ludwig Maximillian University, Munich, and Leibniz Supercomputing Centre, Germany
Henryk Krawczyk	Gdańsk University of Technology, Poland
Piotr Krzyżanowski	University of Warsaw, Poland
Krzysztof Kurowski	PSNC, Poznań, Poland
Jan Kwiatkowski	Wrocław University of Technology, Poland
Jakub Kurzak	University of Tennessee, USA
Giulliano Laccetti	University of Naples Federico II, Italy
Marco Lapegna	University of Naples Federico II, Italy
Alexey Lastovetsky	University College Dublin, Ireland
Joao Lourenco	University Nova of Lisbon, Portugal
Tze Meng Low	Carnegie Mellon University, USA
Hatem Ltaief	KAUST, Saudi Arabia
Emilio Luque	Universitat Autonoma de Barcelona, Spain
Vyacheslav I. Maksimov	Ural Branch, Russian Academy of Sciences, Russia
Victor E. Malyshkin	Siberian Branch, Russian Academy of Sciences, Russia
Pierre Manneback	University of Mons, Belgium
Tomas Margalef	Universitat Autonoma de Barcelona, Spain
Svetozar Margenov	Bulgarian Academy of Sciences, Sofia
Ami Marowka	Bar-Ilan University, Israel
Ricardo Morla	INESC Porto, Portugal
Norbert Meyer	PSNC, Poznań, Poland
Jarek Nabrzyski	University of Notre Dame, USA
Raymond Namyst	University of Bordeaux and Inria, France

Contents – Part II

The 6th Workshop on Language-Based Parallel Programming Models (WLPP 2015)

The 5th Workshop on Performance Evaluation of Parallel Applications on Large-Scale Systems

Workshop on Parallel Computational Biology (PBC 2015)

Workshop on Applications of Parallel Computation in Industry and Engineering

Minisymposium on HPC Applications in Physical Sciences

Workshop on Complex Collective Systems

Special Session on Algorithms, Methodologies and Frameworks
for HPC in Geosciences and Weather Prediction

Contents – Part I

Parallel Non-numerical Algorithms

Minisymposium on GPU Computing

Special Session on Efficient Algorithms for Problems with Matrix and Tensor Decompositions

The Third Workshop on Models, Algorithms, and Methodologies for Hierarchical Parallelism in New HPC Systems

Virtualizing CUDA Enabled GPGPUs
on ARM Clusters

Raffaele Montella[1]([✉]), Giulio Giunta[1], Giuliano Laccetti[2], Marco Lapegna[2],
Carlo Palmieri[1], Carmine Ferraro[1], and Valentina Pelliccia[1]

[1] Department of Science and Technologies, University of Napoli Parthenope,
Centro Direzionale di Napoli, Isola C4, 80143 Naples, Italy
{raffaele.montella,giulio.giunta,carlo.palmieri,carmine.ferraro,
valentina.pelliccia}@uniparthenope.it
[2] Department of Mathematics and Applications, University of Naples Federico II,
Complesso Universitario Monte S. Angelo, Via Cintia, Naples, Italy
{giuliano.laccetti,marco.lapegna}@unina.it

Abstract. The acceleration of inexpensive ARM-based computing
nodes with high-end CUDA enabled GPGPUs hosted on x86 64 machines
using the GVirtuS general-purpose virtualization service is a novel approach to hierarchical parallelism. In this paper we draw the vision
of a possible hierarchical remote workload distribution among different devices. Preliminary, but promising, performance evaluation data
suggests that the developed technology is suitable for real world
applications.

Keywords: GPGPU · Virtualisation · Remoting · Cloud computing ·
ARM · CUDA · HPC

1 Introduction

The High Performance Cloud Computing is now offering an outstanding elastic infrastructure providing the performances required by most e-science applications. Public, private, hybrid and campus clouds are production level reali-
ties managing virtual clusters instanced on cloud infrastructures in a relatively
straightforward fashion [10]. This issue impacts on science democratization in
a scenario where the scientific computing massively relies on general purpose
graphics processing units (GPGPUs) to accelerate data parallel computing tasks
especially using NVIDIA CUDA framework [7].

The virtualization currently provided by popular open source hypervisors
(XEN, KVM, Virtual Box) does not allow software based transparent use of
accelerators as CUDA based GPUs while VMWare and XEN support GPU
on the basis of hardware virtualization provided natively by NVIDIA GRID
devices [8].

In this paper we present the updated component GVirtuS (Generic Virtualization Service) as results in GPGPUs software based transparent virtualization

R. Wyrzykowski et al. (Eds.): PPAM 2015, Part II, LNCS 9574, pp. 3–14, 2016.
DOI: 10.1007/978-3-319-32152-3_1

and remoting with the use of NVIDIA CUDA as main aim [4]. In the latest GVirtuS incarnation we enforced the architecture independence making it working with both CUDA and OpenCL on Intel and ARM architecture.

While GVirtuS is a transparent and VMM independent framework to allows an instanced virtual machine to access GPUs implementing various communicator components (TCP/IP, VMCI for VMware, VMSocket for KVM) to connect the front- end in guestOS and back-end in hostOS, rCUDA [2], GViM [6] and vCUDA [15] are three recent research projects on CUDA virtualization in GPU clusters and virtual machines as GPGPU library. They all use an approach similar to GVirtuS.

The rest of the paper is organized in the following way: the Sect. 2 is about how GVirtuS works on different architectures; the Sect. 3 is dedicated to the design and technical issues about the latest version of GVirtuS; the Sect. 4 deals with the scenarios and prototypal applications; the Sect. 5 shows the preliminary evaluation results; the Sect. 6 is about the conclusions and the future directions of this promising research.

2   GVirtuS on Heterogeneous Architectures

Different application fields motivate an ARM port of GVirtuS such as, but not limited to Sensor as a Service [1], High Performance Internet of Things (HPIoT) [9] and High Performance Cloud Computing (HPCC) [3].

In order to fit the GPGPU/x86 64/ARM application into our generic virtualization system we mapped the back-end on the x86 machine directly connected to the GPU based accelerator device and the front-end on the ARM board(s) using the GVirtuS tcp/ip based communicator. GVirtuS as NVIDIA CUDA remoting and virtualization tool achieve good results in terms of performances and system transparency [11].

The CUDA applications are executed on the ARM board through the GVirtuS front-end. Thanks to the GVirtuS architecture, the front-end is the only component needed on the guest side. This component acts as a transparent virtualization tool giving to a simple and inexpensive ARM board the illusion to be directly connected to a high-end CUDA enabled GPGPU device or devices (Fig. 1).

The computing nodes of a regular old-style cluster behave as input/output nodes for ARM based inexpensive sub-clusters. In this way the amount of heat produced decreases while the high computing power demanding applications have to be refactored in order to fit this new heterogenic approach. Thanks to GVirtuS, these devices are seen by each of the ARM based sub-cluster computing nodes as directly connected to them in a transparent way. This vision permits to gain more computing power reducing the expensive, power hungry and heat producing x86_64 based computing nodes. In the same way, this approach increases the parallelism at the sub-cluster level and, last but not the least, unchain the high-end GPGPU power to ARM based computing nodes [12].

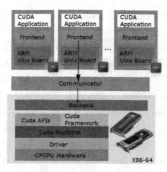

Fig. 1. The GVirtuS architecture independent from computing architecture (ARM, x86_64) and acceleration model (CUDA, OpenCL).

3 Design and Technical Issues

In GVirtuS we used the classic split-driver approach based on front-end, communicator and back-end components.

The front-end is a module that uses the driver APIs supported by the platform. The interposer library provides the familiar driver API abstraction to the guest application. It collects the request parameters from the application and passes them to the back-end driver, converting the driver API call into a corresponding frontend driver call. When a callback is received from the frontend driver, it delivers the response messages to the application. In GVirtuS the front-end is deployed on the virtual machine instance and it is implemented as a stub library.

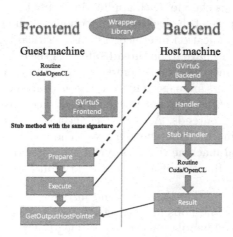

Fig. 2. The GVirtuS approach to the split driver model.

The communicator maps the request parameters from the shared ring and converts them into driver calls to the underlying wrapper library. Once the driver call returns, the backend passes the response on the shared ring and notifies the guest domains. The wrapper library converts the request parameters from the backend into actual driver API calls to be invoked on the hardware. It also relays the response messages back to the backend. The driver API is the vendor provided API for the device. The back-end is a component serving frontend requests through the direct access to the driver of the physical device. This component is implemented as a server application waiting for connections and responding to the requests submitted by frontends. In an environment requiring shared resource the back-end must offer a form of resource multiplexing. Another source of complexity is the need to manage multithreading at the guest application level (Fig. 2).

The CUDA APIs version and GVirtuS are strictly dependent by each other because the nature of the transparent virtualization and remoting. The use of CUDA 6.5 APIs is strongly motivated by the following issues:

- The library design was no longer fitting with the split driver approach leveraged by GVirtuS and other similar products changed after CUDA 3.x;
- CUDA applications can be compiled directly on the ARM board with the installation of ad hoc libraries available from NVIDIA;
- CUDA is strictly proprietary and no open source.

We target on two main goals: (1) provide a fully transparent virtualization/remoting solution; (2) reduce the overhead of virtualization and remoting so that the performance of the virtualized solution is as close as possible to the one of the bare metal execution. The frontend library allows the development of the frontend component and contains a single class, called Frontend. There is only one instance of this class for each application using the virtualized resource. This instance is in charge of the backend connection, which is based on an implementation of a Communication interface. This is a critical issue especially when the virtualized resources have to be thread-safe as in the case of GPUs providing CUDA support. The methods implemented in this class support request preparation, input parameters management, request execution, error checking and output data recovery. The backend is executed on the host machine. It waits for connections from frontends. As a new connection is incoming it spawns a new process for serving the frontend requests. The CUDA enabled application running on the virtual machine requests services to the virtualized device using the stub library. Each function in the stub library follows these steps:

1. Obtains a reference to the single frontend instance;
2. Uses Frontend class methods for setting the parameters;
3. Invokes the Frontend handler method specifying the remote procedure name;
4. Checks the remote procedure call results and handles output data.

In order to implement the NVIDIA CUDA stack split-driver using GVirtuS a developer has to implement the Frontend, Backend and the Handler subclasses.

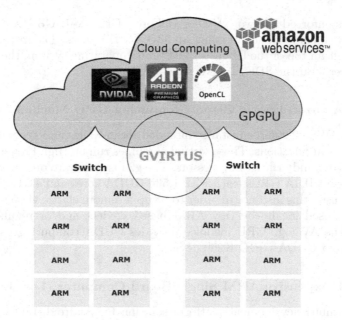

Fig. 3. Sharing high-end GPU accelerated devices hosed by x86_64 machines among inexpensive Beowulf clusters based on ARM

For CUDA runtime virtualization the Handler is implemented as a collection of functions and a jump table for the specified service. The frontend has been implemented as a dynamic library based on the interface of the original libcudart.so library. As an improvement over the NVIDIA CUDA virtualization offered by gVirtuS, we used the general-purpose virtualization service GVirtuS to provide virtualization support for CUDA, openCL and openGL. The CUDA driver implementation is similar to the CUDA runtime except for the low-level ELF binary management for CUDA kernels. A slightly different strategy has been used for openCL and openGL support. The openCL library provided by nVIDIA is a custom implementation of a public specification [5].

4   Scenarios and Prototypal Applications

Our test setup involves a Maxwell based development workstation; a virtual machine instance powered by a NVIDIA GPU provided by the Amazon Web Service Elastic Cloud Computing, inexpensive ARM based single board computer (SBC) and a cluster made by 3 ARM based high-end SBC.

4.1   The Development Workstation

Workstation Genesis GE-i940 Tesla equipped with an i7- 940 2,93 133 GHz fsb, Quad Core hyper-threaded 8 Mb cache CPU and 12 Gb RAM. The GPU subsystem is enforced by two NVIDIA GeForce Titan X 12 Gb RAM powered by the Maxwell chipset and summing up 5376 CUDA cores. The testing system has been built on top of the Ubuntu 14.04 Linux operating system, the NVIDIA CUDA Driver, and the SDK/Toolkit version 6.5.

4.2   The Amazon Elastic Cloud Computing GPU Machine

We used an AWS g2.2xlarge instance intended for graphics and general-purpose GPU compute applications. Those virtual machines run on High Frequency Intel Xeon E5-2670 (Sandy Bridge) Processors. The g2 instances are provided by high-performance NVIDIA GPUs, each with 1,536 CUDA cores and 4 GB of memory (Fig. 3). We used this instance in order to setup a remote elastic virtual GPGPU environment used locally by tiny ARM based devices and/or regular x86_64 machines. The AWS g2 GPU instance provides CUDA computing capabilities 2.x thanks to a CUDA GRID K250 device.

4.3   The Inexpensive ARM Single Board Computer (Low-End)

UDOO is a multi development platform solution for Android, Linux, Arduino and Google ADK 2012. The board has been designed to provide a flexible environment that allows exploring the new frontiers of the Internet of Things. We used an UDOO Quad single computer board equipped by a Freescale i.MX 6 ARM Cortex-A9 CPU Quad core 1 GHz with a custom version of Ubuntu 12.04 Linux as GVirtuS consumer. This board is supported by 1 GB DDR3 RAM and Gigabit Ethernet. The UDOO Quad single board computer has an integrated GPU capabilities, but is not CUDA enabled.

4.4   The Cluster Based on High-End ARM Single Board Computer (High-End)

In order to face with a real next generation high performance computing scenario, we setup an experimental cluster made by 3 NVIDIA Jetson TK1 computing nodes connected by a dedicated Gigabit Ethernet network to the developing workstation mimic an accelerator server. Each computing node relies on 4-PLUS-1 Cortex A15 r3 CPU architecture that delivers higher performance and is more power efficient than the previous generation and a Kepler GPU architecture that utilizes 192 CUDA cores to deliver advanced graphics capabilities, GPU computing with NVIDIA CUDA 6.x support, breakthrough power efficiency and performance for the next generation of gaming and GPU-accelerated computing applications.

4.5   The Benchmark Experiment Setup

Our main goal is to access a remote CUDA GPGPU on an ARM powered SBC in order to provide or expand its CUDA capabilities. The following scenarios were tested:

– Backend on X86/Frontend on X86;
– Backend on AWS X86/Frontend on X86;
– Backend on AWS X86/Frontend on low-end ARM 32bit;
– Backend X86/Frontend on high-end ARM 32bit cluster.

Matrix multiplication is an implementation of matrix multiplication that does take advantage of shared memory. In this implementation, each thread block is responsible for computing one square sub-matrix Csub of C and each thread within the block is responsible for computing one element of Csub. Csub is equal to the product of two rectangular matrices: the sub-matrix of A of dimension (A.width, block_size) that has the same row indices as Csub, and the sub-matrix of B of dimension (block_size, A.width) that has the same column indices as Csub. In order to fit into the device's resources, these two rectangular matrices are divided into as many square matrices of dimension block_size as necessary and Csub is computed as the sum of the products of these square matrices. Each of these products is performed by first loading the two corresponding square matrices from global memory to shared memory with one thread loading one element of each matrix, and then by having each thread compute one element of the product. Each thread accumulates the result of each of these products into a register and once done writes the result to global memory.

Vector addition is a very basic sample that implements element by element vector addition. Used to test the simplest cuda functionalities.

Sorting networks implements bitonic sort and odd-even merge sort, algorithms belonging to the class of sorting networks. While generally subefficient on large sequences compared to algorithms with better asymptotic algorithmic complexity (i.e. merge sort or radix sort), may be the algorithms of choice for sorting batches of short- or mid-sized arrays.

MPI Matrix multiplication is an implementation of matrix multiplication that does take advantage of distributed memory. In order to test GVirtuS on TK1 we chose to use a simple Matrix Multiply program that use MPI, to spawn the processes and the data on the available machines, and CUDA to do the math. The algorithm run as follows:

– The process with rank 0 initializes the matrices, coordinates the work and collects the results. It is never involved in calculation;
– The matrix B is entirely sent to the entire worker processes;
– The rows of the matrix A are fairly divided between the worker processes.
– The worker processes do the entire needed math;
– The process with rank 0 collects from all the workers the portion of the matrix C calculated.

The timing reported is referred to the whole process, considering the time needed for the data passing and the data collecting.

5   Preliminary Evaluations

After running several tests, we can for sure assert the effectiveness of the designed infrastructure. We can evaluate the overhead introduced by GVirtuS faced with the chance to run CUDA code on no CUDA enabled devices. Mainly, he bottle-neck is the communications overhead due to the use of the TCP communicator. This results in poor performances especially stressed out when the GPU remoting is done outside the local dedicated network where the overhead is acceptable. For more details about performances you can look at the following tables.

5.1 GPGPU Virtualization

The development workstation hosts a virtual machine running inside the user space provided by the hypervisor. In this context we are neglecting the optimization of the communicator component focusing on the interoperability of the TCP/IP channel. The Table 1 reports the results of running both the frontend and the backend on the same 64-bit machine and the same setup in a 32-bit environment demonstrating there are no performance issues related with that. This test is needed because both ARM configurations we used are 32-bit.

Table 1. Backend on X86/Frontend on X86.

Test	Without GVirtuS (64-bit)	With GVirtuS (64-bit)	Without GVirtuS (32-bit)	With GVirtuS (32-bit)
MatrixMul	0.092 s	0.139 s	0.098 s	0.149 s
Vector addition	0.059 s	0.063 s	0.057 s	0.067 s
Sorting networks	8.539 s	8.787 s	8.482 s	8.676 s

5.2 Elastic Remote GPGPU Sharing

We explored a scenario in which GPUs are hosted elastically on a public cloud in an infrastructure as a service fashion. We setup the backend on a g2.x2large AWS EC2 instance.

Table 2. Backend on AWS X86/Frontend on X86.

Test	Without GVirtuS	With GVirtuS
MatrixMul	38.236 s	0.098 s
Vector addition	3.298 s	0.057 s
Sorting networks	144.6 s	8.482 s

The Table 2 represents the results of our test suite executed on a local machine sharing CUDA enabled GPUs available on the cloud. This is a test used to

Table 3. Backend on AWS X86/Frontend on ARM 32bit.

Test	Without GVirtuS	With GVirtuS
MatrixMul	N/A	50.61 s
Vector addition	N/A	3.368 s
Sorting networks	N/A	257.5 s

demonstrate the feasibility of GPU remoting on an elastic resource. Is trivial the wall clock time increase strongly using GPU remoting instead of local resources, but the aim is to demonstrate it is working.

The Table 3 is about the same tests, but executed locally on an ARM based single board computer. In this case, without the GPU remoting it could be possible no test running on the ARM machine.

5.3   High-End ARM GPU Cluster

In this experiment we used the MPI Matrix multiplication program in order to investigate about the behavior of GVirtuS in a scenario where a x86 machine is used as an accelerator node of a high-end ARM based cluster. In our setup each computing node is provided by an on-board K20A NVIDIA CUDA enabled GPU with 192 cores, while the accelerator node is powered by a couple of NVIDIA Titan X.

Table 4. ARM cluster CUDA accelerated locally and on X86 - Time.

Number of processes	Without GVirtuS $1600 \times 1600 \times 800$	With GVirtuS $1600 \times 1600 \times 800$	Without GVirtuS $3200 \times 3200 \times 1600$	With GVirtuS $3200 \times 3200 \times 1600$
1	4294 s	830 s	7022 s	2378 s
2	2377 s	883 s	4033 s	2290 s
4	1663 s	1390 s	6440 s	3220 s
8	2037 s	2584 s	6648 s	4731 s

We have performed this benchmark with two problem size: $1600 \times 1600 \times 800$ and $3200 \times 3200 \times 1600$. The experiment compares the performance of the on-board GPU and GVirtuS remoted one on both problems size. The ARM based cluster is build on 3 nodes each provided by 4 CPU cores. The MPI Matrix multiplication program uses MPI, but is not OpenMP enabled, so we performed runs using up to 8 MPI process considering the results for one and two processes as is, while runs for four and eight processes are to be considered good for speculation and for planning the next experiments.

Table 5. ARM cluster CUDA accelerated locally and on X86 - Performance in GFLOPS.

Number of processes	Without GVirtuS 1600 × 1600 × 800	With GVirtuS 1600 × 1600× 800	Without GVirtuS 3200 × 3200 × 1600	With GVirtuS 3200 × 3200× 1600
1	0.95	4.94	4.67	13.78
2	1.72	4.64	8.13	14.31
4	2.46	2.95	5.09	10.18
8	2.01	1.59	4.93	6.93

The Table 4 represents the results, as wall clock time, of our benchmarks run with local and remoted CUDA acceleration for a number of MPI processes as 1, 2, 4 and 8. The Table 5 represents the same results as performance in GFLOPS.

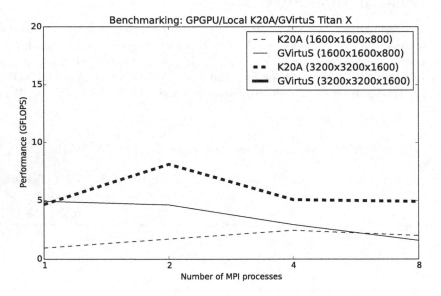

Fig. 4. MPI Matrix multiplication performance with on board K20 GPU and GVirtuS remoted NVIDIA Titan X.

The results in Fig. 4 demonstrate the use of GVirtuS remoted CUDA acceleration is convenient especially when the problems size increase: the weight of the latency due to the communication decrease, as expected. The overall performances are improved by the MPI parallel approach when the CUDA is used locally, but the limited amount of node memory and number of nodes prevented to investigate more in this direction. When the number of MPI processes increases over the 2 the benchmarks are no more suitable for classic parallel programming efficiency and speedup analysis, but could be useful for some speculations about GVirtuS and its use in GPU remoting. The implementation of a new version of the Matrix multiplication enabled for both distributed and shared memory could provide a better performance test for this kind of applications especially when GVirtuS will fully support the multithreading.

6   Conclusions and Future Directions

The most challenging result achieved by our work described in this paper is the implementation of a base tool unchaining the development of really distributed and heterogenic hardware architectures and software applications. The experiments we performed demonstrate how is convenient the path we followed as trailblazer in the hunt for the next big thing in the off the shelf commodity high performance computing clusters. We setup a sub-cluster made by high-end ARM based boards provided by multicore ARM 32-bit CPUs and high bandwidth network interfaces experiencing important improvements from the ARM side, but even a better scalability because a more performing communication. We setup a x86_64 machine and a low-end ARM SBC transparently accelerated by a GVirtuS back-end running on an Amazon Elastic Computer Cloud GPU instance. These experiments demonstrate the use of GVirtuS for GPU remoting and multiplexing. The target of this research is the provisioning of a full production software environment for advanced earth system simulations and analysis based on science gateways, workflow engines and high performance cloud computing [13] giving a support for the next generation of scientific dissemination tools [14].

Acknowledgement. This research was supported in part by the Grant Agreement number: 644312 RAPID H2020-ICT-2014/H2020-ICT-2014-1Heterogeneous Secure Multi-level Remote Acceleration Service for Low-Power Integrated Systems and Devices.

References

1. Di Lauro R., Lucarelli, F., Montella, R.: SIaaS-sensing instrument as a service using cloud computing to turn physical instrument into ubiquitous service. In: 2012 IEEE 10th International Symposium on Parallel and Distributed Processing with Applications (ISPA), pp. 861–862. IEEE (2012)
2. Duato, J., Pena, A.J., Silla, F., Mayo, R., Quintana-Ort, E.S.: rCUDA: reducing the number of GPU-based accelerators in high performance clusters. In: 2010 International Conference on High Performance Computing and Simulation (HPCS), pp. 224–231. IEEE, June 2010
3. Foster, I., Zhao, Y., Raicu, I., Lu, S.: Cloud computing and grid computing 360-degree compared. In: Grid Computing Environments Workshop, GCE 2008, pp. 1–10. IEEE, November 2008

4. Giunta, G., Montella, R., Agrillo, G., Coviello, G.: A GPGPU transparent virtualization component for high performance computing clouds. In: D'Ambra, P., Guarracino, M., Talia, D. (eds.) Euro-Par 2010, Part I. LNCS, vol. 6271, pp. 379–391. Springer, Heidelberg (2010)
5. Giunta, G., Montella, R., Laccetti, G., Isaila, F., Blas, J.G.: A GPU accelerated high performance cloud computing infrastructure for grid computing based virtual environmental laboratory. In: Constantinescu, Z. (ed.) Advances in Grid Computing, pp. 35–43. InTech (2011). ISBN: 978-953-307-301-9
6. Gupta, V., Gavrilovska, A., Schwan, K., Kharche, H., Tolia, N., Talwar, V., Ranganathan, P.: GViM: GPU-accelerated virtual machines. In: Proceedings of the 3rd ACM Workshop on System-level Virtualization for High Performance Computing, pp. 17–24. ACM, March 2009
7. Yang, C.T., Huang, C.L., Lin, C.F.: Hybrid CUDA, OpenMP, and MPI parallel programming on multicore GPU clusters. Comput. Phys. Commun. **182**(1), 266–269 (2011)
8. Younge, A.J., Walters, J.P., Crago, S., Fox, G.C.: Evaluating GPU passthrough in Xen for high performance cloud computing. In: Parallel & Distributed Processing Symposium Workshops (IPDPSW), 2014 IEEE International, pp. 852–859. IEEE (2014)
9. Laccetti, G., Montella, R., Palmieri, C., Pelliccia, V.: The high performance internet of things: using GVirtuS to share high-end GPUs with ARM based cluster computing nodes. In: Wyrzykowski, R., Dongarra, J., Karczewski, K., Waśniewski, J. (eds.) PPAM 2013, Part I. LNCS, vol. 8384, pp. 734–744. Springer, Heidelberg (2014)
10. Montella, R., Foster, I.: Using hybrid grid/cloud computing technologies for environmental data elastic storage, processing, and provisioning. In: Furht, B., Escalante, A. (eds.) Handbook of Cloud Computing, pp. 595–618. Springer, Heidelberg (2010)
11. Montella, R., Coviello, G., Giunta, G., Laccetti, G., Isaila, F., Blas, J.G.: A generalpurpose virtualization service for HPC on cloud computing: an application to GPUs. In: Wyrzykowski, R., Dongarra, J., Karczewski, K., Waśniewski, J. (eds.) PPAM 2011, Part I. LNCS, vol. 7203, pp. 740–749. Springer, Heidelberg (2012)
12. Montella, R., Giunta, G., Laccetti, G.: Virtualizing high-end GPGPUs on ARM clusters for the next generation of high performance cloud computing. Cluster Comput. **17**(1), 139–152 (2014)
13. Montella, R., Kelly, D., Xiong, W., Brizius, A., Elliott, J., Madduri, R., Maheshwari, K., Porter, C., Vilter, P., Wilde, M., Zhang, M., Foster, I.: FACE-IT: a science gateway for food security research. In: Concurrency and Computation: Practice and Experience (2015). doi:10.1002/cpe.3540
14. Pham, Q., Malik, T., Foster, I., Di Lauro, R., Montella, R.: SOLE: linking research papers with science objects. In: Groth, P., Frew, J. (eds.) IPAW 2012. LNCS, vol. 7525, pp. 203–208. Springer, Heidelberg (2012)
15. Shi, L., Chen, H., Sun, J., Li, K.: vCUDA: GPU-accelerated high-performance computing in virtual machines. IEEE Trans. Comput. **61**(6), 804–816 (2012)

A Distributed Hash Table for Shared Memory

Wytse Oortwijn$^{(\boxtimes)}$, Tom van Dijk, and Jaco van de Pol

Formal Methods and Tools, Department of EEMCS, University of Twente,
P.O.-box 217, 7500 AE Enschede, The Netherlands
{w.h.m.oortwijn,t.vandijk,j.c.vandepol}@utwente.nl

Abstract. Distributed algorithms for graph searching require a high-performance CPU-efficient hash table that supports `find-or-put`. This operation either inserts data or indicates that it has already been added before. This paper focuses on the design and evaluation of such a hash table, targeting supercomputers. The latency of `find-or-put` is minimized by using one-sided RDMA operations. These operations are overlapped as much as possible to reduce waiting times for roundtrips. In contrast to existing work, we use linear probing and argue that this requires less roundtrips. The hash table is implemented in UPC. A peak-throughput of 114.9 million op/s is reached on an Infiniband cluster. With a load-factor of 0.9, `find-or-put` can be performed in 4.5 μs on average. The hash table performance remains very high, even under high loads.

Keywords: Distributed hash table · High-performance computing · Partitioned global address space · Remote direct memory access

1 Introduction

A *hash table* is a popular data structure for storing maps and sets, since storing and retrieving data can be done with amortised time complexity $\mathcal{O}(1)$ [2]. A *distributed hash table* is a hash table that is distributed over a number of workstations, connected via a high-performance network. This has the advantage that more memory is available, at the cost of slower accesses due to network latency and bandwidth limitations. In High Performance Computing (HPC) it is desirable to have a fast and scalable distributed hash table, as it enables many distributed algorithms to be implemented efficiently.

Nowadays high-performance networking hardware like Infiniband [7] is available. Infiniband supports Remote Direct Memory Access (RDMA), which allows computers to directly access the memory of other machines without invoking their CPUs. Moreover, RDMA supports zero-copy networking, meaning that no memcopies are performed [14]. Experimental results show that one-sided RDMA is an order of magnitude faster compared to standard Ethernet hardware [10]. Furthermore, scaling along high-performance Infiniband hardware is comparable in price to scaling along standard Ethernet hardware [10]. In this paper, we target supercomputers, i.e. many-core machines connected via Infiniband.

The Partitioned Global Address Space (PGAS) programming model combines the shared and distributed memory models. Each process hosts a local

© Springer International Publishing Switzerland 2016
R. Wyrzykowski et al. (Eds.): PPAM 2015, Part II, LNCS 9574, pp. 15–24, 2016.
DOI: 10.1007/978-3-319-32152-3_2

block of memory. The PGAS abstraction combines all local memory blocks into a single global address space, thereby providing a global view on the memory. PGAS can make use of RDMA if used in a distributed setting [5]. In that case, machine-local accesses to the global address space are handled via standard memory operations, and remote accesses are handled via *one-sided* RDMA. Several PGAS implementations provide support for RDMA, including OpenSHMEM [1] and UPC [4]. We use UPC, since it supports asynchronous memory operations.

Our goal is to implement a distributed hash table for the PGAS abstraction that supports a single operation, namely find-or-put, that either inserts data when it has not been inserted already or indicates that the data has been added before. If necessary, find-or-put could easily be split into two operations find and insert. Furthermore, the hash table should require minimal memory overhead, should be CPU-efficient, and find-or-put should have minimal latency.

Our motivation for designing such a hash table is its use in distributed symbolic verification (e.g. model checking), which only requires a find-or-put operation and garbage collection in a stop-the-world scenario. Garbage collection is, however, omitted in the design of find-or-put presented in this paper. We tried to minimize the number of roundtrips required by find-or-put while keeping the hash table CPU-efficient by not relying on memory polling. Many existing implementations are, however, either CPU-intensive [9] or require more roundtrips [10,15], which motivated this research. We use linear probing and argue that this scheme requires less roundtrips compared to alternative hashing schemes. Furthermore, the design of find-or-put is more widely applicable to any sort of memory-intensive application requiring a hash table, of which distributed model checking is merely an example.

Previous work includes Pilaf [10], a key-value store that employs RDMA. Pilaf uses an optimised version of Cuckoo hashing to reduce the number of roundtrips. In Pilaf, lookups are performed via RDMA reads, but inserts are handled by the server. Nessie [15] is a hash table that uses Cuckoo hashing and RDMA both for lookups and inserts. HERD [9] is a key-value store that only uses one-sided RDMA writes and ignores the CPU bypassing features of RDMA to achieve higher throughput. FaRM [3] is a distributed computing platform that exposes the memory of all machines in a cluster as a shared address space. A hash table is built on top of FaRM that uses a variant of Hopscotch hashing.

This paper is organised as follows. Different hashing strategies are compared in Sect. 2 and we argue that linear probing requires the least number of roundtrips. Section 3 discusses the design of find-or-put. Section 4 shows the experimental evaluation of find-or-put, covering hash table efficiency with respect to latency, throughput, and the required number of roundtrips. Finally, our conclusions are summarised in Sect. 5.

2   Preliminaries

To achieve best performance, it is critical to minimize the number of RDMA roundtrips performed by find-or-put when targeting remote memory. This is because the throughput of the hash table is limited to the throughput of the RDMA devices. Also, the waiting times for roundtrips contribute to the latency of `find-or-put`. In this section some notation is given, followed by a number of hashing strategies and their efficiencies with respect to the number of roundtrips.

2.1 Notation

A hash table $T = \langle b_0, \ldots, b_{n-1} \rangle$ consists of a sequence of buckets b_i usually implemented as a standard array. We denote the *load-factor* of T by $\alpha = \frac{m}{n}$, where m is the number of elements inserted in T and n the total number of buckets. A hash function $h : U \to R$ maps data from some universe $U = \{0,1\}^w$ to a range of keys $R = \{0, \ldots, r - 1\}$. Hash tables use hash functions to map words $x \in U$ to buckets $b_{h(x)}$ by letting $r \leq n$. Let $x \in U$ be some word. Then we write $x \in b_i$ if bucket b_i contains x, and otherwise $x \notin b_i$. We write $x \in T$ if there is some $0 \leq i < n$ for which $x \in b_i$, and otherwise $x \notin T$. For some $x, y \in U$ with $x \neq y$ it may happen that $h(x) = h(y)$. This is called a *hash collision*. A hash function $h : U \to R$ is called a *universal hash function* if $\Pr[h(x) = h(y)] \leq \frac{1}{n}$ for every pair of words $x, y \in U$.

2.2 Hashing Strategies

Ideally only a single roundtrip is ever needed both for finding and inserting data. This can only be achieved when hash collisions do not occur, but in practice they occur frequently. HERD only needs one roundtrip for every operation [9], but at the cost of CPU efficiency, because every machine continuously polls for incoming requests. We aim to retain CPU efficiency to keep the hash table usable in combination with other high-performance distributed algorithms.

Chained hashing is a hashing scheme which implements buckets as linked lists. Insertions take $\mathcal{O}(1)$ time, but lookups may take $\Theta(m)$ in the worst case. It can be shown that lookups require $\Theta(1 + \alpha)$ time on average when a universal hash function is used [2]. Although constant, the average number of roundtrips required for an insert is thus more than one. Furthermore, maintaining linked lists brings memory overhead due to storing pointers.

Cuckoo hashing [11] is an open address hashing scheme that achieves constant lookup time and expected constant insertion time. Cuckoo uses $k \geq 2$ independent hashing functions $h_1, \ldots, h_k$ and maintains the invariant that, for every $x \in T$, it holds that $x \in b_{h_i(x)}$ for exactly one $1 \leq i \leq k$. Lookups thus require at most k roundtrips, but inserts may require more when all k buckets are occupied. In that case, a relocation scheme is applied, which may not only require many extra roundtrips, but also requires a locking mechanism, which is particularly expensive in a distributed environment. A variant on Cuckoo hashing, named *bucketized Cuckoo hashing* [13], enables buckets to contain multiple data elements, which linearly reduces the number of required roundtrips.

Hopscotch hashing [6] also has constant lookup time and expected constant insertion time. In Hopscotch every bucket belongs to a fixed-sized *neighbourhood*. Lookups only require a single roundtrip, since neighbourhoods are consecutive blocks of memory. However, inserts may require more roundtrips when the neighbourhood is full. In that case, buckets are relocated, which may require many roundtrips and expensive locking mechanisms.

Linear probing examines a number of *consecutive* buckets when finding or inserting data. Multiple buckets, which we refer to as *chunks*, can thus be obtained with a single roundtrip. When there is a hash collision, linear probing continues its search for an empty bucket in the current chunk, and requests additional consecutive chunks if necessary. We expect chunks retrievals to require less roundtrips than applying a relocation scheme, like done in Hopscotch. Other probing schemes, like *quadratic probing* and *double hashing*, require more roundtrips, since they examine buckets that are nonconsecutive in memory.

Cache-line-aware linear probing is proposed by Laarman et al. [12] in the context of NUMA machines. Linear probing is performed on cache lines, which the authors call *walking-the-line*, followed by double hashing to improve data distribution. Van Dijk et al. [16] use a probe sequence similar to walking-the-line to implement `find-or-put`, used for multi-core symbolic verification.

3   Design and Implementation

In this section the hash table structure and the design of `find-or-put` are discussed. We expect linear probing to require less roundtrips than both Cuckoo hashing and Hopscotch hashing, due to the absence of expensive relocation mechanisms. We also expect that minimising the number of roundtrips is key to increased performance, since the throughput of the hash table is directly limited by the throughput of the RDMA devices. This motivates the use of linear probing in the implementation of `find-or-put`. Unlike [12], we only use linear probing, since it reduces latency compared to quadratic probing, at the cost of possible clustering. We did not observe serious clustering issues, but if clustering occurs, quadratic probing can still be used, at the cost of slightly higher latencies.

The latency of `find-or-put` depends on the waiting time for roundtrips to remote memory (which is also shown in Sect. 4). We aim to minimize the waiting times by overlapping roundtrips as much as possible, using asynchronous memory operations. Furthermore, the number of roundtrips required by `find-or-put` is linearly reduced by querying for *chunks* instead of individual buckets. We use constant values C to denote the *chunk size* and M to denote the maximum number of chunks that `find-or-put` takes into account. Figure 1 shows the design of `find-or-put`. Design considerations are given in the following sections.

3.1   Memory Layout

In our implementation, each bucket is 64 bits in size. The first bit is used as a flag to denote bucket occupation and the remaining 63 bits are used to store data. When inserting data, the occupation bit is set via a `cas` operation to prevent expensive locking mechanisms. If the hash table needs to support the storage of data elements larger than 63 bits, a separate shared data array could be used.

```
1  def find-or-put(data):
2    h ← hash(data)
3    s₀ ← query-chunk(0, h)
4    for i ← 0 to M - 1:
5      if i < M - 1
6        s_{i+1} ← query-chunk(i + 1, h)
7      sync(s_i)
8      for j ← 0 to C - 1:
9        if ¬occupied(p_{(i,j)})
10         addr ← (h + iC + j) mod kn
11         b ← new-bucket(data)
12         val ← cas(b_a, p_{(i,j)}, b)
13         if val = p_{(i,j)}
14           return inserted
15         elif data(val) = data
16           return found
17       elif data(p_{(i,j)}) = data
18         return found
19   return full
```

```
1  def query-chunk(i, h):
2    start ← (h + iC) mod kn
3    end ← (h + (i + 1)C - 1) mod kn
4    if end < start
5      return split(start, end)
6    else
7      S ← ⟨b_{start} ··· b_{end}⟩
8      P ← ⟨p_{(i,0)} ··· p_{(i,C-1)}⟩
9      return memget-async(S, P)
```

```
1  def split(start, end):
2    S₁ ← ⟨b_{start} ··· b_{kn-1}⟩
3    S₂ ← ⟨b₀ ··· b_{end}⟩
4    P₁ ← ⟨p_{(i,0)} ··· b_{(i,|S₁|-1)}⟩
5    P₂ ← ⟨p_{(i,|S₁|)} ··· p_{(i,C-1)}⟩
6    s₁ ← memget-async(S₁, P₁)
7    s₂ ← memget-async(S₂, P₂)
8    return ⟨s₁, s₂⟩
```

Fig. 1. The implementation of find-or-put, as well as the implementation of query-chunk and split, which are used by find-or-put to query on the ith chunk.

The data elements are then stored in the data array and the corresponding indices are indexed and stored in the hash table. In that case, an extra roundtrip is required by find-or-put to access the data array.

The atomic $\text{cas}(B, c, v)$ operation compares the content of a shared memory location B to a value c. If they match, v is written to B and the *former* value at location B is returned by cas. Otherwise, B is unchanged and its contents are returned. The $\text{occupied}(b)$ operation simply checks if the occupation bit of a bucket b is set and the $\text{new-bucket}(d)$ operation creates a new bucket with its occupation bit set to true and containing d as data.

Assuming that the hash table is used by n processes $t_1, \ldots, t_n$, we allocate a *shared* table $T = \langle b_0, \ldots, b_{kn-1} \rangle$ of buckets, such that each process owns k buckets. In addition, we allocate two-dimensional arrays $P_i = \langle p_{(0,0)}, \ldots, p_{(M-1,C-1)} \rangle$ on every process t_i in *private* memory, which we use as local buffers. The arrays P_i are furthermore cache line aligned. This minimizes the number of cache misses when iterating over P_i, thus reducing the number of data fetches from main-memory. Cache lines are typically 64 bytes in size, so 8 buckets fit on a single cache line. To optimally use cache line alignment we choose C to be a multiple of 8.

3.2 Querying for Chunks

In Fig. 1, when some process t_j queries a chunk, it transfers C buckets from the shared array T into P_j, so that t_j can examine the buckets locally. Because linear probing is used, several *consecutive* chunks might be requested.

The `query-chunk` operation is used to query the ith consecutive chunk and the `sync` operation is used to *synchronize* on the query, that is, waiting for its completion.

It may happen that $end < start$ (line 4 of `query-chunk`), in which case the chunk wraps around the kn-sized array T because of the modulo operator (line 2 and line 3). Then the query has to be split into two, as the chunk spans over two nonconsecutive blocks of shared memory. This is done by the `split` operation.

The `memget`(S, P) operation is supported by many PGAS implementations and transfers a block of shared memory S into a block of private memory P owned by the executing process. Then `memget-async` is a *non-blocking* version of `memget`, as it does not block further execution of the program while waiting for the roundtrip to finish. Instead, `memget-async` returns a handle that can be used by `sync`, which is a blocking operation used to synchronize on the roundtrip. This allows work to be performed in between calls to `memget-async` and `sync`. The `query-chunk` operation itself returns one or two handles, and `sync` can be used to synchronize on them.

3.3   Designing `find-or-put`

In Fig. 1, the call `find-or-put`(d) returns `found` when $d \in T$ before the call and returns `inserted` when $d \notin T$ before the call. Finally, `full` is returned when $d \notin T$ and d could not be inserted in any of the MC examined buckets.

The algorithm first requests the first chunk and, if needed, tries a maximum of $M-1$ more chunks before returning `full`. Before calling `sync`(s_i) on line 7, the next chunk is requested by calling `query-chunk`$(i + 1, d)$ on line 6. This causes the queries to overlap, which reduces the blocking times for synchronization on line 7 and thereby reduces the latency of `find-or-put`.

By iterating over a chunk, if a bucket $p_{(i,j)}$ is empty, `find-or-put` tries to write *data* to the bucket b_a in shared memory via a `cas` operation (line 12). The *former* value of b_a is returned by `cas` (line 12), which is enough to check if `cas` succeeded (line 13). In this case, `inserted` is returned, otherwise the bucket has been occupied by another process in the time between the calls to `query-chunk` and `cas`. It may happen that *data* is inserted at that bucket, hence the check at line 15. If not, the algorithm returns to line 8 to try the next bucket. If $p_{(i,j)}$ *is* occupied, `find-or-put` checks if *data* $\in p_{(i,j)}$ (line 17). In that case, `found` is returned, otherwise the next iteration is tried.

4   Experimental Evaluation

We implemented `find-or-put` in Berkeley UPC, version 2.20.2, and evaluated its performance by measuring the latency and throughput under various configurations. We compiled the implementation using the Berkeley UPC compiler and gcc version 4.8.2, with the options `upcc -network=mxm -O -opt`. All experiments have been performed on the DAS-5 cluster [8], using up to 48 nodes, each running CentOS 7.1.1503 with kernel version 3.10.0. Every machine has 16 CPU

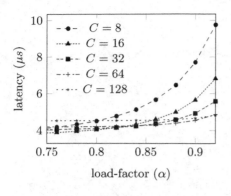
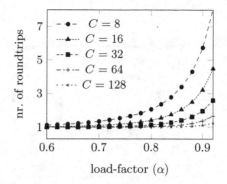

Fig. 2. The left plot shows the average latency of `find-or-put` (in microseconds) and the right plot shows the average number of roundtrips performed by `find-or-put` under different load-factors (α) and chunk sizes (C). Both plots show empirical data.

Table 1. The left table shows the latencies (in μs) and average number of roundtrips (Rt.) required by `find-or-put` to find a suitable bucket under various load-factors (α). The right table shows the total throughput ($\times 10^6$) of `find-or-put` under a *mixed* workload using various machines and processes per machine (Procs/M).

α	$C = 8$ Lat.	Rt.	$C = 16$ Lat.	Rt.	$C = 32$ Lat.	Rt.	$C = 64$ Lat.	Rt.	$C = 128$ Lat.	Rt.
0.5	3.69	1.0	3.71	1.0	3.99	1.0	4.17	1.0	4.50	1.0
0.6	3.74	1.1	3.72	1.0	4.00	1.0	4.18	1.0	4.50	1.0
0.7	3.90	1.3	3.78	1.1	4.00	1.0	4.18	1.0	4.51	1.0
0.8	4.50	2.1	4.00	1.4	4.09	1.1	4.20	1.0	4.52	1.0
0.9	7.70	5.7	5.64	3.2	4.92	2.0	4.54	1.4	4.66	1.1

Procs/M	Machines 1	2	32	48
1	189.46	1.28	8.51	11.73
2	326.36	2.21	16.09	22.05
4	709.52	3.83	28.76	37.82
8	898.34	6.18	49.41	63.36
16	-	10.17	81.55	114.85

cores, 64 GB internal memory and is connected via a high-performance 48 Gb/s Mellanox Infiniband network. All experiments have been repeated at least three times and the average measurements have been taken into account.

4.1   Latency of `find-or-put`

We measured the latency of `find-or-put` using various chunk sizes while increasing the load-factor α. This is done by creating two processes on two different machines, thereby employing the Infiniband network. Both processes maintain a 1 GB portion of the hash table. The first process inserts a sequence of unique integers until α reaches 0.92, which appears to be our limit. The hash table started to return `full` when using 8-sized chunks and having $\alpha > 0.92$. The average latencies and the number of roundtrips have been measured at intervals of 0.02, with respect to α. These measurements are shown in Fig. 2 and Table 1.

The differences between latencies are very small for $\alpha \leq 0.5$, no matter the chunk size. For $\alpha = 0.5$, the average latency when using 64-sized chunks is 13 % higher compared to 8-sized chunks (shown in Table 1). However, the average

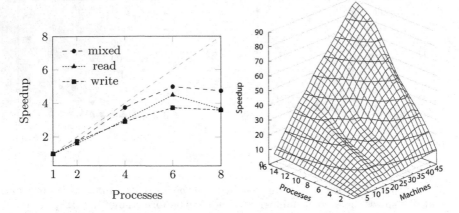

Fig. 3. Both plots show the average speedups with respect to the total throughput of `find-or-put`. The left plot shows the speedup with 1 machine, scaling from 1 to 16 processes. The right plot shows the speedup when scaling from 2 to 48 machines (i.e. using the Infiniband network) and scaling from 1 to 16 processes per machine.

latencies increase vastly for $C \geq 64$. For example, compared to 8-sized chunks, the latency is already 22 % higher for $C = 128$.

Moreover, the average latencies also increase vastly for $\alpha > 0.5$ when a low chunk size is used. By having small chunk sizes, more roundtrips are required by `find-or-put` to find the intended bucket, especially when $\alpha \geq 0.8$. By using a larger chunk size, higher load-factors are supported at the cost of slightly higher latencies. The average number of roundtrips directly influences the average latencies, which shows the importance of minimizing the number of roundtrips.

4.2 Throughput of `find-or-put`

The throughput of the hash table has been measured in terms of `find-or-put` operations per second (ops/sec). We scaled the number of machines from 1 to 48 and the number of processes per machine from 1 to 16. Each process owns a 1 GB portion of the hash table and inserts a total of 10^7 random integers. Three different workloads have been taken into account, namely:

- Mixed: 50 % finds and 50 % inserts
- Read-intensive: 80 % finds and 20 % inserts
- Write-intensive: 20 % finds and 80 % inserts

To get the proper find/insert ratio, each workload uses a different strategy to select the random integers. We used $C = 32$ and $M = 32$ in every configuration.

A subset of the measurements is shown in Table 1, and Fig. 3 shows speedups with respect to the *total* throughput, that is, the total sum of the throughputs obtained by all participating processes. In Fig. 3, the local speedups (left) are calculated relative to single-threaded runs. The remote speedups (right) are calculated relative to 2 machines, each having 1 process, thereby taking the Infiniband network into account. Only throughputs under a *mixed* workload are presented, because the other workloads show very similar behaviour.

By comparing local throughput (i.e. using one machine) with remote throughput (i.e. using at least two machines), we observed a performance drop of several orders of magnitude. The local throughput reaches a peak of 8.98×10^8 ops/s. By using a mixed workload, the local throughput is up to 88 times higher than the peak-throughput obtained with two machines. A remote peak-throughput of 11.5×10^7 is reached, which is still 7.8 times lower than the local peak-throughput.

The local throughput reaches a speedup of 5x with 8 processes (see Fig. 3) under a mixed workload. We observed a vast decrease in local speedup when more than 8 processes were used. However, when we use the Infiniband network, the performance remains stable, even when more than 8 processes per machine are used. The remote throughput reaches a speedup of 90x (with 48 machines, each having 16 processes) compared to 2 machines, each having 1 process. Compared to the single-threaded runs, a speedup of 0.61x is reached with 48 machines. This is expected; single-machine runs have better data-locality, as they do not use the network. Nonetheless, the entire memory of every participating machine can be fully utilized while maintaining good time efficiency.

4.3   Roundtrips Required by `find-or-put`

The average number of probes required by Pilaf during a key lookup in 3-way Cuckoo hashing with $\alpha = 0.75$ is 1.6 [10]. Nessie requires more roundtrips, since it uses 2-way Cuckoo hashing, which increases the chance on hash collisions compared to 3-way Cuckoo hashing. Our design requires only 1.04 probes on average for $C = 32$ and 1.006 probes for $C = 64$. Compared to Pilaf, this is an improvement of 53 % with 32-sized chunks.

Regarding the number of inserts, Pilaf is more efficient, as all inserts are handled by the server, at the cost of CPU efficiency. As part of the insertion procedure, a lookup must be performed to find an empty bucket. After that, the insert can be performed via `cas`, thereby requiring one extra roundtrip, in addition to the lookup operation. Therefore, our inserts are also more efficient than Nessie's inserts.

5   Conclusion

To build an efficient hash table for shared memory it is critical to minimize the number of roundtrips, because their waiting times contribute to higher latencies. The number of roundtrips is limited by the throughput of the RDMA devices. Lowering the number of roundtrips may directly increase the throughput.

Linear probing requires less roundtrips than Cuckoo hashing and Hopscotch hashing due to chunk retrievals, asynchronous queries, and the absence of relocations. Experimental evaluation shows that `find-or-put` can be performed in $4.5\,\mu s$ on average with a load-factor of 0.9 for $C = 64$. This shows that the hash table performance remains very high, even when the load-factor gets big. Furthermore, the entire memory of all participating machines can be used.

Table 1 shows that, in most cases, only one call to `query-chunk` would be enough for `find-or-put` to find a suitable bucket, especially for small values of α and large values of C. As future work, it would be interesting to dynamically determine the value of C to reduce the number of roundtrips. Moreover, we plan to use the hash table in a bigger framework for symbolic verification.

References

1. Chapman, B., Curtis, T., Pophale, S., Poole, S., Kuehn, J., Koelbel, C., Smith, L.: Introducing OpenSHMEM: SHMEM for the PGAS community. In: Fourth Conference on Partitioned Global Address Space Programming Model. ACM (2010)
2. Cormen, T.H., Leiserson, C.E., Rivest, R.L., Stein, C.: Introduction to Algorithms, 3rd edn. MIT press, Cambridge (2009)
3. Dragojevi, A., Narayanan, D., Hodson, O., Castro, M.: FaRM: Fast remote memory. In: 11th USENIX Conference on Networked Systems Design and Implementation, NSDI, vol. 14 (2014)
4. El-Ghazawi, T., Smith, L.: UPC: Unified Parallel C. In: ACM/IEEE Conference on Supercomputing. ACM (2006)
5. Farreras, M., Almasi, G., Cascaval, C., Cortes, T.: Scalable RDMA performance in PGAS languages. In: Parallel and Distributed Processing, pp. 1–12. IEEE (2009)
6. Herlihy, M.P., Shavit, N.N., Tzafrir, M.: Hopscotch hashing. In: Taubenfeld, G. (ed.) DISC 2008. LNCS, vol. 5218, pp. 350–364. Springer, Heidelberg (2008)
7. InfiniBand Trade Association: Accessed 9 May 2015. http://www.infinibandta.org
8. The Distributed ASCI Supercomputer 5 (2015). http://www.cs.vu.nl/das5
9. Kalia, A., Kaminsky, M., Andersen, D.G.: Using RDMA efficiently for key-value services. In: ACM Conference on SIGCOMM, pp. 295–306. ACM (2014)
10. Mitchell, C., Geng, Y., Li, J.: Using one-sided RDMA reads to build a fast, CPU-efficient key-value store. In: USENIX Annual Technical Conference, pp. 103–114 (2013)
11. Pagh, R., Rodler, F.F.: Cuckoo hashing. J. Algorithms **51**(2), 122–144 (2004)
12. Laarman, A., van de Pol, J., Weber, M.: Boosting multi-core reachability performance with shared hash tables. In: Conference on Formal Methods in Computer-Aided Design, FMCAD, pp. 247–256 (2010)
13. Ross, K.A.: Efficient hash probes on modern processors. In: IEEE 23rd International Conference on Data Engineering, pp. 1297–1301. IEEE (2007)
14. Rumble, S.M., Ongaro, D., Stutsman, R., Rosenblum, M., Ousterhout, J.K.: Its time for low latency. In: HotOS (2011)
15. Szepesi, T., Wong, B., Cassell, B., Brecht, T.: Designing a low-latency cuckoo hash table for write-intensive workloads using RDMA. In: First International Workshop on Rack-scale Computing (2014)
16. van Dijk, T., van de Pol, J.: Sylvan: Multi-core decision diagrams. In: Baier, C., Tinelli, C. (eds.) TACAS 2015. LNCS, vol. 9035, pp. 677–691. Springer, Heidelberg (2015)

Mathematical Approach to the Performance Evaluation of Matrix Multiply Algorithm

Luisa D'Amore[1(✉)], Valeria Mele[1], Giuliano Laccetti[1], and Almerico Murli[1,2]

[1] Department of Mathematics and Applications "Renato Caccioppoli",
University of Naples Federico II, Via Cintia, 80126 Naples, Italy
{luisa.damore,valeria.mele,giuliano.laccetti,almerico.murli}@unina.it
[2] SPACI, University of Naples Federico II, Via Cintia, 80126 Naples, Italy
http://matematica.dip.unina.it

Abstract. Matrix multiplication (MM) is a computationally-intensive operation in many algorithms used in scientific computations. Not only one of the kernels in numerical linear algebra, the problem of matrix multiplication is also fundamental for almost all matrix problems such as least square and eigenvalues problem. The performance analysis of the MM needs to be re-evaluated to find out the best-practice algorithm on novel architectures. This motivated the analysis which is presented in this article and which is carried out by means of the new modelling framework that the authors have already introduced (L. D'Amore et al. *On a Mathematical Approach for Analyzing Parallel Algorithms*, 2015). The model exploits the knowledge of the algorithm and the multilevel parallelism of the target architecture and it could help the researchers for designing optimized MM implementations.

Keywords: Matrix-matrix multiply · Performance analysis · Multilevel paralllelism

1 Introduction

The authors proposed a performance model for analysing parallel algorithms. The model assumes that the (parallel) algorithm is represented as a set of operators related to each other according to a rule of dependence. Furthermore, the model has a parameterized formulation intended to exploit the different characteristics of the computing machines such as reconfigurable hardware devices [13].

Here we consider the matrix multiplication (MM) algorithm and we apply the performance model. The algorithm is simple and has not any ambition of optimization (many efforts are spent in the field of linear algebra and recent examples can be found in [1,9,11,12]), instead, our aim is to discuss how easily some implementation choices could be addressed giving rise to different performance results. The focus is on the "opportunity" of implementing the algorithm in hybrid distributed/shared memory computing environments, obtaining the most important information *before* the implementation. The implementations

© Springer International Publishing Switzerland 2016
R. Wyrzykowski et al. (Eds.): PPAM 2015, Part II, LNCS 9574, pp. 25–34, 2016.
DOI: 10.1007/978-3-319-32152-3_3

of MM algorithm will be composed from multiplications with sub matrices. The general MM algorithm can be decomposed into multiple calls to matrix multiplication. These themselves can be decomposed into multiple calls to inner-kernels. The aim now is to understand how these lowest level kernels can attain high performance, then so will the MM algorithm. This paper attempts to describe how to apply the performance model that the authors have developed so as to make it accessible to a broad audience.

2   Matrix Multiplication

Given two matrices $A, B \in \Re^{n \times n}$ and the computational problem

$$\mathcal{B}_{n^2} \equiv MM_{n \times n} := A \cdot B, \tag{1}$$

we introduce the sub problems $matmul^i_{\frac{n}{3} \times \frac{n}{3}}$, for $i = 0, \ldots, 26$ which are defined as follows:

$$\mathcal{B}_{\frac{n}{3} \times \frac{n}{3}} \equiv matmul^i_{\frac{n}{3} \times \frac{n}{3}} := C_i + A_i \cdot B_i, \tag{2}$$

with $A_i \in \Re^{\frac{n}{3} \times \frac{n}{3}}$, $B_i \in \Re^{\frac{n}{3} \times \frac{n}{3}}$ and $C_i \in \Re^{\frac{n}{3} \times \frac{n}{3}}$ blocks of A, B and C, respectively. Finally, we introduce the decomposition

$$D_{27}(MM_{n \times n}) := \{matmul^i_{\frac{n}{3} \times \frac{n}{3}}\}_{0 \leq i < 27}. \tag{3}$$

From (1)–(3), the decomposition matrix is:

$$M_D = \begin{bmatrix} matmul^0_{\frac{n}{3} \times \frac{n}{3}} & matmul^1_{\frac{n}{3} \times \frac{n}{3}} & matmul^2_{\frac{n}{3} \times \frac{n}{3}} & \cdots & matmul^8_{\frac{n}{3} \times \frac{n}{3}} \\ matmul^9_{\frac{n}{3} \times \frac{n}{3}} & matmul^{10}_{\frac{n}{3} \times \frac{n}{3}} & matmul^{11}_{\frac{n}{3} \times \frac{n}{3}} & \cdots & matmul^{17}_{\frac{n}{3} \times \frac{n}{3}} \\ matmul^{18}_{\frac{n}{3} \times \frac{n}{3}} & matmul^{19}_{\frac{n}{3} \times \frac{n}{3}} & matmul^{20}_{\frac{n}{3} \times \frac{n}{3}} & \cdots & matmul^{26}_{\frac{n}{3} \times \frac{n}{3}} \end{bmatrix} \tag{4}$$

The set $D_{27}(MM_{n \times n})$ is made of 27 subproblems $matmul^{(i+j+k)}_{\frac{n}{3} \times \frac{n}{3}} \in D_{27}$, and the problem $MM_{n \times n}$ has concurrence degree $r_D = 9$ and dependence degree $c_D = 3$.

Suppose that the computing environment can be represented by means of the machine $\mathcal{M}_{1,1}$ which has

- $P = 1$,
- $Op_{\mathcal{M}_{1,1}} = \{\otimes, \ldots\}$ where $\otimes :=$ matrix-matrix multiply,
- $L = 2$ two memory levels,
- $rmem_i$ (read) and $wmem_j$ (write) as memory accesses operators on blocks of size $\frac{n}{3} \times \frac{n}{3}$,
- $tmem_1 := tblock_{mem}$,
- for each $\otimes$, 1 read (before the execution) and 1 write (after the execution) are needed.

According to D_{27}, the sequential algorithm $A_{D_{27}, \mathcal{M}_{1,1}}$ on $\mathcal{M}_{1,1}$ is made of the 27 operators $\otimes$ corresponding to the 27 sub-problems. The execution matrix corresponding to $A_{D_{27},M_{1,1}}$ on $\mathcal{M}_{1,1}$ has $r_E = 27$ rows and only one column, i.e. $c_E = 1$. It is the following matrix:

$$
M_E = \begin{bmatrix} \otimes_0 \\ \otimes_1 \\ \vdots \\ \otimes_{26} \end{bmatrix} \tag{5}
$$

while the memory matrix $AM_{A_{D_{27},\mathcal{M}_{1,1}}}$ has $r_{MEM} = 52$ rows and $c_{MEM} = 1$ column, and it can be described in the following way:

$$
AM_{A_{D_{27},\mathcal{M}_{1,1}}} = \begin{bmatrix} rmem_0(\cdot) \\ wmem_0(\cdot) \\ rmem_1(\cdot) \\ wmem_1(\cdot) \\ \vdots \\ rmem_{26} \\ wmem_{26} \end{bmatrix} \tag{6}
$$

The execution time of algorithm $A_{D_{27},M_{1,1}}$ is

$$
T(A_{D_{27},M_{1,1}}) = r_E \cdot T_r \tag{7}
$$

where T_r is the execution time of the row r of the matrix given in (5). It is equal to the execution time of the $\otimes$ operator (since they are all the same). Let $C(\otimes)$ denote the complexity of $\otimes$ operator, then (7) becomes:

$$
T(A_{D_{27},\mathcal{M}_{1,1}}) = 27 \cdot C(\otimes) \cdot tcalc \tag{8}
$$

The memory access time of the software corresponding to $A_{D_{27},M_{1,1}}$, is

$$
T_M(SW(A_{D_{27},\mathcal{M}_{1,1}}(A))) = r_{mem_1} \cdot tblock_{mem} = 54 \cdot tblock_{mem}, \tag{9}
$$

and its execution time is

$$
\begin{aligned}
T(SW(A_{D_{27},\mathcal{M}_{1,1}})) &= T(A_{D_{27},\mathcal{M}_1}) + T_M(SW(A_{D_{27},\mathcal{M}_{1,1}})) \\
&= 27 \cdot C(\otimes) \cdot tcalc + 54 \cdot tblock_{mem}
\end{aligned} \tag{10}
$$

2.1 The Algorithm at the First Level of Decomposition

We consider the machine $\mathcal{M}_{9,9}$ such that

- $P = 9$ (which we call nodes), which are organized in a 3×3 logical grid,
- $Op_{\mathcal{M}_{9,9}} = \{\otimes, ...\}$ where $\otimes$ = matrix-matrix multiply,
- $L = 3$ (two memory levels plus one level for communications),
- $trans_i$ denotes the memory access operator which moves a block of size $\frac{n}{3} \times \frac{n}{3}$ in time $tblock_{com}$[1],

---

[1] Note that typically $tblock_{com} \gg tblock_{mem}$.

– each node can transfer a single block concurrently, that is the machine can transfer 9 blocks at the same time.
– for a broadcast step, each node performs a transfer (one send, other eight receive).
– for a rolling step, each node performs two transfers (send and receive one block).

Starting, each node has a $\frac{n}{3} \times \frac{n}{3}$ block of each matrix. If $matmul(p \cdot i)$ is the subproblem $matmul_{\frac{n}{3} \times \frac{n}{3}}^{p \cdot i} \in D_{27}$, the algorithm $A_{D_{27}, \mathcal{M}_{9,9}}$ is the following (i.e. the so called Broadcast Multiply Rolling (BMR) Algorithm [10]) (Fig. 1):

```
for p=1 to 3
    nodes on the i-th grid diagonal broadcast their A block
        to all its grid row;
    the node i solves  matmul(i*p);
    if p<3, each node sends its B block to the upper one
        on the same column of the grid (rolling).
endfor
```

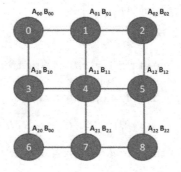

Fig. 1. The starting matrices blocks distribution among the nodes.

The execution matrix of $A_{D_{27}, \mathcal{M}_{9,9}}$ is

$$M_E = \begin{bmatrix} \otimes_0 & \otimes_1 & \otimes_2 & \otimes_3 & \otimes_4 & \otimes_5 & \otimes_6 & \otimes_7 & \otimes_8 \\ \otimes_9 & \otimes_{10} & \otimes_{11} & \otimes_{12} & \otimes_{13} & \otimes_{14} & \otimes_{15} & \otimes_{16} & \otimes_{17} \\ \otimes_{18} & \otimes_{19} & \otimes_{20} & \otimes_{21} & \otimes_{22} & \otimes_{23} & \otimes_{24} & \otimes_{25} & \otimes_{26} \end{bmatrix} \quad (11)$$

and it is perfectly parallel. The memory matrix is

$$AM_{A_{D_{27}, \mathcal{M}_{9,9}}} = \begin{bmatrix} trans_0(\cdot) & trans_1(\cdot) & trans_2(\cdot) & \dots & trans_8(\cdot) \\ trans_9(\cdot) & trans_{10}(\cdot) & trans_{11}(\cdot) & \dots & trans_{17}(\cdot) \\ trans_{18}(\cdot) & trans_{19}(\cdot) & trans_{20}(\cdot) & \dots & trans_{26}(\cdot) \\ trans_{27}(\cdot) & trans_{28}(\cdot) & trans_{29}(\cdot) & \dots & trans_{35}(\cdot) \\ trans_{36}(\cdot) & trans_{37}(\cdot) & trans_{38}(\cdot) & \dots & trans_{44}(\cdot) \\ trans_{45}(\cdot) & trans_{46}(\cdot) & trans_{47}(\cdot) & \dots & trans_{53}(\cdot) \\ trans_{54}(\cdot) & trans_{55}(\cdot) & trans_{56}(\cdot) & \dots & trans_{62}(\cdot) \end{bmatrix} \quad (12)$$

The execution time of each row of M_E, is the execution time of the $\otimes$ operator. If $r_E = 3$ is the number of rows of $E_{A_{D_{27}, \mathcal{M}_{9,9}}}$, the execution time of the BMR algorithm $A_{D_{27}, \mathcal{M}_{9,9}}$ is

$$T(A_{D_{27}, \mathcal{M}_{9,9}}) = r_E \cdot T_r = 3 \cdot C(\otimes) \cdot tcalc \tag{13}$$

Since $r_{mem} = 7$, the memory access time of the software $SW(A_{D_{27}, \mathcal{M}_{9,9}})$ is

$$T_M(SW(A_{D_{27}, \mathcal{M}_{9,9}})) = r_{mem_9} \cdot tblock_{com} = 7 \cdot tblock_{com} \tag{14}$$

and its execution time is

$$\begin{aligned} T(SW(A_{D_{27}, \mathcal{M}_{9,9}})) &= T(A_{D_{27}, \mathcal{M}_{9,9}}) + T_M(SW(A_{D_{27}, \mathcal{M}_{9,9}})) \\ &= 3 \cdot C(\otimes) \cdot tcalc + 7 \cdot tblock_{com}. \end{aligned} \tag{15}$$

Finally, the speed up of the software $SW(A_{D_{27}, \mathcal{M}_{9,9}})$ is

$$Sp(SW(A_{D_{27}, \mathcal{M}_{9,9}})) = \frac{T(SW(A_{D_{27}, \mathcal{M}_{1,1}}))}{T(SW(A_{D_{27}, \mathcal{M}_{9,9}}))} = \frac{26 \cdot C(\otimes) \cdot tcalc + 52 \cdot tblock_{mem}}{3 \cdot C(\otimes) \cdot tcalc + 7 \cdot tblock_{com}} \tag{16}$$

2.2 The Sequential Algorithm at the Second Level of Decomposition

Consider the subproblem $matmul^i_{\frac{n}{3} \times \frac{n}{3}}$ and the decomposition

$$D'_{\frac{n}{3}-1} = \{matvec^i_{\frac{n}{3} \times \frac{n}{3}}\}_{0 \leq i < (\frac{n}{3}-1)} \tag{17}$$

where

$$\begin{aligned} matvec^i_{\frac{n}{3} \times \frac{n}{3}} &:= \text{multiply of a block } A_i \text{ of } \frac{n}{3} \times \frac{n}{3} \\ &\text{elements and a vector } B_i \text{ of } \frac{n}{3} \text{ elements.} \end{aligned} \tag{18}$$

All the subproblems are independent, so the decomposition matrix of $matmul^i_{\frac{n}{3} \times \frac{n}{3}}$ is

$$M_{D'_{\frac{n}{3}-1}} = \left[matvec^0_{\frac{n}{3} \times \frac{n}{3}} \quad matvec^1_{\frac{n}{3} \times \frac{n}{3}} \quad \cdots \quad matvec^{\frac{n}{3}-1}_{\frac{n}{3} \times \frac{n}{3}} \right] \tag{19}$$

and $matmul^i_{\frac{n}{3} \times \frac{n}{3}}$ has concurrence degree $\frac{n}{3}$ and dependence degree 1.

Let us introduce the machine $\mathcal{M}'_{1,1}$ corresponding to a generic node of $\mathcal{M}_{9,9}$. Suppose that $\mathcal{M}'_{1,1}$ is such that

- $P = 1$,
- $Op_{\mathcal{M}'_{1,1}} = \{\boxtimes, ...\}$ where $\boxtimes$ = matrix-vector multiply,
- $L = 2$,
- $rmemv_i$ (read) or $wmemv_j$ (write) denote the memory accesses operators moving a vector of size $\frac{n}{3}$ in time $tmem := tvec_{mem}$[2].

---

[2] Typically $tvec_{mem} \leq tblock_{mem}$.

Since all the subproblems must be solved one after another, the execution matrix of $A_{D'_{\frac{n}{3}-1},\mathcal{M}'_{1,1}}$ is

$$
M_E = \begin{bmatrix} \boxtimes_0 \\ \boxtimes_1 \\ \vdots \\ \boxtimes_{\frac{n}{3}-1} \end{bmatrix} \tag{20}
$$

Since we assume that for the execution of each operator, it is required one read (before the execution) and one write (after the execution) of a vector of size $\frac{n}{3}+1$, the memory matrix has

$$
r_{mem,D'_{\frac{n}{3}-1}} = \left(\frac{n}{3}+2\right) \cdot \frac{n}{3}
$$

rows. The execution time of each row of the matrix in (20) is the execution time of the $\boxtimes$ operator. If

$$
r_{E,D'_{\frac{n}{3}-1}} = \frac{n}{3}
$$

is the number of rows of $E_{A_{D'_{\frac{n}{3}-1},\mathcal{M}'_{1,1}}}$, the execution time of the algorithm $A_{D'_{\frac{n}{3}-1},\mathcal{M}'_{1,1}}$ is

$$
\begin{aligned}
T(A_{D'_{\frac{n}{3}-1},\mathcal{M}'_{1,1}}) &= C(\otimes) \cdot tcalc = r_{E1D'_{\frac{n}{3}-1}} \cdot T_r \\
&= \frac{n}{3} \cdot C(\boxtimes) \cdot tcalc = \frac{n}{3} \cdot 2 \cdot \left(\frac{n}{3}\right)^2 \cdot tcalc \\
&= 2 \cdot \left(\frac{n}{3}\right)^3 \cdot tcalc
\end{aligned} \tag{21}
$$

and the memory access time of the software $SW(A_{D'_{\frac{n}{3}-1},\mathcal{M}'_{1,1}})$ is

$$
T_M(SW(A_{D'_{\frac{n}{3}-1},\mathcal{M}'_{1,1}})) = r_{mem,D'_{\frac{n}{3}-1}} \cdot tvec_{com} = \left(\frac{n}{3}+2\right) \cdot \frac{n}{3} \cdot tvec_{mem}. \tag{22}
$$

Finally, the execution time of the software $SW(A_{D'_{\frac{n}{3}-1},\mathcal{M}'_{1,1}})$ is

$$
\begin{aligned}
T(SW(A_{D'_{\frac{n}{3}-1}},\mathcal{M}'_{1,1})) &= T(A_{D'_{\frac{n}{3}-1},\mathcal{M}'_{1,1}}) + T_M(SW(A_{D'_{\frac{n}{3}-1},\mathcal{M}'_{1,1}})) \\
&= 2 \cdot \left(\frac{n}{3}\right)^3 \cdot tcalc + \left(\frac{n}{3}+2\right) \cdot \frac{n}{3} \cdot tvec_{mem}
\end{aligned} \tag{23}
$$

2.3   The Parallel Algorithm at the Second Level of Decomposition

We consider the machine $\mathcal{M}'_{1.8}$ made of 8 cores/threads for each node of $\mathcal{M}_{9,9}$. Let us assume that $\mathcal{M}'_{1.8}$ is such that

- $P = 8$,
- $Op_{\mathcal{M}'_{1.8}} = \{\boxtimes, ...\}$ where $\boxtimes$ = matrix-vector multiply,
- $L = 2$,

– $rmemv_i$ (read) or $wmemv_j$ (write) denote the memory access operators on a vector of $\frac{n}{3}$ elements concurrently in time $tvec_{mem}$ between the memory levels. Note that $tvec_{mem} \leq tblock_{mem}$.

Then, if $matvec(t \cdot i)$ denotes subproblem $matvec_{\frac{n}{3} \times \frac{n}{3}}^{t \cdot i} \in D'_{\frac{n}{3}-1}$, we get the Multi Thread Matrix multiply Algorithm $A_{D'_{\frac{n}{3}-1}, \mathcal{M}'_{1 \cdot 8}}$:

```
for i:=1 to n/(9*8)
  each thread t solves matvec(t*i)
endfor
```

The first 8 of the $\frac{n}{3}$ subproblems can be solved independently by the 8 cores, and so on until they are all completed. Hence, the execution matrix of the algorithm $A_{D'_{\frac{n}{3}-1}, \mathcal{M}'_{1 \cdot 8}}$ has $r_E = \frac{n}{3 \cdot 8} = \frac{n}{24}$ rows and if we assume that $\frac{n}{3}$ is a multiple of 8^3, the algorithm is perfectly parallel.

Assuming that for the execution of each operator, it is required to read (before the execution) and to write (after the execution) the vector of size $\frac{n}{3} + 1$ and that the cores can transfer their vectors concurrently, that is the machine can concurrently transfer 8 vectors, the memory matrix $AM_{A_{D'_{\frac{n}{3}-1}, \mathcal{M}_{1 \cdot 8}}}$ has

$$r_{mem, D'_{\frac{n}{3}-1}} = \left(\frac{n}{3} + 2\right) \cdot \frac{n}{24}$$

rows. The execution time of each row of the execution matrix is the execution time of the $\boxtimes$ operator. If $r_E = \frac{n}{24}$ is the number of rows of $E_{A_{D'_{\frac{n}{3}-1}, \mathcal{M}'_{1 \cdot 8}}}$, the execution time of the algorithm $A_{D'_{\frac{n}{3}-1}, \mathcal{M}'_{1 \cdot 8}}$ is

$$T(A_{D'_{\frac{n}{3}-1}, \mathcal{M}'_{1 \cdot 8}}) = r_E \cdot T_r = \frac{n}{24} \cdot C(\boxtimes) \cdot tcalc = \frac{n}{24} \cdot 2 \cdot \left(\frac{n}{3}\right)^2 \cdot tcalc. \quad (24)$$

If we denote by $r_{mem, D'_{\frac{n}{3}-1}} = \left(\frac{n}{3} + 2\right) \cdot \frac{n}{24}$ the number of rows of the memory access matrix of the algorithm $A_{D'_{\frac{n}{3}-1}, \mathcal{M}_{1 \cdot 8}}$, the memory access time of the software $SW(A_{D'_{\frac{n}{3}-1}, \mathcal{M}_{1 \cdot 8}})$ we are going to implement is

$$T_M(SW(A_{D'_{\frac{n}{3}-1}, \mathcal{M}_{1 \cdot 8}})) = r_{mem, D'_{\frac{n}{3}-1}} \cdot tvec_{mem} = \left(\frac{n}{3} + 2\right) \cdot \frac{n}{24} \cdot tvec_{mem} \quad (25)$$

and the execution time of the software $SW(A_{D'_{\frac{n}{3}-1}, \mathcal{M}_{1 \cdot 8}})$ is

$$\begin{aligned} T(SW(A_{D'_{\frac{n}{3}-1}, \mathcal{M}_{1 \cdot 8}})) &= T(A_{D'_{\frac{n}{3}-1}, \mathcal{M}'_{8,8}}) + T_M(SW(A_{D'_{\frac{n}{3}-1}, \mathcal{M}_{8,8}})) \\ &= \frac{n}{24} \cdot 2 \cdot \left(\frac{n}{3}\right)^2 \cdot tcalc + \left(\frac{n}{3} + 2\right) \cdot \frac{n}{24} \cdot tvec_{mem}. \end{aligned} \quad (26)$$

---

[3] There is no loss of generality.

Finally, the speed up is

$$
Sp(SW(A_{D'_{\frac{n}{3}-1},\mathcal{M}'_{1\cdot8}})) = \frac{T(SW(A_{D'_{\frac{n}{3}-1},\mathcal{M}'_{1,1}}))}{T(SW(A_{D'_{\frac{n}{3}-1},\mathcal{M}_{1\cdot8}}))}
$$

$$
= \frac{2 \cdot \left(\frac{n}{3}\right)^3 \cdot tcalc + \left(\frac{n}{3}+2\right) \cdot \frac{n}{3} \cdot tvec_{mem}}{\frac{n}{24} \cdot 2 \cdot \left(\frac{n}{3}\right)^2 \cdot tcalc + \left(\frac{n}{3}+2\right) \cdot \frac{n}{24} \cdot tvec_{mem}} > 1 \tag{27}
$$

Let $A'_{D_{\frac{n}{3}-1},\mathcal{M}_{9\cdot8}}$ denote the algorithm that uses 9 nodes and 8 cores per node. We get the following expression of the speed up the algorithm that uses 1 level of parallelism in $M_{9,9}$

$$
Sp(SW(A_{D_{27},\mathcal{M}_{9,9}})) = \frac{T(SW(A_{D_{27},\mathcal{M}_{1,1}}))}{T(SW(A_{D_{27},\mathcal{M}_{9,9}}))}
$$

$$
= \frac{26 \cdot C(\otimes) \cdot tcalc + 52 \cdot tblock_{mem}}{3 \cdot C(\boxtimes) \cdot tcalc + 7 \cdot tblock_{com}} \tag{28}
$$

which should be compared to the speed up of the algorithm that uses 2 levels of parallelism in $M_{9\cdot8}$

$$
Sp(SW(A_{D'_{\frac{n}{3}-1},\mathcal{M}_{9\cdot8}})) = \frac{T(SW(A_{D_{27},\mathcal{M}_{1,1}})}{T(SWA_{D'_{\frac{n}{3}-1},\mathcal{M}_{9\cdot8}})}
$$

$$
= \frac{26 \cdot \left(\frac{n}{3} \cdot C(\boxtimes) \cdot tcalc + \left(\frac{n}{3}+2\right) \cdot tvec_{mem}\right) + 52 \cdot tblock_{mem}}{3 \cdot \frac{n}{24} \cdot \left(C(\boxtimes) \cdot tcalc + \left(\frac{n}{3}+2\right) \cdot tvec_{mem}\right) + 7 \cdot tblock_{com}} \tag{29}
$$

By specializing the parameters we can estimate the performance gain that we get using two levels of parallelism instead of one.

3   Conclusion

Matrix multiplication is one of the fundamental kernels in numerical linear algebra, for almost all matrix problems such as least square problem eigenvalue problem and data assimilation problem [5–8,14]. Future designs of microprocessors and large HPC systems will be heterogeneous in nature, relying on the integration of two major types of components. On the first hand, multi/many-cores CPU technology have been developed and the number of cores will continue to escalate because of the desire to pack more and more components on a chip while avoiding the power wall, instruction level parallelism wall, and the memory wall. On the other hand special purpose hardware and accelerators, especially Graphics Processing Units (GPUs) are in commodity production, and have outpaced standard CPUs in floating point performance in recent years, and have become as easy, if not easier to program than multi-core CPUs. Finally, reconfigurable architectures such as Field programmable Gate Arrays (FPGAs) offer several parameters such as operating frequency, precision, amount of memory, number of computation units, etc. These parameters define a large design space that must be explored to find efficient solutions.

To address this scenario, it is undoubted that performance analysis of MM algorithm should be re-evaluated to find out the best-practice algorithm on novel architectures. This motivated the work to investigate the performance of the standard MM algorithm, by means of the new modelling framework that the authors have introduced.

This paper attempts to describe how to apply the performance model that the authors have developed so as to make it accessible to a broad audience. The model exploits the knowledge of the algorithm and the target architecture and it could help the researchers for designing optimized implementations on emerging computing architectures, such as that one developed in [3,4].

References

1. Ballard, G., Demmel, J., Holtz, O., Schwartz, O.: Minimizing communication in numerical linear algebra. SIAM J. Matrix Anal. Appl. **32**(3), 866–901 (2011)
2. Cuomo, S., D'Amore, L., Murli, A., Rizzardi, M.: Computation of the inverse Laplace transform based on a collocation method which uses only real values. J. Comput. Appl. Math. **198**(1), 98–115 (2007)
3. D'Amore, L., Laccetti, G., Romano, D., Scotti, G., Murli, A.: Towards a parallel component in a GPU-CUDA environment: a case study with the L-BFGS Harwell routine. Int. J. Comput. Math. **92**(1), 59–76 (2015)
4. D'Amore, L., Casaburi, D., Galletti, A., Marcellino, L., Murli, A.: Integration of emerging computer technologies for an efficient image sequences analysis. Integr. Comput.-Aided Eng. **18**(4), 365–378 (2011)
5. D'Amore, L., Murli, A.: Regularization of a Fourier series method for the Laplace transform inversion with real data. Inverse Prob. **18**(4), 1185–1205 (2002)
6. D'Amore, L., Arcucci, R., Carracciuolo, L., Murli, A.: DD-OceanVar: a domain decomposition fully parallel data assimilation software in mediterranean sea. Procedia Comput. Sci. **18**, 1235–1244 (2013)
7. D'Amore, L., Arcucci, R., Carracciuolo, L., Murli, A.: A scalable approach to variational data assimilation. J. Sci. Comput. **61**, 239–257 (2014)
8. D'Amore, L., Arcucci, R., Marcellino, L., Murli, A.: HPC computation issues of the incremental 3D variational data assimilation scheme in OceanVar software. J. Numer. Anal. Ind. Appl. Math. **7**(3–4), 91–105 (2012)
9. Demmel, J., Eliahu, D., Fox, A., Kamil, S., Lipshitz, B., Schwartz, O., Spillinger, O.: Communication-optimal parallel recursive rectangular matrix multiplication. In: Proceedings of the IEEE 27th International Symposium on Parallel and Distributed Processing (IPDPS 2013), pp. 261–272. IEEE Computer Society, Washington, D.C. (2013)
10. Fox, G., Otto, S., Hey, A.: Matrix algorithms on a hypercube I: matrix multiplication. Parallel Comput. **3**(5), 17–31 (1987)
11. Gunnels, J.A., Henry, G.M., van de Geijn, R.A.: A family of high-performance matrix multiplication algorithms. In: Alexandrov, V.N., Dongarra, J., Juliano, B.A., Renner, R.S., Tan, C.J.K. (eds.) ICCS-ComputSci 2001. LNCS, vol. 2073, pp. 51–60. Springer, Heidelberg (2001)

12. Gunnels, J.A., Gustavson, F.G., Henry, G.M., van de Geijn, R.A.: FLAME: formal linear algebra methods environment. ACM Trans. Math. Softw. **27**(4), 422–455 (2001)
13. Kuon, I., Tessier, R., Rose, J.: FPGA architecture: survey and challenges. Found. Trends Electron. Des. Autom. **2**(2), 135–253 (2007)
14. Murli, A., Cuomo, S., D'Amore, L., Galletti, A.: Numerical regularization of a real inversion formula based on the Laplace transform's eigenfunction expansion of the inverse function. Inverse Prob. **23**(2), 713–731 (2007)

How to Mitigate Node Failures in Hybrid Parallel Applications

Maciej Szpindler[✉]

Interdisciplinary Centre for Mathematical and Computational Modelling,
University of Warsaw, Warszawa, Poland
m.szpindler@icm.edu.pl

Abstract. This paper describes approach to distributed node failure
detection and communicator recovery in MPI applications with dynamic
resource allocation. Failure detection is based on a recent proposal for
user-level mitigation. The aim of this paper is to identify distributed and
scalable approach for node failures detection and mitigation. Failed MPI
communication recovery is realized with experimental implementation
for MPI level resource allocation. Re-allocation of resources is used to
replace failed node and enable application continuation with a full performance. Experimental results and performance of proposed techniques
are discussed for schematic application scenarios.

Keywords: Message passing · Fault tolerance · Resource allocation ·
Dynamic execution

1 Introduction

Recent advances in HPC systems design result in increase of node level parallelism. One can expect this trend will continue up to developing substantial
multi-element processing units in a form of many-core hyper-threaded computing nodes with a dozens of cores. No matter which model of software parallelism
is exploited, the case of fault tolerance is significant for applications reliability
and handling of hardware failures.

The most popular model that ensures both high performance and scalability
on systems composed of large shared memory nodes is the hybrid parallelism.
Usually the latter term refers to at least two levels of different parallelisation
techniques coupled together. On the top level, preferred technique is message
passing and distributed memory model such as MPI. On the lower level, different shared memory models usually provide better scalability for a range of
applications classes. Popular choice there are OpenMP or other threading models. Such a combination of inter- and intra-node computing techniques is referred
to as hybrid parallelism.

In the case of the MPI as a choice for the highest level of parallelisation
technique, fault tolerance is widely studied area still not yet standardized.
A number of approaches have been explored in this connection. Both library

© Springer International Publishing Switzerland 2016
R. Wyrzykowski et al. (Eds.): PPAM 2015, Part II, LNCS 9574, pp. 35–44, 2016.
DOI: 10.1007/978-3-319-32152-3_4

specific implementation [6] and MPI functionality extension approaches [7] have been proposed until now without successful adoption in a form of standardized definition. A recent proposal of fault tolerance primitives called User Level Failure Mitigation (ULFM) [3] has attracted wide recognition.

Nested parallelism on the intra-node communication level is supported within MPI model. At least two choices are possible there: either multi-process approach provided by MPI-3 shared memory windows [8] or multi-threading implemented inside MPI processes with chosen threading library. For both of these choices it is usually practical to use dedicated MPI communicator that allows intra-node synchronization. There are also advanced developments on extending this idea to a dynamic endpoints communicators [5] that will be probably included into a future version of the MPI standard.

For all the realizations of intra-node parallelism any type of failure result in a damage of associated communicator. Moreover, usually any serious hardware failure is actually resulting in whole node failure and loss of communication, no matter what the scale of the system is. It is expected that for larger systems with massive inter-node parallelism hardware failures will occur more often comparing to application lifetime. For either single multi-threaded process or multi-process execution on the failed node, the associated intra-node communicator is doomed to failure.

This paper presents basic schemes for failed communicators recovery and reconstruction that enable hybrid parallel application to mitigate node failures. Section 2 gives summary on the distributed detection of intra-node communicator failures, Sect. 3 describe reconstruction approach with a use of dynamic resource allocation. Section 4 contains an analysis of the experiments on the proposed techniques for node failure mitigation and a discussion of the performance for the proposed approach. The key contributions of the described work are the following:

– study on the distributed node failure detection using currently available implementations of the MPI user level failure mitigation approach,
– application of the dynamic resource allocation from the MPI level for failed node reconstruction,
– experiments on the performance and scalability of the proposed techniques.

2 Detecting Node Failures

2.1 User-Level Failure Mitigation Model

The MPI standard [9] defines basic abstraction for handling failures. The default approach is to use MPI_ERRORS_ARE_FATAL error handler. In this case all application processes are immediately terminated if any type of failure occurs. This is also a common choice for most of the legacy MPI codes that are in fact no fault-tolerant. Another approach, supported by MPI, is to use MPI_ERRORS_RETURN handler which gives possibility to post some process local operation before application is terminated. ULMF model is extending the latter approach, enabling application to continue its execution after the failure.

ULFM is a set of functions extending MPI API functionality with primitives for handling process failures explicitly in the application code. It is designed to provide mechanisms for failure detection, notification, propagation and communication recovery. Details of this MPI extension are described in [3]. While this enables MPI application to detect process failures and mitigate them, reconstruction and recovering application consistency are not a part of the extension and remain user responsibility.

Process failures are indicated with specific return codes of MPI communication routines. Either MPI_SUCCESS or MPI_ERR_PROC_FAILED error codes are returned for completion or failure respectively. If global knowledge of failure is required, already started communication can be revoked to assert consistency, raising MPI_COMM_REVOKED error code on all active communication. Functions for local failure acknowledgement and collective agreement on the group of survived processes are provided with the ULMF model.

ULFM extensions are partly implemented in the MPICH project (as for beginning of 2015, version 3.2 pre-release) and in a specific OpenMPI branch (version 1.7ft) dedicated for fault-tolerance studies. First analyses of the ULFM model performance and limitations have already appeared in [4] and real MPI applications with fault tolerance implemented with ULFM have been studied in [1].

Since ULFM model seems likely to be adopted, it is worth to target hybrid parallel applications using this approach. Node failure mitigation is addressed in this paper.

2.2   Intra-node Communicators

MPI model uses abstract communicator construct to represent a group of processes and their interactions. It provides elegant way of separating different communication scopes for collective communication. Also it is an abstraction that allows to express different communication schemes with groups and virtual topologies. More complicated communication designs can be described with either intra-communicator for a single group of processes or inter-communicator for separating two distinct groups of processes participating in the communication (referred to as local group and remote group).

It is practical to express nested parallelism in the hybrid MPI applications with dedicated communicators. This encapsulates intra-node communication and synchronization. It allows separation between intra- and inter-node communication which may overlap. Also it enables fine-grain synchronization depending on the application design. As a result communication costs may be reduced and eventually message exchange optimized on the MPI internal level. MPI provides convenient functionality with MPI_COMM_SPIT_TYPE which partitions global communicator into a disjoint subgroups of given type. The only standardized type is MPI_COMM_TYPE_SHARED which returns groups associated with shared memory nodes. This exposes intra-node shared memory regions for local processes.

This approach is natural for two level parallelism with MPI+MPI model that is composed of MPI communication across the nodes and MPI shared memory windows within the node. It was showed that such model of hybrid parallelism is beneficial for some application classes [8].

Other hybrid parallel applications based on MPI+X approach (where X denotes some other parallel programming model, e.g. OpenMP) may also require logical separation of intra- and inter-node communication. This is quite common approach for reducing total number of MPI messages and its exchange rate.

2.3   Node Failure Detection

In this paper dedicated intra-node communicator is considered. While computing node fails, respective communicator disappears and fault-tolerant application needs to handle with corrupted communicator. With a choice of ULFM model for mitigating node failures, one must decide on detection technique. Two distinct approaches that aims distributed detection are described in this paper. Distributed method is defined as not involving all processes participating in the MPI_COMM_WORLD communicator (global communicator).

First approach relies on the MPI inter-communicators. In this case each of the intra-node communicators has its counterpart communicator acting as remote group of processes. This scheme is depicted on Fig. 1. Local group and its remote neighbour form an inter-communicator. This seemingly complicated construct allows detection of node failures locally. Broken node and associated processes group are identified with a use of ULMF detection function. Unfortunately, inter-communicators were not fully supported by ULFM implementations at the time of this research.

Latter approach does not involve inter-communicators. The most straight-forward way of detecting failed processes is to test MPI_COMM_WORLD. This kind of process failures detection is not scalable while all processes are involved. More distributed attempt is proposed with a special communicators structure. Each inter-node shared memory communicator delegates one "leader" process. These processes participate in "leaders" communicator. This special communicator allows to connect processes between distinct nodes as shown on Fig. 2.

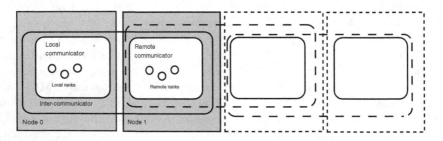

Fig. 1. Pairing local and remote node communicators in inter-communicator for local notification.

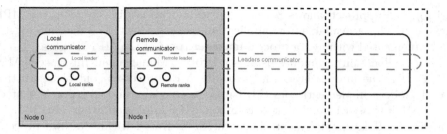

Fig. 2. Leaders communicator for inter-node communication and notification.

Members of the "leaders" communicator notify failures locally. Node failure is detected and group of failed processes is identified without involving global operations on the MPI_COMM_WORLD level.

3 Dynamic Reconstruction

3.1 Communicator Reconstruction

When the failed communicator is identified, reconstruction is possible. The choice of reconstruction approach depends on application type and user requirements. If communicator needs to be recreated to continue execution, then new processes are spawned. Spawning means dynamic creation of processes in the MPI terminology. Spawned processes eventually build new communicator to swap and restore failed one. Such approach introduces significant overheads due to process spawning as discussed in [3]. It is also required that application would support restore of lost data of the failed communicator member processes. At least two choices are considered in previous studies: either using checkpoints to dump application state in a selected points of execution or to replicate node private data on different remote node. Both choices require significant changes on the application level and these are discussed in [1].

3.2 Dynamic Resource Allocation

Another essential issue concerning node-communicator reconstruction is resource utilization. If performance degradation or increased node memory load are not acceptable, over-subscription of processes on the remaining set of nodes is a bad solution. Recreated processes need to be spawned on a new node. New resource need to be granted to application. This is usually not immediate nor possible with a general purpose HPC systems that execute many user jobs simultaneously.

One of the possible solution is to use dynamic resource allocation. It was showed that resizing node allocation is possible and basic implementation was presented for the hydra process manager of the MPICH library and Slurm resource management infrastructure. The details of this work are described in [10].

Proposed approach allows resizing Slurm allocation directly from the MPI spawn call. This is available from application code as depicted on diagram Fig. 3. It is implemented with hydra process manager (part of the MPICH library) and the Slurm allocation techniques using the Process Management Interface (PMI) API. PMI is the interim layer that provides MPI processes control [2]. In the case of modern implementations, MPI process spawning model is implemented with PMI infrastructure. For two common MPI implementations, MPICH is providing PMI layer implementation tightly integrated within its own process manager called *hydra* while OpenMPI has similar approach with closely related project called *PMIx*.

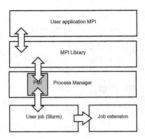

Fig. 3. Dynamic allocation scheme with process and resource managing layers.

Three modes of resource allocation were implemented and provided support for different applications requirements. Immediate allocation mode provides access to the resources only if currently available. It raises an error in the other case. Immediate mode was implemented using native Slurm request features. Non-blocking mode gives immediate return to execution after the allocation request. It was intended to use a helper thread to track allocation status. Blocking mode returns only if resources are successfully allocated. It is using Slurm blocking request. While blocking and non-blocking modes depend on external conditions and availability of the resources were not addressed in experiments discussed in the next section.

4   Experimental Results

In this section experimental results are described. Synthetic application was implemented to test the performance of proposed node failure detection and reconstruction approach. It focuses on node failures in case of hybrid parallelism. Application kernel is a two level reduction with a local operation over node's shared memory and a global MPI reduce operation across nodes. If global reduction raises fault error, failed node is detected and associated communicator is re-created. This schematic kernel aims to reproduce nested parallelism and it's typical communication pattern. Reconstruction of the failed communicator allocates new node dynamically with a use of described resource allocation technique. Experiments were performed using beta release if the MPI library which was the only choice available supporting ULFM extension. More complex communication schemes were considered to behave unstable and schematic application was selected as reliable test at this stage.

Two types of experiments was executed. One type addressed absolute performance of the proposed node failure detection and application reconstruction. Node failure detection overhead was analysed using high precision timer. Performance of the dynamic reconstruction of failed nodes was measured with a focus on dynamic allocation time and process spawning time. Other type of experiments tested relative cost of application reconstruction against the cost of application restart including the cost of resource allocation and application re-initialization.

4.1   Absolute Performance

Performance of the described node failure detection without usage of inter-communicators was addressed. Time overhead introduced by the proposed detection scheme was measured. Time cost versus a number of participating nodes was studied. Averaged results are shown on Fig. 4. Time measurements was based on a CPU cycles. The choice of the time measure was motivated by insufficient precision of the MPI_Wtime function.

Detection scheme was tested for up to 24 nodes running from 4 to 16 local processes using MPI_COMM_TYPE_SHARED sub-communicators. It was found that scalability is limited more by a number of processes per node that by the actual number of nodes. This exposes limitations of the remote process interactions used in a detection scheme.

Cost of the dynamic process allocation used in reconstruction is shown on Fig. 5. Time spend waiting for reallocation of failed nodes was compared to the process spawning cost. As expected spawning new processes were associated with overheads [3]. Dynamic allocation implementation was corrected and time was

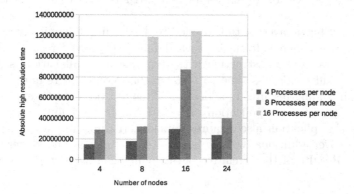

Fig. 4. Relative time spent in detection phase in the case of schematic hybrid parallel application.

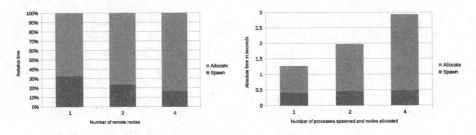

Fig. 5. Left: relative cost of the spawn and allocate operations for increasing number of nodes. Right: Time cost in seconds of spawning processes and allocating additional nodes.

greatly improved comparing to the previous results [10]. Nevertheless significant overheads were observed for dynamic allocation of nodes, due to user job resizing which involves many, possibly slow, system components. Experimental results were collected using "immediate" allocation mode. Presented time measurements are averaged over the series of experimental runs. Despite using a pool of reserved nodes for experiments, results still tend to be highly biased by the internal Slurm allocation procedures.

4.2 Relative Performance

To demonstrate practicality of the discussed approach, cost of the detection and dynamic reconstruction of failed nodes was compared to cost of re-scheduling and re-initialization of the schematic mini-application. Overall approach should also contain full application state recovery, including state of failed node's memory. It can use memory image cached on the remote node that is periodically updated which obviously introduces significant memory footprint and synchronization overhead. Other choices are possible but were not addressed in the described work. Instead of studying application specific state recovery that is discussed in [4], neglected costs of job re-scheduling and MPI related re-initialization were addressed.

Experiment tested average time needed to detect and dynamically re-allocate resources in case when half of nodes used failed. Collected results show that despite its obstacles, reconstruction with dynamic node allocation is practical approach. It still needs less time to recover than complete re-initialization of application including resource re-allocation. This test does not take into account time required to recover application to a state before the failure. Obviously re-scheduling of application also require new job creation, in case of scheduling system. Moreover additional waiting is required if nodes are no longer available to the user. Results of these relative performance comparison are summarized on Fig. 6.

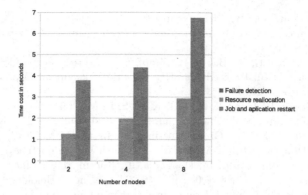

Fig. 6. Relative costs of detection and dynamic allocation versus cost of complete job re-initialization.

5 Summary and Future Work

Node failure detection and associated application reconstruction is important issue in case of hybrid parallelism. In this paper distributed approach for node failure detection is proposed and possible implementation choices with ULFM extension of the MPI standard discussed. This is the attempt to enable scalable and fault-tolerant applications with a hybrid parallelism. Performance of the proposed approach was invalided and experimental tests are discussed. Limitations and performance issues were identified. This work is related to unstable and experimental implementation of ULFM extension and other more scalable approaches are still available. Possible choices are described in this paper and are easy to apply with more refined and stable implementations.

Another contribution of this paper to the discussion on fault-tolerant MPI applications is proposal for communicator reconstruction involving dynamic resources allocation. It is demonstrated as practical alternative for application restart in case of node failures. Implementation of the proposed mechanism is described and experimental results included. Identified limitations are related to immediate allocation and need to be addressed with better Slurm integration. The case of non-blocking allocation requests and pending for resources still need to be refined to provide more capabilities and integrity.

References

1. Ali, M.M., Southern, J., Strazdins, P., Harding, B.: Application level fault recovery: using fault-tolerant open MPI in a PDE solver. In: 2014 IEEE International, Parallel & Distributed Processing Symposium Workshops (IPDPSW), pp. 1169–1178. IEEE (2014)
2. Balaji, P., Buntinas, D., Goodell, D., Gropp, W., Krishna, J., Lusk, E., Thakur, R.: PMI: a scalable parallel process-management interface for extreme-scale systems. In: Keller, R., Gabriel, E., Resch, M., Dongarra, J. (eds.) EuroMPI 2010. LNCS, vol. 6305, pp. 31–41. Springer, Heidelberg (2010)

3. Bland, W., Bouteiller, A., Herault, T., Hursey, J., Bosilca, G., Dongarra, J.J.: An evaluation of user-level failure mitigation support in MPI. Computing **95**(12), 1171–1184 (2013)
4. Bland, W., Raffenetti, K., Balaji, P.: Simplifying the recovery model of user-level failure mitigation. In: Proceedings of the 2014 Workshop on Exascale MPI, pp. 20–25. IEEE Press (2014)
5. Dinan, J., Balaji, P., Goodell, D., Miller, D., Snir, M., Thakur, R.: Enabling MPI interoperability through flexible communication endpoints. In: Proceedings of the 20th European MPI Users' Group Meeting, pp. 13–18. ACM (2013)
6. Fagg, G.E., Dongarra, J.: FT-MPI: fault tolerant MPI, supporting dynamic applications in a dynamic world. In: Dongarra, J., Kacsuk, P., Podhorszki, N. (eds.) PVM/MPI 2000. LNCS, vol. 1908, pp. 346–353. Springer, Heidelberg (2000)
7. Gropp, W., Lusk, E.: Fault tolerance in message passing interface programs. Int. J. High Perform. Comput. Appl. **18**(3), 363–372 (2004)
8. Hoefler, T., Dinan, J., Buntinas, D., Balaji, P., Barrett, B., Brightwell, R., Gropp, W., Kale, V., Thakur, R.: MPI+MPI: a new hybrid approach to parallel programming with MPI plus shared memory. Computing **95**(12), 1121–1136 (2013)
9. MPI Forum: MPI: A Message-Passing Interface Standard. Version 3.0, 21 September 2012. http://www.mpi-forum.org
10. Szpindler, M.: Enabling adaptive, fault-tolerant MPI applications with dynamic resource allocation. In: Proceedings of the 3rd International Conference on Exascale Applications and Software (2015)

A Scalable Numerical Algorithm for Solving Tikhonov Regularization Problems

Rosella Arcucci[1,2,4], Luisa D'Amore[1,2(✉)], Simone Celestino[1],
Giuliano Laccetti[1], and Almerico Murli[2,3]

[1] University of Naples Federico II, Naples, Italy
`luisa.damore@unina.it`
[2] Euro Mediterranean Centre on Climate Change (CMCC), Lecce, Italy
[3] SPACI, Naples, Italy
[4] Imperial College London, London, UK

Abstract. We present a numerical algorithm for solving large scale Tikhonov Regularization problems. The approach we consider introduces a splitting of the regularization functional which uses a domain decomposition, a partitioning of the solution and modified regularization functionals on each sub domain. We perform a feasibility analysis in terms of the algorithm and software scalability, to this end we use the scale-up factor which measures the performance gain in terms of time complexity reduction. We verify the reliability of the approach on a consistent test case (the Data Assimilation problem for oceanographic models).

Keywords: Tikhonov regularization · Large scale inverse problems · Parallel algorithm · Data assimilation

1 Introduction and Motivation

The solution of large scale inverse and ill posed problems arises in a variety of applications, such as those in the earth/climate science, including earth observation (remote sensing) and data assimilation [8,14], or those arising in image analysis, including medical imaging, astronomical imaging and restoration of digital films [2,4,5,9,10,15]. A straightforward solution of such problems is meaningless because the computed solution would be dominated by errors. Therefore some regularization must be employed. In this paper we focus on the standard Tikhonov Regularization (TR) method [16]. The efficient solution of TR problems critically depends on suitable numerical algorithms. Several strategies have been proposed in the literature. Basically, the approaches are based on the Conjugate Gradient iterative method, or on the Singular Value Decomposition. However, because of their formulation, these approaches are intrinsically sequential and none of them is able to address in an acceptable computational time large scale applications. For such simulations we need to address methods which allow us to reduce the problem to a finite sequence of sub problems of a more manageable size, perhaps without sacrificing the accuracy of the computed solution. Indeed, we need to employ scalable parallel algorithms.

© Springer International Publishing Switzerland 2016
R. Wyrzykowski et al. (Eds.): PPAM 2015, Part II, LNCS 9574, pp. 45–54, 2016.
DOI: 10.1007/978-3-319-32152-3_5

Here, scalability refers to the capability of the algorithm to:

- exploit performance of emerging computing architectures in order to get a solution in a suitable acceptable time (strong scaling),
- use additional computational resources effectively to solve increasingly larger problems (weak scaling).

In the present work we introduce a computational model which starts from a decomposition of the global domain into sub domains. On these sub domains we define local regularization functionals such that the minimum of the global regularization functional can be obtained by collecting the minimum of each local functional. The (global) problem is decomposed into (local) sub problems in such a way. The resulted algorithm consists of several copies of the original one, each one requiring approximately the same amount of computations on each sub domain and an exchange of boundary conditions between adjacent sub domains. The data is flowing across the surfaces, the so called surface-to-volume effect is produced.

A research collaboration between us, the Argonne National Laboratory in Chicago, the Imperial College London, the University of California Santa Cruz, and the Barcelona Supercomputing Center, within the H2020-MSCA-RISE-2015 Project NASDAC (iNnovative Approaches for Scalable Data Assimilation in oCeanography) give us the opportunity to work on variational Data Assimilation (DA) in Oceanographic Models [7,9]. Then we applied this approach to the (DA) inverse problem which is ill posed and variational approaches used for solving it are essentially derived from the TR formulation.

2   Preliminary Concepts

Here we introduce some notations we use in the next sections. For more details see [6].

Definition 1 (The Inverse problem). *Given the linear operators $\mathbf{M} \in \Re^{N \times N}$ and $\mathbf{H} \in \Re^{S \times N}$, and the vector $\mathbf{v} \in \Re^{S \times 1}$, where $N >> S$. Assume that $\mathbf{H}$ is highly ill conditioned. To compute $\mathbf{u} : \Omega \mapsto \Re^{N \times 1}$ such that*

$$\mathbf{v} = \mathbf{H}[\mathbf{u}] \qquad (1)$$

subject to the constraint $\mathbf{u} = \mathbf{u}^{\mathbf{M}}$ where $\mathbf{u}^{\mathbf{M}} = \mathbf{M}[\mathbf{u}]$.   ♠

The TR approach provides the approximation $\mathbf{u}(\lambda)$ of $\mathbf{u}$, where λ is the regularization parameter, as follows [13]

Definition 2 (The TR problem). *To compute*

$$\mathbf{u}(\lambda) = argmin_{\mathbf{u}} J(\mathbf{u}) \qquad (2)$$

where

$$J(\mathbf{u}) = \|\mathbf{H}\mathbf{u} - \mathbf{v}\|_{\mathbf{R}}^2 + \lambda \|\mathbf{u} - \mathbf{u}^{\mathbf{M}}\|_{\mathbf{B}}^2, \qquad (3)$$

is the TR problem of (1) and, where $\|\cdot\|_{\mathbf{B}}$ and $\|\cdot\|_{\mathbf{R}}$ denote the weighted norms with respect to the error covariance matrices $\mathbf{B}$ and $\mathbf{R}$ and λ is the regularization parameter. ♠

Definition 3 (Domain Decomposition). *Let*

$$\Omega = \bigcup_{i=1}^{p} \Omega_i \quad \Omega_i \subset \Re^3 \tag{4}$$

be the decomposition of $\Omega \subset \Re^3$ where $\Omega_i \subset \Re^3$ are such that $\Omega_i \cap \Omega_j = \Omega_{ij} \neq \emptyset$ when the subdomains are adjacent. ♠

Starting from a decomposition of the domain Ω, we now introduce the *local* TR functionals. A local TR functional, which describes the local problems on each sub-domain Ω_i, is obtained from the TR functional J in (3), by adding a *local* constraint defined on the overlapping regions in Ω_{ij}. This is in order to enforce the continuity of each solution of the local DA problem onto the overlap region between adjacent domains Ω_i and Ω_j.

Definition 4 (Local TR functional). *Let $\mathbf{H}_i, \mathbf{u}^i, \mathbf{v}^j, (\mathbf{u}^M)^i, \mathbf{R}_i$ and $\mathbf{B}_i$, be the restrictions on Ω_i of $\mathbf{H}, \mathbf{u}, \mathbf{v}$ and $\mathbf{u}^M, \mathbf{R}$ and of $\mathbf{B}$, respectively. Let $\mathbf{u}^j$ be the restriction on Ω_j of $\mathbf{u}$, $\mathbf{B}_{ij}$ be the restriction of $\mathbf{B}$ on the overlapping region Ω_{ij}. Finally, let λ_i and ω_i be the (local) regularization parameters. Then*

$$\mathbf{u}^i(\lambda_i, \omega_i) = argmin_{\mathbf{u}^i} J(\Omega_i, \lambda_i, \omega_i)$$

where

$$J(\Omega_i, \lambda_i, \omega_i) = \|\mathbf{H}_i \mathbf{u}^i - \mathbf{v}^i\|_{\mathbf{R}_i}^2 + \lambda_i \|\mathbf{u}^i - (\mathbf{u}^M)^i\|_{\mathbf{B}_i}^2$$
$$+ \omega_i \|\mathbf{u}^i/\Omega_{ij} - \mathbf{u}^j/\Omega_{ij}\|_{\mathbf{B}_{ij}}^2 \tag{5}$$

is the minimum of the local TR functional $J(\Omega_i, \lambda_i, \omega_i)$. ♠

In [6] the authors proved that

$$\mathbf{u}(\lambda) = \sum_{i=1,p} \mathbf{u}^{EO_i}(\lambda_i, \omega_i), \tag{6}$$

where

$$\mathbf{u}^{EO_i}(\lambda_i, \omega_i) : \Omega_i \mapsto \Omega$$

and

$$\mathbf{u}^{EO_i}(\lambda_i, \omega_i) := \begin{cases} \mathbf{u}^i & on \; \Omega_i \\ 0 & elsewhere \end{cases}$$

This result states that the minimum of J, in (2), can be regarded as a piecewise function obtained by patching together $\mathbf{u}^i$, i.e. the minimum of the operators $J(\Omega_i, \lambda_i, \omega_i)$; it means that, by using the domain decomposition, the global minimum of the operator J can be obtained by patching together the minimum of the *local* functionals $J(\Omega_i, \lambda_i, \omega_i)$.

In the following we refer to the decomposition of TR functional as the DD-TR model.

2.1 The Algorithmic Scalability

Large-scale problems are computationally expensive and their solution requires designing of scalable approaches. Many factors contribute to scalability, including the architecture of the parallel computer and the parallel implementation of the algorithm. However, one important issue is the scalability of the algorithm itself. We use the following measure

Definition 5 (Scalable Algorithm). *If $p \in \mathcal{N}$, and $p > 1$, the algorithm associated to the decomposition given in (4) is*

$$\mathcal{A}(\Omega, p) := \{\mathcal{A}(\Omega_1), \mathcal{A}(\Omega_2), \ldots, \mathcal{A}(\Omega_p)\}$$

where $\mathcal{A}(\Omega_i)$ is the local algorithm on Ω_i. ♠

Definition 6 (Scale up factor). *Let $p_1, p_2 \in \mathcal{N}$ and $p_1 < p_2$. Let $T(\mathcal{A}(\Omega, p_i))$, $i = 1, 2$ denote the time complexity of $\mathcal{A}(\Omega_i, p_i)$, $i = 1, 2$. $\forall \ i \neq j$ we define the (relative) scale up factor of $\mathcal{A}(\Omega, p_2)$, in going from p_1 to p_2, the following ratio:*

$$S_{p_2, p_1}(N) = \frac{T(\mathcal{A}(\Omega, p_1))}{(p_2/p_1)T(\mathcal{A}(\Omega, p_2))}.$$

♠

We observe that:

1. if N is fixed and $p \sim N$ we get the so called *strong scaling*.
2. if $N \to \infty$ and r is kept fixed, then we get the so called *weak scaling*.

3 The Case Study

Let $t \in [0, T]$ denote the time variable. Let $u^{true}(t, x)$ be the evolution state of a predictive system governed by the mathematical model $\mathcal{M}$ with $u^{true}(t_0, x)$, $t_0 = 0$ as initial condition. Here we consider a 3D shallow water model. Let $v(t, x) = \mathcal{H}(u^{true}(t, x))$ denote the observations mapping, where $\mathcal{H}$ is a given nonlinear operator which includes transformations and grid interpolations. According to the real applications of model-based assimilation of observations, we will use the following definition of Data Assimilation (DA) inverse problem [13, 14]. Given

- $D_N(\Omega) = \{x_j\}_{j=1,\ldots,N} \in \Re^N$: a discretization of $\Omega \subset \Re^3$;
- $\mathbf{M}$: a discretization of $\mathcal{M}$;
- $\mathbf{u}_0^{\mathcal{M}} = \{u_0^j\}_{j=1,\ldots,N}^{\mathcal{M}} \equiv \{u(t_0, x_j)\}_{j=1,\ldots,N}^{\mathcal{M}} \in \Re^N$: numerical solution of $\mathcal{M}$ on $D_N(\Omega)$. This is the background estimates, i.e. the initial states at time t_0; it is assumed to be known, usually provided by a previous forecast.
- $\mathbf{u}^{\mathcal{M}} = \{u^j\}_{j=1,\ldots,N} \equiv \{u(x_j)\}_{j=1,\ldots,N} \in \Re^N$: numerical solution of $\mathbf{M}$ on $D_N(\Omega)$;
- $\mathbf{u}^{true} = \{u(x_j)^{true}\}_{j=1,\ldots,N}$: the vector values of the reference solution of $\mathcal{M}$ computed on $D_N(\Omega)$ at t fixed;

- $\mathbf{v} = \{v(y_j)\}_{j=1,\ldots,nobs}$: the vector values of the observations on $D_N(\Omega)$;
- $\mathcal{H}(x) \simeq \mathcal{H}(z) + \mathbf{H}(x - z)$: where $\mathbf{H} \in \Re^{N \times nobs}$ is the matrix obtained by the first order approximation of the Jacobian of $\mathcal{H}$ and $nobs \ll N$;
- $\mathbf{R}$ and $\mathbf{B}$ the covariance matrices of the errors on the observations $\mathbf{v}$ and on the system state $\mathbf{u}^M$, respectively. These matrices are symmetric and positive definite (see [6] for details).

We assume that $N = n_x \times x_y \times n_z$ and $n_x = n_y = n$ while $n_z = 3$. Since the unknown vectors are the fluid height or depth, and the two-dimensional fluid velocity fields, the problem size is $N = 3n^2$. $\mathbf{H}$ is assumed to be a piecewise linear interpolation operator whose coefficients are computed using the points of model domain nearest the observation values. We assume $\mathbf{u}^{true}$ be the solution of $\mathcal{M}$ as given in [1]. Observation values are randomly chosen among the values of $\mathbf{u}^{true}$.

Definition 7 (The DA Inverse problem). *Let $\mathbf{u}^{DA}$ be the solution of:*

$$\mathbf{v} = \mathbf{H}[\mathbf{u}^{DA}]$$

subject to the constraint:

$$\mathbf{u}^{DA} = \mathbf{u}^M.$$ ♠

DA is an ill posed inverse problem [14]. The local DD-TR operator, defined on a subdomain Ω_i, is (see Eq. (5), with $\lambda_i = \omega_i = 1$)):

$$J_i(\mathbf{u}^i) = (\mathbf{H}_i \mathbf{u}^i - \mathbf{v}^i)^T \mathbf{R}_i (\mathbf{H}_i \mathbf{u}^i - \mathbf{v}^i) + (\mathbf{u}^i - (\mathbf{u}^M)^i)^T \mathbf{B}_i (\mathbf{u}^i - (\mathbf{u}^M)^i)$$
$$+ (\mathbf{u}^i - \mathbf{u}^j)^T \mathbf{B}_{ij} (\mathbf{u}^i - \mathbf{u}^i). \tag{7}$$

In [3,7] the authors provided the reliability of DD-TR model for DA problem. In this paper we present results of an implementation of the model on two different computing architectures. We evaluate the efficiency of these implementations by analysing the strong and weak scalability of the algorithm by using the scale up factor defined in Sect. 2.1.

4   The DD-TR Algorithm on Two Reference Computing Architectures

In this paper, our testbed is a distributed computing environment composed of computational resources, located in the University of Naples Federico II campus, connected by local-area network. More precisely, the testbed is made of:

- A1: a 288 CPU-multicore architecture made of distributed memory blades each one with computing elements sharing the same local memory for a total of 3456 cores.
- A2: a GPU+CPU architecture made of the 512 threads NVIDIA Tesla connected to a quad-core CPU.

If *nproc* denotes the number of processing elements of the reference architectures, we have $nproc = 64$ for A1, and $nproc = \#$ threads-blocks, for A2. We assume a 2D uniform decomposition of $D_N(\Omega)$ along the (x, y)-axis, that is the x-axis is divided by s and the y-axis by q then, the size of each subdomain $D_N(\Omega_i)$ is $r = nloc_x \times nloc_y \times nloc_z$ where:

$$nloc_x = \frac{n_x}{s} + 2o_x \ , \ nloc_y = \frac{n_y}{q} + 2o_y \ , \ nloc_z = n_z. \tag{8}$$

These dimensions include the overlapping $(2o_x \times 2o_y)$.

We use the LBFGS method for computing the minimum of DD-TR functionals [11,17]. Then, following result specifies the scale up factor of algorithm $\mathcal{A}(D_N(\Omega), p)$ in our case study [6]:

Proposition 1. *If the time complexity of $\mathcal{A}(D_N(\Omega), 1)$ is $T(N) = O(f(N))$ flops, on a problem of size N, where $f(N) \in \Pi_3$, the scale up factor of the algorithm $\mathcal{A}(D_N(\Omega), p)$ is*

$$S_{p,1}(N) = \alpha(r, p)\, p^2. \tag{9}$$

Remark: Let t_{flop} denote the unitary time required by one floating point operation. As a result, the execution time needed to algorithm $\mathcal{A}(D_N(N), 1)$ for performing $T(N)$ floating point operations, is

$$T_{flop}(N) = T(N) \times t_{flop}.$$

Multiplying and dividing the (9) by t_{flop} we get

$$\alpha(r, p)p^2 = \frac{T_{flop}(N)}{p T_{flop}(N/p)}. \tag{10}$$

Finally, we give the following

Definition 8. *Let $\frac{S}{V} := \frac{T_{oh}(N/p)}{T_{flop}(N/p)}$ denote the surface-to-volume ratio. It is a measure of the amount of data exchange (proportional to surface area of domain) per unit operation (proportional to volume of domain).* ♠

In [12] authors define $T^{nproc}(N)$, the execution time of $\mathcal{A}(N, p)$, as given by time for computation plus an overhead which is given by synchronization, memory accesses and communication time also.

$$T^{nproc}(N) := T_{flop}^{nproc}(N) + T_{oh}^{nproc}(N)$$

where

- A1: $T_{flop}^{nproc}(N)$ is computing time required for the execution of $T(N)$ floating point operations; $T_{oh}^{nproc}(N)$ is overhead time of $T(N)$ data which includes communications among CPU processors.
- A2: $T_{flop}^{nproc}(N) := T^{CPU}(N) + T^{GPU}(N)$, where
  - $T^{CPU}(N)$ is the CPU execution time for the execution of $T(N)$ floating point operations,
  - $T^{GPU}(N)$ is the GPU execution time for the execution of $T(N)$ floating point operations.

and $T_{oh}^{nproc}(N)$ includes the communications time between host (CPU) and device (GPU) and time for memories accesses.

Here we assume that

$$T^{GPU}(N) := T_{flop}^{GPU}(N) + T_{mem}^{GPU}(N), \tag{11}$$

where $T_{mem}^{GPU}(N)$ is the time for global and local memories transfers into the device (GPU) and $T_{flop}^{GPU}(N)$ is the computing time required for execution of floating point operations.

Finally, for A2, $T_{oh}^{nproc}(N) \equiv T_{mem}^{GPU}(N)$, since the communications between host (CPU) and device (GPU) in the algorithm we implement occur just at the begin and the end for I/O transfers and, for this reason, it can be neglected in our considerations.

4.1   Discussion

Table 1 shows results obtained for $\mathcal{A}(\Omega, p)$ on A1 for a problem size $O(10^6)$ and $O(10^7)$ by using $nproc = p$ and Table 2 shows execution time of the algorithm $\mathcal{A}(\Omega, p)$ running on A2 for a problem size $O(10^7)$ by using # thread-blocks= $2p$.

In Table 2, $T^{CPU}(N)$ is execution time that CPU needs for building data. These data are transferred just once as well as output data so we have that $T_{oh}^{GPU}(N)$ is reduced to the time of I/O transfer. For this reasons we evaluate the performance of DD-TR implementation on GPU by analysing $T^{GPU}(N)$. $T_{oh}(N)$ can be estimates by dividing D_N, which is size of processed data espressed in GB by the bandwidth value B_W which is the rate of data transfer espressed in GB/seconds: $T_{oh}(N) := \frac{D_N}{B_W}$ secs .

We have $D_N = 3.7\,\text{GB}$ which gives $T_{oh} \simeq 3.7/208\,\text{s} \simeq 0.017\,\text{s}$. Our considerations will focus on values of $T_{flop}^{GPU}(N)$ reported in Table 3. We now discuss the software scalability as shown in Tables 1 and 3. To this end, we introduce

$$s_{nproc}^{loc} := \frac{T_{flop}(N/p)}{T_{nproc}(N/p)}, \tag{12}$$

which denotes the speed up of the (local) algorithm $\mathcal{A}(D_N(\Omega_i), N/p)$ for solving the local problem on subdomain $D_N(\Omega_i)$. Let us express the measured scale up

Table 1. Results on A1: Execution time and *scale up* factor of $\mathcal{A}(\Omega,p)$ for different values of $N = 3n^2$ and $nproc = 2p$.

n	$nproc$	$T^{nproc}(N)$	Measured $S_{nproc,8}$	$S_{nproc,8}$
$O(10^6)$	8	2.0545e+02	1.0	1
	16	6.3316e+01	3.25	4
	32	2.0005e+01	10.27	16
	64	8.7835e+00	23.39	64
n	$nproc$	$T^{nproc}(N)$	Measured $S_{nproc,16}$	$S_{nproc,16}$
$O(10^7)$	8	–	–	–
	16	3.9091e+03	1.0	1
	32	9.9952e+02	3.91	4
	64	2.7584e+02	14.17	16

Table 2. Execution time of algorithm $\mathcal{A}(\Omega,p)$ running on A2 for a problem size $O(10^7)$ and $nproc = \#thread - blocks$.

N	p	$nproc$	$T^{GPU}(N)$
$O(10^7)$	1	2	0.144
	2	4	0.044
	4	8	0.025
	8	16	0.024

factor in terms of s_{nproc}^{loc}. We have:

$$S_{1,nproc}^{measured} := \frac{T_{flop}(N)}{p \cdot (T_{flop}(r_i) + T_{oh}(N/p))}. \tag{13}$$

From the (12) and the (13) it follows that

$$S_{1,nproc}^{measured} = \frac{T_{flop}(N)}{\frac{pT_{flop}(N/p)}{s_{nproc}^{loc}} + pT_{oh}(N/p)} = \frac{s_{nproc}^{loc}\frac{T_{flop}(N)}{pT_{flop}(N/p)}}{1 + \frac{s_{nproc}^{loc}T_{oh}(N/p)}{T_{flop}(N/p)}}. \tag{14}$$

Table 3. Results on A2: Values of T_{flop}^{GPU} and measured *scale up* factor compared with theoretical once.

N	p	$T_{flop}^{GPU}(N)$	Measured $S_{nproc,2}$	$S_{nproc,2}$
$O(10^7)$	1	0.127	-	-
	2	0.027	4.7	4
	4	0.008	15.9	8
	8	0.007	18.1	16

As we need to guarantee that the so-called *surface-to-volume* effect on each local DA problem is produced [2, 4, 9, 10], we assume:

$$0 \le \frac{S}{V} < 1 - \frac{1}{s_{nproc}^{loc}} < 1.$$

Let

$$\alpha := \frac{s_{nproc}^{loc}}{1 + \frac{s_{nproc}^{loc} T_{oh}(N/p)}{T_{flop}(N/p)}} = \frac{s_{nproc}^{loc}}{1 + s_{nproc}^{loc} \frac{S}{V}},$$

from (14) it comes out that

$$S_{1,nproc}^{measured} = \alpha S_{1,nproc}.$$

Finally, it holds that

(i) if $s_{nproc}^{loc} = 1$ then

$$\alpha < 1 \Leftrightarrow S_{nproc,1}^{measured} < S_{nproc,1};$$

(ii) if $s_{nproc}^{loc} > 1$ then

$$\alpha > 1 \Leftrightarrow S_{nproc,1}^{measured} > S_{nproc,1};$$

(iii) if $s_{nproc}^{loc} = p$ then

$$1 < \alpha < p \Rightarrow S_{nproc,1}^{measured} < p S_{nproc,1};$$

Hence, we may conclude that if

$$s_{nproc}^{loc} \in]1, p] \Rightarrow S_{nproc}^{measured} \in]S_{nproc,1}, p\, S_{nproc,1}[.$$

It is worth noting that in our experiments, in A1, local DA problems are sequentially solved, then

$$s_{nproc}^{loc} = 1$$

while in A2, local DA problems have been concurrently solved on the GPU device, so

$$s_{nproc}^{loc} > 1$$

Thus the above analysis validates the experimental results both in terms of strong and weak scaling.

Acknowledgments. This work has been realised thanks to the use of the SCoPE computing infrastructure at the University of Naples, also in the framework of PON *"Rete di Calcolo per SuperB e le altre applicazioni"* (ReCaS) project. This work was developed within the research activity of the H2020-MSCA-RISE-2016 NASDAC Project N. 691184.

References

1. Moler, C.: Experiments with MATLAB (2011)
2. Antonelli, L., Carracciuolo, L., Ceccarelli, M., D'Amore, L., Murli, A.: Total variation regularization for edge preserving 3D SPECT imaging in high performance computing environments. In: Sloot, P.M.A., Tan, C.J.K., Dongarra, J., Hoekstra, A.G. (eds.) ICCS-ComputSci 2002, Part II. LNCS, vol. 2330, p. 171. Springer, Heidelberg (2002)
3. Arcucci, R., D'Amore, L., Carracciuolo, L.: On the problem-decomposition of scalable 4D-Var data assimilation models. In: International Conference on High Performance Computing & Simulation (HPCS 2015), pp. 589–594 (2015)
4. Carracciuolo, L., D'Amore, L., Murli, A.: Towards a parallel component for imaging in PETSc programming environment: a case study in 3-D echocardiography. Parallel Comp. **32**(1), 67–83 (2006)
5. D'Amore, L., Murli, A.: Regularization of a Fourier series method for the Laplace transform inversion with real data. Inverse Prob. **18**(4), 1185–1205 (2002)
6. D'Amore, L., Arcucci, R., Carracciuolo, L., Murli, A.: A scalable approach to variational data assimilation. J. Sci. Comput. **61**, 239–257 (2014)
7. D'Amore, L., Arcucci, R., Carracciuolo, L., Murli, A.: DD-OceanVar: a domain decomposition fully parallel data assimilation software in mediterranean sea -. Procedia Computer Science **18**, 1235–1244 (2013)
8. D'Amore, L., Arcucci, R., Marcellino, L., Murli, A.: A parallel three-dimensional variational data assimilation scheme. AIP Conf. Proc. **1389**(1), 1829–1831 (2011)
9. D'Amore, L., Arcucci, R., Marcellino, L., Murli, A.: HPC computation issues of the incremental 3D variational data assimilation scheme in OceanVar software. J. Numer. Anal. Ind. Appl. Math. **7**(3–4), 91–105 (2012)
10. D'Amore, L., Casaburi, D., Galletti, A., Marcellino, L., Murli, A.: Integration of emerging computer technologies for an efficient image sequences analysis. Integr. Comput. Aided Eng. **18**(4), 365–378 (2011)
11. D'Amore, L., Laccetti, G., Romano, D., Scotti, G., Murli, A.: Towards a parallel component in a GPU–CUDA environment: a case study with the L-BFGS Harwell routine. Int. J. Comput. Math. **92**(1), 59–76 (2015)
12. Flatt, H.P., Kennedy, K.: Performance of parallel processors. Parallel Comput. **12**, 1–20 (1989)
13. Haben, S.A., Lawless, A.S., Nichols, N.K.: Conditioning of the 3DVAR Data Assimilation Problem, Mathematics Report 3/2009. University of Reading, Department of Mathematics (2009)
14. Kalnay, E.: Atmospheric Modeling, Data Assimilation and Predictability. Cambridge University Press, Cambridge (2003)
15. Murli, A., Cuomo, S., D'Amore, L., Galletti, A.: Numerical regularization of a real inversion formula based on the Laplace transform's eigenfunction expansion of the inverse function. Inverse Prob. **23**(2), 713–731 (2007)
16. Tikhonov, A.N., Arsenin, V.Y.: Solutions of Ill-Posed Problems. Wiley, New York (1977)
17. Zhu, C., Byrd, R.H., Lu, P., Nocedal, J.: Algorithm 778: L-BFGS-B: Fortran subroutines for large-scale bound constrained optimization. ACM TOMS **23**(4), 550–560 (1997)

Workshop on Power and Energy Aspects of Computation

Energy Performance Modeling
with TIA and EML

Francisco Almeida[1](✉), Javier Arteaga[1], Vicente Blanco[1],
and Alberto Cabrera[2]

[1] High Performance Computing Group, Universidad de La Laguna, La Laguna, Spain
`{falmeida,jarteagr,Vicente.Blanco}@ull.es`
[2] Instituto Tecnológico y de Energías Renovables, Granadilla, Spain
`acabrera@iter.es`
`http://cap.pcg.ull.es/`

Abstract. Current parallel performance analysis tools are typically based on either measurement or modeling techniques, with little integration between both approaches. Researchers developing predictive models have to build their own validation experiments. Conversely, most application profiling tools do not produce output that can be readily used to generate automatically approximated models.

The Tools for Instrumentation and Analysis (TIA) framework was originally designed to bridge the gap between analytical complexity modeling and performance profiling tools. Through loosely coupled, but well integrated components, TIA provides both profiling on a specified set of metrics for source-annotated regions of parallel code, and analysis facilities to help find and validate an appropriate performance model.

This methodology can also be applied to power performance. In this work, we enhance TIA with energy measurement capabilities through our Energy Measurement Library (EML). We test the augmented framework by performing power performance profiling and analysis tasks for a simple computation.

Keywords: Analytical modeling · Performance analysis · Energy modeling · Power measurement

1 Introduction

Energy performance is currently a key factor in the design of HPC systems and software, with the move towards ultrascale infrastructure requiring significant advancements in computation energy efficiency. Thus, the attention of many researchers of HPC software has turned towards energy performance analysis.

To guide energy consumption optimization decisions, analysis tools and techniques that are both powerful and simple to use are essential. These tools typically fall under one of two approaches: performance measurement and profiling of actual application code, and analytical modeling aiming to predict real performance in many systems.

© Springer International Publishing Switzerland 2016
R. Wyrzykowski et al. (Eds.): PPAM 2015, Part II, LNCS 9574, pp. 57–65, 2016.
DOI: 10.1007/978-3-319-32152-3_6

There is a clear relationship between the modeling and profiling processes: theoretical models need experimental data to validate them, and experimental data often offers insight on an appropriate analytical model. However, many analysis tools strictly focus on profiling and visualization. While this specialization is not without merits in itself, lack of integration with modeling tools can also lead to significant effort for researchers in building their own model validation or model parameter fitting experiments.

Some attempts exist in the direction of an integrated profiling-modeling analysis solution, typically focusing on performance modeling. In this paper, we build on top of the Tools for Instrumentation and Analysis (TIA) framework, extending it to add energy measurement capabilities through our Energy Measurement Libray (EML) component.

The rest of this paper is structured as follows: Sect. 2 gives a brief overview of prior work related to software energy measurement and energy performance analysis. Section 3 explains both the TIA model of analysis and tooling, describing how energy measurement has been integrated through EML. In Sect. 4, we describe the instrumentation of a simple matrix multiplication experiment. The obtained results are used to showcase energy performance capabilities and show the analysis stage in Sect. 5. Lastly, Sect. 6 presents our conclusions and planned future work.

2   Background

Numerous system and application performance models have been developed in literature. Of these, many recent studies deal with energy consumption modeling in computation systems [8,17]. Some also take into account algorithmic energy performance properties [6,13]. In most cases reviewed, validation experimentation or model generation processes are not automatically performed.

The rapidly evolving landscape of energy measurement tools is making the lack of enery-aware profiling tools less of a likely cause. By now, most of the major parallel performance toolsets (such as TAU [15], Paraver [12] or Vampir [10]) have been updated to expose hardware-counter based power measurements (often collected through PAPI [3]). Smaller and more narrowly focused energy-specific APIs, such as PowerAPI [14], pmlib [1] or our Energy Measurement Library (EML) [4], have also been publicly released.

Explicit profiling tool support for the modeling process of parallel code is not as widespread in HPC, although prior work exists in this area. The Performance Analysis and Characterization Environment (PACE) [11] uses a hierarchy of model objects representing applications, subtasks, parallel communication patterns and hardware. The Prophesy system [16] allows for whole-application model development based the idea of composing kernel performance models through a coupling parameter characterizing their interaction.

Another approach can be found in the Dimemas simulator, capable of building simple application behavior and platform component models from Paraver event traces [12]. A recent study [5] defines a method, now integrated in the

Scalasca framework, for building rough performance models (forgoing model accuracy to increase coverage of critical scalability bottleneck testing) from profiling data obtained from the Score-P [7] instrumentation framework.

For this experience, we use a modeling-oriented analysis environment, the Tools for Instrumentation and Analysis (TIA) framework. This choice is driven by both its modular design and our familiarity with the TIA codebase, as we intend to extend the framework using EML to achieve energy modeling capabilities.

3   The TIA Framework

The TIA framework was conceived as a hybrid approach between theoretical complexity analysis and performance profiling techniques [9]. It is based on the previously published CALL system [2], from which TIA inherits its instrumentation system and modular tool structure.

3.1   Model for Performance Analysis

In the original CALL system, analysis started with the researcher deriving theoretical complexity formulas, in terms of both architecture-dependent and architecture-independent parameters. These are embedded in the application source code as CALL annotations (C preprocessor `#pragma cll` directives) for each segment of code to be analyzed (called an individual *experiment* in this model).

TIA implements a similar scheme, while decoupling model development from the annotation and profiling stage. It is no longer required for the user to commit to a complexity formula at annotation time.

The TIA framework is mainly comprised of a source-to-source translator for annotated C programs, closely related to the original `call`, and a R static analysis package (Fig. 1).

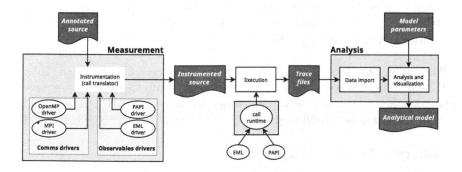

Fig. 1. The TIA model

This annotated source is converted to instrumented code by the translator. When compiled and run, the resulting application then produces trace files for each compute node containing data for each annotated experiment.

3.2 The `call` Translator

The `call` translator takes C source with annotations and replaces them in the emitted program with instrumentation code, including both the framework's own runtime support library (`cll.h`) and any required measurement libraries (such as PAPI).

This translator is extensible through user modules called drivers, of which there are two types: communications and observables drivers.

Communications drivers instruct `call` with knowledge of the underlying communication APIs for message-passing parallel applications. An instance of this would be automatic barrier synchronization of experiments in MPI-based applications.

Observables drivers can interface with both platform native APIs and portable APIs (such as PAPI) to equip `call` with measurement capabilities from different sources. Some of them, like the PAPI driver or an UNIX stdlib-based time driver (UNISTD) come already bundled with the framework.

3.3 The `cll` Analysis Package

The `cll` analysis package for the R environment, developed for TIA as a replacement to CALL's prior `llac` R package [9], includes import functions for the `call` output format, as well as processing and filtering utilities for the experimental data.

Its main feature is the ability to either fit the experimental data to a user-provided model, or try to automatically generate a suitable model as a linear combination of user-provided metrics and parameters.

3.4 Energy Performance Analysis

The proposed model is general enough to apply to energy performance analysis. Additionally, the modular design of the tool allows to extend its profiling capabilities to include energy measurement through an appropriate observables driver.

We implement this driver based on our Energy Measurement Library (EML) [4]. The library has been previously presented as an open-source C abstraction layer over any software energy measurement API. It is designed to be itself extensible to add device support through drivers.

4 Energy Analysis Experiment

In this section, we present a simple experiment as a usage example to illustrate the TIA energy performance analysis process. Our target code is a naive matrix multiplication implementation.

4.1 Experimental Setup

The test is run in a machine with an Intel Xeon E5-2660 processor, which exposes the Intel Running Average Power Limit interface to hardware counter CPU energy measurements. This is one of the metrics currently supported by EML.

4.2 Instrumentation

We first annotate the code with TIA directives:

```
#pragma cll for(N=MIN; N<=MAX; N+=STRIDE)
/* TIA experiment: initialization */
#pragma cll init CLOCK, \
    EML_ENERGY_RAPL = init[0]*N*N
    for(i=0;i<N;i++)
      for(j=0;j<N;j++)
        A(i,j) = B(i,j) = (i == j);
#pragma cll end init

/* TIA experiment: multiplication */
#pragma cll mat CLOCK, \
    EML_ENERGY_RAPL = mat[0]*N*N*N
    for(i = 0; i < N; i++) {
      for(j = 0; j < N; j++) {
        sum = 0;
        for(k = 0; k < N; k++)
          sum += A(i, k) * B(k, j);
        C(i,j) = sum;
      }
    }
#pragma cll end mat
#pragma cll end for
```

Both initialization and multiplication are enclosed in #pragma cll directives delimiting the beginning and start of an experiment, and giving it a name (init, mat). A list is then given indicating desired metrics to be included: in this case, both the standard CLOCK metric (system time) and the EML_ENERGY_RAPL metric (total energy consumption seen by RAPL) provided by the EML driver are recorded.

Finally, the user can provide a complexity formula to be used for the TIA automatic model generation feature, which attempts to find the linear combination of terms that best approximate experiment data gathered for the desired metrics according to a quality criterion. Presently, constants are denoted by the experiment name plus array subscript notation (init[0]), while other terms denote variables. Non-linear terms can be expressed here: in the above example, we are suggesting N^2 (N*N) as a term for the init model, and N^3 (N*N*N) for mat.

Note that it is possible to use measured metrics as part of these formulas by simply representing them as symbolic terms and performing further fitting from the R analysis environment, although this is not yet an automatic process. Later on, this would be further simplified by introducing support to refer to metrics directly in the formula for automatic analysis.

5   Results

Averaging 10 runs of the instrumented code with different matrix sizes, we obtain a trace file with the following energy consumption data (Table 1):

Table 1. RAPL energy consumption data

Size	Energy [J]	
	init	mat
500	0.19	40.02
750	0.23	173.94
1000	0.31	419.80
1250	0.48	954.19
1500	0.66	1777.57
1750	0.89	3290.00
2000	1.14	4651.50

We can now import the trace file in R through the `cll` package. At this stage, `cll` provides a number of data preprocessing functions such as filtering and outlier elimination or joining of multiple trace files.

The main utility is `cll.fit` function. It receives trace data and a list of terms and searches the space of models built as a linear combination of these terms for the best fit. For example, this would be a full search for the `mat` experiment for all models built from the terms N, N^2 (some output has been omitted for brevity):

```
> cll.fit(matdata, type="full",
  var.list=list("N", "N*N"))

[...]

Coefficients:
                Estimate Std. Error t value
(Intercept)    1.753e-01  9.206e-03   19.04
I(N * N * N)   1.338e-10  6.322e-12   21.17

Term Weights:
      weight          terms
1 0.9999998 (Intercept)
2 0.8999677       N*N*N
```

```
3 0.1000321              N*N
4 0.0000000               N

Best Top Models:
     weight                         model
1 0.8999677  EML_ENERGY_RAPL ~ I(N * N * N)
2 0.1000321     EML_ENERGY_RAPL ~ I(N * N)
```

The compute stage has cubic complexity on the problem size, as expected. For more complex programs, the `cll.fit` function can be instructed to consider more factors and other metrics as terms.

Lastly, the package includes functions to produce basic graphical visualizations of predictions and model fit graphs (Fig. 2). More sophisticated representations can take advantage of the plotting functionality in the R environment.

Matrix multiplication fit to model

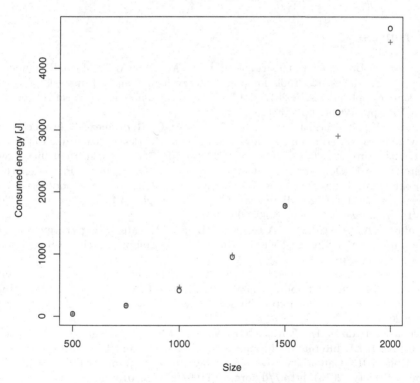

Fig. 2. Generated model (red) from measurements (black) (Color figure online)

6   Conclusions

We have shown that the modular design of both the TIA framework and EML makes them suitable for composition as part of larger analysis frameworks. Additionally, we have presented a simple example application in which the framework can assist in finding a theoretical model, improving experiment platform independence through observables drivers and providing basic visualization facilities.

Future work will be directed towards a production-ready modular analysis framework. This involves streamlined installation and usage processes, improved visualization and statistical capabilities, and a user interface for the analysis package. The EML component is also expected to undergo further API development and improve device support in coming releases.

Acknowledgements. This work was supported by the Spanish Ministry of Education and Science through the TIN2011-24598 project, Spanish CAPAP-H4 network, and NESUS IC1315 COST Action.

References

1. Alonso, P., Badia, R.M., Labarta, J., Barreda, M., Dolz, M.F., Mayo, R., Quintana-Orti, E.S., Reyes, R.: Tools for power-energy modelling and analysis of parallel scientific applications. In: 2012 41st International Conference on Parallel Processing (ICPP), pp. 420–429. IEEE (2012)
2. Blanco, V., González, J.A., León, C., Rodríguez, C., Rodríguez, G.: From complexity analysis to performance analysis. In: Kosch, H., Böszörményi, L., Hellwagner, H. (eds.) Euro-Par 2003. LNCS, vol. 2790, pp. 704–711. Springer, Heidelberg (2003)
3. Browne, S., Dongarra, J., Garner, N., Ho, G., Mucci, P.: A portable programming interface for performance evaluation on modern processors. Int. J. High Perform. Comput. Appl. **14**(3), 189–204 (2000). http://dx.org/10.1177/109434200001400303
4. Cabrera, A., Almeida, F., Arteaga, J., Blanco, V.: Measuring energy consumption using EML (Energy Measurement Library). Computer Science-Research and Development, pp. 1–9 (2014)
5. Calotoiu, A., Hoefler, T., Poke, M., Wolf, F.: Using automated performance modeling to find scalability bugs in complex codes. In: Proceedings of the ACM/IEEE Conference on Supercomputing (SC13), Denver, CO, USA. pp. 1–12. ACM, November 2013
6. Choi, J., Bedard, D., Fowler, R.J., Vuduc, R.W.: A roofline model of energy. In: 27th IEEE International Symposium on Parallel and Distributed Processing, IPDpPS 2013, Cambridge, MA, USA, May 20–24, 2013. pp. 661–672. IEEE Computer Society (2013). http://dx.org/10.1109/IPDPS.2013.77
7. Knüpfer, A., Rössel, C., an Mey, D., Biersdorff, S., Diethelm, K., Eschweiler, D., Geimer, M., Gerndt, M., Lorenz, D., Malony, A., et al.: Score-P: A joint performance measurement run-time infrastructure for Periscope, Scalasca, TAU, and Vampir. In: Tools for High Performance Computing 2011, pp. 79–91. Springer (2012)

8. Lively, C., Wu, X., Taylor, V., Moore, S., Chang, H.C., Cameron, K.: Energy and performance characteristics of different parallel implementations of scientific applications on multicore systems. Int. J. High Perform. Comput. Appl. **25**(3), 342–350 (2011)

9. Martínez, D.R., Cabaleiro, J.C., Pena, T.F., Rivera, F.F., Pérez, V.B.: Accurate analytical performance model of communications in MPI applications. In: 23rd IEEE International Symposium on Parallel and Distributed Processing, IPDpPS 2009, Rome, Italy, May 23–29, 2009. pp. 1–8. IEEE (2009). http://dx.org/10.1109/IPDPS.2009.5161175

10. Nagel, W.E., Arnold, A., Weber, M., Hoppe, H.C., Solchenbach, K.: VAMPIR: Visualization and analysis of MPI resources. Supercomputer **12**, 69–80 (1996)

11. Nudd, G.R., Kerbyson, D.J., Papaefstathiou, E., Perry, S.C., Harper, J.S., Wilcox, D.V.: PACE a toolset for the performance prediction of parallel and distributed systems. Int. J. High Perform. Comput. Appl. **14**(3), 228–251 (2000)

12. Pillet, V., Labarta, J., Cortes, T., Girona, S.: Paraver: A tool to visualize and analyze parallel code. In: Proceedings of WoTUG-18: Transputer and Occam Developments. vol. 44, pp. 17–31, March 1995

13. Roy, S., Rudra, A., Verma, A.: An energy complexity model for algorithms. In: Proceedings of the 4th Conference on Innovations in Theoretical Computer Science. pp. 283–304. ITCS 2013, NY, USA (2013). http://doi.acm.org/10.1145/2422436.2422470

14. Sandia National Laboratories: High performance computing power application programming interface (API) specification (2014). http://powerapi.sandia.gov/

15. Shende, S.S., Malony, A.D.: The TAU parallel performance system. Int. J. High Perform. Comput. Appl. **20**(2), 287–311 (2006)

16. Taylor, V., Wu, X., Stevens, R.: Prophesy: An infrastructure for performance analysis and modeling of parallel and grid applications. ACM SIGMETRICS Perform. Eval. Rev. **30**(4), 13–18 (2003)

17. Wang, H., Chen, Q.: Power estimating model and analysis of general programming on GPU. J. Softw. **7**(5), 1164–1170 (2012)

Considerations of Computational Efficiency
in Volunteer and Cluster Computing

Paweł Czarnul$^{(\boxtimes)}$ and Mariusz Matuszek

Faculty of Electronics, Telecommunications and Informatics,
Gdansk University of Technology, Gdansk, Poland
{pczarnul,mrm}@eti.pg.gda.pl

Abstract. In the paper we focus on analysis of performance and power consumption statistics for two modern environments used for computing – volunteer and cluster based systems. The former integrate computational power donated by volunteers from their own locations, often towards social oriented or targeted initiatives, be it of medical, mathematical or space nature. The latter is meant for high performance computing and is typically installed in a dedicated computing centre. While volunteer systems allow to obtain high computing power, they are not meant for dense computations and do not feature state-of-the-art hardware. Clusters offer best of the best at the cost of high purchase and maintenance cost. In the paper we give computational efficiency statistics for Atlas@Home, Asteroids@Home and BOINC cross-project and compare these to clusters such as Cray XC30, SuperMUC and TRYTON.

Keywords: Performance · Power consumption · Volunteer computing · Cluster computing

1 Introduction

Today, high performance computing can be performed within powerful workstations and servers thanks to multicore CPUs and accelerators such as GPUs or Intel Xeon Phi coprocessors. Reaching towards exaflop performance requires integration of such nodes into systems such as clusters[1] or volunteer based systems. These have reached around 33 (Tianhe-2[2]) and 8 PetaFlop/s (BOINC[3]) performance. However, this comes at a cost of considerable power consumption, either in an HPC center (over 17 MW for Tianhe-2) or in individual volunteers' homes. Measurement and consideration of energy consumption is important both from the cost point of view and negative side effects on the environment. Future clusters are suggested to be developed with an upper bound on power consumption of 20 MW maximum [1].

---

[1] http://www.top500.org.

[2] http://top500.org/system/177999.

[3] http://boinc.berkeley.edu/.

© Springer International Publishing Switzerland 2016
R. Wyrzykowski et al. (Eds.): PPAM 2015, Part II, LNCS 9574, pp. 66–74, 2016.
DOI: 10.1007/978-3-319-32152-3_7

In this paper we focus on volunteer computing and comparison to cluster computing in the aforementioned respects. In volunteer computing energy consumption might be important from the following points of view:

1. individual – which CPUs offer better performance/power ratio?
2. global – possibly in order to provide guidance to individuals willing to donate computing power to specific projects – in this case several performance metrics of a single CPU or GPU need to be considered for various classes of computational code.

2   Related Work

Energy consumption is becoming a concern and is addressed in both cluster and volunteer based systems. For instance in paper [2] the authors proposed theoretical and practical solutions to data partitioning and scheduling in environments that consist of many potentially distributed clusters, each of which may consist of multicore CPUs and GPUs. Various optimizers may be defined, with an example given for minimization of application execution time with an upper bound on the total power consumption of the compute devices used for application execution. In the context of volunteer computing, paper [3] suggests that focus is mostly given towards project with real practical impact, partly because energy consumption has become an important factor.

There exist several practical solutions and systems that perform distributed computations in the volunteer fashion:

1. BOINC[4] [4] is probably the most recognized system of this type of all. In BOINC clients run a dedicated client that fetches parts of work from a central server, sends results back and repeats the process. The system can send a packet to more than one volunteer in order to increase dependability of computations at the cost of processing power. Paper [5] describes scalability of the distribution mechanisms found in BOINC that allow processing of over 23 million tasks per day.
2. Comcute [6,7] can be thought of as an extension of the ideas deployed in BOINC, with two major design and implementation differences:
   - Computations are performed within a web browser on the client side, as opposed to a dedicated client application. This relieves the user from the need for installation and related safety concerns. However, the associated cost is lower performance compared to native code. In Comcute, a special mechanism was implemented for selection of the best performing technology supported on the client side (either JavaScript, Java, Flash etc.).
   - management of computations on the server side is also distributed and is performed by many servers elected for this purpose. This increases reliability of processing for a project that can survive failures of some management servers.

---

[4] http://boinc.berkeley.edu/.

3. WeevilScout prototype framework [8] implements the concept of processing within a browser but is limited to the JavaScript technology.
4. CrowdCL [9] is an open source framework for volunteer computing that enables to perform computations within a webpage. CrowdCL provides both the server and client codes. On the client side, JavaScript, CrowdCLient, KernelContext, WebCL and OpenCL are used to run computations possibly on CPUs or other compute devices.
5. In [10] the authors proposed an architecture and a solution in which the traditional volunteer computing is extended with mobile devices. The latter have become both powerful with multicore processors and energy efficient. The solution extends the traditional distributed computing model with decentralized distribution points. Roles such as task distribution point, task execution point and task distribution and execution point can be assigned to clients. The authors presented both running times and energy consumption of individual clients for an experimental distributed architecture for processing a prediction of a protein structure.

Several works have addressed modeling of computations in volunteer based systems, often adopting probabilistic models. Paper [11] contains models for phases in volunteer processing with distinction of the following stages: distribution, in progress (computations) and validation. Distributions such as Weibull, mixture of Gaussian distributions are suggested. Paper [12] describes computers of Internet users including core count and memory sizes. In paper [13] we presented modeling of computational effort (in credits) in a volunteer system required for processing of a task that requires processing of a certain number of data packets with known computational requirements on a reliable machine. In volunteer computing, the client may be available for a limited time only and if processing of a data packet has not been finished within a time frame, it would need to be repeated by another volunteer. The paper assesses actual overheads of a real volunteer environment based on real BOINC statistics. Energy consumption was also considered in selected works, from various perspectives. In paper [14] the authors propose usage of aggressive volunteer computing in an environment with multi-core computers. As the number of cores is constantly increasing, the authors suggest that aggressive volunteer computing in which computations are assigned to computers already active doing native jobs can bring considerable savings in energy consumption. According to their findings, on average this strategy saved around 50 % energy compared to clusters and 33 % compared to the traditional volunteer approach. In paper [15] the authors analyze a slightly different aspect of volunteer computing related to energy consumption. Namely, if control over volunteers' computers is regained by their owners or users it may result in wasted energy for volunteer jobs. The authors prested an approach using reinforced learning for determination of computers on which jobs should be deployed and demonstrated 30 % to 50 % of energy savings by doing that. In paper [16] the author evaluates impact of CPU throttling on energy consumption when performing jobs in a volunteer system. Specifically, the author compares a dedicated machine with full load, a system in which BOINC restricted CPU load to 20 % and a scenario with fine-grained CPU throttling in which the application goes to sleep after predefined, relatively short intervals. According to experiments this gave savings in energy consumption by around 40 %.

3   Estimating CPU Power Consumption of Volunteer Nodes

To obtain the average expected CPU power consumption of a volunteer node we performed statistical analysis based on data made available by BOINCstats[5]. For analysis we choose two popular Physics and Astrophysics projects Atlas@Home and Asteroids@Home. The Atlas@Home project was started in 2014, while Asteroids@Home begun in 2012. These projects are relatively new. For additional comparison we analyse data from the Boinc combined set, which lists information collected from all projects since the beginning of the BOINC initiative. These lists give us reliable information regarding particular CPU models used in these projects and multiplicity of each CPU model used in a particular project. Furthermore, in order to make our comparison as fair as possible we filtered the raw volunteer data and selected for further analysis only those entries which showed non zero computational activity for the last month, week and day. This allowed us to treat those volunteers similar to clusters which are dedicated computational resources. Apart from the aforementioned filtering we did not discriminate any volunteers.

For each of the aformenetioned CPU models, we used TDP (Thermal Design Power) given by CPU manufacturers as a CPU power consumption metric. While it can be argued that momentary CPU power draw while performing intensive computations may exceed the TDP specification, voluntary computing patterns are typically long-term, where any excessive power draw is guaranteed to activate hardware thermal protection mechanisms which, either by introducing additional wait cycles, limiting clock frequency and/or core voltage will limit the average CPU power consumption to stay within the manufacturer-stated value.

In each case considered, we calculated the expected TDP (ETDP) value from the set of all CPUs listed in a project as follows: let N_{CPU} be a total number of all CPUs considered; let N_T be a number of CPUs of type T in the set; finally let P_T be a TDP value specified for a CPU of type T expressed in Watts. Then, an expected TDP value for a given project p can be calculated as:

$$\mathrm{ETDP_p} = \sum_T \frac{N_T}{N_{CPU}} \cdot P_T \tag{1}$$

In the following sections each considered project is characterised and its ETDP value is given, using data obtained from BOINCStats in January, 2015.

---

[5] http://www.boincstats.com.

3.1 Atlas@Home Expected CPU Power Consumption

The Atlas@Home project, started in 2014, lists 367 different CPUs, for a total sum of 3159. The first place is taken by 77 instances of Six-Core AMD Opteron(tm) Processor 2435 which makes for a total share of 2.4375 %. The second place is held by 73 instances of AMD FX(tm)-8350 Eight-Core Processor and the third by 72 instances of Intel(R) Core(tm) i7-3770 CPU @ 3.40 GHz. The calculated ETDP for this project equals 78.4 W.

3.2 Asteroids@Home Expected CPU Power Consumption

The Asteroids@Home project, started in 2012, lists 1538 different CPU versions. Unlike in Atlas@Home project, where we were able to calculate ETDP across all CPUs, we limit our calculations to the first 300 different CPUs listed, for a total sum of 65920. The first place is taken by 6941 instances of ARMv7 Processor rev 0 which makes for a total share of 10.5 % of a considered CPU subset. The second place is held by 2188 instances of ARMv7 Processor rev 1 and the third by 1156 instances of Intel(R) Core(tm) i7-3770 CPU @ 3.40 GHz. The 300th (final considered) entry is 58 instances of an Intel(R) Core(tm) i5-2467M CPU @ 1.60 GHz. Beyond that point, individual CPU versions listed fall rapidly in instance numbers to low double digit values and we consider them as statistical noise rather than a useful data point for analysis. The calculated ETDP for this project equals 58.9 W.

3.3 BOINC Cross-Project Expected CPU Power Consumption

Analogous to Asteroids@Home, we evaluate only a subset of a total set of all listed CPU versions. Our subset consists of 490 most frequently appearing CPU variants, for a total of 11089271 instances. The first place is held by 587999 instances of Intel(R) Core(tm) i7-3770 CPU @ 3.40 GHz (a 5.3 % of a considered subset). The second place is taken by 461115 instances of Intel(R) Xeon(R) CPU X5355 @ 2.66 GHz (a 4.16 % of a subset) and the third place belongs to Intel(R) Core(tm) i5-3470 CPU @ 3.20 GHz with 436775 instances (3.94 % of a subset). The closing (490th) entry is Intel(R) Pentium(R) CPU G2030 @ 3.00 GHz with 3427 instances (0.031 %). The calculated ETDP for this project equals 72.6 W.

4 Estimating Computational Power of Volunteer Nodes

With the goal of comparing voluntary computing with traditional clusters we had to choose a common performance metric. The first and most obvious choice was a number of floating point operations per second (FLOPS). However, this presented unexpected difficulties. First of all, BOINC project uses its own system of BOINC Credits awarded to a volunteer after computations are performed. There is a mapping between a Boinc Credit and a FLOP[6], however, it appears

---

[6] http://boinc.berkeley.edu/wiki/computation_credit.

that Credits are awarded differently (for different amount of work) between projects and, while there appears to be an effort to provide coefficients[7] to credits of different projects, it is not clear, whether statistics on BOINCStats use normalised credits, credits specific to a project, or a mixture of both. Another big problem with Boinc Credits is, that they do not differentiate between work performed using CPU and work done on a GPU (or another accelerator). This makes Credits data next to useless for any meaningful comparison.

With the above in mind, it was necessary to look for another computational power metric, preferably with data available for most of CPUs listed in the project statistics. We decided to use CPU Mark data available for most processors considered. In very few cases, where the CPU under consideration was not benchmarked, we had to supply extrapolated data.

For each considered project, we calculated the expected CPU Mark (EMARK) value from the set of all CPUs listed in a project as follows: let N_{CPU} be a total number of all CPUs considered; let N_T be a number of CPUs of type T in the set; finally let M_T be a CPU Mark benchmark result specified for a CPU of type T. Then, an expected CPU Mark value for a given project p can be calculated as:

$$\text{EMARK}_\text{p} = \sum_T \frac{N_T}{N_{CPU}} \cdot M_T \tag{2}$$

The CPU sets examined for each considered project were identical with description given in Sect. 3. The benchmark data was obtained from PassMark(R) Software CPU Benchmarks[8]. The resulting EMARK values for all projects are given in Table 1.

Table 1. EMARK values calculated for considered BOINC projects

Atlas@Home	Asteroids@Home	BOINC combined
5984	4085	3286

5   Comparison of Volunteer and Cluster Computational Efficiency

To compare computational efficiency between volunteer computing and traditional clusters, we had to obtain identical metrics for a cluster. Additionally, to guarantee a fair comparison, it was necessary to pick a CPU-only cluster. Using data from November 2014 issue of the Top500 List[9] we selected Cray XC30 listed as the 13th most powerful supercomputer. The Cray XC30 system, located in

---

[7] http://boincstats.com/en/stats/-1/cpcs.
[8] www.cpubenchmark.net.
[9] http://www.top500.org.

United States, is based on Intel(R) Xeon(R) E5-2697v2 12Core @ 2.7 GHz. The
number of cores listed is 225984, which translates to 18832 CPUs. The cluster
uses an Aries interconnect and runs the Cray Linux Environment. The specified TDP value for this CPU is 130 W and the CPUMARK benchmark result is
23549 CPU Marks.

We also decided to have a look at a CPU-only cluster from the November 2012 issue of the Top500 list, which corresponds to the time frame when
Asteroids@Home project was initiated. We selected the 6th entry on the list:
the SuperMUC from Leibnitz Rechenzentrum based on 18432 Intel(R) Xeon(R)
E5-2680 CPUs @ 2.7 GHz. This CPU has a TDP of 130 W and the CPUMARK
benchmark result is 13251 CPU Marks.

For an additional data point we examined TRYTON Supercomputer[10] from
TASK Academic Supercomputing Centre in Gdańsk, Poland. Built in 2015,
TRYTON is based on 2966 Intel(R) Xeon(R) E5-2670 v3 @ 2.3 GHz CPUs.
This CPU has a TDP of 120 W and the benchmark result is 17481 CPU Marks.

With that data available we can calculate a computational efficiency metric
expressed in CPU Marks per Watt of CPU power. The final results are presented
in Table 2.

Table 2. Computational efficiency (in CPU Marks/W) comparison of selected clusters
and BOINC projects

Atlas@Home	Asteroids@Home	BOINC Combined	Cray XC30	SuperMUC	TRYTON
76.3	69.4	45.3	181.2	101.9	145.7

6   Summary and Future Work

We examined computational efficiency (computational power related to CPU
power consumption) of selected BOINC projects, as well as all BOINC projects
combined. We performed the same analysis for three representative clusters.
From the efficiency figures presented in Table 2 it can be seen, that clusters have
computational efficiency advantage over voluntary computing, with an advantage
factor range of 4.0 to 1.3, depending on a particular pair compared. However,
when comparing pairs of similar date of origin, we can see advantage factors
in range of 1.46 – 2.37. This can be explained easily, as clusters are designed
as homogeneous environments built with the then state-of-the-art components,
while volunteer projects group together volunteers with a wider range of heterogeneous computing hardware.

It can be observed, that BOINC Combined has the lowest efficiency and
that the efficiency increases with date of project origin. This is expected, as
the Combined statistics list all processors, also relatively very old ones, which
negatively impacts the results.

---

[10] http://task.gda.pl/kdm/sprzet/tryton/.

In this paper, when estimating power consumption we focused on CPU power consumption using TDP because this data is well documented and available. We are aware that the total power consumption is also influenced by cooling and auxiliary equipment. However, available data is not sufficient for a meaningful comparison. Similarly, extending EMARK with use of accelerators is an interesting future project once enough data is available.

A very interesting future project would be to perform similar efficiency analysis on combined computational resources, taking into consideration both CPU and accelerators. However, at the moment there appears to be not enough data, especially on the volunteer computing side, for such analysis.

Finally, even though efficiency numbers themselves seem to clearly be in favour of traditional clusters, it must not be forgotten, that voluntary computing movement, in addition to the substantial computing power made available, has a very important educational role of disseminating awareness of science and research goals to the wide population.

Acknowledgments. The work was performed within grant "Modeling efficiency, reliability and power consumption of multilevel parallel HPC systems using CPUs and GPUs" sponsored by and covered by funds from the National Science Center in Poland based on decision no DEC-2012/07/B/ST6/01516.

References

1. Dongarra, J.: Overview of high performance computing, SC 2013, UTK Booth talk, Denver, U.S.A (2013). http://www.netlib.org/utk/people/JackDongarra/SLIDES/sc13-UTK.pdf
2. Czarnul, P., Rościszewski, P.: Optimization of execution time under power consumption constraints in a heterogeneous parallel system with GPUs and CPUs. In: Chatterjee, M., Cao, J., Kothapalli, K., Rajsbaum, S. (eds.) ICDCN 2014. LNCS, vol. 8314, pp. 66–80. Springer, Heidelberg (2014)
3. Beberg, A.L., Ensign, D.L., Jayachandran, G., Khaliq, S., Pande, V.S.: Folding@home: lessons from eight years of volunteer distributed computing. In: 8th IEEE International Workshop on High Performance Computational Biology (HiCOMB 2009) in Conjunction with the IEEE International Parallel and Distributed Processing Symposium (IPDpPS 2009) (2009)
4. Anderson, D.P.: Boinc: A system for public-resource computing and storage. In: Proceedings of 5th IEEE/ACM International Workshop on Grid Computing, Pittsburgh, USA (2004)
5. Anderson, D.P., Korpela, E., Walton, R.: High-performance task distribution for volunteer computing. In: Proceedings of the First International Conference on e-Science and Grid Computing, E-SCIENCE 2005, Washington, USA, pp. 196–203. IEEE Computer Society (2005)
6. Czarnul, P., Kuchta, J., Matuszek, M.: Parallel computations in the volunteer – based comcute system. In: Wyrzykowski, R., Dongarra, J., Karczewski, K., Waśniewski, J. (eds.) PPAM 2013, Part I. LNCS, vol. 8384, pp. 261–271. Springer, Heidelberg (2014)

7. Balicki, J., Krawczyk, H., Nawarecki, E. (eds.): Grid and Volunteer Computing. Gdansk University of Technology, Faculty of Electronics, Telecommunication and Informatics Press, Gdansk ISBN: 978-83-60779-17-0 (2012)
8. Cushing, R., Putra, G., Koulouzis, S., Belloum, A., Bubak, M., de Laat, C.: Distributed computing on an ensemble of browsers. Internet Comput. IEEE **17**, 54–61 (2013)
9. MacWilliam, T., Cecka, C.: Crowdcl: Web-based volunteer computing with webcl. In: High Performance Extreme Computing Conference (HPEC 2013), pp. 1–6. IEEE (2013)
10. Funai, C., Tapparello, C., Ba, H., Karaoglu, B., Heinzelman, W.: Extending volunteer computing through mobile ad hoc networking. In: IEEE GLOBECOM Global Communications Conference Exhibition & Industry Forum, Austin, TX, U.S.A (2014)
11. Estrada, T., Taufer, M., Reed, K.: Modeling job lifespan delays in volunteer computing projects. In: 9th IEEE/ACM International Symposium on Cluster Computing and the Grid, CCGRID 2009, pp. 331–338 (2009)
12. Heien, E.M., Kondo, D., Anderson, D.P.: A correlated resource model of internet end hosts. IEEE Trans. Parallel Distrib. Syst. **23**, 977–984 (2012)
13. Czarnul, P., Matuszek, M.: Performance modeling and prediction of real application workload in a volunteer-based system. In: Applications of Information Systems in Engineering and Bioscience, Proceedings of 13th International Conference on SOFTWARE ENGINEERING, PARALLEL and DISTRIBUTED SYSTEMS Conference (SEPADS), Gdansk, Poland, WSEAS, pp. 37–45 (2014). ISBN: 978-960-474-381-0. http://www.wseas.us/e-library/conferences/2014/Gdansk/SEBIO/SEBIO-03.pdf
14. Li, J., Deshpande, A., Srinivasan, J., Ma, X.: Energy and performance impact of aggressive volunteer computing with multi-core computers. In: IEEE International Symposium on Modeling, Analysis Simulation of Computer and Telecommunication Systems, MASCOTS, pp. 1–10 (2009)
15. McGough, A.S., Forshaw, M.: Reduction of wasted energy in a volunteer computing system through reinforcement learning. Sustainable Computing: Informatics and Systems, vol. 4, pp. 262–275. Special Issue on Energy Aware Resource Management and Scheduling (EARMS) (2014)
16. Hanappe, P.: Fine-grained cpu throttling to reduce the energy footprint of volunteer computing. Technical report, Sony Computer Science Laboratory Paris (2012)

Workshop on Scheduling for Parallel Computing (SPC 2015)

Parallel Programs Scheduling
with Architecturally Supported Regions

Łukasz Maśko[1]([✉]) and Marek Tudruj[1,2]

[1] Institute of Computer Science, Polish Academy of Sciences,
Jana Kazimierza 5, 01-248 Warsaw, Poland
{masko,tudruj}@ipipan.waw.pl
[2] Polish-Japanese Academy of Information Technology,
Koszykowa 86, 02-008 Warsaw, Poland

Abstract. Scheduling of programs for hierarchical architectures of Chip Multi-Processor (CMP) modules interconnected by global data networks is the subject of this paper. The CMP modules are of double nature: architecturally specialized modules which execute time-critical computations and standard CMP modules which interconnect the specialized ones. Inside application programs, so called architecturally supported regions are identified meant for efficient execution on dedicated architecturally supported modules. Programs are represented by macro dataflow graphs built of architecturally supported nodes and program glue nodes. The paper proposes a new task scheduling algorithm for programs meant for execution in such CMP-based systems. The algorithm is based on list scheduling with modified ETF (Earliest Task First) heuristics. It is assessed by experiments based on simulation of program execution which shows parallel speedup improvements.

Keywords: Parallel programming · Program graph scheduling · Parallel architectures · Heterogeneity

1 Introduction

Putting a large number of cores inside a processor chip in the Networks on Chip (NoCs), Systems on Chip (SoCs) or Chip Multi-Processors (CMPs) technologies [1,2] sets new challenges in the design of core interconnection networks. Designers have to be conscious of technology limitations, such as power dissipation, wire delays, signal cross talks and silicon area, which make designing large monolithic CMPs problematic. The self-imposing solution are modular hierarchical structures of many CMPs interconnected by an external global network with improved efficiency and scalability. Although already viable in the current chip technology, the ideas of core clustering inside CMPs nor CMPs clustering have not been yet investigated in a mature way.

In globally interconnected systems of CMPs modules some CMPs can be strongly architecturally supported to provide high parallel speedup for some time-critical computations, while other CMPs can remain standard multicore

© Springer International Publishing Switzerland 2016
R. Wyrzykowski et al. (Eds.): PPAM 2015, Part II, LNCS 9574, pp. 77–89, 2016.
DOI: 10.1007/978-3-319-32152-3_8

processors. Usually, the architecturally supported CMPs are more intelligent and more difficult to be designed. Such architecturally supported CMP modules usually impose some particular requirements on program structures. This means at least identification of so called architecturally supported regions in the program code. Special features of programs imply special task scheduling algorithms to optimize program execution including adequate graph representation of programs with architecturally supported region nodes.

Scheduling algorithms have been intensively studied for years. Most of techniques like list scheduling, clustering or evolutionary algorithms focus on homogeneous architectures [3,4,8,9]. Extensive surveys of such scheduling algorithms can be found in [4–6]. There are also works dealing with heterogeneous architectures [7], but heterogeneity there is limited to different speed of processing units. In this paper, we assume a different idea of heterogeneity. The system is built of two classes of globally connected computing units (architecturally supported and standard CMP modules). Consequently, we use a macro data flow graph representation in which program graphs consist of two kinds of nodes: architecturally supported and glue nodes.

Our previous paper [10] presents an improved ETF-based list scheduling algorithm [3] for such program and system assumptions. It aims at obtaining better schedules by taking special attention of the order of in which ready graph nodes are scheduled. For this, program graph nodes are assigned scheduling priorities based on static, topological properties of the graph. They do not take task computing nor communication times into account. The priorities are first assigned to architectural nodes after their division into layers, using an analysis of an architectural task activation graph. These priorities are next propagated to glue nodes to control glue node selection during list scheduling to prevent too early execution of such glue nodes, which are not needed for execution of the topologically nearest architectural nodes.

This paper presents a new scheduling algorithm for the program and system assumptions as above, based on modified ETF heuristics with a different definition of task node priorities in the program graphs. Contrary to the previously proposed algorithm, here the priorities are not defined based exclusively on static topological properties of program graphs, but they are determined dynamically based on simultaneous scheduling of the input program graph and scheduling of an equivalent architecturally supported region activation graph. The paper examines the influence of some structural properties of program graphs on the make-spans of such defined program scheduling method and compares its results to those of standard ETF scheduling.

The paper is composed of 4 main parts. The first part presents general system architectural assumptions, the idea of architecturally-supported program regions, and describes structuring of programs for execution in the assumed system architecture. The second part presents the proposed task scheduling algorithm with architecturally supported regions. The third part introduces the program graph measures used for the selection of the adequate scheduling algorithm. The fourth part presents comparative experimental results obtained by simulated use of the proposed algorithm.

2 Architecture-Supported Regions in Application Programs

The general structure of the assumed parallel multi-CMP system with a global data exchange network is presented in Fig. 1(a). We have two kinds of CMPs in this system: Architectural CMPs – ACMPs, which have architecture optimized for execution of some critical program functions and General-Purpose CMPs – GCMPs, similar to typical commercial multicore processors. A program for such architecture can be logically divided into two types of fragments (see Fig. 1b):

- Architecturally-Supported Regions (ASR), whose execution will be accelerated using ACMPs, which can correspond to subroutines and are treated as graph nodes ("architectural nodes"). An ASR will usually have a parallel internal structure. We assume, that the ASR program graph has already been mapped to cores in an ACMP by a separate special scheduling algorithm. Papers [8,9] describe different kinds of such scheduling algorithms meant for heuristic optimization of exemplary ASR modules for efficient execution of parallel matrix multiplication in the ACMP architecture based on communication on the fly. The architecture is especially efficient for parallel programs featuring strong sharing of processed data.
- The glue code, not showing features for special hardware acceleration, which fills gaps between ASRs and will be executed using a set of GCMPs. The glue code is represented as glue nodes.

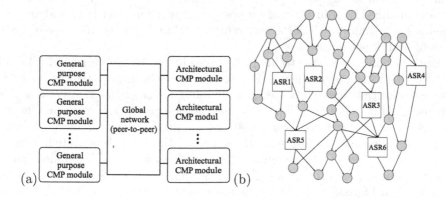

Fig. 1. The general system structure (a) and an application program graph (b)

Formally, a program is described by a macro data flow graph $G = (V, E)$, where V, E are the set of nodes and edges of the graph, respectively. The set V can be divided into two disjoint sets of nodes V_s and V_a, such that $V = V_s \cup V_a$, $V_s \cap V_a = \emptyset$, where V_a contains architectural nodes corresponding to ASRs, while V_s contains glue nodes which exist between nodes from V_a. This division may be determined automatically by a compiler, or manually by a programmer.

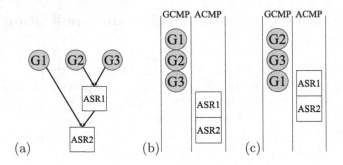

Fig. 2. Motivation for introduction of priorities in the ETF heuristics.

Each node is characterized by its weight representing time needed for execution of instructions included in this node (on ACMP or GCMP, respectively), which is determined automatically by a compiler. Each edge has a weight representing volume of data transmitted with a communication, which such edge represents.

3 Task Scheduling with Architecturally Supported Regions

The assumed multi-CMP architecture requires a proper scheduling algorithm to exploit all its advantageous features. We propose an algorithm, which is derived from the list scheduling technique with the Earliest Task First (ETF) heuristics, but was modified to adjust to the proposed computation model. This algorithm schedules glue nodes to GCMPs and architectural nodes to ACMPs to eliminate stalls of both kinds of resources.

List scheduling is a basic technique for scheduling parallel tasks and ETF [3] is one of the most popular heuristics used for list scheduling. Unfortunately, it has some disadvantages when used in scheduling for heterogeneous systems like that assumed in this paper. The ETF heuristics examine all ready task nodes and selects one with the earliest possible start time. If there are several ready nodes with the same earliest start time, any of them can be selected. It may cause that some nodes would be executed, which should be delayed since their results will be required much "later" in the graph. An example of such situation is depicted in Fig. 2a. Assuming, that the executive system has 1 GCMP with 1 core, 1 ACMP, and weights of all glue nodes G1-G3 are the same, a classical ETF-based list scheduling would give a schedule shown in Fig. 2b. But execution of node G1 should be delayed until nodes G2 and G3 are executed. Then G3 can be executed in parallel with execution of architecturally supported region ASR1, giving a better makespan as in Fig. 2c.

The proposed algorithm aims at minimal program execution time by obtaining permanent loads of ACMP and GCMP modules. The node selection method is modified by special ordering of nodes in the program graph. The nodes are so classified to assure selection in the first place of such ready glue nodes, whose

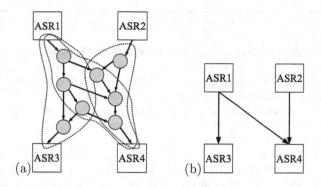

Fig. 3. Conversion of glue subgraphs (a) to edges in RAG (b)

results are required for execution of the topologically nearest ASRs in the graph. The selected glue nodes are scheduled on available processing resources without delaying higher classified graph nodes.

The general algorithm consists of 2 phases. First, the Region Activation Graph (RAG) of the input program graph is created. Then, the program graph is scheduled using list scheduling with modified ETF-based heuristics enriched by an analysis of RAG to set priorities of the ASR and glue nodes.

3.1   Region Activation Graph

A Region Activation Graph (RAG) $RAG_G = (V_a, E')$ is an acyclic unweighted directed graph derived from the original program graph described in previous sections. Nodes in RAG_G correspond to ASR nodes in an initial graph G (the V_a set), while edges depict data dependencies between the ASRs. For two nodes $u, v \in V_a$, an edge $u \rightarrow v \in E'$ exists in RAG_G, if there is a directed path between u and v in G containing only glue nodes. Two ASRs are data-dependent if there exists at least one directed path between them in G with only glue nodes. Two data-dependent ASR nodes are connected in RAG_G with an edge replacing the subgraph consisting of glue nodes and edges on all paths between the two ASRs in G. Subgraphs corresponding to different RAG_G edges may intersect, so some glue nodes can be in more than 2 edges in RAG_G.

Figure 3a presents a part of an input program graph with 4 architectural nodes (ASR1-ASR4) connected with a set of glue nodes. Figure 3b shows a RAG obtained after converting the sets of glue nodes to inter-ASR edges. There are directed paths which connect ASR1 with ASR3 and ASR4 and also directed paths from ASR2 to ASR4. Therefore, we obtain 3 edges in the resulting RAG. There is no directed paths between ASR2 and ASR3 in the input graph, thus, there is no edge between these nodes in RAG.

Algorithm 1. List scheduling algorithm with modified ETF heuristics

1: {Input: a program graph $G = (V, E)$}
2: Determine Region Activation Graph $RAG_G = (V_a, E')$, based on graph G.
3: Let $V_a^R \subseteq V_a$ be the set of nodes without predecessors in RAG_G.
4: Let V_H be the set of glue nodes without predecessors in G (ready nodes), corresponding to edges leading to nodes from V_a^R (high priority ready glue nodes). Let V_L be the set of other ready glue nodes. Let V_A be the set of ready architectural nodes from graph G.
5: **while** $V_H \cup V_L \cup V_A$ is not empty **do**
6:    Find the node $u \in V_A$ (if available) with the earliest possible execution start time. Let p be the index of a free ACMP, on which execution of u is possible.
7:    Find the node $v \in V_H$ (if available) with the earliest execution start time. Let q be the core index, on which execution of v is possible.
8:    Find the node $w \in V_L$ (if available) with earliest execution start time and for which execution ends before execution of the node v found in the previous step may start. If the node v was not found, select any node $w \in V_L$ with the earliest execution start time. Let r be the core index, on which execution of w is possible.
9:    **if** the node u has been found **then**
10:       Schedule u for execution on p^{th} ACMP and remove it from V_A.
11:       Virtually schedule u in RAG_G and remove it from V_a^R.
12:       **for all** descendants u' of u in RAG_G **do**
13:          **if** u' becomes ready in RAG_G **then**
14:             Insert u' in V_a^R. Move from V_L to V_H all nodes corresponding to edges leading to u' in RAG_G.
15:          **end if**
16:       **end for**
17:    **else**
18:       **if** the node w has been found **then**
19:          Schedule the node w for execution on the core r and remove it from V_L.
20:          Insert to V_L all the descendants of the node w, for which all their predecessors have already been scheduled.
21:       **else**
22:          Schedule the node v for execution on the core q and remove it from V_H. Insert to V_H (or V_L) all the descendants of the node v, for which v was the last scheduled predecessors and which correspond to edges leading to ready nodes in RAG_G (other nodes in RAG_G, respectively).
23:       **end if**
24:    **end if**
25: **end while**

3.2 Scheduling Algorithm Based on RAG Topology Analysis

The proposed scheduling algorithm assumes RAG analysis to delay execution of glue nodes not used for execution of the soonest ASR nodes. The algorithm concurrently schedules the initial program graph and its RAG. Classical list scheduling divides all nodes into three sets: already scheduled nodes, nodes which are ready for execution (with all predecessors scheduled) and nodes waiting for completion of their predecessors. In the original ETF heuristics, all ready nodes are examined and one of them is chosen. Based on RAG analysis, we introduce two subsets of ready glue nodes: the high priority nodes needed for execution of the topologically nearest ASRs in the graph and the low priority nodes needed for execution of topologically more distant ASRs. The *topologically nearest* ASR nodes are such, which are also ready for execution in the RAG of the scheduled program graph. At every scheduling step, the glue nodes, which correspond to edges in the RAG leading to currently ready nodes in the RAG, have high priority, while other glue nodes have low priority. If an ASR node is scheduled, we also simulate its assignment to the same computing resources in the RAG of the scheduled graph. As a result of this assignment, the descendants of the scheduled node in the RAG may become ready – then all the low priority ready glue nodes on the ASR incoming edges obtain high priority.

The pseudo-code of the proposed scheduling algorithm is shown as Algorithm 1. It follows list scheduling principles. Each time, when a glue node is to be assigned to a GCMP, first the high priority nodes are considered. Low priority glue nodes are scheduled only when their execution doesn't impede high priority nodes. Such node selection strategy assumes that architectural nodes can be executed as soon as possible on ACMPs. Additionally, GCMPs if free, can execute glue nodes, which are required for further computations.

Time complexity of the algorithm remains polynomial, although with a higher degree than list scheduling with standard ETF heuristics. All the additional steps have polynomial complexity, including for instance computation of a RAG and layers using breadth first graph traversals as well as transfers of ready nodes between V_L and V_H sets in the loop.

4   Graph Metrics for Right Selection of the Scheduling Algorithm

We have compared make-spans obtained for different program graphs. Experiments show, that comparison results depend on features of the graphs in terms of topology, weights of nodes and edges but also on resources available for program execution. In our study we deal with layered program macro data flow graphs, built of node layers and edges for inter-node communication between layers. A layer in such program graph contains all architectural nodes, which have the same depth in the program RAG, plus all the glue nodes, which provide data for these architectural nodes.

In list scheduling of a program graph a node may be scheduled too early, which may lead to an un-optimal use of processor time for other ready nodes. We introduce a metrics, which we call **Cumulated Activation Stride (CAS)** of a program graph to measure the potential of the graph for this non-optimality. The metrics is determined starting with a traversal of a program graph by breadth first search, in which for every node a layer of a deepest architectural region activating this node is determined ($max\_act\_layer(v)$):

1. For each glue node v, determine its layer number, $layer(v)$ (they depend only on architectural nodes of the graph and their dependencies in RAG).

2. For each glue node v, determine $max\_act\_layer(v)$ – the maximal layer number in which this node may be activated, by computing the maximum over the following values, depending on all the predecessors u of node v:

   (a) If v has no predecessors, then $max\_act\_layer(v) = 0$.

   (b) If u is an architectural node, $max\_act\_layer(v) = layer(u) + 1$. It means, that node v should be treated in exactly the same way as glue nodes from layer $layer(u) + 1$, because it is activated within this layer.

   (c) If u is a glue node, then $max\_act\_layer(v) = max\_act\_layer(u)$. The node u can be activated earlier then needed, so we consider its max_act_layer, not its layer number.

After all the glue nodes in the graph are examined, we determine the **Activation Stride** for each glue node v:

$$activation\_stride(v) = layer(v) - max\_act\_layer(v)$$

This value will be non-zero only for nodes, which become ready before architectural nodes preceding their layer are completed.

The **Cumulated Activation Stride** metrics $CAS(G)$ for graph G is defined as the sum of node activation strides multiplied by node weights over all glue nodes, divided it by the product of sum of glue node weights and the maximal layer number in the graph ($Arch(G)$ and $Glue(G)$ correspond to architectural and glue nodes of the graph G, respectively):

$$CAS(G) = \frac{\sum_{v \in Glue(G)} activation\_stride(v) * weight(v)}{max_{u \in Arch(G)} layer(u) * \sum_{v \in Glue(G)} weight(v)}$$

So defined metrics will be equal to 0 if all the glue nodes are activated by architectural nodes, which precede their layers. The maximal value may be obtained for a graph, in which there are no glue nodes in all layers except the last one, and all these nodes are ready at the beginning of the program graph execution (they have no predecessors). Since the maximal stride cannot be greater than the maximum layer number, $CAS(G)$ cannot exceed 1. It also does not change if all the weights in the graph are multiplied by the same constant.

Figure 4 presents an exemplary program graph with layered structure. Each layer is composed as a set of uniform subgraphs containing nodes Aj^i, Bj^i and Mj^i. Nodes Aj^i and Bj^i are glue nodes, while nodes Mj^i are architectural nodes. The long, black edge corresponds to communication between layers, which activates node $B1^{i+2}$. The other activation edge of nodes Bj^i corresponds to read of initial data from shared memory. Node $B1^{i+2}$ in layer $i + 2$ is activated by architectural node from layer $i - 1$, therefore its activation stride equals 3 (layers are computed with respect to architectural nodes, not glue nodes).

5   Experimental Results

To evaluate and compare performance of the presented scheduling algorithm, the following exemplary iterative application program was considered:

```
func benchmark(stride) {
        // Let i be the iteration number
        for i=1 to N pardo
                // Let j be the path number
                for j = 1 to K pardo
                        // select parts of the results of the previous iteration
                        a[i, j] = A(m[i − 1, 1], m[i − 1, 2], ..., m[i − 1, N]);
                        // if i ≤ stride then initial data are read
                        b[i, j] = B(u[i, j], m[i − stride, j]);
                        m[i, j] = M(a[i, j], b[i, j]);
                end for
        end for
}
```

This program corresponds to a computational algorithm, which includes common functions $A()$, $B()$ and $M()$ on elements of square matrices, such as matrix addition and multiplication. We assume, that functions $A()$ and $B()$ have irregular internal structure and are not promising for faster implementation in ACMP modules. Therefore they will be treated as glue nodes in the program graph. $M()$ is a parallel matrix multiplication based on recursive matrix decomposition into quarters. The stride parameter corresponds to the activation stride for computations of $B()$ functions which provide data for $M()$ regions. A vector $u[]$ and matrices $m[0, 1..N]$ are initial parameters of the program used for computation. A single iteration of the outer loop creates a layer of subgraphs (each subgraph corresponds to an iteration of the inner loop), which are mutually independent. Macro dataflow graph of a part of the considered exemplary program for $stride = 3$ is shown in Fig. 4.

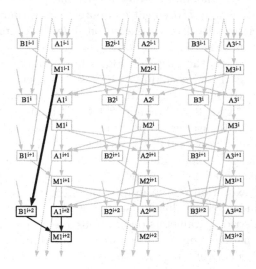

Fig. 4. Exemplary program graph with communication between layers that causes non-zero strides for nodes Bj^i; Mj^i – architectural matrix multiplication nodes.

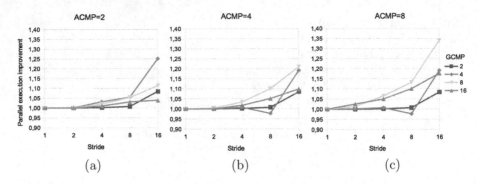

Fig. 5. Parallel execution improvement of the proposed algorithm with 2, 4 and 8 ACMPs and 2, 4, 8 or 16 GCMPs for graphs with activation strides 1, 2, 4, 8 and 16.

All graphs for experiments were generated with $K = 8$ (8 paths) and $N = 16$ (16 layers). For parallel execution of such graph, the maximal number of 8 ACMP and 16 GCMP modules is needed. We have considered a set of graphs for a range of values for parameter stride: 1 which corresponds to a graph with no strides, 2, 4, 8 and 16, which corresponds to a graph, in which all the B nodes are ready at the beginning of computations. The graph is uniform due to node weights, which were selected in the arbitrary way: all A and M nodes have weights equal to 8000 units, while B nodes have weights equal to 6500.

The graphs were scheduled for executive systems with a range of ACMP (2, 4 and 8) and GCMP (2, 4, 8 and 16) modules. ACMP modules execute only one ASR node at a time. GCMP modules were assumed to contain 1 computing core. We scheduled the graphs with a standard ETF-based scheduling algorithm and compared it to the schedules obtained with the proposed algorithm. We examined parallel schedule improvement computed as a ratio of execution time obtained by the reference algorithm to execution time of a graph scheduled with the proposed algorithm (Fig. 5).

Experiments show that the proposed algorithm performs in general better than classical ETF-based list scheduling. The results depend on the number of both ACMP and GCMP modules applied. The biggest execution time improvement was equal to 1.34 for 8 ACMP and 8 GCMP modules and graphs with $stride = 16$. The smallest average improvement was obtained for the smallest (2) and the biggest (16) number of GCMP modules. This is due to the fact, that 2 GCMPs are insufficient to prepare data for 16 ACMPs on time (both algorithms are forced to serialize parallel computations), and 16 is the number of GCMPs needed for optimal execution of the graph, therefore the standard algorithm was capable of finding good schedule for such system. For some combinations of ACMP, GCMP and stride parameters one can observe improvements, which are smaller than 1. The parallel improvements are computed in comparison to ETF-based list scheduling algorithm, not the sequential execution. Therefore, it must be considered, how ETF deals with the graph for a given number of resources.

Table 1. Values of $CAS(G)$ for graphs generated for different values of the stride parameter and different relative weights of the B nodes.

Relative weight of B nodes	Activation Stride				
	1	2	4	8	16
20 %	0.000	0.009	0.025	0.049	0.070
40 %	0.000	0.015	0.043	0.086	0.123
60 %	0.000	0.021	0.058	0.115	0.165
80 %	0.000	0.025	0.069	0.139	0.198
100 %	0.000	0.028	0.079	0.158	0.226
120 %	0.000	0.031	0.087	0.173	0.248
140 %	0.000	0.033	0.093	0.187	0.267
160 %	0.000	0.035	0.099	0.199	0.284
180 %	0.000	0.037	0.104	0.209	0.298

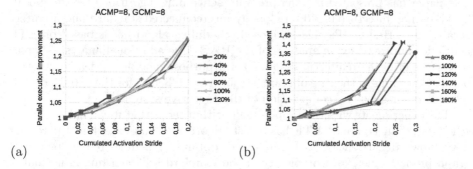

(a)   (b)

Fig. 6. Parallel speedup improvement as a function of the $CAS(G)$ for different relative weights of the B-type nodes.

We assume, that for those configurations, ETF was able to find a solution, which was better than the one found by the proposed algorithm.

We have also examined the relation between the CAS metrics of the graph G and schedule improvement. In the assumed graph, the value of CAS depends on the stride parameter, which influences the topology of a graph, but also on weights of B nodes. We have checked a range of graphs, which differ in weights of B nodes. We have assumed a series of B node weights being a percentage of the B node weight (100 %) in the uniform graph used in the experiment discussed above. Experiments were done for the number of both ACMP and GCMP modules equal to 8. The results for other combinations of ACMP and GCMP numbers show similar tendencies to those shown in the paper. The values of CAS measures of examined graphs are shown in Table 1.

Figure 6 shows correspondance of the parallel improvement to the CAS metrics of graphs. Improvements for the graphs with CAS metrics equal to 0 ($stride = 1$) are the smallest in general and equal to 1. In such graphs all the glue nodes are activated in their layers and therefore the standard ETF-based algorithm has no chance to make a wrong scheduling decision. With the increase of the CAS metrics we can observe a better improvement obtained by the proposed algorithm. Graphs with higher CAS contain more nodes that can be scheduled too early, when compared to their layer. Also, these nodes are heavier, therefore, they have bigger impact on the overall schedule. It makes such graphs harder to be correctly scheduled – especially with the standard ETF scheduling. Due to a different way of handling of ready nodes, the proposed algorithm shows much better resistance to such situations. Execution of the questionable nodes is delayed, which allows faster start of architectural nodes from previous layers and better resource use, leading to better schedules. The best improvements were noticed for graphs with the biggest CAS value ($stride = 16$).

6   Conclusions

The paper has presented parallel program scheduling algorithms for the modular system architecture based on globally interconnected standard and architecturally supported CMPs. The proposed scheduling algorithm is based on ETF heuristics improved by an analysis of the RAG. The additional analysis enables better use of both architectural and general purpose modules. It leads to better parallel speedups in the case of graph structures "difficult" for the standard ETF schedulers for adequate composition of the executive system.

The experiments with the proposed algorithm show, that it can deliver better schedules than standard ETF-based list algorithm. The experimental results have shown dependencies of the quality of obtained schedules on the proposed graph property metrics. Complexity of the standard ETF scheduling is smaller than complexity of the presented improved algorithm, therefore for graphs with small CAS values it is enough to use the standard ETF algorithm – the schedules are the same or very close to the schedules obtained with the improved algorithm, but they may be computed faster. For graphs with high values of CAS, it is profitable to use a better, although more complicated algorithm we propose.

References

1. Owens, J.D., et al.: Research challenges for on-chip interconnection networks. IEEE MICRO **27**, 96–108 (2007)
2. Kundu, S., Peh, L.S.: On-chip interconnects for multicores. IEEE MICRO **25**, 3–5 (2007)
3. Hwang, J.-J., Chow, Y.-C., Anger, F.D., Lee, C.-Y.: Scheduling precedence graphs in systems with interprocessor communication times. SIAM J. Comput. **18**(2), 244–257 (1989)
4. Yu-Kwong, K., Ishfaq, A.: Benchmarking and comparison of the task graph scheduling algorithms. J. Parallel Distrib. Comput. **59**, 381–422 (1999)
5. Sinnen, O.: Task Scheduling for Parallel Systems. Wiley, England (2007)

6. Błażewicz, J., Ecker, K.H., Pesch, E., Schmidt, G., Weglarz, J.: Handbook on Scheduling. International Handbooks on Information Systems. Springer, Heidelberg (2007)
7. Topcuoglu, H., Hariri, S., Min-You, W.: Performance-effective, low-complexity task scheduling for heterogeneous computing. IEEE Trans. Parallel, Distrib. Syst. 13(3), 260–274
8. Masko, Ł., Dutot, P.F., Mounié, G., Trystram, D., Tudruj, M.: Scheduling moldable tasks for dynamic SMP clusters in SoC technology. In: Wyrzykowski, R., Dongarra, J., Meyer, N., Waśniewski, J. (eds.) PPAM 2005. LNCS, vol. 3911, pp. 879–887. Springer, Heidelberg (2006)
9. Maśko, Ł., Tudruj, M.: Task scheduling for SoC-Based dynamic SMP clusters with communication on the fly. In: 7th International Symposium on Parallel and Distributed Computing, ISPDC, pp. 99–10. IEEE CS (2008)
10. Tudruj, M., Maśko, Ł.: Scheduling parallel programs based on architecture–supported regions. In: Wyrzykowski, R., Dongarra, J., Karczewski, K., Waśniewski, J. (eds.) PPAM 2011, Part II. LNCS, vol. 7204, pp. 51–60. Springer, Heidelberg (2012)

Adaptive Multi-level Workflow Scheduling with Uncertain Task Estimates

Tomasz Dziok, Kamil Figiela, and Maciej Malawski[✉]

Department of Computer Science, AGH University of Science and Technology,
Al. Mickiewicza 30, 30-059 Kraków, Poland
{kfigiela,malawski}@agh.edu.pl

Abstract. Scheduling of scientific workflows in IaaS clouds with pay-per-use pricing model and multiple types of virtual machines is an important challenge. Most static scheduling algorithms assume that the estimates of task runtimes are known in advance, while in reality the actual runtime may vary. To address this problem, we propose an adaptive scheduling algorithm for deadline constrained workflows consisting of multiple levels. The algorithm produces a global approximate plan for the whole workflow in a first phase, and a local detailed schedule for the current level of the workflow. By applying this procedure iteratively after each level completes, the algorithm is able to adjust to the runtime variation. For each phase we propose optimization models that are solved using Mixed Integer Programming (MIP) method. The preliminary simulation results using data from Amazon infrastructure, and both synthetic and Montage workflows, show that the adaptive approach has advantages over a static one.

Keywords: Cloud · Workflow · Scheduling · Optimization · Adaptive algorithm

1 Introduction

Scientific workflow is a widely accepted method for automation of complex computational processes on distributed computing infrastructures, including IaaS clouds [7]. When using clouds and their pay-per-use pricing model with multiple types of virtual machine (VM) resources, usually called instances, the problem of scheduling and cost optimization becomes a challenge. The specific problem we address in this paper is that most static scheduling algorithms assume that the estimates of task runtimes are known in advance, while in reality these estimates may be inaccurate. These discrepancies may be a result of inherent uncertainty in performance models of the application, or may be caused by unexpected dynamic behavior of the infrastructure. On the other hand, dynamic scheduling approaches that adapt to such uncertainties cannot be easily used for scheduling

M. Malawski—This work is partially supported by EU FP7-ICT project PaaSage (317715), Polish grant 3033/7PR/2014/2 and AGH grant 11.11.230.124.

© Springer International Publishing Switzerland 2016
R. Wyrzykowski et al. (Eds.): PPAM 2015, Part II, LNCS 9574, pp. 90–100, 2016.
DOI: 10.1007/978-3-319-32152-3_9

under deadline or budget constraints, since meeting a constraint requires some form of advance planning based on estimates.

In this paper, we propose an adaptive scheduling algorithm for deadline constrained workflows that consist of multiple levels. Such levels are present in real scientific workflows and they often have up to 1 000 000 tasks [7,13]. The main idea behind the algorithm is to produce a global approximate plan for the whole workflow in a first phase, and a local detailed schedule for the current level of the workflow. The algorithm is then invoked iteratively after each level completes the execution, in this way being able to adjust to the runtime variation from the estimated execution times. Another advantage of this approach is that we can reduce the complexity of scheduling of the whole workflow by reducing it into two smaller problems that can be solved using Mixed Integer Programming (MIP). The algorithm has been evaluated by simulation using data from Amazon infrastructure and workflows from Pegasus Workflow Gallery [13].

This paper is organized as follows: in Sect. 2 we discuss other scheduling models and algorithms for workflows. Section 3 contains detailed description of the algorithm proposed in this paper, and its illustration on a simple example is given in Sect. 4. In Sect. 5 we outline the optimization models used. Then in Sect. 6 we show results for real workflows. Finally, in Sect. 7 we present conclusions and future work.

2   Related Work

Mathematical programming has been applied to the problem of workflow scheduling in clouds. The model presented in [12] is applied to scheduling small-scale workflows on hybrid clouds using time discretization. Large-scale bag-of-task applications on hybrid clouds are addressed in [4]. The cloud bursting scenario described in [3], where a private cloud is combined with a public one, also addresses workflows. None of these approaches addresses the problem of inaccurate estimates of actual task runtimes.

Adaptive approach is known from engineering systems [1]. Dynamic algorithms for workflow scheduling in clouds have been proposed e.g. in [17], where they assume the dynamic stream of workflows. In [9] the goal is to minimize makespan and monetary cost, assuming an auction model, which differs from our approach where we assume a cloud pricing model of Amazon EC2.

In our earlier work [14], we also used the MIP approach to schedule multi-level workflows, but the dynamic nature of cloud is not considered. We have also analyzed the impact of uncertainties of runtime estimations on the quality of scheduling for bag-of-task in [15] and workflow ensembles in [16], with the conclusion that these uncertainties cannot be always neglected.

Task estimation for workflow scheduling is a non-trivial problem, but several approaches exist,

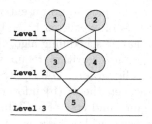

Fig. 1. DAG example

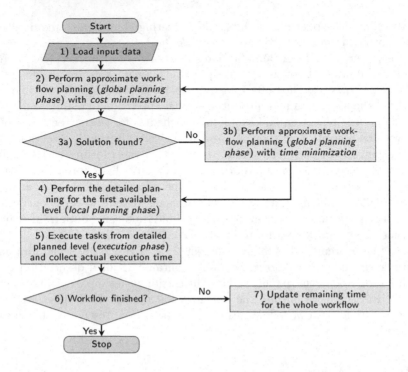

Fig. 2. High level flow of scheduling algorithm.

e.g. those based on stochastic modeling and workflow reductions [5]. It is also possible to create performance models to estimate workflow execution time using application and system parameters, as proposed in [18]. The error of these estimates is less than 20 % for most cases, which gives a hint on the size of possible uncertainties.

3   Adaptive Scheduling Algorithm

Our algorithm provides an adaptive method for optimizing cost of workflow execution in IaaS clouds, under a deadline constraint. We assume that the workflow tasks can be divided by their levels, where a level of a task is a length of the longest path from an entry node. Tasks from one level can have different estimates of execution time. It can be considered as a hybrid between static and dynamic scheduling algorithms.

The algorithm requires: (a) workflow (see Fig. 1) represented as directed acyclic graph (DAG), where nodes represent tasks and edges dependencies between them; (b) information about available infrastructure, i.e. the performance and cost of available VM instance types; and (c) global deadline for the whole workflow. We assume that (a) all tasks in each level are independent and can be executed in parallel on multiple VMs; (b) each VM has price per hour and a performance metric called CCU (which is a result of a benchmark, as in Cloud Harmony Compute Units [6]); (c) each task has estimated size which is execution time on a VM with performance of 1 CCU; (d) tasks in one level could have different estimated size; and (e) execution time of a task on given VM is inversely proportional to VM performance expressed in CCU.

The objective of the algorithm is to minimize the execution cost under a deadline constraint. The algorithm is run before each level of tasks begins its execution. Each time it consists of two phases. In the first *global planning phase*, the algorithm uses an approximation that tasks in each level are uniform, and finds assignments between the tasks and VMs for the whole workflow. In the second *local planning phase*, a detailed plan is prepared for the closest level of individual tasks. After a level completes, the algorithm takes into account the real execution time of already completed tasks, and based on that updates the remaining time. Thanks to that it is able to adjust to differences between an estimated and actual execution time.

The algorithm is shown in Fig. 2, and consists of the following steps.

1. First, the information about workflow, available infrastructure (list of VMs) and global deadline are loaded.
2. In this step (*global planning phase*) algorithm assigns VMs to levels. For each level, we calculate average estimated task execution time and we pass it as input. The aim of this optimization is to find assignments between VMs and levels with minimal total cost under global deadline. As a result, the algorithm returns information which VMs are assigned to each level and also how many tasks should be executed on each VM. It also returns estimated execution time and cost for levels.
3. If the solver does not find a solution, the optimization is run again without deadline constraint, but with time minimization as an objective. This may be the case when the deadline is too short. We then fallback to minimization of deadline overrun and we ignore the cost objective.
4. Next, we perform *local planning phase* that assigns individual tasks to VMs in the current level. It uses the results from step 2 as an input: VMs assigned to this level and number of tasks which should be executed on each VM. The objective of optimization is to minimize the total execution time. Total cost is not taken into account, because the VMs are already chosen and the estimated execution time for each one is known – so the cost does not change. As a result the algorithm returns information on which VM task will be executed.
5. Then we execute tasks on VMs assigned in *local planning phase* and collect the actual task execution time. Tasks may be executed on real VMs instances or in a cloud simulator (which allows to test many scenarios easily).
6. The algorithm finishes if there are no remaining levels to be scheduled.
7. We update remaining total time with actual execution time and perform planning for remaining part of the workflow, repeating process from step 2.

4   Illustrative Example

To illustrate the operation of our algorithm, we prepared an example using the simple workflow from Fig. 1. The input is provided in Table 1. The workflow consists of 3 levels, so the algorithm is executed in three iterations, as shown in Table 2. The resulting execution times and costs in all iterations are presented and commented in Fig. 3.

Table 1. Example input to the algorithm: estimated task sizes, VM performance and costs for the workflow shown in Fig. 1. We assume the global deadline is 15.

TaskID	T1	T2	T3	T4	T5	VM ID	Performance (CCU)	Cost per Time Unit
Est. Task Size	22	18	10	10	20	A	5	10
						B	10	25

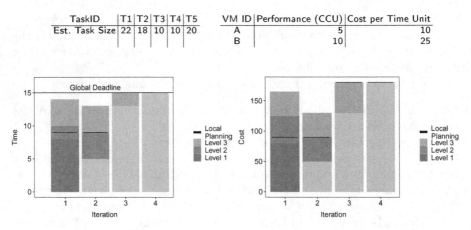

Fig. 3. Execution time and cost of the algorithm, shown level by level. In the first iteration, the global planning phase estimates the completion time of level 1 is 8 (purple bar) and the local planning estimates it to be 9 (solid line). In iteration 2, it turns out that the level L1 finished at time 5 (grey bar). Both global and local planning for level 2 (red bar and solid line) predict the finish time for time 9. The actual execution of level 2 completes in time 13 (grey bar), so in iteration 3 both global and local phases plan the execution of level 3 (orange bar) to complete just within the deadline). The execution in iteration 4 shows that the level 3 actually completed as planned (Color figure online).

5   Optimization Models

We use three optimization models in the algorithm: the first one for global planning phase, the second one in the case when deadline cannot be met, and the third one for the local planning. Since the domain is discrete, each model belongs to a mixed-integer programming (MIP) class. In all three models we assume for simplicity that VMs start immediately and have no latency. Thanks to that the problems are solved quicker. On the other hand, we assume that all possible delays are included in the error of estimates, which is taken into account in step 7 of the algorithm. Here we outline the main features of the models, and for the details we refer to the source code in the public repository [8].

Table 2. Planning and execution flow for illustrative example. The assignments of tasks to VMs change whenever the actual execution time differs from the estimated one.

Level	# Tasks	Avg. Task Exec. Time	Planned Time	Cost	Assigned VMs (number of tasks)		Task	VM	Planned Time	Cost	Actual Time	Cost
L1	2	20	8	80	A (2)		T1	A	5	50	3	30
L2	2	10	2	45	A (1), B (1)		T2	A	4	40	2	20
L3	1	20	4	40	A (1)							

(a) 1st iteration, *global planning* (left): estimated cost of executing the workflow is 165 and the total time is 14. Algorithm plans to use all the available time, selects cheaper instance and minimizes the total cost. *Local planning* and execution (right): task runtimes from level L1 were overestimated, the remaining time is 10.

Level	# Tasks	Avg. Task Exec. Time	Planned Time	Cost	Assigned VMs (number of tasks)		Task	VM	Planned Time	Cost	Actual Time	Cost
L2	2	10	4	40	A (2)		T3	A	2	20	4	40
L3	1	20	4	40	A (1)		T4	A	2	20	4	40

(b) 2nd iteration, *global planning* (left): Due to more time than expected, the algorithm assigned all the tasks to the cheapest VM – to minimize the total cost. *Local planning* and execution (right): tasks from level L2 were underestimated. Remaining time is 2, current total cost is 130.

Level	# Tasks	Avg. Task Exec. Time	Planned Time	Cost	Assigned VMs (number of tasks)		Task	VM	Planned Time	Cost	Actual Time	Cost
L3	1	20	2	50	B (1)		T5	B	2	50	2	50

(c) 3rd iteration, *global planning* (left): the algorithm selects the more powerful VM B to keep the deadline constraint. *Local planning* and execution (right): total time is 15, total cost is 180 and the workflow completed in the given deadline.

Model used in *global planning phase* assigns VMs and sub-deadlines to each level, but instead of scheduling individual tasks, it uses an approximation of average task runtimes. For each level, it calculates an average task size, and based on this, an estimated cost of executing its tasks on a given VM. As a result, it is known which VMs should be used for each level and how many tasks should be executed on selected VM. The objective is to minimize total cost of the whole workflow execution.

Input to this approximate planning is defined with the following data: m is number of VMs, n is number of levels, V is a set of VMs, L is a set of levels, d is global deadline, L_l is number of tasks in level l, $T_{l,v}^a$ is average estimated execution time of task from level l on VM v, p_v is cost of running VM v for one time unit, $C_{l,v} = p_v T_{l,v}^a$ is average estimated cost of executing task from level l on VM v.

The search space is defined with the following variables: $A_{l,v}$ is binary matrix which tells if VM v will execute at least one task from level l, $Q_{l,v}$ is integer matrix which tells how many tasks from level l will be executed on VM v, T_l^e is vector of real numbers which stores execution time for level l (estimated sub-deadlines), $T_{l,v}^v$ is matrix which stores execution time for VM v on level l. $A_{l,v}$ is used as an auxiliary variable to simplify defining constraints.

The objective is to minimize total cost: Minimize: $\sum_{l}^{L} \sum_{v}^{V} C_{l,v} * Q_{l,v}$. We constrain the search space to keep the total execution time below the deadline, to divide the deadline into sub-deadlines and to enforce them, and to ensure that all the tasks from all the levels are executed.

Model used in *global planning phase* when deadline cannot be met is used when searching for solution using the first model fails. It can happen e.g. when real execution time of previous level takes much more time than expected. Comparing to the previous model, the algorithm ignores global deadline constraint and the objective function minimizes total time of workflow execution:

Minimize: $\sum_{l}^{L} T_l^e$.

Model used in *local planning phase* assigns VMs to each task from a single level. The goal is to minimize time of level execution, which is equal to the time of the longest working VM. The input to this optimization problem is defined with the following data: m is number of VMs, k is number of tasks in current level, K is a set of tasks, V is a set of VMs (only VMs assigned to current level – results from *global planning phase*), $T_{k,v}^e$ is an estimated execution time of task k on VM v, N_v is a number of tasks which will be executed on VM v (results from *global planning phase*).

Search space is defined with the following variables: $A_{k,v}$ is binary matrix which tells if task k will be executed on VM v, T_v^r is vector of real numbers which tells how long does each VM v work, w is helper variable which stores the longest working time for VMs from V.

The objective is to minimize time of the longest working VM: Minimize: $max(T_v^r | v \in V)$ that is implemented as Minimize: w. We constrain the search space to ensure that all the tasks are executed, to assign given number of tasks on each VM, and to assign the correct value to w which is the longest working VM.

Implementation of Algorithm and Models. Optimization models are implemented in CMPL modeling language [19]. As a solver we use CBC [11]. Input data is loaded from DAG files (workflows) and JSON files (infrastructure). The simulator which executes the tasks and introduces the runtime variations is implemented in Java. Source code (including optimization models) is available in the repository [8].

6   Evaluation Using Synthetic Workflows

For evaluation of the algorithm we implemented a simple simulator. Its goal is to execute one level of tasks on the assigned VMs and to introduce the runtime variation of task execution times to simulate the behavior of the real infrastructure.

We present here the results of our adaptive algorithm obtained using Montage workflow [13] representing astronomical image processing, consisting of 5000 tasks. As estimates of task sizes we used data from the logs of our earlier runs performed on Amazon EC2 [10]. We used the m3.large as a reference VM type and for performance estimation of other instance types we used the ECU value as provided by Amazon [2]. As the error of estimates we introduced a normal distribution with the standard deviation of 0.25. Since the real Montage workflow consists of very small tasks (having execution time in the order of seconds), we artificially extended them by multiplying their execution time by 3600. The deadline was set to 3500 time units (hours).

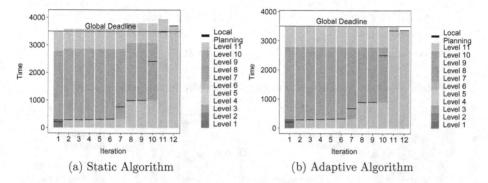

(a) Static Algorithm   (b) Adaptive Algorithm

Fig. 4. Execution time plot for Montage 5000 workflow with random errors of estimates.

We compared our adaptive algorithm to its static scheduling variant as a baseline. The static scheduling works in the same way as our algorithm, but it plans all the levels in advance. This means it does not update the global and local planning phases after execution of each level, so it does not adjust to the runtime variations.

Figure 4 shows the results of the static and adaptive scheduling algorithms, presented in the same convention as in the illustrative example (Fig. 3). We can observe that in the plot (b) the adaptive algorithm adjusts to the actual execution time after each level, while the static algorithm (a) does not, which leads to the deadline overrun.

Figure 5 presents how the completion time and total cost depend on the varying estimation error μ. The errors were generated using the normal distribution with the standard deviation of 0.25 and the mean of μ, with μ from -0.25 (overestimation) to 0.25 (underestimation). In plot (a) we observe that our adaptive algorithm succeeds to meet the deadline in more cases than the static algorithm. Even for the largest error ($\mu = 0.25$) the deadline overrun is only 5 %, while for the static algorithm it is over 25 %. On the other hand, plot (b) shows that the adaptation costs more, i.e. in most cases the cost is higher for adaptive algorithm, but never more than by 5 %. This is explained by the need to choose more expensive VMs to complete the workflow before the deadline.

In addition to Montage, we tested our algorithms using other workflows from the gallery [13]. Generally, we observed similar behavior as in the case of Montage. Sample results are shown in Fig. 6, where overestimation and underestimation represent error distribution shifted by -0.25 and 0.25, respectively. Relative

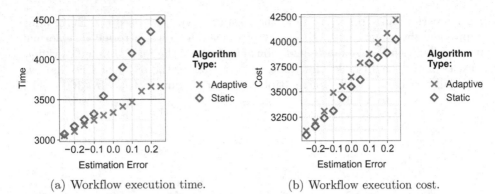

(a) Workflow execution time.   (b) Workflow execution cost.

Fig. 5. Workflow execution time/cost depending on the estimation error.

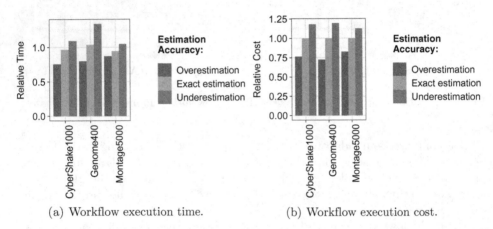

(a) Workflow execution time.   (b) Workflow execution cost.

Fig. 6. Plots with normalized execution time/cost for other workflows.

execution time is normalized to the deadline, while the relative cost is normalized to the cost of execution with exact estimates (errors with $\mu = 0$ and standard deviation of 0.25). The deadline overrun for large errors is caused by the fact that when the task runtimes are underestimated in the final level, the algorithm cannot adjust to them. Improving the algorithm would require adding a learning capability to predict the estimation error based on previous levels, which will be the subject of future work.

7   Conclusions and Future Work

In this paper we presented the adaptive algorithm for scheduling workflows in clouds with inaccurate estimates of run times. The preliminary evaluation results have shown that the implemented algorithm works as designed, and is able to meet the given deadline while minimizing the cost.

The algorithm adapts to the actual situation at runtime: when tasks execute quicker than estimated – the algorithm selects slower (and cheaper) VMs, and minimizes the total cost. When tasks execute slower than estimated – the algorithm selects faster (and more expensive) VMs, which increases total cost, but allows not exceeding the deadline for the whole workflow. When deadline is exceeded (or it is not possible to plan execution under deadline) then the algorithm minimizes the total time regardless of cost. When estimated execution time for tasks from the same levels has a big variation, then there are visible differences between estimated time in *global planning phase* and *local planning phase*. When execution of tasks is longer is final levels (which is the worst case scenario) then the total cost increases, but this is general problem for all adaptive scheduling algorithms.

During implementation and evaluation we found out a few ways that could be enhanced in future work. They include improvement of pricing in optimization models by e.g. reusing already assigned VMs, extending models with data transfer time and cost, or splitting levels with many tasks on smaller ones on 'logic' independent levels. It would be also interesting to improve task estimation (i.e. take into account multi-core CPUs) or use machine learning in estimating task execution time. After more systematic testing, we plan to use this algorithm as a part of engine to executing workflows in computing clouds.

References

1. Abdelzaher, T., Diao, Y., Hellerstein, J.L., Lu, C., Zhu, X.: Introduction to control theory and its application to computing systems. In: Liu, Z., Xia, C.H. (eds.) Performance Modeling and Engineering, pp. 185–215. Springer, Heidelberg (2008)
2. Amazon: AWS pricing (2015). http://aws.amazon.com/ec2/pricing/
3. Bittencourt, L.F., Madeira, E.R.M.: Hcoc: A cost optimization algorithm for workflow scheduling in hybrid clouds. J. Internet Serv. Appl. **2**(3), 207–227 (2011)
4. den Bossche, R.V., Vanmechelen, K., Broeckhove, J.: Online cost-efficient scheduling of deadline-constrained workloads on hybrid clouds. Future Gener. Comput. Syst. **29**(4), 973–985 (2013)
5. Chirkin, A.M., Belloum, A.S.Z., Kovalchuk, S.V., Makkes, M.X.: Execution time estimation for workflow scheduling. In: 2014 9th Workshop on Workflows in Support of Large-Scale Science, pp. 1–10. IEEE, November 2014
6. CloudHarmony: What is ECU? CPU benchmarking in Cloud (2010). http://blog.cloudharmony.com/2010/05/what-is-ecu-cpu-benchmarking-in-cloud.html
7. Deelman, E., et al.: Pegasus, a workflow management system for science automation. Future Gener. Comput. Syst. **46**, 17–35 (2015)
8. Dziok, T.: Repository with optimization models (2015). https://bitbucket.org/tdziok/mgr-cloudplanner
9. Fard, H.M., Prodan, R., Fahringer, T.: A truthful dynamic workflow scheduling mechanism for commercial multicloud environments. IEEE Trans. Parallel Distrib. Syst. **24**(6), 1203–1212 (2013)
10. Figiela, K., Malawski, M.: Modeling, optimization and performance evaluation of scientific workflows in clouds. In: 2014 IEEE Fourth International Conference on Big Data and Cloud Computing, p. 280. IEEE, December 2014

11. Forrest, J.: Cbc (coin-or branch and cut) open-source mixed integer programming-solver (2012). https://projects.coin-or.org/Cbc
12. Genez, T.A.L., Bittencourt, L.F., Madeira, E.R.M.: Using time discretization to schedule scientific workflows in multiple cloud providers. In: 2013 IEEE Sixth International Conference on Cloud Computing, pp. 123–130. IEEE, June 2013
13. Juve, G., Chervenak, A., Deelman, E., Bharathi, S., Mehta, G., Vahi, K.: Characterizing and profiling scientific workflows. Future Gener. Comput. Syst. **29**(3), 682–692 (2013)
14. Malawski, M., Figiela, K., Bubak, M., Deelman, E., Nabrzyski, J.: Scheduling Multilevel Deadline-Constrained Scientific Workflows on Clouds Based on Cost Optimization. Scientific Programming, New York (2015)
15. Malawski, M., Figiela, K., Nabrzyski, J.: Cost minimization for computational applications on hybrid cloud infrastructures. Future Gener. Comput. Syst. **29**(7), 1786–1794 (2013)
16. Malawski, M., Juve, G., Deelman, E., Nabrzyski, J.: Algorithms for cost and deadline-constrained provisioning for scientific workflow ensembles in IaaS clouds. Future Gener. Comput. Syst. **48**, 1–18 (2015)
17. Mao, M., Humphrey, M.: Auto-scaling to minimize cost and meet application deadlines in cloud workflows. In: SC 2011. SC 2011, ACM, Seattle, Washington (2011)
18. Pietri, I., Juve, G., Deelman, E., Sakellariou, R.: A performance model to estimate execution time of scientific workflows on the cloud. In: Proceedings of the 9th Workshop on Workflows in Support of Large-Scale Science, pp. 11–19. WORKS 2014, IEEE Press, Piscataway, NJ, USA (2014)
19. Steglich, M.: CMPL (Coin mathematical programming language) (2015). https://projects.coin-or.org/Cmpl

Accelerating the Min-Min Heuristic

Martín Pedemonte[⊠], Pablo Ezzatti, and Álvaro Martín

Instituto de Computación, Universidad de la República,
11.300 Montevideo, Uruguay
{mpedemon,pezzatti,almartin}@fing.edu.uy

Abstract. Min-Min is a classical heuristic for scheduling tasks to heterogeneous computational resources, which has been applied either directly or as part of more sophisticated heuristics. The time complexity of the direct implementation of Min-Min is $O(mn^2)$ for scheduling n tasks on m machines. This has motivated the use of simpler heuristics and parallel implementations of Min-Min for the sake of acceptable runtimes in large scenarios. Recently, we have proposed an efficient algorithm for computing Min-Min, whose time complexity is $O(mn)$. In this work, we study mult-many core versions of this new algorithm. The experimental evaluation of our proposal shows important runtime reductions compared to the sequential version.

Keywords: Min-Min heuristic · Parallel implementation · Heterogeneous computing

1 Introduction

The performance of a *distributed heterogeneous computing (HC)* platform depends, to great extent, on the scheduling algorithm applied to assign tasks to computing resources. Since the number of tasks and resources in, for example, any Grid-like computational system, may be very large, and minimizing the overall execution time for a set of tasks is an NP-hard problem [6,7], several heuristic methods have been proposed and analyzed in the literature [2,13,26]. Of course, the overall performance depends on both the performance of the scheduling algorithm, which determines the time required to compute a schedule, and the quality of the schedule, which determines the time required to complete the execution of the set of tasks. *Min-Min* is a classical greedy heuristic [11, Algorithm D] that usually yields good schedules in moderate execution times [13] for small-medium scenarios. For this reason, Min-Min has been widely used as the base for more sophisticated scheduling algorithms [2,8,18,20,23,25,27], e.g., to generate initial solutions for population-based heuristics [24].

A direct implementation of Min-Min requires $O(mn^2)$ operations [11] to schedule n tasks on m machines, which makes impracticable its application when the number of tasks is large. For this reason, with the aim of reducing the computation time, some authors have proposed other heuristics [4,23] loosely inspired

© Springer International Publishing Switzerland 2016
R. Wyrzykowski et al. (Eds.): PPAM 2015, Part II, LNCS 9574, pp. 101–110, 2016.
DOI: 10.1007/978-3-319-32152-3_10

on Min-Min, and other authors have explored the use of parallel computing techniques, including multi-core implementations [19] and GPU-based implementations [3,19]. Recently, we have proposed an efficient sequential implementation of Min-Min that requires $O(mn \log n)$ operations[1] [5] and yields a great performance improvement in practice. A similar algorithm, based on a heap data structure, was independently presented in [21]. Similar ideas have also been proposed in [8] for a different problem setting.

In this paper, we study high performance computing (HPC) implementations of Min-Min based on this new sequential algorithm. Particularly, we propose three different parallel implementations, two shared memory variants and a hybrid CPU-GPU version, and we perform an experimental evaluation in order to understand the benefits of each proposal. The results demonstrate that HPC techniques yield important runtime reductions, which makes the construction of very large schedules in very short times possible (e.g., scheduling 128 K tasks on 4 K machines takes a few seconds). Beyond improving the performance in direct applications of Min-Min, these results encourage the use of this heuristic as part of more sophisticated scheduling schemes for large scenarios.

The rest of the paper is structured as follows. In Sect. 2, we introduce the task scheduling problem, the classical direct implementation and the recently proposed two-stage implementation of Min-Min. Then, in Sect. 3, we propose three different two-stage parallel implementations. These proposals are then evaluated experimentally in Sect. 4. Finally, we discuss some conclusions and future work in Sect. 5.

2   The Min-Min Scheduling Heuristic

Consider a set M of m machines and a set T of n tasks. For $J \in T$ and $i \in M$, we denote by $E(i, J)$ the *expected time to compute (ETC)* task J on machine i. A *schedule* S is a mapping that assigns a machine to each task, i.e., $S(J)$ denotes the machine where task J is assigned to be executed on. Given an ETC for all tasks and machines, the *makespan* of S, denoted $\hat{f}(S)$, is the estimated time required to execute *all* tasks according to schedule S, i.e.,

$$\hat{f}(S) = \max_{i \in M} \left\{ \sum_{J:S(J)=i} E(i, J) \right\}. \tag{1}$$

Since minimizing $\hat{f}(S)$ is computationally unaffordable for large problem instances, schedules are usually obtained in practice following a heuristic approach.

Algorithm 1 implements the Min-Min heuristic [11]. The algorithm iterates through a loop in Step 5 assigning, in each iteration, a task J_{min} to a machine i_{min} (Step 23). The loop is repeated while the variable U, which maintains the set of tasks that have not been assigned yet, is nonempty. The variable

---

[1] The time complexity is $O(mn)$ for fixed numeric precision implementations.

$t(i)$ maintains, for each machine i, the sum of the execution time of all tasks assigned so far to i. In each iteration of the loop in Step 5, J_{min} and i_{min} are selected such that the completion time of J_{min} on i_{min}, which is given by $t(i_{min}) + E(i_{min}, J_{min})$, is minimized among all machines in M and all tasks in U. This selection is formulated as a two-step minimization, implemented through two nested loops. The inner loop in Step 9, selects a task $\hat{J}$ that minimizes the completion time for a fixed machine i, where the minimization is over the set U of non-assigned tasks. The outer loop in Step 7 iterates over all machines $i \in M$, and selects a machine for which the completion time of the task $\hat{J}$ determined in the inner loop is minimum. As observed in [11, Theorem 1], it is readily verified that Algorithm 1 requires $\Theta(mn^2)$ operations.

> **input** : A set T, a set M, and the ETC $E(i, J)$ for all $J \in T, i \in M$
>
> 1 **foreach** $i \in M$ **do**
> 2 | $t(i) = 0$
> 3 **end**
> 4 Set $U = T$
> 5 **while** $U \neq \emptyset$ **do**
> 6 $minCT = +\infty$
> 7   **foreach** $i \in M$ **do**
> 8 $iMinCT = +\infty$
> 9   **foreach** $J \in U$ **do**
> 10 $cTime = t(i) + E(i, J)$
> 11   **if** $cTime < iMinCT$ **then**
> 12 $iMinCT = cTime$
> 13 $\hat{J} = J$
> 14   **end**
> 15   **end**
> 16   **if** $iMinCT < minCT$ **then**
> 17 $minCT = iMinCT$
> 18 $J_{min} = \hat{J}$
> 19 $i_{min} = i$
> 20   **end**
> 21   **end**
> 22 $t(i_{min}) = minCT$
> 23 $S(J_{min}) = i_{min}$
> 24 $U = U \setminus \{i_{min}\}$
> 25 **end**

Algorithm 1. Min-min heuristic.

2.1 An Efficient Sequential Algorithm

Notice that, for each machine i, Algorithm 1 selects tasks for execution on i in nondecreasing order of expected time to compute. Thus, the two-step minimization performed in Algorithm 1 to obtain J_{min} and i_{min} is simplified if, for each machine, the set T of tasks is sorted in that order. This idea is exploited in Algorithm 2, which we have proposed recently in [5]. For the sake of concreteness, we assume that tasks and machines are enumerated as $J = \{1 \ldots n\}$ and $M = \{1 \ldots m\}$, respectively, and the ETC is given by an $n \times m$ matrix E. We follow, loosely, Matlab style notation and denote by $E(i,:)$ the i-th row of E, which contains the ETC of all tasks for a fixed machine i.

Algorithm 2 consists of two stages. In Stage 1, the loop in Step 1 iterates over the set of machines and, for each $i \in M$, sorts, in Step 2, the elements of $E(i,:)$ in nondecreasing order. Simultaneously, Step 2 also computes a permutation $Perm(i,:)$ of the set $\{1 \ldots n\}$ such that $Perm(i,j)$, $1 \leq j \leq n$, is the task whose expected time to compute occupies position j in the sorted version of $E(i,:)$. In addition, for each machine i, a variable $n(i)$ maintains the smallest index within $Perm(i,:)$ such that the task $Perm(i,n(i))$ has not been assigned yet, and a variable $t(i)$ maintains the sum of the expected time to compute of all tasks assigned so far to i. In Stage 2, the loop in Step 7 follows essentially the same steps as Algorithm 1, except that, since tasks have been sorted, the loop in Step 9 of Algorithm 1, which selects a task $\hat{J}$ that minimizes the completion time, reduces to picking the first non-assigned task in the sorted array of tasks. This is performed by the loop in Step 10 of Algorithm 2.

The time complexity of Algorithm 2 depends on the implementation of Step 2. In the usual situation in which ETC matrix entries are represented with a fixed numeric precision, Algorithm 2 requires $O(mn)$ operations [5] using a radix exchange sort algorithm [10]; it requires $O(mn \log n)$ operations otherwise.

3   Parallel Implementations of Min-Min

In this section we present three variants of parallel implementations of Algorithm 2. Firstly, we describe a straightforward Matlab implementation, which just exploits parallel support of natively implemented Matlab routines. In Sect. 3.2 we describe a highly optimized C/C++ shared memory parallel implementation, and in Sect. 3.3, we discuss a hybrid CPU-GPU version.

3.1   A Simple Parallel Implementation

Our first parallel implementation of the Min-Min heuristic follows a simple yet efficient strategy, which consists in using the multithreaded sort function of Matlab [14]. Sort, as many other functions in Matlab, can automatically run on multiple computational threads in a multicore machine. This naive approach has the advantage that it is extremely simple and it does not require any parallel programming skills. This version will be referred to as *pM*.

3.2   An Optimized Shared Memory Parallel Implementation

In this subsection we describe *pSM*, a parallel shared memory implementation of Algorithm 2, implemented in C using the well known OpenMP API. Both stages of the algorithm are parallelized in *pSM*.

input : A set T, a set M, and the ETC $E(i, J)$ for all $J \in T, i \in M$
// Stage 1
1 **foreach** $i \in M$ **do**
2     $[E(i, :), Perm(i, :)] = sort(E(i, :))$
3     $t(i) = 0$
4     $n(i) = 1$
5 **end**
// Stage 2
6 Set $U = T$
7 **while** $U \neq \emptyset$ **do**
8     $minCT = +\infty$
9   **foreach** $i \in M$ **do**
10   **while** $Perm(i, n(i)) \notin U$ **do**
11           $n(i) = n(i) + 1$
12   **end**
13        $cTime = t(i) + E(i, n(i))$
14   **if** $cTime <= minCT$ **then**
15           $minCT = cTime$
16           $J_{min} = Perm(i, n(i))$
17           $i_{min} = i$
18   **end**
19   **end**
20     $t(i_{min}) = minCT$
21     $S(J_{min}) = i_{min}$
22     $U = U \setminus \{i_{min}\}$
23 **end**

Algorithm 2. An efficient implementation of Min-Min.

In Stage 1, since the set of tasks can be sorted independently for each machine, each of the m calls to sort in Step 2 can be computed in parallel. Since the distribution of the ETC of tasks for each machine is unknown a priori, we assume that independent sorts are equally balanced. For this reason, we use a static mapping between the chunks that need to be computed and the executing threads. Step 2 of Algorithm 2 is implemented using a sort routine adapted from the source code publicly available authored by Michael Herf [9], which implements the radix exchange sort algorithm [10] for floating point numbers.

In Stage 2, the selection of a machine i_{min} and task J_{min} with minimum completion time (Step 9) can also be parallelized; an optimal algorithm requires $m/p + \log \log p + O(1)$ comparisons on p processors [22]. In our implementation we use the reduction parallel pattern [15], where each thread performs a reduction independently on some subset of the set of machines M, and then the results are synchronized choosing the machine and task with the overall minimum completion time.

3.3 A Hybrid CPU-GPU Implementation

In this subsection, we explore the use of Graphics Processing Units (GPUs) to accelerate the computation of the Min-Min heuristic.

The architecture of GPUs is designed following the principle of devoting more transistors to computation than traditional CPUs. As a consequence, current GPUs have a large number of small cores and are usually referred to as *many-cores* processors. Based on the CUDA platform [17] for GPU programming, GPUs can be viewed as a set of shared memory multicore processors. When a CUDA kernel is called, a large number of threads are generated on the GPU (modern GPUs can execute thousands of threads in parallel). The group of all the threads generated by a kernel invocation is divided for their execution into warps that are the basic scheduling units in CUDA and consist of 32 consecutive threads. A warp executes one common instruction at a time; when threads of a warp have to execute different instructions (which is known as thread divergence), the execution is serialised.

In order to design an algorithm that runs efficiently on GPU, we should consider the following aspects: avoid thread divergence to fully exploit GPU resources, minimize data transfers between main memory and GPU since they are expensive operations, and make coalesced accesses to the GPU global memory since this reduces the number of memory transactions and thus increases the instruction throughput (warp accesses to global memory can be coalesced if they refer to addresses in the same segment of memory).

Stage 1 of Algorithm 2 is well suited to GPU, as there exist efficient algorithms to compute many sorts concurrently in a GPU. On the other hand, Stage 2 is not well suited to GPU implementation. Even though this stage is a reduction, the accesses to global memory are often not coalesced since each thread is processing the tasks of a different machine, and each thread runs out of work rapidly, wasting the potential of the platform.

For these reasons, we designed a hybrid CPU-GPU implementation that computes Stage 1 in the GPU and Stage 2 in the CPU. This version, which we refer to as *pCPU+GPU*, is implemented using the CUDA programming language. Step 2 of Algorithm 2 is implemented using the segmented sort routine authored by Sean Baxter [1], which uses merge sort [12] and the odd-even transposition algorithm [16] for sorting networks.

In *pCPU+GPU*, the ETC matrix and the original *Perm* matrix are completely transferred from main memory to the GPU, then the tasks are sorted for each machine in the GPU, and finally the updated *Perm* matrix is transferred back to main memory in order to proceed to the execution of Stage 2 in CPU.

4 Experimental Evaluation

The experimental platform is a PC with a 4 cores Intel i7-3770 processor at 3.40 GHz, 16 GB of RAM, using the CentOS Linux 6.3 operating system, connected to a Tesla K20c with 2496 CUDA cores at 705 MHz and 5 GB of RAM.

Table 1. Runtime in seconds (mean ± std) and speedup of pM.

Scenario	Single Thread Runtime	Four Threads	
		Runtime	Speedup
8192 × 256	0.252 ± 2.5e-3	0.169 ± 1.8e-3	1.49×
16384 × 512	1.093 ± 1.9e-3	0.716 ± 8.1e-3	1.53×
32768 × 1024	4.673 ± 1.3e-2	3.022 ± 8.7e-3	1.55×
65536 × 2048	23.373 ± 2.5e-1	16.457 ± 2.8e-1	1.42×
131072 × 4096	100.529 ± 9.4e-1	75.298 ± 1.2e0	1.34×

Table 2. Runtime in seconds (mean ± std) and speedup of pSM.

Scenario	Single Thread Runtime	Two Threads		Four Threads	
		Runtime	Speedup	Runtime	Speedup
8192 × 256	0.043 ± 6.9e-4	0.026 ± 7.2e-4	1.64×	0.021 ± 4.7e-4	1.99×
16384 × 512	0.148 ± 9.1e-4	0.082 ± 9.4e-4	1.80×	0.061 ± 1.1e-3	2.41×
32768 × 1024	0.770 ± 1.2e-2	0.381 ± 3.5e-3	2.02×	0.239 ± 2.1e-2	3.22×
65536 × 2048	4.364 ± 2.2e-2	2.277 ± 2.8e-2	1.92×	1.173 ± 2.4e-2	3.72×
131072 × 4096	20.740± 1.3e-1	10.512± 6.3e-2	1.97×	5.376 ± 1.1e-1	3.86×

The test set was designed to evaluate the performance of the parallel implementations of Algorithm 2 in several scenarios, with different number of tasks and machines. It consists of 5 scenarios, each composed of 20 instances of the same size. The number of tasks in the different scenarios ranges from 8192 to 131072, and the number of machines ranges from 256 to 4096 (8192 × 256, 16384 × 512, 32768 × 1024, 65536 × 2048 and 131072 × 4096). All the results reported are the average over the 20 different instances of each scenario on independent runs.

The instances for the three smallest scenarios were taken from the repository publicly available at http://par-cga-sched.gforge.uni.lu/instances/etc/. The instances for the two largest scenarios were created using the generator publicly available at http://www.fing.edu.uy/inco/grupos/cecal/hpc/HCSP.

In the first place, we analyze the performance of pM. Table 1 presents the execution time in seconds of pM running with one and four threads. The table also includes the speedup of the parallel four threaded execution over the sequential execution. The maximum speedup value obtained is 1.55, for the scenarios with 32768 tasks and 1024 machines, which is not close to a linear scalability (4×). These results are not surprising since pM only computes in parallel the first stage of Algorithm 2. Although the improvement in performance of the parallel execution is modest in magnitude, it is an interesting result from a practical point of view since pM involves no parallelization effort.

Let us now analyze the performance of pSM. Table 2 presents the runtime in seconds rounded to three figures of pSM executing with one, two and four threads. The table also includes the speedup of the parallel executions over the

Table 3. Runtime in seconds (mean ± std) and speedup of *pCPU+GPU* for Stage 1.

Scenario	pSM	pCPU+GPU	Speedup	pCPU+GPU	Speedup
	Sort	Sort	Sort	Total	Total
8192 × 256	0.006± 0.0e1	0.003 ± 4.7e-4	2.00×	0.009 ± 7.5e-4	0.64×
16384 × 512	0.026 ± 5.1e-4	0.014 ± 4.1e-4	1.86×	0.035 ± 1.2e-3	0.74×
32768 × 1024	0.119 ± 1.7e-3	0.060 ± 5.0e-4	1.98×	0.137 ± 1.3e-3	0.87×
65536 × 2048	0.805 ± 2.3e-2	0.260 ± 4.9e-4	3.10×	0.567 ± 3.6e-3	1.42×
131072 × 4096	3.218± 1.0e-1	1.139± 9.4e-4	2.86×	2.369 ± 2.0e-3	1.36×

sequential execution. The maximum speedup value of the two threaded execution is 2.02× for the scenarios with 32768 tasks and 1024 machines, and remains stable for larger scenarios. On the other hand, the maximum speedup value (3.86×) of the four threaded execution is obtained for the largest scenario, which is close to a linear speedup. The higher speedup values are obtained when solving instances from larger scenarios, showing that larger scenarios allow *pSM* to better profit from the parallel computation of the threads. These results indicate that Algorithm 2 is well suited to parallel computing.

Finally, we examine the performance of *pCPU+GPU*. Table 3 presents the runtime of the sort routine of the *pCPU+GPU* implementation and the total execution time for Stage 1 on the GPU, including the data transfers from main memory to GPU and in the opposite direction. The table also includes the speedups over the *pSM* execution with four threads of the sort routine.

For the settings of this experiment, we observe that the total speedup increases with the instance size except for the largest scenario. The explanation for the slight degradation in relative performance with respect to the 65536×2048 scenario is that, due to GPU memory restrictions, the computation of Stage 1 of Algorithm 2 for the largest scenario needs to be divided in two parts. Indeed, every instance requires 4 GB of RAM to store the ETC and permutation matrices, which, together with the additional memory required by the sort library, makes impossible to store the whole data in the 5 GB of RAM of the Tesla K20c.[2]

Table 3 shows that the GPU outperforms the CPU for sorting the tasks in all cases and, in the largest two scenarios, the difference in performance even compensates the transfer overhead, yielding a global speedup greater than unity. These results demonstrate that involving GPUs for solving Stage 1 of Algorithm 2 can be highly profitable (notice from Tables 2 and 3 that Stage 1 consumes more than half of the total time in *pSM* for these two scenarios). In practice this stage can be solved concurrently in CPU and GPU by partitioning the rows of the ETC matrix in two sets, and sorting each set in a different type of processing unit. A partition that yields the best load balance depends on the specific hardware characteristics of the CPU, the GPU, and the transfer bus.

---

[2] The real available free memory is 4.92 GB.

5   Conclusions and Future Work

In this article we have studied, from an empirical perspective, the application of parallel techniques to accelerate recent implementations of Min-Min.

The parallel implementations proposed in this work achieve important runtime reductions. Specifically, both *pSM* and *pCPU+GPU* are able to solve instances with 65536 tasks and 2048 machines in about one second, and instances with 131072 tasks and 4096 machines in about five seconds. These improvements allow using Min-Min for the scheduling of very large scenarios with a negligible overhead and, additionally, help to include Min-Min as a part of more sophisticated scheduling strategies.

As part of future research, the results in this paper could be complemented with a model capable of estimating the performance of the aforementioned HPC variants of Min-Min on different platforms, as a function of the problem instance size and certain hardware specifications, such us size and performance of cache memories, RAM memory, and the GPU features. Another line of future work is to address the scheduling problem in a dynamic environment using Min-Min. Specifically, after Stage 1 is completed, the first tasks of the schedule can start executing as soon as they are assigned in the first iterations of Stage 2. When new tasks arrive in the meanwhile, they can be efficiently inserted if the tasks are maintained in sorted lists. On the other hand, if a new machine becomes available, the set of pending tasks need to be sorted for the new machine. The small execution times reported in this paper make this scheme faceable.

Acknowledgment. The authors acknowledge support from Programa de Desarrollo de las Ciencias Básicas, Uruguay, and Sistema Nacional de Investigadores, Uruguay.

References

1. Baxter, S.: Modern gpu. http://nvlabs.github.io/moderngpu/. Accessed April 2015
2. Braun, T.D., Siegel, H.J., Beck, N., Bölöni, L.L., Maheswaran, M., Reuther, A.I., Robertson, J.P., Theys, M.D., Yao, B., Hensgen, D., Freund, R.F.: A comparison of eleven static heuristics for mapping a class of independent tasks onto heterogeneous distributed computing systems. J. Parallel Distrib. Comput. **61**(6), 810–837 (2001)
3. Canabé, M., Nesmachnow, S.: Parallel implementations of the MinMin heterogeneous computing scheduler in GPU. In: V Latin American Symposium on High Performance Computing. HPCLatam (2012). www.clei.cl/cleiej/papers/v15i3p8.pdf
4. Diaz, C.O., Guzek, M., Pecero, J.E., Danoy, G., Bouvry, P., Khan, S.U.: Energy-aware fast scheduling heuristics in heterogeneous computing systems. In: Proceedings of the 2011 International Conference on High Performance Computing & Simulation (HPCS 2011), pp. 478–484 (2011)
5. Ezzatti, P., Pedemonte, M., Martín, A.: An efficient implementation of the min-min heuristic. Comput. Oper. Res. **40**(11), 2670–2676 (2013)
6. Fernandez-Baca, D.: Allocating modules to processors in a distributed system. IEEE Trans. Softw. Eng. **15**(11), 1427–1436 (1989)
7. Garey, M.R., Johnson, D.S.: Computers and Intractability: A Guide to the Theory of NP-Completeness. W. H. Freeman & Co., New York (1979)

8. Giersch, A., Robert, Y., Vivien, F.: Scheduling tasks sharing files on heterogeneous master-slave platforms. J. Syst. Archit. **52**(2), 88–104 (2006)
9. Herf, M.: Radix tricks. http://stereopsis.com/radix.html. Accessed April 2015
10. Hildebrandt, P., Isbitz, H.: Radix exchange-an internal sorting method for digital computers. J. ACM **6**(2), 156–163 (1959)
11. Ibarra, O.H., Kim, C.E.: Heuristic algorithms for scheduling independent tasks on nonidentical processors. J. ACM **24**(2), 280–289 (1977)
12. Knuth, D.E.: The Art of Computer Programming, Sorting and Searching, vol. 3, 2nd edn. Addison Wesley Longman Publishing Co., Inc., Redwood City (1998)
13. Luo, P., Lü, K., Shi, Z.: A revisit of fast greedy heuristics for mapping a class of independent tasks onto heterogeneous computing systems. J. Parallel Distrib. Comput. **67**(6), 695–714 (2007)
14. Mathworks: Multicore matlab. http://www.mathworks.com/discovery/multicore-matlab.html. Accessed April 2015
15. McCool, M.D., Robison, A.D., Reinders, J.: Structured Parallel Programming: Patterns for Efficient Computation. Morgan Kaufmann, Burlington (2012)
16. Haberman, N.: Parallel neighbor sort (or the glory of the induction principle). Technical report 2087, Computer Science Department, Carnegie Mellon University (1972)
17. Nvidia Corporation: CUDA C Programming Guide Version 5.5. Nvidia Corporation (2013)
18. Pinel, F., Dorronsoro, B., Pecero, J., Bouvry, P., Khan, S.: A two-phase heuristic for the energy-efficient scheduling of independent tasks on computational grids. Cluster Comput. **16**, 1–13 (2012)
19. Pinel, F., Dorronsoro, B., Bouvry, P.: Solving very large instances of the scheduling of independent tasks problem on the GPU. J. Parallel Distrib. Comput. **73**(1), 101–110 (2013)
20. Ritchie, G., Levine, J.: A fast, effective local search for scheduling independent jobs in heterogeneous computing environments. In: PLANSIG 2003: Proceedings of the 22nd Workshop of the UK Planning and Scheduling Special Interest Group, pp. 178–183, December 2003
21. Tabak, E., Cambazoglu, B., Aykanat, C.: Improving the performance of independent task assignment heuristics minmin, maxmin and sufferage. IEEE Trans. Parallel Distrib. Syst. **25**(5), 1244–1256 (2013)
22. Valiant, L.G.: Parallelism in comparison problems. SIAM J. Comput. **4**(3), 348–355 (1975)
23. Wu, M.Y., Shu, W., Zhang, H.: Segmented min-min: A static mapping algorithm for meta-tasks on heterogeneous computing systems. In: Proceedings of the 9th Heterogeneous Computing Workshop, HCW 2000, pp. 375–385. IEEE Computer Society, Washington, DC (2000)
24. Xhafa, F., Abraham, A.: Computational models and heuristic methods for grid scheduling problems. Future Gener. Comput. Syst. **26**(4), 608–621 (2010)
25. Xhafa, F., Alba, E., Dorronsoro, B., Duran, B.: Efficient batch job scheduling in grids using cellular memetic algorithms. J. Math. Model. Algorithms **7**, 217–236 (2008)
26. Xhafa, F., Barolli, L., Durresi, A.: Batch mode scheduling in grid systems. Int. J. Web Grid Serv. **3**(1), 19–37 (2007)
27. Xhafa, F., Carretero, J., Dorronsoro, B., Alba, E.: A TABU search algorithm for scheduling independent jobs in computational grids. Comput. Artif. Intell. **28**(2), 237–250 (2009)

Divisible Loads Scheduling in Hierarchical Memory Systems with Time and Energy Constraints

Maciej Drozdowski[✉] and Jędrzej M. Marszałkowski

Institute of Computing Science, Poznań University of Technology,
Piotrowo 2, 60-965 Poznań, Poland
{Maciej.Drozdowski,Jedrzej.Marszalkowski}@cs.put.poznan.pl

Abstract. In this paper we consider scheduling distributed divisible computations in systems with hierarchical memory for energy and time performance criteria. Hierarchical memory allows to conduct computations on big data sets using out-of-core processing instead of coercing application data fit into core storage. However, out-of-core computations are more costly both in time and energy. A model for scheduling divisible loads under time and energy criteria is introduced. Two types of scheduling algorithms are proposed and evaluated: a single-installment algorithm which builds optimum schedules but may use out-of-core storage, and a set of multi-installment algorithms which use limited memory but require more communications.

Keywords: Scheduling · Divisible loads · Hierarchical memory · Energy efficiency · Performance evaluation

1 Introduction

Providing electricity and bearing its cost has become a key element in designing and running big data centers and supercomputing installations [9]. Dissipating heat generated in computations is currently one of the limitations to the further growth of the CPU speeds [8]. Hence, energy efficiency is a very active research area and recent advantages in this field are closely analyzed [14].

In this paper we study the trade-off between time performance and energy cost in processing divisible loads on systems with hierarchical memory. Divisible loads are data-parallel applications which can be divided into parts of arbitrary sizes, and the parts can be processed independently in parallel. Divisible load theory (DLT) has been proposed in [1,4] to analyze performance of distributed computations and schedule them accordingly. Thus, DLT provides methods of scheduling and analyzing performance of a broad class of distributed applications operating on big data volumes [2,3,5,12]. Contemporary computer systems have hierarchical memory organization. At the top of the hierarchy CPU registers have the shortest access time, but they are scarcest. Processor caches establish the next level of memory hierarchy. Main memory, here by convention referred

© Springer International Publishing Switzerland 2016
R. Wyrzykowski et al. (Eds.): PPAM 2015, Part II, LNCS 9574, pp. 111–120, 2016.
DOI: 10.1007/978-3-319-32152-3_11

to as RAM, has much bigger size but is again slower. The following levels of memory hierarchy are based on external and networked storage: HDDs, NAS, tapes, optical media, etc. In this study we reduce the above hierarchy to just two types of memory: *core* comprising registers, caches, RAM, and *out-of-core* memory comprising all types of external storage. This partitioning has a practical motivation. On the one hand, sizes of data (load) processed in big data applications far exceed size of CPU registers and caches. Hence, to a great extent, core is transparent for a developer of such applications. On the other hand, core accesses are managed by hardware, while out-of-core memory is accessed via software wrappers (virtual memory, (networked) file systems), and consequently, it is by orders of magnitude slower. Due to the limited core size a developer must undertake steps to fit data in core. Contrarily, out-of-core storage offers nearly unlimited storage but requires use of virtual memory or dedicated data management subsystem [11]. Consequently, on-core and out-of-core computations have different character both in the development and in performance.

Systems with hierarchical memory have been analyzed in DLT [7]. Energy may be considered a special type of cost in DLT. Scheduling with monetary cost has been considered in [13]. Energy in processing divisible loads on flat memory systems has been subject of [6]. In this paper we combine nonlinear energy consumption and computing time models specific for systems with hierarchical memory. We analyze two types of solutions: a single-installment method which sends load to processors once and multi-installment algorithms which send the load in many iterations. Since the problem is bicriterial the trade-off between time and energy will be analyzed.

Further organization of this work is as follows. In the next section we formulate the scheduling problem and provide timing and energy use models. In Sect. 3 algorithms solving the problem are proposed. Section 4 is dedicated to evaluation of the proposed methods. Section 5 summarizes results of this work.

2   Problem Formulation

It is assumed that computations are performed in a single-level tree system with root M_0 (a.k.a. master, server, originator) and machines (computers,processors) $M_1, \ldots, M_m$ at the leaves. The machines can be in one of four states: (1) idle - consuming power P^I, (2) starting - which takes time S and power P^S, (3) networking - using power P^N, (4) computing. Busy-waiting is considered the same as networking state. Initially volume V of load is held by M_0, M_0 is in the networking state, $M_1, \ldots, M_m$ are idle. M_0 activates $M_1, \ldots, M_m$ which takes energy SP^S on each machine. Next load V is distributed in parts to machines $M_1, \ldots, M_m$. Transferring α units of load to M_i takes time αC, where C is communication rate (in seconds per byte). M_0 sends the load to processors one after the other, i.e. load is distributed to slaves in the sequential manner. M_0 activates M_is just-in-time which means that completion of the starting procedure coincides with the beginning of receiving of the load to process. For simplicity of exposition we assume that the time of returning results from M_is to M_0 is very short compared to the whole schedule length T and can be neglected (this can be easily relaxed in DLT [3,5,12]). The duration and energy cost of sending the waking signal is negligible and starting some machine M_j can be performed in parallel with some other machine M_i communicating with M_0.

The time and energy of computations on load α depend on the load size, precisely whether the load part fits in the main memory [7,10]. It is assumed that the time of computing on load of size α is determined by a piecewise-linear function $\tau(\alpha) = \max\{a_1\alpha, a_2\alpha + b_2\}$. The first component of τ corresponds with computations in core with speed $1/a_1$, the second component represents out-of-core computations. Function τ has two properties: $\tau(0) = 0$ and $\tau(\rho) = a_1\rho = a_2\rho + b_2$, where ρ is the size of main memory (RAM) available to the application beyond which system starts using out-of-core memory. The energy consumed in the computations is determined by an analogous function $\varepsilon(\alpha) = \max\{k_1\alpha, k_2\alpha + l_2\}$ satisfying conditions $\varepsilon(0) = 0, \varepsilon(\rho) = k_1\rho = k_2\rho + l_2$. The problem considered here consists in constructing a schedule of minimum length T and energy E. Since this problem is effectively bicriterial we will be solving energy E minimization problem under constrained schedule length T.

3   Solution Methods

In this section we propose two strategies of load distribution. The first sends the load to machines once. Consequently, load parts can be big and out-of-core processing may be unavoidable. The second, iteratively distributes load chunks of small size in multiple communications.

3.1   Optimum Single-Installment

A schedule for the current method is shown in Fig. 1a. In the schedule M_0 busy-waits S units of time for M_1 initiation, then load V is distributed in parts $\alpha_1, \ldots, \alpha_m$ to machines $M_1, \ldots, M_m$, respectively. M_0 communicates continuously for time $C\sum_{i=1}^{m} \alpha_i = CV$ and switches off. Thus, in the schedule of length T, M_0 consumes energy

$$E_0 = P^N(CV + S) + P^I(T - CV - S).$$

Machine M_i remains idle until time $C\sum_{i=1}^{i-1} \alpha_i$ (where $\sum_{i=1}^{0} \alpha_i = 0$), starts in time S, receives its part of load in time $C\alpha_i$, computes it in time $\tau(\alpha_i)$ and switches off. Let us denote by $t_i = \tau(\alpha_i)$ the time of computations on M_i and by $e_i = \varepsilon(\alpha_i)$ the energy consumed in these computations. The duration of idle intervals on machine M_i is $T - S - C\alpha_i - t_i$. The energy consumed by M_i is

$$E_i = P^S S + P^N C\alpha_i + e_i + P^I(T - S - C\alpha_i - t_i)$$

The problem of minimizing energy consumption E under limited schedule length T can be formulated as a linear program:

$$\min \sum_{i=0}^{m} E_i \tag{1}$$

$$S + C \sum_{j=1}^{i} \alpha_j + t_i \leq T \quad i = 1, \ldots, m \tag{2}$$

$$\max\{a_1\alpha_i, a_2\alpha_i + b_2\} = t_i \quad i = 1, \ldots, m \tag{3}$$

$$\max\{k_1\alpha_i, k_2\alpha_i + l_2\} = e_i \quad i = 1, \ldots, m \tag{4}$$

$$\sum_{i=1}^{m} \alpha_i = V \tag{5}$$

$$\alpha_i, t_i, e_i \geq 0 \quad i = 1, \ldots, m \tag{6}$$

In the above formulation inequality (2) guarantees feasibility of the schedule on each processor. Constraints (3), (4) instantiate functions $\tau(\alpha), \varepsilon(\alpha)$. We present constraints (3), (4) in a simplified form which is accepted by contemporary solvers (e.g. CPLEX), but it can be implemented in any LP solver by splitting the max function into two inequalities and adding cost of exceeding constraints. By Eq. (5) all work is executed.

3.2   Multi-Installment Methods

In the next three algorithms M_0 sends load chunks of equal size α. Actual methods of calculating α for each specific algorithm will be given in the following. The sequence of communications to $M_1, \ldots, M_m$ is repeated iteratively until exhausting the load. The number of communications may be indivisible by m and the size α^f of the last sent chunk may be smaller than α. It is assumed that computations on each of the machines $M_1, \ldots, M_m$ last longer than sending the load to the remaining $m - 1$ machines. This imposes a requirement that $(m-1)C\alpha \leq \tau(\alpha)$ which can be reformulated as $m \leq a_1/C + 1$ for $\alpha \leq \rho$ and $m \leq a_2/C + 1 + b_2/(C\alpha)$ for $\alpha > \rho$. Thus, the number of processors which can be effectively exploited is limited and it is bigger when slower out-of-core processing takes place. Now we derive schedule length T and energy E used when chunks of size α are applied. For simplicity of exposition let $m > 1$.

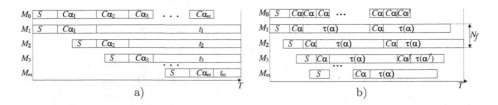

Fig. 1. (a) Single-installment schedule. (b) Multi-installment schedule.

The number of complete distribution iterations in which each of m machines obtain load α is $N_o = \lfloor \frac{V}{\alpha m} \rfloor$. The number of chunks of size α in the last incomplete iteration is $N_f = \lfloor (V - N_o m\alpha)/\alpha \rfloor$. Size of the last chunk is $\alpha^f = V - (mN_o + N_f)\alpha$. Then, the schedule length is (cf. Fig. 1b):

$$T = S + N_o(C\alpha + \tau(\alpha)) + \begin{cases} N_f C\alpha + \max\{\alpha^f C + \tau(\alpha^f), \tau(\alpha)\} & N_f > 0 \\ \max\{(m-1)C\alpha, \alpha^f C + \tau(\alpha^f)\} & N_f = 0 \end{cases}$$

Deriving energy consumption requires calculating idle time, computing and communication durations on $M_1, \ldots, M_m$. At the start of the schedule M_i is idle until time $C(i-1)\alpha$. Thus, total energy used before machines activation is $E_1^I = P^I \sum_{i=1}^m (i-1)C\alpha = P^I(m-1)m/2C\alpha$. Starting m machines consumes $E^S = P^S mS$ units of energy. Energy consumed on $M_1, \ldots, M_m$ in the computations and communications is $E^R = (N_o m + N_f)(P^N C\alpha + \varepsilon(\alpha)) + P^N C\alpha^f + \varepsilon(\alpha^f)$.

Let us assume that $\alpha^f C + \tau(\alpha^f) < \tau(\alpha)$, i.e., the schedule ends on the last machine receiving a chunk of size α (see Fig. 1b). The idle time on $M_i \in \{M_1, \ldots, M_{N_f}\}$ is $(N_f - i)C\alpha$, on M_{N_f+1} it is $\tau(\alpha) - C\alpha^f - \tau(\alpha^f)$, and on $M_i \in \{M_{N_f+2}, \ldots, M_m\}$ it is $\tau(\alpha) - (i - N_f - 1)C\alpha$. Thus, total idle time on $M_1, \ldots, M_m$ at the end of the schedule is

$$I = \sum_{i=1}^{N_f} (N_f - i)C\alpha + \tau(\alpha) - C\alpha^f - \tau(\alpha^f) + \sum_{i=N_f+2}^m (\tau(\alpha) - (i - N_f - 1)C\alpha).$$

Suppose that $\alpha^f C + \tau(\alpha^f) \geq \tau(\alpha)$, which means that M_{N_f+1} has no idle time. Idle time on machines $M_i \in \{M_1, \ldots, M_{N_f}\}$ is $(N_f - i)C\alpha + \tau(\alpha^f) + C\alpha^f - \tau(\alpha)$ and on $M_i \in \{M_{N_f+2}, \ldots, M_m\}$ it is $\tau(\alpha^f) + C\alpha^f - (i - N_f - 1)C\alpha$. Hence, total idle time on $M_1, \ldots, M_m$ at the end of the schedule is

$$I = \sum_{i=1}^{N_f} ((N_f - i)C\alpha + \tau(\alpha^f) + C\alpha^f - \tau(\alpha)) + \sum_{i=N_f+2}^m (\tau(\alpha^f) + C\alpha^f - (i - N_f - 1)C\alpha).$$

Energy wasted in idle waiting at the end of the schedule is $E_2^I = P^I I$.

It remains to calculate the energy consumed by the originator. M_0 starts in networking state and then it is continuously communicating or busy-waiting until distributing the last piece of work. The idle time on M_0 is $\max\{\tau(\alpha) - C(\alpha^f), \tau(\alpha^f)\}$. Hence, the energy consumed on M_0 is $E_0 = P^N T + (P^I - P^N)\max\{\tau(\alpha) - C(\alpha^f), \tau(\alpha^f)\}$. Finally, total energy consumed by the methods using load chunks of fixed size α is

$$E = E_1^I + E^S + E^R + E_2^I + E_0.$$

Below we outline multi-installment scheduling algorithms with their specific ways of defining load chunk sizes α.

Simple Static Chunk (SSC) algorithm assumes that load chunk sizes are equal to the size of available RAM memory, i.e. $\alpha_{SSC} = \rho$. Thus, SSC avoids using out-of-core memory. A disadvantage of simple static chunk algorithm are the final outstanding load chunks. It means that if $q_1 = \lceil V/(\rho m) \rceil \neq \lfloor V/(\rho m) \rfloor = q_2$ then in the last iteration of load distribution many processors may remain idle.

Static Chunk with Underload (SCU) algorithm assumes $\alpha_{SCU} = V/(q_1 m)$. Thus, algorithm SCU sends load chunks of size at most ρ and avoids out-of-core processing at the cost of one more iteration.

Static Chunk with Overload (SCO) attempts to round the number of communication iterations down, at the cost of possibly using out-of-core processing. Hence, in SCO size of the load chunk is $\alpha_{SCO} = V/(m \max\{1, q_2\})$.

Guided Self-Scheduling (GSS) algorithm adapts the idea of the classic loop scheduling algorithm [5]. Let V' be the size of load remaining on M_0. Chunk sizes are calculated as $\alpha_{GSS} = \min\{V', \max\{1, \min\{V'/m, \rho\}\}\}$. For $V > \rho$, the algorithm starts with load chunk sizes of RAM size. When $V' < \rho$, GSS gradually decreases chunk sizes and thus minimizes the spread of machine completion times. GSS does not send load chunk sizes smaller than some fixed size which is denoted here as 1 by convention. This can be a result of data structures representing the solved problem or some size which sufficiently amortizes fixed overheads in processing one chunk. For $V \gg m\rho$ the maximum number of usable processors in GSS is the same as in the previous algorithms because initial load chunks have size ρ. However, if $V < m\rho$ GSS uses chunks smaller than ρ, chunk sizes decrease and communications are getting shorter. In such a situation GSS is able to start more machines than SSC, SCU, SCO without entailing idle time on $M_1, \ldots, M_m$.

4   Performance Comparison

In this section we will analyze performance as consumed energy E vs schedule length T. We will also analyze sensitivity of the algorithms to changing problem size V and system parameters. Note, that only the single-installment (SI) method is capable of changing energy consumption E with changing T. A study of the E vs T trade-off computed by SI can be found in [10]. The multi-installment methods do not offer such a trade-off and the only parameter which can impact E and T, given computing system and problem size, is the number of machines m. Hence, in the following figures we study impact of m on E and T. In order to compare SI against multi-installment methods, the shortest schedules on the given number of machines m will be used for SI method. Unless stated to be otherwise the system and application parameters were the following: $V = 10\,\text{GB}$, $a_1 = 0.082\,\text{s/MB}$, $a_2 = 2.366\,\text{s/MB}$, $b_2 = -2274.9\,\text{s}$, $k_1 = 13\,\text{J/MB}$, $k_2 = 294\,\text{J/MB}$, $l_2 = -280\,\text{kJ}$, $C = 7.8\,\text{ms/MB}$, $S = 10\,\text{s}$, $P^I = 14\,\text{W}$, $P^N = 91\,\text{W}$, $P^S = 101\,\text{W}$, $\rho = 996\,\text{MB}$. It can be verified that processing out-of-core is roughly 28 times slower per MB than processing on-core. The energy consumption per MB is roughly 23 times higher out-of-core. Communication rate C corresponds with communication speed of $\approx 1\,\text{Gb}$ per second. P^I, S, P^S represent a very light-weight system which quite effectively switches from hibernation to the running state. The size of RAM accessible for storing data is $\rho = 996\,\text{MB}$. These values have been measured in a real system, for an application consisting in searching for patterns in a big data file [6, 10].

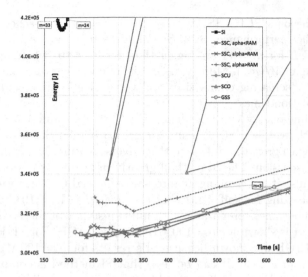

Fig. 2. Time-energy diagram for the default system.

We start with a time-energy chart in Fig. 2 for the above reference parameters, to introduce the phenomena guiding performance. The dependencies are shown only partially for better visibility (but will be shown in their entirety in the next figure). It can be observed that with growing m not only T decreases but so does used energy E. Hence, the smallest m is shown on the right-hand-side of the chart. Energy performance is ruled by the following effects. On the one hand, growing number of machines shortens the schedule and the root M_0 is using less energy. On the other hand, adding machines incurs energy cost. As a result, it can be observed that energy first decreases with shortening of the schedule, but then is starts to increase. This phenomenon can be observed in the following figures. The shortest schedules are built by the single installment method (SI, in the upper-left corner), but using $m = 24$ and more machines has big cost in energy needed to start them. At these values of m it is possible to fit the whole load V in core memories. Note that SI has apparent energy use minimum at $m \approx 28$. Big irregularities in time and energy can be observed in SCO. Since V is not always divisible by $m\rho$ and rounding chunk sizes up results in various values of the difference between α and ρ, even small excesses of chunk sizes above ρ escalate time and energy consumption. Consequently, SCO has big irregularity in performance and should be avoided. Results for the simple static chunk (SSC) algorithm are shown for three chunk sizes: 680 MB, 996 MB, 998 MB, where $\rho = 996$ MB. It can be seen that even small increase of the chunk size beyond ρ has bad impact on the energy use. Chunks smaller than ρ have advantage of shorter waiting time at the start of the schedule and better load balance at its end. Hence, a small dominance of SSC with $\alpha < \rho$ for the maximum usable number of machines. For the given parameters the maximum number of processors which can be applied without idle time is $m = 11$. Static chunk with underload (SCU), SSC with $\alpha < \rho$ and guided-self-scheduling (GSS) have very similar performance. Still, SCU suffers from minor irregularities in performance (T, E for $m = 11$ are bigger than for $m = 10$) which are results of uneven rounding of $V/(m\rho)$. Moreover, GSS is able to construct slightly shorter schedule due to decreasing chunk sizes and consequently smaller dispersion of processor completion times.

In Fig. 3a time-energy chart is shown for $V = 10\,\text{G}$ and $V = 100\,\text{G}$. The static chunk with overload (SCO) manifests great irregularities because T, E are not monotonic with growing m. Due to this adverse feature SCO will be omitted in the further discussion. The SI method greatly improves its performance with growing m because it is becoming able to shift the load from the out-of-core to the on-core processing for sufficiently big m. Finally, at $V = 10\,\text{G}$ and $m > 11$ its performance becomes comparable with multi-installment methods. In Fig. 3b time-energy chart is shown for $\rho = 100\,\text{MB}$ and $\rho = 10\,\text{GB}$. For SI dependencies for $\rho = 1\,\text{GB}, 10\,\text{GB}$ are shown because SI's results for $\rho = 100\,\text{MB}$ are out of the range shown in Fig. 3b. It can be seen that SI method is competitive with the remaining algorithms only if the load is stored in core. What is more, under such circumstances SI is able to build the best energy schedules (lower-left part of the chart). SI is capable of constructing shorter schedules, but it activates new machines which brings energy costs bigger than in the other methods. SSC method for $\rho = 10\,\text{G}$ uses just one load chunk, schedule length T is constant, and adding each new machine only increases energy costs. Surprisingly, energy performance of the multi-installment methods for small $\rho = 100\,\text{MB}$ is better than for $\rho = 10\,\text{GB}$ because small load chunks reduce initial and final idle times. It can be also observed that GSS for $\rho = 10\,\text{GB}$ is capable of constructing shorter schedules than other multi-installment methods because by shrinking chunk sizes it is able to avoid idle times on processors and still activate more of them, though using more energy. Both GSS and SI approach the minimum

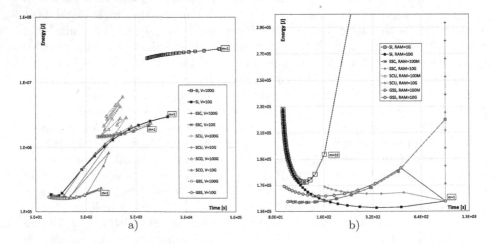

Fig. 3. Time-energy dependence (a) for $V = 10\,\text{G}$ and $V = 100\,\text{G}$. (b) for varying ρ.

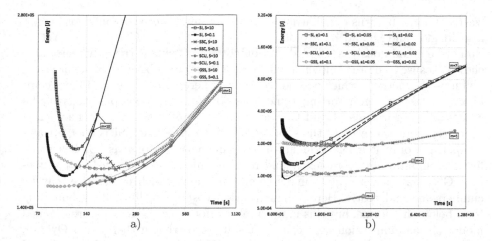

Fig. 4. Time-energy dependence (a) for $S = 10$ s and $S = 0.1$ s, (b) for changing a_1.

schedule length determined by communication time: $S + CV$. However, GSS is more energy-efficient.

In Fig. 4a time-energy relation is shown for two values of the startup time $S = 0.1$ s and $S = 10$ s. Two effects of reducing startup time can be observed. The schedules get shorter roughly by the startup time of the first processor, and energy consumption is decreased by the amount of energy saved in the startup of the machines. In Fig. 4b impact of changing processing rate a_1 is analyzed. The value of a_1 can be changed by designing a faster algorithm to solve the considered problem. Assuming, that this new application runs on the same computer, also k_1 must decrease proportionally. Three values of a_1 are shown: $a_1 = 0.1, 0.05, 0.02$ which corresponds with an algorithm twice and five time faster. The number of processors which can be activated by algorithms SSC, SCU decreases with increasing processing speed (a_1 decreases). Hence, this number decreases from $m = 13$ machines for $a_1 = 0.1$ to $m = 3$ for $a_1 = 0.02$. Though time- and energy-performance of all multi-installment algorithms is similar, GSS algorithm has an advantage of using more machines than SSC, SCU and consequently building shorter schedules though at higher energy costs. The SI method is able to construct schedules of comparable length but by using more machines and energy. The advantage in energy of multi-installment methods over SI grows with decreasing a_1 (i.e. increasing speed).

5   Conclusions

In this paper time- and energy-performance of scheduling algorithms for divisible computations in systems with hierarchical memory has been studied. The time- and energy-performance is determined by: (i) size of load chunks which regulates on-/out-of-core processing, (ii) number of usable processors which decide on minimum schedule length, (iii) amount of idle time which rule wasted energy. It turns out that intensive use of out-of-core computations is not a good idea and should be avoided as demonstrated by SCO method. Yet, it cannot be unanimously concluded that on-core processing is the only reasonable choice because in more complex applications the results obtained in small pieces still must be merged (which was not considered here). Hence, in such more complex applications, e.g. in sorting, some degree of out-of-core computations maybe acceptable. Algorithms SI and GSS are able to employ the biggest number of processors and hence build the shortest schedules. However, SI is energetically competitive only if whole load fits in core memories. Moreover, SI has much higher computational complexity and requires information on system parameters. GSS may perform quite many communications which in practice may be cumbersome and costly. Here communication costs were limited by bounding from below chunk size, but this may cripple performance. Thus, a more detailed model of communication cost can be a subject of the further work. Overall, GSS algorithm can be recommended as a good compromise of performance and implementation simplicity.

References

1. Agrawal, R., Jagadish, H.V.: Partitioning techniques for large-grained parallelism. IEEE Trans. Comput. **37**, 1627–1634 (1988)
2. Berlińska, J., Drozdowski, M.: Scheduling divisible MapReduce computations. J. Parallel Distrib. Comput. **71**, 450–459 (2011)
3. Bharadwaj, V., Ghose, D., Mani, V., Robertazzi, T.: Scheduling Divisible Loads in Parallel and Distributed Systems. IEEE Computer Society Press, Los Alamitos (1996)
4. Cheng, Y.-C., Robertazzi, T.G.: Distributed computation with communication delay. IEEE Trans. Aerosp. Electron. Syst. **24**, 700–712 (1988)
5. Drozdowski, M.: Scheduling for Parallel Processing. Springer, London (2009)
6. Drozdowski, M., Marszałkowski, J.M., Marszałkowski, J.: Energy trade-offs analysis using equal-energy maps. Future Gener. Comput. Syst. **36**, 311–321 (2014)
7. Drozdowski, M., Wolniewicz, P.: Out-of-core divisible load processing. IEEE Trans. Parallel Distrib. Syst. **14**, 1048–1056 (2003)
8. Fuller, S.H., Millett, L.I.: Computing performance: game over or next level? Computer **41**, 31–38 (2011)
9. Katz, R.H.: Tech titans building boom. IEEE Spectr. **46**(INT), 36–49 (2009). http://www.spectrum.ieee.org/feb09/7327
10. Marszałkowski, J.M., Drozdowski, M., Marszałkowski, J.: Time and energy performance of parallel systems with hierarchical memory. J. Grid Comput. (2015, accepted). doi:10.1007/s10723-015-9345-8
11. Mills, R.T., Yue, C., Stathopoulos, A., Nikolopoulos, D.S.: Runtime and programming support for memory adaptation in scientific applications via local disk and remote memory. J. Grid Comput. **5**, 213–234 (2007)
12. Robertazzi, T.: Ten reasons to use divisible load theory. IEEE Comput. **36**, 63–68 (2003)
13. Sohn, J., Robertazzi, T.G., Luryi, S.: Optimizing computing costs using divisible load analysis. IEEE Trans. Parallel Distrib. Syst. **9**, 225–234 (1998)
14. The Green 500, November 2014. http://www.green500.org/

The 6th Workshop on Language-Based Parallel Programming Models (WLPP 2015)

Extending Gustafson-Barsis's Law
for Dual-Architecture Computing

Ami Marowka[✉]

Parallel Research Lab, Jerusalem, Israel
amimar2@yahoo.com

Abstract. This study has investigated how *scaled* performance is affected by the energy constraints imposed on dual-architecture processors. Theoretical models were developed to extend the Gustafson-Barsis Law by accounting for energy limitations before examining the three processing modes available to hybrid processors: symmetric, asymmetric, and simultaneous asymmetric. Analysis shows that by choosing the optimal chip configuration, energy efficiency and energy savings can be increased considerably.

Keywords: Energy efficiency · Gustafson-Barsis's Law · Hybrid architecture · Performance per watt · Modeling techniques

1 Introduction

The major challenge that microprocessor designers will face in the coming decade is not just power, but also energy efficiency. Although Moore's Law [1] continues to offer solutions with more transistors, power budgets limit our ability to use them. However, there are promising solutions such as heterogeneous many-core architectures that will provide higher performance at lower energy requirements and reduced leakage. Recent research shows that integrated CPU-GPU processors have the potential to deliver more energy efficient computations, which is encouraging chip manufacturers to reconsider the benefits of heterogeneous parallel computing [3–8]. Chip manufacturers such as Intel, NIVIDIA, and AMD have already announced such architectures, i.e., Intel Sandy Bridge, AMD's Fusion APUs, and NVIDIA's Project Denver.

Despite some criticisms [9,10], Amdahl's law [11] and Gustafson-Barsis's Law [12] are still relevant at the dawn of a heterogeneous many-core computing era. Both laws are simple analytical models that help developers to evaluate the actual speedup that can be achieved using a parallel program. They represent two points of view that are not contradictory, but rather complement each other. However, neither of these laws is perfect. Amdahl's Law and Gustafson-Barsis's Law do not account for overheads associated with the creation/destruction of processes/threads and with maintaining cache coherence. Neither do they account for other types of serial tasks such as identification of critical sections, synchronization, lock management, and load balancing.

© Springer International Publishing Switzerland 2016
R. Wyrzykowski et al. (Eds.): PPAM 2015, Part II, LNCS 9574, pp. 123–132, 2016.
DOI: 10.1007/978-3-319-32152-3_12

Furthermore, the future relevance of the laws requires their extension by the inclusion of constraints and architectural trends demanded by modern multiprocessor chips. In [13] we extended Amdahl's law according to the work of Woo and Lee [2] and applied it to the case of a hybrid CPU-GPU multi-core processor. In this work we repeat on our previous study, but this time we extend the Gustafson-Barsis's Law. The main contributions of this paper are as follows:

- To define and formulate two metrics: speedup and performance per watt.
- Using the above metrics, to evaluate the energy efficiency and scalability of three processing schemes available for heterogeneous computing: symmetric, asymmetric and simultaneous asymmetric.
- For each processing scheme, to examine how performance and power are affected by different chip configurations.
- Finally, to analyze and compare the outcomes of the three analytical models and to show how considerable energy savings can be achieved by choosing the optimal chip configuration.

2   Symmetric Processors

In this section we reformulate Gustafson-Barsis's Law to capture the necessary changes imposed by power constraints. We start with the traditional definition of a symmetric multi-core processor and continue by applying energy constraints to the equations following the method of Woo and Lee [2].

2.1   Symmetric Speedup

Gustafson-Barsis's Law begins with a parallel computation and estimates how much faster the parallel computation is than the same computation executing on a single core. Gustafson argues that, as processor power increases, the size of the problem set also tends to increase. This is why the speedup determined by Gustafson-Barsis's Law, also called *scaled speedup*, is the time required by a parallel computation divided into the time hypothetically required to solve the same problem on a single core.

According to the Gustafson-Barsis's Law, a typical program has a serial portion that cannot be parallelized (and therefore can be executed only by a single core) and a parallel portion that can be parallelized (and therefore can be executed by any number of cores in the processor). Let the parallel execution time of the program be normalized to 1, and let the serial and parallel portions be denoted by s and p respectively. Then the following equation concisely describes the law:

$$Speedup_s = s + (1 - s) \cdot c = c + (1 - c) \cdot s \tag{1}$$

where c is the number of cores and s is the fraction of a program's execution time that is spend in serial code ($0 \leq s \leq 1$).

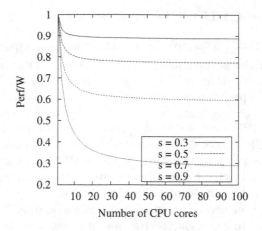

Fig. 1. Performance per watt as a function of the number of CPU cores of a symmetric multi-core processor when $k_c = 0.3$

2.2   Symmetric Performance per Watt

To model power consumption in realistic scenarios, we introduce the variable k_c to represent the fraction of power a single CPU core consumes in its idle state ($0 \leq k_c \leq 1$). In the case of a symmetric processor, one core is active during the sequential computation and consumes a power of 1, while the remaining $(c-1)$ CPU cores consume $(c-1)k_c$. During the sequential computation period, the processor consumes a power of $1+(c-1)k_c$. Thus, during the parallel computation time period, c CPU cores consume c power. It requires s and $(1-s)$ to execute the sequential and parallel codes, respectively, so the formula for the average power consumption W_s of a symmetric processor is as follows.

$$W_s = \frac{s \cdot \{1 + (c-1) \cdot k_c\} + (1-s) \cdot c}{s + (1-s)} \tag{2}$$

Next, we define the *performance per watt (Perf/W)* metric to represent the amount of performance that can be obtained from 1 watt of power. *Perf/W* is basically the reciprocal of energy. The *Perf/W* of a single CPU core execution is 1, so the *Perf/W_s* achievable for a symmetric processor is formulated as follows.

$$\frac{Perf}{W_s} = \frac{Speedup_s}{W_s} = \frac{c + (1-c) \cdot s}{s \cdot \{1 + (c-1) \cdot k_c\} + (1-s) \cdot c} \tag{3}$$

Figure 1 plots the performance per watt for a symmetric multi-core processor as modeled by Eq. (3), showing that the performance per watt decreases rapidly for a small number of cores. However, as the number of cores increases, so does the problem size, and the inherently serial portion becomes much smaller as a proportion of the overall problem. Therefore, the performance per watt remains almost constant as the number of cores increases and reflects the assumption that the execution time remains fixed.

3 Asymmetric CPU-GPU Processors

In this section, an asymmetric CPU-GPU processor where CPU and GPU cores are **integrated on the same die and share the same memory space and power budget** will be referred to as a *hybrid processor*.

We assume that a program's execution time can be composed of a time period where the program runs sequentially (s), a time period where the program runs in parallel on the CPU cores (α), and a time period where the program runs in parallel on the GPU cores $(1 - \alpha)$. **Note that in this case it is assumed that the program runs in parallel on the CPU cores *or* on the GPU cores, but not on both at the same time. Simultaneous asymmetric processing will be the topic of the next section.**

To model the power consumption of an asymmetric processor we introduce another variable, k_g, to represent the fraction of power a single GPU core consumes in its idle state $(0 \leq k_g \leq 1)$. We introduce two further variables, α and β, to model the performance difference between a CPU core and a GPU core. The first variable represents the fraction of a program's execution time that is parallelized on the CPU cores $(0 \leq \alpha \leq 1)$, while the second variable represents a GPU core's performance normalized to that of a CPU core $(0 \leq \beta)$. For example, comparing the performance of a single CPU core (Intel Core-i7-960 multicore processor) against the performance of a single GPU core (NVIDIA GTX 280 GPU processor) yields values of β between 0.4 and 1.2.

We assume that one CPU core in an active state consumes a power of 1 and the *power budget (PB)* of a processor is 100. Thus, $g = (PB-c)/w_g$ is the number of the GPU cores embedded in the processor where variable w_g represents the active GPU core's power consumption relative to that of an active CPU core $(0 \leq w_g)$.

3.1 Asymmetric Speedup

Now, if the sequential code of the program is executed on a single CPU core the following equation represents the theoretical achievable *asymmetric speedup* $(speedup_a)$.

$$Speedup_a = s + N \cdot (1 - s) \cdot \{\alpha \cdot c + \frac{(1 - \alpha) \cdot g}{\beta}\} \qquad (4)$$

where N is the number of hybrid processors. Each hybrid processor contains c CPU cores and g GPU cores.

3.2 Asymmetric Performance per Watt

To model the power consumption of an asymmetric processor we assume that during the sequential computation phase, one CPU core is in active state and the amount of power it consumes is 1, the $c-1$ idle CPU cores consume $(c-1)k_c$ and the g idle GPU cores consume $g \cdot w_g \cdot k_g$. During the parallel computation

s = 0.3

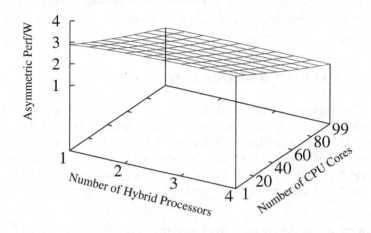

Fig. 2. Asymmetric perf/W as a function of the number of hybrid processors and various CPU-GPU chip configurations for $s = 0.3, w_g = 0.25, \alpha = 0.5, k_c = 0.3$, $k_g = 0.2$ and $\beta = 1.0$.

on the CPU cores, the CPU cores consume c and the g idle GPU cores consume $g \cdot w_g \cdot k_g$. During the parallel computation on the GPU cores, the GPU cores consume $g \cdot w_g$ and the idle CPU cores consume $c \cdot k_c$.

Let P_s, P_c, and P_g denote the power consumption during the sequential, CPU, and GPU processing phases, respectively.

$$P_s = s \cdot \{1 + (c - 1) \cdot k_c + g \cdot w_g \cdot k_g\}$$
$$P_c = \alpha \cdot (1 - s) \cdot \{c + g \cdot w_g \cdot k_g\}$$
$$P_g = (1 - \alpha) \cdot (1 - s) \cdot \{g \cdot w_g + c \cdot k_c\}$$

It requires time $(1 - p)$ to perform the sequential computation, and times $\alpha \cdot p$ and $(1 - \alpha) \cdot p$ to perform the parallel computations on the CPU and GPU, respectively, so the average power consumption W_a of an asymmetric processor is as follows.

$$W_a = P_s + P_c + P_g \tag{5}$$

Consequently, $Perf/W_a$ of N asymmetric processors is expressed as

$$\frac{Perf}{W_a} = \frac{s + N \cdot (1 - s) \cdot \{\alpha \cdot c + \frac{(1-\alpha) \cdot g}{\beta}\}}{P_s + N \cdot (P_c + P_g)} \tag{6}$$

Figure 2 shows the performance per watt of an asymmetric processor for $s = 0.3$ as a function of the number of hybrid processors and as a function of CPU cores within each hybrid processor. It can be seen that the $Perf/W_a$ decreases slowly with the increase in the number of hybrid processors, as expected, and decreases faster as the number of the CPU cores increases. Furthermore, the optimal $Perf/W_a$ is obtained for a chip configuration of 1 CPU core and 396 GPU cores.

4   CPU-GPU Simultaneous Processing

In the previous analysis we assumed that a program's execution time is divided into three phases as follows: a *sequential phase* where one core is active, a *CPU phase* where the parallelized code is executed by the CPU cores, and a *GPU phase* where the parallelized code is executed by the GPU cores. However, the aim of hybrid CPU-GPU computing is to divide the program while allowing the CPU and the GPU will execute their codes simultaneously.

4.1   Simultaneous Asymmetric Speedup

We conduct our analysis assuming that the CPU's execution time overlaps with the GPU's execution time. Such an overlap occurs when the CPU's execution time $\alpha \cdot p \cdot c$ equals the GPU's execution time $\frac{(1-\alpha)\cdot p\cdot g}{\beta}$. Let α' denote the value of α that applies to this equality:

$$\alpha' = \frac{g}{g + c \cdot \beta}$$

We assume that the sequential code of the program is executed on a single CPU core. Thus, the following equation represents the theoretical achievable *simultaneous asymmetric speedup (speedup$_{sa}$)*:

$$Speedup_{sa} = s + N \cdot (1 - s) \cdot \{\alpha' \cdot c\}$$
$$= s + N \cdot (1 - s) \cdot \{\frac{(1 - \alpha') \cdot g}{\beta}\} \tag{7}$$

where N is the number of hybrid processors. Each hybrid processor contains c CPU cores and g GPU cores.

4.2   Simultaneous Asymmetric Perf/W

To model the power consumption of an asymmetric processor in a simultaneous processing mode, we assume that one core is active during the sequential computation and consumes a power of 1, while the remaining $c - 1$ idle CPU cores consume $(c - 1)k_c$ and g idle GPU cores consume $g \cdot w_g \cdot k_g$. Thus, during the parallel computation time period, c active CPU cores consume c and g active GPU cores consume $g \cdot w_g$. It requires $(1 - p)$ to execute sequential code and $\alpha' \cdot p$ to execute the parallel codes on the CPU and the GPU simultaneously,

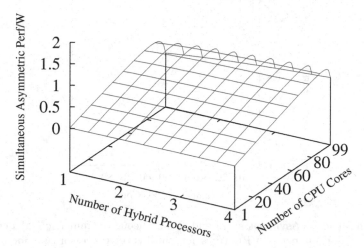

Fig. 3. Simultaneous Asymmetric Perf/W as a function of the number of hybrid processors and various CPU-GPU chip configurations for $s = 0.3, w_g = 0.25, k_c = 0.3$, $k_g = 0.2$ and $\beta = 1.0$.

so the average power consumption of an asymmetric processor in a simultaneous processing mode is

$$W_{sa} = P_s + P_c + P_g \tag{8}$$

where

$$P_s = s \cdot \{1 + (c-1) \cdot k_c + g \cdot w_g \cdot k_g\}$$
$$P_c + P_g = \alpha' \cdot (1-s) \cdot \{c + g \cdot w_g\}$$

Consequently, $Perf/W_{sa}$ of N asymmetric processors in a simultaneous processing mode is expressed as follows.

$$\frac{Perf}{W_{sa}} = \frac{s + N \cdot (1-s) \cdot \{\alpha' \cdot c\}}{P_s + N \cdot (P_c + P_g)} \tag{9}$$

Figure 3 shows the performance per watt of an asymmetric processor, as modeled by Eq. (9), for $s = 0.3$ as a function of the number of hybrid processors and as a function of CPU cores within each hybrid processor. It can be observed that the $Perf/W_{sa}$ slightly decreases with the increase in the number of hybrid processors. When the performance of the CPU cores dominates, the graph increases rapidly as the number of CPU cores increases (and the number of GPU cores is decreases). Then, it reaches the point beyond which the performance per watt decreases very rapidly because the dominance of the GPU cores is negligible.

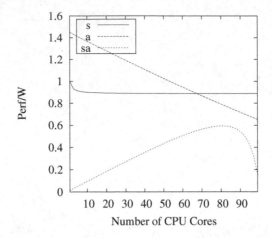

Fig. 4. Symmetric (s) Asymmetric (a) and Simultaneous Asymmetric (sa) Perf/W as a function of the number of CPU cores for one hybrid processor and for $s = 0.3$, $w_g = 0.25, \alpha = 0.5$ and $\beta = 2.0$.

5   Synthesis

Figure 4 shows the performance per watt of the three processing schemes that were studied in this research (symmetric (s), asymmetric (a), and simultaneous asymmetric (sa)) and how they are affected by chip configuration. First, it can be observed that the chip configuration has no effect on $Perf/W$ while processing in symmetric mode, as can be expected. In simultaneous asymmetric processing mode, $Perf/W$ improves with increasing number of CPU cores until it reaches peak performance for a chip configuration of approximately 85 CPU cores and 60 GPU cores. Beyond this point, $Perf/W$ decreases rapidly to a point where the contribution of the GPU cores is negligible. On the other hand, in asymmetric processing mode, a chip configuration consisting of a single CPU core yields an optimal performance per watt, and any attempt to increase the number of CPU cores in the chip organization leads to a significant decrease in performance per watt.

6   Related Work

Hill and Marty [14] studied the implications of Amdahl's law on multi-core hardware resources and proposed the design of future chips based on the overall chip performance rather than core efficiencies. The major assumption in that model was that a chip is composed of many basic cores and their resources can be combined dynamically to create a more powerful core with higher sequential performance. Using Amdahl's law, they showed that asymmetric multi-core chips designed with one fat core and many thin cores exhibited better performance than symmetric multi-core chip designs.

Woo and Lee [2] developed a many-core performance per energy analytical model that revisited Amdahl's Law. Using their model the authors investigated the energy efficiency of three architecture configurations. The first architecture studied contained multi-superscalar cores, the second architecture contained many simplified and energy efficient cores, and the third architecture was an asymmetric configuration of one superscalar core and many simplified energy efficient cores. The evaluation results showed that under restricted power budget conditions the asymmetric configuration usually exhibited better performance per watt. The energy consumption was reduced linearly as the performance was improved with parallelization scales. Furthermore, improving the parallelization efficiency by load balancing among processors increased the efficiency of power consumption and increased the battery life.

Sun and Chen [15] studied the scalability of multi-core processors and reached more optimistic conclusions compared with the analysis conducted by Hill and Marty [14]. The authors suggested that the fixed-size assumption of Amdahl's law was unrealistic and that the fixed-time and memory-bounded models might better reflect real world applications. They presented extensions of these models for multi-core architectures and showed that there was no upper bound on the scalability of multi-core architectures. However, the authors suggested that the major problem limiting multi-core scalability is the memory data access delay and they called for more research to resolve this memory-wall problem.

Esmaeilzadeh et al. [16] performed a systematic and comprehensive study to estimate the performance gains from the next five multi-core generations. Accurate predictions require the integration of as many factors as possible. Thus, the study included: power, frequency and area limits; device, core and multi-core scaling; chip organization; chip topologies (symmetric, asymmetric, dynamic, and fused); and benchmark profiles. They constructed models based on pessimistic and optimistic forecasts, and observations of previous works with data from 150 processors. The conclusions were not encouraging.

7 Conclusions

The analysis of three analytical models of symmetric, asymmetric, and simultaneous asymmetric processing using two performance metrics with regard to various chip configurations suggest that future many-core processors should be a priori designed to include one or a few fat cores alongside many efficient thin cores to support energy efficient hardware platforms. Moreover, to achieve optimal scalability and energy savings, a dynamic configuration mechanism is required for identifying and implementing the optimal chip organization.

References

1. Moore, G.: Cramming more components onto integrated circuits. Electronics **38**(8), 114–117 (1965)
2. Woo, D.H., Lee, H.S.: Extending Amdahl's law for energy-efficient computing in the many-core era. IEEE Comput. **38**(11), 32–38 (2005)
3. Kumar, R., et al.: Heterogeneous chip multiprocessors. IEEE Comput. **38**(11), 32–38 (2005)
4. Mantor, M.: Entering the Golden Age of Heterogeneous Computing. C-DAC PEEP 2008. http://ati.amd.com/technology/streamcomputing/IUCAA_Pune_PEEP_2008.pdf
5. Kogge, P., et al.: ExaScale Computing Study: Technology Challenges in Achieving Exascale Systems. DARPA, Washington, D.C (2008)
6. Fuller, S.H., Millett, L.I.: Computing performance: game over or next level? IEEE Comput. **44**(1), 31–38 (2011)
7. Borkar, S.: Thousand core chips: a technology perspective. In: Proceedings of 44th Design Automation Conference (DAC 2007), pp. 746–749. ACM Press (2007)
8. Marowka, A.: Back to thin-core massively parallel processors. IEEE Comput. **44**(12), 49–54 (2011)
9. Hillis, D.: The Pattern on the Stone: The Simple Ideas that Make Computers Work. Basic Books, New York (1998)
10. Shi, Y.: Reevaluating Amdahl's Law and Gustafson's Law (1996). http://www.cis.temple.edu/shi/docs/amdahl/amdahl.html
11. Amdahl, G.M.: Validity of the single-processor approach to achieving large-scale computing capabilities. In: Proceeidngs of American Federation of Information Processing Societies Conference, pp. 483–485. AFIPS Press (1967)
12. Gustafson, J.L.: Reevaluating Amdahl's law. Comm. ACM **31**, 532–533 (1988)
13. Marowka, A.: Analytical modeling of energy efficiency in heterogeneous processors. Comput. Electr. Eng. J. **39**(8), 2566–2578 (2013). Elsevier press
14. Hill, M.D., Marty, M.R.: Amdahl's law in the multicore era. IEEE Comput. **41**, 33–38 (2008)
15. Sun, X.-H., Chen, Y.: Reevaluating Amdahl's law in the multicore era. J. Parallel Distrib. Comput. **70**, 183–188 (2010)
16. Esmaeilzadeh, H., Blem, E., St. Amant, R., Sankaralingam, K., Burger, D.C.: Dark silicon and the end of multicore scaling. In: Proceeding of 38th International Symposium on Computer Architecture (ISCA), pp. 365–376, June 2011

Free Scheduling of Tiles Based on the Transitive Closure of Dependence Graphs

Wlodzimierz Bielecki[✉], Marek Palkowski, and Tomasz Klimek

Faculty of Computer Science and Information Systems,
West Pomeranian University of Technology in Szczecin,
Zolnierska 49, 71210 Szczecin, Poland
{wbielecki,mpalkowski}@wi.zut.edu.pl
http://www.wi.zut.edu.pl

Abstract. A novel approach to form the free schedule of tiles comprising statement instances of the program loop nest is presented. Forming both valid tiles and free scheduling are based on the transitive closure of loop nest dependence graphs. Under the free schedule, tiles are executed as soon as their operands are available. To describe and implement the approach, loop dependences are presented in the form of tuple relations. A discussed algorithm is implemented in the open source TRACO compiler. Experimental results exposing the effectiveness of the introduced algorithm and speed-up of parallel programs, produced by means of this algorithm, are discussed.

Keywords: Loop nest tiling · Transitive closure · Dependence graphs · Coarse-grained parallelism · Free scheduling

1 Introduction

Tiling is a very important iteration reordering transformation for both improving data locality and extracting loop parallelism. Loop tiling for improving locality groups loop statement instances in a loop iteration space into smaller blocks (tiles) allowing reuse when the block fits in local memory. On the basis of a valid schedule of tiles, parallel coarse-grained code can be generated.

To our best knowledge, well-known tiling techniques are based on linear or affine transformations of program loop nests [6,9,10,13,20]. In paper [5], we describe the limitations of affine transformations and present how the free-scheduling of loop nest statement instances can be formed by means of the transitive closure of program dependence graphs. In this paper, we demonstrate how the approach, presented in our paper [5], can be adapted to form the free-scheduling of valid tiles. To generate both valid tiles and free-scheduling, we apply the transitive closure of dependence graphs. The proposed approach allows generation of parallel tiled code even when there does not exist an affine transformation allowing for producing a fully permutable loop nest. This approach is a result of a combination of the polyhedral model and the iteration space slicing framework.

© Springer International Publishing Switzerland 2016
R. Wyrzykowski et al. (Eds.): PPAM 2015, Part II, LNCS 9574, pp. 133–142, 2016.
DOI: 10.1007/978-3-319-32152-3_13

2   Background

A considered approach uses the dependence analysis proposed by Pugh and Wonnacott [16] where dependences are represented by dependence relations. Dependences of a loop nest are described by dependence relations with constraints presented by means of the Presburger arithmetic.

A dependence relation is a tuple relation of the form [*input list*]→[*output list*]: *formula*, where *input list* and *output list* are the lists of variables and/or expressions used to describe input and output tuples and *formula* describes the constraints imposed upon *input list* and *output list* and it is a Presburger formula built of constraints represented with algebraic expressions and using logical and existential operators. A dependence relation is a mathematical representation of a data dependence graph whose vertices correspond to loop statement instances while edges connect dependent instances. The input and output tuples of a relation represent dependence sources and destinations, respectively; the relation constraints point out instances which are dependent.

Standard operations on relations and sets are used, such as intersection ($\cap$), union ($\cup$), difference ($-$), domain (dom R), range (ran R), relation application ($S' = R(S) : e' \in S'$ iff exists e s.t. $e \to e' \in R, e \in S$). In detail, the description of these operations is presented in papers [11,16].

The positive transitive closure for a given relation R, R^+, is defined as follows [11]: $R^+ = \{e \to e' : e \to e' \in R \vee \exists e''s.t.\ e \to e'' \in R \wedge e'' \to e' \in R^+\}$. It describes which vertices e' in a dependence graph (represented by relation R) are connected directly or transitively with vertex e.

Transitive closure, R^*, is defined as follows [12]: $R^* = R^+ \cup I$, where I the identity relation. It describes the same connections in a dependence graph (represented by R) that R^+ does plus connections of each vertex with itself.

The composition of given relations $R_1 = \{x_1 \to y_1|f_1(x_1,y_1)\}$ and $R_2 = \{x_2 \to y_2|f_2(x_2,y_2)\}$, is defined as follows [11]: $R_1 \circ R_2 = \{x \to y|\exists z\ s.t.\ f_1(z,y) \wedge f_2(x,z)\}$.

3   Finding Free Scheduling

The algorithm, presented in our paper [5], allows us to generate fine-grained parallel code based on the free schedule representing time partitions; all statement instances of a time partition can be executed in parallel, while partitions are enumerated sequentially. The free schedule function is defined as follows.

Definition 1 [7,8]. The *free schedule* is the function that assigns discrete time of execution to each loop nest statement instance as soon as its operands are available, that is, it is mapping $\sigma : LD \to \mathbb{Z}$ such that

$$\sigma(p) = \begin{cases} 0\ if\ there\ is\ no\ p_1 \in LD\ s.t.\ p_1 \to p \\ 1 + max(\sigma(p_1), \sigma(p_2), ..., \sigma(p_n)); p, p_1, p_2, ..., p_n \in LD; \\ p_1 \to p, p_2 \to p, ..., p_n \to p, \end{cases}$$

where $p, p_1, p_2, ..., p_n$ are loop nest statement instances, LD is the loop nest domain, $p_1 \rightarrow p, p_2 \rightarrow p, ..., p_n \rightarrow p$ mean that the pairs p_1 and p, p_2 and p, ...,p_n and p are dependent, p represents the destination while $p_1, p_2, ..., p_n$ represent the sources of dependences, n is the number of operands of statement instance p (the number of dependences whose destination is statement instance p).

The free schedule is the fastest legal schedule [8]. In paper [5] we presented fine-grained parallelism extraction based on the power k of relation R.

The idea of the algorithm is the following [5]. Given relations $R_1, R_2, ..., R_m$, representing all dependences in a loop nest, we first calculate $R = \bigcup\limits_{i=1}^{m} R_i$ and then R^k, where $R^k = \underbrace{R \circ R \circ ... R}_{k}$, "$\circ$" is the composition operation. Techniques of calculating the power k of relation R are presented in the following publications [12,17] and they are out of the scope of this paper. Let us only note that given transitive closure R^+, we can easily convert it to the power k of R, R^k, and vice versa, for details see [17].

Given set UDS comprising all loop nest statement instances that are ready to execution at time $k = 0$ (Ultimate Dependence Sources), each vertex, represented with the set $S_k = R^k(UDS) - R^+ \circ R^k(UDS)$, is connected in the dependence graph, defined by relation R, with some vertex(ices) represented by set UDS with a path of length k. Hence at time k, all the statement instances belonging to the set S_k can be scheduled for execution and it is guaranteed that k is as few as possible.

4   Loop Nest Tiling Based on the Transitive Closure of Dependence Graphs

In this paper, to generate valid tiled code, we apply the approach presented in paper [4], which is based on the transitive closure of dependence graphs. Next, we briefly present the steps of that approach.

First, we form set $TILE(II, B)$, including iterations belonging to a parametric tile, as follows $TILE(II, B) = \{[I] | B*II + LB \leq I \leq \min(B*(II + 1) + LB - 1, UB)$ AND $II \geq 0\}$, where vectors LB and UB include the lower and upper loop index bounds of the original loop nest, respectively; diagonal matrix B defines the size of a rectangular original tile; elements of vector I represent the original loop nest iterations contained in the tile whose identifier is II; 1 is the vector whose all elements have value 1; here and further on, the notation $x \geq (\leq)y$ where x, y are two vectors in $\mathbb{Z}^n$ corresponds to the component-wise inequality, that is, $x \geq (\leq)y \Longleftrightarrow x_i \geq (\leq)y_i$, i=1,2,...,n.

Next, we build sets $TILE\_LT$ and $TILE\_GT$ that are the unions of all the tiles whose identifiers are lexicographically less and greater than that of $TILE(II, B)$, respectively:

$TILE\_LT = \{[I] \mid$ exists II' s. t. $II' \prec II$ AND $II \geq 0$ AND $B*II+LB \leq UB$ AND $II' \geq 0$ and $B*II' + LB \leq UB$ AND I in $TILE(II', B)\}$,

$TILE\_GT = \{[I] \mid$ exists II' s. t. $II' \succ II$ AND $II \geq 0$ AND $B*II + LB \leq$ UB AND $II' \geq 0$ and $B*II' + LB \leq UB$ AND I in $TILE(II', B)\}$,

where "$\prec$" and "$\succ$" (here and further on) denote the lexicographical relation operators for two vectors. Then, we calculate set

$$TILE\_ITR = TILE - R^!(TILE\_GT),$$

which does not include any invalid dependence target, i.e., it does not include any dependence target whose source is within set $TILE\_GT$. The following set

$$TVLD\_LT = (R^+(TILE\_ITR) \cap TILE\_LT) - R^+(TILE\_GT)$$

includes all the iterations that (i) belong to the tiles whose identifiers are lexicographically less than that of set $TILE\_ITR$, (ii) are the targets of the dependences whose sources are contained in set $TILE\_ITR$, and (iii) are not any target of a dependence whose source belong to set $TILE\_GT$. Target tiles are defined by the following set $TILE\_VLD = TILE\_ITR \cup TVLD\_LT$.

Lastly, we form set $TILE\_VLD\_EXT$ by means of inserting (i) into the first positions of the tuple of set $TILE\_VLD$ elements of vector II: $ii_1, ii_2, ..., ii_d$; (ii) into the constraints of set $TILE\_VLD$ the constraints defining tile identifiers $II \geq 0$ and $B*II + LB \leq UB$. Target code is generated by means of applying any code generator allowing for scanning elements of set $TILE\_VLD\_EXT$ in the lexicographic order, for example, CLooG [1].

5   Free Scheduling for Tiles

The algorithm presented in this paper is a combination of the approaches presented in the two previous sections. First, we generate tiled code as it is described in Sect. 4, then we find free scheduling for tiles of the tiled code. For this purpose, first, we form relation, R_TILE, which describes dependences among tiles as follows

$R\_TILE := \{[II] -> [JJ]$: exist I, J s.t. (II, I) in $TILE\_VLD\_EXT(II)$ AND (JJ, J) in $TILE\_VLD\_EXT_i(JJ)$ AND J in $R(I)\}$,

where II, JJ are the vectors representing tile identifiers; vectors I, J comprise iterations belonging to tiles whose identifiers are II, JJ, respectively.

The following step is to calculate set, UDS, including the tile identifiers which state for tile ultimate dependence sources and/or independent ones as follows: $UDS = II\_SET$ – range (R_TILE), where set $II\_SET = \{[II] \mid II \geq 0$ and $B*II + LB \leq UB\}$ represents all tile identifiers.

Now, we apply the algorithm presented in paper [5] to form free-scheduling for tiles of tiled code. With this purpose, we calculate the transitive closure and power k of relation R_TILE and next calculate set S_k, representing the free schedule, as follows $S_k = R\_TILE^k(UDS) - (R\_TILE^+ \circ R\_TILE^k(UDS))$. Finally, we extend the tuple of set S_k with variable k and variables representing statement instances of a parametric target tile(together with corresponding constraints) and generate code applying any code generator, for example, CLooG to scan iterations within set S_k in the lexicographical order. Algorithm 1 presents the discussed above idea in a formal way.

Algorithm 1. Parallel tiled code generation based on the free schedule

---

Input: A loop nest of depth d; constants $b_1, b_2, ..., b_d$ defining the size of a rectangular original tile, relation R representing all the dependences in the loop nest.

Output: Code enumerating time partitions according to the free schedule, tiles for each time partition (in parallel), and statement instances in each tile.

Method:

1. Apply the algorithm, presented in paper [4] to the original loop nest to generate sets II_SET, $TILE\_VLD$, $TILE\_VLD\_EXT$, and tiled code.
2. Form relation, R_TILE, which describes dependences among tiles but ignores dependences within each tile as follows

   $R\_TILE := \{[II] -> [JJ]$: exist I, J s.t. (II,I) in $TILE\_VLD\_EXT(II)$ AND (JJ,J) in $TILE\_VLD\_EXT(JJ)$ AND J in $R(I)\}$,
   where II, JJ are the vectors representing tile identifiers, $TILE\_VLD\_EXT$ is the set returned by step 1.
3. Calculate set, UDS, including the tile identifiers which state for tile ultimate dependence sources and/or independent ones as follows
   $UDS := II_SET - $ range (R_TILE),
4. Calculate set
   $S_k = R\_TILE^k(UDS) - (R\_TILE^+ \circ R\_TILE^k(UDS))$.
5. Extend set S_k as follows: insert in the first its tuple position symbolic variable k responsible for representing time under the free schedule; insert in the last its tuple positions the elements of set $TILE\_VLD$ returned by step 1; insert into the constraint of set S_k the constraint of set $TILE\_VLD$.
6. Apply to the set, returned by step 5, CLooG [1] and postprocess the code generated by CLooG to get the following code structure

   ```
   seqfor for k  //enumerating time partitions
       parfor Sk //enumerating tile identifiers contained in set Sk
               //formed in step 4  for a given value of k
           seqfor TILE_VLD //enumerating  statement instances comprised in
                   //set TILE_VLD defined by the tile identifiers
                   //represented  by the previous parfor loop
   ```

---

6   Illustrative Example

In this section, we illustrate steps of Algorithm 1 by means of the following loop:

```
for(i=1; i<=6; i++)
 for(j=1; j<=6; j++)
  a[i][j] = a[i+1][j-1];
```

We use the ISL library to carry out operations on relations and sets required by the presented algorithm. A dependence relation, returned by Petit, the Omega project dependence analyzer, is the following

```
R:= {[i,j,v] -> [i',j',v'] : ( i' = 1+i and j' = j-1 and v = 6 and v' = 6
and 1 <= i <= 5 and 2 <= j <= 6 )},
```

where here and further on "6" states for the statement identifier represented via the corresponding line number in the original loop nest.

The algorithm presented in paper [4] returns the following set $TILE\_VLD\_EXT$ representing both tile identifiers and statement instances within each target tile.

```
TILE_VLD_EXT:= { [i0, i1, i2, i3, 6] : i0 >= 0 and i2 >= 1 + 2i0 and
i2 <= 6 and i3 >= 1 + 2i1 and i3 <= 6 and i3 >= 1 and i3 <= 3 + 2i0 +
2i1 - i2; [i0, i1, 2 + 2i0, 2i1, 6] : i0 <= 2 and i0 >= 0 and i1 <= 2
and i1 >= 1; [i0, 2, 2 + 2i0, 6, 6] : i0 <= 2 and i0 >= 0 }.
```

Using relation R and set $TILE\_VLD\_EXT$, we form realtion R_TILE that is of the form below.

```
R_TILE:= { [i0, i1, 6] -> [1 + i0, -1 + i1, 6] : i0 >= 0 and i0 <= 1 and
i1 <= 2 and i1 >= 1; [i0, 2, 6] -> [1 + i0, 2, 6] : i0 <= 1 and i0 >= 0 }.
```

Set UDS is the following $\{[0, \, jj, 6] : jj \leq 2 \text{ and } jj \geq 0\}$.

Using the appropriate functions of the ISL library to calculate relations R_TILE^k and R_TILE^+, we calculate set S_k according to the formula in step 4 of Algorithm 1, and extend set S_k as presented in step 5 of Algorithm 1, to get:

```
Sk:= { [i0, i0, i2, i3, i4, 6] : i3 >= 1 + 2i0 and i4 >= 1 + i2 and
i2 <= 2 and i3 <= 2 + 2i0 and i0 >= 0 and i4 <= 2 + 2i2 and i4 >= 2 +
2i0 + 2i2 - i3 and i0 <= 2 and 2i4 <= 6 + 4i0 + 5i2 - 2i3 }.
```

Finally, we apply to set S_k the GLooG code generator and postprocess the code returned by CLooG to yield the following OpenMP C code.

```
1. for (c0 = 0; c0 <= 2; c0 += 1)
2. #pragma omp parallel for
3.   for (c2 = 0; c2 <= 2; c2 += 1)
4.     for (c3 = 2 * c0 + 1; c3 <= 2 * c0 + 2; c3 += 1)
5.       for (c4 = max(c2 + 1, 2 * c0 + 2 * c2 - c3 + 2);
            c4 <= min(2*c0 + 2 * c2 - c3 + c2/2 + 3, 2 * c2 + 2); c4++)
6.         a[c3][c4]=a[c3+1][c4-1];
```

where line 1 presents the serial *for* loop enumerating time partitions; line 2 represents the two OpenMP directives (*parallel for*) pointing out that the iterations of the *for* loop in line 3 can be executed in parallel; the *for* loops in line 1 and line 3 enumerate tile identifiers, whereas the *for* loops in line 4 and line 5 scan iterations within a tile. Figure 1 presents original tiles, while Fig. 2 shows target tiles returned by the algorithm, presented in paper [4] (depicted by dashed lines), and the three time partitions (k=0, 1, 2) for the illustrative example.

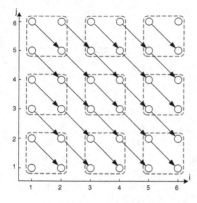

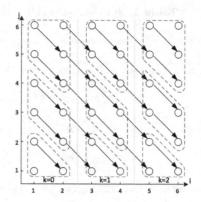

Fig. 1. Original tiles   **Fig. 2.** Target tiles and time partitions

7   Experimental Study

The presented algorithm has been implemented in the optimizing compiler TRA
CO, publicly available at the website http://traco.sourceforge.net. For calculating R^+ and R^k, TRACO uses the corresponding functions of the ISL library
[17]. To evaluate the effectiveness of proposed approach, we have experimented
with NAS Parallel Benchmarks 3.3 (NPB) [14].

From 431 loops of the NAS benchmark suite, Petit is able to analyse 257
loops, and dependences are available in 134 loops (the rest 123 loops do not
expose any dependence). For these 134 loop nests, ISL is able to calculate
R_TILE^k for 58 ones and accordingly TRACO is able to generate parallel tiled
code for those programs. Such a limitation is not the limitation of the algorithm,
it is the limitation of the corresponding ISL function.

To check the performance of parallel tiled code, produced with TRACO,
the following criteria were taken into account for choosing NAS programs: (i) a
loop nest must be computationally intensive (there are many NAS benchmarks
with constant upper bounds of loop indices, hence their parallelization is not
justified), (ii) structures of chosen loops must be different (there are many loops
of a similar structure).

Applying these criteria, we have selected the following five NAS
loops: *BT_rhs_1* (Block Tridiagonal Benchmark), *FT_auxfnct.f2p_2* (Fast
Fourier Transform Benchmark), *UA_diffuse_5*, *UA_setup_16* and *UA_transfer_4*
(Unstructured Adaptive Benchmark).

To carry out experiments, we have used a computer with Intel i5-4670
3.40 GHz processors (Haswell, 2013), 6 MB cache and 8 GB RAM. Source and
target codes of the examined programs are available in http://sourceforge.net/
p/issf/code-0/HEAD/tree/trunk/examples/fstile/.

Table 1 presents execution time and speed-up for the studied loop nests.
Speed-up is the ratio of sequential and parallel program execution times, i.e.,
$S = T(1)/T(P)$, where $T(P)$ is the parallel program execution time on P processors. Speedups were computed against the serial original code execution time.

Table 1. Speed-up of parallel tiled loop nests for 4 CPU cores.

Program	Loop up. bounds	Time of serial run (in seconds)	Block size	Time of parallel run (in seconds)	Speed-up
FT_auxfnct.f2p_2	N1, N2, N3 = 500	6.857	16	0.817	8.393
			32	0.795	8.625
	N1, N2, N3 = 600	13.403	16	1.176	11.397
			32	1.228	10.914
BT_rhs.f2p_1	N1, N2, N3 = 200	2.87	16	0.892	3.217
			32	1.112	2.581
	N1, N2, N3 = 300	10.598	16	2.936	3.610
			32	3.549	2.986
UA_diffuse.f2p_5	N1, N2, N3, N4 = 100	0.444	16	0.209	2.124
			32	0.187	2.374
	N1, N2, N3, N4 = 200	10.875	16	3.85	2.825
			32	3.556	3.058
UA_setup.f2p_16	N1, N2, N3 = 1000	1.325	16	0.662	2.002
			32	0.445	2.978
	N1, N2, N3 = 1100	15.285	16	0.976	15.661
			32	0.746	20.489
UA_transfer.f2p_4	N1, N2, N3 = 700	5.541	16	0.742	7.468
			32	0.745	7.438
	N1, N2, N3 = 1000	22.751	16	1.501	15.157
			32	1.499	15.177

Experiments were carried out for 4 CPUs. Analysing the data in Table 1, we may conclude that for all parallel tiled loops, positive speed-up is achieved. It depends on the problem size defined by loop index upper bounds and a tile size. It is worth to note that for the *FT_auxfnct.f2p_2* and *UA_transfer_4* programs, super-linear speed-up is achieved, i.e., the speed-up is greater than 4 – the number of CPUs used. This phenomenon could be explained by the fact that the data size required by the original program is greater than the cache size when executed sequentially, but could fit nicely in each available cache when executed in parallel, i.e., due to increasing program locality.

8   Related Work

There has been a considerable amount of research into tiling demonstrating how to aggregate a set of loop iterations into tiles with each tile as an atomic macro statement, starting with pioneer paper [10] and those presenting advanced techniques [6,9,19].

One of the most advanced reordering transformation frameworks is based on the polyhedral model. Let us remind that "*Restructuring programs using the polyhedral model is a three steps framework. First, the Program Analysis phase aims at translating high level codes to their polyhedral representation and to provide data dependence analysis based on this representation. Second, some optimizing or parallelizing algorithm uses the analysis to restructure the programs in the polyhedral model. This is the Program Transformation step. Lastly, the Code Generation step returns back from the polyhedral representation to a high level program*" [3].

All above three steps are available in the approach presented in this paper. But there exists the following difference in step 2: in the polyhedral model "*a (sequence of) program transformation(s) is represented by a set of affine functions, one for each statement*" [3] while the presented approach does not find and use any affine function. It applies the transitive closure of a program dependence graph to specific subspaces of the source loop iteration space. At this point of view the program transformation step is rather within the Iteration Space Slicing Framework introduced by Pugh and Rosser [15], where the key step is calculating the transitive closure of a program dependence graph.

Papers [10,18] are a seminal work presenting the theory of tiling techniques based on affine transformations. These papers present techniques consisting of two steps: they first transform the original loop into a fully permutable loop nest, then transform the fully permutable loop nest into tiled code. Loop nests are fully permutable if they can be permuted arbitrarily without altering the semantics of the source program. If a loop nest is fully permutable, it is sufficient to apply a tiling transformation to this loop nest [18].

Papers [2,5] demonstrate how we can extract coarse- and fine-grained parallelism applying different Iteration Space Slicing algorithms, however they do not consider any tiling transformation.

Wonnacott and Strout review implemented and proposed techniques for tiling dense array codes in an attempt to determine whether or not the techniques permit on scalability. They write [19]: "*No implementation was ever released for iteration space slicing*". This permits us to state that TRACO, which implements the algorithm, presented in this paper, is the first compiler where Iteration Space Slicing is applied to produce parallel tiled code based on the free-schedule of tiles.

9 Conclusion

In this paper, we presented a novel approach based on a combination of the Polyhedral Model and the Iteration Space Slicing framework. It allows generation of parallel tiled codes which demonstrate significant speed-up on shared memory machines with multi-core processors. The usage of the free schedule of tiles instead of that of loop nest statement instances allows us to adjust the parallelism grain-size to match the inter-processor communication capabilities of the target architecture. In the future, we plan to present an extended approach allowing for tiling with parallelepiped original tiles.

References

1. Bastoul, C.: Code generation in the polyhedral model is easier than you think. In: PACT 2013 IEEE International Conference on Parallel Architecture and Compilation Techniques, Juan-les-Pins, pp. 7–16, September 2004

2. Beletska, A., Bielecki, W., Cohen, A., Palkowski, M., Siedlecki, K.: Coarse-grained loop parallelization: iteration space slicing vs affine transformations. Parallel Comput. **37**, 479–497 (2011)
3. Benabderrahmane, M.-W., Pouchet, L.-N., Cohen, A., Bastoul, C.: The polyhedral model is more widely applicable than you think. In: Gupta, R. (ed.) CC 2010. LNCS, vol. 6011, pp. 283–303. Springer, Heidelberg (2010). http://dx.doi.org/10.1007/978-3-642-11970-5_16
4. Bielecki, W., Palkowski, M.: Perfectly nested loop tiling transformations based on the transitive closure of the program dependence graph. Soft Comput. Comput. Inf. Sci. **342**, 309–320 (2015)
5. Bielecki, W., Palkowski, M., Klimek, T.: Free scheduling for statement instances of parameterized arbitrarily nested affine loops. Parallel Comput. **38**(9), 518–532 (2012)
6. Bondhugula, U., Hartono, A., Ramanujam, J., Sadayappan, P.: A practical automatic polyhedral parallelizer and locality optimizer. SIGPLAN Not. **43**(6), 101–113 (2008)
7. Darte, A., Khachiyan, L., Robert, Y.: Linear scheduling is nearly optimal. Parallel Process. Lett. **1**(2), 73–81 (1991)
8. Darte, A., Robert, Y., Vivien, F.: Scheduling and Automatic Parallelization. Birkhauser, New York (2000)
9. Griebl, M.: Automatic parallelization of loop programs for distributed memory architectures (2004)
10. Irigoin, F., Triolet, R.: Supernode partitioning. In: Proceedings of the 15th ACM SIGPLAN-SIGACT Symposium on Principles of Programming Languages, POPL 1988, pp. 319–329. ACM, New York (1988)
11. Kelly, W., Maslov, V., Pugh, W., Rosser, E., Shpeisman, T., Wonnacott, D.: The omega library interface guide. Technical report, College Park, MD, USA (1995)
12. Kelly, W., Pugh, W., Rosser, E., Shpeisman, T.: Transitive closure of infinite graphs and its applications. Int. J. Parallel Program. **24**(6), 579–598 (1996)
13. Lim, A., Cheong, G.I., Lam, M.S.: An affine partitioning algorithm to maximize parallelism and minimize communication. In: Proceedings of the 13th ACM SIGARCH International Conference on Supercomputing, pp. 228–237. ACM Press (1999)
14. NAS benchmarks suite (2013). http://www.nas.nasa.gov
15. Pugh, W., Rosser, E.: Iteration space slicing and its application to communication optimization. In: International Conference on Supercomputing, pp. 221–228 (1997)
16. Pugh, W., Wonnacott, D.: An exact method for analysis of value-based array data dependences. In: Banerjee, U., Gelernter, D., Nicolau, A., Padua, D. (eds.) Languages and Compilers for Parallel Computing. LNCS, vol. 768, pp. 546–566. Springer, Heidelberg (1993)
17. Verdoolaege, S.: Integer set library - manual. Technical report (2011). http://www. kotnet.org/~skimo//isl/manual.pdf
18. Wolf, M.E., Lam, M.S.: A loop transformation theory and an algorithm to maximize parallelism. IEEE Trans. Parallel Distrib. Syst. **2**(4), 452–471 (1991)
19. Wonnacott, D.G., Strout, M.M.: On the scalability of loop tiling techniques. In: Proceedings of the 3rd International Workshop on Polyhedral Compilation Techniques (IMPACT), January 2013
20. Xue, J.: On tiling as a loop transformation (1997)

Semiautomatic Acceleration of Sparse Matrix-Vector Product Using OpenACC

Przemysław Stpiczyński[✉]

Institute of Mathematics, Maria Curie–Skłodowska University,
Pl. Marii Curie-Skłodowskiej 1, 20-031 Lublin, Poland
przem@hektor.umcs.lublin.pl

Abstract. The aim of this paper is to show that well known SPARSKIT SpMV routines for *Ellpack-Itpack* and *Jagged Diagonal* formats can be easily and successfully adapted to a hybrid GPU-accelerated computer environment using OpenACC. We formulate general guidelines for simple steps that should be done to transform source codes with irregular data access into efficient OpenACC programs. We also advise how to improve the performance of such programs by tuning data structures to utilize hardware properties of GPUs. Numerical experiments show that our accelerated versions of SPARSKIT SpMV routines achieve the performance comparable with the performance of the corresponding CUSPARSE routines optimized by NVIDIA.

Keywords: Sparse matrices · SpMV · GPUs · OpenACC · CUSPARSE

1 Introduction

Recently, GPU-accelerated computer architectures have become very attractive for achieving high performance execution of scientific applications at low costs [1,2], especially for linear algebra computations [3,4]. Unfortunately, the process of adapting existing software to such new architectures can be difficult. Compute Unified Device Architecture (CUDA) programming interface can be used only for NVIDIA cards, while the use of OpenCL (Open Computing Language [5]) leads to a substantial increase of software complexity.

SPARSKIT is a well known package tool for manipulating and working with sparse matrices [6]. It is a very good example of widely used valuable software packages written in Fortran. Unfortunately, it does not utilize modern computer architectures, especially GPU-accelerated multicore machines. The new implementation of the most important SPARSKIT routines for NVIDIA GPUs has been presented in [7].

Sparse matrix-vector product (SpMV) is a central part of many numerical algorithms [6,8]. There are a lot of papers presenting rather sophisticated techniques for developing SpMV routines that utilize the underlying hardware of GPU-accelerated computers [9–13]. Unfortunately, these methods are rather complicated and usually machine-dependent. However, the results presented

© Springer International Publishing Switzerland 2016
R. Wyrzykowski et al. (Eds.): PPAM 2015, Part II, LNCS 9574, pp. 143–152, 2016.
DOI: 10.1007/978-3-319-32152-3_14

in [14] show that simple SPARSKIT SpMV routines using CSR (Compressed Sparse Row) format [6] can be easily and efficiently adapted to modern multi-core CPU-based architectures. Loops in source codes can be easily parallelized using OpenMP directives [15,16], while the rest of the work can be done by a compiler. Such parallelized SpMV routines achieve the performance comparable with the performance of the SpMV routines available in libraries optimized by hardware vendors (i.e. Intel MKL).

OpenACC is a new standard for accelerated computing [17]. It offers compiler directives for offloading C/C++ and Fortran programs from host to attached accelerator devices. Such simple directives allow to mark regions of source code for automatic acceleration in a vendor-independent manner [18]. However, sometimes it is necessary to apply some high-level transformations of source codes to achieve reasonable performance [19–21]. Paper [22] shows attempts to apply OpenACC for accelerating SpMV. However, the authors consider only some modifications of the CSR format and apply other GPU-specific optimizations (just like communication hiding).

In this paper we show that well known SPARSKIT SpMV routines for *Ellpack-Itpack* (ELL) and *Jagged Diagonal* (JAD) formats [6] can be easily and successfully adapted to a hybrid GPU-accelerated computer environment using OpenACC. We also advise how to improve the performance of such programs by tuning data structures to utilize hardware properties of GPUs applying some high-level transformation of the source code. The paper is structured as follows. Section 2 describes ELL and JAD – two formats which are suitable for GPU-accelerated computations. We show how to apply some basic source code transformations to obtain accelerated versions of SpMV routines. In Sect. 3 we present pJAD - a new format, which allows to outperform SpMV routine for JAD. Section 4 discusses the results of experiments performed for a set of test matrices. We also compare the performance of our OpenACC-accelerated routines with the performance of SpMV for the HYB (ELL/COO) format [23]. Finally, in Sect. 5 we formulate general guidelines for simple steps that should be done to transform irregular source codes into OpenACC programs.

2   SPARSKIT and SpMV Routines

ELL format for sparse matrices assumes the fixed-length rows [24]. A sparse matrix with n rows and at most $ncol$ nonzero elements per row is stored column-wise in two dense arrays of dimension $n \times ncol$ (Fig. 1). The first array contains the values of the nonzero elements, while the second one contains the corresponding column indices.

JAD format removes the assumption on the fixed-length rows [7]. Rows of a matrix are sorted in non-increasing order of the number of nonzero elements per row (Fig. 2). The matrix is stored in three arrays. The first array a contains nonzero elements of the matrix (i.e. jagged diagonals), while the second one (i.e. ja) contains column indices of all nonzeros. Finally, the array ia contains the beginning position of each jagged diagonal. The number of jagged diagonals is

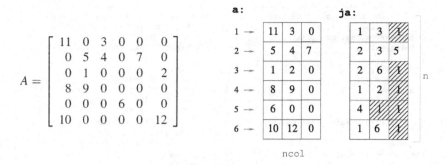

Fig. 1. ELL format for sparse matrices

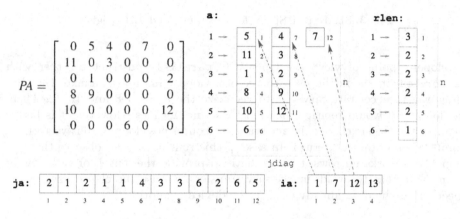

Fig. 2. JAD format for sparse matrices

stored in jdiag. Optionally, we can consider just another array rlen containing lengths of all rows [11]. Elements of this array can be easily calculated (even in parallel) using the following formula

$$\mathbf{rlen}(i) = |\{j : 1 \le j \le \mathbf{jdiag} \wedge \mathbf{ia}(j+1) - \mathbf{ia}(j) \ge i\}|, \quad i = 1, \dots, n. \qquad (1)$$

Figure 3 shows Fortran subroutines which implement SpMV for ELL and JAD. Note that SPARSKIT subroutines amuxe and amuxj were originally written in Fortran 77, but here we present their equivalents written in Fortran 90.

OpenACC provides the parallel construct that launches gangs that will execute in parallel. Gangs may support multiple workers that execute in vector or SIMD mode [17]. This standard also provides several constructs that can be used to specify the scope of data in accelerated parallel regions. It should be noticed that proper data placement and carefully planned data transfers can be crucial for achieving reasonable performance of accelerated programs [19].

In our OpenACC program, a GPU is responsible for performing SpMV while the host program has to read data and initialize computations. The accelerated subroutines accamuxe and accamuxj are presented in Fig. 4. From the

```
1   subroutine amuxe (n,x,y,na,ncol,a,ja)        subroutine amuxj(n,x,y,jdiag,a,ja,ia)
2     real*8 x(n), y(n), a(na,*)                    integer n, jdiag, ja(*), ia(*)
3     integer  n, na, ncol, ja(na,*)               real*8 x(n), y(n), a(*)
4     integer i, j                                 integer i, ii, k1, len, j
5
6     do 1 i=1, n                                  do i=1, n
7       y(i) = 0.0                                   y(i) = 0.0d0
8     end do                                       end do
9     do j=1,ncol                                  do ii=1, jdiag
10      do i=1,n                                     k1 = ia(ii)-1
11        y(i)=y(i)+a(i,j)*x(ja(i,j))                len = ia(ii+1)-k1-1
12      end do                                       do j=1,len
13    end do                                           y(j)=y(j)+a(k1+j)*x(ja(k1+j))
14   end subroutine amuxe                           end do
15                                                 end do
16                                               end subroutine amuxj
```

Fig. 3. SPARSKIT SpMV for ELL (left) and JAD (right)

developer's point of view, the OpenACC `parallel` construct together with `vector_length` should be used to vectorize loops. In case of `amuxe`, the simplest way to accelerate SpMV is to vectorize the loops 6–8 and 10–12. Then, the loop 9–13 would repeat generated kernel `ncol` times. However, it is better to apply the loop exchange. In `accamuxe`, the outermost loop 9–16 is vectorized. Similarly we obtain `accamuxj`. In case of this routine we can observe that the loop 11–19 works on rows, thus we have to provide the length of each row in `rlen`. Note that to avoid unnecessary transfers, we use the clause `present` to specify that the data already exist in the device memory.

```
1   subroutine accamuxe(n,x,y,na,ncol,a,ja)    subroutine accamuxj(n,x,y,jdiag,a,ja,ia,
2     real*8 x(n), y(n), a(na,*)                                               rlen,iperm)
3     integer  n, na, ncol, ja(na,*)           integer n,jdiag,ja(*),ia(*),rlen(*),
4     integer i, j                                                             iperm(*)
5     real*8 t                                  real*8 x(n), y(n), a(*)
6                                               integer i, ii, k1, len, j, k
7   !$acc parallel loop vector_length(128)&     real*8 t
8   !$acc present(x,y,a,ja)
9     do i = 1,n                               !$acc parallel loop vector_length(128)&
10      t=0                                     !$acc present(a,ja,ia,y,x,rlen,iperm)
11      !$acc loop seq                          do i=1, n
12      do j=1,ncol                               t = 0.0d0
13        t = t+a(i,j)*x(ja(i,j))                 !$acc loop seq
14      end do                                    do j=1,rlen(i) !for each within a row
15      y(i)=t                                      k=ia(j)-1+i
16    end do                                        t=t+a(k)*x(ja(k))
17   end subroutine accamuxe                      end do
18                                                y(iperm(i)) = t  !apply permutation
19                                              end do
20                                            end subroutine accamuxj
```

Fig. 4. Accelerated versions of SpMV for ELL (left) and JAD (right)

3   Optimizing SpMV Using pJAD Format

Our version of SpMV for JAD can be further optimized. We can improve memory access by aligning (padding) columns of the arrays a and ja. Thus, in each column we add several zero elements and each column's length should be a multiple of a given *bsize*. Then, each block of threads will have to work on rows of the same length. The number of elements in a and ja will be increased to the size which is bounded by $n_{nz} + jdiag \cdot (bsize - 1)$, where n_{nz} is the number of nonzero elements (Fig. 5). This modified format can be called pJAD (i.e. padded JAD format). Similar modifications have been introduced in [25]. However, Kreutzer et al. consider *bsize* equal to the length of half-warp, what is specific for NVIDIA GPUs. They also assume that threads with a block can be responsible for processing various amount of data. Figure 6 shows the source code of accpamuxj. Note that the array brlen contains the length of each block of rows of a given size *bsize*.

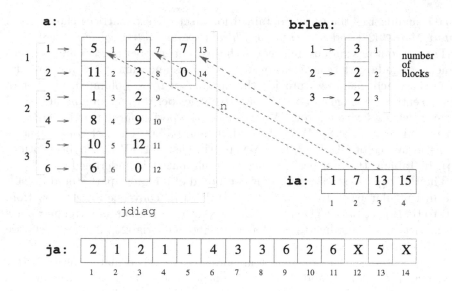

Fig. 5. pJAD format and its data structures

4   Results of Experiments

Our OpenACC implementation of the SpMV routines has been tested on a computer with two Intel Xeon X5650 (6 cores each with hyper- threading, 2.67 GHz, 48 GB RAM) and two NVIDIA Tesla M2050 (448 cores, 3 GB GDDR5 RAM with ECC off), running under Linux under with NVIDIA CUDA Toolkit version 6.5 and PGI Accelerated Server version 15.4, which supports OpenACC [26].

```
1    subroutine accpamuxj(n,x,y,jdiag,a,ja,ia,brlen,iperm,bsize)
2     integer n,jdiag,ja(*),ia(*),brlen(*),iperm(*),bsize
3     real*8 x(n), y(n), a(*)
4     integer i,ii,k1,len,j,k
5     real*8 t
6    !
7     !$acc parallel loop vector_length(128) &
8     !$acc present(a,ja,ia,y,x,brlen,iperm)
9     do i=1,n
10      t = 0.0d0
11      !$acc loop seq
12      do j=1,brlen((i-1)/bsize+1)
13        k=ia(j)-1+i
14        t=t+a(k)*x(ja(k))
15      end do
16      y(iperm(i)) = t
17     end do
18    end subroutine accpamuxj
```

Fig. 6. Accelerated SpMV routine for pJAD format

Table 1 summarizes the results obtained for a set of test matrices chosen from *Matrix Market* [27] and *University of Florida sparse matrix collection* [28].

The set contains various matrices with different number of rows and nonzero elements. The largest *cage15* has over $5 \cdot 10^6$ rows and almost 10^8 nonzero elements. For each matrix we provide the number of rows (columns), number of nonzero entries, average number of nonzero entries per row, maximum number of nonzero elements within a row. We also show the performance (in GFLOPS) of accelerated versions of SpMV for ELL, JAD and pJAD. The last column shows the performance of CUSPARSE SpMV routine using HYB (i.e. hybrid format [23]). In Table 1, the best performance for each matrix is underlined.

The HYB sparse storage format is composed of a regular part stored in ELL and an irregular part stored in COO. CUSPARSE conversion operation from CSR to HYB partitions a given sparse matrix into the regular and irregular parts automatically or according to developer-specified criteria [23]. For our tests, we have chosen the first option.

We can observe that for almost all matrices pJAD format achieves better performance than ELL and JAD. ELL outperforms pJAD only for matrices *cry1000*, *af23560*, *majorbasis*, *ecology2*, *atmosmodl*, where all rows have almost the same number of nonzero elements (i.e. $n_{nz}/n \approx \max_{nz}$) or where the number of nonzero elements is rather big in comparison with the number of rows (i.e. $n \ll n_{nz}$ for *nd24k*). It should be noticed that for some matrices ELL exceeds the memory capacity of Tesla M2050 (*pre2*, *torso1*, *inline_1*). The performance of pJAD is a little bit worse than the performance of HYB, because pJAD format requires re-permutation of the result's entries. For some matrices with $n_{nz}/n \ll \max_{nz}$, pJAD outperforms HYB (i.e. *af23560*, *bcsstk36*, *bbmat*, *cfd1*, *torso1*, *ldoor*). Note that for *cage15*, CUSPARSE routine for conversion from CSR to HYB has failed because memory capacity has been exceeded.

Table 1. Results of experiments for a set of test matrices

Matrix	n	n_{nz}	n_{nz}/n	$\max_{nz}$	ELL	JAD	pJAD	HYB
cry10000	10000	49699	5.0	5	3.99	3.79	3.37	2.11
poisson3Da	13514	352762	26.1	110	2.14	3.40	3.54	4.41
af23560	23560	484256	20.6	21	12.03	11.88	11.94	9.05
g7jac140	41490	565956	13.6	153	1.21	3.23	3.39	4.90
fidapm37	9152	765944	83.7	255	4.97	6.19	6.78	6.84
bcsstk36	23052	1143140	49.6	178	4.34	11.64	13.52	9.21
majorbasis	160000	1750416	10.9	11	15.20	13.78	13.80	14.01
bbmat	38744	1771722	45.7	126	6.03	7.73	8.88	7.34
cfd1	70656	1828364	25.9	33	12.84	12.77	13.21	12.83
ASIC_680ks	682712	2329176	3.4	210	0.28	4.32	4.46	7.54
FEM_3D_thermal2	147900	3489300	23.6	27	13.09	13.13	14.95	14.98
parabolic_fem	525825	3674625	7.0	7	9.80	9.33	9.83	11.58
ecology2	999999	4995991	5.0	5	13.83	12.00	12.84	15.76
pre2	659033	5959282	9.0	628	—	6.59	7.32	8.74
boneS01	127224	6715152	52.8	81	9.96	9.51	11.43	12.58
torso1	116158	8516500	73.3	3263	—	10.70	11.73	6.96
thermal2	1228045	8580313	7.0	11	5.92	4.80	5.03	8.65
atmosmodl	1489752	10319760	6.9	7	13.67	12.42	12.68	15.55
bmw3_2	227362	11288630	49.7	336	2.16	11.58	14.17	16.08
af_shell8	504855	17588875	34.8	40	13.56	14.69	17.59	19.48
cage14	1505785	27130349	18.0	41	6.40	9.48	11.05	12.79
nd24k	72000	28715634	398.8	520	11.46	4.23	4.52	12.52
inline_1	503712	36816342	73.1	843	—	9.89	11.74	12.26
ldoor	952203	46522475	48.9	77	9.05	12.52	15.43	14.68
cage15	5154859	99199551	19.2	47	5.86	9.12	10.49	—

5   Conclusions and Future Work

We have shown that well known SPARSKIT SpMV routines for ELL and JAD
formats can be easily and successfully adapted to a hybrid GPU-accelerated
computer environment using OpenACC. Such routines achieve reasonable performance. Further improvements can be obtained by introducing the new data
formats for sparse matrices to utilize specific GPU hardware properties. Numerical experiments have justified that the performance of our optimized SpMV
routines is comparable with the performance of the routine provided by the
vendor. We have also discussed when the use of considered formats would be
profitable. We believe the use of OpenACC and accelerated Fortran routines

can be attractive for people who prefer to develop applications using high-level directive programming techniques instead of complicated CUSPARSE API.

The general guidelines for semiautomatic acceleration of irregular codes using OpenACC can be summarized as follows:

1. Define regions where data should exist on accelerators. Try to reduce transfers between host and accelerators.
2. Try to vectorize outermost loops within your code. Vectorized loops should have sufficient computational intensity, namely the ratio of the number of computational operations to the number of memory operations should be greater than one.
3. If necessary, apply loop exchange and inform the compiler that loops are safe to parallelize using the `independent` clause in OpenACC `loop` constructs.
4. Try to keep threads within gangs (or *blocks* in terms of CUDA) working on the same amount of data.
5. The best performance occurs when coalesced memory access takes place [29, 30]. Threads within gangs should operate on contiguous data blocks.
6. Tune your data structures by aligning data in arrays. It can be done by data structure padding.

In the future, we plan to implement some other important routines from SPARSKIT, especially well-known solvers for sparse systems of linear equations. We also plan to implement multi-GPU support using OpenACC and OpenMP [31]. The full package with the software will soon be available for the community.

References

1. Nickolls, J., Buck, I., Garland, M., Skadron, K.: Scalable parallel programming with CUDA. ACM Queue **6**, 40–53 (2008)
2. Leist, A., Playne, D.P., Hawick, K.A.: Exploiting graphical processing units for data-parallel scientific applications. Concurrency Comput. Pract. Experience **21**, 2400–2437 (2009)
3. Agullo, E., Demmel, J., Dongarra, J., Hadri, B., Kurzak, J., Langou, J., Ltaief, H., Luszczek, P., Tomov, S.: Numerical linear algebra on emerging architectures: the PLASMA and MAGMA projects. J. Phys. Conf. Ser. **180**, 012037 (2009)
4. Nath, R., Tomov, S., Dongarra, J.: Accelerating GPU kernels for dense linear algebra. In: Palma, J.M.L.M., Daydé, M., Marques, O., Lopes, J.C. (eds.) VECPAR 2010. LNCS, vol. 6449, pp. 83–92. Springer, Heidelberg (2011)
5. Kowalik, J.S., Puzniakowski, T.: Using OpenCL - Programming Massively Parallel Computers. Advances in Parallel Computing, vol. 21. IOS Press, Amsterdam (2012)
6. Saad, Y.: Iterative Methods for Sparse Linear Systems. SIAM, Philadelphia (2003)
7. Li, R., Saad, Y.: GPU-accelerated preconditioned iterative linear solvers. J. Supercomputing **63**, 443–466 (2013)
8. Helfenstein, R., Koko, J.: Parallel preconditioned conjugate gradient algorithm on GPU. J. Comput. Appl. Math. **236**, 3584–3590 (2012)
9. Feng, X., Jin, H., Zheng, R., Shao, Z., Zhu, L.: A segment-based sparse matrix-vector multiplication on CUDA. Concurrency Comput. Pract. Experience **26**, 271–286 (2014)

10. Pichel, J.C., Lorenzo, J.A., Rivera, F.F., Heras, D.B., Pena, T.F.: Using sampled information: is it enough for the sparse matrix-vector product locality optimization? Concurrency Comput. Practi. Experience **26**, 98–117 (2014)
11. Vázquez, F., López, G.O., Fernández, J., Garzón, E.M.: Improving the performance of the sparse matrix vector product with GPUs. In: 10th IEEE International Conference on Computer and Information Technology, CIT 2010, Bradford, West Yorkshire, UK, 29 June-1 July 2010, pp. 1146–1151 (2010)
12. Williams, S., Oliker, L., Vuduc, R.W., Shalf, J., Yelick, K.A., Demmel, J.: Optimization of sparse matrix-vector multiplication on emerging multicore platforms. Parallel Comput. **35**, 178–194 (2009)
13. Matam, K.K., Kothapalli, K.: Accelerating sparse matrix vector multiplication in iterative methods using GPU. In: International Conference on Parallel Processing, ICPP 2011, Taipei, Taiwan, 13–16 September 2011, pp. 612–621 (2011)
14. Bylina, B., Bylina, J., Stpiczyński, P., Szałkowski, D.: Performance analysis of multicore and multinodal implementation of SpMV operation. In: Proceedings of the Federated Conference on Computer Science and Information Systems, 7–10 September 2014, Warsaw, Poland, pp. 575–582. IEEE Computer Society Press (2014)
15. Chandra, R., Dagum, L., Kohr, D., Maydan, D., McDonald, J., Menon, R.: Parallel Programming in OpenMP. Morgan Kaufmann Publishers, San Francisco (2001)
16. Marowka, A.: Parallel computing on any desktop. Commun. ACM **50**, 74–78 (2007)
17. OpenACC: The OpenACC Application Programming Interface (2013). http://www.openacc.org
18. Sabne, A., Sakdhnagool, P., Lee, S., Vetter, J.S.: Evaluating performance portability of OpenACC. In: Brodman, J., Tu, P. (eds.) LCPC 2014. LNCS, vol. 8967, pp. 51–66. Springer, Heidelberg (2015)
19. Wang, C., Xu, R., Chandrasekaran, S., Chapman, B.M., Hernandez, O.R.: A validation testsuite for OpenACC 1.0. In: 2014 IEEE International Parallel & Distributed Processing Symposium Workshops, Phoenix, AZ, USA, 19–23 May 2014, pp. 1407–1416 (2014)
20. Reyes, R., López-Rodríguez, I., Fumero, J.J., de Sande, F.: A preliminary evaluation of OpenACC implementations. J. Supercomputing **65**, 1063–1075 (2013)
21. Eberl, H.J., Sudarsan, R.: OpenACC parallelisation for diffusion problems, applied to temperature distribution on a honeycomb around the bee brood: a worked example using BiCGSTAB. In: Wyrzykowski, R., Dongarra, J., Karczewski, K., Waśniewski, J. (eds.) PPAM 2013, Part II. LNCS, vol. 8385, pp. 311–321. Springer, Heidelberg (2014)
22. Fegerlund, O.A., Kitayama, T., Hashimoto, G., Okuda, H.: Effect of GPU communication-hiding for SpMV using OpenACC. In: Proceedings of the 5th International Conference on Computational Methods (ICCM 2014) (2014)
23. NVIDIA: CUDA CUSPARSE Library. NVIDIA Corporation (2015). http://www.nvidia.com/
24. Grimes, R., Kincaid, D., Young, D.: ITPACK 2.0 users guide. Technical report CNA-150, Center for Numerical Analysis, University of Texas (1979)
25. Kreutzer, M., Hager, G., Wellein, G., Fehske, H., Basermann, A., Bishop, A.R.: Sparse matrix-vector multiplication on GPGPU clusters: a new storage format and a scalable implementation. In: 26th IEEE International Parallel and Distributed Processing Symposium Workshops & PhD Forum, IPDpPS 2012, Shanghai, China, 21–25 May 2012, pp. 1696–1702 (2012)

26. Wolfe, M.: Implementing the PGI accelerator model. In: Kaeli, D.R., Leeser, M. (eds.) Proceedings of 3rd Workshop on General Purpose Processing on Graphics Processing Units, GPGpPU 2010, Pittsburgh, Pennsylvania, USA, 14 March 2010. ACM International Conference Proceeding Series, vol. 425, pp. 43–50. ACM (2010)

27. Boisvert, R.F., Pozo, R., Remington, K.A., Barrett, R.F., Dongarra, J.: Matrix market: a web resource for test matrix collections. In: Boisvert, R.F. (ed.) Quality of Numerical Software - Assessment and Enhancement, Proceedings of the IFIP TC2/WG2.5 Working Conference on the Quality of Numerical Software, Assessment and Enhancement, Oxford, UK, 8–12 July 1996. IFIP Conference Proceedings, vol. 76, pp. 125–137. Chapman & Hall (1997)

28. Davis, T.A., Hu, Y.: The University of Florida sparse matrix collection. ACM Trans. Math. Softw. **38**, 1–25 (2011)

29. NVIDIA Corporation: CUDA Programming Guide. NVIDIA Corporation (2015). http://www.nvidia.com/

30. NVIDIA: CUDA C Best Practices Guide. NVIDIA Corporation (2015). http://www.nvidia.com/

31. Xu, R., Chandrasekaran, S., Chapman, B.M.: Exploring programming multi-GPUs using OpenMP and OpenACC-based hybrid model. In: 2013 IEEE International Symposium on Parallel & Distributed Processing, Workshops and Phd Forum, Cambridge, MA, USA, 20–24 May 2013, pp. 1169–1176 (2013)

Multi-threaded Construction of Neighbour Lists for Particle Systems in OpenMP

Rene Halver[1] and Godehard Sutmann[1,2(✉)]

[1] Jülich Supercomputing Centre (JSC), Institute for Advanced Simulation (IAS), Forschungszentrum Jülich (JSC), 52425 Jülich, Germany
[2] ICAMS, Ruhr-University Bochum, 44801 Bochum, Germany
g.sutmann@fz-juelich.de

Abstract. The construction of neighbour lists based on the linked cell method is investigated in the context of particle simulation methods within the OpenMP shared memory programming model. Various implementations are studied which avoid memory collisions and race conditions. Performance and optimisation considerations are made along with run time behaviour and memory requirements. Performance models are proposed, which reproduce the measured runtime behaviour and which provide insight into the performance dependence on specific system parameters. Benchmarks are performed for different implementations on a number of multi-core architectures and thread numbers up to 240 are considered on the Xeon Phi architecture in the SMT mode, so that performance can be studied for a large number of threads working concurrently on the construction of linked cells on a shared memory partition.

1 Introduction

Particle simulation methods, e.g. Molecular Dynamics (MD) [4], Smoothed Particle Hydrodynamics [6] or element free methods [7], are nowadays applied to large scale systems, composed of millions or even billions of particles. For systems, composed of particles interacting via short range potentials or local propagation rules large scale simulations can be executed on massively parallel computers applying domain decomposition schemes [8,10]. This reduces the calculation of interactions to finite spatial regions administrated by either single CPUs, nodes or many-core accelerators. For short range interactions it is usually sufficient to communicate with adjacent processors so that the total number of communications is constant. Domain decomposition is therefore a suitable method to design and implement algorithms which scale to a large number of processors. However, scaling starts to saturate if the surface area of a domain and the related effort in communication of data between neighbour processors become as time consuming as the work within the volume part of a domain. This usually occurs for strong scaling problems, where the number of processors is increased for a given problem size.

For programming models, based on a distributed memory paradigm, like MPI. a compute node which is composed of a number of cores of $\mathcal{O}(10)$ or

© Springer International Publishing Switzerland 2016
R. Wyrzykowski et al. (Eds.): PPAM 2015, Part II, LNCS 9574, pp. 153–165, 2016.
DOI: 10.1007/978-3-319-32152-3_15

even $\mathcal{O}(100)$ administers the same number of computational domains, which all have to communicate explicitly their data. Although the data exchange between cores which have access to the same memory can be implemented efficiently, the ratio of total surface area to total domain volume increases proportionally with the number of subdivisions of cartesian directions. Therefore, in order to avoid explicit communication between domains administered on the same compute node, hybrid programming models are an attractive alternative. While the number of domains per node is reduced to one and communication between compute nodes is performed by MPI, a shared memory programming model like OpenMP [1] can be applied to exploit multi-core parallelism on the nodes. Until now there it is no clear decision, whether pure distributed memory models like MPI or a hybrid implementation like MPI plus OpenMP are most successful. However, modern multi-core architectures implement simultaneous multi-threading (SMT) features, which allow to allocate more threads than available compute cores while still benefitting from an efficiency gain due to parallel pipeline execution on the cores, which makes hybrid programming models indispensable for exploiting maximum performance offered by modern computers.

To calculate interactions between particles most efficiently and to avoid a quadratical complexity with the number of degrees of freedom, neighbour list techniques have been developed which reduce the evaluation between particles to linear complexity. It is the construction of neighbour lists which can be a bottleneck in the parallel performance in a hybrid scheme or a single node OpenMP implementation. One reason is that for every particle i located on the processor the local environment of particles j is stored into a data structure, which is later on used for a fast evaluation of interactions. Since particles most often perform mutual movements and undergo diffusion, the location in physical space gets uncorrelated with the position in memory over time, and consequently race conditions can occur when different threads access the memory locations in the list arrays (as it is the case for linked-cell lists [9]) or have to respect sequential ordering (as it is for Verlet lists [11]).

There is a number of works, focusing on either parallel sorting under OpenMP [12] or the analysis and performance of linked cell methods [9,11]. However, to our best knowledge there is no in depth discussion about the implementation and performance analysis of linked cell methods under OpenMP.

In the following we will consider the implementation of linked-cell lists in more detail and especially compare different implementations for a shared memory programming model based on OpenMP with a special focus on the avoidance of race conditions due to simultaneous memory access by different threads at the same time. We will consider implementations based on OpenMP pragmas `critical` and `atomic`, which ensure non-conflicting memory access for execution blocks or single statements but which often lead to a quasi-serialisation of program execution or might even end in wrong execution sequences, as will be discussed later on. As an alternative multi-array implementations, `lock` statements and their hybrid implementation are studied, which provide a large improvement of scalability up to 240 threads/node. Benchmarks are conducted on a number of architectures are tested in order to check general trends of the findings.

2 Linked-Cell Lists

2.1 Method

When the range of interaction between particles is limited in space by a distance R_c, called the cutoff radius, a common sorting method is to subdivide the physical space into equal cells of length $L_c \geq R_c$. This ensures that a particle i, located in a cell, which is characterised by a cartesian index triplet $\{ix, iy, iz\}$ has an effective search space for interaction partners in the index range $\{ix \pm 1, iy \pm 1, iz \pm 1\}$, which gives a maximum of 27 cells to be accessed. If symmetry considerations between particle interactions can be made (like Newton's principle of action-counteraction in MD simulations), the search space can be further reduced to the local cell, where a particle is located plus 13 neighbour cells. In principle one can have a larger number of sub-cells for each particle in order to approximate better the spherical shape of the interaction range. However, if the size of a cell is chosen too small, the number of cells, located within the search volume around a particle i, increases and might exceed the number of particles, which actually interact with i, so that during a search a large number of cells is empty resulting in both an unfavourable computational and a memory access overhead, which eventually slows down the simulation. In practical cases, most often the cell size is in the order of the cutoff radius, whereas an optimal size of cells was found in different studies [11] to be close to $R_c/2$. The main procedure in the linked-cell list consists in sorting the particles into cells in an ordered way (cmp. Fig. 1). Algorithm 1 shows a pseudo-code for the procedure, which also demonstrates that the complexity of the method is $\mathcal{O}(N)$. The array of particles is scanned in an ordered way (although in principle also arbitrary sequences would be allowed) and the first particle which is sorted into the cell links to the null-index (or any other index, which is outside the index space of the particle set). Each following particle in the same cell is linked then to the index of the particle sorted last into the same cell.

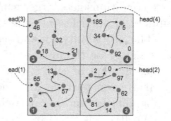

Fig. 1. Schematic of linked-cell list construction. Arrows indicate how particles are linked together. The head-array stores the index of the latest sorted particle in the cell.

2.2 Race Conditions

Linked-cell lists are efficient for storing particle geometric neighbourhood information in memory which grows linearly with particle number. For the implementation with OpenMP the problem of race conditions arises, if no natural sorting of the particles is already in place. Assuming the general case, where particles with indices $i \in \{1, N\}$ are distributed randomly over a volume Ω, where the linked-cell list is applied. If the the loop over particles is parallelized

---

Algorithm 1. Serial linked-cell algorithm

1: head(1:n_c) ← 0   ▷ initialize cell head
2: **for** (all particles i) **do**   ▷ run over all particles in system or domain
3:     ic ← compute_ic($\mathbf{r}(i), L_c$)   ▷ compute cell index from coordinates $\mathbf{r}$ and cell
    length L_c
4:     list(i) ← head(ic) ▷ update particle list with current entry point of computed
    cell
5:     head(ic) ← i ▷ update entry point of computed cell to index of actual particle
6: **end for**

---

with OpenMP, each thread operates on a different chunk of particles with contiguous index ranges. In the general case the particles are not localised within a defined sub-volume and in general for active volumes of distinct threads p and q it is $\Omega_p \bigcap \Omega_q \neq \emptyset$. This implies that distinct threads partially operate on the same volume elements. In the case where two threads treat particles i and j which are to be sorted into the same linked-cell at the same time, there is high risk for a race-condition and the algorithm gets error prone. The risk for race conditions gets larger with increasing global or local density. Sorting a large number of particles into the same cell, the collision probability increases.

The probability for race-conditions, p_{rc}, can be quantified in the following way. Let N be the number of particles, which are simulated on n_{th} threads and n_c the number of cells into which the system is subdivided.

If a given particle, i_{th_r}, administrated by thread r is considered, then the probability for another particle i_{th_s}, administrated by thread s to be both in the same cell, c_k, is given by $p(i_{th_r}, i_{th_s} \in \Omega_{c_k}) = 1/n_c$.

Furthermore, the probability that $r \neq s$, i.e. that both particles are on different threads, is $p(r \neq s) = (n_{th} - 1)/n_{th}$ and therefore, the probability for a race condition of a particle pair is given by

$$p_{rc} = p(i_{th_r}, i_{th_s} \in \Omega_{c_k}) \times p(r \neq s) \tag{1}$$

$$= \frac{1}{n_c} \frac{n_{th} - 1}{n_{th}} \tag{2}$$

These considerations are made for a particle pair and therefore on average one has to consider N/n_{th} of these operations in every time step for every thread. Since two particles on different threads were considered, the total number of counts has to be multiplied by $n_{th}/2$. These considerations provide the number of particle pairs on different threads in a given loop iteration. If there are n_g statements, accessing global memory out of n_l statements in the loop, then the number of threads being located coincidently at global memory statements is n_g/n_l (neglecting memory latencies and complexity of operations). Therefore, the total number of memory collisions in each time step can be estimated approximately as

$$n_{rc} = p_{rc} \times \frac{N}{n_{th}} \times \frac{n_{th}}{2} \times \frac{n_g}{n_l} \qquad (3)$$

$$= \frac{1}{n_c} \frac{n_{th} - 1}{n_{th}} \times \frac{n_{th}}{2} \times \frac{n_g}{n_l} \qquad (4)$$

$$= \frac{1}{2} \frac{N}{n_c} \frac{n_g}{n_l} (n_{th} - 1) \qquad (5)$$

This result shows that the number of race conditions follows a simple dependence on N, n_c and n_{th}. In order to test this theoretical prediction with measurements, test scenarios were conducted to check the number of race conditions, which are to be expected in realistic scenarios. To this aim, systems were defined with number of particles $N \in \{10^3, 10^8\}$ for different number of cells, into which the particles were sorted via the linked-cell method. The number of cells was modified according to $n_c \in \{10^3, 100^3\}$. The number of threads was varied in the range $n_{th} \in \{1, 240\}$, where for the case $n_{th} = 1$, no race conditions should appear, which therefore served as a control run. Race conditions were identified in the following way: if two particles located in the same cell but managed on different threads are treated at the same time and therefore threads try to access the same memory location at the same time, the information of one particle is overwritten and therefore this particle is not properly sorted into the linked-cell structure. If the resulting list is read in a second step and particle information is reconstructed a number of particles is not stored in the list due to race-conditions. Therefore, the number of lost particles is a measure for the number of race-conditions. We note that the probabilistic model, suggested below is a lower limit for the number of race conditions since multiple memory collisions or other side effects due to collisions are not taken into account. In a comparison with the measured numbers for race conditions, we find a very good agreement between model and numerical experiment (cmp. Fig. 2. The theoretical model can be understood as an upper limit for the number of race conditions, which is in very good agreement with measurements.

As a consequence of the predictions and measurements, correctness of operations has to be ensured by either synchronisation operations between the threads or by ensuring fully asynchronous or disjunct operations of threads on the memory. Therefore, an implementation of the linked-cell algorithm under OpenMP programming model has to respect the fact that multiple threads can work on overlapping volume elements. The avoidance of risk can be achieved either by data structures or OpenMP commands or the combination of both. In the following we will implement and compare the following scenarios: application of array copies, plain use of OpenMP locks, locks with caches in synchronous and asynchronous mode and the combination of locks and array copies.

3   Parallel Implementations

3.1   OpenMP with Critical and Atomic

The constructs, offered by OpenMP to access memory exclusively by one thread, are the *critical* environment and the *atomic* and *lock* statements. We have

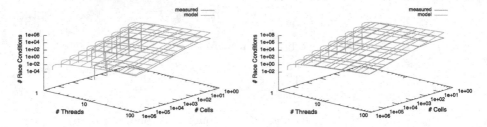

Fig. 2. Number of race conditions for the linked cell algorithm as a function of number of cells, n_c, and number of threads, n_{th}, when no synchronisation steps between threads are considered. Compared are measurements on JUQUEEN with the probabilistic model, Eq. 5 for $N = 10^4$ (left) and $N = 10^6$ (right).

implemented all three of them but have found that only the lock-statement offers an acceptable solution to achieve both avoidance of race conditions and scalability. The difference between the three constructs is obvious: *critical* is the strongest statement to enforce memory thread-safety. The whole block which is inside the critical region is executed exclusively on a single thread, blocking all the other threads to avoid possible race conditions. If a loop is protected by a critical statement this implies a serialisation and has the obvious consequence that the implementation is not scalable. Even worse, the overhead by administrating the critical regions by the threads adds on top and results, in general, in a longer execution time than running on a single thread without *critical*.

Atomic ensures exclusive access of a memory location on a single thread. Compared to the serial implementation this mainly involves the introduction of atomic statements "`#pragma omp atomic write`" before the head- and list-items in lines 1, 5 and 1, 4 in Algorithm 1, respectively. However, if there are multiple data accesses in a loop which depend on each other, a finite probability exists to get asynchronous execution between the threads within a loop. With respect to the serial version of the linked cell algorithm (Algorithm 1) this occurs when one thread is updating the **head** item (cmp. Algorithms 1 and 5) immediately before an other thread is mapping the **list** item onto **head**. If particles are located on different threads but geometrically located within the same cell, this will result in mismatches of list-entries where finally particles are lost from the list. Depending on the parameters of the simulation setup (number of particles N, number of cells n_c and number of threads n_{th}) the number of such collisions does vary but in every tested simulation scenario we found that a small percentage of particles was lost from the system due to race conditions, which, as a consequence, declassifies this method from practical use for the construction of linked cell list.

3.2 Copies of Arrays

The safest way to avoid race conditions on a multi-threaded architecture is to assure that every thread is operating on its own memory space. Since the linked cell algorithm has the dual characteristics of locality in cell-index space but non-locality in particle-index space, the cell array has to be copied for each thread to ensure that an OpenMP paralleled do-loop over particles does not collide in memory with another thread if particle positions fall into the same cell. For small number of cells and threads this method might be still appropriate and manageable. But for large number of threads the danger exists that the allocation of the cell structure consumes a big portion of memory space potentially leading to memory problems and consequently to limitations of system size. Compared to the serial implementation, the copied version requires a minor extension of the algorithm (cmp. Algorithm 2). Since information is now stored for the same cell on different threads, this information has to be combined in a final step, since for further processing of the linked cell scheme (e.g. in the force routine), parallelisation is to be explored on a cell level, where the number of cells is distributed with its complete information over the threads. Therefore, not only the last particle found within a cell on a thread has to be stored (t_last(th)), but also the first one (t_first(th)), in order to efficiently concatenate the list structure without running over all list entries. Considering two threads with information of a given cell, ic. The particles are linked together on each thread via the list-array. If, on any thread, list(i) $\rightarrow$ 0 points to zero, the list is finished on this thread. A concatenation of information from two threads into a new array can be easily performed via two operations: (1) head(ic) $\leftarrow$ t_last(th2), (2) list(t_first(th2)) $\leftarrow$ t_last(th1). Since the first operation has only to be performed once for each cell, this implies that if information about particles in a given cell is distributed over n_{th} threads, $(n_{th}+1)$ operations are necessary to concatenate the linked cell structure for every cell. For large number of cells this does not only imply a memory bottleneck for the n_{th} copies of the cell structure, but also $(n_{th}+1) \times n_c$ operations are necessary to concatenate the list array. As shown in Algorithm 2, this operation can be parallelised over cells, so that every thread has to perform $(n_{th}+1)n_c/n_{th}$ operations.

3.3   OpenMP Locks

Extension of the serial linked cell algorithm invokes the extension via setting "OMP_SET_LOCK(locks(ic))" before statement Algorithm 1 line 4 and "OMP_UNSET_LOCK(locks(ic))" after statement Algorithm 1 line 5. The array locks takes care on the simultaneously locking of ic if it is a multidimensional array. To model the performance of the locked linked-cell implementation, one can consider the following steps. The creation of locks will first of all induce an overhead, since a memory location has to be exclusively tagged by a thread and all other threads have to check whether a memory address, which is to be accessed by a given thread is locked or free to write. This locking and unlocking of addresses is measured as τ_l. Since the loop runs over all particles, the total overhead time for the lock-operation is given as $T_l = N\tau_l$. Compared to the serial execution time $T_s = N\tau_s$, this will lead to an execution time on a single thread of $T_L(1) = T_l + T_s$ or on n_{th} threads as $T_L(n_{th}) = (T_l + T_s)/n_{th} = (1+\alpha)T_s/n_{th}$, where $\alpha = \tau_l/\tau_s$ was introduced. On the other hand it was shown before that the

Algorithm 2. Copied linked-cell algorithm

1: t_head(1:n_c) ← 0; t_last(1:n_c) ← 0 ▷ initialize arrays for first and last particle in a cell on each thread
2: #pragma OMP PARALLEL PRIVATE(i,ic,tid)
3: tid ← OMP_GET_THREAD_NUM()
4: **for** (all particles i on tid) **do** ▷ run over all particles administrated by thread tid
5:    ic ← compute_ic($\mathbf{r}(i), L_c$)   ▷ compute cell index from coordinates $\mathbf{r}$ and cell length L_c
6:    list(i) ← t_head(tid, ic)   ▷ map cell entry index to list
7:    **if** (t_head(tid, ic) = 0) t_last(tid, ic) ← i ▷ store initial particle of the cell in t_last
8:       t_head(tid, ic) ← i ▷ over write cell entry index with current particle index
9: **end for**
10: #pragma OMP END PARALLEL DO
11:
12: #pragma OMP PARALLEL DO PRIVATE(ic,tid,sid,last)
13: **for** (all cells ic) **do**   ▷ combine information from all threads
14:    **for** (all threads $n_{th} - 1 : 0 : -1$) **do** ▷ run over all particles in reverse order from $n_{th} - 1$ to 0
15:       **if** (t_head(tid, ic) ≠ 0) **then** ▷ if cell information is not empty on thread tid
16:          head(ic) ← t_head(tid, ic)
17:          last ← t_last(tid, ic)
18:          **for** (sid ← $tid - 1 : 0 : -1$) **do**   ▷ concatenate lists from different threads
19:             **if** (t_head(sid, ic) ≠ 0) **then** ▷ if cell information is not empty on thread sid
20:                list(last) ← t_head(sid, ic) ▷ end of list is updated with head of thread sid
21:                last ← t_last(sid, ic)   ▷ set pointer to the last element of cell ic
22:             **end if**
23:          **end for**
24:          exit thread loop
25:       **end if**
26:    **end for**
27: **end for**
28: #pragma OMP END PARALLEL DO

number of race conditions, n_{rc} (Eq. 5). as function of threads, cells and particles can be considerable. If we consider the effect of a lock to prevent the occurrence of race conditions, we can assume in a first approximation that this is the number of threads which will be cumulated in the locks and induce some additional waiting time until the locks are released. Since the access time of a thread on a locked memory location may occur at any instance during the locked state, the average waiting time for a thread will be $\tau_l/2$. Therefore, the performance and

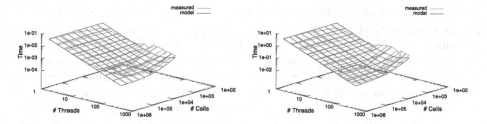

Fig. 3. Performance on the Xeon Phi for synchronisation between threads with locks. Compared are results for measured data with the probabilistic model, Eq. 7, for $N = 10^4$ (left) and $N = 10^6$ (right).

scalability of the locked version of linked cell list can be expressed as

$$T_L(N, n_c, n_{th}|\alpha) = N\frac{\tau_l + \tau_s}{n_{th}} + n_{rc}\frac{\tau_l}{2} \tag{6}$$

$$= N\tau_s\left(\frac{1+\alpha}{n_{th}} + \frac{\alpha}{4}\frac{n_{th}-1}{n_c}\right) \tag{7}$$

It is obvious that the first term in Eq. 7 presents a scalable contribution which is inversely proportional to the number of threads. However, the second term will contribute to a saturation, which is mainly dependent on the number of cells. As we have seen before a high collision probability in memory is given, when many particles are located in a cell operated by a large number of threads. Therefore, from this representation it might be even expected that the performance can be slowed down with increasing number of threads. In Fig. 3 we compare performance results for the lock variant of the linked cell algorithm on a Xeon-Phi processor with 60 cores, which allows according to the four-way SMT usage of the cores the application of 240 threads. The comparison is made for particle systems of $N = 10^4$ and 10^6 particles. Although not perfect, the main characteristics of the performance model are found for the Xeon-Phi, especially the slowing down behaviour in performance for large number of threads and small number of cells. Furthermore, it is seen that the qualitative behaviour for the cases is well reproduced and that the results differ mainly throughout the scaling factor $N\tau_s$ introducing the dependence on the number of particles. For large number of threads and large number of cells a deviation gets more apparent for the case $N = 10^4$, where the execution time is increased relative to the model prediction. We attribute this behaviour to memory access, where random memory addresses (distribution of particle indices) has a relatively larger scatter on the same size of grid cells and thread numbers.

3.4 Combination of Copies and Locks

Buffered copy-lock: As was shown before, setting and unsetting the lock introduces a considerable overhead to the execution time. Therefore, reducing the number of locks is desirable, which can be achieved by an extension of the simple copy-lock version. Each thread allocates a private memory segment of size n_b, which is smaller than the complete replication of a cell structure and which serves as a buffer to store information of the particle index and the corresponding cell, where it is located. To reduce memory collision probability between threads during storage of buffer information, the global grid structure is copied n_{cp} times. If the buffer is filled, the thread sets a lock on one of these copies of the grid array, which is not locked by another thread, constructs the linked list information from the buffer and continues with filling the buffer anew. This is repeated, until the particles, attributed to the thread are fully treated. The free parameters, which have to be specified are the size of the buffer and the number of copies of the grid structure. When all threads have finished, the number of grid copies are concatenated to a single linked cell structure, similar to the copy-variant.

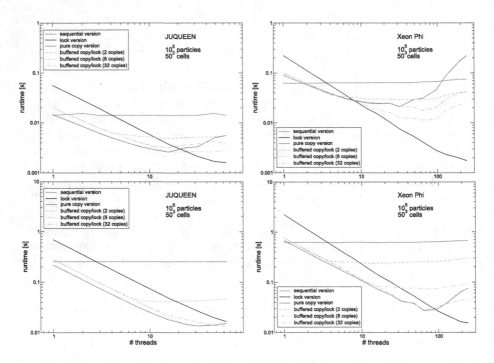

Fig. 4. Scalability comparison between JUQUEEN (left) and Xeon Phi (right) for different test scenarios for variations in particle number and densities, $\rho = N/n_c$ in the system. From top to bottom: (i) $N = 10^5$, $n_c = 125000$, $\rho = 0.8$; (ii) $N = 10^6$, $n_c = 125000$, $\rho = 8$. Ordinates of JUQUEEN and Xeon Phi have the same range to allow direct comparison.

4   Discussion

We have studied the scalability of the linked cell method on two different architectures, the IBM BlueGene/Q (JUQUEEN [2]) with 16 cores/node and a Xeon Phi 5110P with 60 compute cores, which is mounted to an Intel Xeon E5-2650 CPU [3]. Starting point of the study was for both machines a sequential version of the linked-cell algorithm, Algorithm 1. One goal of the work was to test the scalability of the algorithm in a multi-core environment where many threads concurrently work on the same problem accessing the same shared memory. Employing a 4-way SMT mode on each architecture, comparisons could be made on JUQUEEN for up to 64 threads (16 cores/node) and on Xeon Phi for up to 240 threads. The algorithmic problem in the study shows up in the construction of the lists, where an index ordered array in memory is accessed, i.e. particle positions, but where results of the sorting are stored, in general with random memory access. Depending on the size of the array, it is not kept in cache and therefore a risk of random memory accesses is given leading easily to memory bandwidth limited performance (memory bandwidth is 6.4 GB/s/core on JUQUEEN and 320 GB/s in total on Xeon Phi). In addition there is an indirect memory addressing,characteristic for linked cell algorithm, which makes it difficult to vectorize. A method was proposed in Reference [5], where the main loop was split into two parts of which one could be vectorized. This does not avoid the main problem of the linked cell method and therefore, we have chosen not to concentrate on a possible vectorized implementation. Possibly the Xeon Phi might benefit to some extend from a partial vectorisation, but the main features, as observed in the present study are likely to be conserved. Here we concentrated on different thread safe implementations showing very different scalability properties (see Fig. 4). As a starting point the number of potential collisions in memory in a multi-threaded environment was demonstrated for a simple model which followed rather well the actual findings from simulations. As standard implementations within OpenMP standard the `critical`, `atomic` and `lock` statements are offered. It could be shown that while `critical` leads to a serialisation of the execution and `atomic` does not prevent race conditions in the algorithm, only `lock` is able to map the algorithm on a thread safe implementation. As was observed for both architectures, `lock` induces some considerable overhead and leads to considerable longer execution times ($\approx$ 5 times) on a single thread. A faster implementation is given by a copy version, where each thread is working on its own local copy of the grid structure. Although this was observed to give fastest results for large number of particles and cells at moderate number of threads, this implementation comes soon to its end, because of its large memory demand. Especially in the SMT mode, where more threads are invoked than physical cores, strong saturation is observed, which might be due to cache coherency across the threads, i.e. SMT is strongly hindered if after changes in an array location, updates on the other threads have to be done. This situation is also observed for the buffered copy-variants, where the number of copies is varied between 2 and 32 in the studies. Although locks have to be set for every copy the performance is improved for large number of particles and cells since the number of locks is inversely reduced by the number of entries n_b in the buffer, which is $n_b = 2 \times 10^4$ for all cases studied here. Although a strong performance degradation is observed for the different copy versions for small number of cells and particles, it is interesting to observe a reverse behaviour , i.e. small number of copies shows the best runtime behaviour for copy versions. It is, however, a matter of fact that the induced overhead, induced in the concatenation step of the copied grid arrays, leads to a worse performance than the sequential implementation. We have to stress here that a bad performance is always to be expected for the case where the number of particles gets smaller than the number of cells, leading to overhead in the copy versions when iterating over empty cells in the concatenation step. Also, from Algorithm 2 it is clear that a formal complexity of $\mathcal{O}(n_{th}^2)$ is induced, although for a small density of particles there will be never a quadratic execution (if $N/n_c < n_{th}$). For large densities, however, the quadratic behaviour will appear as a pre factor, limiting scalability for large n_{th}.

It has to be mentioned that all algorithms, which were considered in the present study, show a degraded efficiency with respect to the single thread performance. For the case of $N = 10^6$ particles and $n_c = 50^3$ cells the efficiency is reduced to $\approx 11\%$ and $\approx 17\%$ on JUQUEEN and Xeon Phi, respectively. This can be explained either with too much work in copying data and gathering information between the threads, enhanced for large number of threads, or degradation is due to a too large prefactor in OpenMP specific operations, e.g. the lock statement. Considering, however, the efficiency of an algorithm compared to itself, the scaling behaviour is not bad. E.g., the algorithm including locks for thread synchronisation shows an efficiency of 70 % (95 %) and 60 % (82 %) on JUQUEEN and Xeon Phi, where numbers in brackets refer to efficiencies before entering hyperthreading mode.

From the present study we can conclude that the application of the `lock` statement gives acceptable results, whereas efficiency degradation is always present due to the considerable overhead introduced by setting and unsetting the locks. Therefore, for small number of threads and $N/n_c \geq 1$ copy variants are an attractive implementation for the small n_{th}, in several cases up to the number of threads where SMT mode gets important. There are other ideas, introducing asynchronous models for locking arrays and for partially vectorising the loops, which is in progress and will be communicated in future.

References

1. http://openmp.org/wp/openmp-specifications/
2. http://www.fz-juelich.de/ias/jsc/EN/Expertise/Supercomputers/JUQUEEN/JUQUEEN_node.html
3. http://www.fz-juelich.de/ias/jsc/EN/Research/HPCTechnology/ClusterComputing/JUROPA-3/JUROPA-3_node.html
4. Frenkel, D., Smit, B.: Understanding Molecular Simulation: From Algorithms to Applications. Academic Press, San Diego (2002)
5. Grest, G.S., Dünweg, B., Kremer, K.: Vectorized link cell Fortran code for molecular dynamics siulations for large number of particles. Comput. Phys. Comm. **55**, 269–285 (1989)

6. Hoover, W.G.: Smooth Particle Applied Mechanics. World Scientific, Singapore (2006)
7. Li, S., Liu, W.: Meshfree and particle methods and their applications. Appl. Mech. Rev. **55**, 1–36 (2002)
8. Plimpton, S.: Fast parallel algorithms for short range molecular dynamics. J. Comput. Phys. **117**, 1 (1995)
9. Rapaport, D.: The Art of Molecular Dynamics Simulation. Cambridge University Press, Cambridge (2001)
10. Sutmann, G.: Classical molecular dynamics. In: Grotendorst, J., Marx, D., Muramatsu, A. (eds.) Quantum Simulations of Many-body Systems: From Theory to Algorithm, vol. 10, pp. 211–254. NIC, Jülich (2002)
11. Sutmann, G., Stegailov, V.: Optimization of neighbor list techniques in liquid matter simulations. J. Mol. Liq. **125**, 197–203 (2006)
12. Zurek, D., Pietron, M., Wielgosz, M., Wiatr, K.: The comparison of different sorting algorithms implemented on different hardware platforms. Comput. Sci. **14**, 679–691 (2013)

NumCIL and Bohrium: High Productivity and High Performance

Kenneth Skovhede$^{(\boxtimes)}$ and Simon Andreas Frimann Lund

Niels Bohr Institute, University of Copenhagen, Copenhagen, Denmark
{skovhede,safl}@nbi.ku.dk

Abstract. In this paper, we explore the mapping of the NumCIL C#
vector library where operations are offloaded to the Bohrium runtime
system and evaluate the performance gains. By using a feature-rich language, such as C#, we argue that productivity can be increased. The
use of the Bohrium runtime system allows all vector operations written
in C# to be executed efficiently on multi-core systems.

We evaluate the presented design through a setup that targets a 32
core machine. The evaluation includes well-known benchmark applications, such as *Black Sholes*, *5-point stencil*, *Shallow Water*, and *N-body*.

Keywords: C# · NumCIL · Bohrium · High performance · High productivity · Vector programming · Array programming

1 Introduction

We have previously introduced the NumCIL library [13] for performing linear
algebra in C#, using an approach known as vector programming, array programming or collection programming [12]. In such an approach, the programmer
writes high-level operations on multidimensional vectors rather than looping over
the individual elements. One of the primary benefits of such an approach is that
it leaves the program more readable because it is more of a description of what
should be done, rather than how it should be done. This approach can greatly
speed up the development cycle, as the developer can focus on the structure of
compact expressions, rather than explicitly specify details such as loop indicies.

The Bohrium runtime system [10] is a related project aiming to deliver architecture specific optimizations. In Bohrium, a program will use the C or C++
interface to describe multidimensional vectors and request various operations on
these. The execution of these operations is deferred until the program requires
access to the result. This lazy evaluation approach enables the Bohrium runtime to collect a number of scheduled instructions and perform optimizations on
these. The optimizations are an ongoing research project.

Since Bohrium uses a common intermediate representation of the scheduled
operations, it is possible to apply different optimization strategies to different
execution targets. The Bohrium intermediate representation also enables execution of Bohrium bytecode on multi-core CPU's, GPGPU's and even cluster
setups.

© Springer International Publishing Switzerland 2016
R. Wyrzykowski et al. (Eds.): PPAM 2015, Part II, LNCS 9574, pp. 166–175, 2016.
DOI: 10.1007/978-3-319-32152-3_16

In this article, we only evaluate the performance using a multi-core CPU. A more detailed description of the Bohrium system is available in *Kristensen et al.* [10].

By adding an extension to the NumCIL library, the vector operations expressed in C# can be forwarded to the Bohrium runtime system. This enables the programmer to have a rapid development cycle, without even having Bohrium installed. Once the program is tested for correctness, the unmodified program can then be executed with Bohrium support, such that all vector operations are executed with an efficient multi-core implementation.

2   Related Work

The array programming approach is in widespread use over a number of different programming languages, including Ada [5], CoArray Fortran [8], Chapel [3], NumPy [9] and numerous others. The NumPy approach differs in that it has no explicit support in Python but is implemented using Pythonic constructs in such a way that it seems *natural* to Python programmers. This approach means that nothing needs to change, in the Python programmers toolchain, to take advantage of the array programming found in NumPy. This non-intrusive approach with a natural language integration is the inspiration for the NumCIL library.

The idea of using language features to add support for vector programming instead of modifying the language is also found in the C++ libraries Armadillo [11] and Blitz++ [16]. The Armadillo library leverages existing linear algebra systems to achieve high performance but does so at template instantiation time, rather than at runtime.

The RyuJIT [4] compiler adds support for smaller vectors by converting vector operations to SIMD instructions. This approach helps in handling memory access and accelerates the execution time, but does require changes to the runtime system and does not offer any features for larger arrays. The RuyJIT is scheduled to ship with Microsoft's .Net framework 5 [4]. The Mono runtime [17] offers the Mono.Simd library with similar capabilities, implemented as a library with special support from the runtime [18].

The ideas for providing an intermediate representation of the requested operations, and performing optimizations on this, are also found in the, now discontinued, Intel Array Building Blocks (ArBB) project [7]. The ArBB system relies on a special compiler and an extended C++ syntax to describe computational kernels. When executing a batch of instructions, a number of optimization techniques are applied, such as removal of scratch memory, loop fusion, etc.

The Bohrium runtime system [10] is similar to ArBB and Chapel, in that the programmer uses vectors and describes *what* should be done, rather than *how* it is done. Internally this is achieved by means of a vector-oriented byte-code, i.e. simple instructions for a pseudo vector processing system. This abstraction allows Bohrium to be programming language agnostic, and is used to express a flat C API. With this API, it is possible to support a number of programming languages, such as Python, C++ and C#, in which the developer uses some array-library to interact with Bohrium.

The programming model used by Bohrium and NumCIL is very similar to the one found in NumPy [9], for which there also exists a Bohrium interface. In that sense, NumCIL fills the same role as NumPy, by providing an abstraction for interacting with Bohrium.

3   Implementation

The NumCIL library consists of three main item types: Multidimensional views, data storage and operators. The views are applied to the data storage to select a subset of the flat data storage, and project it into multiple dimensions, using offset, stride and skip values. Applied operators affect only the subset of the data that view projects, which greatly reduces the need for copying data into appropriately sized containers. The implementation of the multidimensional views found in NumCIL are compatible with NumPy's ndarrays [9] and also the Bohrium data views.

The primary design goal for the Bohrium extension to NumCIL has been to allow a non-intrusive addition. This allows code already written and tested with NumCIL to use the Bohrium runtime system without any changes. The non-intrusive design is achieved by hooking into the **DataAccessor** class, which is normally a simple wrapper for an array. By replacing the NumCIL factory instance that produces **DataAccessor** items, it becomes possible to provide Bohrium enabled data accessors.

Table 1 shows a simple multidimensional program written with Num-CIL. It illustrates how a flat array can be projected into multiple dimensions, and how the data can be *broadcasted* into larger dimensions. The program can be executed in Bohrium, simply by adding the statement **NumCIL.Bohrium.Utility.Activate();** prior to running the code.

If the program in Table 1 is executed with Bohrium loaded, the variable "a" will not be allocated until it is needed in the very last line. In that very last line, the allocation, multiplication, addition and summation is executed in Bohrium as a single instruction batch. Depending on the Garbage Collector, the batch may or may not contain instructions to deallocate the memory as well.

When a Bohrium enabled data accessor is created, it can be created with or without existing data. If there is no existing data, as with "a", an empty array is allocated by the Bohrium system and a handle for this is maintained by the data accessor. If existing data is already present, as with "c", the data accessor behaves as a non-Bohrium enabled data accessor facilitating access to the array data. This ensures that data is always kept where it is already allocated and not copied needlessly.

When an operation is applied to a multidimensional view that is referencing a Bohrium enabled data accessor, such as the multiplication, the views involved are created in Bohrium and an instruction matching the requested operation is emitted to the Bohrium runtime system. However, emitting the operation does

Table 1. A simple vector program with NumCIL

C# code	Resulting data
```using NumCIL.Float32;```	
...	
```var a = Generate.Range(3);```	[0, 1, 2]
```var b = a[Range.All, Range.NewAxis];```	[[0], [1], [2]]
```var data = new float[] { 2, 3 };```	[2, 3]
```var c = new NdArray(data);```	[2, 3]
```var d =```	[[0, 0], [2, 3], [4, 6]] =
b	[[0 ,0], [1, 1], [2, 2]]
*	*
c;	[[2, 3], [2, 3], [2, 3]]
```Console.WriteLine((d```	[[0, 0], [2, 3], [4, 6]]
+ 1)	+ [[1, 1], [1, 1], [1, 1]]
.Sum());	= [[1, 1], [3, 4], [5, 7]] = 15

nothing more than adding the operations to the current batch. Since the CLR is using a garbage collected approach, there is a chance that the GC will run before the operations are executed. If the GC runs, it can reuse the memory occupied by non-referenced items, and it may also choose to move existing data to a new location, and thus invalidating a pointer to the data. This problem is exacerbated by the introduction of temporary storage when compiling a composite statement as shown in Table 2.

Table 2. A composite expression and the equivalent single expression version

Composite expression	Expansion to single expressions

```var e = Generate.Range(10);```
```var f = ((e + 10) * e) - 1;```
	```var e = Generate.Range(10);```
	```var t0 = e + 10;```
	```var t1 = t0 * e;```
	```var f = t1 - 1;```

All of the temporary variables shown in Table 2 will be short lived and eliminated when the GC runs. In order to avoid issues with the GC, it is possible to Pin the memory when obtaining a pointer to the data. As long as the pointer is Pinned, the GC will not attempt to move or reuse the data. Since multiple multidimensional views may point to the same data, as with "a" and "b", a reference counting scheme is used to defer the Unpinning until the last reference is out of scope. This further ensures that data is not copied but used where it is located, with minimal overhead.

When a Bohrium enabled data accessor is created without existing data, only the view data is initialized, and the data storage is kept uninitialized. When the operations eventually execute, the Bohrium runtime will allocate only the needed data. This allows for using more memory than what is physically available on the machine with no side effects.

When data is requested by the CIL, i.e. for the summation operation which returns a scalar, all pending operations need to execute to ensure that the data observed by the application is seeing the expected results. This is accomplished by performing a Bohrium sync command on the target data, and then requesting a flush of all pending instructions.

In the case where the data being requested is not backed by a CIL array, an extra instruction is inserted that will copy the data allocated by Bohrium into a freshly created CIL array. This copy operation is done prior to the sync and flush commands, such that the intermediate storage can easily be eliminated by the Bohrium runtime system, thus allowing the results to be written directly to the CIL array.

If the user is only requesting a single element from the data, the entire data stays in the Bohrium allocated memory region, and only the requested element is copied into a CIL variable. This greatly reduces memory usage if only single elements are requested in a large array, such as when reading only the border values. If the user is writing to a single element in data that is not backed by a CIL array, all pending operations are flushed before writing the element directly into the memory region allocated by Bohrium. Table 3 shows the different states the DataAccessor goes through.

Table 3. State flow for a Bohrium enabled NumCIL DataAccessor

Created with		Used in operation		Access element		Access array
No data	$\rightarrow$	create bh handle	$\rightarrow$ read from pointer $\rightarrow$	flush		create new array emit copy free bh handle flush convert to CIL return array
CIL array	$\rightarrow$	pin create bh handle	$\rightarrow$	flush unpin read array	$\rightarrow$	flush unpin return array

4   Results

To evaluate the performance of the library, we have implemented a number of computational cores for classic simulations. The benchmarks are all implemented in C# and run using Mono 3.2.8 on Ubuntu 14.04.02. In order to provide a reasonable baseline, the benchmarks are also implemented with NumPy 1.8.2

and executed with Python 2.7.6. The hardware platform has two AMD Opteron 6272 CPUs with a total of 32 cores and 128 GB DDR3 memory with 4 memory busses. GCC version 4.8.2 was used to compile the Bohrium runtime. Source code is available for the Bohrium and NumCIL packages [15], as well as for the benchmarks [14].

Various options were used when executing the C# benchmarks. The basic Managed mode is using only C# and CIL functionality. The Unsafe configuration, utilizes the option to bypass array bounds checks within the CIL runtime, by accessing the data through memory pointers, with so-called unsafe code. The Unsafe configuration does not appear to influence the execution times on Mono, but is shown here as it does have an effect on the Microsoft .Net runtime [13]. When executing the benchmarks with Bohrium enabled, the number of utilized threads are varied to give an indication of the scalability.

The NumPy versions execute faster than the C# versions of the same code in general. There are two main reasons for this. Firstly, the NumPy implementation is written mostly in C, which means that none of the Python overhead is present. Secondly, the Mono JIT compiler does not perform various optimizations, such as efficient function inlining. When using the Microsoft .Net runtime, the execution times are roughly half of the Mono results, and approximately 20 % faster than the NumPy code [13]. On the Windows platform, we would expect NumCIL by itself to perform roughly 50 % faster than the reported Mono results, but when coupled with Bohrium, only metadata is handled by the .Net runtime, and thus the obtained execution times would be the same.

4.1 General Observations

The speedup does not exceed a factor of 4, even when using 32 cores. This limitation stems from the current execution mode in Bohrium, where each operation is executed individually. This approach has the effect that each operation will read all memory inputs and write all memory outputs for each operation, even if the inputs or outputs are needed for other operations. As the inputs and outputs are vectors, the caches are not utilized, effectively limiting the output to the bandwidth of the memory system.

This issue, and many other performance issues, can be mitigated through a technique known as *loop fusion*, where loop traversals are transposed, such that less memory access is required. Even though these optimizations are not yet implemented in Bohrium, we still see speedups. Once these optimizations are fully implemented in Bohrium, the NumCIL library will automatically perform even better.

4.2 Black-Scholes Model

The Black-Scholes model is a financial method for estimating the price of stock options [1]. It can be considered an embarrassingly parallel computation kernel, similar to Monte-Carlo π, but with a heavier computational workload. As shown in Fig. 1, the performance gains from the Bohrium runtime are fairly low, due to the current configuration not being able to efficiently fuse the operations, causing a high load on the memory system. As the memory system is saturated, adding execution units does not improve performance.

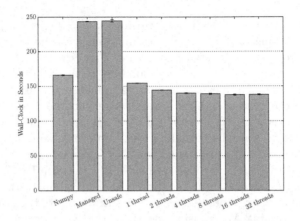

Fig. 1. BlackScholes 3200000, 36 iterations

4.3   Heat Equation

The Heat Equation benchmark is implemented as a 5-point stencil and simulates thermal dispersion in a material. The stencil is applied ten times to a 5000×5000 element array of single precision floating point numbers. The computation is simple additions, using multiple parallel accesses to the same memory. Even though the Mono implementation has some drawbacks and performs significantly slower than NumPy, the Bohrium runtime can re-use memory allocations, which allows for significant speedups [6]. Despite the low computational complexity, the Bohrium runtime can speed up execution when using all cores Fig. 2.

4.4   n-Body Simulation

The n-body simulation is implemented in a naive manner, yielding a $O = N^2$ complexity. For each time-step, the forces of all bodies on all bodies are computed, and their velocities and positions are updated. The Mono runtime slightly outperforms the NumPy version for this benchmark. When the Bohrium runtime is activated, it is capable of memory re-use and runs over twice as fast on a single core, with speedup on up to 16 threads Fig. 3.

4.5   Shallow Water

The Shallow Water simulation [2] is performed on a grid of 5000 by 5000 single precision numbers, over ten discrete timesteps, simulating water movements.

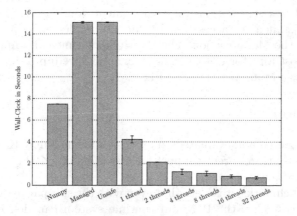

Fig. 2. HeatEquation 5000 × 5000, 10 iterations

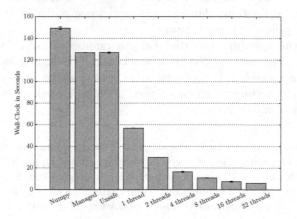

Fig. 3. nBody 5000, 10 iterations

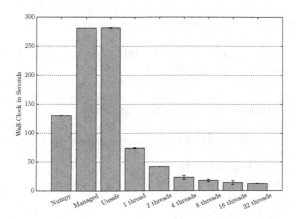

Fig. 4. Shallow water 5000 × 5000, 10 iterations

Many independent computations on each element dominate the computations, yielding irregular memory accesses. The NumPy implementation is more than twice as fast as the Mono version. The Bohrium runtime can improve this even further, by almost a factor of two, yielding a total speedup of four times, compared to the basic Mono performance Fig. 4.

5   Conclusion

We have implemented and evaluated an extension to the NumCIL library, which enables completely transparent support for execution of existing programs with the Bohrium runtime system.

From the benchmarks, it is clear that even with the sub-par performance from the Mono JIT compiler, the Bohrium runtime system can deliver substantial speedups.

Given the high-level language features in C#, it is clear that the NumCIL library can be used for rapid development, and when paired with the Bohrium runtime, it also yields high performance.

Even with the speedups reported here, a number of additional optimizations are being developed for the Bohrium runtime, including loop fusion and NUMA-aware memory handling. Once these optimizations are implemented in Bohrium, the loosely coupled approach used in NumCIL will automatically give even greater performance boosts.

Acknowledgment. This research was supported by grant number 131-2014-5 from Innovation Fund Denmark. This research has been partially supported by the Danish Strategic Research Council, Program Committee for Strategic Growth Technologies, for the research center 'HIPERFIT: Functional High Performance Computing for Financial Information Technology' (hiperfit.dk) under contract number 10-092299.

References

1. Black, F., Scholes, M.: The pricing of options and corporate liabilities. J. Polit. Econ. **81**, 637–654 (1973)
2. Burkardt, J.: Shallow water equations. http://people.sc.fsu.edu/~jburkardt/m_src/shallow_water_2d/. Accessed May 2015
3. Chamberlain, B., Vetter, J.S.: An introduction to chapel: cray cascade high productivity language. In: AHPCRC DARPA Parallel Global Address Space (PGAS) Programming Models Conference, Minneapolis (2005)
4. Frei, K.: RyuJIT CTP3: How to use SIMD. http://blogs.msdn.com/b/clrcodegeneration/archive/2014/04/03/ryujit-ctp3-how-to-use-simd.aspx. Accessed May 2015
5. Ichbiah, J.D., Krieg-Brueckner, B., Wichmann, B.A., Barnes, J.G., Roubine, O., Heliard, J.C.: Rationale for the design of the ada programming language. ACM Sigplan Not. **14**(6b), 1–261 (1979)
6. Lund, S.A., Skovhede, K., Kristensen, M.R.B., Vinter, B.: Doubling the performance of Python/NumPy with less than 100 SLOC. In: IEEE International Conference on Performance, Computing and Communications (2013)

7. Newburn, C.J., So, B., Liu, Z., McCool, M., Ghuloum, A., Toit, S.D., Wang, Z.G., Du, Z.H., Chen, Y., Wu, G., et al.: Intel's array building blocks: a retargetable, dynamic compiler and embedded language. In: 2011 9th Annual IEEE/ACM International Symposium on Code Generation and Optimization (CGO), pp. 224–235. IEEE (2011)
8. Numrich, R.W., Reid, J.: Co-array fortran for parallel programming. In: ACM Sigplan Fortran Forum, vol. 17, pp. 1–31. ACM (1998)
9. Oliphant, T.E.: Python for scientific computing. Comput. Sci. Eng. **9**(3), 10–20 (2007). http://scitation.aip.org/content/aip/journal/cise/9/3/10.1109/MCSE.2007.58
10. Kristensen, M.R.B., Lund, S.A.F., Blum, T., Skovhede, K., Vinter, B.: Bohrium: a virtual machine approach to portable parallelism. In: 2014 IEEE 28th International Parallel and Distributed Processing Symposium Workshops & Ph.D. Forum (IPDPSW). IEEE (2014)
11. Sanderson, C.: Armadillo: C++ linear algebra library. http://arma.sourceforge.net/. Accessed May 2015
12. Sipelstein, J.M., Blelloch, G.E.: Collection-oriented languages. Proc. IEEE **79**(4), 504–523 (1991)
13. Skovhede, K., Vinter, B.: Numcil: numeric operations in the common intermediate language. J. Next Gener. Inf. Technol. **4**(1), 9–18 (2013)
14. Team, B.: Benchpress source code. https://github.com/bh107/benchpress. Accessed May 2015; used revision 349ce5c1a69bb723a76783f7720c6ff0874519af
15. Team, B.: Bohrium source code. https://github.com/bh107/bohrium. Accessed May 2015; used revision 6f27c1fb3ae46c9b2541ba6d15b44e4a02e2cb01
16. Veldhuizen, T., Cummings, J.: Armadillo: C++ linear algebra library. http://blitz.sourceforge.net/. Accessed May 2015
17. Xamarin: Mono: Cross platform, open source .net framework. http://www.mono-project.com/. Accessed May 2015
18. Xamarin : Mono.simd namespace: hardware accelerated simd-based primitives. http://api.xamarin.com/index.aspx?link=N3AMono.Simd. Accessed May 2015

Parallel Ant Brood Graph Partitioning in Julia

Jose Juan Mijares Chan, Yuyin Mao, Ying Ying Liu,
Parimala Thulasiraman$^{(\boxtimes)}$, and Ruppa K. Thulasiram

University of Manitoba, Winnipeg, Canada
{thulasir,tulsi}@cs.umanitoba.ca

Abstract. Many big data applications are usually categorized as irregular. Irregular problems feature unpredictable and unstructured properties in terms of the program flow, data access pattern and typically use pointer-based data structures such as graphs. The problems are data, compute and communication intensive in nature. The algorithms are therefore designed and implemented on high performance architectures. The first stage of the parallel algorithm design is data partitioning. In this stage, the data is sub-divided into equally sized disjoint elements such that the communication volume among the processors is minimized. If the data is represented as a graph, it can be stated as the graph partitioning problem, which is NP-hard. In this work, we consider the meta-heuristic, ant brooding algorithm based on larval sorting by ants to solve the graph partitioning problem. The parallel ant brooding algorithm is implemented on a cluster using MIT's Julia language. We test the parallel algorithm on different benchmark and synthetic graphs. We compare our Julia parallel implementation with Julia sequential and C sequential implementations. We found that the performance of Julia is comparable to C with good scalability, and the parallel Julia implementation achieves speedup greater than 1 for a synthetic graph with 200 vertices and 1000 edges.

Keywords: Julia language · Ant brooding · Graph partitioning

1 Introduction

Many big data applications such as social, biological and complex networks, are usually categorized as irregular. Irregular problems [1] feature very unpredictable/unstructured properties for their program flow and data access patterns. These problems typically use pointer-based data structures such as graphs. They are data/communication intensive. Therefore, algorithms for these problems are designed and implemented on high performance computing architectures.

The first stage in designing algorithms for such problems on a parallel computer is data partitioning. The purpose of this stage is to partition the data among the processors such that the data locality is maximized and communication among processors is minimized. When the data structure is represented as a graph, the problem can be stated as a *graph partitioning* problem.

© Springer International Publishing Switzerland 2016
R. Wyrzykowski et al. (Eds.): PPAM 2015, Part II, LNCS 9574, pp. 176–185, 2016.
DOI: 10.1007/978-3-319-32152-3_17

Given a graph $G = (V, E)$ where V is the set of vertices and E is the set of edges, the k-way graph partitioning problem [2] is solved by partitioning the graph into k nonempty disjoint subsets of vertices such that the number of edges connecting two partitions is minimized. The graph partitioning problem is NP-hard. Therefore, heuristics [3] have been proposed to find approximate solutions to this problem.

Well-known sequential partitioning algorithms include Kernighan-Lin algorithm [4], spectral bisection method [5], and k-means algorithm [6]. An alternative method for solving the graph partitioning problem, is to transform the problem into identifying the "natural" clusters by constructing a bijective mapping between the graph vertices and points in a geometric space [7]. The clustering approach is invaluable when k is not known in advance.

In recent decades, meta-heuristics to solve graph partitioning/clustering problems in various applications have been considered. Meta-heuristics are general algorithmic frameworks that are designed to solve complex optimization problems [8]. In the literature, ant colony optimization [9], particle swarm optimization [10] and hybrid heuristics [11] have been considered.

In this paper, we focus on the ant brooding algorithm based on larval sorting by ants [7,12,13] for the graph partitioning problem from a graph clustering approach. We develop a parallel ant brooding algorithm and implement it in Julia parallel programming language. The paper is organized as follows: In the next section, we discuss the ant brooding algorithm followed by the parallel algorithm (Sect. 3) and its implementation in Julia. Sections 4 and 5 provide evaluations and results, respectively. And finally, Sect. 6 concludes the paper.

2 The Ant Brood Sorting Algorithm

Ant brooding was proposed by Deneubourg $et\ al.$ [12] as a distributed sorting algorithm to cluster robots. In the algorithm, the ant-like robots move randomly deciding whether to pick up or drop off objects based on the fraction of nearby points occupied by objects of the same type in their limited memory. Deneubourg $et\ al.$'s model was extended by Lumer and Faieta introducing a dissimilarity measure [13]. In this new approach, the simulation evolves in discrete time steps. At each step, a randomly selected ant can either pick or drop an object at its current location based on the following probabilities function.

$$P_{pick}(i) = \left(\frac{k_p}{k_p + f(i)}\right)^2 \tag{1}$$

$$P_{drop}(i) = \begin{cases} 2f(i) & \text{if } f(i) < k_d \\ 1 & \text{otherwise} \end{cases} \tag{2}$$

where k_p and k_d are set to educated guesses that allow the customization of the probability of pick up, P_{pick}, and drop, P_{drop}. $f(i)$ is the local density estimator of the object i and the ant's current neighborhood. The local density estimator is defined by

$$f\left(i\right) = \begin{cases} \frac{1}{l}\sum_j \left(1 - \frac{d(i,j)}{\alpha}\right) & \text{if } f > 0 \\ 0 & \text{otherwise} \end{cases} \tag{3}$$

The dissimilarity, $d\left(i,j\right)$, between the object i and current neighbor object j is scaled by a constant $\alpha \in (0,1]$. Then, it is normalized by l representing the neighborhood size (for example in a 3×3 area, $l = 8$). The dissimilarity measure is extended from Euclidean distance to the graph dissimilarity measure by Kuntz et $al.$ [7] in solving graph partitioning problem using ant brooding. The dissimilarity between two objects, now represented as vertices, depend only on their respective neighborhood relationships.

Let $N\left(v_i\right)$ be the set of vertices adjacent to v_i and including v_i: $N\left(v_i\right) = \{v_j \in V; \left(v_i, v_j\right) \in E\} \cup \{v_i\}$. Then the dissimilarity matrix can be expressed as

$$d(v_i, v_j) = \frac{|N\left(v_i\right) \triangle N\left(v_j\right)|}{|N\left(v_i\right)| + |N\left(v_j\right)|} \tag{4}$$

where the $\triangle$ operator is the symmetric difference (union minus intersection), while the $|\cdot|$ is the set cardinality operator.

Handl et $al.$ [14] compared the performance of the ant brooding clustering algorithm with k-means, agglomerative hierarchical clustering and one-dimensional self-organizing maps, concluding that the strengths of the algorithm include scalability, the capability to work with any kind of data that can be described in terms of symmetric dissimilarities, as well as, automatically determine the number of clusters without assumption on its shape.

Inspired by the decentralized manner ants cluster objects, our approach maps each ant to a thread on a parallel architecture. Ants work concurrently and independently to solve the problem.

3   Parallel Ant Brooding Algorithm in Julia

3.1   Julia Language

Julia [15] has been in development since 2009. Currently, it is an open source project that was started by the Massachusetts Institute of Technology and rapidly has become an emerging programming language alternative to R, Matlab, Octave, Python, and SciLab, or even C, C++ and Fortran.

Julia has gained enormous popularity since it combines the high-level programming style, with a high performance dynamic programming language and a familiar syntax, similar to Python, Matlab and R. It is dynamically typed, supporting polymorphism, metaprogramming, recursivity, user-defined parametric types, and multiple dispatch, among others [16]. For parallel processing, Julia provides a variety of specialized primitives and macros. For the distributed memory programming, the primitive $remote$ $call$ executes instructions on a remote processor, while the $remote$ $references$ are used to refer to an object stored in remotely. As well, distributed array data structures, *pmap and DArray*, and shared arrays (in an experimental stage) for system shared memory, like *SharedArray*, are provided to facilitate the development. Julia also provides a group of primitives for scheduling tasks, general parallel computing set of instructions, a cluster management inter-phase, besides a vectorization set of instructions.

3.2   Modifications to the Basic Ant Brooding Model

Relaxed Drop Behavior. A major issue with the basic model [13] is that ants easily pick up objects but do not drop them after many iterations. This issue may not be obvious in the sequential implementation because the clustering is formed gradually by ants. In parallel implementation, however, all the ants move simultaneously on the grid, and it is less likely for high-similarity neighborhoods to emerge gradually. In order to encourage the drop behavior of ants, we modified the drop function as the following:

$$P_{drop}(i) = \begin{cases} 1 & \text{if } f(i) > 1 \\ f(i)^2 & \text{otherwise} \end{cases} \tag{5}$$

More Intelligent Ants. Instead of moving the ants one cell at a time on the grid, we assume that the ants are more intelligent so they are able to perceive a neighborhood size l that is larger than or equal to 8, the default size. In addition, the program maintains a stack of free vertices and allows the ants to jump to the positions of these vertices directly after they become unloaded. Loaded ants continue to walk step-wise for exploration. This modification greatly improves the speed of the algorithm without affecting the solution quality.

3.3   Data Structures

The Lumer and Faieta algorithm [13] relies on the sequence of how ants are selected to randomly walk, pick or drop objects, therefore, the variables holding the position of objects or vertices are accessible only by one ant at a time. In contrast, the parallel implementation requires a shared data structure that holds information about the objects or vertices randomly distributed over the search space, denoted as $object_{coordinates}$. Also, each ant must store, in its local memory, its current coordinates and the object it is carrying, denoted as $ant_{coordinates}$ and ant_{carry} respectively. The data structure holding the vertices and edges of the graph, G, is passed to local memory as it is frequently accessed.

The passing of data structures is benefited from the "pass-by-sharing" [17] feature of the Julia language, which means that values are not copied when they are passed by functions, instead, pointers that refer to the passed values are passed to the caller function.

Algorithm 1. Asynchronous Parallel Ant Brooding Algorithm

Input: N_{ants}, G, dim, l, P_{procs}, k_p, α, and N_{iter}
Output: $object_{coordinates}$ after N_{iter} iterations
1 *Initialization*;
2 Assign positions of random objects to N_{ants} ants;
3 Distribute ants over P_{procs}

4 **for** $iter \leftarrow 1$ **to** N_{iter} **do**
5 | **foreach** p *in* P_{procs} **do**
6 | | **foreach** *ant,* a_k *owned by* p **do**
7 | | | map (move to next position);
8 | | | map (
9 | | | **if** a_k *is unloaded and* $ant_{coordinates}$ *is dissimilar and occupied* **then**
10 | | | | pickup object according to Eq. 1
11 | | | **end**
12 | | | **else**
13 | | | | **if** a_k *is loaded and* $ant_{coordinates}$ *is empty and carrying object*
       | | | | *is similar* **then**
14 | | | | | drop object according to Eq. 5
15 | | | | **end**
16 | | | **end**
17 | | |)
18 | | **end**
19 | **end**
20 **end**

3.4   Program Flow

Let N_{ants} denote the number of ants. The core of the parallel implementation is the load distribution among the ants, where in most cases the kth-ant, a_k, receives the $(N_{ants})^{-1}$ part of the workload. Let P_{procs} be the pool of available processors. Each processor is assigned a given number of ants. We assume $N_{ants} \geq P_{procs}$. The ants iterate for N_{iter} iterations. We designed an asynchronous parallel algorithm (Algorithm 1) that passes the iteration control to each of the ants, leaving the ants to run freely at different speeds on the algorithm. The algorithm is initialized with the number of ants N_{ants}, a graph G, a grid with dimension dim, the neighborhood size $l \geq 8$, the pool of available processors, P_{procs}, the variables, k_p and α, that govern the decisions of the ants, and finally the number of iterations, N_{iter}. The first step in the algorithm is to assign positions of random objects to the N_{ants} ants (line 2). Then, all ants are distributed among the P_{procs} processors.

Each ant at each processor performs the instructions from step 7 to 15 in a loop for N_{iter} iterations. The behavior of the ant can be summarized into three actions: move to a new position, pickup objects (vertices), or drop objects. Due to the randomness of the objects' position, the load on each ant is not known in advance. We use Julia's *pmap* function to achieve dynamic scheduling. We also use *@simd* macro for vectorization that exploits SIMD instructions when executing loops.

4   Evaluations

4.1   Cluster Retrieval

The output of ant brooding is the coordinates of the vertices on the grid, without clustering labels. Although the layout may reveal obvious structures to human eyes, we need to translate this implicit information to explicit membership labels for the computer. Since ant brooding improves the clustering quality by changing the geometric positions of the objects, an Euclidean distance based cluster retrieval algorithm will render good result. For simplicity, we use k-means [6] together with our modified ant brooding algorithm for the cluster retrieval.

4.2   Metrics

To evaluate the clustering quality, we introduce the mean intra-cluster distance (MICD) to measure the cluster "compactness", but in order to evaluate the effect of all the clusters we use its sum of squares, $SS\,(MICD)$. This metric should tend to minimize ideally if the clustering is of good quality

$$SS\,(MICD) = \sum_n MICD_n^2 = \sum_n \left(\sum_{v_i \in C_n} \frac{\|v_i - \mu_n\|}{|C_n|} \right)^2 \tag{6}$$

where $\|\cdot\|$ is the Euclidean distance, and C_n is the n cluster of vertices with centroid in μ_n. To measure "sparsity" among clusters, we use the sum of squares of the inter-clusters distance (SSICD). This metric should tend to maximize ideally if the cluster is of good quality. That is,

$$SSICD = \sum_i \sum_j \|\mu_i - \mu_j\|^2 \tag{7}$$

The ratio $SSICD/SS(MICD)$ normalizes the two metrics. A greater value of this ratio indicates a better clustering quality overall. We measure the solution quality of ant brooding by comparing the ratio $SSICD/SS(MICD)$ for the clusters retrieved by k-means before and after ant brooding.

For performance, we compare the run time of Julia parallel implementation with Julia sequential and C sequential implementations.

4.3   Experiments

We perform all of our experiments on a cluster with 12 CPUs (Intel Xeon CPU E5-2430 v2 @ 2.50 GHz), 4 GB system memory and IvyBridge. communication network. To test the robustness of the algorithm, we evaluate our parallel ant brooding algorithm on three common benchmark graphs for clustering, including karate club [18] (34 nodes, 78 edges, 2 clusters), dolphins [19] (62 nodes, 159 edges, 2 clusters), football [20] (115 nodes, 613 edges, a hierarchical community structure), and a recursive matrix (R-MAT) graph [21] (200 nodes, 1000 edges), which is a synthetic graph that mimics the power-law distribution [22] of real world networks.

The selection of ant brooding parameters is based on the heuristic by Handl *et al.* [14]. Let $|V|$ denote the number of vertices on the graph, The number of ants N_{ants} is roughly in the order of $0.1 * |V|$, the grid dimension dim is in the order of $\sqrt[2]{10 * |V|}$, number of iterations N_{iter} is in the order of $2000 * |V|$. It is also suggested [7] that a value that is close to 1 for the scaling constant α leads to visual separation of clusters on a grid. We experiment on the input graphs with different combinations of the above parameters, as well as other user defined parameters k_p and l. The best parameter settings are shown in Table 1. In each iteration, one ant makes one move within a small neighbourhood l, but the complexity of calculating dissimilarity in this neighbourhood is $O(|V|)$, the computational complexity of our parallel algorithm is therefore $N_{iter} * N_{ants} * O(|V|)/P_{procs}$. By substituting N_{iter} with $2000 * |V|$ and N_{ants} with $0.1 * |V|$, we get $O(|V|)^3/P_{procs}$, where P_{procs} is not a constant.

Table 1. Ant brooding parameter settings

Graph	N_{iter}	N_{ants}	dim	l	k_p	α
Karate	100000	4	20	48	0.3	1.0
Dolphins	100000	6	25	48	0.3	1.0
Football	100000	10	30	48	0.3	1.0
R-MAT	10000	20	50	80	0.3	1.0

5   Results

We present the visual display in Fig. 1 for the input graphs on the Cartesian plane before and after ant brooding. In all cases, ant brooding turns a random projection of vertices into a placement of noticeable clusters. The algorithm correctly identifies the two clusters in karate club and dolphins. The football graph has a hierarchical clustering structure [20]. The algorithm clearly separates the two clusters on the top level and the two lower level clusters shown on the left side of the grid. Fine level clustering result can also be observed on a higher resolution of the grid. For R-MAT graph, the algorithm is able to display clusters with different sizes, and the one vertex in the upper middle of the grid represents an outliner that does not belong to any other clusters.

We fine tune the results obtained from ant brooding by applying a k-means algorithm, where k is chosen such that it produces the highest $SSICD/SS(MICD)$ ratio. We assume that k represents the number of processors on a parallel machine. Table 2 confirms that our algorithm improves $SSICD/SS(MICD)$ of the clustering on all the graphs.

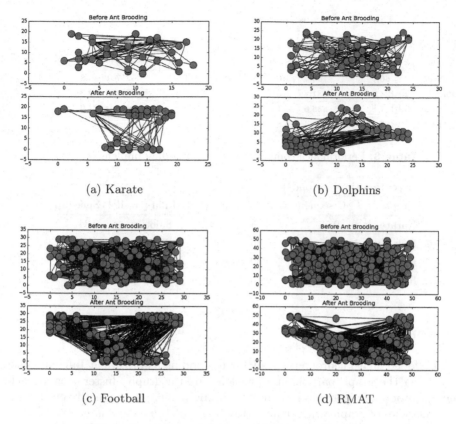

(a) Karate

(b) Dolphins

(c) Football

(d) RMAT

Fig. 1. Graph Layout before and after Ant Brooding

In terms of performance, we implement the algorithm in sequential and in parallel. We implemented the algorithm sequentially in both C and Julia to test the efficiency of the two languages. As shown in Table 3, for karate and dolphins, the execution time of *Julia sequential* is several folds of that of *C sequential*, and the same observation for *Julia parallel* versus *Julia sequential*. This is expected because the language overhead still occupies a large portion of the total execution time. When the graph size gets larger, the performance of *Julia sequential* and *Julia parallel* gets consistently better compared to *C sequential* that scales almost linearly with the graph size. *Julia parallel* is able to achieve speedup greater than 1 for the R-MAT graph with 200 nodes and 1000 edges. The reader may notice that in average, the executions of R-MAT on all three programs are faster than those of football graph, although the R-MAT graph is bigger. This is because the executions of R-MAT uses less iterations to achieve good solution quality (refer to Table 1). The cluster retrieval is considered as a post-processing step of ant brooding algorithm. Therefore, its execution time is not included in the performance measurement of ant brooding.

Table 2. k-means clustering before and after ant brooding

		Before ant brooding			After ant brooding		
Graph	k	SS(MICD)	SSICD	SSICD/SS(MICD)	SS(MICD)	SSICD	SSICD/SS(MICD)
Karate	2	66.7	105.4	1.6	44.6	257.0	5.8
Dolphins	2	124.9	181.8	1.5	136.6	314.2	2.3
Football	3	177.8	836.6	4.7	76.3	1666.9	21.9
R-MAT	6	433.4	11621.4	26.8	320.0	13493.5	42.2

Table 3. Performance comparison of ant brooding implementations

Execution time (seconds)					
Graph	k	C sequential	Julia sequential	Julia parallel	Speedup
Karate	2	0.57	2.75	8.66	0.31
Dolphins	2	2.51	5.46	8.90	0.61
Football	3	10.69	9.59	15.8	0.60
R-MAT	6	9.10	9.70	8.53	1.13

6   Conclusions

We developed an asynchronous parallel ant brooding algorithm in Julia language for solving the graph partitioning problem from a graph clustering approach. The ant brooding algorithm is a good mapping technique to identify clusters on different types of graphs with small intra-cluster distances and large inter-cluster distance. We show that our parallel Julia implementation has comparable speed to C with good scalability, and successfully achieves speedup for the R-MAT graph with 200 nodes and 1000 edges. Meanwhile, parameter tuning remains a challenge to generalize ant brooding algorithm to work with any types of graphs. For future work, we will experiment with larger graphs and improve our modified algorithm to accommodate this large data size.

References

1. Thulasiraman, P.: Irregular Computations on Fine-grain Multithreaded Architecture. Lambert Publishing Company (2009)
2. Karypis, G., Kumar, V.: Multilevel k-way partitioning scheme for irregular graphs. J. Parallel Distrib. Comput. **48**(1), 96–129 (1998)
3. Bui, T.N., Moon, B.R.: Genetic algorithm and graph partitioning. IEEE Trans. Comput. **45**(7), 841–855 (1996)
4. Kernighan, W.B., Lin, S.: An efficient heuristic procedure for partitioning graphs. Bell Syst. Tech. J. **49**(2), 291–307 (1970)
5. Pothen, A., Simon, H.D., Liou, K.-P.: Partitioning sparse matrices with eigenvectors of graphs. SIAM J. Matrix Anal. Appl. **11**(3), 430–452 (1990)
6. Hendrickson, B., Kolda, T.G.: Graph partitioning models for parallel computing. Parallel Comput. **26**, 1519–1534 (2000)

7. Kuntz, P., Snyers, D., Layzell, P.: A stochastic heuristic for visualising graph clusters in a bi-dimensional space prior to partitioning. J. Heuristics **5**(3), 327–351 (1999)
8. Bianchi, L., Dorigo, M., Gambardella, L.M., Gutjahr, W.J.: A survey on metaheuristics for stochastic combinatorial optimization. Nat. Comput. Int. J. **8**(2), 239–287 (2009)
9. Tashkova, K., Korošec, P., Šilc, J.: A distributed multilevel ant-colony algorithm for the multi-way graph partitioning. Int. J. Bio-Inspired Comput. **3**(5), 286–296 (2011)
10. Picarougne, F., Azzag, H., Venturini, G., Guinot, C.: A new approach of data clustering using a flock of agents. Evol. Comput. **15**(3), 345–367 (2007)
11. Soper, A.J., Walshaw, C., Cross, M.: A combined evolutionary search and multilevel optimisation approach to graph-partitioning. J. Global Optim. **29**(2), 225–241 (2004)
12. Deneubourg, J.-L., Goss, S., Franks, N., Sendova-Franks, A., Detrain, C., Chrétien, L.: The dynamics of collective sorting robot-like ants and ant-like robots. In: Proceedings of the First International Conference on Simulation of Adaptive Behavior on From Animals to Animats, pp. 356–363 (1991)
13. Lumer, E.D., Faieta, B.: Diversity, adaptation in populations of clustering ants. In: Proceedings of the Third International Conference on Simulation of Adaptive Behavior: From Animals to Animats 3, pp. 501–508. MIT Press (1994)
14. Handl, J., Knowles, J., Dorigo, M.: Ant-based clustering: a comparative study of its relative performance with respect to k-means, average link and id-som. In: Proceedings of the Third International Conference on Hybrid Intelligent Systems. IOS Press (2003)
15. The julia language. http://julialang.org/. Accessed on 05 October 2015
16. Bezanson, J., Chen, J., Karpinski, S., Shah, V., Edelman, A.: Array operators using multiple dispatch: a design methodology for array implementations in dynamic languages. In: Proceedings of ACM SIGPLAN International Workshop on Libraries, Languages, and Compilers for Array Programming, p. 56. ACM (2014)
17. Julia package listing. http://pkg.julialang.org/. Accessed on 05 October 2015
18. Zachary, W.W.: An information flow model for conflict and fission in small groups. J. Anthropol. Res. **33**, 452–473 (1977)
19. Lusseau, D., Schneider, K., Boisseau, O.J., Haase, P., Slooten, E., Dawson, S.M.: The bottlenose dolphin community of doubtful sound features a large proportion of long-lasting associations. Behav. Ecol. Sociobiol. **54**(4), 396–405 (2003)
20. Girvan, M., Newman, M.E.J.: Community structure in social and biological networks. Proc. Nat. Acad. Sci. **99**(12), 7821–7826 (2002)
21. Chakrabarti, D., Zhan, Y., Faloutsos, C.: R-MAT: a recursive model for graph mining. In: SDM, vol. 4, pp. 442–446. SIAM (2004)
22. Strogatz, H.S.: Exploring complex networks. Nature **410**(6825), 268–276 (2001)

The 5th Workshop on Performance Evaluation of Parallel Applications on Large-Scale Systems

Scalability Model Based on the Concept of Granularity

Jan Kwiatkowski[1]($\boxtimes$) and Lukasz Olech[2]

[1] Department of Informatics, Faculty of Computer Science and Management,
Wroclaw University of Technology,
Wybrzeze Wyspianskiego 27, 50-370 Wroclaw, Poland
jan.kwiatkowski@pwr.edu.pl
[2] Department of Artificial Inteligence, Faculty of Computer Science
and Management, Wroclaw University of Technology,
Wybrzeze Wyspianskiego 27, 50-370 Wroclaw, Poland
lukasz.olech@pwr.edu.pl

Abstract. In the recent years it can be observed increasing popularity of parallel processing using multi-core processors, local clusters, GPU and others. Moreover, currently one of the main requirements the IT users is the reduction of maintaining cost of the computer infrastructure. It causes that the performance evaluation of the parallel applications becomes one of the most important problem. Then obtained results allows efficient use of available resources. In traditional methods of performance evaluation the results are based on wall-clock time measurements. This approach requires consecutive application executions and includes a time-consuming data analysis. In the paper an alternative approach is proposed. The decomposition of parallel application execution time onto computation time and overheads related to parallel execution is use to calculate the granularity of application and then determine its efficiency. Finally the application scalability can be evaluates.

Keywords: Parallel processing · Scalability of parallel application · Granularity concept

1 Introduction

In the recent years there has been rapid development of new technologies related to the evolution of the technical possibilities offered by computer hardware - increasing calculation speed, decreasing communication time, increasing bandwidth communications, etc. Moreover it can be observed increasing popularity of parallel processing by using multi-core processors, clusters, GPU and others. Equally important as the evolution of the information systems are changes of the requirements of the IT users. Increasingly, the basic requirement of the IT users are not systems, offering improved processing speed, but ones that will reduce the cost of maintaining infrastructure. It causes that performance evaluation constitutes an intrinsic part of every application development process.

© Springer International Publishing Switzerland 2016
R. Wyrzykowski et al. (Eds.): PPAM 2015, Part II, LNCS 9574, pp. 189–198, 2016.
DOI: 10.1007/978-3-319-32152-3_18

In parallel programming the goal of the design process is not to optimise a single metrics, a good design has to take into consideration memory requirements, communication cost, efficiency, implementation cost, and others. Therefore performance evaluation of parallel programs is very important for the development of efficient parallel applications.

In the paper [1] three categories of performance metrics have been proposed. The first are speedup metrics that show how faster results can be obtain when using some number of processing units comparing with using only one processing unit. The second one are efficiency metrics that determine the percentage of CPU utilization during parallel program execution. And finally scalability, which say how application behaves when increasing the number of available processing units and/or the size of the problem being solved. In the paper all of these metrics will be used for performance evaluation of parallel application.

In general the performance analysis can be carried out analytically or through experiments. The paper focusses on the second approach. Independently on the used measurement method during experimental performance evaluation of parallel programs is the need to measure the run time of sequential and parallel programs, which is time consuming. In the paper the method, which overcomes above problem is proposed. Basing on the concept of granularity and decomposition of the parallel application execution time onto the computation time and the overhead time presented in [2,3] we show that by measurement only wall-clock time and computation time it is possible to evaluate the performance of parallel programs. The paper extends previous one by presentations results of experiments performed up to 4096 processing units (cores) and by scalability analysis.

The paper is organised as follows. Section 2 briefly describes different performance metrics and two main approaches to scalability analysis - strong and weak scalability. How granularity can be used in performance evaluation is presented in Sect. 3. The next section illustrates the experimental results obtained during evaluation of two parallel algorithms, strong and weak scalability are considered. Finally, Sect. 5 outlines the work and discusses ongoing work.

2   Performance Metrics and Scalability Analysis

During performance evaluation of parallel applications different metrics are used [4]. The first one is the parallel run time ($t_{runtime}$). It is the time from the moment when computation starts to the moment when the last processor finishes its execution and is composed of three different times: computation time (t_{comp}) is the time spent on performing computation by all processors, communication time (t_{comm}) is the time spent on sending and receiving messages by all processors and idle time (t_{idle}) is when processors stay idle. The next commonly used metric is speedup, which captures the relative benefit of solving a given problem using a parallel system. There exist different speedup definitions. Generally the speedup (S) is defined as the ratio of the time needed to solve the problem on a single processor to the time required to solve the same problem on a parallel system with p processors. Theoretically, speedup cannot exceed the number of processors used during program execution, however, different speedup anomalies can be observed [5]. Both above mentioned performance metrics do not take into account the utilisation of processors in the parallel system. While executing a parallel algorithm processors spend some time on communicating and some processors can be idle. Then the efficiency (E) of a parallel program is defined as a ratio of speedup to the number of used processors. In the ideal parallel system the efficiency is equal to one but in practice efficiency is between zero and one, however because of different speedup anomalies, it can be even greater than one.

The last performance metrics is scalability of the parallel system. It can be considered in different ways, we can use it for hardware, algorithms, data bases, execution environment, etc. One can say that currently it is one of the most important performance metrics. In general it can be say that it is a metrics, which consider the "system" capacity to increase speedup in proportion to the number of available processors. There are a lot of approaches to modelling the scalability, for example by using so called isoefficiency analysis [4], Universal Scalability Model proposed by the Neil Gunther [6], H-isoefficiency function [7] and others.

One can find two different approaches to way in which scalability is defined [8]. The first one based on Amdahl law (1) is called strong scalability. The strong scalability is also called scalability with a fixed size of the problem, it means that our goal is to minimize the program execution time by using more processing units. It means that we can say that system is scalable when increasing number of processing units are used effectively. For example, when the number of processing units equals 8 and the speedup received equals 8, too, then we have excellent scalability. This approach is the pessimist because of indicates a bounded speedup.

$$Speedup(n) = \frac{T(1)}{T(n)} = \frac{1}{(1-p) + \frac{p}{n}} \tag{1}$$

where n denotes the number of processing units, p denotes the non-scaled fraction of the application parallel part and $T(1)$, $T(n)$ execution time at 1 and n processors respectively.

The second one is weak scalability that based on Gustafson law (2). The week scalability is also called the scalability with variable problem size, when the problem size increased at the time when the number of processing units increased (the input is fixed for each processor). We say that a system is scalable when the efficiency (execution time)is the same for increasing the number of processors and the size of the problem [4]. This approach is the optimistic because of indicates an unlimited speedup.

$$Speedup(n) = \frac{T(1)}{T(n)} = 1 + (n-1) * p^* \tag{2}$$

where n denotes number processing units, p^* denotes the scaled fraction of the application parallel part and $T(1)$, $T(n)$ execution time at 1 and n processors respectively.

3   Using Granularity for Performance Analysis

In general the granularity of a parallel computer is defined as a ratio of the time required for a basic communication operation to the time required for a basic computation operation. Let's define the granularity of the parallel algorithm similarly as the ratio of the amount of computation to the amount of communication within a parallel algorithm execution $(G = T_{comp}/T_{comm})$. Above definition can be used for calculating the granularity of a single process executed during program execution on each processor as well as for the whole program by using total communication and computation times of all program processes. Then let's use the overhead function, which is a function of problem size and the number of processors and is defined as follows [4]:

$$T_o(W, p) = p * T_p - W \tag{3}$$

where W denotes the problem size, T_p denotes time of parallel program execution and p is the number of used processors.

The problem size can be defined as the number of basic computation operations required to solve the problem using the best serial algorithm. Let us assume that a basic computation operation takes one unit of time. Thus the problem size is equal to the time of performing the best serial algorithm on a serial computer. Based on the above assumptions after rewriting the Eq. (3) we obtain the following expression for parallel run time:

$$T_p = \frac{W + T_o(W, p)}{p} \tag{4}$$

Recalling that the parallel run time consists of computation time, communication time and idle time, let's assume that the main overhead of parallel program execution is communication time. The total communication time is equal to the sum of the communication time of all performed communication steps. Assuming that the distribution of data among processors is equal then the communication time can be calculated using equation $T_{total\_comm} = p * T_{comm}$. Note that the above is true when the distribution of work between processors and their performance is equal. Similarly, the computation time is the sum of the time spent by all processors performing computation. Then the problem size W is equal to $p * T_{comp}$. Therefore the expression for the efficiency takes the form:

$$\mathcal{E} = \frac{1}{1 + \frac{T_{comm}}{T_{comp}}} = \frac{1}{1 + \frac{1}{G}} = \frac{G}{G + 1} \tag{5}$$

It means that using the concept of granularity we can calculate the efficiency and speedup of parallel algorithms. Concluding above consideration it is possible to evaluate a parallel application using such metrics as efficiency, speedup and scalability by measuring only the computation and wall-clock times during execution of parallel version of a program on a parallel computer. Deeper presentation of the above discussion can be find in [2].

4   Case Studies

To confirm the usefulness of the theoretical analysis presented in the previous sections the series of experiments were performed. During the experiments two different algorithms were used: K-means and Monte Carlo method (calculation of Pi number). The tests were executed on the BEM cluster at Wroclaw Centre for Networking and Supercomputing (720 homogeneous nodes (2 procesors) Intel Xeon E5 2670 v3). For both algorithms the strong scalability was checked and weak scalability was check only for K-means algorithm.

To avoid the execution time anomalies [2] the experiments were performed for data sizes sufficiently larger than CPU cache size and smaller than the main memory limits for strong scalability analysis and for weak scalability analysis the problem size increased proportionally to the number of used processors. Because the experiments were performed in a multi-user environment the execution times depended on computer load, therefore the presented results are the averages from the series of 10 identical experiments performed. Moreover the results of measurement lying in the distance above 1.5 interquartile range of the whole series were treated as erroneous and omitted, and the measurement was repeated. To evaluate the accuracy of the new method the relative error defined as $\frac{S-\mathcal{S}}{S}$ where S is the actual speedup and $\mathcal{S}$ is the estimated one has been used. Moreover because of way in which different times have been measured for speedup calculation instead of granularity isogranularity defined as ($G_{iso} = t_{comp}/t_{overhead}$) was used.

K-means is one of the algorithms that is used for solving the clustering problem [9]. It classifies a given data set into defined fixed number of clusters k (predefined). In the first algorithm's step so called centroids for each cluster should be chosen - one for each cluster. These centroids can be defined in random way however the better choice is to place them as much as possible far away from each other. In the next step all points from the data set are assign to the nearest centroid. After completion of this step the new centroids for each cluster are calculated using the means metrics for the created clusters. Then we repeated the second step using these new centroids. The process is continue as long as the differences between coordinates of new and old centroids are satisfied. Alternatively the process can be finished after predefined number of iterations.

The above algorithm was parallelized in the following simple way. The chosen processor reads input data, and then distributes them to other processors. Each processor received N/p data, when p is a number of available processors and N is the number of input data. Then each processor generates the appropriate number of centroids and exchanges information about them with other processors. After completion of above step each processor has information about all the centroids and performs the second step of the sequential algorithm. In the next step each processor calculates the data necessary to calculate new centroids (the number of point in each cluster and sums of points coordinates) and exchange this information with other processors. Then the new centroids are calculated, in parallel by all processors (execution replication), and again the process returns to the second step of sequential algorithm. The algorithm ends when the stop criterion is met. Then the chosen processor collects clustering results from other processors and merge them. Description of Monte Carlo method is skiped because of the common knowledge about it and the lack of space.

Below the results of experiments performed to check the strong scalability of k-means algorithm and Monte Carlo methods are presented. In the experiments performed for K-means algorithm different data set sizes were used. The number of generated clusters was 1024 during all tests. Moreover different hardware configurations by means different number of cores from each processor were used. Received results are presented on Figs. 1, 2, 3 and 4.

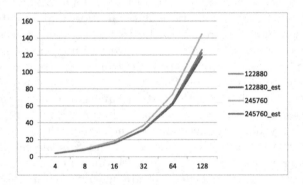

Fig. 1. K-means algorithm speedup and estimated speedup - 2 cores at each node

The first test was performed using 2 cores from 2, 4, 8, 16, 32 and 64 processors, its results are presented on Fig. 1. As can be seen the actual speedup and estimated speedup are very close, however when the size of data set is equal 245760 there are large differences between actual and estimated speedup and the precise relative error is even over 16 % when using 128 cores, for other cases is less than 5 %.

In the second test 4, 8, 16, 32, 64, 128 and 512 processing units (cores), four from each processor were used. Results of this test are presented on Fig. 2. As previously can be seen that the actual speedup and estimated speedup are very close. In general the precise relative error was less then 2 %, however for problem size 1966080 was slightly larger when using 512 cores.

In the third test 8, 16, 32, 64, 128, 256, 512 and 2048 processing units (cores), eight from each processor were used. Results of this test are presented on Fig. 3. In this test results are really satisfied, the precise relative error were between 1 % and 2 %.

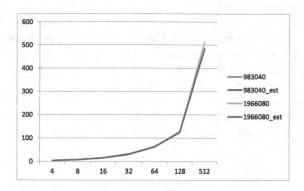

Fig. 2. K-means algorithm speedup and estimated speedup - 4 cores at each node

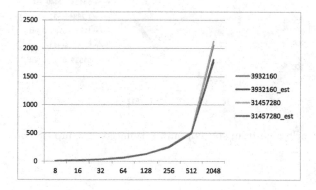

Fig. 3. K-means algorithm speedup and estimated speedup - 8 cores at each node

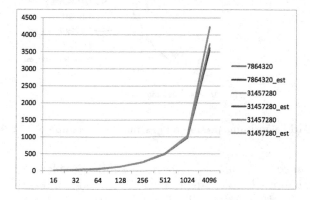

Fig. 4. K-means algorithm speedup and estimated speedup - 16 cores at each node

In the last test performed for K-means algorithm 16, 32, 64, 128, 256, 512, 1024 and 4096 processing units (cores), sixteen from each processor were used. Results of this test are presented on Fig. 4. As during the previous tests the results were very good, the precise relative error values were between 0,2 % and 6 %, only for problem size equals 7864320 for 4096 processing unites was larger, close to 15 %.

In the experiments performed for Monte Carlo method different data set sizes were used. Moreover different hardware configurations by means different number of cores from each processor were used. Received results are presented on Figs. 5 and 6.

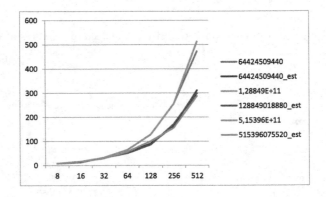

Fig. 5. Monte Carlo algorithm speedup and estimated speedup - 8 cores at each node

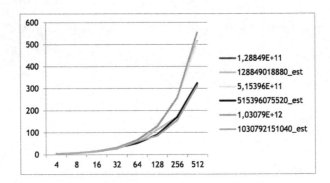

Fig. 6. Monte Carlo algorithm speedup and estimated speedup - 16 cores at each node

Results obtained for Monte Carlo methods similarly as for k-means algorithm are very promising, the shape of diagrams are very close and the precise relative error was not larger than 5 % in all cases. Considering the strong scalability we can conclude that results of experiments show that both algorithms are scalable in the limits of defined by the limits of performed tests.

4.1 Experimental Results - Weak Scalability

During test related to checking weak scalability for k-means algorithm 8, 16, 32, 64, 128, 256, and 512 processing units (cores) randomly chosen have been used. Problem sizes were from 122880 to 31457280 to satisfy requirements that during program execution each processing unit should used the same amount of data. Typically weak scalability is presented as a diagram using scaling efficiency. In the paper we present it in different way proposed in the paper [10] by presenting execution time and speedup in the tables (Tables 1 and 2).

Table 1. Execution time of parallel k-means algorithm

Problem size	T_1	T_8	T_{16}	T_{32}	T_{64}	T_{128}	T_{256}	T_{512}
122880	2,918	0,364	0,182	0,091	0,046	0,023	0,0064	0,0047
983040	178,132	22,240	11,120	5,659	2,779	1,388	0,698	0,345
1966080	708,634	88,522	44,262	23,137	11,063	5,5308	2,554	1,385
3932160	2831,905	318,146	176,702	88,353	44,175	22,079	10,240	5,517
7864320	11337,77	1416,032	812,677	353,901	176,960	88,462	38,918	21,949
15728640	45318,33	5660,593	2830,199	1624,234	707,452	353,600	176,745	88,342
31457280	181417,6	22653,32	11325,1	5661,966	3249,23	1414,826	706,802	353,248

Table 2. Speedup of parallel k-means algorithm based on Gustafson's model

Problem size	S_8	S_{16}	S_{32}	S_{64}	S_{128}	S_{256}	S_{512}
122880	7,946	15,744	31,297	61,302	117,931	70,688	0,0047
983040	7,948	15,822	31,431	62,546	122,945	240,221	408,803
1966080	7,972	15,868	31,232	62,909	125,105	81,220	483,412
3932160	7,958	15,843	31,630	63,024	125,784	100,921	495,631
7864320	7,982	15,837	31,672	62,941	125,793	118,578	475,919
15728640	7,913	15,808	31,536	63,008	125,728	250,092	498,543
31457280	7,930	15,883	31,582	55,060	125,540	125,300	499,833

From the Table 1, we can observe that for a problem size 983040 the run time on 8 procesors equeals 22,24 s, then when 32 procesors are used and problem size is increased to 1966080, the run time is very close 23,13 s. Similarly for 128 processors and problem size equeals 3932160 the run time is 22,07 s. Therefore we can conclude that the speedup is scaling from 7,94 to 125,78 for workload from 983040 to 3932160 when 128 instead 8 processors are available.

5 Conclusions and Future Work

In the paper the new way of scalability evaluation of parallel application is proposed. Utilizing the separate measurements of wall-clock time and CPU time, it offers the possibility to estimate the application speedup and efficiency using only the measurement for a single, parallel execution. For the method to be successful it requires only the readily available data, without the need of installation of additional software or application modifications. The experiments performed proved that the estimation accuracy is sensitive to the simplifying assumption taken. For all analysed algorithms the results obtained are similar: the shape of diagrams is similar and the value of speedup is close. In the future works a broader class of algorithms will be taken into consideration, as well as improving the way of weak scalability evaluation will be considered.

Acknowledgments. Calculations have been carried out using resources providing by Wroclaw Centre for Networking and Supercomputing (http://wcss.pl), grant No. 266.

References

1. Jogalekar, P., Woodside, M.: Evaluating the scalability of distributed systems. IEEE Trans. Parallel Distrib. Syst. **11**(6), 589–603 (2000)
2. Kwiatkowski, J.: Parallel applications performance evaluation using the concept of granularity. In: Wyrzykowski, R., Dongarra, J., Karczewski, K., Waśniewski, J. (eds.) PPAM 2013, Part II. LNCS, vol. 8385, pp. 215–224. Springer, Heidelberg (2014)
3. Kwiatkowski, J., Pawlik, M., Konieczny, D.: Comparison of execution time decomposition methods for performance evaluation. In: Wyrzykowski, R., Dongarra, J., Karczewski, K., Wasniewski, J. (eds.) PPAM 2007. LNCS, vol. 4967, pp. 1160–1169. Springer, Heidelberg (2008)
4. Grama, A.Y., Gupta, A., Kumar, V.: Isoefficiency: measuring the scalability of parallel algorithms and architectures. IEEE Parallel Distrib. Technol. **1**, 12–21 (1993)
5. Kwiatkowski, J., Pawlik, M., Konieczny, D.: Parallel program execution anomalies. In: Proceedings of First International Multiconference on Computer Science and Information, Wisla, Poland (2006)
6. Gunther, N.J., Puglia, P., Tomasette, K.: Hadoop superlinear scalability: the perpetual motion of parallel performance. Commun. ACM **58**(4), 46–55 (2015)
7. Bosque, J.L., Robles, O.D., Toharia, P., Pastor, L.: H-Isoefficiency: scalability metric for heterogeneous systems. In: Proceedings of the 10th International Conference on Computational and Mathematical Methods in Science and Engineering, CMMSE 2010, pp. 240–250 (2010)
8. Shoukourian, H., Wilde, T., Auweter, A., Bode, A.: Predicting the energy and power consumption of strong and weak scaling HPC applications. Supercomput. Front. Innovations **1**(2), 20–41 (2014)
9. MacQueen, J.B.: Some methods for classifcation and analysis of multivariate observations. In: Proceedings of the Fifth Symposium on Math, Statistics, and Probability, pp. 281–297. University of California Press, Berkeley (1967)
10. Kartawidjaja, M.A.: Analyzing scalability of parallel matrix multiplication using dusd. Asian J. Inf. Technol. **9**(2), 78–84 (2010). ISSN: 1682–3915

Performance and Power-Aware Modeling of MPI Applications for Cluster Computing

Jerzy Proficz$^{(\boxtimes)}$ and Paweł Czarnul$^{(\boxtimes)}$

Academic Computer Center TASK, Faculty of Electronics,
Telecommunications and Informatics, Gdańsk University of Technology,
Gdańsk, Poland
jerp@task.gda.pl, pczarnul@eti.pg.gda.pl

Abstract. The paper presents modeling of performance and power consumption when running parallel applications on modern cluster-based systems. The model includes basic so-called blocks representing either computations or communication. The latter includes both point-to-point and collective communication. Real measurements were performed using MPI applications and routines run on three different clusters with both Infiniband and Gigabit Ethernet interconnects. Regression allowed to obtain specific coefficients for particular systems, all modeled with the same formulas. The model has been incorporated into the MERPSYS environment for modeling parallel applications and simulation of execution on large-scale cluster and volunteer based systems. Using specific application and system models, MERPSYS allows to predict application execution time, reliability and power consumption of resources used during computations. Consequently, the proposed models for computational and communication blocks are of utmost importance for the environment.

Keywords: Performance model · Energy consumption · Cluster computing · MPI

1 Introduction

Modern parallel systems have increased in sizes considerably in recent years. The most powerful cluster on the TOP500 list – Tianhe-2 features over 3 million cores, offers over 33 PFlop/s performance but at over 17 MWatts of power consumption[1]. It should be noted that growth in performance of such parallel systems stems from incorporation of more and more computational cores into the system. At the same time, such large clusters, due to a large number of components, are prone to failures. This may effectively impact execution times of parallel applications due to necessary checkpoints and restarts in order to continue from the last consistent application state. For instance, for Sequoia the reported failure rate reaches 1.25 per day [1]. Consequently, it is of utmost

---

[1] www.top500.org.

© Springer International Publishing Switzerland 2016
R. Wyrzykowski et al. (Eds.): PPAM 2015, Part II, LNCS 9574, pp. 199–209, 2016.
DOI: 10.1007/978-3-319-32152-3_19

importance for developers to be able to assess application running times and speed-ups taking into account not only the design and bottlenecks in the application but also potential failures of such large scale cluster systems. On one hand, if we assume a given input data size for an application then speed-up will be dependent on the ratio of computations to communication, synchronization and specific optimization techniques such as overlapping communication and computations, piggybacking etc. On the other hand, hardware parameters such as CPU, GPU performance as well as latency and bandwidth of an interconnect will also impact the speed-up. The aforementioned failure possibilities will further limit performance because if a failure occurs then the application will need to be restarted either from scratch or from the last saved checkpoint. Consequently, a question should be asked what would be the optimal number of CPUs in order to minimize the application running time given this constraints. Furthermore, power-aware metrics are considered nowadays apart from performance only. For instance, one of considered optimization goals is minimization of application running time with a constraint on the total power consumption used by computing devices [2]. Consequently, a good model for parallel applications in a cluster based environment is still crucial for optimization, especially if it addresses power-aware aspects along with performance. Our main contribution is to provide a model related to the cluster performance, power consumption and reliability estimations. We designed the set of formulas working exactly for our simulation tool, we tested them and tuned for specific real environments.

2   Background and Related Work

A cluster model, or more generally, a hardware model needs to be introduced in every simulator of running an application in a parallel environment.

GSSIM [3] provides a configurable solution, where it is possible to use the default mode, where computation times are simply calculated as a linear function of the processor clock, and communication times are analytically solved according to the used network devices, their latencies and bandwidths. The power consumption is based on the three schemes: constant where a device always consumes the same power, the resource based, usually using values for idle and full utilization of the resource and the application related where the exact values need to be provided by the user. The model considered in this work is more focused on the cluster environment, thus it concerns such operations like disk data transfers or HyperThreading out-of-the box, without additional user configuration.

There are two main analytical models of the communication behavior in the network of computation nodes, the one proposed by Hockney in [4] and LogP [5]. The former assumes the time of the message passing between nodes equals $L + m/B$, where L is a latency of the network, B is a network bandwidth and m is a message size. The latter assumes the message delivery time equals $L + 2o$, where L is latency of the network, o is an overhead and an additional parameter describing the modeled system: P – the number of nodes communicating each other. However, it also assumes that the next message cannot be sent during the gap time, denoted by g, thus the network can carry messages $\lfloor L/g \rfloor$ at the time.

While the gap parameter in the LogP model reflects the network contention, it is suited for short messages only. Therefore, many variants were proposed. The LogGP model [6] introduces an additional parameter: G – gap per byte, thus the time of sending m bytes between two nodes can be presented as $o + (m - 1)G + o$. The PLogP model [7] introduces additional dependency of the gap and overhead parameters on the message size, and additionally distinguishes overhead for sending and receiving the message: $g(m)$, $o_s(m)$ and $o_r(m)$.

The HLoGP model [8] provides support for heterogeneous environment introducing the parameter matrices instead of scalar model parameters, e.g. $L = \{L_{11}, \ldots, L_{MM}\}$, where M is a number of the nodes (which all can be different from each other) and an additional vector reflecting the differences between computational power of the nodes, i.e. $P = \{P_1, \ldots, P_M\}$. Similarly the computational power of the processor/cores is also considered in MLogP model [9] where multicore processor architecture is taken into account. Finally Log_nP models [10] enable the hierarchical performance analysis for layered systems, including the impact of the memory and middleware on distributed communication.

3 Simulation Environment for Parallel Applications Running on Cluster-Based Systems

Within project "Modeling efficiency, reliability and power consumption of multilevel parallel HPC systems using CPUs and GPUs" sponsored by and covered by funds from the National Science Center in Poland we created an environment for simulation of parallel applications run on large-scale cluster, grid and volunteer based systems. For an application run for a particular input data size on a given system, the environment returns the following:

1. Application execution time.
2. Success/failure of the application – potential hardware failures have become a concern in large scale parallel systems [1] because of the number of components.
3. Energy consumed during execution of the application thanks to considering power consumption of devices such as CPUs, GPUs and network interconnects.

The distributed architecture of the system comprises the following components:

1. A client-side system and application editor as well as a simulation panel:
   (a) System model editor (Fig. 1). A user creates a model of the system by selecting predefined computational components such as CPUs/GPUs which are interconnected using predefined network types such as WANs, LANs or buses within each node. The system model can be defined at multiple levels starting at the top from WAN through LAN up to the node/machine level. For each particular computational or network type

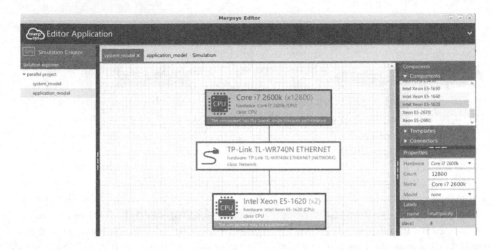

Fig. 1. System model editor – an exemplary system

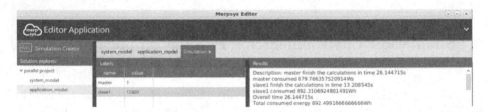

Fig. 2. Sample results

component the user selects one hardware component with specification
stored in a database. This includes single, double floating point performance for CPUs/GPUs, power consumption etc. The user assigns labels
to particular computational components. Specifically, if a CPU or a GPU
is assigned a label e.g. "master" with cardinality 12 this means that
up to 12 processes or threads marked "master" in the application can be
run there. It is the scheduler described next that will decide how many
processes or threads with such a label will be launched on such a computational component. An exemplary system modeling an environment with
Intel Xeon E5-1620 CPUs for master and i7 2600k for slave processes is
shown in Fig. 1.

(b) Application editor. A user writes code of a parallel application using a
special Java type meta-language which uses a generalization of message
passing paradigms such as MPI. The application includes codes of various processes/threads each marked with a distinct label such as "master",
"slaveX", "slaveY" etc. The code consists of computational or communication blocks and can contain any basic Java constructs such as for,
while loops, conditional instructions etc. Computational blocks take as
input data size, a function that determines the number of operations vs

the input data size, optional software stack and optimizations. Communication blocks include point-to-point, barrier, broadcast, scatter, gather similar to MPI.

(c) Simulation panel (Fig. 2). The panel allows definition of the number of processes or threads with given labels and optionally a number of variables for which values would be available from within the application code. The panel allows starting a simulation and display of results. Figure 2 shows exemplary results for running a parallel application in the MERPSYS environment.

2. System server that acts as a proxy between users and simulators. It launches and manages several simulations on a cluster in parallel. Upon termination of a simulation, the client application displays results to the user.

3. Scheduler – an application that decides where the required number of processes or threads should be launched considering slots available in the system model.

4. Simulator – an application that simulates execution of the aforementioned application on a large scale system. For each distinct label defined within the application, the simulator starts a separate thread that simulates a given number of processes/threads of this type. Proper scaling is used for both computations and communication using the given number of processes/threads. This allows simulation of thousands of processes/threads running the same code using one simulation thread. The simulator offers two advantages over running a real application: it allows consideration of application and system sizes for which a real application could not be run due to resource limitations (such as the limited RAM size), simulation time for an application can be much shorter than the running time of a real application – this is possible because of encapsulation of computations within computational blocks.

5. Hardware database which stores information about various hardware components such as specific CPUs, GPUs, interconnects within nodes (such as PCI) and among nodes (LANs, WANs etc.).

Consequently, it is important for the simulator to have detailed formulas for CPUs, GPUs and interconnects for correct prediction of execution times and energy consumption of particular blocks of code, either representing computations or communication among processes or threads of a parallel application. Additionally, a rough estimate on successful execution of an application can be derived that uses the number of the nodes involved in computations.

4 Model Formulas

Our model is based on the statistical approach. The measurements realized during the experiments reflecting each simulated block were manually compared to the commonly used formulas (e.g. in [11] for the group communication) and the closest approximation was chosen. In general we used the expert knowledge which reflects the internals of the MPI implementation and the cluster design.

We used the hardware specific implementation (e.g. Infiniband for MVAPICH) thus using of the already proposed models [11] directly could be misleading. For regression we used two different error measurements i.e.: (i) mean percentage error (MPE) in cases where the differences for small absolute value were important (e.g. sending of short messages) in the model, and (ii) a traditional mean squared error for others.

Table 1 presents the formulas of the proposed model. Their parameters are split into two groups:

1. Input parameters that need to be provided for the application model, i.e.: the number of instructions to be executed: h, the number of the active threads executed on a particular node: p_{th}, input data size: d and the number of the nodes involved in computations: P.
2. Parameters related to a specific environment, constant during the execution of an application on the modeled cluster, and dependent on the cluster hardware, software (e.g. the operating system and/or the used message library implementation) and their configuration, e.g. the number of cores in the CPU. Their values were provided directly (like the mentioned number of cores) or by the regression (for less obvious ones). Table 3 contains a complete list of the parameters.

Table 1. Model formulas

Modeled block	Execution time
Computation block (O_x)	$t_{ox} = \begin{cases} T_{min}h & \text{if } p_{th} \leq P_{low} \\ T_{low}h & \text{if } p_{th} \in (P_{low}, P_{hi}) \\ (T_{hi} + K_{hi}p_{th})h & \text{if } p_{th} > P_{hi} \end{cases}$
Communication peer-to-peer (C_{p2p})	$t_{p2p} = T_{p2p} + K_{p2p}\lceil d/D_{tu}\rceil D_{tu}$
Communication broadcast (C_{bcast})	$t_{bcast} = T_{bcast} + K_{bcast}\lceil d/D_{tu}\rceil D_{tu}log(P)$
Communication scatter (C_{scat})	$t_{scat} = T_{scat} + K_{scat}\lceil d/D_{tu}\rceil D_{tu}\frac{log(P)}{P}$
Communication gather (C_{gath})	$t_{gath} = T_{gath} + K_{gath}\lceil d/D_{tu}\rceil D_{tu}\frac{log(P)}{P}$
Communication all-to-all (C_{a2a})	$t_{a2a} = T_{a2a} + K_{a2a}\lceil d/D_{tu}\rceil D_{tu}P$
Communication barrier (C_{bar})	$t_{bar} = T_{bar} + K_{bar}log(P)$
Read block from a network disk (R_{disk})	$t_{rdisk} = T_{rdisk} + K_{rdisk}\lceil d/D_{tu}\rceil D_{tu}$
Write block to a network disk (W_{disk})	$t_{wdisk} = T_{wdisk} + K_{wdisk}\lceil d/D_{tu}\rceil D_{tu}$
Modeled block	Power consumption
Any number of blocks on a single node	$pw = \begin{cases} PW_{low} + KW_{low} \times p_{th} & \text{if } p_{th} \leq P_{low} \\ PW_{hi} + KW_{hi} \times p_{th} & \text{if } p_{th} \in (P_{low}, P_{hi}) \\ PW_{max} & \text{if } p_{th} > P_{hi} \end{cases}$
Modeled block	Probability of the correct execution
Any number of blocks on a set of nodes	$s(\Delta t, P) = e^{-\lambda \Delta t P}$

A computation block O_x is a basic block which represents data processing in the cluster and grid environment. We assume the most typical arithmetical or comparison operations performed on data. Such a block can be performed sequentially or in parallel using different cores of the nodes processors.

We assumed that the time of parallel computation of a single instruction is constant (T_{min}) as long as the number of threads is lower than the number of cores, thus in this case the total time depends only of the number of executed instructions: $t_{ox} = T_{min}h$. We assume a similar constant time (T_{low}) for HyperThreading however, in this case this time is longer: $T_{low} > T_{min}$, thus $t_{ox} = T_{low}h$. Finally, for the number of threads exceeding the number of (virtual) cores, we assumed the time of the single instruction increases linearly with the number of the threads: $t_{ox} = (T_{hi} + K_{hi}p_{th})h$. Figure 3a presents the regression results with a comparison to the real measurements for an exemplary computation block of 11 million instructions executed by a single thread.

We distinguish a number of communication blocks, including direct peer-to-peer and group operations. The time for the former was assumed to be linear to the size of the message, and for the latter we relied mainly on the analytical models (e.g. provided in [11]). For both types of formulas we introduced an adjustment related to the (maximum) data transfer unit. Figures 4 and 5 present the regression results in comparison to the real measurements for peer-to-peer, barrier and broadcast blocks. Similarly, blocks related to the network disk I/O operations were modeled, and appropriate formulas were proposed. Due to their characteristics, we assumed the same time complexity as for peer-to-peer communication.

During the experiments, we observed the electrical power load being strongly dependent on the number of the active threads rather than on the type of operation (a specific computational block). Thus we proposed a model, in which the power consumed at the moment (expressed in Watts), depends on the number of used processor cores, which are directly involved in processing. For regression, we assumed segmented and linear increase of power consumption. However, we introduced a distinction between the power consumed by active threads assigned to the real and logical (HyperThreading) cores. Obviously, after passing a certain threshold additional, active threads do not introduce additional increase in power consumption. Figure 3b presents the power regression results in comparison to the real measurements.

5   Model Parameter Regression Results

The model parameters were derived for three different cluster environments: Galera+ (all parameters including power), Galera (performance parameters only) clusters, located in the Academic Computer Centre – TASK, and a KASK cluster (performance parameters only) located at the Department of Computer Architecture, Faculty of Electronics, Telecommunications and Informatics. All these machines are located at Gdańsk University of Technology. The clusters work under a Linux operating system and the measurements were performed for the MVAPICH v1.8 MPI implementation.

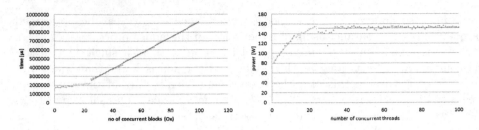

Fig. 3. Computation block regressions vs real measurements for Galera+: (a) execution time, (b) power consumption

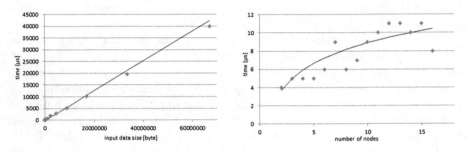

Fig. 4. Execution time regressions versus real measurements for Galera+: (a) peer-to-peer, (b) barrier

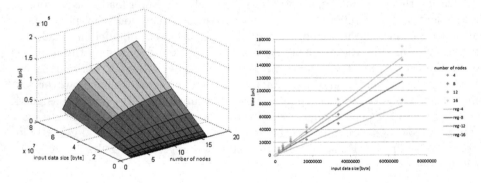

Fig. 5. Broadcast execution time regression for Galera+: (a) 3D view (b) mapping to 2D versus real measurements

The Galera+ cluster consists of 192 computation nodes. Every node is equipped with two Intel Xeon 2.27 GHz multicore processor units, with 6 physical and 12 logical (HyperThreading) cores each, 16 GB RAM, and network interface cards. The cluster uses two interconnection networks: (i) Infiniband QDR (40 Gbps) and (ii) GB Ethernet, supported by respective network switches. Additionally there is a 500 TB disk array exposed to the nodes using a Glustre remote file system. The Galera cluster consists of 672 computation nodes. Every node is equipped with two Intel Xeon 2.33 GHz multicore processor units, with 4 cores each, 8-32 GB RAM, and network interface cards. The cluster uses two interconnection networks: (i) Infiniband DDR (20 Gbps) and (ii) GB Ethernet, supported by respective network switches. Additionally the 500 TB disk array is exposed to the nodes using a Lustre remote file system.

Table 2. Pearson coefficient squared (R^2) calculated for the regression formulas versus the real measurements

Formula	R^2 Galera+	R^2 Galera	R^2 KASK
t_{ox}	0.9993	0.9999	0.9999
t_{p2p}	0.9996	0.9998	0.9996
t_{bcast}	0.9902	0.9688	0.9102
t_{scat}	0.9643	0.9176	0.9668
t_{gath}	0.9620	0.9374	0.9823
t_{a2a}	0.9296	0.7858	0.8384
t_{bar}	0.7324	0.9223	0.8527
t_{rdisk}	0.9999	0.9971	0.9976
t_{wdisk}	0.9988	0.9999	0.9999
pw	0.9062	—	—

The KASK cluster consists of 10 computation nodes. Every node is equipped with two Intel Xeon 2.8 GHz multicore processor units, with 2 physical and 4 logical (HyperThreading) computation cores each, 4 GB RAM, and network interface cards. The cluster uses two interconnection networks: (i) Infiniband (10 Gbps) and (ii) GB Ethernet, supported by the corresponding network switches. Additionally there is a 4 TB disk array exposed to the nodes using the NFS file system. The regression was performed numerically using the gathered power and time measurements. The final results for three different clusters are presented in Table 3. Figures 3, 4 and 5 present the charts with comparison of some regression results to the real measurements and Table 2 provides the evaluation with the Pearson coefficient squared (R^2) for all estimated formulas.

6   Summary and Future Works

In the paper, we presented modeling of parallel processing in a cluster environment that includes equations representing execution times of computational and communication blocks, power consumption of such blocks as well as a simple estimate on reliability of computations. Furthermore, coefficients for these computational and communication equations were found for three clusters and power consumption and reliability indicated for Galera+ cluster located at Academic Computer Center, Gdańsk, Poland. These equations constitute an integral part of an environment that allows simulation of parallel applications running on

Table 3. Model instance parameters for Galera+, Galera and KASK clusters

Param	Galera+	Galera	KASK	Param	Galera+	Galera	KASK
P_{low}	12	8	4	T_{a2a} [μs]	7.5	18.3	10.1
P_{hi}	24	—	8	K_{a2a} [μs/B]	0.00012	0.00012	0.00048
T_{min} [μs]	0.00165	0.0019	0.0033	T_{bar} [μs]	1.4	-1.2	-1.8
T_{low} [μs]	0.00194	—	0.0034	K_{bar} [μs]	7.5	20.6	22.6
T_{hi} [μs]	3.29e-04	1.16e-04	2.06e-04	T_{rdisk} [μs]	5200	1390	7600
K_{hi} [μs]	7.96e-05	2.35e-04	4.14e-04	K_{rdisk} [μs/B]	0.00253	0.0324	0.0290
D_{tu} [B]	2048	2048	2048	T_{wdisk} [μs]	1200	270	1100
T_{p2p} [μs]	3.7	1.3	1.3	K_{wdisk} [μs/B]	0.00474	0.0066	0.0093
K_{p2p} [μs/B]	0.00063	0.00132	0.00181	PW_{low} [W]	78	—	—
T_{bcast} [μs]	1.7	-1.1	-1.04	KW_{low} [W]	4.84	—	—
K_{bcast} [μs/B]	0.00188	0.00199	0.0580	PW_{hi} [W]	109	—	—
T_{scat} [μs]	5.0	1.1	2.9	KW_{hi} [W]	1.90	—	—
K_{scat} [μs/B]	0.00340	0.00820	0.00880	PW_{max} [W]	151	—	—
T_{gath} [μs]	8.1	16.4	10.0	λ	5.03372 e-10	—	—
K_{gath} [μs/B]	0.00346	0.00100	0.00990				

large scale systems and prediction of execution time, potential failures and energy consumed during a run. The environment with its basic components such as a system editor, an application editor and a simulation panel with sample results were presented. The system editor and consequently the simulator use the presented equations for simulation of runs of modeled parallel applications on the three aforementioned clusters. This in turn allows estimation of execution times, power consumption and reliability for various applications and configurations including the number of nodes run on these clusters.

Acknowledgments. The work was performed within grant "Modeling efficiency, reliability and power consumption of multilevel parallel HPC systems using CPUs and GPUs" sponsored by the National Science Center in Poland based on decision no DEC-2012/07/B/ST6/01516.

References

1. Dongarra, J.: Emerging heterogeneous technologies for high performance computing. In: Heterogeneity in Computing Workshop (2013). http://www.netlib.org/utk/people/JackDongarra/SLIDES/hcw-0513.pdf
2. Czarnul, P., Rościszewski, P.: Optimization of execution time under power consumption constraints in a heterogeneous parallel system with GPUs and CPUs. In: Chatterjee, M., Cao, J., Kothapalli, K., Rajsbaum, S. (eds.) ICDCN 2014. LNCS, vol. 8314, pp. 66–80. Springer, Heidelberg (2014)
3. Bak, S., Krystek, M., Kurowski, K., Oleksiak, A., Piatek, W., Weglarz, J.: GSSIM - a tool for distributed computing experiments. sci. program. **19**, 231–251 (2011)
4. Hockney, R.W.: The communication challenge for mpp: Intel paragon and meiko cs-2. Parallel Comput. **20**, 389–398 (1994)

5. Culler, D., Karp, R., Patterson, D., Sahay, A., Schauser, K.E., Santos, E., Sub-ramonian, R., von Eicken, T.: Logp: towards a realistic model of parallel computation. In: Proceedings of the Fourth ACM SIGPLAN Symposium on Principles and Practice of Parallel Programming, PPOPP 1993, pp. 1–12. ACM, New York (1993)
6. Alexandrov, A., Ionescu, M.F., Schauser, K.E., Scheiman, C.: Loggp: Incorporating long messages into the logp model—one step closer towards a realistic model for parallel computation. In: Proceedings of the Seventh Annual ACM Symposium on Parallel Algorithms and Architectures, SPAA 1995, pp. 95–105. ACM, New York (1995)
7. Kielmann, T., Bal, H.E., Verstoep, K.: Fast measurement of LogP parameters for message passing platforms. In: Rolim, J.D.P. (ed.) IPDPS-WS 2000. LNCS, vol. 1800, pp. 1176–1183. Springer, Heidelberg (2000)
8. Bosque, J.L., Perez, L.P.: Hloggp: a new parallel computational model for heterogeneous clusters. In: CCGRID, pp. 403–410. IEEE Computer Society (2004)
9. Chui, C.K.: The logp and mlogp models for parallel image processing with multi-core microprocessor. In: Proceedings of the 2010 Symposium on Information and Communication Technology, SoICT 2010, pp. 23–27. ACM, New York (2010)
10. Cameron, K.W., Ge, R., Sun, X.: $\log_n p$ and $\log_3 p$: accurate analytical models of point-to-point communication in distributed systems. IEEE Trans. Comput. **56**, 314–327 (2007)
11. Pjesivac-Grbović, J., Fagg, G.E., Angskun, T., Bosilca, G., Dongarra, J.J.: Mpi collective algorithm selection and quadtree encoding. In: Proceedings of the 13th European PVM/MPI Users' Group Meeting, Bonn, Germany (2006)

Running Time Prediction
for Web Search Queries

Oscar Rojas[1,2], Veronica Gil-Costa[1,2(✉)], and Mauricio Marin[1,2]

[1] CITIAPS, DIINF, University of Santiago, Santiago, Chile
ggvcosta@gmail.com
[2] Center for Biotechnology and Bioengineering, Santiago, Chile

Abstract. Large scale Web search engines have to process thousands of
queries per second and each query has to be solved within a fraction of
a second. To achieve this goal, search engines rely on sophisticated services capable of processing large amounts of data. One of these services
is the search service (or index service) which is in charge of computing
the top-k document results for user queries. Predicting in advance the
response time of queries has practical applications in efficient adminis-
tration of hardware resources assigned to query processing. In this paper,
we propose and evaluate a query running time prediction algorithm that
is based on a discrete Fourier transform which models the index as a
collection of signals to obtain patterns. Results show that our approach
performs at least as effectively as well-known prediction algorithms in
the literature, while significantly improving computational efficiency.

Keywords: WAND · Inverted files · Multi-threading

1 Introduction

Large scale Web search engines are complex systems composed by several services deployed on different clusters of processors interconnected by a high speed
communication network. In particular, the so-called index service where it is necessary to determine the K most "suitable" Web documents for a given query, is
the most time consuming task. This service uses a document-similarity-to-query
function which running time depends on the query contents and the size of the
Web sample (usually huge), kept indexed in the distributed memory cluster of
processors supporting the index service. When similarity calculation functions
are used in tandem with cost-saving strategies such as the WAND [1] or the
Block-Max WAND [5], the cost is not linear on the dataset size. The WAND
strategy allows to skip documents with no chance of being part of the top K documents, making the running time cost unpredictable and widely variable across
queries. Additionally, unbalance is introduced among the distributed processors
and resources may be under-utilized.

In this context, running time prediction algorithms represent a non-trivial
problem which can be used to decide at run time the number of resources to be

© Springer International Publishing Switzerland 2016
R. Wyrzykowski et al. (Eds.): PPAM 2015, Part II, LNCS 9574, pp. 210–220, 2016.
DOI: 10.1007/978-3-319-32152-3_20

assigned to each incoming query under unstable conditions such as high query traffic. In this paper, we propose a method suitable for this the Block-Max WAND query evaluation strategy operating on an index service for a Web search engine. The paper describes the method and compares it against a state of the art method for the same problem [2] on a multi-thread environment.

The underlying data structure in the index service is the so called inverted file, composed of a table of terms and for each term there is a posting list of Web documents where the term appears and its in-document term frequency. The similarity function scans the posting lists associated with the query terms to determine the documents that are the most similar ones to the query.

Our prediction method uses the discrete Fourier transform (DFT) to obtain a spectrum of the posting lists of terms stored in the inverted file. This process is performed off-line. The information obtained with the DFT (which is represented as a vector with six descriptors) is used to feed a feed-forward neural network with back-propagation which estimates the query running time on-line. Figure 1(a) shows a general scheme of the proposed query running time prediction method. The DFT has been previously used in other context like patter recognition, data mining [10], etc. In the Web search engine domain, it has been used as a scoring method to order the relevance of documents when related to a specific query [7]. However, to the best of our knowledge, the novelty of our proposal comes from the application of the DFT to significantly reduce the number of dataset descriptors required to train the machine learning model.

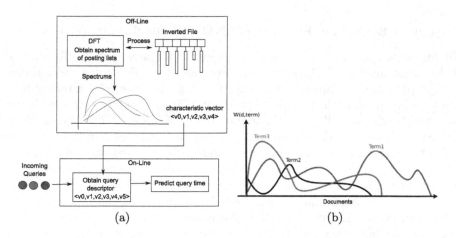

Fig. 1. (a) General scheme. (b) Distribution of the scores of posting lists.

The remaining of this paper is as follows. Section 2 describes the Block-Max WAND used in this work and related works. Section 3 presents our proposed dynamic query time prediction algorithm. Section 4 presents our experiment results and Sect. 5 presents conclusions.

2   Background and Related Work

Large scale Web Search Engines (WSE) have to manage huge quantities of documents while achieving the goal of effectively answering user queries within a fraction of a second. WSEs are usually built as a collection of services hosted in large clusters of multi-core processors wherein each service is deployed on a set of processors supporting multi-threading. Services are software components executing operations such as (a) calculation of the top-k documents that best match an user query; (b) construction of the result Web page for queries; (c) advertising related to query terms; (d) query suggestions, among other operations.

One important bottleneck in WSE is the service in charge of computing the top-k documents for user queries, named search service (or index service), where a ranking algorithm is executed on an inverted index (or inverted file) which is a data structure used by all well-known WSEs. This index enables the fast determination of the documents that contain the query terms and contains data to calculate document scores for ranking. The index is composed of a vocabulary table and a set of posting lists. The vocabulary table contains the set of relevant terms found in the document collection. Each of these terms is associated with a posting list which contains the document identifiers where the term appears in the collection along with data used to assign a score to the document. To solve a query, it is necessary to get from the posting lists the set of documents associated with the query terms and then to perform a ranking of these documents in order to select the top-k documents as the query answer [1,3].

2.1   Query Evaluation Process: WAND and Block-Max WAND

Ranking algorithms return the top-k documents for user queries. To quickly process large inverted lists, these algorithms use dynamic pruning techniques to avoid processing complete lists. Some ranking algorithms for inverted lists have been proposed in the technical literature [1,5]. In this paper we use on the current state-of-the-art WAND [1] and its variant BM-WAND [5], which achieve significant benefits [5]. They use a pointer movement strategy based on pivoting to skip many documents that would be evaluated by an exhaustive algorithm.

The WAND algorithm assumes a single threaded processor containing an inverted index, which is usually kept in compressed format. The algorithm process each query by looking for query terms in the inverted index and retrieving each posting list. Documents referenced from the intersection of the posting lists allow to answer conjunctive queries (AND bag of word query) and documents retrieved at least from one posting list allow to answer disjunctive queries (OR bag of word query). It uses a standard docID sorted index and it is based on two levels. In the first level, some potential documents are selected as results using an approximate evaluation. Then, in the second level those potential documents are fully evaluated (e.g. using the BM25 or vector model) to obtain their scores.

Let us consider an additive document scoring model for document ranking, i.e. for query q and document d we have $\text{Score}(d, q) = \sum_{t \in q \cap d} w(d, t)$, where $w(d, t)$ represents the score for term t in d. Each term t is associated with an upper bound UB$_t$ which corresponds to its maximum contribution to any document score in the collection. The currently found top-k results are stored in a heap, which is initially empty. The document with least score is located in the root. The root score provides a threshold value which is used to decide the full score evaluation of the remaining documents in the posting lists associated with the query terms. The admission of a new document in the heap is produced when the WAND operator is true, i.e. when the score of the new document is greater than the score of the document with the minimum score stored in the heap. If the heap is full, the document with the minimum score is replaced, updating the value of the threshold. Documents with a score smaller than the threshold score in the heap are skipped. This scheme allows skipping many documents that would have been evaluated by an exhaustive algorithm.

The Block-Max WAND (BM-WAND) algorithm is proposed in [5]. It uses compressed posting lists organized in blocks. Each block stores the upper bound (Block max) for the documents inside that block in uncompressed form, thus enabling to skip large parts of the posting lists by skipping blocks. This reduces the cost of the WAND algorithm but does not guarantee correctness because some relevant documents could be lost. To solve this problem, the authors propose a new algorithm that moves forward and backwards in the posting lists to ensure that no documents are missed. Independently, the same idea was presented in [3].

2.2   Query Time Prediction Algorithms

One may think that queries terms with larger posting lists are more expensive to process. However, the cost of processing a query cannot be directly related to the posting list lengths of its terms when dynamic pruning techniques are used, because many documents can be skipped. In this context and under different user query bursts, query time prediction algorithms can be useful to determine which resources are going to be allocated to a given query. In [4] a performance query predictor is proposed. It is based on the relative entropy between a query language model and the corresponding collection language model.

A query efficiency predictor is proposed in [2] for the WAND algorithm. The algorithm is designed for a distributed search engine. This work shows that there is a strong correlation between the distribution of postings in the query terms and the response time of the query. Recently, the work in [6] detects the most relevant parameters used in [2] and propose to optimize memory usage.

The work in [9] aims to achieve a minimum query response time when query traffic is high, by adjusting the value of k (top-k document results) and the threshold used by the WAND algorithm which increases the aggressiveness of the pruning. In other words, it reduces the number of document retrieved for each user query. The algorithm is configured to prune more or less aggressively, depending on the expected duration of the query. The value of k is also estimated in [8]. However, in this work the effectiveness of the search engine is not compromised. The algorithm ensures the top-k document results.

3 Proposed Query Time Prediction Algorithm

The score distribution $w(t, d)$, the location of documents representing the upper bounds in posting lists and the length of the posting lists, varies from term to term. Figure 1(b) shows the score distribution of the posting lists of three terms. The x-axis shows the documents sorted in ascending order by their identifiers, and the y-axis shows the score $w(t, d)$. A good query representation combines different features that allow establishing a mathematical relationship between the time required to process the query and the information of the inverted index. We propose using the DFT to represent the main characteristics of posting lists of query terms, such as the power spectral density of posting lists among others. This characteristics will be used later to predict query response times.

Given a query q containing the terms t_l with $l \geq 1$, where each term has a posting list L_t containing pairs $< d, w(d, t) >$ where d is the document identifier and $w(d, t)$ is the score of the term in the document (e.g. the frequency of occurrence of the term t in the document d), our method works as follows. We use information regarding the frequency spectrum of density functions Φ_t obtained from the posting lists of the terms $t_l \in q$, and also considers the information related to the spectrum of frequency of the processing time $T(t_l, k)$ for each term t_l required to retrieve the top-k document results. The spectrum of frequencies is obtained with the discrete Fourier transform DFT. In addition, we use: (a) the size of each posting list $s_t = |L_t|$ (i.e. the number of documents where the term appears), (b) the processing time for $T(t, 10)$, $T(t, 100)$, $T(t, 1000)$ and for $T(t, 10000)$, and (c) the threshold value for the top-k document. Then, we describe each term with a five dimension characteristic vector $\psi :< \psi_0, \psi_1, \psi_2, \psi_3, \psi_4 >$.

The density function X_{DFT} of the posting lists of the term t_l, describes the search space Ω_t of the posting list L_t. The X_{DFT} of the processing times functions $T(t, k)$ describes the differences of the times required to process the posting list of a term t with different k values. In practice, the values of $X_{DFT}[u]$ are the u-th coefficients of Fourier and express the frequency content of a function or a signal. In this analysis, the DFT of Φ_t, can be considered as a characterization of the distribution of the values $w(d, t)$, and therefore it can be seen as a function of bulk density in the frequency domain.

We use the spectral power density of X_{DFT} over $w(d, t)$, because it represents the cost of processing the signal in the frequency domain. It shows how the power is scattered as a function of the frequency $F = 1/10$, which is the minimum frequency (or fundamental frequency) of the DFT. The fundamental frequency $F = 1/10$ describes the density of posting lists by using the convolution of the broader sinusoidal signal. Thus, it describes well the posting lists that have a higher density. We also use the magnitude of the spectrum of the fundamental frequency $F = 1/4$ of the DFT for the posting lists and for the processing times obtained for each term $T(t, k)$, which describes the difference between the processing times as the value of k increases in a quadratic way. Table 1 summarizes the descriptors used in our predictor.

PSD of Φ_t of Frequency $1/10$: ψ_0 is the Power Spectral Density (PSD) of the DFT of Φ_t in the fundamental frequency $F = 1/10$. The calculation is $|X_{DFT}[u]\{\Phi_t\}|^2, u = 1$. Φ_t (Eq. 1) is a vector containing the cumulative sums on Φ_G (Eq. 2) of scores $w(d,t)$ of each document $d \in L_t$ inside 10 intervals I_j. Each j-th interval I is equi-spaced at the rate of $\#Postings/10$ items. If there are empty intervals, the cumulative sum is zero in those positions. Each value of $\Phi_{t,i}$ is obtained with Eq. 3.

$$\Phi_t = <\Phi_{t,1}, \Phi_{t,2}, ..., \Phi_{t,i}, ...\Phi_{t,10}> \tag{1}$$

$$\Phi_G = max\{\sum_{d\in L_t} w(d,t)\}, t \in V \tag{2}$$

$$\Phi_{t,i} = \frac{1}{\Phi_G}\sum_{d\in I_i} w(d,t) \tag{3}$$

DFT Magnitude of Rank-Score of Frequency $1/4$: ψ_1 is the magnitude of the frequency spectrum of the DFT in the fundamental frequency $F = 1/4$ of the distribution of cumulative density of the documents scores from $k = 1$ to $k = \{10, 100, 1000, 10000\}$.

#Postings: ψ_2 is the number of documents where the term appears.

DFT Magnitude of Processing Times: ψ_3 is the magnitude of the frequency spectrum of the DFT obtained for the vector containing the processing times $T(t, k)$ of a term t at frequency $T = 1/4$. The vector elements are $< T(t, 10), T(t, 100), T(t, 1000), T(t, 10000) >$. $T(t, k)$ is the processing time required to retrieve the top-k documents results for the term t.

Threshold: ψ_4 score value (threshold) for the $k - th$ document (top-k). If the list has less than k documents, then $\psi_4 = 0$.

To predict the query response time, we compute the query descriptor Ψ_q as a six dimension vector $< x_0, x_1, x_2, x_3, x_4, x_5 >$ as follows. For each term $t \in q$, we add the corresponding descriptors $t_{\psi_0}, t_{\psi_1}, t_{\psi_2}$ and t_{ψ_3} of each term in q, so we compute an initial query vector with dimension four. Then, we include two

Table 1. Elements of the term descriptors.

Descriptor for term t
1. ψ_0 : PSD of Φ_t at a frequency $1/10$.
2. ψ_1 : DFT magnitude of Rank-Score at a frequency $1/4$.
3. ψ_2 : $\#Postings$
4. ψ_3 : DFT magnitude for the processing times.
5. ψ_4 : score value for the $k - th$ document in the term list.
Additional descriptor $\mathcal{S}()$
a. Sum: $x_i = \sum_{t\in q} t_{\psi_i}$ for i=0,1,2,3
b. Maximum: $x_4 = max_{t\in q}\{t_{\psi_1}\}$, $x_5 = max_{t\in q}\{t_{\psi_4}\}$

additional descriptors computed as the $max\{t_{\psi_2}\}$ and $\max\{t_{\psi_4}\}$ for each $t \in q$. All vectors ψ_t are calculated off-line, while Ψ_q is obtained on-line.

For a given query q, the descriptors $< x_0, x_1, x_3 >$ represent the sum of integrals obtained with the DFT (for a different feature) and the descriptor x_2 represents the sum of documents, which gives an approximation to the search space of q. We do not compute the sum of ψ_4 for each term of the query, because it is a lower bound of the score of the top-k and if there are several term lists with high scores, the sum of those scores will increase the value of ψ_4 and it will lose its characteristic of lower bound. We also use the maximum values of ψ_1 and ψ_4 that are *minimum bounds*.

All descriptors using the DFT, are based on the use the fundamental frequency F that depends on the period of P of the input signal. That is, the distributions of $w(d, t)$ with period of $P = 10$, the distributions of the cumulative density with period of $P = 4$, and the distributions of processing time with period of $P = 4$ where the fundamental frequency is $F = 1/P$. We use the fundamental frequency to quantify how is the distribution of the high values of the input signal. In our case, high values of the magnitudes of the DFT represent a higher processing list costs, either in the spatial domain as in the time domain.

4   Experiment Results

4.1   Experimental Setup

Experiments were conducted on a Intel Processor Core i7-3820. We used a 50.2 million document corpus TREC ClueWeb09 (category B)[1]. We index this corpus using the Terrier IR platform[2]. We select the first 14,289 queries from the TRECMillion Query Track 2009. We use a feed-forward neural network with back-propagation with six input neurons, one for each of the six dimension characteristic vector ψ, and one output. We trained the neural network with 60 % of the queries and we used 40 % for validation. We conducted experiments with $1, 5, 10, 25$ and 50 neurons in the hidden layer. We used the transfer functions log-Sigmoid in the hidden layer and transfer function lineal in the output layer.

We pruned the index to keep only data related to the terms of the query log. We evaluated the prediction algorithms (our proposal and the Baseline approach according to [2], which uses vectors with 42 descriptors to represent queries) over two strategies to efficiently organize multi-threading in search engine processors [8]. We refer to a situation in which (1) multiple threads are assigned to a given query and each thread keeps a local heap data structure to hold its current local top-k results to then determine the global top-k results for the query (LBM-WAND), and (2) one shared heap is kept for each active query to hold its current global top-k results and the assigned threads access it in a concurrent manner to process the respective query (SBM-WAND). We set $k = 1000$.

---

[1]   http://www.lemurproject.org/clueweb09.php/.
[2]   http://terrier.org/.

4.2   Accuracy Evaluation

Table 2 shows results with different number of neurons in the hidden layer to determine how much the precision of the algorithms can be improved with a greater numbers of neurons. Running time is measured in seconds. r is the Pearson's correlation, and the root of the mean square error computed as: $RMSE = \sqrt{(\sum(x_i - y_i)^2/N)}$, where N is the number of queries used in the experiment, x_i is the response time for q_i and y_i is the predicted query time.

Table 2. Accuracy evaluation obtained with three query time prediction algorithms. L stands for Local heap and S for shared heap.

	Neurons of the Hidden Layer									
	1		5		10		25		50	
	r	RMSE	r	RMSE	r	RMSE	r	RMSE	r	RMSE
Sequential BM-WAND										
Baseline	**0,898**	**0,049**	0,923	0,043	0,908	0,047	0,862	0,059	0,844	0,063
Proposed	0,886	0,052	**0,930**	**0,041**	**0,928**	**0,042**	**0,925**	**0,043**	**0,919**	**0,044**
LBM-WAND / SBM-WAND										
2 thread										
Baseline-L	**0,912**	**0,047**	0,932	0,042	0,925	0,044	0,894	0,052	0,876	0,058
Proposed-L	0,895	0,051	**0,944**	**0,038**	**0,940**	**0,039**	**0,941**	**0,039**	**0,918**	**0,046**
Baseline-S	**0,897**	**0,051**	0,901	0,050	0,891	0,054	0,861	0,062	0,813	0,074
Proposed-S	0,880	0,055	**0,926**	**0,043**	**0,929**	**0,043**	**0,906**	**0,049**	**0,903**	**0,050**
4 thread										
Baseline-L	**0,920**	**0,044**	0,945	0,037	0,928	0,043	0,921	0,045	0,884	0,055
Proposed-L	0,902	0,048	**0,948**	**0,035**	**0,951**	**0,034**	**0,949**	**0,035**	**0,928**	**0,042**
Baseline-S	**0,897**	**0,052**	0,905	0,050	0,901	0,051	0,865	0,061	0,853	0,064
Proposed-S	0,875	0,056	**0,926**	**0,044**	**0,927**	**0,044**	**0,925**	**0,044**	**0,918**	**0,047**
8 thread										
Baseline-L	**0,939**	**0,035**	0,955	0,031	0,954	0,031	0,948	0,034	0,922	0,042
Proposed-L	0,927	0,039	**0,956**	**0,030**	**0,963**	**0,028**	**0,967**	**0,026**	**0,954**	**0,031**
Baseline-S	**0,913**	**0,050**	0,936	0,043	0,927	0,046	0,902	0,053	0,891	0,057
Proposed-S	0,892	0,055	**0,938**	**0,042**	**0,938**	**0,042**	**0,925**	**0,047**	**0,919**	**0,048**

Results show that a good prediction can be achieved with 5 neurons. From this point on, results do not improve significantly. With more threads, the prediction is more accurate, because processing times reported by each thread tends to be lower (each thread access to a small portion of the index which reduces the average query processing time). Also, the Pearson correlation between real and estimated query times is greater than 0.8 with all approaches. Thus, there is a positive correlation between the real and the estimated query response time.

Results reported with the LBM-WAND and the SBM-WAND algorithms show that our proposal presents good query time estimations with errors

Table 3. Average query response times in second reported by the real execution, and the query times estimated by our proposal and the Baseline approach.

	5 neurons of Hidden Layer								
	Real		Baseline				Proposed		
Threads	MQT	max_t	MQT	max_t	EM	max_e	MQT	max_t	EM	max_e
1T	0,133	1,062	0,133	0,839	**0,025**	0,376	0,134	0,909	**0,025**	**0,370**
2T-L	0,091	0,668	0,091	0,506	0,016	0,283	0,091	0,492	**0,014**	**0,256**
2T-S	0,075	0,586	0,076	0,422	0,015	0,489	0,075	0,364	**0,014**	**0,222**
4T-L	0,062	0,424	0,062	0,356	0,009	0,191	0,061	0,353	**0,008**	**0,148**
4T-S	0,043	0,323	0,043	0,263	**0,008**	0,305	0,043	0,195	**0,008**	**0,132**
8T-L	0,057	0,406	0,056	0,403	0,007	0,204	0,056	0,296	**0,006**	**0,196**
8T-S	0,036	0,256	0,035	0,240	**0,006**	0,104	0,034	0,206	**0,006**	**0,093**

values close to the ones reported by the Baseline. The neuronal network tends to improve the precision of the prediction algorithms as more threads. With 8 threads the Pearson correlation is increased in average by 3,6 % for LBM-WAND and by 1,1 % for SBM-WAND. The RMSE of LBM-WAND (baseline and proposed) decreases by 37 % in average. Therefore, the LBM-WAND approach tends to be more effective in terms of time prediction.

Table 3 shows the query processing times in seconds obtained with a real execution and the query times predicted by our proposal and the Baseline approach using five neurons in the hidden layer of the neuronal network. Both prediction algorithms were executed with the LBM-WAND and the SLB-WAND parallel algorithms. MQT is the average time required to process queries. max_t is the maximum computation time achieved in the experiments. max_e is the maximum observed error of the prediction. The average mean error of the prediction is computed as $EM = (\sum |x_i - y_i|)/N$, where N is the number of queries used in the experiment, x_i is the response time for q_i and y_i is the predicted time.

As expected query response time is reduced with more threads, because each thread processes a smaller portion of the inverted index. In particular, the SBM-WAND approach with 8 threads reports an average query response time of 0.036 which reduces by 73 % the time reported by the sequential algorithm (0.133 vs 0.036). The maximum query response time is reduced by 75 % (1.062 vs. 0.256) and the EM is reduced by 76 % (0.025 vs. 0.006). In general, our proposal reports the same EM as the baseline approach, but with lower maximum errors. These results are achieved when using at least five neurons in the hidden layer.

4.3   Performance Evaluation

Table 4 shows running times in nanoseconds required to build the query vector and the time required to predict the query processing time using 2 and 5 terms and using 5, 10, 25 and 50 neurons in the hidden layer of the neuronal network.

Table 4. Running time in seconds reported by our proposal and the Baseline approach.

	Vector Construction		Neuronal Network			
Strategy	2 terms	5 terms	5 Neurons	10 Neurons	25 Neurons	50 Neurons
Baseline	22129 ns	31429 ns	6844 ns	9150 ns	14108 ns	23397 ns
Proposed	**2095 ns**	**2376 ns**	**4610 ns**	**5238 ns**	**6984 ns**	**8894 ns**

Table 5. Amount of memory required to build the vector for terms and for queries.

	Vector for Terms	Vector for Queries
Baseline	92 Bytes	213 Bytes
Proposed	45 Bytes	46 Bytes

Results show that our proposal is capable of reducing by 91 % the time required to build the query vector using two terms in each query (22129 vs. 2095) and 94 % in the case of using 5 terms per query (31429 vs. 2376). With our proposed prediction algorithm, the time required to predict the query response time is 33 % lower than the time reported by the Baseline approach when using a hidden layer with 5 neurons. With 50 neurons, our proposal reduces by 62 % the time required to predict a query processing time.

Table 5 shows the memory required to build the vector for term of the inverted list and the vectors used to describe the queries. Our proposal reduces by 51 % the amount of memory used to store the terms descriptors and reduces by 78 % the amount of memory required for the query vectors. Therefore, our proposed algorithm drastically reduces on-line execution times and also the memory space required to manage the vector descriptors.

5   Conclusion

We proposed a query running time predictor for the BM-WAND algorithms based on two components: (1) a discrete Fourier transform (DFT), and (2) a feed-forward neural network with back-propagation. The DFT is used to obtain values for characteristics of the posting lists associated with the query terms. These characteristics are used to train a neuronal network which is used to predict the query execution time. We evaluate our proposed prediction algorithm using two multi-threading query processing approaches. Results show that, with both SBM-WAND and LBM-WAND parallel query processing approaches, our proposal is capable of estimating the query execution time with a mean error close to the one reported by the Baseline approach [2], with the benefit that our proposal dramatically reduces the processing time per query and memory space.

In the near future we plan to evaluate the use of our predictor as part of a scheduler devised to assign one or more threads to solve any given query.

Acknowledgments. This research was partially funded by Basal funds FB0001, Conicyt, Chile; PMI USA 1204 and PICT 2014-1146.

References

1. Broder, A.Z., Carmel, D., Herscovici, M., Soffer, A., Zien, J.Y.: Efficient query evaluation using a two-level retrieval process. In: CIKM, pp. 426–434 (2003)
2. Macdonald, N.T.C., Ounis, I.: Learning to predict response times for online query scheduling. In: SIGIR, pp. 621–630 (2012)
3. Chakrabarti, K., Chaudhuri, S., Ganti, V.: Interval-based pruning for top-k processing over compressed lists. In: ICDE, pp. 709–720 (2011)
4. Cronen-Townsend, S., Zhou, Y., Croft, W.B.: Predicting query performance. In: SIGIR, pp. 299–306 (2002)
5. Ding, S., Suel, T.: Faster top-k document retrieval using block-max indexes. In: SIGIR, pp. 993–1002 (2011)
6. Kim, S., He, Y., Hwang, S., Elnikety, S., Choi, S.: Delayed-dynamic-selective (DDS) prediction for reducing extreme tail latency in web search. In: WSDM, pp. 7–16 (2015)
7. Park, L., Ramamohanarao, K., Palaniswami, M.: Fourier domain scoring: a novel document ranking method. TKDE **16**(5), 529–539 (2004)
8. Rojas, O., Gil-Costa, V., Marin, M.: Efficient parallel block-max wand algorithm. In: Wolf, F., Mohr, B., an Mey, D. (eds.) Euro-Par 2013. LNCS, vol. 8097, pp. 394–405. Springer, Heidelberg (2013)
9. Tonellotto, N., Macdonald, C., Ounis, I.: Efficient and effective retrieval using selective pruning. In: WSDM, pp. 63–72 (2013)
10. Warren, T.: Clustering of time series data-a survey. JPR **38**(11), 1857–1874 (2005)

The Performance Evaluation
of the Java Implementation of Graph500

Magdalena Ryczkowska[1], Marek Nowicki[1], and Piotr Bala[2(✉)]

[1] Faculty of Mathematics and Computer Science,
Nicolaus Copernicus University, Chopina 12/18, 87-100 Torun, Poland
{gdama,faramir}@mat.umk.pl
[2] Interdisciplinary Centre for Mathematical and Computational Modeling,
University of Warsaw, Pawinskiego 5a, 02-106 Warsaw, Poland
bala@icm.edu.pl

Abstract. Graph-based computations are used in many applications. Increasing size of analyzed data and its complexity make graph analysis a challenging task. In this paper we present performance evaluation of Java implementation of Graph500 benchmark. It has been developed with the help of the PCJ (Parallel Computations in Java) library for parallel and distributed computations in Java. PCJ is based on a PGAS (Partitioned Global Address Space) programming paradigm, where all communication details such as threads or network programming are hidden. In this paper, we present Java implementation details of first and second kernel from Graph500 benchmark. The results are compared with the existing MPI implementations of Graph500 benchmark, showing good scalability of PCJ library.

Keywords: High performance computing · Graph processing · PGAS · Parallel and distributed computation · Performance evaluation · Parallel graph algorithms · Java

1   Introduction

Many computational problems can be formulated in the terms of graphs. Graphs are very convenient when talking about relations in any context. That is why they are commonly used in many scientific fields, for example in biology (to model protein interactions or a food chain), in sociology (for social network analysis), WWW mining, analysis of networks or data transfer processing.

Since size of analyzed problems increases, fast analysis and short time for getting solution for the large graph is becoming more and more important [1]. Graphs of interest often consist of millions of vertices and processing of them is not an easy task. From the computational point of view graph processing requires intensive integer calculations and is different from traditional CPU intensive floating point operations measured with, for example, LINPACK benchmark. The important challenge is large memory demand to store and process graphs

© Springer International Publishing Switzerland 2016
R. Wyrzykowski et al. (Eds.): PPAM 2015, Part II, LNCS 9574, pp. 221–230, 2016.
DOI: 10.1007/978-3-319-32152-3_21

and in the case of parallel processing, huge amount of communication and synchronization. In order to analyze supercomputers powers across the world in the context of graph problems the Graph500 benchmark has been created [2].

Exploitation of graphs is ubiquitous, therefore new solutions, languages or libraries are still being invented to improve performance and make programming easy end efficient. Building systems that process vast amounts of data has been made simpler by the introduction of the MapReduce framework [3], and its opensource implementation Hadoop [4]. These systems offer automatic scalability to extreme volumes of data, automatic fault-tolerance, and a simple programming interface based around implementing a set of functions. However, it has been recognized [5] that these systems are not always suitable when processing data in the form of a large graph. Therefore dedicated tools for graph processing has been developed. A good example are open source Graphlab [6], Ligra [7] or Graph Processing System [8] as well as proprietary one such as Pregel [5].

Most of the tools for graph processing is using traditional programming languages such as C/C++. However the growing adoption of Java as programming language for the data analytics opens requirement for new scalable solutions. The parallel execution in Java is based on the Thread class or fork-join framework. Recent addition to the Java parallelization capabilities is the Java Concurrency Package introduced in Java SE5 and improved in Java SE6. All these features can be used within single Java Virtual Machine which limits parallelization capabilities to the single shared memory node.

One of the new programming tools for the parallel computing in Java is PCJ library [9]. PCJ addresses increasing demand for easy and efficient tools to parallelize Java applications. The library is based on the PGAS (Partitioned Global Address Space) model [10] which allows programmer to view a distributed memory system as a global address space. In PGAS model communication details are hidden improving ease of programming. PGAS languages use one-sided communication which allows for access to the remote memory without involving threads on the remote nodes. Unlike existing solutions based on the MPI or RMI, the PCJ is written as pure Java library and does not use external libraries nor JNI technology. The performance of the PCJ library has been already tested based on the standard microbenchmarks such as ping-pong, broadcast and barrier showing good scalability and performance up to thousands of cores [11].[1]

In this paper we present parallelization of the problem of a large graph traversing. Our solution is based on the Java language and is parallelized with the PCJ library. We have implemented Graph500 benchmark and evaluated its performance. The obtained results are compared to the standard MPI implementation of the Graph500. The approach used in PCJ implementation of the first kernel of Graph500 is different than those used in reference codes. The idea of the second kernel algorithm is based on reference 'MPI simple' implementation and has been adopted to PGAS Java library and one-sided communication.

---

[1] PCJ has been appreciated as an efficient way of programming parallel applications, receiving HPC Challenge Class 2 Best Productivity Award [12].

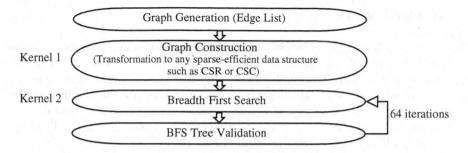

Fig. 1. The Graph500 benchmark execution flow.

The rest of the paper is organized as follows: in the Sect. 2 we describe Graph500 benchmark. Next section describes related work. Section 4 describes basic features of the PCJ library. The parallelization of the Graph500 benchmark with the PCJ is described in the Sect. 5. The following section contains performance evaluation of the solution. Paper is concluded by the conclusions and remarks on future work.

2 Graph500 Benchmark

Graph500 was developed to help to analyze and evaluate performance of the computers and computer clusters in the context of graph algorithms [2].

The flow of tasks in Graph500 is illustrated in Fig. 1. The benchmark provides Graph Generator which constructs a graph in a form of an edge list. Each edge is undirected and is represented with endpoints given in the tuple as start-vertex and end-vertex. The Graph500 Generator adopts a Kronecker Graph model [13] that simulates sparse, real networks with small diameter, which are scale-free and their degree distribution follows a power law.

The size of a graph is specified by the following parameters: $SCALE$ (the logarithm base two of the number of vertices) and $edgefactor$ (the ratio of the graph's edge count to its vertex count), which means that the total number of vertices is $N = 2^{SCALE}$ and the number of edges equals $M = edgefactor \cdot N$.

The benchmark has two computational kernels that are timed and included in the performance information:

1. The first kernel transforms the edge tuples list obtained from the Graph Generator and constructs an undirected graph in a format usable by second kernel.
2. The second kernel performs a breadth-first search (BFS) of the graph from a randomly chosen source vertex.

No subsequent modifications are permitted between kernels computations. After both kernels have finished, there is a validation phase to check if the BFS result is correct. There are 64 iterations of kernel 2 together with validation tests.

3   Related Work

Graph500 benchmark provides several implementations: GNU Octave (may be Matlab[TM] compatible), sequential, OpenMP, Cray XMT and a couple of implementations using MPI. The implementations differ in partitioning of a graph (1D horizontal or 1D vertical) and sparse graph format used in the first kernel (CSR or CSC). CSR (Compressed Sparse Row) is an efficient representation of sparse graphs, where graph is stored in two one-dimensional arrays. The first array holds all nonzero entries (end-points of the edges) of sparse adjacency matrix reading rows in top-to-bottom order. The second array stores offsets which indicate the start vertex of the edges. CSC (Compressed Sparse Column) is similar except that nonzero entries are read by columns in left-to-right order and the second array stores offsets which indicate the end vertex of the edges. All the Graph500 kernel 2 reference MPI implementations are based on a level-synchronized breath-first search. The difference between them is in the way of keeping current queue while doing BFS. In the two MPI implementations: 'replicated-csr' and 'replicated-csc' current queue is replicated across all the nodes, while in simple and one-sided implementations every task hold its own current queue. The types of communication between processors and the partitioning of the graph among processors have impact on performance [14]. In PCJ kernel 1 algorithm we present new approach, in which threads does not need to merge computed outcomes. Unlike in MPI implementation, in our solution after initial computations every thread is able to create its own part of CSR outcome. The PCJ kernel 2 idea is based on 'MPI simple' reference implementation and has been adopted to PGAS Java library and one-sided communication. The BFS which is a crucial part of Graph 500 benchmark and is a key part of many algorithms has been widely studied [15].

The problem of graph processing has been studied in PGAS languages recently [16]. Authors present fast PGAS implementation of graph algorithms for the connected components and minimum spanning tree problems. With additional algorithmic and PGAS specific optimizations, authors achieved significant speedups over both the best sequential implementation and the best single-node SMP implementation for large, sparse graphs with more than a billion edges.

There is also several papers describing graph traversal in PGAS UPC language [17,18], where authors present fast PGAS connected components algorithms and an abstraction of Queues used to communication between tasks using buffering. This implementation outperforms UPC-threads.

The PGAS X10 language has been used for creation of X-Pregel [19] a graph processing system based on Google's Computing Pregel model.

4   PCJ Library

There have been many solutions designed for parallel computations based on Java language. Among them we can mention Java Grande, Titanium, Parallel Java or ProActive. Most of these solutions have performance problems or they are difficult to use because they contain extensions to the Java Language and are based on the translation of Java code to C. Currently there is lack of easy and efficient tools designed for parallel and distributed computations in Java.

PCJ is a library for parallel and distributed computations in Java. It is based on the PGAS (Partitioned Global Address Space) paradigm [10]. In this approach all communication details such as threads or network programming are hidden for the programmer, what allows to develop distributed applications easily. Unlike other implementations, PCJ does not need to use dedicated compiler to preprocess code nor defines new language constructs. It has a form of Java library which can be used without any modifications of the language. Programs developed with PCJ can be run on the distributed systems with different JVM running on the nodes.

5   Graph500 in PCJ

In this section we present the implementation details of the Graph500 benchmark implemented in Java with the PCJ library. The implementation of first and second kernel has been prepared.

5.1   Kernel 1

The aim of the first kernel is to transform the graph given in the form of edge tuples list (the graph is undirected) to more optimized data structure that is usable by second kernel [20]. The Graph500 specification does not impose specific data structure. Since the graph is static and it does not change during computations. In PCJ implementation of Graph500 benchmark we chose to transform the edge list to the CSR format.

The arrangement of vertices and edges of a graph on distributed memory system is realized by 1D partitioning, which can be easily visualized by one-dimensional decomposition of the adjacency matrix of the graph. All vertices and edges of the original graphs are partitioned, so each processor owns N/p vertices and its incident edges (p is a number of processors and N is a number of vertices in a graph). In a case of undirected graphs all edges are stored twice.

In this representation all adjacent vertices of vertex are stored in a continuous portion of memory. Adjacent vertices of vertex v_i are next to adjacent vertices of vertex v_{i+1}. The offsets array stores the start point of each contiguous vertex adjacency block. The data organization in this form is sparse-efficient, because the accumulated storage of the distributed data structure would have the same order as the storage that is needed for exactly the same data but on a single machine. Unlike in MPI code, the data gathering and merging is not necessary, because CSR representation for the original graph is just simple continuous rewrite of outcomes from all threads.

The first step of the algorithm is to find maximum vertex identifier in a graph edge tuples list. Every process looks through different chunk of edge tuples and finds its maximum. Later the reduction is made and global maximum vertex is found. Having maximum vertex, processors compute the range of owned vertices which is based on 1D partitioning. Graph vertices are distributed in the way that every vertex is owned by exactly one task. Every task chooses only those edges that are incident to its owned vertices. Later all tasks create the CSR representation of a part of a graph, taking into consideration only owned vertices and their incident edges. For a specific vertex, at first the number of all its incident edges is computed (offsets) and later all endpoints are put into the proper place of array holding edges.

Algorithm 1. Creation of CSR graph representation from a list of edge tuples.

Input: Undirected graph G(V, E) in the form of list of edge tuples L
Output: CSR representation of the graph G
1: **for** all processors **in parallel do**
2:    localMax ← FIND-MAX(Lp) ▷ Lp is a chunk of L processed by processor p
3: **if** processor 0
4:    globalMax ← REDUCE(localMax)
5:    BROADCAST(globalMax)
6: **for** all processors **in parallel do**
7:    FIND-RANGE(globalMax) ▷ range of vertices the processor owns
8:    COMPUTE-CSR(L) ▷ CSR computation for owned vertices

5.2   Kernel 2

The second kernel performs a breadth-first search traversal of a given graph. The BFS starts from a distinguished vertex taken from the Graph Generator. The source vertex is randomly sampled from the vertices in the graph. To avoid trivial searches, only vertices with degrees at least one not counting self-loops are taken.

PCJ implementation of the second kernel is also based on 1D vertex partitioning of the graph. After the first kernel, original graph is kept in distributed CSR format. Every thread holds its own subset of vertices together with adjacent edges. The implementation is based on level-synchronous BFS strategy. This means that all vertices at a level k form vertex s are visited before vertices at distance $k + 1$ (Fig. 2). The distance between two vertices is the shortest path connecting those vertices. In PCJ implementation when vertices at level k are discovering vertices at level k + 1 a benign race condition is allowed as it does not change the correctness of BFS result. It does not matter which vertex is considered a predecessor in BFS result tree as far as the outcome is correct. The PCJ algorithm uses the same idea as in MPI Simple reference implementation.

The BFS implementation uses two queues. At any time, current frontier queue (CQ) keeps local vertices owned by specific task at the current level. The next frontier queue (NQ) holds local vertices that are within one vertex away from the current level and should be processed at the next level. The predecessor array, which keeps information about parent vertices in BFS result tree is distributed among different tasks. Every PCJ thread maintains the information about vertices he owns. Only the vertex owner can identify if the vertex has already been visited or not. When the adjacent vertex does not belong to the task then the notification needs to be send to its corresponding owner PCJ thread. This leads to all-to-all communication. After visiting all of the vertices at each level, the current frontier and next frontier queues are swapped.

The first step of the algorithm is to find task that owns source vertex. The task marks start vertex as visited and adds it to local CQ. Later, search starts in while loop. For all vertices in CQ all adjacent vertices needs to be visited. There are two different situations relating to visiting adjacent vertices. When a newly visited vertex is owned by the running task it simply becomes visited and is put into the next queue. When the adjacent vertex is owned by a different thread then the owner task needs to be notified that its vertex should be visited. The owner thread checks if the vertex was visited. If no, it marks the vertex as visited and puts it to the next queue, otherwise it is ignored. Vertices can be added to queues multiple times and in different order. To mitigate the overhead connected with sending large number of small communications, message coalescing has been used. Multiple messages about adjacent vertices, destined to the same remote task are grouped together and send in one put. At the end of each level all tasks synchronize to check if there are any new vertices that need to be visited in the next level. If NQ is empty for all tasks the BFS is complete. Otherwise, CQ and NQ are swapped and search continues to the next level.

6   Results

All tests have been carried out on the distributed memory cluster with nodes build of 28 cores (on two sockets) Intel(R) Xeon(R) CPU E5-2697 v3 @ 2.60 GHz with 125 GB RAM. The nodes are connected within InfiniBand. We used 64-bit JVM Oracle Java 8 and OpenMPI version 1.6.5 with gcc compiler.

Sample graphs in the form of edge tuples list and source vertices for BFS implementation (kernel 2) used in performance tests has been generated from Graph Generator of the Graph 500 benchmark. The performance has been tested on graphs of SCALE $= 25$ and SCALE $= 26$ with edgefactor $= 16$. Among others, the performance is compared using TEPS (Traversed Edges per Second) metric.

We show performance of our solution and compare it with native MPI reference implementations. Figure 3 presents overall time and TEPS of Kernel 1 computations for PCJ implementation compared with native reference MPI implementations. The scaling for graphs of SCALE 25 and 26 is similar. Weaker PCJ result in TEPS may be connected with the algorithm because PCJ implementation while constructing CSR representation checks all edges to pick vertices owned by particular node (see Algorithm 1, line 8). In this case the communication is not needed, which keeps the algorithm simple. On the other hand, the MPI implementation uses merge instead, which gives better performance results.

The PCJ implementation tests have been carried out running 4 or 16 processes per node. Figure 4 shows that PCJ implementation results behave similar in case of one node (16 ppn) and 4 nodes (4 ppn). As PCJ algorithm is based

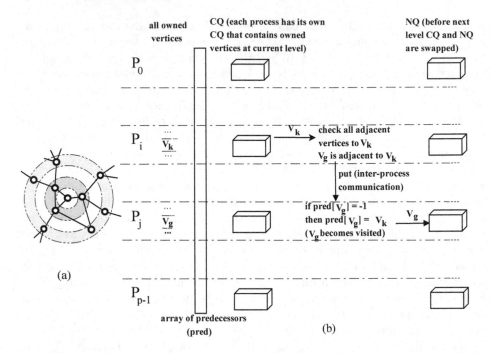

(a)

(b)

Fig. 2. Undirected graph with source vertex as central point together with vertices in first and second level (a). BFS schematic implementation (b).

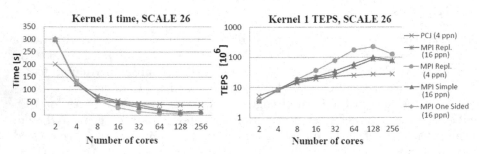

Fig. 3. Kernel 1 overall execution time together with TEPS performance for the SCALE 25 and 26. The PCJ running 4 PCJ threads per node is presented together with the selected MPI implementations running 4 or 16 processes per node (marked as 4 ppn and 16 ppn respectively).

on the same idea as MPI simple code we compare this two implementations. The PCJ implementation is slower and scales up to 4 or 8 threads. Since Kernel 2 parallel processing requires significant amount of communication between threads while exchanging information about visited vertices, this reflects lower performance of the PCJ communication compare to MPI.

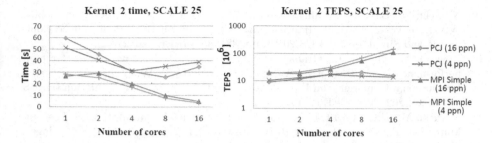

Fig. 4. Kernel 2 overall execution time together with TEPS performance for the SCALE 25. The PCJ running 4 or 16 PCJ threads per node is presented together with the simple MPI implementation running 4 or 16 processes per node (marked as 4 ppn and 16 ppn respectively). As we show performance results up to 16 cores the tests have been carried out using 4 nodes (for 4 ppn) and 1 node (for 16 ppn).

7 Conclusions and Future Work

In this paper we evaluated PCJ library for graph problems implementing Graph 500 benchmark. The PCJ library shows good performance for Kernel 1, however performance for Kernel 2 leaves space for further improvements. The work on the reason for such behavior has to be performed to check the source of the problem which can be located in the algorithm design or in the PCJ communication while using InfiniBand interconnect. The second reason is suggested by the comparison of communication microbenchmarks for MPI and PCJ.

Acknowledgements. This work has been performed using the PL-Grid infrastructure. Partial support from CHIST-ERA consortium is acknowledged.

References

1. Lumsdaine, A., Gregor, D., Hendrickson, B., Berry, J.: Challenges in parallel graph processing. Parallel Process. Lett. **17**(01), 5–20 (2007)
2. Murphy, R.C., Wheeler, K.B., Barrett, B.W., Ang, J.A.: Introducing the Graph 500. Cray Users Group (CUG) (2010)
3. Dean, J., Ghemawat, S.: MapReduce: simplified data processing on large clusters. Commun. ACM **51**(1), 107–113 (2008)
4. Hadoop Home Page. https://hadoop.apache.org/. Accessed 06 November 2015
5. Malewicz, G., Austern, M.H., Bik, A.J.C., Dehnert, J.C., Horn, I., Leiser, N., Czajkowski, G.: Pregel: a system for large-scale graph processing. In: SIGMOD (2011)
6. Low, Y., Gonzalez, J., Kyrola, A., Bickson, D., Guestrin, C., Hellerstein, J.M.: GraphLab: a new framework for parallel machine learning. In: UAI (2010)
7. Shun, J., Blelloch, G.E.: Ligra: a lightweight graph processing framework for shared memory. ACM SIGPLAN Not. **48**(8), 135–146 (2013)
8. Salihoglu, S., Widom, J.: GPS. A graph processing system. In: Proceedings of the 25th International Conference on Scientific and Statistical Database Management, p. 22. ACM (2013)

9. Nowicki, M., Bala, P.: Parallel computations in Java with PCJ library. In: Smari, W.W., Zeljkovic, V. (eds.) International Conference on High Performance Computing and Simulation (HPCS 2012), pp. 381–387. IEEE (2012)

10. Nowicki, M., Gorski, L., Grabarczyk, P., Bala, P.: PCJ - Java library for high performance computing in PGAS model. In: Smari, W.W., Zeljkovic, V. (eds.) International Conference on High Performance Computing and Simulation (HPCS 2014), pp. 202–209. IEEE (2014)

11. Nowicki, M., Bała, P.: PCJ - new approach for parallel computations in java. In: Manninen, P., Öster, P. (eds.) PARA. LNCS, vol. 7782, pp. 115–125. Springer, Heidelberg (2013)

12. Nowicki, M., Gorski, L., Bala, P.: HPC challenge PCJ benchmarks. In: The International Conference for High Performance Computing, Networking, Storage and Analysis, SC 2014, New Orleans (2014)

13. Leskovec, J., Chakrabarti, D., Kleinberg, J., Faloutsos, C., Ghahramani, Z.: Kronecker graphs: an approach to modeling networks. J. Mach. Learn. Res. 11, 985–1042 (2010)

14. Suzumura, T., Ueno, K., Sato, H., Fujisawa, K., Matsuoka, S.: Performance characteristics of Graph500 on large-scale distributed environment. In: IEEE International Symposium on Workload Characterization (IISWC 2011), pp. 149–158. IEEE (2011)

15. Amer, A., Lu, H., Balaji, P., Matsuoka, S.: MPI+ threads applications at scale: a case study with parallel breadth-first search. In: 2nd Workshop on Parallel Programming Model for the Masses. IEEE (2015)

16. Cong, G., Almasi, G., Saraswat, V.: Fast PGAS connected components algorithms. In: Proceedings of the Third Conference on Partitioned Global Address Space Programing Models, p. 13. ACM (2009)

17. Jose, J., Potluri, S., Luo, M., Sur, S., Panda, D.: UPC queues for scalable graph travelsals: design and evaluation on infiniband clusters. In: Conference on PGAS Programming Models (2011)

18. Cong, G., Almasi, G., Saraswat, V.: Fast PGAS implementation of distributed graph algorithms. In: Proceedings of the ACM/IEEE International Conference for High Performance Computing, Networking, Storage and Analysis, pp. 1–11. IEEE Computer Society (2010)

19. Bao, N.T., Suzumura, T.: Toward highly scalable pregel-based graph processing platform with X10. In: Proceedings of the 22nd International Conference on World Wide Web Companion, pp. 501–508. International World Wide Web Conferences Steering Committee (2013)

20. Ryczkowska, M.: Evaluating PCJ library for graph problems - Graph500 in PCJ. In: Smari, W.W., Zeljkovic, V. (eds.) Internationa Conference on High Performance Computing and Simulation (HPCS 2014), pp. 1005–1007. IEEE (2014)

Workshop on Parallel Computational Biology (PBC 2015)

Performance Analysis of a Parallel, Multi-node Pipeline for DNA Sequencing

Dries Decap[1,5(✉)], Joke Reumers[2,5], Charlotte Herzeel[3,5],
Pascal Costanza[4,5], and Jan Fostier[1,5]

[1] Department of Information Technology, Ghent University - iMinds,
Gaston Crommenlaan 8 bus 201, 9050 Ghent, Belgium
`dries.decap@intec.ugent.be`
[2] Janssen Research & Development, A Division of Janssen Pharmaceutica N.V.,
2340 Beerse, Belgium
[3] Imec, Kapeldreef 75, 3001 Leuven, Belgium
[4] Intel Corporation, Brussels, Belgium
[5] ExaScience Life Lab, Kapeldreef 75, 3001 Leuven, Belgium

Abstract. Post-sequencing DNA analysis typically consists of read
mapping followed by variant calling and is very time-consuming, even on
a multi-core machine. Recently, we proposed Halvade, a parallel, multi-
node implementation of a DNA sequencing pipeline according to the
GATK Best Practices recommendations. The MapReduce programming
model is used to distribute the workload among different workers. In this
paper, we study the impact of different hardware configurations on the
performance of Halvade. Benchmarks indicate that especially the lack of
good multithreading capabilities in the existing tools (BWA, SAMtools,
Picard, GATK) cause suboptimal scaling behavior. We demonstrate that
it is possible to circumvent this bottleneck by using multiprocessing on
high-memory machines rather than using multithreading. Using a 15-
node cluster with 360 CPU cores in total, this results in a runtime of
1 h 31 min. Compared to a single-threaded runtime of ~12 days, this
corresponds to an overall parallel efficiency of 53 %.

Keywords: DNA sequencing · MapReduce · Hadoop · Cloudera · Distributed file systems

1 Introduction

Post-sequencing DNA analysis typically consists of the alignment of reads to a
reference genome ('read mapping') followed by the identification of differences
between the reference genome and the aligned reads ('variant calling'). For both
tasks, numerous tools have been described in literature. Recently, the Broad
Institute has proposed the Best Practices recommendations [1] for a DNA variant
calling pipeline based on BWA [2] for read alignment, SAMtools [3]/Picard [4]
for data preprocessing and GATK [5,6] for variant calling. Especially for whole-
genome datasets, this pipeline is very time consuming with a single-core runtime of ~12 days to process the NA12878 dataset (Illumina Platinum genomes,

© Springer International Publishing Switzerland 2016
R. Wyrzykowski et al. (Eds.): PPAM 2015, Part II, LNCS 9574, pp. 233–242, 2016.
DOI: 10.1007/978-3-319-32152-3_22

1.5 billion paired-end reads, 100 bp, 50-fold coverage, human genome). Even when enabling multithreading support in the individual tools, the execution time for this dataset is still ~5 days on a 24-core machine (dual socket Intel Xeon E5-2695 v2 @ 2.40 GHz), indicative of a poor scaling behavior.

To deal with this bottleneck, we recently proposed Halvade [7], a parallel, multi-node framework in which a variant calling pipeline has been implemented according to the GATK Best Practices recommendations. Halvade relies on the MapReduce programming model [8] to run multiple instances of existing tools (BWA, SAMtools/Picard, GATK) in parallel both across and within nodes on subsets of the data. Halvade is based on the simple observation that read mapping is parallel by read (i.e., aligning a certain read does not depend on the alignment of other reads) while variant calling is parallel by genomic region (i.e., variant calling in a certain genomic region does not depend on variant calling in other genomic regions). During the map phase, BWA is used to align reads to a reference genome in parallel, whereas data preprocessing (SAMtools/Picard) and variant calling (GATK) are handled during the reduce phase by operating on different genomic regions in parallel. In between the map and reduce step, the aligned reads are sorted according to genomic position using the MapReduce sorting functionality. For details about the implementation of Halvade and the tools involved we refer to [7].

In [7], it was demonstrated that Halvade strongly reduces the runtime: on a 15-node cluster, each node containing 24 CPU cores and 64 GB of RAM, the NA12878 is processed in 2 h 39 min. Additionally, it was shown that the multi-node parallel efficiency of Halvade is excellent (around 90 %), which means that the runtime is significantly reduced by using 15 nodes compared to using only a single node. However, significant performance loss can still be observed *within* each node. This can be seen from the overall performance: with a runtime of 2 h 39 min using 360 CPU cores (15 nodes × 24 cores/node), a speedup of ~108 is obtained compared to a single-threaded runtime of ~12 days. This corresponds to an overall parallel efficiency of about 30 %, suggesting the presence of certain performance bottlenecks. Understanding the performance of a sequencing pipeline is a non-trivial matter. Certain components in the pipeline are very compute-intensive (e.g. read alignment) whereas other components (e.g. data preprocessing) are mostly data-intensive. Therefore, certain tools might be CPU bound whereas others might be limited by I/O bandwidth. In order to better understand the influence of hardware configuration on the performance of sequencing pipelines, we have set up a range of benchmarks in order to identify possible bottlenecks. Specifically, in this paper, we study the influence on the total runtime of the amount of available RAM, the presence of NUMA domains, the type of network interconnection, the use of solid-state disks versus hard-disk drives and finally, the use of a distributed vs. centralized file system. We demonstrate that the use of high-memory machines and NUMA optimizations can further reduce the overall runtime whereas other hardware aspects have only limited influence. Ultimately, this allows us to process the entire NA12878 dataset in 1 h 31 min, yielding an overall parallel efficiency of 53 %.

Halvade is written in Java using the Hadoop MapReduce 2.0 API. The source is available at http://bioinformatics.intec.ugent.be/halvade under GPL license.

2   Dataset and Tool Versions

In all benchmarks variant calling was performed on a whole-genome DNA sequencing dataset (NA12878, human genome, Illumina Platinum Genomes) or a subset thereof. The full dataset consists of 1.5 billion 100 bp paired-end reads (50-fold coverage) stored in two 43 GB compressed (gzip) FASTQ files.

For these benchmarks, GATK version 3.1.1, BWA version 0.7.12-r1044, BED-Tools version 2.17.0, elPrep version 1.0 [9], SAMtools version 0.1.19 and Picard version 1.112 were used. The dbSNP [10] database and human genome reference found in the GATK hg19 resource bundle [11] were used.

3   Single Node Benchmarks

As the runtime of the complete NA12878 dataset on a single node is impractically high, all benchmarks in this section were performed on a representative subset of 131 million paired-end reads (about 9 % of the total number of reads). Benchmarks in this section were run on a single 24-core node (dual Intel E5-2680v3 @ 2.50 GHz) with 512 GB of RAM.

3.1   Influence of the Number of Tasks per Node

When running Halvade, the number of parallel tasks (mappers/reducers) per node can have a big influence on performance. The number of tasks per node corresponds to the number of instances of the individual tools (BWA, GATK, etc.) that are being run in parallel on a machine. One scenario is to run only a single task and to use the multithreading functionality of the tools to make use of the available cores. An alternative scenario is to run multiple tasks in parallel on the same node, each task then using only a fraction of the available cores. Because of suboptimal multithreading scalability of certain individual tools, the choice in number of tasks can have a big impact on runtime. This is illustrated in Table 1 where the runtime is shown for three scenarios: (i) 1 task using 24 cores for multithreading; (ii) 4 tasks each using 6 cores for multithreading and (iii) 24 tasks without multithreading. The sequential runtime (single core) of the pipeline is ~30.5 h. When allowing the individual tools to run 24 threads on the same machine, the runtime reduces to ~16.5 h, resulting in a very low parallel efficiency of only 7.7 %. This poor scaling can be observed in both map and reduce phase, but is especially pronounced in the reduce phase. It is caused partly by the lack of multithreading support in some of the tools used, e.g. BWA sampe and Picard. However, even the modules of GATK that do support multithreading exhibit poor scaling behavior. When moving from multithreading to multitasking as supported by Halvade, runtimes decrease significantly.

Table 1. Runtime and parallel efficiency as a function of the number of tasks per node.

	Map phase		Reduce phase		Total	
	Runtime	Efficiency	Runtime	Efficiency	Runtime	Efficiency
Single-threaded	14 h 50 min	n/a	15 h 38 min	n/a	30 h 28 min	n/a
1 task × 24 threads	4 h 28 min	13.84 %	12 h 3 min	5.41 %	16 h 31 min	7.69 %
4 tasks × 6 threads	1 h 21 min	45.78 %	3 h 6 min	21.01 %	4 h 27 min	28.53 %
24 tasks × 1 threads	47 min	78.80 %	55 min	71.06 %	1 h 42 min	74.67 %

Using 4 tasks with 6 threads each, runtime reduces to ∼4.5 h. When using 24 tasks without multithreading a runtime of only 1 h 42 min is obtained, corresponding to a parallel efficiency of 74.7 %. We observed an increased CPU utilization during pipeline execution when using 24 parallel tasks compared to using multithreading in 1 task.

On this type of node, optimal runtime is achieved when using a maximum number of tasks without multithreading. However, this is only possible because the node provides a sufficient amount of RAM (512 GB in this case). Tests indicate that certain GATK modules require almost 16 GB of RAM. Therefore, the maximum number of tasks might be limited by the memory that is provided by a node.

3.2 Influence of the Presence of NUMA Domains

Many recent systems make use of non-uniform memory access (NUMA) domains. Each NUMA domain contains a number of CPU cores and part of the RAM. Cores have faster access to memory that resides in the same NUMA domain ('local' access) and slower access to memory that is outside this domain ('remote' access). Files on disk that are accessed by a tool are typically buffered in memory by the Linux operating system. If different processes are accessing the same file, this buffered copy of (part of) the file can be located in a different NUMA domain than that of the core accessing it. If sufficient memory is available we can make distinct copies of the reference file to each of the NUMA domains and as such speed up the file access and seek times. We implemented this idea through the use of wrappers around certain Java calls. In this wrapper the NUMA domain of the assigned cores are determined and a copy is made for that domain on local scratch if it was not yet created. This way each domain has its own local copy which will be cached in the different NUMA domains.

Using 24 tasks on a single node and the entire NA12878 dataset, Fig. 1 shows the runtime of the different components (summed over all 24 tasks) of the pipeline with and without the use of the wrappers. For most components, the influence is only marginal with the ScoreRecalibrator module from GATK being a notable exception. In that particular case, a reduction in runtime of 45 % can be observed when using the wrappers. This is a process where a dbSNP database file (roughly 10 GB) is intensively used to generate recalibration tables. In this case, the improved NUMA data locality considerably improves runtime.

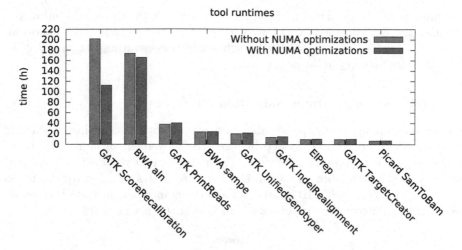

Fig. 1. Comparison of the runtime (summed over all 24 parallel tasks) for each individual tool/module used in Halvade with and without optimized NUMA locality.

4 Multi-node Benchmarks

4.1 Influence of the Use of Solid State Disks

Many tools within Halvade rely on local disk I/O (scratch). This includes reading the reference genome and accessing the dbSNP database as well as writing and reading intermediate data generated by the different GATK modules as well as BWA-aln and BWA-sampe. We tested the performance difference between using solid state drives (SSD) and regular hard disk drives (HDD). Test results indicate only minimal differences in runtime. This is due to the relatively low overall disk usage during the execution of the Halvade job. The disk I/O volume was measured in intervals of one minute and converted to MB/s (see Fig. 2). With the exception of a peak during sorting phase, the disk I/O is well below 100 MB/s (averaged over

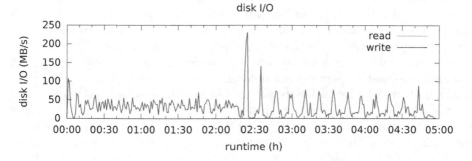

Fig. 2. Disk I/O (scratch) observed on a worker node. Note that almost no data is actually being read from disk as data are still cached in memory.

one minute) which is well within the range of modern HDDs. During the entire job, volumes read from disk were very low, leading us to the conclusion that almost all data written to local disk was cached in memory by the operating system and again accessed from memory in the next step.

4.2 Influence of the Interconnection Network

In between map and reduce phase, aligned reads are sorted according to genomic position. This parallel sorting step involves the movement of large volumes of data over the interconnection network. The network I/O volume was measured in intervals of one minute and again converted to MB/s (see Fig. 3). Again, as network I/O is below 100 MB/s, almost no performance benefit was observed by using an Infiniband interconnect over a 10 Gbit Ethernet network.

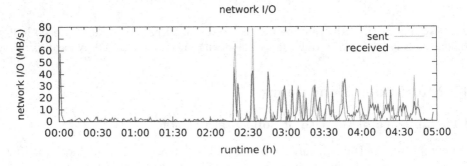

Fig. 3. Network I/O observed on a single worker node.

4.3 Influence of the File System

Traditionally, MapReduce relies on the Hadoop Distributed File System (HDFS) to read input and write final output data. In that case, data is stored on the local disks of the worker nodes in a distributed fashion. Alternatively, centralized file systems such as IBM's Generalized Parallel File System (GPFS) or the Intel Enterprise Edition for Lustre software can be used. In that case, data is stored on separate data nodes and transferred to the worker nodes through an interconnection network. As the pipeline is rather compute-intensive, all three systems were able to provide data to the worker nodes at a sufficiently high rate, hence almost no performance difference was observed. However, the use of Intel's Hadoop Adapter for Lustre included in Intel Enterprise Edition for Lustre software has two advantages. First, it decreases the time spent during the sort & shuffle phase compared with HDFS/GPFS. Second, Lustre uses less memory on the worker nodes. This can be important on nodes with limited memory capacity. For instance, on nodes equipped with 64 GB of RAM running 4 Halvade tasks, we noticed that certain reduce tasks failed because of memory shortage. The cause of this is the difference in coverage over the different genomic regions and thus some tasks will have more reads to process. These reduce tasks had to be rescheduled causing an increase in runtime. On a 7-node cluster, the use of Intel's Hadoop Adapter for Lustre included in Intel Enterprise Edition for Lustre software decreased the runtime from 5 h 27 min (using HDFS) to 4 h 48 min on the same cluster.

5   Benchmark of NA12878 Dataset on a 15-Node Cluster

Halvade was used to process the complete NA12878 dataset on a 15-node cluster, each node containing 24 CPU cores (dual-socket Intel E5-2680v3 @ 2.50 GHz) with 512 GB of RAM and three solid-state drivers (SSD) of 400 GB in RAID 0 to store intermediate data (local scratch). The nodes are interconnected through an Infiniband network and access a GPFS storage through a second Infiniband network. Note that Lustre was not available on this cluster. Cloudera CDH 5.3 is deployed as a Hadoop distribution by HanythingOnDemand [12]. Halvade was configured to use 24 tasks per node, hence up to 360 tasks (24 tasks × 15 nodes) were run in parallel. NUMA optimizations were in place. On this cluster, Halvade completed read alignment and variant calling of the NA12878 dataset in 1 h 31 min. Compared to a single-threaded runtime of ∼12 days, this represents and overall speedup of a factor of ∼190 or a parallel efficiency of 53 %.

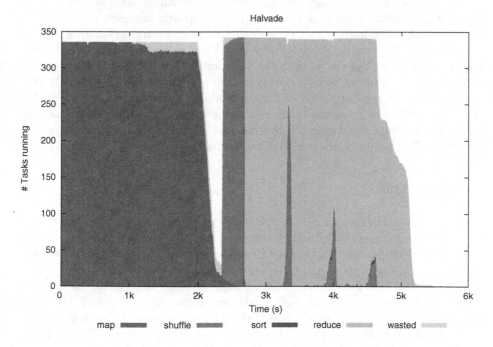

Fig. 4. Execution of the NA12878 dataset on a 15-node cluster. Each node runs 24 tasks in parallel. Note that certain tasks are used by MapReduce for task tracking and scheduling purposes.

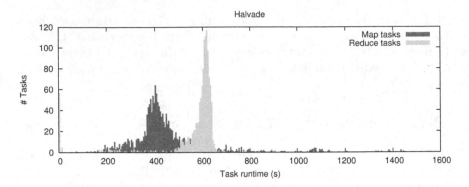

Fig. 5. Distribution of runtime of the different tasks. Each map task (1569 in total) consists of aligning a chunk of ~60 MB of the input FASTQ file to the reference genome. Each reduce task (1303 in total) involves data preprocessing and variant calling in a genomic region of about ~2.3 Mbp.

We can now compare this result to previously reported results in [7]. A runtime of 2 h 39 min was reported on a comparable 15-node cluster, however, in that case the nodes were equipped with only 64 GB of memory. Therefore, it was optimal to run only 4 parallel tasks per node instead of 24 causing significant loss of efficiency within each node. On the other hand, running 360 tasks in parallel significantly increases the task scheduling overhead and makes it more difficult for the MapReduce framework to evenly distribute the workload among the different tasks. This can be clearly observed in Fig. 4 where a non-negligible load imbalance can be observed in both the map and reduce phase. The underlying cause for this is a rather large variation in task execution time (see Fig. 5). Ultimately, with the current status of multithreading performance in the available tools, it is still best to use as many tasks on a node as possible. Note that the newer BWA-mem (also supported by Halvade) already features much improved multithreading performance over BWA-aln/sampe.

6   Conclusion

We investigated the impact of different hardware configurations on the runtime of Halvade, a parallel, multi-node framework that implements a variant calling pipeline according to the GATK Best Practices recommendations. Halvade relies on BWA for read mapping and GATK for variant calling.

Even though Halvade is primarily intended to allow for a multi-node parallelization of sequencing pipelines, Halvade can be used to significantly speed up post-sequencing analysis on a single node. This is because the overall parallel efficiency of the individual tools is very low: a speedup of less than 2 is observed when moving from single-threaded execution to multithreaded execution on a 24-core machine. Part of this poor scaling behavior can be explained by the fact that BWA-sampe and Picard do not support multithreading, however, most of the GATK modules involved in the pipeline also do not exhibit good scaling behavior. This scaling behaviour has also been observed in several other mapping tools in [13]. By using Halvade on high-memory nodes, multithreading can be replaced by multitasking. The latter is far more efficient, which has also been shown in [14], and a speedup of ~18 is obtained on a 24-core machine.

Additionally, having much memory in a system allows to hold a copy of buffered files in each of the NUMA domains. As such, CPU cores have access to a copy in the local NUMA domain, thus avoiding remote memory access. For the GATK ScoreRecalibrator module, this improves the runtime by nearly a factor of two.

Other hardware aspects, such as local disk speed (solid state drives vs. regular hard disk drives), speed of interconnection network (Infiniband vs. Ethernet networks) or file system (HDFS vs. GPFS) have only a minor influence on overall runtime. Even though a typical whole-genome dataset involves hundreds of GB of input data and a multiple thereof of intermediate data, the sequencing pipeline is mostly compute-intensive and hence, runtime is mostly influenced by the compute capacity of a node, rather than I/O speed.

Finally, Intel Enterprise Edition for Lustre software was investigated. The use of Intel's Hadoop Adapter for Lustre included in Intel Enterprise Edition for Lustre software simplifies the shuffle & sort which leads to better performance. Additionally, Lustre uses less memory which can be important when high-memory machines are not available.

With all optimizations in place, Halvade is able to complete read alignment and variant calling of the complete NA12878 dataset in 1 h and 31 min on a 15-node cluster, each node containing 24 CPU cores and 512 GB of RAM. Compared to a single-threaded runtime of ~12 days for this pipeline, this represents an overall speedup of a factor of ~190 or a parallel efficiency of 53 %.

Acknowledgments. This work is funded by Intel, Janssen Pharmaceutica and by the Institute of the Promotion of Innovation through Science and Technology in Flanders (IWT). The computational resources (Stevin Supercomputer Infrastructure) and services used in this work were provided by the VSC (Flemish Supercomputer Center), funded by Ghent University, the Hercules Foundation and the Flemish Government department EWI. Special thanks goes to Stijn De Weirdt for his assistance with the Java wrappers to improve NUMA locality. Benchmarks on Lustre were run at the Intel Big Data Lab, Swindon, UK. We acknowledge the support of Ghent University (Multidisciplinary Research Partnership Bioinformatics: From Nucleotides to Networks).

References

1. Van der Auwera, G.A., Carneiro, M.O., Hartl, C., Poplin, R., del Angel, G., Levy-Moonshine, A., Jordan, T., Shakir, K., Roazen, D., Thibault, J., Banks, E., Garimella, K.V., Altshuler, D., Gabriel, S., DePristo, M.A.: From FastQ data to high-confidence variant calls: the genome analysis toolkit best practices pipeline. Curr. Protoc. Bioinformat. **43**, 11.10.1–11.10.33 (2013)

2. Li, H., Durbin, R.: Fast and accurate short read alignment with Burrows-Wheeler transform. Bioinformatics **25**, 1754–1760 (2009)
3. Li, H., Handsaker, B., Wysoker, A., Fennell, T., Ruan, J., Homer, N., Marth, G., Abecasis, G., Durbin, R.: The sequence alignment/map format and SAMtools. Bioinformatics **25**, 2078–2079 (2009)
4. Picard. http://broadinstitute.github.io/picard/
5. McKenna, A., Hanna, M., Banks, E., Sivachenko, A., Cibulskis, K., Kernytsky, A., Garimella, K., Altshuler, D., Gabriel, S., Daly, M., DePristo, M.A.: The genome analysis toolkit: a MapReduce framework for analyzing next-generation DNA sequencing data. Genome Res. **20**, 1297–1303 (2010)
6. Depristo, M.A., Banks, E., Poplin, R., Garimella, K.V., Maquire, J.R., Hartl, C., Philippakis, A.A., del Angel, G., Rivas, M.A., Hanna, M., McKenna, A., Fennell, T.J., Kernytsky, A.M., Sivachenko, A.Y., Cibulskis, K., Gabriel, S.B., Altshuler, D., Daly, M.J.: A framework for variation discovery and genotyping using next-generation DNA sequencing data. Nat. Genet. **43**, 491–498 (2011)
7. Decap, D., Reumers, J., Herzeel, C., Costanza, P., Fostier, J.: Halvade: scalable sequence analysis with MapReduce. Bioinformatics **31**, 2482–2488 (2015)
8. Dean, J., Ghemawat, S.: MapReduce: simplified data processing on large clusters. Commun. ACM **51**, 107–113 (2008)
9. elPrep. http://github.com/exascience/elprep
10. Sherry, S.T., Ward, M.H., Kholodov, M., Phan, L., Smigielsky, E.M., Sirotkin, K.: dbSNP: the NCBI database of genetic variation. Nucleic Acids Res. **29**(1), 308–311 (2001)
11. GATK resource bundle. ftp://gsapubftp-anonymous@ftp.broadinstitute.org/bundle/2.8/hg19
12. HanythingOnDemand. https://github.com/hpcugent/hanythingondemand
13. Hatem, A., Bozda, D., Toland, A.E., Catalyurek, V.: Benchmarking short sequence mapping tools. BMC Bioinform. **14**, 184 (2013)
14. Kutlu, M., Agrawal, G.: PAGE: a framework for easy parallelization of genomic applications. In: IPDPS (2014)

Parallelising the Computation of Minimal Absent Words

Carl Barton[1], Alice Heliou[2,3], Laurent Mouchard[4], and Solon P. Pissis[5]($\boxtimes$)

[1] The Blizard Institute, Barts and The London School of Medicine and Dentistry,
Queen Mary University of London, London, UK
c.barton@qmul.ac.uk

[2] Inria Saclay-Île de France, AMIB, Bâtiment Alan Turing, Palaiseau, France

[3] Laboratoire d'Informatique de l'École Polytechnique (LIX),
CNRS UMR 7161, Palaiseau, France
alice.heliou@polytechnique.org

[4] University of Rouen, LITIS EA 4108, TIBS, Rouen, France
laurent.mouchard@univ-rouen.fr

[5] Department of Informatics, King's College London, London, UK
solon.pissis@kcl.ac.uk

Abstract. An *absent word* of a word y of length n is a word that does not occur in y. It is a *minimal absent word* if all its proper factors occur in y. Minimal absent words have been computed in genomes of organisms from all domains of life; their computation also provides a fast alternative for measuring approximation in sequence comparison. There exists an $\mathcal{O}(n)$-time and $\mathcal{O}(n)$-space algorithm for computing all minimal absent words on a fixed-sized alphabet based on the construction of suffix array (Barton *et al.*, 2014). An implementation of this algorithm was also provided by the authors and is currently the fastest available. In this article, we present a new $\mathcal{O}(n)$-time and $\mathcal{O}(n)$-space algorithm for computing all minimal absent words; it has the desirable property that, given the indexing data structure at hand, the computation of minimal absent words can be executed in parallel. Experimental results show that a multiprocessing implementation of this algorithm can accelerate the overall computation by more than a factor of two compared to state-of-the-art approaches. By excluding the indexing data structure construction time, we show that the implementation achieves near-optimal speed-ups.

Keywords: Algorithms on strings · Absent words · Suffix array

1 Introduction

Sequence comparison is an important step in many tasks in bioinformatics. It is fundamental in many applications; from phylogenies reconstruction to the reconstruction of genomes. Traditional algorithms for measuring approximation in sequence comparison are based on the notions of distance or of similarity between sequences, which are generally computed through sequence alignment

© Springer International Publishing Switzerland 2016
R. Wyrzykowski et al. (Eds.): PPAM 2015, Part II, LNCS 9574, pp. 243–253, 2016.
DOI: 10.1007/978-3-319-32152-3_23

techniques. An issue with using alignment techniques is that they are computationally expensive, requiring quadratic time in the length of the sequences—a truly sub-quadratic algorithm for this problem seems highly unlikely [1]. This has led to increased research into *alignment free* techniques [10].

Whole-genome alignments prove computationally intensive and have little biological significance. Hence standard notions for sequence comparison are gradually being complemented and in some cases replaced by alternative ones that refer either implicitly or explicitly to the composition of sequences in terms of their constituent patterns. One such notion is based on comparing the words that are absent in each sequence. A word is an *absent word* of some sequence if it does not occur in the sequence. Absent words represent a type of *negative information*: information about what does not occur in the sequence. For instance, considering the words which occur in one sequence but do not in another can be used to detect mutations or other biologically significant events [17].

Given a sequence of length n, the number of absent words of length at most n is exponential in n. However, the number of certain classes of absent words is only linear in n. A *minimal absent word* of a sequence is an absent word whose proper factors all occur in the sequence. Notice that minimal and *shortest absent words* [18] are not the same; minimal absent words are a superset of shortest absent words [15]. An upper bound on the number of minimal absent words is known to be $\mathcal{O}(\sigma n)$ [6,13], where σ is the size of the alphabet. This suggests that it may be possible to compare sequences in time proportional to their lengths, for a fixed-sized alphabet, instead of proportional to the product of their lengths [10].

Recently, there has been a number of studies on the biological significance of absent words in various species. The most comprehensive study on the significance of absent words is probably [2]; in this, the authors suggest that the deficit of certain subsets of absent words in vertebrates may be explained by the hypermutability of the genome. It was later found in [9] that the compositional biases observed in vertebrates in [2] are not uniform throughout different sets of minimal absent words. Moreover, the analyses in [9] support the hypothesis that minimal absent words are inherited through a common ancestor, in addition to lineage-specific inheritance, only in vertebrates. In [8], the minimal absent words in four human genomes were computed, and it was shown that, as expected, intra-species variations in minimal absent words were lower than inter-species variations. Very recently, in [17], it was shown that there exist three minimal words in the *Ebola* virus genomes which are absent from human genome. The authors suggest that the identification of such species-specific sequences may prove to be useful for the development of both diagnosis and therapeutics.

From an algorithmic perspective, an $\mathcal{O}(n)$-time and $\mathcal{O}(n)$-space algorithm for computing all minimal absent words on a fixed-sized alphabet based on the construction of suffix automata was presented in [6]. An alternative $\mathcal{O}(n)$-time solution for finding minimal absent words of length at most ℓ, such that $\ell = \mathcal{O}(1)$, based on the construction of tries of bounded-length factors was presented in [5]. A drawback of these approaches, in practical terms, is that the construction of suffix automata (or of tries) often have a large memory footprint. Hence, an important problem was to be able to compute minimal absent words with more memory-efficient data structures (cf. [4]).

The computation of minimal absent words based on the construction of suffix arrays was considered in [15]; although this algorithm has a linear-time performance in practice, the worst-case time complexity is $\mathcal{O}(n^2)$. The first $\mathcal{O}(n)$-time and $\mathcal{O}(n)$-space suffix-array-based algorithm was recently presented in [3] to bridge this unpleasant gap. An implementation of this algorithm is currently, and to the best of our knowledge, the fastest available for the computation of minimal absent words. With the continuous efforts in whole-genome sequencing, the computation of minimal absent words remains the main bottleneck in analysing a large set of large genomes [8,9,17]. Hence due to the large amounts of data being produced, it is desirable to further engineer this computation.

Our Contribution. In this article, our contribution is threefold: (a) We present a new $\mathcal{O}(n)$-time and $\mathcal{O}(n)$-space algorithm for computing all minimal absent words on a fixed-sized alphabet; (b) We show that this algorithm has the desirable property that, given the relevant indexing data structure at hand, the computation of minimal absent words can be executed in parallel; and (c) We make available an implementation of this algorithm for shared-memory multiprocessing programming. Experimental results, using real and synthetic data, show that the *overall* computation is accelerated by more than a factor of two compared to the state of the art. By excluding the indexing data structure construction time, we show that the implementation achieves *near-optimal* speed-ups. This is important as engineering further the involved indexing data structure construction is an ongoing research topic [16], which is beyond the scope of this article.

2   Definitions and Notation

To provide an overview of our result and algorithm, we begin with a few definitions from [3]. Let $y = y[0]y[1] .. y[n-1]$ be a *word* of *length* $n = |y|$ over a finite ordered *alphabet* Σ of size $\sigma = |\Sigma| = \mathcal{O}(1)$. We denote by $y[i .. j] = y[i] .. y[j]$ the *factor* of y that starts at position i and ends at position j and by ε the *empty word*, word of length 0. We recall that a prefix of y is a factor that starts at position 0 ($y[0 .. j]$) and a suffix is a factor that ends at position $n-1$ ($y[i .. n-1]$), and that a factor of y is a *proper* factor if it is not the empty word or y itself.

Let x be a word of length $0 < m \leq n$. We say that there exists an *occurrence* of x in y, or, more simply, that x *occurs in* y, when x is a factor of y. Every occurrence of x can be characterised by a starting position in y. Thus we say that x occurs at the *starting position* i in y when $x = y[i .. i+m-1]$. Oppositely, we say that the word x is an *absent word* of y if it does not occur in y. The absent word x, $m \geq 2$, of y is *minimal* if and only if all its proper factors occur in y.

We denote by SA the *suffix array* of y, that is the array of length n of the starting positions of all sorted suffixes of y, i.e. for all $1 \leq r < n$, we have $y[\mathsf{SA}[r-1] .. n-1] < y[\mathsf{SA}[r] .. n-1]$ [12]. Let $\mathsf{lcp}(r, s)$ denote the length of the longest common prefix of the words $y[\mathsf{SA}[r]\mathinner{.\,.}n-1]$ and $y[\mathsf{SA}[s]\mathinner{.\,.}n-1]$, for all $0 \leq r,s < n$, and 0 otherwise. We denote by LCP the *longest common prefix* array of y defined by $\mathsf{LCP}[r] = \mathsf{lcp}(r-1,r)$, for all $1 \leq r < n$, and $\mathsf{LCP}[0] = 0$. SA [14] and LCP [7] of y can be computed in time and space $\mathcal{O}(n)$.

In this article, we consider the following problem.

MINIMALABSENTWORDS
Input: a word y on Σ of length n
Output: all tuples $< a, (i,j) >$, such that word x, defined by $x[0] = a, a \in \Sigma$, and $x[1\mathinner{.\,.}m-1] = y[i\mathinner{.\,.}j]$, $m \geq 2$, is a minimal absent word of y

---

3   Algorithm pMAW

In this section, we present algorithm pMAW, a new $\mathcal{O}(n)$-time and $\mathcal{O}(n)$-space algorithm for computing all minimal absent words of a word of length n using arrays SA and LCP. We first start by explaining some useful properties from [15] we use in algorithm pMAW. Then we present our algorithm in detail, and, finally, we show how it can be adapted for parallel computing.

3.1   Useful Properties

A minimal absent word $x[0\mathinner{.\,.}m-1]$ of a word $y[0\mathinner{.\,.}n-1]$ is an absent word whose proper factors all occur in y; *equivalently*, both the longest proper suffix and prefix of x occur in y.

Definition 1. *A* repeated pair *in a word $\boldsymbol{y}$ is a tuple $< \boldsymbol{i,j,w} >$ such that word $\boldsymbol{w}$ occurs in $\boldsymbol{y}$ at starting positions $\boldsymbol{i}$ and $\boldsymbol{j}$. A repeated pair is* right *(resp.* left*) maximal, if $\boldsymbol{y}[\boldsymbol{i}+|\boldsymbol{w}|] \neq \boldsymbol{y}[\boldsymbol{j}+|\boldsymbol{w}|]$ (resp. $\boldsymbol{y}[\boldsymbol{i}-1] \neq \boldsymbol{y}[\boldsymbol{j}-1]$) A repeated pair is maximal if it is left maximal and right maximal.*

Lemma 1 ([15]). *If $\boldsymbol{awb}$ is a minimal absent word of a word $\boldsymbol{y}$, where $\boldsymbol{a}$ and $\boldsymbol{b}$ are letters and $\boldsymbol{w}$ a word, then there exist two positions $\boldsymbol{i}$ and $\boldsymbol{j}$ such that $< \boldsymbol{i,j,w} >$ is a maximal repeated pair in $\boldsymbol{y}$.*

By Lemma 1, we can exhaustively compute minimal absent words by examining all the maximal repeated pairs. To compute maximal repeated pairs, we consider all right maximal repeated pairs and check the letters that occur just before.

Definition 2. *Given the LCP array of a word of length $\boldsymbol{n}$, we say that interval $[\boldsymbol{i,j}]$, $0 \leq \boldsymbol{i} < \boldsymbol{j} \leq \boldsymbol{n}-1$, is an* LCP-interval *of* LCP-depth $\boldsymbol{d}$ *if*

- $\mathsf{LCP}[\boldsymbol{i}] < \boldsymbol{d}$, *and* $\boldsymbol{j} = \boldsymbol{n}-1$ *or* $\mathsf{LCP}[\boldsymbol{j}+1] < \boldsymbol{d}$
- $\mathsf{LCP}[\boldsymbol{k}] \geq \boldsymbol{d}$, *for all* $\boldsymbol{i} < \boldsymbol{k} \leq \boldsymbol{j}$
- $\mathsf{LCP}[\boldsymbol{k}] = \boldsymbol{d}$, *for at least one* $\boldsymbol{k}, \boldsymbol{i} < \boldsymbol{k} \leq \boldsymbol{j}$.

Right maximal repeated pairs are given by the suffix array with the notion of LCP-interval. Indeed if positions i and j are in an LCP-interval of depth d then $< i,j,y[\mathsf{SA}[i]\mathinner{.\,.}\mathsf{SA}[i]+d-1] >$ is a right maximal repeated pair. Analogously, if $< i,j,w >$ is a right maximal repeated pair then i and j are in the same LCP-interval of depth $|w|$.

3.2   Computation of Minimal Absent Words

For the rest of this section we denote minimal absent words by maws. We first pre-compute SA, LCP, and a bit-vector v such that $v[i] = 1$ if and only if LCP$[i]$ is a local maximum. We use rank and select data structures and denote by MaxRank(k) the operation giving the number of 1's in $[0 : k)$ and by MaxSelect(k) the operation giving the position of the kth 1. The following function presents maws computation for a given interval $[k_1, k_2)$ of SA and LCP.

Function *ComputeMaws* (k_1, k_2, y, SA, LCP, MaxRank, MaxSelect)

   SetLetter$\leftarrow\emptyset$; LifoPos.push(0); LifoSet.push(SetLetter);

   foreach $t \in [\text{MaxRank}(k_1) + 1 : \text{MaxRank}(k_2)]$ **do**

     $i\leftarrow$MaxSelect(t); *left*$\leftarrow i - 1$; *right*$\leftarrow i + 1$;

     pos$\leftarrow$LifoPos.top(); *lpos*$\leftarrow$LCP[*pos*]; SetLetter$\leftarrow\emptyset$;

     while *1* **do**

        while *pos* > 0 and LCP$[i] <$ *lpos* **do**

           we pop from LifoPos the positions with an LCP value equal to *lpos*; we pop their set of letters from LifoSet; we have visited the whole LCP-interval of depth *lpos*, so we infer maws using these sets and SetLetter; we update *left* and *right*; *pos*$\leftarrow$LifoPos.top(); *lpos*$\leftarrow$LCP[*pos*];

        if LCP$[i] > \max($LCP$[\textit{left}], LCP[\textit{right}], \textit{lpos})$ **then**

           we have visited the whole LCP-interval of depth LCP$[i]$, so we infer maws with SetLetter, $y[\text{SA}[i]-1]$, and $y[\text{SA}[\textit{left}]-1]$;

        SetLetter$\leftarrow$SetLetter $\cup \{y[\text{SA}[i] - 1]\}$;

        if LCP$[\textit{left}] = LCP[i]$ *or* LCP$[\textit{right}] = LCP[i]$ **then**

           LifoPos.push(i); LifoSet.push(SetLetter);

           we push onto LifoPos all the successive neighbours of interval (*left*,*right*) with an LCP value equal to LCP$[i]$; for each of them we push onto LifoSet the letter preceding their corresponding suffix; we update *left* and *right*;

        if LCP$[\textit{right}] \leq LCP[\textit{left}] < LCP[i]$ **then** $i\leftarrow$*left*; *left*$\leftarrow i - 1$;

        else if LCP$[\textit{right}] > LCP[i]$ **then** we push onto stacks the positions skipped and their corresponding set of letters;

        break;

       else $i\leftarrow$*right*; *right*$\leftarrow i + 1$

If i is a local maximum in the LCP array, then $[i - 1, i]$ is the LCP-interval of LCP-depth LCP$[i]$ that contains i. Consequently our idea is to start the computation at the first local maximum of the LCP array and to visit the surrounding positions in decreasing order of their LCP value. In this process we keep in the array SetLetter the set of letters that occur before the repeated factor. When we

j	LCP	$y[SA[j]-1]$	suffixes $\overset{8}{\underset{j}{\longrightarrow}}$LCP
$k-1$	11	T	w A T T T $\cdots$
k	8	A	w C A A G $\cdots$
$k+1$	9	G	w C C A A $\cdots$
$k+2$	9	A	w C G C T $\cdots$
$k+3$	10	A	w C G T A $\cdots$
$k+4$	11	A	w C G T T $\cdots$
$k+5$	9	T	w C T A C $\cdots$
$k+6$	10	A	w C T G C $\cdots$
$k+7$	8	T	w G C G G $\cdots$

step	i	left	right	SetLetter	Inferred maws and action on stacks
1	$k+4$	$k+3$	$k+5$	$\emptyset$	
2	$k+3$	$k+2$	$k+5$	$\{$A$\}$	
3	$k+2$	$k+1$	$k+5$	$\{$A$\}$	we push $k+2$, $k+1$, and $k+5$ onto LifoPos; we push SetLetter,
	k		$k+6$		$\{$G$\}$, and $\{$T$\}$ onto LifoSet
4	$k+6$	$k+5$	$k+7$	$\emptyset$	we infer 2 maws: AwCTA, TwCTG
5	$k+5$	$k+4$	$k+7$	$\{$A$\}$	$k+5$ is already in LifoPos
6	$k+7$	$k+4$	$k+8$	$\{$A,T$\}$	we pop $k+5$, $k+1$, and $k+2$ from LifoPos and $\{$T$\}$, $\{$G$\}$, $\{$A$\}$ from LifoSet
	k			$\{$A,G,T$\}$	we infer 7 maws: GwCA, TwCA, AwCC, TwCC, GwCG, TwCG, GwCT

Fig. 1. Illustration of the algorithm step by step for the interval $[k, k+7]$. The example is taken from the *Lactobacillus casei* genome (Accession #: NC010999). $w = $ TCTGAGCG is a common prefix of the considered suffixes and $k = 2,554,910$.

reach a local minimum we store its position on the SA array in the stack LifoPos, and the current array SetLetter in the stack LifoSet. We will analyse them once we have visited their whole LCP-interval. In this way, we consider each maximal repeated pair and infer from them the whole set of maws using Lemma 1. An example of this function is illustrated in Fig. 1. Contrary to MAW [3], the previous linear-time algorithm, in pMAW we do not consider our data structures globally; we rather consider each LCP-interval *independently*. This important property will allow us to use parallel computations, as shown in Sect. 3.3.

Overall Complexity. We use arrays SA and LCP, which can be computed in time and space $\mathcal{O}(n)$ [7,14]. There also exists a representation which uses $n+o(n)$ bits of storage space and supports rank and select on a bit-vector of size n in constant time [11]. We also use two stacks, LifoPos and LifoSet, where we push and pop $\mathcal{O}(n)$ elements, each containing at most σ integers. Thus the whole algorithm requires time and space $\mathcal{O}(\sigma n)$. We obtain the following result.

Theorem 1. *Algorithm pMAW solves problem* MINIMALABSENTWORDS *in time and space $\mathcal{O}(n)$.*

The *advantages* of pMAW over existing works are as follows. It is (provably) linear-time in the worst case as opposed to the one in [15]. Contrary to the linear-time algorithm in [3], we explicitly compute the LCP-intervals. For a given depth, LCP-intervals have no overlap, therefore we can consider them independently.

3.3 Parallelisation Scheme

Lemma 2. *Let y be a word of length n over an alphabet of size σ and let ℓ be the length of the shortest minimal absent word of y. Then the following hold:*

- *For all $k \in [0, \ell-2]$, $|\{s \in [0, n-1] : \mathsf{LCP}[s] = k\}| = (\sigma-1)\sigma^k + 1$;*
- *For all $k \in [\ell-1, n-1]$, $|\{s \in [0, n-1] : \mathsf{LCP}[s] = k\}| < (\sigma-1)\sigma^k + 1$.*

Proof. Let $k \in [0, n-1]$, we denote by $s_0, \ldots, s_{m-1}$, ordered increasingly, the m elements of the set $\{s \in [0, n-1] : \mathsf{LCP}[s] = k\}$. For all $i \in [0, m-1]$, we have

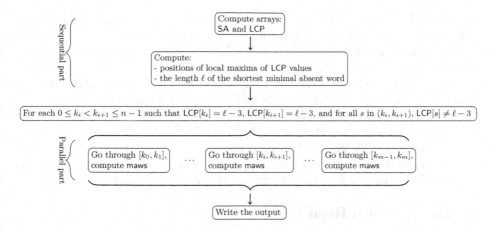

Fig. 2. Overview of Algorithm pMAW

$y[\mathsf{SA}[s_i - 1] .. \mathsf{SA}[s_i - 1] + k - 1] = y[\mathsf{SA}[s_i] .. \mathsf{SA}[s_i] + k - 1]$ and $y[\mathsf{SA}[s_i - 1] + k] < y[\mathsf{SA}[s_i] + k]$. We consider the pair (s_i, s_{i+1}) with $i \in [0, m - 2]$, there are two cases:

- $\mathsf{lcp}(s_i, s_{i+1}) = k$, so $y[\mathsf{SA}[s_i] .. \mathsf{SA}[s_i] + k - 1] = y[\mathsf{SA}[s_{i+1}] .. \mathsf{SA}[s_{i+1}] + k - 1]$ and $y[\mathsf{SA}[s_i - 1] + k] < y[\mathsf{SA}[s_i] + k] \leq y[\mathsf{SA}[s_{i+1} - 1] + k] < y[\mathsf{SA}[s_{i+1}] + k]$. The alphabet is of size σ; this can happen at most $\sigma - 2$ times consecutively.
- $\mathsf{lcp}(s_i, s_{i+1}) < k$, so $y[\mathsf{SA}[s_i] .. \mathsf{SA}[s_i] + k - 1] < y[\mathsf{SA}[s_{i+1}] .. \mathsf{SA}[s_{i+1}] + k - 1]$. There are σ^k different words of length k; this can happen at most $\sigma^k - 1$ times.

In the first case, we have an additional sub-case, when $\mathsf{SA}[s_i - 1] + k = n$. Then $y[\mathsf{SA}[s_i - 1] + k]$ is not a letter of the alphabet Σ, so we have one more position with an LCP value equal to k. Thus, there are at most $(\sigma - 1)\sigma^k$ pairs (s_i, s_{i+1}), so there are at most $(\sigma - 1)\sigma^k + 1$ positions with an LCP value equal to k.

The equality holds if and only if all the words of length $k + 1$ appear in y, so only if $k < \ell' - 1$ where ℓ' is the length of the shortest absent word. A minimal absent word is an absent word so $\ell \geq \ell'$. Let x be a shortest absent word, then all its proper factors occur in y because they are smaller than x, so x is a minimal absent word. Therefore $\ell = \ell'$, the equality holds if and only if $k \in [0, \ell - 2]$. □

By Lemma 2, the length ℓ of the shortest minimal absent word of some word of length n satisfies: $\ell - 1 = \min\{k \geq 0 : |\{s \in [0, n - 1] : \mathsf{LCP}[s] = k\}| < (\sigma - 1)\sigma^k + 1\}$. As the alphabet is of size σ, there are σ^k distinct words of length k, but a word y of length n has exactly $n + 1 - k$ factors of length k. Thus, if $\sigma^k > n + 1 - k$ there are absent words of size k in y. Consequently we have $\ell \leq \log_\sigma(n + 1 - \ell) < \log_\sigma(n)$. Thus, we compute ℓ, the length of the shortest minimal absent word, in one pass over the LCP array by counting the number of positions having an LCP value equal to d, for all $d \in [0, \lfloor \log_\sigma(n) \rfloor]$.

According to Lemma 1 we can ignore positions having an LCP value lower than $\ell - 2$ when computing minimal absent words. Hence, we focus on LCP-intervals of LCP-depth above or equal to $\ell - 2$: they are sufficient to exhaustively compute the set of minimal absent words. Consequently we compute the set of positions k_i with i in $[0, (\sigma - 1)\sigma^{\ell-3}]$ such that $\mathsf{LCP}[k_i] = \ell - 3$. $[0, k_0), [k_0, k_1), \ldots, [k_{m-1}, k_m), [k_m, n)$, with $m = (\sigma - 1)\sigma^{\ell-3}$, is a partition of $[0, n - 1]$. This partition is such that, every LCP-interval of LCP-depth above or equal to $\ell - 2$ is entirely included in one of the sub-intervals $[k_i, k_{i+1})$.

Therefore we can consider each one of these sub-intervals *independently*, and thus parallelise the computation of minimal absent words. In each sub-interval we go through the SA and LCP arrays starting at the first (from left to right) local maximum and going down until we reach a local minimum, as described in Sect. 3.2. For an overview of the algorithm pMAW inspect Fig. 2.

4   Experimental Results

We implemented algorithm pMAW as a programme to compute all minimal absent words of a given sequence. The programme was implemented in the C programming language, using Open Multi-Processing (OpenMP) API for shared-memory multiprocessing programming, and developed under GNU/Linux operating system. It takes as input arguments a file in (Multi) FASTA format and the minimal and maximal length of minimal absent words to be outputted; and then produces a file with all minimal absent words of length within this range as output. There are additional input parameters; for example, the number t of available processing elements. The implementation is distributed under the GNU General Public License (GPL), and it is available at http://github.com/solonas13/maw, which is set up for maintaining the source code and the man-page documentation. The experiments were conducted on a Desktop PC using 1 to 16 cores of 2 Intel Xeon E5-2670V2 Ten-Core CPUs at 2.50 GHz and 256 GB of main memory under 64-bit GNU/Linux.

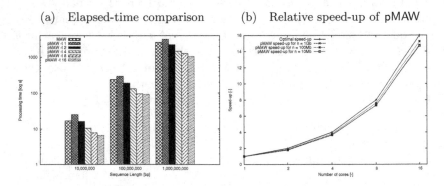

(a)   Elapsed-time comparison   (b)   Relative speed-up of pMAW

Fig. 3. Elapsed-time comparison of pMAW and MAW and relative speed-up of pMAW for computing minimal absent words using synthetic DNA sequences

To evaluate the efficiency of our implementation, we compared it against the corresponding performance of MAW [3], which is currently the fastest available implementation for computing minimal absent words. We generated three random sequences of length 10 Mbp, 100 Mbp, and 1 Gbp, respectively, by using a uniform frequency distribution of letters of the DNA alphabet. We computed all minimal absent words of length at most 20 for each sequence. We considered both the $5' \rightarrow 3'$ and the $3' \rightarrow 5'$ DNA strands. Figure 3a depicts elapsed-time comparisons of pMAW and MAW, *including* the sequential part of the algorithm. pMAW becomes the fastest in *all* cases when $t \geq 2$ accelerating the computation by more than a factor of two when $t = 16$. Notice that the y-axis is on logarithmic scale. The measured relative speed-up of pMAW is illustrated in Fig. 3b. The relative speed-up was calculated as the ratio of the runtime of pMAW on 1 core to the runtime of pMAW on t cores, excluding the sequential part of the algorithm. The results highlight the *excellent* scalability of pMAW when the letters have a uniform frequency distribution in the sequence. In this case, pMAW achieves near-optimal speed-ups, confirming our theoretical findings.

To further evaluate the efficiency of our implementation, we compared it against the corresponding performance of MAW using real data. We considered the genomes of *Homo sapiens* and *Mus musculus*, obtained from the NCBI database (ftp://ftp.ncbi.nih.gov/genomes/). We computed all minimal absent words of length at most 20 of the complete sequence of the *Homo sapiens* $(2,937,639,113\,\text{bp})$ and *Mus musculus* $(2,647,521,431\,\text{bp})$ genomes—ignoring unknown bases. We considered both the $5' \rightarrow 3'$ and the $3' \rightarrow 5'$ DNA strands. Figure 4a depicts elapsed-time comparisons of pMAW and MAW, *including* the sequential part of the algorithm. pMAW becomes the fastest in *all* cases when $t \geq 2$ accelerating the computation by more than a factor of two when $t = 16$. Notice that the y-axis is on logarithmic scale. The measured relative speed-up of pMAW is illustrated in Fig. 4b. The relative speed-up was calculated as the ratio of the runtime of pMAW on 1 core to the runtime of pMAW on t cores, excluding the sequential part of the algorithm. The results highlight the *good* scalability of pMAW with real data. The computation is accelerated by a factor of 10 when $t = 16$. The maximum allocated memory was 137 GB for both programmes.

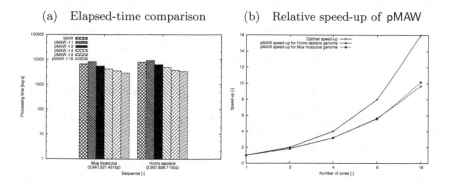

(a)   Elapsed-time comparison   (b)   Relative speed-up of pMAW

Fig. 4. Elapsed-time comparison of pMAW and MAW and relative speed-up of pMAW for computing minimal absent words using real DNA sequences

5 Final Remarks

The importance of our contribution here is underlined by the fact that any parallel algorithms for the construction of the involved indexing data structure can be used *directly* to replace the sequential part of the algorithm proposed here (see Fig. 2). This would result in a *fully* parallel algorithm for the computation of minimal absent words. Our immediate target is to investigate the performance of such an algorithm by using the parallel algorithms presented in [16] for constructing the suffix array and the longest common prefix array.

References

1. Abboud, A., Williams, V.V., Weimann, O.: Consequences of faster alignment of sequences. In: Esparza, J., Fraigniaud, P., Husfeldt, T., Koutsoupias, E. (eds.) ICALP 2014. LNCS, vol. 8572, pp. 39–51. Springer, Heidelberg (2014)
2. Acquisti, C., Poste, G., Curtiss, D., Kumar, S.: Nullomers: really a matter of natural selection? PLoS One **2**(10), e1022 (2007)
3. Barton, C., Heliou, A., Mouchard, L., Pissis, S.P.: Linear-time computation of minimal absent words using suffix array. BMC Bioinform. **15**, 388 (2014)
4. Belazzougui, D., Cunial, F., Kärkkäinen, J., Mäkinen, V.: Versatile succinct representations of the bidirectional Burrows-Wheeler transform. In: Bodlaender, H.L., Italiano, G.F. (eds.) ESA 2013. LNCS, vol. 8125, pp. 133–144. Springer, Heidelberg (2013)
5. Chairungsee, S., Crochemore, M.: Using minimal absent words to build phylogeny. Theoret. Comput. Sci. **450**, 109–116 (2012)
6. Crochemore, M., Mignosi, F., Restivo, A.: Automata and forbidden words. Inf. Process. Lett. **67**, 111–117 (1998)
7. Fischer, J.: Inducing the LCP-array. In: Dehne, F., Iacono, J., Sack, J.-R. (eds.) WADS 2011. LNCS, vol. 6844, pp. 374–385. Springer, Heidelberg (2011)
8. Garcia, S.P., Pinho, A.J.: Minimal absent words in four human genome assemblies. PLoS One **6**(12), e29344 (2011)
9. Garcia, S.P., Pinho, O.J., Rodrigues, J., Bastos, C.A.C., Ferreira, G.P.J.S.: Minimal absent words in prokaryotic and eukaryotic genomes. PLoS One **6**, e16065 (2011)
10. Haubold, B., Pierstorff, N., Möller, F., Wiehe, T.: Genome comparison without alignment using shortest unique substrings. BMC Bioinform. **6**, 123 (2005)
11. Jacobson, G.: Space-efficient static trees and graphs. In: 30th SFCS 1989, pp. 549–554. IEEE Computer Society (1989)
12. Manber, U., Myers, E.W.: Suffix arrays: a new method for on-line string searches. SIAM J. Comput. **22**(5), 935–948 (1993)
13. Mignosi, F., Restivo, A., Sciortino, M.: Words and forbidden factors. Theoret. Comput. Sci. **273**(1–2), 99–117 (2002)
14. Nong, G., Zhang, S., Chan, W.H.: Linear suffix array construction by almost pure induced-sorting. In: DCC 2009, pp. 193–202. IEEE Computer Society (2009)
15. Pinho, A.J., Ferreira, P.J.S.G., Garcia, S.P., Rodrigues, J.M.: On finding minimal absent words. BMC Bioinformatics **10** (2009)
16. Shun, J.: Fast parallel computation of longest common prefixes. In: SC 2014, pp. 387–398. IEEE Computer Society (2014)

17. Silva, R.M., Pratas, D., Castro, L., Pinho, A.J., Ferreira, P.J.S.G.: Three minimal sequences found in Ebola virus genomes and absent from human DNA. Bioinformatics **31**(15), 2421–2425 (2015)
18. Wu, Z.D., Jiang, T., Su, W.J.: Efficient computation of shortest absent words in a genomic sequence. Inf. Process. Lett. **110**(14–15), 596–601 (2010)

Accelerating 3D Protein Structure Similarity Searching on Microsoft Azure Cloud with Local Replicas of Macromolecular Data

Dariusz Mrozek[✉], Tomasz Kutyła, and Bożena Małysiak-Mrozek

Institute of Informatics, Akademicka 16, 44-100 Gliwice, Poland
dariusz.mrozek@polsl.pl

Abstract. Searching similarities among 3D protein structures deposited in macromolecular data repositories, like Protein Data Bank, is one of the time-consuming processes performed in structural bioinformatics. When performed in one-to-many or many-to-many model, the process requires increased computational resources. Moreover, exponential growth of protein structures in the Protein Data Bank causes the necessity to prepare computer systems to be able to deal with such huge volumes of data. Cloud computing provides both, theoretically infinite computational resources and a great possibility of scaling systems out and up. In this paper, we show how 3D protein structure similarity searching can be scaled out on Microsoft Azure cloud and performed by a loosely coupled, many-task computing system with local replicas of macromolecular data.

Keywords: Bioinformatics · Proteins · 3D protein structures · Similarity searching · Alignment · Superposition · Cloud computing · Parallel computing · Parallel systems · Scalability · Microsoft Azure

1 Introduction

3D protein structures (Fig. 1) allow to understand the functions of proteins at a molecular level. Since protein structures exhibit high conservation in the evolution of organisms [3], they have a unique feature of becoming indicators of cellular functions, which can be used while identifying functions of newly discovered proteins. Even if protein sequences diverged significantly during the evolution, comparison of 3D protein structures and finding structural similarities and common substructures allow to draw conclusions on functional similarity of proteins from various, sometimes evolutionary distant organisms. This emphasizes the great importance of 3D protein structure similarity searching for scientific domains, such as structural bioinformatics, systems biology, and indirectly, molecular modeling [14].

There have been a number of methods for protein structure similarity searching developed in the last decades, including VAST [2], DALI [5], LOCK2 [19],

---

This project was supported by Microsoft Research in USA within Microsoft Azure for Research Award granted for the Cloud4Psi project.

R. Wyrzykowski et al. (Eds.): PPAM 2015, Part II, LNCS 9574, pp. 254–265, 2016.
DOI: 10.1007/978-3-319-32152-3_24

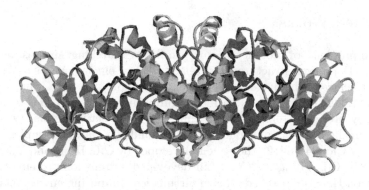

Fig. 1. Crystal structure of *Titin kinase*, a part of the muscle protein titin, which is essential in the temporal and spatial control of the assembly of the highly ordered sarcomeres (contractile units) of striated muscle. The molecule has two chains, each of which is more than three hundred amino acids (residues) long and, therefore, several thousand atoms. PDB entry 1TKI [7] displayed as cartoons.

FATCAT [22], CE [20], FAST [24], TM-align [23], MICAN [9], CASSERT [11, 13], and others. However, despite of advances made in this area in recent years, protein structure similarity searching is still a time-consuming and computationally intensive problem, especially when performed in one-to-many or many-to-many model. This is caused by several factors, including (1) protein structure complexity (hundreds of amino acids, and therefore, thousands of atoms), (2) computational complexity of algorithms for 3D structure similarity searching, (3) the exponential growth of the number of 3D structures in macromolecular data repositories, such as the Protein Data Bank (PDB) [1] - as of Oct 15, 2015 there were 112,968 structures in the PDB.

These factors cause that the scientific community develop new methods for 3D structure similarity searching and look for computing platforms that would allow to scale computations, complete similarity searches much faster, and handle the growing amount of macromolecular data. Cloud computing, which is a computing model that allows a convenient, on-demand network access to a shared pool of configurable computing resources [8], provides such a kind of scalable, high-performance computational platforms. Cloud platforms, such as Microsoft Azure, Amazon EC2, or Google App Engine, can be particularly beneficial for institutions that need to quickly gain access to a computer system which has a higher than average computing power.

The idea of Cloud computing that provides access to configurable computing resources (such as networks, servers, storage, applications, and others) as a service also became very interesting for scientific community. Scientists can now provision computing resources from cloud providers without having to build entire computing infrastructure within their own institutions. This paper presents an example of implementation of frequently performed process on the Cloud. In this paper, we show how Microsoft Azure public cloud can be utilized to scale out protein 3D structure similarity searching on many compute nodes.

2   Related Works

Mentioned problems of protein structure similarity searching, such as high complexity of 3D protein structures and relatively low efficiency of existing algorithms in the face of the dynamic growth of macromolecular data, caused several trials to parallelize the process on many computing nodes in order to increase its efficiency. Recent attempts to speed up protein structure similarity searches benefit from the concepts of Cloud computing and Big Data. Examples of these attempts are Cloud4Psi [12], the system developed by C.L. Hung and Y.L. Lin [6] and PH2 [4]. C.L. Hung and Y.L. Lin developed system for protein structure alignment on Hadoop installed on own virtualized computing environment. The system employs DALI and VAST algorithms for protein structure alignment and authors developed their own refinement method to reduce the RMSD of the original alignments. The system uses MapReduce computing paradigm to complete the task. The PH2 system developed by S. Hazelhurst allows to store PDB files in a replicated way on the Hadoop Distributed File System, and then, enables formulation of SQL queries concerning various features of 3D protein structures. PH2 also makes use of the MapReduce model.

On the other hand, Cloud4Psi for 3D protein structure similarity searching, which was reported in our previous works [12], is a system that is fully based on the concepts of Cloud computing model. The system uses computational resources of Microsoft Azure public cloud and provides dynamic scaling in response to increasing demands. The theoretical model of the system was described in [10]. Early tests reported in [12] have shown advantages of using Clouds in scaling similarity searches against a repository of protein structures. In this paper, we will show how the same process is performed in a modified architecture with local replicas of processed data and different scheduling schemes.

3   Microsoft Azure Cloud Platform

Microsoft Azure is Microsoft's cloud platform that delivers services for building scalable web-based applications. Microsoft Azure allows developing, deploying and managing applications and services through a network of data centers located in various countries throughout the world. Microsoft Azure is a public cloud, which means that the infrastructure of the cloud is available for public use and is owned by Microsoft selling cloud services. Microsoft Azure provides computing resources in a virtualized form, including processing power, RAM, storage space and appropriate bandwidth for transferring data over the network, within Infrastructure as a Service (IaaS) service model. Moreover, within Platform as a Service model, Azure also delivers a platform and dedicated cloud service programming model for developing applications that should work in the cloud. Basic tier of the Microsoft Azure platform provides five classes of virtual machines (compute units): ExtraSmall, Small, Medium, Large, ExtraLarge. They differ with the number of cores possessed, CPU/core speed and amount of memory delivered, and efficiency of I/O channel. These compute units can be used while building custom cloud-based applications. Detailed features of available compute units are listed in [21].

Microsoft Azure programming model provides an abstraction for building cloud applications. Each cloud application is defined in terms of component roles that implement the logic of the application. There are two types of roles that can be used to implement the logic of the application: Web roles that provide a web based front-end for the cloud service, and Worker roles used for background processing, scalable computations, long running or intermittent tasks. Microsoft Azure also provides a rich set of data services for various storage scenarios. These data services enable storing, modifying and reporting on data in Microsoft Azure: BLOBs that allow to store unstructured text or binary data (video, audio and images), Tables that can store large amounts of unstructured non-relational (NoSQL) data, Azure SQL Database for storing large amounts of relational data, and others. An important element of the Azure programming model are messaging mechanisms. Messaging mechanisms allow for effective communication between components and processes running in the whole cloud service. Queues are a general-purpose technology that can be used for messaging in a wide variety of scenarios, including: communication between web and worker roles in multi-tier Azure applications (like in our project), communication between on-premises applications and Azure hosted applications in hybrid solutions, communication between components of distributed applications running on-premises in different organizations or departments of an organization.

4 Architecture of the System

We have designed and developed system for 3D protein structure similarity searching, called CloudPSR (Cloud-based Protein Similarity Searching Runner), working on the Microsoft Azure Cloud. The system has been developed based on the Microsoft Azure programming model, including different types of roles for computational purposes, messaging for communication, and various storage components. General architecture of the system is presented in Fig. 2.

Outside users interact with CloudPSR through the Web role. The Web role provides a web-based graphical user interface (GUI), which is used to specify parameters of the search process and query protein structure. The system shall do pairwise comparison of the query protein structure to structures in the repository located in the Storage BLOB. The pairwise comparison involves alignment and superimposition of the query structure specified by a user and candidate structure from the repository. Each such a comparison is independent, so it is quite easy to parallelize the search on many compute units that are provisioned in the Cloud. When the user starts the execution of the search process, the Web role divides the *search job* into a number of smaller *search tasks* and sends these tasks to the Input queue. The search job assumes comparison of a specified query protein structure to all structures in the repository. Search tasks assume doing these comparisons against a part of the repository. How large the part is depends on the scheduling scheme that is used. In our system, we implemented and tested two schemes.

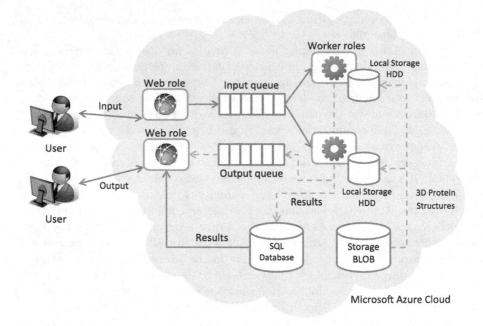

Fig. 2. General architecture of the CloudPSR system for 3D protein structure similarity searching with local replicas of macromolecular data.

In the first scheduling scheme $(S1)$ the whole repository is divided into n parts, where n is equal to the number of Worker roles performing structure comparisons. Therefore, size of the search task construed as the number of candidate protein structures is calculated as follows:

$$size(task) = \frac{size(repository)}{n}. \tag{1}$$

In this scheduling scheme each Worker role is invoked only once for a collection of protein structures from the repository, and withing this execution compares many candidate protein structures, one by one.

In the second scheduling scheme $(S2)$ each Worker role performs a single comparison of the query structure to only one candidate from the repository. Therefore, size of the search task construed as the number of candidate protein structures:

$$size(task) = 1. \tag{2}$$

In this scheduling scheme each Worker role is invoked m times:

$$m = \frac{size(repository)}{n}, \tag{3}$$

and withing each execution compares only one candidate protein structure from the repository. Both scheduling schemes are symbolically presented in Fig. 3.

In both cases, the process is implemented as a parametric sweep. Search task descriptors with sweep parameters determining the range of molecules to be processed by a Worker role in a single execution are encoded as messages and placed in the Input queue by the Web role. Worker roles consume these tasks and perform similarity searches according to the obtained scheme. Queues provide message delivery on First In, First Out (FIFO) basis. Messages are typically consumed and processed by instances of the Worker role in the order in which they were added to the queue. Each message is received and processed by only one instance of the Worker role, except if there is a failure of one of the Workers. In such cases, the failed task is performed by a different Worker. Queues also introduce temporal decoupling of system components. Web roles and Worker roles do not have to communicate directly or synchronously, because messages containing descriptors of search tasks are stored durably in the Input queue. Moreover, Web role does not have to wait for any reply from the Worker roles in order to continue to process user's search requests and generate messages.

Macromolecular data (repository of protein structures) are stored in a dedicated container of Storage BLOB. Protein structures that are needed by particular Workers are temporarily replicated to local hard drives (Local Storage HDD) according to the needs of a particular Worker role and according to assumed scheduling scheme. Results are returned through the Output queue, and then, presented to the user by the Web role. Additionally, they can be collected in the Azure SQL Database, also located in the Cloud, for aggregated statistics and future reuse. In scheduling scheme $S1$ Workers replicate large parts of the repository (according to $size(task)$) before the alignment phase and return/save results after finishing calculations for the whole part of repository assigned to them (see Fig. 3). In scheduling scheme $S2$ single structures that are currently needed for the alignment are replicated before each single alignment occurs, and results are returned/saved directly after each alignment. In Fig. 3 both, data replication and saving results phases for scheduling scheme $S2$ are parts of the alignment steps for each protein P_i. Protein structure alignment is performed with use of one of the popular algorithms. In the system, we have implemented the newest versions of FATCAT algorithm reported in [18].

5   Algorithms for Protein Structure Similarity Searching

Through many years the scientific community designed and developed many algorithms for 3D protein structure similarity searching. These algorithms rely on various representative features of protein structures and have different effectiveness. They also vary in quality of results returned. In the presented system we have implemented jFATCAT rigid and jFATCAT flexible algorithms [18]. These are new, enhanced versions of the Flexible structure AlignmenT by Chaining Aligned fragment pairs allowing Twists (FATCAT) [22]. FATCAT has good reputation among researchers and is publicly available through the Protein Data Bank (PDB) website for those, who want to search for structural neighbors. Moreover, the algorithm is used for pre-calculated all-to-all 3D-structure comparisons for the whole PDB that are updated on a weekly basis [17].

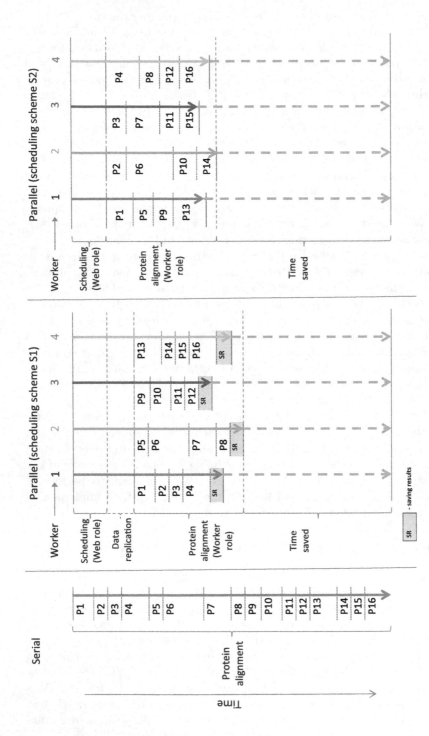

Fig. 3. Execution of the search process against repository of proteins P1-P16: serial and parallel on four Worker roles with parametric sweep according to scheduling schemes $S1$ and $S2$. In scheduling scheme $S2$ data replication and saving results phases are parts of each single alignment step.

FATCAT works on the basis of matching protein structures using Aligned Fragment Pairs (AFPs) representing parts of protein structures that fit to each other. It represents protein structures by means of local geometry, rather than global features such as orientation of secondary structures and overall topology. The algorithm constructs the alignment path that shows which parts of protein structures can be treated as identical or similar. Moreover, FATCAT flexible eliminates drawbacks of many existing methods by treating proteins as flexible structures, not rigid bodies. It allows to enter twists in protein structures while matching their fragments, which leads to finding new regions reflecting structural similarity.

6   Experimental Results

Presented cloud-based architecture with implemented scheduling schemes was tested in order to verify its performance and scalability. Tests of CloudPSR were performed in Microsoft Azure cloud with the use of two Small-sized Web roles and 1 to sixteen Small-sized Worker roles. Small-sized compute units for Web and Worker roles were equipped with 1 CPU core, 1.6 GHz, RAM 1.75 GB, local HDD 224 GB.

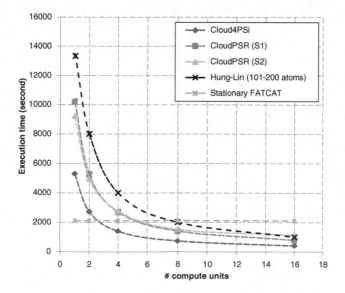

Fig. 4. Dependency between execution time and the number of worker roles/compute units for CloudPSR with both scheduling schemes $S1$ and $S2$, and for competitive systems, when scaling from 1 to sixteen compute units. Horizontal line shows the execution time for stationary, serial version of jFATCAT.

In Fig. 4 we can observe how the execution time changed with the growing number of compute units. Tests were performed by comparing 100 random pairs of protein structures from macromolecular data repository containing 1,000 sample protein structures taken from the Protein Data Bank. Scaling the system horizontally from 1 to sixteen Worker roles allowed to decrease the execution time from 2 h and 50 min to 13 min and 13 s for scheduling scheme $S1$, and from 2 h and 34 min to 19 min for scheduling scheme $S2$. Comparing CloudPSR to other systems working in the Cloud we can see that it is faster than the system developed by Hung and Lin [6], and slower than Cloud4Psi which has slightly different architecture. We can also see that the stationary version of jFATCAT, tested on a PC workstation with Intel Core i7-4700 CPU and 4GB RAM, is faster then CloudPSR (and also other cloud-based systems) for a small number of compute units engaged in computations, e.g. one to four units. CloudPSR with both scheduling schemes catches up the performance of the stationary, serial version of jFATCAT somewhere between four and six compute units. Cloud4Psi becomes more efficient then the serial version much earlier, between two and four units, and Hung an Lin system, which implements different alignment method, much later, between six and eight compute units (but closer to eight compute units). However, it should be noted that compute capabilities of Cloud virtual machines used in tests were much worse than those of the PC workstation.

In Fig. 5 we can observe n-fold speedups as a function of the number of Worker roles for CloudPSR with both scheduling schemes $S1$ and $S2$ and for other cloud-based systems. When scaling the system horizontally from 1 to sixteen Worker

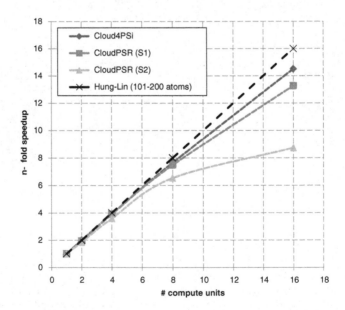

Fig. 5. Dependency between n-fold speedup and the number of Worker roles/compute units for CloudPSR with both scheduling schemes $S1$ and $S2$, and for other, competitive systems, when scaling from 1 to sixteen compute units. Dashed line represents the extrapolation of the speedup obtained by Hung and Lin.

roles CloudPSR gained the acceleration ratio at the level of 14.30 for scheduling scheme $S1$ and at the level of 9.62 for scheduling scheme $S2$, which is slightly worse than the speedup gained by Cloud4Psi (14.50) and Hung and Lin's system (16.00, extrapolated based on reported results). While the acceleration achieved by CloudPSR for scheduling scheme $S1$ is quite stable, it looses the dynamics for the scheduling scheme $S2$ above eight Worker roles. When working according to scheduling scheme $S2$, Worker roles have to perform the same preparation steps, like decoding a message with search task descriptor and replicating small peace of macromolecular data to the Local Storage, before they start a pairwise structure comparison. They also have to execute some post-processing operations, like storing results of the structure comparison in the Azure SQL Database, after they finish the comparison and structural alignment. This proves that multiple executions for single candidate structures from the repository are less efficient than processing protein structures in larger packages. The latter strategy is also employed in Cloud4Psi, hence its execution time is much better. On the other hand, Hung and Lin compare proteins in pairs, one by one, without grouping them into larger packages. As a consequence, the execution time of their system is worse, even though they obtain better speedup.

7   Discussion and Concluding Remarks

Development of system architectures that possess the ability to adjust to the increasing amount of work, such as the architecture of the CloudPSR presented in the paper, is very important in the face of the dynamical growth of macro-molecular data in the Protein Data Bank and other repositories for protein structures. The value of such systems manifests especially in one-to-many similarity searches run by users and many-to-many similarity searches performed for pre-calculated comparisons for future re-uses.

CloudPSR represents scalable, cloud-based solution for 3D protein structure similarity searching, alignment, and superposition. As opposed to the systems, like PH2 [4] and the one developed by C.L. Hung and Y.L. Lin [6], CloudPSR does not rely on Hadoop and MapReduce computing model. Moreover, by focusing on 3D protein structure similarity searching it enables more complex calculations for protein structures than the PH2 system, which is one of the first implementations of protein structure exploration on Hadoop clusters. In this regard, the aim of the CloudPSR is convergent with the system developed by C.L. Hung and Y.L. Lin, although both systems use different algorithms for the same purpose. The architecture of CloudPSR has been designed for Microsoft Azure public cloud using the role-based and queue-based model. From the viewpoint of architecture, CloudPSR is more similar to its predecessor - Cloud4Psi [10,12], also developed for Microsoft Azure cloud. However, architecture of the Cloud4Psi contained additional Manager role for scheduling purposes and it worked without local replicas of macromolecular data and according to the scheduling scheme that divided the whole repository into packages. Architectural and scheduling solutions let CloudPSR reach quite good acceleration ratio of 14.30 on sixteen

Worker roles for scheduling scheme $S1$, which is only slightly worse than 14.50 reached by Cloud4Psi.

Performance tests have proved that CloudPSR enables shortening of the search time significantly. The system with the presented architecture and scheduling scheme $S1$ achieved good scalability, which is important for planning computations in the future. Finally, results of our experiments give a great promise that by using Cloud computing paradigm we can scale scientific computations on many computing nodes and compensate constant, massive influx of data. Future works will be focused on the implementation of other scheduling techniques and more sophisticated methods for controlling the activity of Worker roles, e.g., methods that are based on artificial immune systems [15,16]. Such bio-inspired techniques are widely used to solve many optimization problems. We believe that it will be a correct step toward the coordination of distributed systems containing a large number of compute units.

Acknowledgements. We would like to thank Microsoft Research for providing us with free access to the computational resources of the Microsoft Azure cloud under the Microsoft Azure for Research Award program. Further development of the system will be carried out by the Cloud4Proteins non-profit, scientific group (http://www.zti.aei. polsl.pl/w3/dmrozek/science/cloud4proteins.htm).

References

1. Berman, H., et al.: The protein data bank. Nucleic Acids Res. **28**, 235–242 (2000)
2. Gibrat, J., Madej, T., Bryant, S.: Surprising similarities in structure comparison. Curr. Opin. Struct. Biol. **6**(3), 377–385 (1996)
3. Gu, J., Bourne, P.: Structural Bioinformatics (Methods of Biochemical Analysis), 2nd edn. Wiley-Blackwell, Hoboken (2009)
4. Hazelhurst, S.: PH2: an Hadoop-based framework for mining structural properties from the PDB database. In: Proceedings of the 2010 Annual Research Conference of the South African Institute of Computer Scientists and Information Technologists, pp. 104–112 (2010)
5. Holm, L., Kaariainen, S., Rosenstrom, P., Schenkel, A.: Searching protein structure databases with DaliLite v. 3. Bioinformatics **24**, 2780–2781 (2008)
6. Hung, C.L., Lin, Y.L.: Implementation of a parallel protein structure alignment service on cloud. Int. J. Genomics **439681**, 1–8 (2013)
7. Mayans, O., van der Ven, P., Wilm, M., Mues, A., Young, P., Wilmanns, M., Gautel, M.: Structural basis for activation of the titin kinase domain during myofibrillogenesis. Nature **395**(6705), 863–869 (1998)
8. Mell, P., Grance, T.: The NIST definition of Cloud Computing. Special Publication, pp. 800–145 (2015). http://csrc.nist.gov/publications/nistpubs/800-145/SP800-145.pdf (Accessed 7th May 2015)
9. Minami, S., Sawada, K., Chikenji, G.: MICAN: a protein structure alignment algorithm that can handle multiple-chains, inverse alignments, Ca only models, alternative alignments, and non-sequential alignments. BMC Bioinform. **14**(24), 1–22 (2013)
10. Mrozek, D.: High-Performance Computational Solutions in Protein Bioinformatics. Springer, Heidelberg (2014)

11. Mrozek, D., Małysiak-Mrozek, B.: CASSERT: a two-phase alignment algorithm for matching 3D structures of proteins. In: Kwiecień, A., Gaj, P., Stera, P. (eds.) CN 2013. CCIS, vol. 370, pp. 334–343. Springer, Heidelberg (2013)
12. Mrozek, D., Małysiak-Mrozek, B., Kłapciński, A.: Cloud4Psi: cloud computing for 3D protein structure similarity searching. Bioinformatics **30**(19), 2822–2825 (2014)
13. Mrozek, D., Brożek, M., Małysiak-Mrozek, B.: Parallel implementation of 3D protein structure similarity searches using a GPU and the CUDA. J. Mol. Model. **20**(2), 2067 (2014). http://dx.doi.org/10.1007/s00894-014-2067-1
14. Mrozek, D., Gosk, P., Małysiak-Mrozek, B.: Scaling Ab initio predictions of 3D protein structures in Microsoft Azure cloud. J. Grid Comput. **13**(4), 561–585 (2015). http://dx.doi.org/10.1007/s10723-015-9353-8
15. Poteralski, A.: Optimization of mechanical structures using artificial immune algorithm. In: Kozielski, S., Mrozek, D., Kasprowski, P., Małysiak-Mrozek, B. (eds.) BDAS 2014. CCIS, vol. 424, pp. 280–289. Springer, Heidelberg (2014)
16. Poteralski, A., Szczepanik, M., Ptaszny, J., Kuś, W., Burczyński, T.: Hybrid artificial immune system in identification of room acoustic properties. Inverse Prob. Sci. Eng. **21**(6), 957–967 (2013)
17. Prlić, A., Bliven, S., Rose, P., et al.: Pre-calculated protein structure alignments at the RCSB PDB website. Bioinformatics **26**, 2983–2985 (2010)
18. Prlić, A., Yates, A., Bliven, S., et al.: BioJava: an open-source framework for bioinformatics in 2012. Bioinformatics **28**, 2693–2695 (2012)
19. Shapiro, J., Brutlag, D.: FoldMiner and LOCK2: protein structure comparison and motif discovery on the web. Nucleic Acids Res. **32**, 536–541 (2004)
20. Shindyalov, I., Bourne, P.: Protein structure alignment by incremental combinatorial extension (CE) of the optimal path. Protein Eng. **11**(9), 739–747 (1998)
21. Virtual Machine and Cloud Service Sizes for Azure (2015). https://msdn.microsoft.com/library/azure/dn197896.aspx (Accessed 7th May 2015)
22. Ye, Y., Godzik, A.: Flexible structure alignment by chaining aligned fragment pairs allowing twists. Bioinformatics **19**(2), 246–255 (2003)
23. Zhang, Y., Skolnick, J.: TM-align: a protein structure alignment algorithm based on the TM-score. Nucleic Acids Res. **33**(7), 2302–2309 (2005)
24. Zhu, J., Weng, Z.: FAST: a novel protein structure alignment algorithm. Proteins **58**, 618–627 (2005)

Workshop on Applications of Parallel Computation in Industry and Engineering

Modeling and Simulations of Edge-Emitting Broad-Area Semiconductor Lasers and Amplifiers

Mindaugas Radziunas[✉]

Weierstrass Institute, Mohrenstrasse 39, 10117 Berlin, Germany
Mindaugas.Radziunas@wias-berlin.de
http://www.wias-berlin.de

Abstract. A (2+1)-dimensional partial differential equation model describing spatial-lateral dynamics of edge-emitting broad-area semiconductor devices is considered. A numerical scheme based on a split-step Fourier method is implemented on a parallel computing cluster. Numerical integration of the model equations is used for optimizing of existing devices with respect to the emitted beam quality, as well as for creating and testing of novel device design concepts.

Keywords: Traveling wave model · Numerical scheme · Simulations · Parallel computations · MPI · Semiconductor device · Broad area · Beam quality improvement

1 Introduction

High power high brightness edge-emitting broad-area semiconductor (BAS) lasers and optical amplifiers are compact, efficient and reliable light sources playing a crucial role in different laser technologies, such as material processing, precision metrology, medical applications, nonlinear optics and sensor technology. BAS lasers and amplifiers have a relatively simple geometry [see Fig. 1(a)] allowing an efficient energy pumping through a broad electric contact on the top of the device and can operate at high power (tens of Watts) regimes.

However, BAS devices have one serious drawback: operated at high power, they suffer from a low beam quality due to simultaneous irregular contributions of different lateral and longitudinal optical modes. As a result, the emitted optical beam is irregular, has undesirable broad optical spectra, and large divergence. Thus, a quality improvement of the beam amplified in BAS amplifiers or generated by BAS lasers is a critical issue of the modern semiconductor laser technology.

Seeking to understand the dynamics of BAS devices, to suggest improvements of existing devices or to propose novel device design concepts we do a

M. Radziunas—This work was supported by EU FP7 ITN PROPHET, Grant No. 264687 and by the Einstein Center for Mathematics Berlin under project D-OT2.

variety of related tasks. We perform modeling at different levels of complexity, do mathematical analysis of the hierarchy of models, create and implement efficient and robust numerical algorithms, and make numerical integration of the model equations. Typically, all these steps are done within research projects in cooperation with developers of the devices.

2   Mathematical Modeling and Numerical Algorithm

The dynamics of BAS devices can be described in different ways. The most comprehensive approach resolving the spatio-temporal evolution of full semi-conductor equations self-consistently coupled to the optical fields is given by 3 (space) +1 (time)-dimensional nonlinear PDEs. Since the *height* of the active zone where the optical beam is generated and amplified (y dimension) is considerably smaller than the longitudinal (z) and lateral (x) dimensions of a typical BAS device [see Fig. 1(a)], a significant simplification can be achieved by averaging over the vertical direction and by describing certain effects phenomenologically. The resulting (2+1)-dimensional dynamical traveling wave (TW) model [1] can be resolved numerically orders of magnitudes faster allowing for parameter studies in an acceptable time.

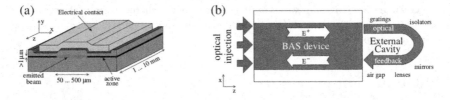

Fig. 1. (a): Schematic diagram of a BAS device. (b): Simplified representation of the BAS device, as considered by the (2+1)-dimensional TW model (Color figure online).

2.1   Basic (2+1)-Dimensional TW Model

The simplest version of the TW model is a degenerate system of second order PDEs for the slowly varying complex amplitudes of the counter-propagating optical fields, $E(z,x,t) = (E^+, E^-)^T$ [see white arrows in Fig. 1(b)], nonlinearly coupled to a rate equation for the real carrier density distribution $N(z,x,t)$. It accounts for the diffraction of fields and diffusion of carriers in the lateral direction, whereas spatially non-homogeneous device parameters capture the geometrical design of the device. The normalized TW model reads as

$$\tfrac{\partial}{\partial t}E = \left[\left(\begin{smallmatrix}-1 & 0 \\ 0 & 1\end{smallmatrix}\right)\tfrac{\partial}{\partial z} - \tfrac{i}{2}\tfrac{\partial^2}{\partial x^2}\right]E + \left[B(N,\|E\|^2) - (\alpha + i\delta)\mathcal{I}\right]E + F_{sp},$$

$$\tfrac{1}{\mu}\tfrac{\partial}{\partial t}N = D\tfrac{\partial^2}{\partial x^2}N + I(z,x) - R(N) - 2\Re e\left[E^{*T}B(N,\|E^\pm\|^2)E\right],$$

where μ is small, α, δ, F_{sp}, D, I, and $R(N) = AN + BN^2 + CN^3$ represent the field losses, the built-in contrast of the refractive index, the spontaneous emission noise, the carrier diffusion, the injected current density, and the spontaneous recombination of carriers, respectively. The complex matrix B models the carrier and photon density dependent semiconductor material gain, $G(N, \|E^\pm\|^2)$, the carrier-induced changes of the refractive index, $\tilde{n}(N)$, as well as the distributed coupling of the counter-propagating fields $\kappa^\pm$:

$$B_{11} = B_{22} = \frac{G(N, \|E\|^2)}{2} + i\tilde{n}(N), \qquad B_{12} = -i\kappa^-, \qquad B_{21} = -i\kappa^+.$$

Here, for example,

$$G(N, \|E\|^2) = \frac{g' \log\left(\max(N, N_*)\right)}{1 + \varepsilon\|E\|^2}, \qquad \tilde{n}(N) = \sigma\sqrt{N}, \qquad \kappa^+ = \kappa^- \in \mathbf{R},$$

where g', σ, ε, and N_* are the differential gain, the refractive index scaling, the nonlinear gain compression parameters, and the small positive carrier density used to determine an appropriate cut-off of the logarithmic gain function.

In general, this model should be considered in the (laterally) unbounded region $Q = Q_{z,x} \times (0, T]$, where $Q_{z,x} = \{(z, x) : (z, x) \in (0, L) \times \mathbf{R}\}$ is the spatial domain, L represents the length of the device, x is the coordinate of the unbounded lateral axis of the device, and T defines the length of the time interval where we perform the integration. Far from the active zone, $|x| \gg 1$, the optical fields and carriers usually are well damped. Thus, in our numerical simulations we truncate the lateral domain at $x = -X$ and $x = X$ so that the truncated domain $Q^t_{z,x} = \{(z, x) : (z, x) \in (0, L) \times [-X, X]\}$ [large rectangular in Fig. 1(b)] contains the considered BAS device [red area in the same figure]. Next, we assume either periodic boundary conditions [2,3] or mixed Dirichlet (for the carrier densities)/approximate transparent (for the field functions) boundary conditions [4].

The boundary conditions for the optical fields at the longitudinal edges of the device, $z = 0$ and $z = L$, account for reflections of the counter-propagating fields and optional injection of external optical beams, $a_{0,L}(x, t)$:

$$E^+(0, x, t) = r_0 E^-(0, x, t) + a_0(x, t), \quad E^-(L, x, t) = r_L E^+(L, x, t) + a_L(x, t),$$

with r_0 and r_L denoting the complex field reflectivity parameters at the laser facets.

2.2 Modifications of the TW Model

The *basic* TW model described above can be reduced to lower dimensional systems, allowing a more detailed analysis, understanding and control of specific dynamical effects. Different types of model reduction and analysis were discussed in Refs. [1,5–8]. On the other hand, different extensions of the basic TW model allow to achieve a more precise description of various relevant properties of BAS devices.

First of all, an introduction of the couple of linear equations for induced polarization functions $P^+(z, x, t)$ and $P^-(z, x, t)$ enables modeling of nontrivial material gain dependence on the lasing frequency [9]:

$$B_{\text{new}} = B - \mathcal{ID}, \qquad \mathcal{D}E^{\pm} := \overline{g}\left(E^{\pm} - P^{\pm}\right), \qquad \frac{\partial}{\partial t}P^{\pm} = i\overline{w}P^{\pm} + \overline{\gamma}\left(E^{\pm} - P^{\pm}\right).$$

Here, the parameters $\overline{g}$, $\overline{w}$, and $\overline{\gamma}$ define the Lorentzian fit of the gain profile and denote the amplitude, the central frequency, and the half width at half maximum of this Lorentzian, respectively.

Another modification is related to the heating of the BAS device by the injected current. It is known, that the gain and refractive index change functions are depending on the local temperature of semiconductor material. A proper coupling of the TW model with the full heat transport equation and the numerical resolution of this extended model, however, is a challenging task due to different time scales. Whereas the typical time scale of the thermal diffusion in semiconductors is measured in microseconds, the carrier and the photon lifetimes are given in nanoseconds and picoseconds, respectively. Thus, in order to simulate the impact of the changing heating to the dynamics of BA lasers in a reasonable time, we propose to use the following parametric approach. Namely, in Refs. [1,7] we have proposed to model injection current induced heating by the linear nonlocal dependence of the refractive index change and the gain peak frequency shift on the inhomogeneous injection $I(x, z)$:

$$\delta_{\text{new}}(z, x) = \delta(z, x) + \iint c_T(z, x, \tilde{z}, \tilde{x})I(\tilde{z}, \tilde{x})d\tilde{z}d\tilde{x},$$

$$\overline{w}_{\text{new}}(z, x) = \overline{w}(z, x) + \iint \nu_T(z, x, \tilde{z}, \tilde{x})I(\tilde{z}, \tilde{x})d\tilde{z}d\tilde{x}.$$

Here, thermal factors c_T and ν_T describe local and nonlocal crosstalk thermal effects in BAS devices with a single or several electrical contacts. This simple model with the properly defined [7] contact-wise constant coefficients c_T and ν_T has allowed a proper theoretical reproduction of the state jumping behavior with tuning of the injected currents [1].

Another useful extension of the basic TW model can be performed when simulating an emitted field propagation through the external cavity (EC) and its re-injection to the BAS device [see thick green arrow in Fig. 1(b)]. In the presence of the optical feedback from the EC, the optical injection function $a_L(x, t)$ in the longitudinal boundary conditions should be replaced by the corresponding (delayed) feedback term. The form of this term depends on the different components within the EC as well as on the field propagation time along the EC.

For example, in the case of a simple EC composed of the collimating lens and the flat mirror located perpendicularly to the optical axis of the BAS device, the re-injection term can be given by a simple delayed term

$$a_L(x, t) = t_L^2 \sqrt{R_{ec}}\, e^{i\varphi_{ec}} E^+\left(L, x, t - 2d_{ec}/c_0\right).$$

Here, $t_L = \sqrt{1 - |r_L|^2}$ is the field amplitude transmission through the right facet of the laser, R_{ec} and φ_{ec} are the field intensity reflection and phase change in the EC, whereas d_{ec} is the distance from the center of the right facet of the BAS diode to the external reflector.

When the collimating lens is absent, and the reflector or the diffractive grating is located at the small angle α_{EC} to the optical axis, the feedback term turns to be more complicated [10]:

$$a_L(x,t) \approx t_L^2 \sqrt{\tfrac{-i}{2d_{ec}\lambda_0}}\ \mathcal{F} \int\limits_{x'\in \mathbf{R}} E^+ \left(L, x', t - 2d_{ec}/c_0\right) e^{-ik_0\rho(x,x')} dx'.$$

Here, $\rho(x, x')$ is the *shortest* distance between two lateral points x' and x at the diode facet that the light takes to travel via the (infinitely broad) external reflector, whereas the operator $\mathcal{F}$ accounts for the spectral filtering by the external grating.

Another external cavity including the lens, the refractive grating located at the angle α_{ec} to the optical axis, and the small reflecting aperture was investigated experimentally in Ref. [11]. The corresponding feedback term in this case can be written as

$$a_L(x,t) \approx \tfrac{-r_g^2 T_L^2}{2\pi} \int_{\mathbf{R}} \chi \left(\tfrac{\lambda_0 f \cot \alpha_{ec}}{2\pi c_0} \omega - x \right) \int_{\mathbf{R}} E^+(L, x, t') e^{-i\omega t'}\, dt'\, e^{i\omega(t - 8f/c_0)}\, d\omega,$$

where ω denotes the relative optical frequency of the field, f is the focal distance of the lense, r_g and $\chi(x)$ are the field amplitude reflections at the grating and the aperture (the step-function $\chi(x)$ is non-vanishing if only x belongs to the aperture).

2.3   Performance of the Parallel Numerical Algorithm

Precise dynamic simulations of long and broad devices and tuning/optimization of the model parameters require huge process time and memory resources. A proper resolution of rapidly oscillating fields in typical BAS devices in a sufficiently large optical frequency range requires a fine space ($10^6 - 10^7$ mesh points) and time (up to 10^6 points for typical 5 ns transients) discretization. Dynamic simulations of such devices can easily take several days or even weeks on a single processor. Some speedup of computations is achieved by using problem-dependent variable grid steps [4]. However, for extended parameter studies with the numerical integration times up to 1000 ns parallel computers and parallel solvers have to be employed.

For the numerical integration of the TW model, we use either a split-step fast Fourier transform based numerical method [2] or a full finite difference scheme [4]. The method of domain decomposition is used to parallelize the sequential algorithm. Namely, the numerical mesh of the full problem defined by N_x lateral and N_z longitudinal uniform discretization steps is splitted along the longitudinal z-direction into K (K: number of processors) non-overlapping rectangular subgrids of the similar size $N_x \times N_{z,j}$, $j = 1, \ldots, K$, $N_{z,j} \approx \text{ceil}(N_z/K)$ [2].

Exemplary simulations of three test problems on the parallel cluster of computers (see Fig. 2) show a good scaling of the algorithm [2]. For example, the simulations performed on 32 processors give a speedup factor of 25. That is, the simulations requiring two weeks of process time on a single processor computer can be efficiently performed over a single night. For a larger number of processes, the relative time needed for communications between them grows and implies a saturation of the speedup (see an increasing deviation of the test results from the ideal speedup in Fig. 2). More details on the performance and scalability of the parallel algorithm can be found in Ref. [2].

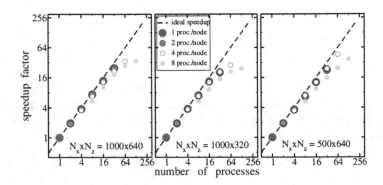

Fig. 2. Speedup of computations in multi-process simulations of three test problems defined on spatial meshes with $N_x \times N_z = 1000 \times 640$, 1000×320, and 500×640 points in lateral (x) and longitudinal (z) directions. Bullets of different color indicate tests with 1, 2, 4 or 8 processes used on each node (Color figure online).

3   Application: Suppression of Mode Jumps in MOPAs

The TW model and our numerical algorithms were successfully used for simulations of different BAS lasers and amplifiers, also showing good agreement with experimental observations [1]. In many cases, our simulations have helped to improve the design of the existing devices.

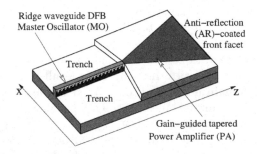

Fig. 3. Schematic representation of Master Oscillator Power Amplifier (MOPA).

For example, the master oscillator (MO) tapered power-amplifier (PA) laser shown in Fig. 3 was analyzed theoretically and experimentally in [1,12]. The narrow waveguide of the distributed feedback (DFB) MO generates a stable stationary optical field determined by a single *transversal* mode, which later is amplified in the tapered PA part of the device. An ideal MOPA laser should be able to maintain a good quality of the emitted beam. The operation of realistic MOPA devices, however, is spoiled by the amplification of the spontaneous emission in the PA, by the small separation of the MO and PA electrical contacts, and by the residual field reflectivity at the PA facet of the device.

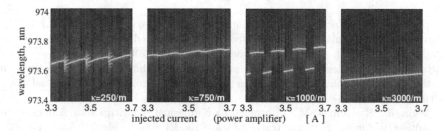

Fig. 4. Simulated optical spectra of DFB MOPA devices with different DFB field coupling coefficients κ for increased injected current. More than three days of parallel 32-processor computations on $N_x \times N_z = 400 \times 800$ spatial mesh with $\sim 2 \cdot 10^7$ time steps were required for calculation of the data represented by each panel.

In Ref. [1] we have analyzed how this residual reflectivity and thermally induced changes of the refractive index imply experimentally observable unwanted switchings between operating states determined by adjacent *longitudinal* optical modes. We have found that these bifurcations are due to the changing phase relations of complex forward- and back-propagating fields at the interface of the MO and PA parts of the device. Simulations of a typical state-jumping behavior with increasing injected current is shown in the left panel of Fig. 4. In the theoretical paper [12] we have demonstrated that a proper choice of the field coupling parameter within the DFB MO part of the device makes it less sensitive to the optical feedback, leading to a stabilization of the laser emission (see second and fourth panels of Fig. 4).

4   Conclusions

In conclusion, we have presented several modifications of (2+1)-dimensional Traveling Wave model used to describe the nonlinear dynamics of broad-area edge-emitting semiconductor lasers and discussed implementation and performance of corresponding numerical algorithms on the parallel cluster of computers. We have found, that a speedup factor of typical problem simulations performed on 32 processors is around 25. For a larger number of processors, the saturation of this speedup factor is observed. Finally, we have presented an example of practical optimization simulations of Master Oscillator Power Amplifier semiconductor laser. Here, 32-processor parallel computations of a single numerical continuation diagram with the change of parameter took more than three days. Thus, without parallelization of the numerical algorithm, an efficient study of laser parameters in a reasonable time would not be possible.

Acknowledgments. This work was supported by EU FP7 ITN PROPHET, Grant No. 264687 and by the Einstein Center for Mathematics Berlin under project D-OT2.

References

1. Spreemann, M., Lichtner, M., Radziunas, M., Bandelow, U., Wenzel, H.: Measurement and simulation of distributed-feedback tapered master-oscillators power-amplifiers. IEEE J. Quantum Electron. **45**, 609–616 (2009)
2. Radziunas, M., Čiegis, R.: Effective numerical algorithm for simulations of beam stabilization in broad area semiconductor lasers and amplifiers. Math. Model. Anal. **19**, 627–646 (2014)
3. Čiegis, R., Radziunas, M., Lichtner, M.: Numerical algorithms for simulation of multisection lasers by using traveling wave model. Math. Model. Anal. **13**, 327–348 (2008)
4. Čiegis, R., Radziunas, M., Lichtner, M.: Effective numerical integration of traveling wave model for edge-emitting broad-area semiconductor lasers and amplifiers. Math. Model. Anal. **15**, 409–430 (2010)
5. Radziunas, M., Botey, M., Herrero, R., Staliunas, K.: Intrinsic beam shaping mechanism in spatially modulated broad area semiconductor amplifiers. Appl. Phys. Lett. **103**, 132101 (2013)
6. Radziunas, M., Herrero, R., Botey, M., Staliunas, K.: Far field narrowing in spatially modulated broad area edge-emitting semiconductor amplifiers. J. Opt. Soc. Am. B **32**, 993–1000 (2015)
7. Radziunas, M., Tronciu, V.Z., Bandelow, U., Lichtner, M., Spreemann, M., Wenzel, H.: Mode transitions in distributed-feedback tapered master-oscillator power-amplifier. Opt. Quantum Electron. **40**, 1103–1109 (2008)
8. Pimenov, A., Tronciu, V.Z., Bandelow, U., Vladimirov, A.G.: Dynamical regimes of a multistripe laser array with external off-axis feedback. J. Opt. Soc. Am. B **30**(6), 1606–1613 (2013)
9. Bandelow, U., Radziunas, M., Sieber, J., Wolfrum, M.: Impact of gain dispersion on the spatio-temporal dynamics of multisection lasers. IEEE J. Quantum Electron. **37**, 183–188 (2001)
10. Jechow, A., Lichtner, M., Menzel, R., Radziunas, M., Skoczowsky, D., Vladimirov, A.: Stripe-array diode-laser in an off-axis external cavity: theory and experiment. Opt. Express **17**, 19599–19604 (2009)
11. Zink, C., Jechow, A., Heuer, A., Menzel, R.: Multi-wavelength operation of a single broad area diode laser by spectral beam combining. IEEE Photonics. Technol. Lett. **26**(3), 253–256 (2014)
12. Tronciu, V.Z., Lichtner, M., Radziunas, M., Bandelow, U., Wenzel, H.: Improving the stability of distributed-feedback tapered master-oscillator power-amplifiers. Opt. Quantum Electron. **41**, 531–537 (2009)

Application of the Parallel INMOST Platform to Subsurface Flow and Transport Modelling

Igor Konshin[1,2]($\boxtimes$), Ivan Kapyrin[1,3], Kirill Nikitin[1,3], and Kirill Terekhov[1,4]

[1] Institute of Numerical Mathematics of the Russian Academy of Sciences,
Moscow 119333, Russia
igor.konshin@gmail.com
[2] Dorodnicyn Computing Centre of the Russian Academy of Sciences,
Moscow 119333, Russia
[3] Nuclear Safety Institute of the Russian Academy of Sciences,
Moscow 115191, Russia
[4] Stanford University, Stanford, CA 94305, USA

Abstract. INMOST (Integrated Numerical Modelling and Object-oriented Supercomputing Technologies) is a tool for supercomputer simulations characterized by a maximum generality of supported computational meshes, distributed data structure flexibility and cost-effectiveness, as well as crossplatform portability. INMOST is a software platform for developing parallel numerical models on general meshes. User guides, online documentation, and the open-source code of the library is available at http://www.inmost.org.

To demonstrate the power and efficiency of the specified technology the solutions of subsurface flow and transport problems was considered. The efficiency of the parallel solution of the multiphase flow model was shown for up to several thousands of cores. Real-life examples of advective–diffusive–dispersive transport with sorption and decay modeling as well as a reactive transport problem were also considered.

Keywords: Numerical modelling · Software platform · Distributed meshes · Subsurface flow and transport

1 Introduction

The solution of industrial boundary-value problems requires high quality approximation and discretization of the problem. In addition to the high accuracy discretization it is necessary to use the general unstructured meshes that fit the problem geometry. On the other hand, to increase the approximation accuracy the usage of very large dimension meshes is required. It results in exploiting of the distributed computations on a modern parallel computers. A developer requires a tool that helps operating with distributed mesh data.

In general, the boundary-value problems solution consist of the following stages:

– mesh generation;

© Springer International Publishing Switzerland 2016
R. Wyrzykowski et al. (Eds.): PPAM 2015, Part II, LNCS 9574, pp. 277–286, 2016.
DOI: 10.1007/978-3-319-32152-3_26

- distribution of mesh data to processors;
- problem discretization;
- assembling linear system;
- solution of linear system;
- visualization of initial data and solution results.

A lot of existing softwares (STK, FMDB, MOAB, MSTK, OpenFOAM, Salome and some others) try to operate with distributed mesh data. While solving a particular problem we were not able to find a library that completely satisfies our requirements. Some of the libraries does not support operations with arbitrary polygon elements, some are insufficiently reliable and their realizations are not effective, several libraries support only one layer of ghost cells, some libraries does not support mesh modification during modelling, there is no possibility (or some difficulties) with incorporating of user approximation schemes, or there is no code portability between different computer platforms (Windows, Linux). That is why a decision to develop our own mesh platform INMOST was done [3–5,9].

2   Algorithmic Specification of INMOST Kernel

Mesh platform should support the following mesh elements:

- Vertex, which contains information on the position in the space;
- Edge, which consists of 2 or more vertices;
- Face, polygon in general case, which is based on a set of edges;
- Cell, polyhedron in general case, which is based on a set of faces.

These mesh elements suppose the following hierarchy (see Fig. 1):

$$\text{Cell} \Rightarrow \text{Face} \Rightarrow \text{Edge} \Rightarrow \text{Vertex}.$$

On the other hand the same hierarchy can be presented in the reverse order (see Fig. 2).

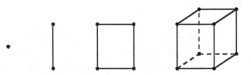

Fig. 1. Basic mesh elements: Vertex, Edge, Face, Cell.

In addition, the mesh elements can be specified by their dimension (see Fig. 3):

- (0D) Node: Vertex;

Fig. 2. Elements composition: Cell ⇒ Face ⇒ Edge ⇒ Vertex.

- (1D) Edge: Line;
- (2D) Face: Triangle, Quad, Polygon;
- (3D) Cell: Tetrahedron, Hexahedron, Prism, Pyramid, Polyhedron.

One of the most important mesh function used on discretization stage is a search of neighboring elements:

$$
\begin{array}{ccc}
\text{Cell} & \Leftrightarrow & \text{Face} \\
\updownarrow & & \updownarrow \\
\text{Vertex} & \Leftrightarrow & \text{Edge}
\end{array}
$$

Not only the above mentioned connections but a complete set of element connections can be applied as well.

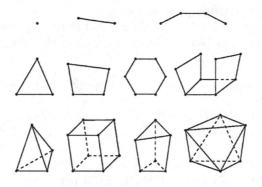

Fig. 3. Elements types.

Except for elements description to operate with mesh elements data is also required. INMOST mesh platform gives such an opportunity: Elements – to store the mesh configuration; Data – to store information in the mesh elements (data types can be dense or sparse; integer, float, or binary; a single value or an array of values); Tags – to connect the mesh data to the elements. To realize the above mentioned opportunity the mesh functions should operate with: Data, Tags, Elements, Set of elements, and Mesh.

But in spite of the wide set of the mesh functions, INMOST is not a mesh generator although such a generator can be written on its base (see example GridGen in [3]).

Except for the mentioned features, a user can require the following operations to handle distributed mesh data:

- Distribute mesh between processors;
- Specify ghost elements;
- Store data for elements in tags;
- Exchange tag data for ghost elements;
- Generate problem matrix from distributed data;
- Call parallel linear solver for distributed matrix;
- Perform global operations (min, max, sum, etc.);
- Save mesh data in a parallel format file (.pvtk, .pmf).

All the functions above are implemented in INMOST.

To distribute, redistribute, and balance the mesh data external packages ParMetis and Zoltan can used as an internal INMOST function. The solution of distributed linear systems generated during discretization can be performed by PETSc, Trilinos, or by a set of internal linear solvers in the same interface.

3   Numerical Experiments

3.1   Two- and Three-Phase Black-Oil Modelling

We consider parallel two- and three-phase black-oil models with the nonlinear monotone flux approximation presented in [7].

The first numerical test for a parallel version of three-phase black-oil model was performed on the BlueGene/P cluster located in the Moscow State University and two parts of the INM cluster:

- BG/P system consists of relatively slow PowerPC 450 (850 MHz) cores with 2 GB RAM each.
- The first part of the INM cluster (INM-1) consists of nodes with two quad-core Intel Xeon X5355 (2.66 GHz) or Intel Xeon E5462 (2.80 GHz) processors and 8 GB RAM per node.
- The second part of the INM cluster (INM-2) consists of nodes with two six-core Intel Xeon X5650 (2.67 GHz) and 24 GB RAM per node.

The problem set-up is the following. The square region contains two wells in the opposite corners: one injector and one producer with given bottom hole pressures.

In our parallel simulation, we use parallel grid generation. At the first stage the computational domain is split into subdomains which are distributed between available cores. At the second stage each core constructs a local grid inside the associated subdomain and exchanges ghost cells with neighbours. Only one layer of ghost cells is sufficient due to the compact stencil of discrete operators. Grid partitioning example is shown on Fig. 4.

The total grid dimensions are $128 \times 128 \times 16$ which gives us total of 304 192 nodes (cells + boundary entities).

Linear systems were solved with the PETSc package. The chosen solver is BCG iterations combined with the additive Schwarz preconditioner and ILU0 preconditioners in subdomains.

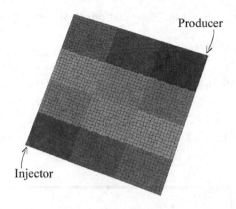

Fig. 4. Grid partitioning example.

Table 1. Relative speed-up of simulation, BG/P.

#cores	Nodes/core	#lit	t_{init}	t_{sol}	speed-up
8	38 024	71 024	68.8 s	28 549 s	1x
16	19 012	71 042	37.2 s	14 471 s	1.97x
32	9 506	71 648	19.6 s	7 464 s	3.82x
64	4 753	72 174	10.5 s	3 874 s	7.36x
128	2 377	73 806	5.9 s	2 059 s	13.86x

Table 1 shows the results of the parallel experiment on BG/P for 200 days simulation. One can see good relative speed-up for up to 128 cores (2.4 k nodes per core). The number of total nonlinear iterations is 648 and does not depend on the number of cores. The number of linear iterations increases slightly as #cores grows, while the initialization and computation times decrease almost linearly. We note that the BG/P system has fast connection with relatively slow computational cores.

Table 2 shows the results for INM-1 and INM-2 which have much faster cores than BG/P. As expected, the relative speed-up is lower albeit satisfactory: up to 11x for 8-to-128 cores on INM-1 and 4.5x for 8-to-64 cores on INM-2.

Figure 5 presents the relative speed-up of the parallel computation which in case of BG/P cluster is close to the ideal linear speed-up. Figure 6 shows the diagram with computational times on three clusters.

The presented results demonstrate good quality of the developed parallel data structure and algorithms, although we use the third-party PETSc linear solver which also can be improved.

The second experiment deals with two-phase flow model on a massively parallel BG/P system with up to 8192 cores. Problem setup and grid construction method is similar to the first test case. We consider 50 days simulation on 0.9 million cells nonorthogonal hexahedral grid (1.8 million unknowns).

Table 2. Relative speed-up of simulation, INM-1 and INM-2 clusters.

#cores	INM-1			INM-2		
	t_{init}	t_{sol}	speed-up	t_{init}	t_{sol}	speed-up
8	9.9 s	12 506 s	1x	6.2 s	4 909 s	1x
16	5.2 s	6 182 s	2.02x	3.8 s	2 980 s	1.65x
32	3.0 s	3 756 s	3.33x	2.4 s	1 957 s	2.51x
64	1.7 s	1 926 s	6.49x	2.0 s	1 092 s	4.50x
128	1.0 s	1 131 s	11.06x	–	–	–

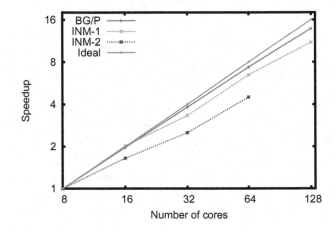

Fig. 5. Relative speed-up, BG/P, INM-1 and INM-2.

Table 3 presents number of linear and nonlinear iteration, initialization, grid generation and total simulation times of the parallel experiment for two-phase flow model. The reference results are taken for 512 cores run. One can see that the total simulation time decreases, yet there is almost no speed-up for 1024-to-2048 and 4096-to-8192 pairs (see Fig. 7, left). This is explained by the reduction of the subproblem sizes and the sharp increase of the number of the linear iterations in these pairs (see Fig. 7, right).

The performed parallel experiments show that the INMOST platform is useable for parallel simulations even with severely low unknowns per core numbers (down to 220 in the last test) and allows to achieve good scalability of the numerical model with minimal changes of the serial code.

3.2   Groundwater Flow and Contaminant Transport Modelling

The platform may be efficiently applied for complex multidimensional problems demanding high-performance simulation of various coupled processes. An example is the groundwater flow and contaminant transport in porous media modellings [6]. One can observe the general trends in the development of

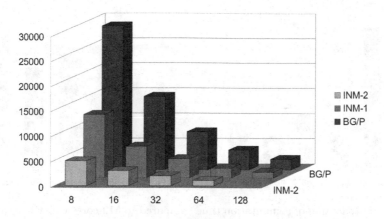

Fig. 6. Solution times for parallel computation, BG/P, INM-1 and INM-2.

Table 3. Relative speed-up of the two-phase flow simulation, BG/P.

#cores	#nonlit	#lit	t_{init}		t_{grid} t_{sol}	speed-up
512	151	247 919	2.69	1.19	1478.7	1x
1024	151	190 099	1.80	0.70	559.0	2.64x
2048	150	333 369	1.34	0.49	536.2	2.76x
4096	148	291 533	1.53	0.41	296.7	4.98x
8192	147	402 742	1.91	0.46	296.0	5.0x

hydrogeological modelling codes. From the numerical point of view these are the use of unstructured adaptive grids, the development of discretizations suitable for this type of grids and parallelization (examples are the MODFLOW-USG code [8] and the ASCEM project of US DOE [2].

The means for mesh and data storage, matrix and vector assembly implemented in the platform were used to create the models of the following processes on unstructured grids:

- saturated and unsaturated groundwater flow;
- advective–diffusive–dispersive transport with sorption and decay;
- reactive transport;
- density-driven flow.

On Fig. 8 we show the application of the code to the safety assessment of a surface radioactive waste disposal facility. The problem features an adaptive mesh composed of triangular prisms, and quite a specific geological structure featuring 10 layers with heterogeneous hydraulic conductivity tensor. Combined with Qt and VTK libraries the INMOST platform allows to organize a full workflow for hydrogeological modelling: creation of a geological model, model data setting, grid generation, numerical flow and transport modelling with the ability to run in parallel, and finally the visualization of results. The broad

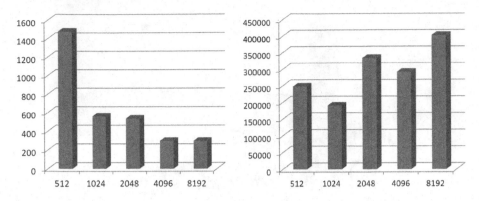

Fig. 7. Left: reduction of computation time compared to 512-cores experiment. Right: total number of linear iterations for simulation.

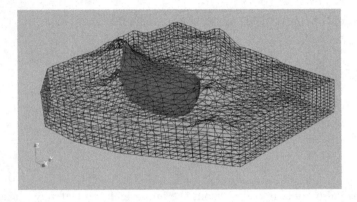

Fig. 8. Modelling the evolution of a contaminant plume from a surface waste disposal in a realistic geological media.

options of the platform to attach data to mesh entities is widely used to impose the boundary conditions, express the heterogeneity of the domain properties and take into account various objects affecting the flow, namely wells, drains, rivers and lake, pollution sources.

Another application is the reactive transport modelling using a combination of a domestic flow and transport code with a third-party chemical code PHREEQC [1]. On Fig. 9 the results of five-spot test case are shown. The domain is a parallelepiped $[-100; 100] \times [-100; 100] \times [-5; 5]$ (in meters). Four injection wells are located close to the corners of the domain: points $(-95; -95)$, $(95; -95)$, $(95; 95)$, $(-95; 95)$ in the X-Y plane with the well screen in the range $(-2; 2)$ meters along the Z-axis. A production well is located in the middle of the domain. The injection intensity is $10 \, \mathrm{m}^3/\mathrm{day}$ for each injection well, the production rate is $40 \, \mathrm{m}^3/\mathrm{day}$. The boundaries are impervious. The porous media is homogeneous and contains an ionic exchanger X. The injected liquids have an equal concentration of strontium $3.3 \cdot 10^{-9} \, \mathrm{Mol/l}$ while the concentration of

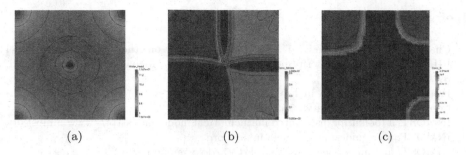

(a)   (b)   (c)

Fig. 9. Modelling groundwater flow and reactive transport of solute containing sodium nitrate and strontium in the presence of ionic exchanger on the rock ($T = 3000$ days). (a) Water head; (b) Nitrate concentration in groundwater; (c) Strontium concentration in groundwater.

sodium nitrate varies from zero (lower left injection well) to 0.6 Mol/l (upper left injection well) with a step 0.2 Mol/l (thus having 0.2 Mol/l in the lower right and 0.4 in the upper right injection well). The initial solution in the media also contains K, Cl, Ca, Mg, S, C in equilibrium with the media. The results of the modelling are in good agreement with observed experimental data showing a decrease in the strontium attenuation caused by the presence of sodium nitrate in the injected solution. On Fig. 9 one can see that the higher the sodium nitrate concentration will be, the less will strontium be sorbed causing a quicker pollution propagation. Note that in sequential mode the modelling of this problem for 1000 days on a very coarse mesh containing 1600 hexahedral cells took around 3 h on an Intel CoreI7 machine substantiating a strong need in parallelization.

Conclusions

In conclusion we would like to formulate the major benefits of INMOST:

- Cross-platform code;
- Supports parallel mesh generation;
- Supports various input/output mesh formats (.gmsh, .vtk, .pvtk, .gmv, internal .pmf);
- Mesh can be distributed and redistributed in parallel (works with Zoltan, Parmetis and internal partitioner);
- Full set of mesh elements;
- Supports element markers and tag data of different types (integer, double, byte, element);
- Basis for parallel grid modification (is under development now).

Acknowledgements. This work has been supported in part by RFBR grants 14-01-00830, 15-35-20991.

References

1. Charlton, S.R., Parkhurst, D.L.: Modules based on the geochemical model PHREEQC for use in scripting and programming languages. Comput. Geosci. **37**(10), 1653–1663 (2011)
2. Freedman, V.L., et al.: A high-performance workflow system for subsurface simulation. Environ. Model. Softw. **55**, 176–189 (2014)
3. INMOST: a toolkit for distributed mathematical modelling. http://www.inmost.org
4. INMOST: user guides. http://wiki.inmost.org
5. INMOST: an online documentation. http://doxygen.inmost.org
6. Kapyrin, I., Vassilevski, Y., Utkin, S.: Concept of the design and application of the GeRa numerical code for radioactive waste disposal safety assessment. Voprosy Atomnoi Nauki i Tekhniki. Ser.: Math. Model. Phys. Process Iss. **4**, 44–54 (2014)
7. Nikitin, K.D., Terekhov, K.M., Vassilevski, Y.V.: A monotone nonlinear finite volume method for diffusion equations and multiphase flows. Comput. Geosci. **18**(3), 311–324 (2014)
8. Panday, S., Langevin, C.D., Niswonger, R.G., Ibaraki, M., Hughes, J.D.: MODFLOW-USG version 1: An unstructured grid version of MODFLOW for simulating groundwater flow and tightly coupled processes using a control volume finite-difference formulation: U.S. Geological Survey Techniques and Methods, Book 6, chap. A45, p. 66 (2013)
9. Vassilevski, Y.V., Konshin, I.N., Kopytov, G.V., Terekhov, K.M.: INMOST - a software platform and a graphical environment for development of parallel numerical models on general meshes. Moscow State University Publishing, Moscow, p. 144 (2013)

Parallel Procedure Based on the Swarm Intelligence for Solving the Two-Dimensional Inverse Problem of Binary Alloy Solidification

Edyta Hetmaniok[✉], Damian Słota, and Adam Zielonka

Institute of Mathematics, Silesian University of Technology,
Kaszubska 23, 44-100 Gliwice, Poland
{edyta.hetmaniok,damian.slota,adam.zielonka}@polsl.pl

Abstract. In the paper an application of Ant Colony Optimization algorithm for solving the two-dimensional inverse solidification problem of binary alloy is presented. Aim of the considered problem lies in reconstruction of the boundary condition on the basis of temperature values measured in selected points of the cast. Presented approach is grounded on two procedures: the finite difference method with application of the generalized alternating phase truncation method and the parallelized Ant Colony Optimization algorithm serving for minimization of a functional representing the important part of the procedure.

Keywords: Swarm intelligence · Parallel ACO algorithm · Solidification · Binary alloy

1 Introduction

With regard to the increasing level of complexity of the technical problems and the increasing time needed to carry out the required calculations, the parallel computations appeared to be very efficient. Parallel computing means that the calculations are executed simultaneously [1], basing on the assumption that the discussed large problem can be divided into smaller tasks solved concurrently. One may distinguish several forms of parallel computing, that is the bit-level, instruction level, data and task parallelism. To employ the parallel computing the appropriate algorithm is needed, possible to parallelize.

Ant Colony Optimization (ACO) algorithm is this kind of algorithm, since its structure enables to involve the parallel computations. ACO algorithm is a swarm intelligence algorithm [2], grounds of which have been taken from the observations of ants exploring the environment in search for the source of food. During this search the ant marks the traversed path by using a chemical substance, called the pheromone, stimulating other members of the swarm to follow the path. In this way the pheromone trace on the most attractive and promising trails is intensified and on the long, leading to nowhere trails it is evaporated. Such system of collective actions of simple individuals belonging to the swarm

© Springer International Publishing Switzerland 2016
R. Wyrzykowski et al. (Eds.): PPAM 2015, Part II, LNCS 9574, pp. 287–297, 2016.
DOI: 10.1007/978-3-319-32152-3_27

creates some kind of artificial intelligence and is imitated in the algorithm by the artificial "ants" representing the possible solutions of discussed problem dispersed in the investigated region and condensing their presence around the best located solutions [3–5]. Authors of the current paper have already contributed in developing the technical applications of ACO algorithm by using it in solving the selected heat conduction problems [6,7]. Problems investigated in these exemplary references have been considered in one-dimensional space in dependance on time, therefore the classical version of ACO algorithm was enough for our needs. In the current paper we deal with a problem considered in two-dimensional space and changing with time which results in significant prolongation of time needed to execute the calculations. Therefore we decided to parallelize the ACO algorithm which was possible, since in the procedure each swarm member is created independently, so all the required computations can be carried out simultaneously for all the created individuals.

Problem taken up in this paper lies in solving the inverse solidification problem, more specifically, in reconstructing the heat transfer coefficient on boundary of the region, together with the temperature distribution in the entire region. The reconstruction will be carried out on the basis of temperature measurements read in selected points of the domain, on the way of minimizing the functional expressing differences between the calculated and measured values of temperature with the aid of ACO algorithm, whereas the process of alloy solidification will be described by the model called as the solidification in temperature interval [8,9]. The used model, taking into account only the temperature distribution, is based on the heat conduction equation with the enclosed source element including the latent heat of fusion and the volume contribution of solid phase. Procedure of minimizing the mentioned functional requires the solution of direct problem which will be realized by applying the finite difference method combined with the generalized alternating phase truncation method [10–12].

2   Governing Equations

Two-dimensional domain $\overline{\Omega} = [0, b] \times [0, d]$, shown in Fig. 1 for the selected moment of time, is occupied by a solidifying material. Region $\overline{\Omega}$ is divided into three subregions taken by the liquid and solid phase, separated by the intermediate two-phase zone (called the mushy zone) and the boundary $\Omega = \overline{\Omega} \times [0, t^*]$ is divided into five parts like it is defined in the figure.

Inside considered domain Ω the function T, describing the distribution of temperature, satisfies the heat conduction equation

$$C \varrho \frac{\partial}{\partial t} T(x, y, t) = \lambda \nabla^2 T(x, y, t), \tag{1}$$

where C denotes the substitute thermal capacity and ϱ and λ are, respectively, the mass density and thermal conductivity, whereas t describes the time variable and x, y refer to the spatial locations. Thus, solving of this problem consists in

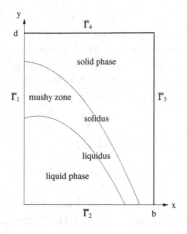

$$\Gamma_0 = \{(x, y, 0); \ x \in [0, b], \ y \in [0, d]\}$$
$$\Gamma_1 = \{(0, y, t); \ y \in [0, d], \ t \in [0, t^*]\}$$
$$\Gamma_2 = \{(x, 0, t); \ x \in [0, b], \ t \in [0, t^*]\}$$
$$\Gamma_3 = \{(b, y, t); \ y \in [0, d], \ t \in [0, t^*]\}$$
$$\Gamma_4 = \{(x, d, t); \ x \in [0, b], \ t \in [0, t^*]\}$$

Fig. 1. Domain of the problem for the selected moment of time $\bar{t}$ $(\bar{\Gamma}_i = \Gamma_i \cap \{\bar{t}\})$ and the general definition of boundaries

determination of function T fulfilling Eq. (1) as well as the initial condition on boundary Γ_0:

$$T(x, y, 0) = T_0, \tag{2}$$

where T_0 denotes the initial temperature, the homogeneous condition of the second kind on boundaries Γ_1 and Γ_2:

$$-\lambda \frac{\partial}{\partial n} T(x, y, t) = 0 \tag{3}$$

and condition of the third kind on boundaries Γ_3 and Γ_4:

$$-\lambda \frac{\partial}{\partial n} T(x, y, t) = \alpha(x, y, t) \left(T(x, y, t) - T_\infty \right), \tag{4}$$

where α denotes the heat transfer coefficient, T_∞ expresses the ambient temperature.

The substitute thermal capacity is a parameter appearing while modeling the solidification process by using the one domain approach. It varies in dependence on temperature and is defined as

$$C = \begin{cases} c_l & T > T_L, \\ c_{mz} + \dfrac{L}{T_L - T_S} & T \in [T_S, T_L], \\ c_s & T < T_S, \end{cases} \tag{5}$$

where c_l, c_{mz} and c_s denote, respectively, the specific heat of liquid phase, mushy zone and solid phase, L describes the latent heat of fusion and T_L and T_S refer to the liquidus and solidus temperatures. Values of density and the thermal conductivity coefficient in Eq. (1) are also variable in dependence on temperature

$$\varrho = \begin{cases} \varrho_l & T > T_L, \\ \varrho_{mz} & T \in [T_S, T_L], \\ \varrho_s & T < T_S, \end{cases} \qquad \lambda = \begin{cases} \lambda_l & T > T_L. \\ \lambda_{mz} & T \in [T_S, T_L]. \\ \lambda_s & T < T_S. \end{cases} \tag{6}$$

Thanks to the above expression of parameters C, ϱ and λ as dependent on temperature, Eq. (1) describes the distribution of temperature in the entire domain, that is in the liquid phase, two-phase (mushy) zone and in the solid phase.

Goal of the investigated inverse problem is to determine the temperature distribution in considered region Ω and to identify the heat transfer coefficient on boundaries Γ_3 and Γ_4 in case when the values of temperature in selected points $(x_i, t_j) \in \Omega$ are given

$$T(x_i, y_i, t_j) = U_{ij}, \qquad i = 1, 2, \ldots, N_1, \quad j = 1, 2, \ldots, N_2, \tag{7}$$

where N_1 denotes the number of sensors and N_2 refers to the number of measurements read from each sensor.

For some assumed fixed form of heat transfer coefficient α problem (1)–(4) turns into the direct problem which can be solved by using one of the known methods dedicated for this kind of inverse problem. In this case we decided to apply the finite-difference method supplied by the generalized alternating phase truncation method [10, 11]. In this method the temperature is replaced by the enthalpy and the calculations are executed in three stages – in each stage the domain is reduced to one phase (the liquid phase, mushy phase and solid phase, respectively), so in every step of calculations the problem is replaced by three one-phase direct solidification problems solved with the aid of finite-difference method.

Thus, after solving problem (1)–(4) for the given form of heat transfer coefficient we are able to find the values of temperature $T_{ij} = T(x_i, t_j)$ corresponding to the assumed form of coefficient α. Next, by using the calculated temperatures T_{ij} and the measured temperatures U_{ij} we construct a functional expressing the error of approximate solution in the following way

$$J(\alpha) = \sum_{i=1}^{N_1} \sum_{j=1}^{N_2} \left(T_{ij} - U_{ij}\right)^2. \tag{8}$$

On the way of minimizing functional (8), realized with the aid of ACO algorithm, we are able to find the values of sought heat transfer coefficient such that the reconstructed values of temperature will be as close as possible to the values of measurements.

3   Ant Colony Optimization Algorithm

The Ant Colony Optimization is an algorithm of heuristic nature belonging to the group of artificial intelligence algorithms imitating the behavior of the swarm of ants sharing the information about traversed paths by leaving on the ground a chemical substance, called pheromone, excreted by most of the ant species. To mimic this specific way of communication between ants we use the following algorithm, in which the vectors $\mathbf{x}$ of investigated region play the role of ants forced to gather around the solution considered as the best (see also [3, 6, 7]).

Pseudo-code of ACO algorithm:

1. Input data:

   $J(\mathbf{x})$ – minimized function, $\mathbf{x} = (x_1, \ldots, x_n) \in D$, where D is a domain of the problem; m – number of ants in one population; I – number of iterations; β – narrowing parameter.

2. Random selection of the initial ants location: $\mathbf{x}^k = (x_1^k, \ldots, x_n^k)$, where $\mathbf{x}^k \in D$, $k = 1, 2, \ldots, m$.

3. Determination of the best located ant $\mathbf{x}^{best}$ in the initial ants population, that is the ant for which the minimized function J takes the lowest value.

for $i = 1 \to I$ **do**

  for $j = 1 \to I^2$ **do**

4. Parallel updating of the ants locations:

   – parallel random selection of vector $\mathbf{dx}^k$, $k = 1, \ldots, m$, such that

   $$-\beta_i \leq dx_t^k \leq \beta_i;$$

   – parallel creation of the new ants population:

   $$\mathbf{x}^k = \mathbf{x}^{best} + \mathbf{dx}^k, \quad k = 1, 2, \ldots, m;$$

   – parallel determination of values $J(\mathbf{x}^k)$, $k = 1, 2, \ldots, m$.

5. Determination of the best located ant $\mathbf{x}^{best}$ in the current ant population.

end for

6. Narrowing of the ants dislocations range according to relation $\beta_{i+1} = 0.1\beta_i$.

end for

Step 6 of the above procedure, in which the range of ants dislocations is reduced, simulates the process of evaporation of the pheromone trail. Parameter β is a measure of concentration of the ants around the best located one, symbolizing the path leading to the source of food. Decreasing of this parameter forces the ants to gather around the best location. And the way of generating the new population of ants, such that each new ant is created independently, allows to parallelize the procedure. Since the calculation of the value of function (8) requires to solve the associated direct problem, we may execute the calculations independently for each tested solution, which about nine times decreases the working time of the procedure. Another element, which should be taken into account, is the heuristic character of the ACO algorithm causing that the obtained result is the best, but the best one achieved in this specific run. Every other execution of the procedure can give slightly different solution. It is not any limitation of this kind of algorithms, it is just a specificity of the heuristic algorithms. However, to ensure the best result the procedure must be repeated some number of times and the best of received solution will be accepted as the best one.

4   Computational Example

Considered inverse solidification problem consists in identification of heat transfer coefficient occurring in boundary condition (4) defined on boundaries Γ_3 and Γ_4 (see Fig. 1). Function α, describing this coefficient, will be sought as the step function in the following form (where $t_1 = 40$, $t_2 = 100$ [s]):

$$\alpha(x,y,t) = \begin{cases} \alpha_1 & \text{for } t \leq t_1 \wedge x \in [0,b] \wedge y = d, \\ \alpha_3 & \text{for } t \in (t_1, t_2] \wedge x \in [0,b] \wedge y = d, \\ \alpha_5 & \text{for } t > t_2 \wedge x \in [0,b] \wedge y = d, \\ \alpha_2 & \text{for } t \leq t_1 \wedge y \in [0,d] \wedge x = b, \\ \alpha_4 & \text{for } t \in (t_1, t_2] \wedge y \in [0,d] \wedge x = b, \\ \alpha_6 & \text{for } t > t_2 \wedge y \in [0,d] \wedge x = b. \end{cases} \tag{9}$$

Solidification process carried out within domain Ω according to Eqs. (1)–(4) is defined by the following values of material parameters:

$b = d = 0.08$ [m], $t^* = 400$ [s], $\lambda_l = 54$ [W/(m K)], $\lambda_s = 30$ [W/(m K)], $c_l = 840$ [J/(kg K)], $c_s = 668$ [J/(kg K)], $\varrho_l = 7000$ [kg/m^3], $\varrho_s = 7500$ [kg/m^3], $L = 272000$ [J/kg], $T_\infty = 50$ [K] and $T_0 = 1803$ [K].

The exact values of the sought heat transfer coefficient are known and are the following

$$\alpha_1 = 1200, \qquad \alpha_3 = 250, \qquad \alpha_5 = 500,$$
$$\alpha_2 = 800, \qquad \alpha_4 = 800, \qquad \alpha_6 = 250 \, [\text{W}/(\text{m}^2\,\text{K})].$$

Thanks to this information we can compare the reconstructed values of coefficients α_i with the exact ones and to evaluate the precision of obtained results.

The sought values were reconstructed on the basis of measurements of temperature made on four thermocouples ($N_1 = 4$) located 4 mm away from boundary of the region. The readings of temperature were taken in two rounds, at every 1 and 4 s, which gives us 400 and 100 measurement values ($N_2 = 400$ or $N_2 = 100$, respectively). For investigating the precision and stability of obtained approximate values of heat transfer coefficient we performed the calculations for the exact as well as for the burdened input data. The burdened input data were perturbed by the 1 % and 2 % random error of normal distribution.

The generalized alternating phase truncation method supported by the finite difference method, applied for solving the direct problem associated with the discussed inverse problem, was realized for the discretization grid of steps $\Delta x = b/100$ and $\Delta y = d/100$ and for the time step $\Delta t = 0.02$. In our computational experiment the same procedure was used for determining the values treated later as the measurements, but to avoid the inverse crime in this case we took the grid of different density. Parameters of the ACO algorithms used for minimizing functional (8), that is the number of ants in one populations and the number of external iterations were equal to $m = 18$ and $I = 6$, respectively. Such values of parameters were the result of our test calculations performed for some well-known benchmark functions. We also took into account in that issue our experiences gained while working on our previous papers (for example on papers [6,7]). According to the literature [3] the decreasing coefficient of the narrowing parameter β at the beginning of the procedure was taken as 0.1 but, basing on our experience, starting from the second iteration we decided to change this value to 0.2. We did it to shorten the time of calculations because thanks to this change the solutions gather faster around the best one. Another way for shortening the execution time is the idea of parallelizing the ACO algorithm. Thanks to this each run of the procedure took about 36 h, whereas the execution time before this modification was about 330 h. It is a very reasonable time for the task of discussed kind solved in two-dimensional space, especially with regard to the fact that for computations we used the system of ordinary PC computers with processor Intel Core i7-3930K 3.2 GHz. In this six physical core processor with twelve logical cores we used in calculations ten logical cores thanks to the intel's hyperthreading technology. Another element, which should be taken into account, is the heuristic nature of ACO algorithm, therefore, to avoid the uncertainty of obtained results, in each case of initial data perturbation and the number of measurements the computations were made eight times.

In Figs. 2 and 3 there are presented two collections of figures. The first one shows the values of the objective functional (8) minimized by using the ACO algorithm together with the mean and maximal relative errors in reconstructing all six values α_i in dependance on the iteration number (within each external iteration I there is executed I^2 internal iterations, therefore $I = 6$ means in fact $6^3 = 216$ iterations) obtained for measurements taken at every 4 s and exact input data. The second figure shows similar results only in the worst case of input data, that is for measurements read with the same frequency but for 2 % input data. Values of minimized functional suppose to decrease which can be observed in both figures – for the exact input data the values approach to zero, as expected, for the burdened input data the values also decrease, however not to zero because of the perturbation. Consequence of the decreasing functional (8) is that the mean and maximal reconstruction errors decrease as well, which is the aim of the procedure and is visible in both figures.

Figure 4 displays the comparison of mean relative errors in reconstructing all six values α_i received for various noises of input data and both frequencies of measurements. We may see that in all cases of input data the reconstruction errors are lower than the input errors. More detailed analysis of results is collected in Table 1 presenting the reconstructed values of the heat transfer coefficient, relative percentage errors of these reconstructions and standard deviations of results obtained in multiple executions of the procedure for the case of 100 measurements. Low values of errors and of standard deviations as well indicate the precise and stable reconstruction of the sought coefficient. Similar conclusion concerns the reconstruction of temperature. Relative and absolute errors of the temperature reconstruction in all four points of the sensor locations for more rarely read measurements and for various noises of input data are presented in Table 2 and we can see that they are insignificant.

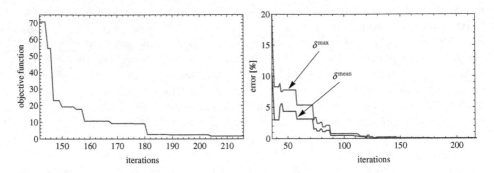

Fig. 2. Values of the objective function (8) of the ACO algorithm (left figure) together with the mean and maximal relative errors in reconstructing all values α_i, $i = 1, ..., 6$, (right figure) in dependance on the iterations number obtained for measurements taken at every 4 s and exact input data

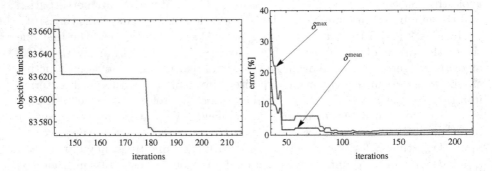

Fig. 3. Values of the objective function (8) of the ACO algorithm (left figure) together with the mean and maximal relative errors in reconstructing all values α_i, $i = 1, ..., 6$, (right figure) in dependance on the iterations number obtained for measurements taken at every 4 s and 2 % perturbation of input data

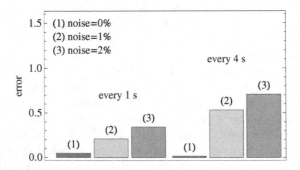

Fig. 4. Mean relative errors in reconstructing all values α_i, $i = 1, ..., 6$, for various noises of input data and various frequencies of measurements

Table 1. Results of the calculations for measurements taken at every 4 s and for various noises of input data ($\overline{\alpha_i}$ – reconstructed values of the heat transfer coefficient, δ_{α_i} – relative percentage error, s_{α_i} – standard deviation)

Noise	i	$\overline{\alpha_i}$	δ_{α_i} [%]	s_{α_i}
0%	1	1199.94	0.005	0.451
	2	800.20	0.025	0.183
	3	249.98	0.010	0.015
	4	800.18	0.022	0.280
	5	499.94	0.012	0.203
	6	250.04	0.015	0.009
1%	1	1203.49	0.291	0.794
	2	797.11	0.361	1.027
	3	250.33	0.130	0.086
	4	788.59	1.426	0.141
	5	503.54	0.708	0.281
	6	250.70	0.282	0.068
2%	1	1192.34	0.638	0.008
	2	811.46	1.433	0.741
	3	249.45	0.222	0.047
	4	803.97	0.496	0.152
	5	494.78	1.043	0.660
	6	248.98	0.408	0.065

Table 2. Relative and absolute errors of the temperature reconstruction in all four points of the sensors locations, for measurements taken at every 4 s and for various noises of input data

Noise	Sensor	δ^{max} [%]	δ^{mean} [%]	Δ^{max} [K]	Δ^{mean} [K]
0 %	1	0.0108	0.0016	0.1877	0.0205
	2	0.0094	0.0011	0.1493	0.0137
	3	0.0217	0.0008	0.3717	0.0095
	4	0.0172	0.0012	0.2948	0.0145
1 %	1	0.1056	0.0409	1.7024	0.4786
	2	0.1236	0.0439	2.1058	0.4770
	3	0.3318	0.0506	5.7017	0.5935
	4	0.4208	0.0504	7.2321	0.6476
2 %	1	0.1165	0.0346	1.5138	0.4041
	2	0.1032	0.0297	1.5269	0.3416
	3	0.3205	0.0122	5.5069	0.1473
	4	0.0613	0.0145	0.7062	0.1792

5 Summary

Goal of the paper was the presentation of the procedure for solving the inverse problem in solidification of the binary alloy. Studied problem was solved thanks to the application of two procedures: the generalized phase truncation method and the Ant Colony Optimization algorithm with a modification consisted in paralleling the computations carried out for the individuals. Numerical verification of the approach showed that in each discussed case of input data the reconstruction errors were comparable or significantly smaller than errors of input data and values of the standard deviations of results were relatively small. The comparison of the selected swarm intelligence algorithms and the genetic algorithm in solving the two-dimensional inverse Stefan problem is investigated in paper [6], where it is shown that the ACO and Artificial Bee Colony algorithms give the results equally precise as the genetic algorithm, but they are faster in working. In paper [7] the comparative analysis of three algorithms (ACO, ABC and Harmony Search algorithm) in solving the inverse Stefan problem is executed which shows that the ant and bee algorithms give similarly precise and stable results in the shortest time. The comparative study of the parallel ACO procedure and selected other algorithms in solving the investigated problem is now in progress.

Acknowledgements. This project has been financed from the funds of the National Science Centre granted on the basis of decision DEC-2011/03/B/ST8/06004.

References

1. Gottlieb, A., Almasi, G.S.: Highly Parallel Computing. Benjamin-Cummings Publishing Co., Menlo Park (1989)
2. Eberhart, R.C., Shi, Y., Kennedy, J.: Swarm Intelligence. Morgan Kaufmann, San Francisco (2001)
3. Toksari, M.D.: Ant colony optimization for finding the global minimum. Appl. Math. Comput. **176**, 308–316 (2006)
4. Dorigo, M., Blumb, C.: Ant colony optimization theory: a survey. Theor. Comput. Sci. **344**, 243–278 (2005)
5. Duan, H.: Ant colony optimization: principle, convergence and application. Adapt. Learn. Optim. **8**, 373–388 (2011)
6. Hetmaniok, E., Słota, D., Zielonka, A.: Using the swarm intelligence algorithms in solution of the two-dimensional inverse Stefan problem. Comput. Math. Appl. **69**(4), 347–361 (2015)
7. Hetmaniok, E., Słota, D., Zielonka, A.: Experimental verification of selected artificial intelligence algorithms used for solving the inverse Stefan problem. Numer. Heat Transf. B. **66**(4), 343–359 (2014)
8. Ionescu, D., Ciobanu, I., Munteanu, S.I., Crisan, A., Monescu, V.: 2D mathematical model for the solidification of alloys within a temperature interval. Metal. Int. **16**(4), 39–44 (2011)
9. Piasecka Belkhayat, A.: Numerical modelling of solidification process using interval boundary element method. Arch. Foundry Eng. **8**(4), 171–176 (2008)

10. Mochnacki, B., Suchy, J.S.: Numerical Methods in Computations of Foundry Processes. PFTA, Cracow (1995)
11. Rogers, J.C.W., Berger, A.E., Ciment, M.: The alternating phase truncation method for numerical solution of a Stefan problem. SIAM J. Numer. Anal. **16**, 563–587 (1979)
12. Słota, D.: Restoring boundary conditions in the solidification of pure metals. Comput. Struct. **89**, 48–54 (2011)

Minisymposium on HPC Applications in Physical Sciences

A Highly Parallelizable Bond Fluctuation Model on the Body-Centered Cubic Lattice

Christoph Jentzsch[1,2], Ron Dockhorn[1,2(✉)], and Jens-Uwe Sommer[1,2]

[1] Leibniz Institute of Polymer Research Dresden, Dresden, Germany
[2] Institute for Theoretical Physics, Technische Universität Dresden,
Dresden, Germany
{jentzsch-christoph,dockhorn,sommer}@ipfdd.de
http://www.ipfdd.de

Abstract. We present a new Monte Carlo method which is based on the original Bond Fluctuation Model (scBFM) for simulating polymeric systems in three dimensions. A body centered cubic lattice is used instead of a simple cubic lattice. This modified Bond Fluctuation Model (bccBFM) fulfills the same requirements as the original scBFM, namely excluded volume and the cut-avoidance of bond vectors. Most remarkably the algorithm allows for a very efficient parallelization. This leads to a performance gain of about two orders of magnitude, when using graphics processor units (GPU). The bccBFM shows universal behavior both for static and dynamic properties and can be used to solve the same problems as the original scBFM, but provides an efficient implementation especially on GPUs.

Keywords: Polymers · Monte Carlo · CUDA · Bond fluctuation model · Parallel algorithm · GPU

1 Introduction

Since the development of the Bond Fluctuation Method (BFM) by Carmesin and Kremer in 1988 [1], and the extension to three dimensions by Deutsch and Binder 1990 [2], the model has served as a basis for numerous computer simulations of polymeric systems such as melts, networks, membranes, dendrimers, copolymers, rings and others [3–9]. It models polymeric systems on a coarse-grained level. Monomers are represented as cubes on a simple cubic lattice as an efficient look-up table for neighbor interactions. Excluded volume interactions were implemented by excluding multiple occupation of single lattice nodes. One of the very remarkable features is the cut-avoidance of bond vectors, meaning that a single chain can not intersect with itself or another chain preserving the local and global topological constraints of the system. Thereby entanglement effects could be properly simulated [2]. This feature is implemented very efficiently by a special set of allowed bond vectors between the monomers [2].

C. Jentzsch and R. Dockhorn contributed equally to this work.

© Springer International Publishing Switzerland 2016
R. Wyrzykowski et al. (Eds.): PPAM 2015, Part II, LNCS 9574, pp. 301–311, 2016.
DOI: 10.1007/978-3-319-32152-3_28

The model is very simple and easy to implement and runs very efficient on modern computer systems. Therefore it can reach large length and time scales.

Here, we present a new version of this model which is based on the body centered cubic lattice. This new model has the same remarkable features as the original model, but allows for a very efficient parallelization giving a performance gain of about two orders of magnitude between CPU and GPU.

The motivation for the development of the new algorithm stems from the recent developments in high performance computing with the need of massively parallel algorithms. Especially GPUs have become a powerful tool in simulating molecular systems [10]. The development of better programming models for GPU [11] has made it possible to make use of this computing power. Our algorithm makes use of the single instruction multiple data (SIMD) [12] model, by performing the same instructions for every monomer. Excluded volume and bond sharing conflicts which arise by moving all monomers in parallel, could be overcome by defining four subsets in which every monomer can be moved in parallel without conflicts. This is not possible for the original BFM (scBFM).

In Sect. 2, the model will be explained in detail. In Sect. 3, we compare the different implementations of the BFM to each other. This is followed by our conclusions and outlook.

2   The Model

The concept of our model is based on the original bond fluctuation model [1,2]. The differences have their origin in the underlying lattice type for modelling the coarse-grained monomers and efficiently implementing the look-up table for short-range non-bonded monomer interactions. Instead of a simple cubic lattice (scBFM), we use a body-centered cubic lattice in this approach (bccBFM). The bcc-lattice can be represented as two interpenetrating sc-lattices with lattice constant a shifted by $(1a, 1a, 1a)$ units as displayed in Fig. 1. The two simple cubic sub-lattices are named even (X) and odd (O) referring to the case that all coordinates of the monomer's midpoint are only even or only odd. As a consequence the used bcc-lattice is twice the size of the sc-lattice, but only half of the nodes need to be accessed within this algorithm.

2.1   Excluded Volume

A single monomer is represented as a cube on one of the two sub-lattices, see Fig. 1. It occupies one node for the midpoint on its lattice and 8 nodes on the other sub-lattice. The length of one edge is two lattice units $(2a)$. On the right-hand side of Fig. 1 we display the model in two dimensions for simplicity. The complete overlap of those cubes/monomers, or all nodes, is forbidden. But they are allowed to "touch" each other meaning that they can occupy 1, 2 or 4 edge nodes together (see Fig. 1). There is no algorithmic advantage in the bccBFM-algorithm to populate all 9 nodes explicitly, instead only the one interior node of the cube is sufficient. In the following, we restrict our explanation only to this

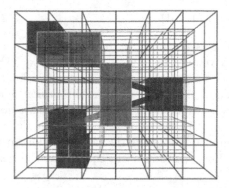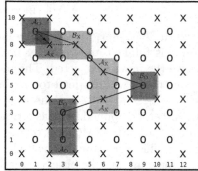

Fig. 1. Sketch of a polymer in the new bccBFM in three dimensions (left) and projected to two dimensions (right) for simplicity. We have only drawn the accessible nodes within the bcc-lattice (O as odd sub-lattice and X as even sub-lattice) and highlight the four subsets $\mathcal{A}_O, \mathcal{A}_X, \mathcal{B}_O$, and $\mathcal{B}_X$ for parallelization. The monomer in the top left corner is allowed to move to the new position (green) and thereby changing its position from the odd (blue) to the even (red) sub-lattice (Color figure online).

center node. Thereby, the minimal distance due to excluded volume constraint between the centers of mass of two monomers is two lattice units.

2.2 Set of Bond Vectors

The bond vector between connected monomers has to be chosen out of the predefined set:

$$\mathbf{B}_{bccBFM} = \mathbf{P}_{\pm} \begin{pmatrix} 2 \\ 0 \\ 0 \end{pmatrix} \cup \mathbf{P}_{\pm} \begin{pmatrix} 2 \\ 2 \\ 0 \end{pmatrix} \cup \mathbf{P}_{\pm} \begin{pmatrix} 2 \\ 2 \\ 2 \end{pmatrix} \cup \mathbf{P}_{\pm} \begin{pmatrix} 3 \\ 1 \\ 1 \end{pmatrix}, \tag{1}$$

where $\mathbf{P}_{\pm}$ stands for all permutations and sign combinations of a triple. We classify those bond vectors in a set of even and odd vectors. Even bond vectors connect monomers on the same sub-lattice, whereas odd vectors connect monomers between different sub-lattices. There are 26 different even and 24 different odd bond vectors. The bond length can vary between $b_{min} = 2$, $b = \sqrt{8}$, $b = \sqrt{11}$, and $b_{max} = \sqrt{12}$. The intersecting and crossing of strands formed by connected monomers out of this set is forbidden, thereby entanglement interactions are taken into account. The cut-avoidance preserves local topological constraints and therefore the global topology, which has been proven by the algorithm of Trautenberg et al. [13].

2.3 Algorithm

Sequential. During the elementary step of the Monte Carlo procedure, one monomer is randomly chosen. The monomer displacement is randomly chosen by one out of 8 diagonal move directions

$$\mathbf{M}_{\mathrm{bccBFM}} \in \mathbf{P}_{\pm}(1,1,1) \qquad\qquad (2)$$

with length $\|\mathbf{M}_{\mathrm{bccBFM}}\| = \sqrt{3}$. The move is accepted, if the excluded volume constraint is fulfilled and the new bond vectors belong to the set in Eq. (1). Otherwise the move is rejected. The move vector $\mathbf{M}_{\mathrm{bccBFM}}$ in context with the given constraints will not lead to any bond crossings during the simulation. For long or short-ranged interactions between the monomers the Metropolis algorithm [14] can be applied. As basic time unit in the simulation model we define one Monte Carlo-Step (MCS) as one attempted Monte Carlo move per monomer in average. This algorithm is repeated as long as necessary to equilibrate the system and for sampling the observables.

Parallel. The bccBFM can be efficiently parallelized without additional rules as necessary for the scBFM [15]. In order to parallelize the simulation we divide all monomers into two subsets $\mathcal{A}$ and $\mathcal{B}$, where every monomer is only connected to monomers of the other subset. Monomers of the same subset are not directly connected. Each of those two subsets can be divided again into two subsets, one subset where all monomers are on the even lattice (X) and another one with all monomers on the odd lattice (O) labeling the position of the monomer's interior node to be on the X or O-lattice. For parallelization all monomers are classified to belong to one of the four subsets $\mathcal{A}_{\mathrm{O}}$, $\mathcal{A}_{\mathrm{X}}$, $\mathcal{B}_{\mathrm{O}}$, and $\mathcal{B}_{\mathrm{X}}$ (for illustration see Fig. 1). During the simulation we randomly choose one of those subsets and attempt to move every monomer in parallel with the same conditions as in the sequential algorithm. This includes excluded volume, restricted bond vectors and - optionally - Metropolis criteria to include thermal interactions. This classification into four subsets guarantees implicitly that all bonds are in the allowed set and excluded volume constraints are fulfilled after performing the accepted moves in parallel. The instructions for a single monomer move applies for every monomer within the subset $\mathcal{A}_{\mathrm{O}}$, $\mathcal{A}_{\mathrm{X}}$, $\mathcal{B}_{\mathrm{O}}$, or $\mathcal{B}_{\mathrm{X}}$ and can be implemented on a SIMD (Single Instruction Multiple Data) machine such as a GPU as a efficient parallelized algorithm. We use the same time unit, MCS, as in the sequential version to be the mean number of attempted moves per monomer.

3   Results

In this section, we test the sequential algorithm (bccBFM-CPU) and parallel algorithm (bccBFM-GPU) and compare their results. For this we use the well studied systems of polymer melts and single polymer chains [16–18]. We show that the bccBFM shows universal behavior both for static and dynamic properties of polymers.

3.1   Static Properties

Single linear chains consisting of $N = 16$ to $N = 1024$ monomers were simulated in a cubic simulation box with length $256\,a$ under periodic boundary conditions.

The radius of gyration R_g^2 was calculated as follows:

$$R_g^2 = \frac{1}{N}\langle\sum_{i=1}^{N}(\boldsymbol{r}_i - \boldsymbol{r}_{\text{COM}})^2\rangle, \qquad (3)$$

where N is the number of monomers of the chain, $\boldsymbol{r}_i$ is the position of the ith monomer, $\boldsymbol{r}_{\text{COM}} = 1/N\sum_{i=1}^{N}\boldsymbol{r}_i$ is the center of mass of the chain, and $\langle\ldots\rangle$ denotes the ensemble average. The results are displayed in Fig. 2.

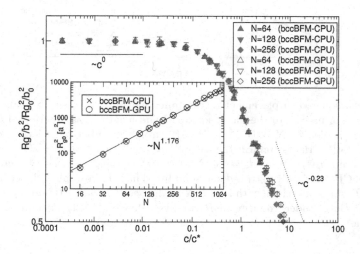

Fig. 2. The chain extension as function of the concentration c normalized by the overlap concentration $c^* = N/\frac{4\pi}{3}(R_{g0}^2/b_0^2)^{3/2}$ in the bccBFM on CPU (filled red) and GPU (open blue). Both variants confirm the theoretical prediction of dense system ($\sim c^{-0.23}$, dotted line) and the diluted case ($\sim c^0$, solid line) for linear chains with length $N = 64, 128, 256$ (from top to bottom: triangle up, triangle down, diamonds). The concentration dependent radius of gyration R_g^2 divided by its average squared bond length b^2 is normalized by the value in the highly diluted case of a single chain R_{g0}^2/b_0^2. No further rescaling between the data in the bccBFM on CPU and GPU have to be applied. Inset: Radius of gyration R_{g0}^2 in lattice units a for single chains with length N in the bccBFM on CPU (crosses) and GPU (circles). Both variants show the expected scaling with a slope of $2\nu \simeq 1.176$ without further rescaling (solid line) (Color figure online).

The bccBFM show the (expected) deviation [19] for shorter chains ($N < 32$) on the scaling relation for single self-avoiding chains $R_{g0} \propto N^\nu$ [16]. In the limit of long chains both models reach the theoretical prediction of the exponent $\nu \approx 0.588$ [20]. The average bond length for highly diluted systems, such as single chains, is $\sqrt{\langle b_0^2\rangle}_{\text{bccBFM}} \simeq 3.1$ in the parallel and sequential implementation.

In order to compare the scaling in the bccBFM, polymer solutions of various concentrations c have been simulated. We normalized R_g^2/b^2 with it's

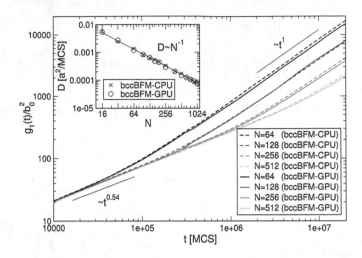

Fig. 3. Mean squared displacement of the inner monomer of a single chain in good solvent $g_1(t)$ normalized to the mean squared bond length b_0^2 as function of time t of single chain with length $N = 64, 128, 256, 512$ (from top to bottom: black, red, green, and cyan). Both variants confirm the theoretical prediction $g_1(t) \sim t^{2\nu/(1+2\nu)} \sim t^{0.54}$ for $t \ll \tau_R$ and $g_1(t) \sim t$ for $t \gg \tau_R$. No further rescaling between the data in the bccBFM on CPU and GPU (stroked and solid lines) have to be applied. Inset: Diffusion coefficient D of a single chain with length N in the bccBFM on CPU and GPU (crosses and circles). Both models confirm the Rouse-like behavior $D \sim 1/N$ (solid line) (Color figure online).

value for the highly diluted case R_{g0}^2/b_0^2 (a single chain) and plotted it versus the concentration normalized by the dimensionless overlap concentration $c^* = N/\frac{4\pi}{3}(R_{g0}^2/b_0^2)^{3/2}$ (see Fig. 2). In the diluted and the dense case the scaling relation $f(c/c^*) \sim (c/c^*)^m$ with scaling exponent m [16] should hold

$$\frac{R_g^2/b^2}{R_{g0}^2/b_0^2} \sim f\left(\frac{c}{c^*}\right) \sim \begin{cases} c^0 & \text{if } c \ll c^* \\ c^{-\frac{2\nu-1}{3\nu-1}} \sim c^{-0.23} & \text{if } c \gg c^* \end{cases}. \tag{4}$$

Both variants bccBFM-CPU and bccBFM-GPU show the same universal behavior, and confirm the scaling relation.

3.2   Dynamic Properties

In order to describe the dynamics of the simulation model, the diffusion properties of polymer chains have been inspected. We start with the discussion of the mean squared displacement (MSD) of the center of mass of the single chain as function of time t which is defined as

$$g_3(t) = \langle [\mathbf{r}_{\text{COM}}(t) - \mathbf{r}_{\text{COM}}(0)]^2 \rangle. \tag{5}$$

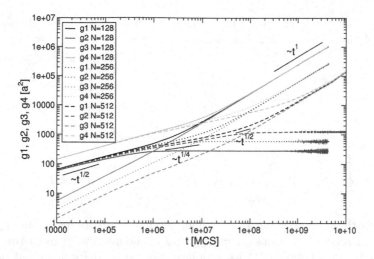

Fig. 4. Mean squared displacement of the inner monomer (black) $g_1(t)$, the inner monomer relative to the center of mass (red) $g_2(t)$, the center of mass (green) $g_3(t)$, and the end monomer (cyan) $g_4(t)$ for melts of linear chains with $N = 128$, 256, and 512 (solid, dashed and stroked lines) at $c = 0.25$ as function of time t simulated with the bccBFM on GPU. All data confirm the theoretical predictions for dynamics, but only chains with length $N = 512$ show the onset of the reptation dynamics (Color figure online).

In the limiting case for a freely diffusing polymer chain the MSD $g_3(t)$ has a linear dependence with time $g_3(t) = 6Dt$. The diffusion constant D was extracted for single chains of different lengths N in the bccBFM (see Inset Fig. 3). The diffusion constant D scales with the chain length N as $D \propto 1/N$. This is the theoretical prediction of the Rouse-like [21] behavior as expected in the bccBFM. Next, we analyze the mean squared displacement of the single monomers. We define the mean squared displacement of the inner monomer of the chain $g_1(t)$, the same displacement relative to the chain's center of mass $g_2(t)$, and $g_4(t)$ as mean squared displacement of the end monomer of the chain as

$$g_1(t) = \langle [\boldsymbol{r}_{N/2}(t) - \boldsymbol{r}_{N/2}(0)]^2 \rangle$$
$$g_2(t) = \langle ([\boldsymbol{r}_{N/2}(t) - \boldsymbol{r}_{\mathrm{COM}}(t)] - [\boldsymbol{r}_{N/2}(0) - \boldsymbol{r}_{\mathrm{COM}}(0)])^2 \rangle. \qquad (6)$$
$$g_4(t) = \langle [\boldsymbol{r}_{N}(t) - \boldsymbol{r}_{N}(0)]^2 \rangle$$

For the short times the inner modes dominate the dynamic behavior and chains in good solvent obey $g_1 \approx g_2 \sim t^{2\nu/(1+2\nu)} \sim t^{0.54}$ [22]. For times larger than the relaxation time τ_R the dynamics of the chain is purely diffusive: $g_1 \approx 6Dt$. In Fig. 3 we compare the MSD $g_1(t)$ of a single chain between CPU and GPU. Both variants confirm the theoretical predictions. No rescaling between the data of the bccBFM on CPU and GPU has to be applied since the parallelization does not require additional rules.

In the framework of the reptation theory [16,17] one expect the following scaling relations [16–18,23,24] for dense polymer system:

$$g_1(t) \sim g_4(t) \sim \begin{cases} t^{1/2} & \text{if } \tau_0 < t < \tau_e \\ t^{1/4} & \text{if } \tau_e < t < \tau_R \\ t^{1/2} & \text{if } \tau_R < t < \tau_d \\ t^1 & \text{if } t > \tau_d \end{cases}$$

$$g_3(t) \sim \begin{cases} t^1 & \text{if } t < \tau_e \\ t^{1/2} & \text{if } \tau_e < t < \tau_R, \\ t^1 & \text{if } t > \tau_R \end{cases} \quad g_2(t) \sim \begin{cases} t^{1/2} & \text{if } \tau_0 < t < \tau_e \\ t^{1/4} & \text{if } \tau_e < t < \tau_R \\ t^0 & \text{if } t > \tau_R \end{cases} \quad . \tag{7}$$

Here, τ_0 is the monomer relaxation time, the constraint induced entanglement time τ_e, the relaxation (Rouse) time τ_R, and τ_d the disentanglement time of the tube. For testing the bccBFM we simulating polymer melts made out of linear chains with length N at a volume fraction $c = 0.25$ in cubic box with length $256\,a$ over several decades in time on the GPU. There are no data on CPU available due to the long time scale needed for equilibration and statistics. In Fig. 4 the different MSD for chain length $N = 128$, 256, and 512 as function of time is shown. All data confirm the theoretical predictions for the diffuse and sub-diffusive behavior. The melt with $N = 512$ shows the regimes for repetation dynamics. Even longer chains are necessary to inspect this behavior in more detail.

In summary, we find a very good agreement between the results of the sequential (CPU) and the parallel (GPU) algorithm. There is no shift on the time scales in the dynamical behavior nor on the static properties. The bccBFM appears as a powerful tool for simulation of dense polymeric systems.

3.3   Performance Optimizations

A challenging task is the comparison of the performance between both variants of the bccBFM on CPU and GPU due to difficulties in the implementation and technical setup. Therefore, we restrict ourselves to the general explanation on techniques used on the GPU for gaining this significant speed-up. As useful comparable quantity we report the measured performance $p = t_{sim} \cdot N \cdot n/t_{real}$ in attempted moves per second (att. moves/s) as the real-time t_{real} (in seconds) needed to perform a simulation of $N \cdot n$ monomeric units for t_{sim} simulation steps (in MCS). Our testing system consists of a melt of $n = 4096$ linear chains with length $N = 128$ under periodic boundary conditions in a cubic simulation box $V = 256^3$ at concentration $c = 0.25$. The sequential variant on CPU achieves a performance of roughly $p \simeq 8 \cdot 10^6$ att. moves/s on an Intel® Xeon® CPU E5620 @ 2.40 GHz (12 MB cache) under usual code optimization. For programming the algorithm in C/C++ on GPUs the parallel computing CUDA®-framework (Compute Unified Device Architecture-Release 4.2) by NVIDIA [11] was used on a GeForce® GTX® 580 with 512 CUDA Cores. The parallelization of the application is realized by parallel execution units naming threads in an array of blocks in a grid by kernel calls (for details see [25–27]). One thread will be associated as elementary parallel movement of a monomeric unit checking excluded volume, bond vector constraints, and Metropolis criterion and applying or rejecting the invoking movement. The bond and lattice conflicts arising from the parallel movement will implicitly be avoided by splitting all monomers in the four subsets $\mathcal{A}_O$, $\mathcal{A}_X$, $\mathcal{B}_O$, and $\mathcal{B}_X$. Instead of checking one node in move direction, the vicinity of 7 lattice nodes of the same sub-lattice (X or O) has to be considered as very efficient parallelization strategy leading to a performance of $p \simeq 300 \cdot 10^6$ att. moves/s without innovative code optimization. A further improvement can be done by using CUDA's Textures1D [28,29] for accessing the lattice look-up as linearized and buffered array yielding $p \simeq 440 \cdot 10^6$ att. moves/s. Checking the lattice occupation requires only the same sub-lattice, therefore, we can re-address the linearized lattice with space filling Z-order-curve [30]. This calculation is very demanding but enhances the caching of the physical vicinity on the GPU [28,29] to provide a performance of $p \simeq 580 \cdot 10^6$ att. moves/s. Further improvements by grouping the monomer position to four consecutive `int32` for coalesced memory load and avoidance of unused threads in the kernel launch yield a performance of $p \simeq 720 \cdot 10^6$ att. moves/s. A very specific improvement is the reduction of the overall information needed for simulating the system. The monomer position and connectivity can be merged onto one `int32` reducing the overall data transfer in the calculation. These improvements lead to a performance of $p \simeq 1050 \cdot 10^6$ att. moves/s on the GPU for the bccBFM instead of $p \simeq 8 \cdot 10^6$ att. moves/s for the sequential CPU version. In summary, the highly optimized implementation of the bccBFM on the GPU runs with a performance of two orders of magnitude faster as on the CPU.

4 Conclusion

We have introduced a highly parallelizable variant of the Bond Fluctuation Model based on the body-centered cubic lattice. The underlying lattice serves as an efficient look-up table for neighbor interactions and the subdivision into four disjunct groups implicitly resolves bond and lattice conflicts and thereby facilitates the parallelization. We have shown that the bccBFM shows universal behavior both for static and dynamic properties for single polymer chains and under dense conditions, and is found to be in agreement with theoretical predictions. In summary, a performance gain of two orders of magnitude for the parallel algorithm can be realized as compared to sequential implementations. The bccBFM allows to explore new length and time scales in polymeric systems by using cost efficient and powerful GPUs. This has been proven by applying it to large scale polymer brushes for a rather large parameter space [31].

Acknowledgments. This work was supported by the Deutsche Forschungsgemeinschaft (DFG) SO-277/8-1. We thank the Center for Information Services and High Performance Computing (ZIH) at TU Dresden for generous allocations of GPU time. We thank Anne Herrmann for implementing the Trautenberg test for the bccBFM and Marco Werner for fruitful discussions (all Leibniz Institute of Polymer Research Dresden).

References

1. Carmesin, I., Kremer, K.: The bond fluctuation method: a new effective algorithm for the dynamics of polymers in all spatial dimensions. Macromolecules **21**, 2819–2823 (1988)
2. Deutsch, H.P., Binder, K.: Interdiffusion and self-diffusion in polymer mixtures: a Monte Carlo study. J. Chem. Phys. **94**, 2294–2304 (1991)
3. Sommer, J.-U., Lay, S.: Topological structure and nonaffine swelling of bimodal polymer networks. Macromolecules **35**, 9832–9843 (2002)
4. Werner, M., Sommer, J.-U.: Polymer-decorated tethered membranes under good- and poor-solvent conditions. Eur. Phys. J. E **31**, 383–392 (2010)
5. Di Cecca, A., Freire, J.J.: Monte Carlo simulation of star polymer systems with the bond fluctuation model. Macromolecules **35**, 2851–2858 (2002)
6. Subramanian, G., Shanbhag, S.: Conformational free energy of melts of ring-linear polymer blends. Phys. Rev. E Stat. Nonlin. Soft Matter Phys. **80**, 041806 (2009)
7. Lang, M., Sommer, J.-U.: Analysis of entanglement length and segmental order parameter in polymer networks. Phys. Rev. Lett. **104**, 177801 (2010)
8. Nedelcu, S., Sommer, J.-U.: Single chain dynamics in polymer networks: a Monte Carlo study. J. Chem. Phys. **130**, 204902 (2009)
9. Nedelcu, S., Sommer, J.-U.: Single-chain dynamics in frozen polymer networks. Rheol. Acta **49**, 485–494 (2010)
10. Anderson, J.A., Lorenz, C.D., Travesset, A.: General purpose molecular dynamics simulations fully implemented on graphics processing units. J. Comput. Phys. **227**, 5342–5359 (2008)
11. Nickolls, J., Buck, I., Garland, M., Skadron, K.: Scalable parallel programming with CUDA. ACM Queue **6**, 40–53 (2008)
12. Flynn, M.J.: Some computer organizations and their effectiveness. IEEE Trans. Comput. **C-21**, 948–960 (1972)
13. Trautenberg, H.L., Hölzl, T., Göritz, D.: Evidence for the absence of bond-crossing in the three-dimensional bond fluctuation model. Comput. Theor. Polym. Sci. **6**, 135–141 (1996)
14. Metropolis, N., Rosenbluth, A.W., Rosenbluth, M.N., Teller, A.H., Teller, E.: Equation of state calculations by fast computing machines. J. Chem. Phys. **21**, 1087 (1953)
15. Nedelcu, S., Werner, M., Lang, M., Sommer, J.-U.: GPU implementations of the bond fluctuation model. J. Comput. Phys. **231**, 2811–2824 (2012)
16. de Gennes, P.G.: Scaling Concepts in Polymer Physics. Cornell University Press, Ithaca (1979)
17. Doi, M., Edwards, S.F.: The Theory of Polymer Dynamics. Clarendon Press, Oxford (1987)
18. Rubinstein, M., Colby, R.: Polymer Physics. Oxford University Press, Oxford, New York (2003)

19. Wittkop, M., Kreitmeier, S., Göritz, D.: Swelling of subchains of a single polymer chain with excluded volume in two and three dimensions: a Monte Carlo study. Macromolecules **29**, 4754–4758 (1996)
20. Le Guillou, J.C., Zinn-Justin, J.: Critical exponents from field-theory. Phys. Rev. B **21**, 3976–3998 (1980)
21. Rouse, P.E.: A theory of the linear viscoelastic properties of dilute solutions of coiling polymers. J. Chem. Phys. **21**, 1272–1280 (1953)
22. Kremer, K., Binder, K.: Dynamics of polymer chains confined into tubes: scaling theory and Monte Carlo simulations. J. Chem. Phys. **81**, 6381–6394 (1984)
23. Paul, W., Binder, K., Heermann, D.W., Kremer, K.: Dynamics of polymer solutions and melts. Reptation predictions and scaling of relaxation times. J. Chem. Phys. **95**, 7726–7740 (1991)
24. Kremer, K., Grest, G.S.: Dynamics of entangled linear polymer melts: a molecular-dynamics simulation. J. Chem. Phys. **92**, 5057–5086 (1990)
25. Kirk, D.B., Hwu, W.-M.W.: Programming Massively Parallel Processors: A Hands-on Approach. Morgan Kaufmann Publishers Inc., San Francisco (2010)
26. Sanders, J., Kandrot, E.: CUDA by Example: An Introduction to General-Purpose GPU Programming. Addison-Wesley Professional, Boston (2011)
27. Wilt, N.: The CUDA Handbook: A Comprehensive Guide to GPU Programming. Addison-Wesley Professional, Upper Saddle River, New Jersey (2013)
28. NVIDIA: CUDA Toolkit Documentation. https://docs.nvidia.com/cuda/index.html
29. NVIDIA: CUDA Best Practices Guide. https://docs.nvidia.com/cuda/cuda-c-best-practices-guide/index.html
30. Morton, G.M.: A Computer Oriented Geodetic Data Base; and a New Technique in File Sequencing. IBM Germany Scientific Symposium Series (1966)
31. Jentzsch, C., Sommer, J.-U.: Polymer brushes in explicit poor solvents studied using a new variant of the bond fluctuation model. J. Chem. Phys. **141**, 104908 (2014)

Genetic Algorithm and Exact Diagonalization Approach for Molecular Nanomagnets Modelling

Michał Antkowiak[✉], Łukasz Kucharski, and Grzegorz Kamieniarz

Faculty of Physics, Adam Mickiewicz University,
ul. Umultowska 85, 61-614 Poznań, Poland
antekm@amu.edu.pl

Abstract. We combined the genetic algorithm search procedure and exact diagonalization method to obtain the fitting system with two-level parallelism and optimally balanced workload which was implemented in the HPC environment. Applying the system to the experimental magnetic susceptibility data of Cr_8Ni molecule we obtained the non-uniform exchange couplings parameters for more general models and we achieved not only better agreement with experiment but we also demonstrated that the values known in literature are systematically overestimated.

Keywords: Molecular nanomagnets · Genetic algorithms · Exact diagonalization

1 Introduction

Molecular nanomagnets based on transition metal ions have been very intensively investigated [7]. Their popularity is mostly due to the fact that quantum phenomena characteristic for a single molecule (like, e.g., quantum tunnelling or step like field dependence of magnetization) can be observed in bulk samples. It is possible because nanomolecules are magnetically shielded from each other by organic ligands and the dominant interactions are those within the molecule. There are also expectations that this kind of materials may find application in quantum computing [2,8,13,18] and information storage [14].

A large family of molecular nanomagnets comprises ring-shaped molecules. Most of them contain even number of antiferromagnetically interacting ions. Only recently the first odd membered antiferromagnetic molecules have been reported [1,3,4,10,11,19]. They are especially interesting because of magnetic frustration which is expected to appear in this kind of materials.

The molecule $(C_2H_{11})_2NH_2Cr_8NiF_9[O_2CC(CH_3)_3]_{18}$ (Cr_8Ni in short), synthesized by the Winpenny group [4,5], belongs to the chromium rings family [15]. It was obtained by doping the ion of nickel into the ring of eight ions of chromium (Cr_8) and represents an exemplary frustrated nanomagnet.

Antiferromagnetic interactions between the nearest chromium ions and between chromium and nickel ions (respectively $J = 16\,\mathrm{K}$ and $J_1 = 70\,\mathrm{K}$) were

© Springer International Publishing Switzerland 2016
R. Wyrzykowski et al. (Eds.): PPAM 2015, Part II, LNCS 9574, pp. 312–320, 2016.
DOI: 10.1007/978-3-319-32152-3_29

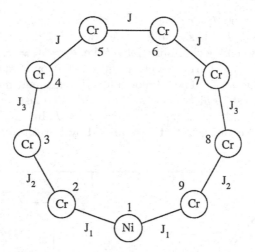

Fig. 1. The model of Cr$_8$Ni molecule. The integers denote the positions of magnetic ions within a ring.

initially obtained by fitting the magnetic susceptibility χ [4,5]. In the calculations equal value of the factor $g = 2$ was assumed for both types of ions and the single-ion anisotropy was neglected. Parameters showing better fit to the susceptibility curve were proposed by Furukawa et al. [6] ($J = 14.7\,\mathrm{K}$, $J_1 = 85\,\mathrm{K}$, $D_{Cr} = -0.42\,\mathrm{K}$, $D_{Ni} = -4.9\,\mathrm{K}$). This time the realistic values of g was adopted, 1.98 for chromium and 2.20 for nickel, and the anisotropy was included. However, surprising deviations from the experimental results in the intermediate temperature range have survived which motivated our study towards more general modeling and advanced optimization techniques needed for accurate data fitting.

2 Microscopic Models

The model of Cr$_8$Ni molecule with an increased number of couplings is presented in Fig. 1 and described by the following Heisenberg Hamiltonian:

$$\mathcal{H} = \sum_{j=4}^{6} J\boldsymbol{S}_j \cdot \boldsymbol{S}_{j+1} + J_1\boldsymbol{S}_1 \cdot (\boldsymbol{S}_2 + \boldsymbol{S}_9) + \sum_{j=2}^{3} J_j(\boldsymbol{S}_j \cdot \boldsymbol{S}_{j+1} + \boldsymbol{S}_{10-j} \cdot \boldsymbol{S}_{11-j})$$

$$+ \sum_{j=2}^{9} \left(D_{Cr}(S_j^z)^2 - g_{Cr}\mu_B \boldsymbol{B} \cdot \boldsymbol{S}_j \right) + D_{Ni}(S_1^z)^2 - g_{Ni}\mu_B \boldsymbol{B} \cdot \boldsymbol{S}_1. \tag{1}$$

We consider models with varying degrees of differentiation of exchange integrals between chromium ions (Fig. 1). For all models the value of the coupling between the chromium and nickel ions is denoted by J_1, the value of the exchange integral between the chromium ions closest to the nickel ion is denoted by J_2 and the next in the order by J_3 (see Fig. 1). Other couplings between the chromium ions are denoted by J. The basic model, which we call 2J, is characterized by the same couplings between the chromium ions J, i.e. $J_2 = J_3 = J$. Models proposed by Cador et al. [4,5] and Furukawa et al. [6] denoted by C and F, respectively, are variants of the model 2J. The 3J model has the exchange integral J_2 between chromium ions nearest to nickel ions and $J_3 = J$. The last proposed model 4J is the most general and corresponds to $J_2 \neq J_3 \neq J$. In our calculations the g values are fixed and coincide with those quoted by Furukawa et al. [6].

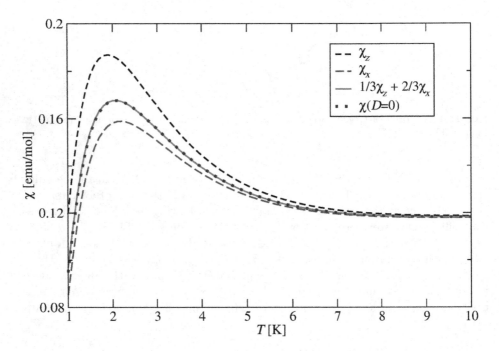

Fig. 2. Magnetic susceptibility χ_x, χ_z and averaged susceptibility (green curve) calculated for Cr_6Ni with parameters of model F compared to susceptibility $\chi(D=0)$ of corresponding isotropic model ($D_{Cr} = D_{Ni} = 0$) (Color figure online).

The experimental data were obtained on a polycrystalline sample, therefore it is important to calculate the susceptibility in both the z and x direction. However, the calculations of the magnetic properties while the magnetic field is applied in x direction would be too much time and memory consuming, making our simulations impossible. Our calculations for systems with a reduced number of chromium ions (Cr_6Ni, Cr_4Ni) showed a negligible impact of anisotropy on the averaged magnetic susceptibility due to the relatively low value of anisotropy for chromium and nickel. In the range of temperature for which experimental results have been obtained for Cr_8Ni [4,5] the susceptibility curve of Cr_6Ni theoretical ring for zero anisotropy practically coincides with that calculated for a non-zero anisotropy and averaged using the formula $\chi = \frac{1}{3}\chi_z + \frac{2}{3}\chi_x$, as shown on

Fig. 2. It presents the susceptibility curves in the x and z directions, the curve representing their averaged values and the curve for vanishing anisotropy. Using this feature we could omit the x direction calculations and fully exploit the exact diagonalization method described in the subsequent section.

3 Computational Methods

To obtain the precise values of the magnetic properties of the model we use the exact diagonalization (ED) technique. The results obtained by this method are numerically accurate, but a major constraint and challenge is the exponential increase of the size of the matrix defined by $(2S+1)^n$, where n stands for the size of the system. It is very helpful to exploit fully symmetry of a given compound. If the magnetic field is oriented along the z axis, the Hamiltonian takes a quasi-diagonal form in the basis formed by eigenvectors of the total spin projection S^z and can be divided into a number of submatrices labelled by quantum number M and the symmetry of the eigenstates.

We used the MPI [21] library to paralellize the processes of the diagonalization of separate submatrices. For the most efficient use of computing time of all processes we implemented the *Longest Processing Time* algorithm [9]. In the final version of our code we applied ScaLAPACK library [20] which not only accelerates the diagonalization process, but also allows to paralellize the diagonalization of a single submatrix over all the computational cores at a single node with shared memory.

For the fitting-based determination of the model parameters we used Genetic Algorithm (GA) approach. It is proven to be useful when applied to a similar problem of a smaller computational scale [12] or a formation of "classical atoms" [17]. Fitting process comes down to an optimization problem, the domain at which GA excels. To express the problem in terms of GA, the main points are definition of specimen and fitness function. Specimen coding is a method of translating a proposed solution into a specimen which is processed by GA operators. In our case the specimen was a vector of the parameters of a given model such as (J_1, J), (J_1, J_2, J) or (J_1, J_2, J_3, J). The fitness function defines how well is the specimen fitted to the environment. The higher this value is, the better solution a given specimen represents. GA naturally seeks the maximum of the fitness function which should also be positively defined.

We have chosen Mean Squared Error (MSE) as a function to minimize which is easy to be transformed into the fitness function defined as an inversion of MSE. To compute MSE we use Eq. (2), where e_i is the i-th experimental value of a physical quantity and c_i is a computed value using the model and parameters encoded in the specimen:

$$\text{MSE} = \frac{1}{N} \sum_{i=1}^{N} (c_i - e_i)^2. \tag{2}$$

GA processes a set of N_s specimens called a population. Each iteration of GA is called a generation and consists of few steps during which specimens are chosen to participate in the next generation, to exchange the genetic information and to be modified and evaluated again. Based on the values of fitness function the specimens introduce their copies into the temporary population which is partitioned into pairs of parents. The purpose of selection is to ensure that better fitted specimens will have more children. The method of selection we chose is a standard proportional roulette method.

The exchange of information is realized by the crossover operator which mimics production of the offspring from two parent specimens. The method we choose is a standard arithmetic crossover procedure [16]. The pairs of parents are chosen randomly from the parents pool without returning. During this step the temporary population effectively becomes a children population.

Random modification of a specimen (i.e., a set of the model parameters forming a vector) is carried by the mutation operator. Its purpose is to introduce diversity into the gene pool of the whole population. Our mutation method is not a standard one. Each specimen has a fixed chance p_{ms} to be chosen as the subject of mutation procedure. During the mutation procedure, each gene (i.e., the component of a given vector) has a fixed chance p_{mo} to be mutated. Such a two-phase approach allowed us to shape better the distribution of mutation. The values of the corresponding parameters varied between successive runs from 0.1 to 0.5. Individual gene mutation can be described by the following formula

$$g' = g \cdot (\alpha_1 \cdot c_1 + 1) + \alpha_2 \cdot c_2, \tag{3}$$

where g and g' are the values of the gene before and after the mutation respectively, α_1 and α_2 are random real numbers chosen with linear probability from $[-1, 1]$ range. The constant value of C_1 describes the magnitude of the relative value change. In our implementation $c_1 = 0.1$ which means that the value g could be changed by 10 % at most. The second term in the sum is the absolute change in value. It is needed to allow mutation to escape the region of near-zero values for g. Since we had some knowledge about the parameters, we could choose it arbitrarily so that it would not dominate expected values but still allowed to effectively escape near-zero values. We decided that $c_2 = 0.1$.

The mutated children become the new population which is evaluated before starting the next iteration. The phase of the specimen evaluation is the most time consuming part of the calculations in a given step, as it requires computing the eigenvalues and eigenvectors from the physical model for each set of parameters represented by a specimen. However, this step requires no exchange of information between specimens so that this stage can be easily parallelized and effectively executed in the HPC infrastructure.

4   HPC Implementation of the Fitting System

We used the *Huygens* supercomputer to perform the large scale calculations. Its nodes contained 16 IBM Power6 dual core processors per node. The queue system allowed to use maximum nine nodes and the longest walltime was 48 h. As our calculations were planned to last not less than a week we decided to run the comparatively not time demanding GA part of the software on our local cluster, while both memory and time consuming ED part had to be run on *Huygens*. In each iteration of GA a number of parameter sets (specimens) were generated, which were then sent to *Huygens* as an input of ED program. After performing the diagonalization and calculating the susceptibility, results were send back to the local cluster allowing the GA to create next generation.

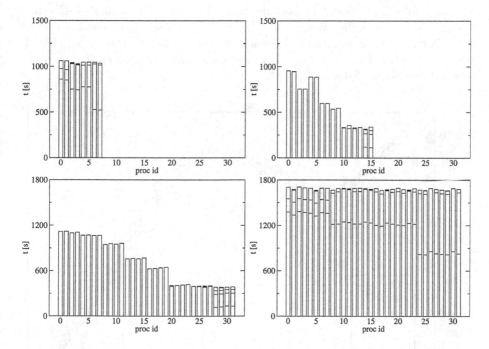

Fig. 3. Computing time balance. Upper row presents the allocation of tasks using one set of parameters for eight and 16 processes. In the lower row all 32 processes are used to compute tasks of two and four sets of parameters. Using four sets of parameters on 32 processes causes balanced work load on the node.

In principle we would be able to use three-level parallelization: specimens in population, blocks in matrix representation and SMP parallelization of math kernels to solve the eigenvalue problem for a given block. However, the maximal allowed size of computing resources per job was too small to utilize such approach efficiently.

For the most optimal use of computing time on *Huygens* we performed tests of work balance on single node. One set of parameters could be effectively parallelized on eight cores only and four sets were needed to equally load all 32 cores (see Fig. 3). Using nine nodes we could perform effective simulations for 36 sets of parameters at a time. Two sets of parameters were needed for each specimen and the best specimen in generation remained unchanged, therefore we adjusted the population size to 19. Each GA run consisted of 1000 generations.

5   The Susceptibility Fits and Conclusions

We confirmed the temperature dependence of the susceptibility found for the original model C [4,5], denoted C_1 in Fig. 4. However, its variant C_2 with realistic g values is inaccurate. Our model $2J$ improved the quality of fit for the higher temperatures range only compared to the C and F predictions (inset in Fig. 4) but significant improvement in the quality of fit we obtained using the $3J$ model (red curve in Fig. 4). Model $4J$ is characterized by an even smaller MSE value, however the susceptibility values are very close to those calculated for the $3J$ model. For the sake of the clarity we do not present the $4J$ model results on Fig. 4. The parameters obtained in this work as well as the parameters of previously proposed models are listed in Table 1.

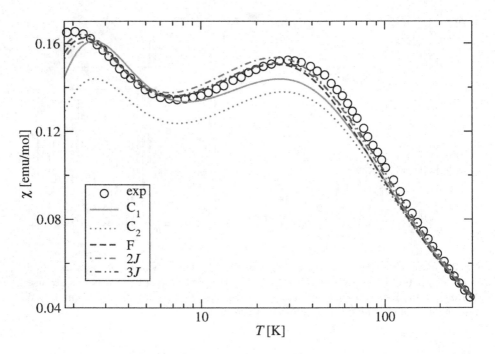

Fig. 4. Temperature dependence of susceptibility for different models of Cr_8Ni molecule. Curves are specified in the legend. Symbols represent experimental data (log-linear scale) (Color figure online).

We have not tried to establish neither the role of different couplings for the quality of the fits nor the uncertainty in their values. However, it is clear from the data in Table 1 that the values of couplings previously calculated are overestimated. In particular, the value of the microscopic model parameter J_1 should be substantially reduced. This outcome provides an evidence for uncertainties of the parameter-fit estimates inherent to the optimization procedure which should be very carefully performed.

Table 1. Comparison between the parameters of the Cr_8Ni molecule models: C_1 – Cador model [4,5], C_2 – Cador model with realistic g_{Cr} and g_{Ni} values, F – Furukawa model [6], $2J$, $3J$, $4J$ – models proposed in this paper.

Model	J [K]	J_3 [K]	J_2 [K]	J_1 [K]	g_{Cr}	g_{Ni}
C_1	16.00	J	J	70.00	2.00	2.00
C_2	16.00	J	J	70.00	1.98	2.20
F	14.70	J	J	85.00	1.98	2.20
$2J$	14.38	J	J	64.76	1.98	2.20
$3J$	14.89	J	14.15	48.66	1.98	2.20
$4J$	13.55	15.23	15.75	46.20	1.98	2.20

In conclusion, we have established that the non-uniform coupling models improve the susceptibility fit to experiment and predict lower values of magnetic couplings than those determined in literature. This was accomplished performing highly parallelized and optimally balanced extensive calculations in the HPC environment as well as optimizing the quantitative measure of the fit by using the genetic algorithm based procedure.

Acknowledgments. The calculations were performed on computer facilities granted by PSNC (Poznań, Poland) as well as within DECI programme by the PRACE-3IP (FP7/2007–2013) under grant agreement no RI-312763 in Nicosia (Cyprus). Support from the Polish MNiSW through the grant No. N519 579138 is also acknowledged.

References

1. Antkowiak, M., Kozłowski, P., Kamieniarz, G., Timco, G., Tuna, F., Winpenny, R.: Detection of ground states in frustrated molecular rings by in-field local magnetization profiles. Phys. Rev. B **87**, 184430 (2013)
2. Ardavan, A., Rival, O., Morton, J., Blundell, S., Tyryshkin, A., Timco, G., Winpenny, R.: Will spin-relaxation times in molecular magnets permit quantum information processing? Phys. Rev. Lett. **98**, 057201 (2007)
3. Baker, M., Timco, G., Piligkos, S., Mathieson, J., Mutka, H., Tuna, F., Kozłowski, P., Antkowiak, M., Guidi, T., Gupta, T., Rath, H., Woolfson, R., Kamieniarz, G., Pritchard, R., Weihe, H., Cronin, L., Rajaraman, G., Collison, D., McInnes, E., Winpenny, R.: A classification of spin frustration in molecular magnets from a physical study of large odd-numbered-metal, odd electron rings. Proc. Natl. Acad. Sci. USA **109**(47), 19113–19118 (2012)
4. Cador, O., Gatteschi, D., Sessoli, R., Barra, A.L., Timco, G., Winpenny, R.: Spin frustration effects in an oddmembered antiferromagnetic ring and the magnetic Möbius strip. J. Magn. Magn. Mater. **290**(291), 55–60 (2005)
5. Cador, O., Gatteschi, D., Sessoli, R., Larsen, F., Overgaad, J., Barra, A.L., Teat, S., Timco, G., Winpenny, R.: The magnetic Möbius strip: synthesis, structure, and magnetic studies of odd-numbered antiferromagnetically coupled wheels. Angew. Chem. Int. Ed. **43**, 5196–5200 (2004)

6. Furukawa, Y., Kiuchi, K., Kumagai, K., Ajiro, Y., Narumi, Y., Iwaki, M., Kindo, K., Bianchi, A., Carretta, S., Santini, P., Borsa, F., Timco, G., Winpenny, R.: Evidence of spin singlet ground state in the frustrated antiferromagnetic ring Cr_8Ni. Phys. Rev. B **79**, 134416 (2009)
7. Gatteschi, D., Sessoli, R., Villain, J.: Molecular Nanomagnets. Oxford University Press, Oxford (2006)
8. Georgeot, B., Mila, F.: Chirality of triangular antiferromagnetic clusters as qubit. Phys. Rev. Lett. **104**, 200502 (2010)
9. Graham, R.: Bounds of multiprocessing timing anomalies. SIAM J. Appl. Math. **17**, 416–429 (1969)
10. Hoshino, N., Nakano, M., Nojiri, H., Wernsdorfer, W., Oshio, H.: Templating odd numbered magnetic rings: oxovanadium heptagons sandwiched by β-cyclodextrins. J. Am. Chem. Soc. **131**, 15100 (2009)
11. Kamieniarz, G., Florek, W., Antkowiak, M.: Universal sequence of ground states validating the classification of frustration in antiferromagnetic rings with a single bond defect. Phys. Rev. B **92**, 140411 (2015)
12. Kucharski, L., Kamieniarz, G., Antkowiak, M., Drzewiński, A.: Single-ion anisotropy estimates for the rhenium (IV-based) molecular magnets: modeling and simulations studies. J. Phys. Soc. Jpn. **83**, 064702 (2014)
13. Lehmann, J., Gaita-Ariño, A., Coronado, E., Loss, D.: Spin qubits with electrically gated polyoxometalate molecules. Nat. Nanotechnol. **2**, 312 (2007)
14. Mannini, M., Pineider, F., Sainctavit, P., Danieli, C., Otero, E., Sciancalepore, C., Talarico, A., Arrio, M.A., Cornia, A., Gatteschi, D., Sessoli, R.: Magnetic memory of a single-molecule quantum magnet wired to a gold surface. Nat. Mater. **8**, 194 (2009)
15. McInnes, E.J.L., Piligkos, S., Timco, G., Winpenny, R.: Studies of chromium cages and wheels. Coord. Chem. Rev. **249**, 2577–2590 (2005)
16. Michalewicz, Z.: Genetic Algorithms + Data Structures = Evolution Programs. Springer, Heidelberg (1996)
17. Sobczak, P., Kucharski, L., Kamieniarz, G.: Genetic algorithm approach to calculation of geometric configurations of 2D clusters of uniformly charged classical particles. Comp. Phys. Comm. **182**, 1900 (2011)
18. Timco, G., Carretta, S., Troiani, F., Tuna, F., Pritchard, R., Muryn, C., McInnes, E., Ghirri, A., Candini, A., Santini, P., Amoretti, G., Affronte, M., Winpenny, R.: Engineering the coupling between molecular spin qubits by coordination chemistry. Nat. Nanotechnol. **4**, 173–178 (2009)
19. Yao, H., Wang, J., Ma, Y., Waldmann, O., Du, W., Song, Y., Li, Y., Zheng, L., Decurtins, S., Xin, X.: An iron(III) phosphonate cluster containing a nonanuclear ring. Chem. Commun. **16**, 1745–1747 (2006)
20. ScaLAPACK – Scalable Linear Algebra PACKage. http://www.netlib.org/scalapack/
21. The Message Passing Interface (MPI) standard. http://www.mcs.anl.gov/research/projects/mpi/

Augmented Symmetry Approach to the DFT Simulations of the Chromium-Based Rings

Michał Wojciechowski[1(✉)], Bartosz Brzostowski[1], and Grzegorz Kamieniarz[2]

[1] Institute of Physics, University of Zielona Góra,
ul. Prof. Szafrana 4a, 65-516 Zielona Góra, Poland
20000339@stud.uz.zgora.pl
[2] Faculty of Physics, Adam Mickiewicz University,
ul. Umultowska 85, 61-614 Poznań, Poland

Abstract. We suggest an augmented symmetry approach to reduce the computational complexity of the DFT electronic structure calculations based on the Wien2k package and to extend its applicability to the studies of heterogeneous ring-shape molecular nanomagnets. The approach is tested for the reference chromium-based rings Cr_8, Cr_7Cd and Cr_7Ni, and a good agreement with the results of the previous standard studies is reached with the substantial gain in the computing time.

Keywords: Molecular magnets · Chromium nanorings · Density functional theory

1 Introduction

Scientific investigations of magnetic systems have been focused on single-molecule complexes, specifically a subgroup of these materials known as molecular rings [1–16]. These types of compounds consist of paramagnetic core and organic ligand shell. They are ideal for investigating magnetic properties of spin coupled systems, which depend mostly on transitional metals embedded in the molecule. Unfortunately these systems are extremely difficult to analyse theoretically as their studies are time and resource consuming, especially using the Wien2k package [17–19]. Thus there is a need for decreasing computational complexity. So far several solutions to this problem have been proposed such as different models [11–13] or using different computational packets (SIESTA [5], NWChem [4], Gaussian [20]). In this paper we show that the investigation of the Cr_7M family of molecular rings using the WIEN2k package and augmented symmetry leads to a substantial gain in the computing time without a loss of a quality of the physical results.

2 Studied Molecules and Computational Details

The molecules that were the object of our study, are $Cr_8F_8(Piv)_{16}$, $Cr_7CdF_8(Piv)_{16}$ and $Cr_7NiF_8(Piv)_{16}$ [1,2] to which for short we refer as Cr_8,

© Springer International Publishing Switzerland 2016
R. Wyrzykowski et al. (Eds.): PPAM 2015, Part II, LNCS 9574, pp. 321–331, 2016.
DOI: 10.1007/978-3-319-32152-3_30

Cr_7Cd and Cr_7Ni. The Piv group is pivalic acid - trimethyl acetic acid $CO_2C(CH_3)_3$. The original synthesized homonuclear Cr_8 molecule consists of eight chromium atoms arranged in a ring. They lay almost in a single layer. Each pair of Cr atoms is connected with each other by a single fluorine bridge oriented inside a ring and two pivalic groups that span outside. For heteronuclear derivatives one Cr atom is substituted by Cd or Ni. The doped molecule is presented in Fig. 1.

For computational reasons the molecules here are simplified by a process called "hydrogen saturation" which replaces each methyl group CH_3 of the pivalic group by a single H atom. As a result the $CO_2C(CH_3)_3$ becomes CO_2CH_3. This process is applied once again, producing O_2CH bridge which replaces the whole pivalic group [15,16]. It allows us to reduce significantly the number of atoms from 272 to 80, yet does not affect magnetic properties.

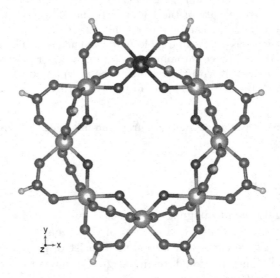

Fig. 1. The structure of the Cr_7M molecule with hydrogen atoms in place of pivalic groups. Chromium is gray, oxygen is red, carbon is brown, fluorine is blue and hydrogen is yellow. Color green represents the substituting atom (Color figure online).

The crystallographic structures of the studied compounds have been deposited in Cambridge Crystallographic Data Centre (CCDC). Records CCDC-164814 and CCDC-164814 through CCDC-191623 contain the supplementary crystallographic data used for this paper. These data can be obtained free of charge at CCDC website: www.ccdc.cam.ac.uk/conts/retrieving.html.

The ab initio calculations are carried using all electron linearized augmented plane wave (LAPW) method [21,22] implemented in WIEN2k computational package [17–19]. The LAPW method uses muffin-tin radii (RMT) approximation [23] to describe the atomic spheres. The values of the RMT parameters which have been chosen for different elements are the same as in [6,16], i.e. 2.40, 1.20,

1.24, 1.00 and 0.83 Bohrs for Cr, F, O, C and H, respectively. The parameters defining the basis set are RKMAX = 2.0 and GMAX = 20. We account for the exchange and correlation effects using the Perdew, Burke and Ernzerhof (PBE) functional [24]. We do not include spin-orbit coupling in our calculations, as it was already shown that the differences in the results are negligible [12,16] and the single-ion anisotropy for the Cr ions is known to be very small [1,25,26].

3 Augmented Symmetry Approach

The Cr_8 molecule posses rotational symmetry. The basic element consists of two, neighboring chromium ions along with accompanying bridges. This 90° wide sliver can be rotated around the axis perpendicular to the plane of the molecule (the z axis) in order to reconstitute the whole ring. Schematic representation of symmetry in Cr_8 molecule is shown in Fig. 2(a). This symmetry can be used in calculations regardles of the specific DFT method as long as the computational software used for research implements such option. For example, WIEN2k uses symmetries to reduce the number of inequivalent atoms in the structure (which is the reason we use it in the first place), while SIESTA package [27] does not. This reduction puts artificial constrains on the system in question. By lowering the number of inequivalent atoms one reduces the number of degrees of freedom. Lower amount of freedom means that less steps have to be made before self-consistency is reached. In turn this leads to faster convergence of calculations.

Unfortunately what we gain in speed we loose in variety of spin configurations we can investigate. As the properties of six chromium ions mimic those of the remaining two, we are left only with four spin configurations: up-up, up-down, down-up and down-down. However, due to symmetry of spin inversion configuration up-up is identical to down-down and up-down to down-up. Thus we are left only with two inequivalent states: antiferromagnetic and ferromagnetic. Furthermore, the reduction of complexity based on internal symmetry of the ring can be used only in case of Cr_8 molecule. For heteronuclear rings substituting single chromium ion with some other element completely removes the symmetry. This limitation applies also for homonuclear odd-number rings like Cr_9 synthesised recently [20].

In order to deal with these problems we have introduced augmented symmetry approach. Here the basic symmetry element is made up from half of the ring. It consists of five, consecutive ions forming an 180° wide arc, and the doped ion M is located on one of the ends of that arc. Now we rotate that arc 180° around an axis going through its ends. That way we reconstitute the whole ring. Schematic representation of this augmented symmetry approach in Cr_7M molecule is shown in Fig. 2(b).

The augmented symmetry approach does not reconstruct the ring ideally. This is due to the arrangement of atoms in the ring along the z axis (perpendicular to the plane of the ring), specifically the metallic ions, which get slightly distorted. Nonetheless these distortions are very small and do not influence properties of the molecule.

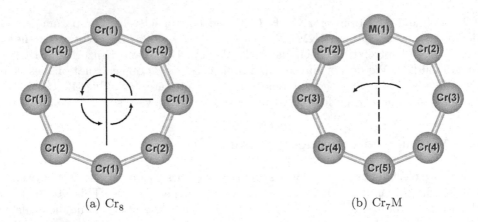

(a) Cr_8 (b) Cr_7M

Fig. 2. Schematic representation of symmetry in Cr_8 molecule and augmented symmetry for Cr_7M molecule. (a) The basic element consists of two, neighboring chromium ions Cr(1) and Cr(2). This 90° wide sliver reconstitutes the whole ring by rotating it around the axis perpendicular to the plane of the molecule by 90°, 180° and 270°. (b) The basic element consists of five, consecutive chromium ions Cr(1), Cr(2), Cr(3), Cr(4) and Cr(5). This 180° wide element reconstitutes the whole ring by rotating it 180° around the axis going through its ends.

The nature of distortion in augmented symmetry is presented in Fig. 3. For original Cr_8 molecule inequivalent chromium atoms lay in two separate layers. These layers are separated along z axis by a distance of 0.15 Å. Since the whole structure is reconstituted by rotation around z axis, the remaining atoms are placed in these two layers (Fig. 3(a)). For Cr_7M molecule in augmented symmetry the third layer is created. Two parts of the rings — basic one and reconstituted — are still identical, but are joined in a slightly different manner (Fig. 3(b)). These differences are not significant. The distance between neighboring layers is 0.15 Å and between outer-most layers is equal to 0.30 Å which is still small compared to the size of the ring (distance between two opposing magnetic ions in the ring is 8.85 Å).

One can try to avoid any geometrical error by reducing the number of symmetry operations and thus allowing for examination of configurations beyond the obvious fully ferromagnetic and fully antiferromagnetic. However this still applies only for the original Cr_8 ring and not for Cr_9 or Cr_7M because the lack of any symmetry in those structures excludes their reduction. In case of WIEN2k this reduction is possible for Cr_8, as long as the symmetry operations that are used constitute one of the existing 230 crystallographic three-dimensional symmetry space groups. Octanuclear magnetic rings that we study consist of 80 atoms, among which eight can be considered magnetic ones. The original Cr_8 molecule belongs to space group No. 75 (P4 unique axis c). Within this group it has 20 nonequivalent atoms and two of them are magnetic which allows for two nonequivalent spin configurations. By reducing the number of symmetries we can change it to space group No. 3 (P2 unique axis c). Within this group

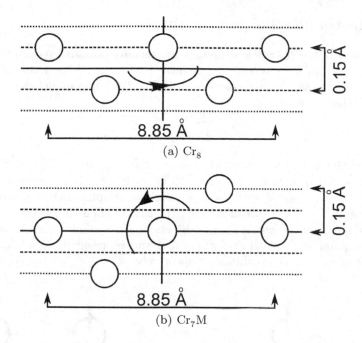

Fig. 3. Schematic representation of distortions in augmented symmetry for Cr_7M molecule in reference to symmetry in Cr_8 molecule.

the molecule has 40 nonequivalent atoms and four of them are magnetic which allows for four nonequivalent spin configurations and no geometrical distortion. The Cr_7M molecules belong to space group No. 1 (P1). Within this group they have 80 nonequivalent atoms. For Cr_7Ni eight of them are magnetic which allows for 72 nonequivalent spin configurations. For Cr_7Cd seven atoms are magnetic which allows for 36 nonequivalent spin configurations. By applying augmented symmetry approach we can reconstruct Cr_7M molecules in space group No. 3 (P2 unique axis b). This however introduces geometrical distortion. Within this group they have 41 nonequivalent atoms. For Cr_7Ni five of them are magnetic which allows for 16 nonequivalent spin configurations and for Cr_7Cd four of them are magnetic which allows for 8 nonequivalent spin configurations.

Some reduction of inequivalent positions can be obtained using chain models [11–13]. Admittedly they reproduce the physical results very well but unfortunately they significantly deform the compound. While augmented symmetry slightly distorts the way two halves of the molecular ring are joined, the chain models remove the ring-like shape of the molecule entirely by replacing it with a one dimensional chain, effectively changing a single molecule into a proper crystal.

4   Extracting Exchange Couplings

The estimates of the magnetic couplings are usually obtained within the standard projected broken symmetry approach (PBS) [28]. Within this approach the magnetic interactions are considered within Ising-like model expressed by Heisenberg Hamiltonian

$$H = \sum_i^N J_i \sigma_i \sigma_{i+1}, \tag{1}$$

where N is the number of magnetic ions in the ring, σ_i is the classical spin variable at site i equal to $\sigma_i = \pm S_i$, which is subject to the periodic boundary condition $\sigma_i = \sigma_{i+N}$ and J_i is the nearest-neighbor coupling between two ions at positions i and $i+1$. The expression under the sum in (1) is the energy of a single interacting pair and the whole sum is the energy of a whole ring. As shown in Fig. 4 these pairs are parallel or anti-parallel and represent the high-spin (HS) and the low-spin (LS) configuration, respectively.

(a) High-spin configuration   (b) Low-spin configuration

Fig. 4. The magnetically interacting pair of neighboring ions with spins S_i and S_{i+1}.

For magnetically interacting pair of neighboring ions with spins S_i and S_{i+1}, the excess energy of the configurations HS and LS considered in Fig. 4 amounts to

$$\Delta E_i = E_i^{HS} - E_i^{LS} = 2 J_i S_i S_{i+1}. \tag{2}$$

If entire configuration consists only from HS pairs we call it the ferromagnetic configuration (FM) and if it consists only from LS pairs than we call it antiferromagnetic (AFM) even if the total spin of molecule is nonzero and can be considered an ferrimagnetic one (as long as consecutive spins are alternating we consider it AFM). The total energy gap ΔE_{TOT} between FM and AFM configurations is given by the sum of the ΔE_i contributions described by (2) and can be expressed as

$$\Delta E_{TOT} = E_{FM} - E_{AFM} = 2 \sum_i^N J_i S_i S_{i+1}. \tag{3}$$

For the homometallic Cr_8 ring all the ions have the same spin $S_i = \frac{3}{2}$ and all the couplings are $J_i = J_{Cr\text{-}Cr}$ (like in Fig. 5(a)) so that (3) yields the coupling

$$J_{Cr\text{-}Cr} = \Delta E_{TOT}/36.$$

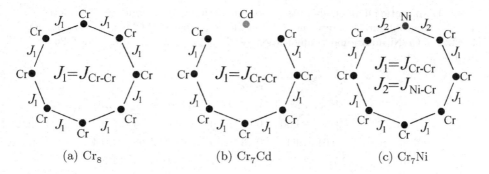

Fig. 5. Schemata of nearest neighbor interactions for Cr_8 (a), Cr_7Cd (b) and Cr_7Ni (c).

For the heterometallic Cr_7Cd ring all seven Cr ions have the same spin $S_i = \frac{3}{2}$ and single Cd ion is non-magnetic. Then Fig. 5(b) and (3) imply the coupling

$$J_{Cr-Cr} = \Delta E_{TOT}/27.$$

In the heterometallic Cr_7Ni ring all the Cr ions have the same spin $S_i = \frac{3}{2}$ and a single Ni ion has spin $S_{Ni} = 1$. We assume that all the couplings between Cr ions are the same and the couplings between Ni and Cr are different (like in Fig. 5(c)). As a consequence, we have two coupling parameters and we need energies of three different spin configurations. To that end we introduce another configuration, which we call AM. In this configuration all Cr-Cr pairs are in LS states and Ni-Cr pairs are in HS states. Following the same logic as before we obtain final J parameters from expressions

$$J_{Cr-Cr} = \frac{E_{FM} - E_{AM}}{27}, \qquad J_{Ni-Cr} = \frac{E_{AM} - E_{AFM}}{6}.$$

5   Results and Discussion

The ground states for the Cr_7M molecules are antiferromagnetic ones with the total magnetic moment $m = 0$ for Cr_8, $m = 3\,\mu_B$ for Cr_7Cd and $m = 1\,\mu_B$ for Cr_7Ni. Magnetic moments obtained for Cr ions are of magnitude 2.71–2.78 μ_B. Magnetic moments obtained for Ni ions are of magnitude 1.50–1.54 μ_B. The values of magnetic moments for the remaining atoms, including Cd, are close to 0. The magnetic moments are strongly localized on magnetic ions. The calculated values are shown in Table 1.

These magnetic moments compare very well with other results, i.e. the ones from [6] or [8]. In [6] results were obtained for Cr_8 exploiting WIEN2k and full structural symmetry as described in Sect. 3. There the magnetic moments for Cr ions are of magnitude 2.68–2.82 μ_B depending on the exchange and correlation functional that was considered. Similar agreement takes place referring to [8]. The authors performed calculations for Cr_7Cd and Cr_7Ni using SIESTA package [27] that does not implement symmetries. The magnetic moments found are of

Table 1. Magnetic moments in units μ_B for magnetic ions in augmented symmetry. For all cases two configurations are shown — antiferromagnetic (AFM) and ferromagnetic (FM). For Cr_7Ni additional configuration (AM) is given. Site numbering follows the one in Fig. 2(b).

	Cr_8		Cr_7Cd		Cr_7Ni		
	AFM	FM	AFM	FM	AFM	FM	AM
Total	0.00	24.00	3.00	21.00	1.00	23.00	5.00
Interstitial	−0.01	1.64	−0.23	1.45	−0.03	1.63	−0.41
M(1)	2.72	2.77	0.00	0.01	−1.50	1.54	1.53
Cr(2)	−2.72	2.76	2.73	2.74	2.70	2.76	2.75
Cr(3)	2.72	2.77	−2.72	2.78	−2.72	2.78	−2.72
Cr(4)	−2.71	2.76	2.72	2.77	2.72	2.77	2.72
Cr(5)	2.72	2.77	−2.72	2.77	−2.73	2.77	−2.72

magnitude 2.86–2.89 μ_B for Cr ions and 1.57 μ_B for Ni ion. For the remaining atoms, including Cd, the estimates reported are close to 0.

As for the interaction parameters J, they were estimated using the standard projected broken symmetry (PBS) approach [28] stated in Sect. 4. For Cr_8 molecule the value of J_{Cr-Cr} parameter is 7.4 meV. For Cr_7Cd molecule the value of J_{Cr-Cr} parameter is 6.4 meV. No interaction between Cr and Cd ions is assumed since Cd has no magnetic moment. For Cr_7Ni molecule the value of J_{Cr-Cr} parameter is 6.6 meV and J_{Ni-Cr} is 4.2 meV.

Again these results compare very well with those known in literature, e.g. in [6,8]. In [6] the value of J_{Cr-Cr} for Cr_8 is 6.3 meV. In [8] the value of J_{Cr-Cr} for Cr_7Cd is 6.9 meV. For Cr_7Ni molecule the value of J_{Cr-Cr} is 6.9 meV and J_{Ni-Cr} between Ni and Cr ions is 5.1 meV. The values of J parameters in [8] differ from those quoted here, because they are calculated using unprojected broken symmetry (UBS) approach [29]. One can easily convert results obtained in PBS and UBS by multiplying them by a certain factor. For Cr rings with Cd and Ni this factor is $\frac{3}{4}$, so $J_{UBS} = \frac{3}{4}J_{PBS}$ and $J_{PBS} = \frac{4}{3}J_{UBS}$.

The physical results were obtained using augmented symmetry approach on a series of different machines. However, to demonstrate the advantages of our approach the numerical calculations were carried out on a single platform with Intel Xeon X5460 processors considering Cr_8 ring only. Three different symmetry groups were considered: the original (full) group No. 75, the augmented group No. 3 and basic group No. 1 (no symmetry imposed). Both execution time and memory consumption were measured for execution of the procedure "lapw0" which implements usage of symmetry. This procedure is responsible for generation of energy potential from electronic density within WIEN2k package. The exact results of this analysis are shown in Table 2. It is clear, that while the augmented symmetry approach is not as effective as using the full symmetry in the system, it still provides significant reduction of computational time and moderate decrease in memory consumption.

Table 2. Execution time t and memory consumption mem of single iteration of "lapw0" procedure for different symmetry groups applied to Cr_8. Results are given in absolute units (hours and minutes for time and gigabytes for memory) as well as in relation to the value of calculations without usage of symmetry. The values quoted are averaged over series of iterations.

Symmetry	t [hh:mm]	t [%]	mem [GB]	mem [%]
None	01:20	100	2.5	100
Augmented	00:25	31	2.1	84
Full	00:09	11	1.6	64

6   Conclusions

We have shown that the augmented symmetry imposed on a system within the Wien2k package not only speeds up calculations but also successfully reproduces the results obtained using actual symmetry in a system [6] or no symmetry at all [8]. Calculated exchange couplings and magnetic moments are in quantitative agreement with other DFT theoretical investigations made for the Cr_7M family [6,8] and carried using SIESTA package and WIEN2k. We expect that our approach will be applicable for other DFT packages exploiting symmetry and will facilitate the quantitative analysis of more complex molecular nanomagnets, too.

Acknowledgments. One of the authors (M.W.) was supported by the European Union scholarship No. DFS.VI.052.4.62.8.2013 from Human Capital Programme, submeasure 8.2.2, priority VIII, funded from EFS and the Lubusz Voivodeship budget. Calculations were performed on computing servers in the Institute of Physics at University of Zielona Góra. We also acknowledge access to the HPC resources in PSNC Poznań (Poland) and those available within the DECI program by PRACE-3IP project (FP7 RI-312763) in Nicosia (Cyprus).

References

1. van Slageren, J., Sessoli, R., Gatteschi, D., Smith, A.A., Helliwell, M., Winpenny, R.E.P., Cornia, A., Barra, A.-L., Jansen, A.G.M., Rentschler, E., Timco, G.: Magnetic anisotropy of the antiferromagnetic ring [Cr_8F_8Piv_16]. Chem. Eur. J. **8**, 277 (2002)
2. Larsen, F.K., McInnes, E.J.L., Mkami, H.E., Overgaard, J., Piligkos, S., Rajaraman, G., Rentschler, E., Smith, A.A., Smith, G.M., Boote, V., Jennings, M., Timco, G.A., Winpenny, R.E.P.: Synthesis and characterization of heterometallic {Cr_7M} wheels. Angew. Chem. Int. Ed. **42**, 101 (2003)
3. Christian, P., Rajaraman, G., Harrison, A., McDouall, J.J.W., Rafterya, J.T., Winpenny, R.E.P.: Structural, magnetic and DFT studies of a hydroxide-bridged {Cr_8} wheel. Dalton Trans. **10**, 1511 (2004)
4. Bellini, V., Affronte, M.: A density-functional study of heterometallic Cr-based molecular rings. J. Phys. Chem. B **114**, 14797 (2010)

5. Brzostowski, B., Ślusarski, T., Kamieniarz, G.: DFT study of octanuclear molecular chromium-based ring using new pseudopotential parameters. Acta Phys. Pol. A. **121**, 1115 (2012)
6. Wojciechowski, M., Brzostowski, B., Kamieniarz, G.: DFT estimation of exchange coupling constant of Cr_8 molecular ring using the hybrid functional B3LYP. Acta Phys. Pol. A. **127**, 407 (2015)
7. Kamieniarz, G., Kozłowski, P., Antkowiak, M., Sobczak, P., Ślusarski, T., Tomecka, D.M., Barasiński, A., Brzostowski, B., Drzewiński, A., Bieńko, A., Mroziński, J.: Anisotropy, geometric structure and frustration effects in molecule-based nanomagnets. Acta Phys. Pol. A. **121**, 992 (2012)
8. Brzostowski, B., Lemański, R., Ślusarski, T., Tomecka, D., Kamieniarz, G.: Chromium-based rings within the DFT and Falicov-Kimball model approach. J. Nanopart. Res. **15**, 1528 (2013)
9. Brzostowski, B., Wojciechowski, M., Lemański, R., Kamieniarz, G., Timco, G.A., Tuna, F., Winpenny, R.E.P.: DFT and Falicov-Kimball model approach to Cr9 molecular ring. Acta Phys. Pol. A. **126**, 270 (2014)
10. Wojciechowski, M., Brzostowski, B., Lemański, R., Kamieniarz, G.: Mapping of the DFT spin configuration energies of Cr8Cd molecular ring onto the energy structure of Falicov-Kimball model. Acta Phys. Pol. A. **127**, 410 (2015)
11. Bellini, V., Tomecka, D.M., Brzostowski, B., Wojciechowski, M., Troiani, F., Manghi, F., Affronte, M.: DFT study of the Cr8 molecular magnet within chain-model approximations. In: Wyrzykowski, R., Dongarra, J., Karczewski, K., Waśniewski, J. (eds.) PPAM 2013, Part II. LNCS, vol. 8385, pp. 428–437. Springer, Heidelberg (2014)
12. Tomecka, D.M., Bellini, V., Troiani, F., Manghi, F., Kamieniarz, G., Affronte, M.: Ab initio study on a chain model of the Cr_8 molecular magnet. Phys. Rev. B **77**, 224401 (2008)
13. Ślusarski, T., Brzostowski, B., Tomecka, D.M., Kamieniarz, G.: Application of the package SIESTA to linear models of a molecular chromium-based ring. Acta Phys. Pol. A. **118**, 967 (2010)
14. Brzostowski, B., Wojciechowski, M., Kamieniarz, G.: Fundamental gaps in Cr_8, Cr_7Ni and Cr_7 molecules. Acta Phys. Pol. A. **126**, 234 (2014)
15. Ślusarski, T., Brzostowski, B., Tomecka, D., Kamieniarz, G.: Electronic structure and magnetic properties of a molecular octanuclear chromium-based ring. J. Nanosci. Nanotechnol. **11**, 9080 (2011)
16. Bellini, V., Olivieri, A., Manghi, F.: Density-functional study of the Cr_8 antiferromagnetic ring. Phys. Rev. B **73**, 184431 (2006)
17. Blaha, P., Schwarz, K., Madsen, G.H.K., Kvasnicka, D., Luitz, J.: WIEN2k: An Augmented Plane Wave + Local Orbitals Program for Calculating Crystal Properties, Karlheinz Schwarz, Techn., Universität Wien (1999)
18. Sjöstedt, E., Nordström, L., Singh, D.J.: An alternative way of linearizing the augmented plane-wave method. Solid State Commun. **114**, 15 (2000)
19. Madsen, G.K.H., Blaha, P., Schwarz, K., Sjöstedt, E., Nordström, L.: Efficient linearization of the augmented plane-wave method. Phys. Rev. B **64**, 195134 (2001)
20. Baker, M.L., Timco, G.A., Piligkos, S., Mathieson, J.S., Mutka, H., Tuna, F., Kozłowski, P., Antkowiak, M., Guidi, T., Gupta, T., Rath, H., Woolfson, R.J., Kamieniarz, G., Pritchard, R.G., Weihe, H., Cronin, L., Rajaraman, G., Collison, D., McInnes, E.J.L., Winpenny, R.E.P.: A classification of spin frustration in molecular magnets from a physical study of large odd-numbered-metal, odd electron rings. Proc. Natl. Acad. Sci. U.S.A. **109**, 19113 (2012)

21. Andersen, O.K.: Linear methods in band theory. Phys. Rev. B **12**, 3060 (1975)
22. Singh, D.J.: Planewaves Pseudopotentials and the LAPW Method. Kluwer Academic Publishers, Boston (1994)
23. Slater, J.C.: Wave functions in a periodic potential. Phys. Rev. **51**, 846 (1937)
24. Perdew, J.P., Burke, K., Ernzerhof, M.: Generalized gradient approximation made simple. Phys. Rev. Lett. **77**, 3865 (1996)
25. Baker, M.L., Guidi, T., Carretta, S., Ollivier, J., Mutka, H., Güdel, H.U., Timco, G.A., McInnes, E.J.L., Amoretti, G., Winpenny, R.E.P., Santini, P.: Spin dynamics of molecular nanomagnets unravelled at atomic scale by four-dimensional inelastic neutron scattering. Nat. Phys. **8**, 906 (2012)
26. Antkowiak, M., Kozłowski, P., Kamieniarz, G., Timco, G.A., Tuna, F., Winpenny, R.E.P.: Detection of ground states in frustrated molecular rings by in-field local magnetization profiles. Phys. Rev. B **87**, 184430 (2013)
27. Soler, J.M., Artacho, E., Gale, J.D., García, A., Junquera, J., Ordejón, P., Sánchez-Portal, D.: The SIESTA method for ab initio order-N materials simulation. J. Phys. Condens. Matter **14**, 2745 (2002)
28. Noodleman, L.J.: Valence bond description of antiferromagnetic coupling in transition metal dimers. Chem. Phys. **74**, 5737 (1981)
29. Ruiz, E., Cano, J., Alvarez, S., Alemany, P.: Broken symmetry approach to calculation of exchange coupling constants for homobinuclear and heterobinuclear transition metal complexes. J. Comput. Chem. **20**, 1391 (1999)

Parallel Monte Carlo Simulations for Spin Models with Distributed Lattice

Szymon Murawski$^{(\boxtimes)}$, Grzegorz Musiał, and Grzegorz Pawłowski

Faculty of Physics, Adam Mickiewicz University, Poznań, Poland
szymon.murawski@gmail.com

Abstract. A method for parallelization of Monte Carlo simulations of 2 d lattice systems is proposed, in which lattice is distributed and processed among parallel processes. Two ways of distributing the lattice are proposed. In the first one the lattice is divided into stripes of equal width, whereas in the second one the lattice is splitted into blocks of minimum perimeter. Communication between processes is handled by remote memory access MPI-2 protocols, but results for standard MPI-1 communication are also presented for comparison. The scalability of proposed method is tested on high performance multicomputers and discussed on the basis of speedup and efficiency.

Keywords: Parallel processing · Distributed processing · One-sided MPI communication · Monte Carlo simulations · Classical spin lattice model · High performance computing

1 Introduction

Over the years Monte Carlo simulations proved to be one of the most important tools not only in physics. Improving the performance of this method is then topic of utmost importance. For lattice models, in which a simulated system is a collection of particles on a two dimensional lattice, one of the most popular methods are cluster algorithms, where the whole cluster is updated instead of single particle. This method however cannot be used for models including interactions with external magnetic field or chemical potential, i.e. simulations under grand canonical ensemble, cannot benefit from cluster updates.

In this work we address the problem of increasing the performance of grand canonical Monte Carlo by dividing the lattice into a processing grid. Each local lattice is then handled by a single process and undergoes a normal Monte Carlo procedure.Each process communicates with the ones processing the neighboring parts of the lattice via MPI interface. By this means we aim to obtain a raise in performance, which allows processing larger lattices.

Computations for larger systems are very important in a Monte Carlo procedure. As we cannot obtain results for macroscopic models directly, we need to extrapolate results obtained for finite lattice sizes to infinity [1–5] or, in case of phase transitions, to undergo finite-size scaling procedure [6]. For these procedures the greater the lattice size, the more accurate the results.

© Springer International Publishing Switzerland 2016
R. Wyrzykowski et al. (Eds.): PPAM 2015, Part II, LNCS 9574, pp. 332–341, 2016.
DOI: 10.1007/978-3-319-32152-3_31

2   The Simulated System

The model we aim to study using procedures described in this paper is the extended Hubbard model with intersite magnetic interaction in the atomic limit on the two dimensional square lattice, described by the following hamiltonian

$$H = U \sum_i n_{i\uparrow} n_{i\downarrow} - 2J \sum_{\langle i,j \rangle} s_i s_j - \mu \sum_i n_i, \tag{1}$$

where U, J are coulomb and magnetic interaction parameters, respectively, $n_{i\uparrow}$ is the number of electrons at site i with spin $\uparrow$, $s_i = n_{i\uparrow} - n_{i\downarrow}$ is the total spin at site i, $n_i = n_{i\uparrow} + n_{i\downarrow}$ is the total number of particles at site i, μ is the chemical potential. $\sum_{\langle \rangle}$ describe summation over the nearest neighbors. The chemical potential part of this Hamiltonian allows for simulations with fluctuating number of particles in the system which means that the grand canonical ensemble is applied.

Our Monte Carlo procedure for this model requires executing three commands in each Monte Carlo step: CREATE, DESTROY and MOVE, responsible for creating, destroying the particle and moving the particle on the lattice [7]. For testing purposes we reduce the complexity of the model by fixing concentration at $n = 1$ and by taking the infinite coulomb potential $U \to \infty$. Thus, we arrive at the system with exactly one particle at each lattice site with the only degree of freedom being the spin flip. This is the so-called Ising model [8], described by the reduced Hamiltonian

$$H = -J^z \sum_{\langle i,j \rangle} s_i s_j. \tag{2}$$

Here each spin can take only two values ± 1. As our system dimension $d = 2$, we can represent it in the computer memory as just zeroes and ones in the two-dimensional table.

3   Simulation Details

Our simulations require executing numerous Monte Carlo steps, each consisting of $N = L \times L$ tries to flip a single spin for fixed values of model parameters with the Gibbs probability

$$P \sim e^{-\frac{\Delta E}{kT}}, \tag{3}$$

where ΔE is the energy difference between the trial state x' and the old state x calculated using Eq. (2). One Monte Carlo step is completed when each lattice site has been visited once. Thus we need to exploit the Metropolis algorithm [9]. By this procedure we are forming a Markov chain of possible states of the system. To avoid correlation with the initial state, the thermalization procedure is applied, what means discarding part (1/4) of the Monte Carlo steps.

The problem of parallel Monte Carlo simulations, especially for the Ising model, was previously analyzed with the use of multispin approach [14–16], as well as with cluster division [17,18]. Division of the lattice into checkerboard domains, active and inactive, proved very successful [19], especially with the use of graphics processing units (GPU), where this method solves the problem of data access bottleneck [20,21]. There were also publications concerning parallel Monte Carlo simulations without lattice distribution and without any modification to Markov Chain ("embarassingly parallel problem") [22], as well as with some modifications [23,24]. However, it should be noted, that the aim of those studies was to increase the precision of simulations, not to reduce the time of computations. Parallel cluster updates algorithms were also presented, both for Wolff [25] and Swendsen-Wang [26,27] methods. A hot topic recently are hardware simulations using field programmable gate arrays (FPGA) [28] that allow for full lattice flip in a single computer cycle. Our work supplements those results by describing a method of parallelization of processing lattice systems, being a large step to study the extended Hubbard model with intersite magnetic interactions in the atomic limit.

4   Communication Within Parallelized Monte Carlo Simulations

Increasing size L of the lattice results in two challenges from computing perspective: longer computing time and greater memory needed for calculations. As time increases proportionally to L^2, one arrives very quickly at maximum lattice size with realistic computing time. To address this problem we divide lattice $L \times L$ among p parallel processes. In the first attempt each process gets a stripe of the lattice of size $L \times L/p$. Each stripe then undergoes a normal Monte Carlo procedure and at the end results from all stripes are accumulated and presented. As number of communication calls is proportional to the perimeter of local lattice, different way of distributing the lattice was also tested in which each part has minimum perimeter. This is accomplished by splitting the lattice to blocks, according to the following function

```
for (int i=floor(sqrt(p));i>0;i--)
  if(p%i==0 && Lx/i>2 && Ly/(p/i)>2)
  {gridX=i;gridY=p/i;break;}
```

where Lx and Ly are original lattice sizes. Moreover, $gridX$ and $gridY$ are number of processes in each dimension of processing grid. For $p = x^2$ this way will split the lattice into squares, for p equals to prime number into stripes, and for other ways into two dimensional blocks. So at the worst case we arrive at the same number of communication calls as with stripe splitting, at best we reduce it significantly (from $2(p+1)/L$ to $4\sqrt{p}/p$). Figure 1 shows the ratio of states on the border to bulk ones versus number of parallel processes and Fig. 2 shows lattice splitting according to block distribution.

The consequence of this approach is a problem arising when a trial site x' happens to be on the border of the local lattice. As energy difference ΔE requires

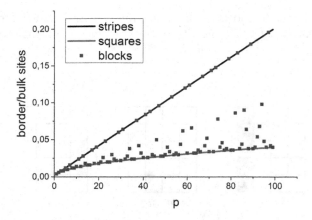

Fig. 1. The ratio of states at the border to all states plotted versus number p of parallel processes taking part in our calculations.

0	1	2
3	4	5
6	7	8

0
1
2
3
4

0	1	2	3
4	5	6	7
8	9	10	11

0	1
2	3
4	5
6	7
8	9

Fig. 2. Lattice splitting for different numbers of parallel processes, numbers indicate rank of the process. Only the leftmost lattice is dividable into squares, all other are rectangles of unequal sides.

information not only of own spin state but also of its neighbors states, MPI communication is necessary.

The standard way of communication using MPI involves both the sender and the receiver taking active part in communication. As we cannot predict which site at what time will be chosen, this implementation of special window in each Monte Carlo step is required, in which all processes communicate with all their neighbors and send all data of states on all borders. This results in drastic decrease of performance and limits the expressiveness of the application, as each send has to match a receive. Using of non-blocking communication does not change the essence of the problem.

To solve this problem we propose to use remote memory access procedures of MPI-2 standard with data of boundary states in MPI memory window [12]. As communication with MPI memory window is non-blocking and one-sided, there is no disruption in calculations due to data transfer between nodes. The MPI memory window is a contiguous memory region available to all processes in specified communication group for read and write operations using MPI_Get(), MPI_Put() and MPI_Accu() functions. Synchronization and coherence of shared data is ensured by calling locking and unlocking memory window for data transfer with `MPI_Win_lock()` and `MPI_Win_unlock()` functions. This process is visualized in Fig. 3.

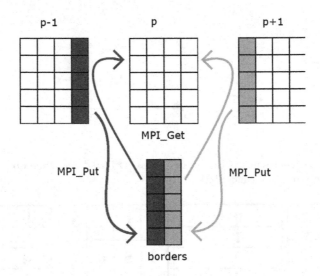

Fig. 3. An example of a piece of a lattice distributed among three processes. Process p keeps information about neighbouring states in separate table **borders** which is put in MPI window for remote memory access. One-sided communication is used to access this table by neighboring processes $p-1, p+1$ and to access the data by process p.

5   Results

Our MC simulations were performed on multicomputer *reef* in the Supercomputing and Networking Center in Poznań built up of 22 nodes with two dual-core Intel Xeon 3 GHz CPUs and of 122 nodes with two quad-core Intel Xeon 2.33 GHz CPUs with OpenMPI and InfiniBand technology of interconnect links. Due to availability of the cores we restricted our calculations to maximum of 20 nodes. Execution times of these simulations for a single process (the number of parallel processes $p = 1$) and stripe distribution were $t_{L=100} \sim 10^2\,s$, $t_{L=1000} \sim 10^3\,s$, $t_{L=10000} \sim 10^4\,s$. For block distribution times were longer by order of four, due to much greater communication overlay.

To evaluate results obtained from different lattice sizes L, different ways of the lattice distributing and different numbers of parallel processes p, we have calculated speedup S defined as $S = t_{seq}/t_{par}$ and efficiency $E = S/p$ [13], where t_{seq} and t_{par} denote sequential and parallel execution times of our program, respectively. For test purposes we drastically reduced the number of generated Monte Carlo steps in our simulations what of course leads to physical results being rather unreliable, but we have obtained reasonable computing times.

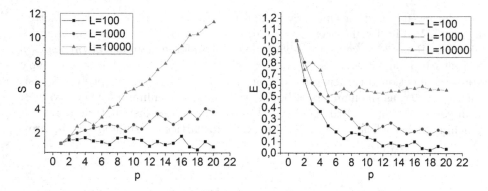

Fig. 4. The dependence of speedup S and efficiency E on the number of parallel processes p for splitting of the lattice into stripes with the use of one-sided communication and remote memory procedures of MPI-2. Different symbols corresponds to different lattice sizes L as explained in the legend box.

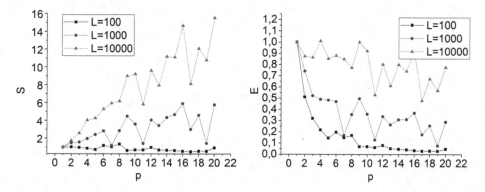

Fig. 5. The dependence of speedup S and efficiency E on number of parallel processes p for splitting lattice into blocks with the use of one-sided communication and remote memory procedures of MPI-2. Different symbols corresponds to different lattice sizes L as explained in the legend box.

Our results with the lattice distributed into stripes are presented in Fig. 4. We observe the constant speedup and the decline of efficiency, as increase of the number of parallel processes is outweighed by the number of spins at borders. Thus, the overlay on communications between p parallel processes (proportional to amount of data to be shared i.e. to $2pL$) evidently is balanced to high degree by the speedup of calculations when one increases the number p of parallel processes. For small lattice sizes ($L = 100$) we observe the saturation of speedup (see Fig. 4), indicating that further increasing the number of parallel processes will not result in better performance. At best, distributing the lattice into stripes for $L = 1000$ will reduce the time needed for calculations by four compared to sequential algorithm. For $L = 10000$ no such saturation is observed, which is very promising for further studies.

The speedup and the efficiency for splitting the lattice into blocks are plotted in Fig. 5. In Fig. 4 some decrease of efficiency is caused by overlay on communication. Relative large speedup variations for different number of processes p are clearly seen, what is the size-effect of the different number of boundary states as mentioned earlier. Greater than usual improvement of speedup is visible for $p = x^2$, as the numbers of boundary states are minimum for those values. Compared to stripe distribution, this kind of the lattice distribution allows for reaching higher efficiency, with local maxima for lattice distributed into squares, i.e. with $p = 4, 9, 16$.

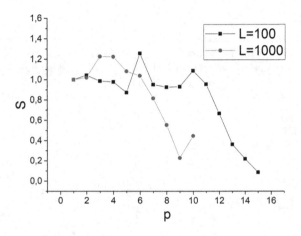

Fig. 6. Speedup S versus number of parallel processes p for distributing lattice into stripes using MPI-1 standard communication procedures.

As the reference point for different communication modes, Fig. 6 shows the speedup calculated for stripe splitting using MPI-1 standard communication, namely `MPI_Send()` and `MPI_Recv()` functions. In this kind of simulations after each Monte Carlo step all processes send two messages containing values on boundary states to their neighbors and then receive two messages from the neighboring sites. To improve performance, communication of odd ID of parallel processes consist of sequence `send, receive, send, receive` while for even IDs the sequence is `receive, send, receive, send`. The speedup is poor compared to one-sided communication with remote memory access procedures of MPI-2 presented in Figs. 4 and 5.

6 Parallel Grand Canonical Monte Carlo Simulations

As distributing the lattice among parallel processes can significantly increase speedup in simulations for Ising-like model, the question arises how to apply it to the model described by Hamiltonian (1). The difference of energy when constructing a new microstate in system described by this model depends only on

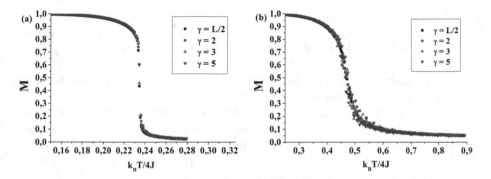

Fig. 7. Magnetization M versus reduced temperature $T/4J$ for Hubbard U-J model, γ describes the maximum allowable distance for particle movement in the Monte Carlo step. Figure 7(a) shows the data for the region of a transition of the first order and 7(b) of the second order. Simulations were done for $L = 20$ using the sequential algorithm.

nearest-neighbor data, just like in the Ising model. As mentioned in Sect. 2, simulations under grand canonical ensemble require three steps: creating, destroying and moving a particle. Creating and destroying of a spin are local updates, handled very similarly to those in Ising model. However, moving a particle is problematic, as it requires information about two different sites on a lattice.

One way to solve this problem would be to introduce an extra layer of communication, where each process asks for a free site on the same or other local lattices. However, this would result in big increase of communication overlay and would be troublesome to implement. Way around this problem would be to restrict movement of particle to local lattice only. Figure 7 shows the influence of maximum allowable movement γ on the quality of the results. For $\gamma = 2$ a particle can move to the next-nearest neighbor site, with $\gamma = L/2$ it can move on the whole lattice. In the region of the first order transition (Fig. 7a) reducing the distance to next-nearest neighbors does not significantly influence on the quality of results. For the second order transition (Fig. 7b) the results change only slightly when varying maximum allowable distance for particle movement.

7   Conclusions

Despite the preliminary character of our results, they prove that distributed processing of the lattice can be successfully performed for spin-lattice models. The gain in performance is significant, especially for large systems, what looks promising for further increasing of the number of parallel processes. The interesting question arises, for how many more computer nodes the overlay on communications is enough balanced by the speedup of calculations to keep realistic time of processing in MC simulations of a spin lattice system.

Comparing different methods of lattice distribution favors splitting the lattice into blocks, as reduced number of boundary states means reduced communication, which outweighs increased communication overlay. We would like to stress out, that the proposed method for increasing the effectiveness of simulations operates on data access layer and keeps all the physics unchanged, in particular the detailed balance. As such applying this method to 3d problems is rather straightforward, as it requires only adding another dimension to the array holding the information of the sites and adding few more rows to the array holding the information of neighboring sites.

We acknowledge the useful tools for effective operations on the distributed lattice: remote memory access, the MPI memory window and one-sided communication [12]. Compared to MPI-2 procedures, cooperative way of sending data of MPI-1 standard does not show improvement of performance, on the contrary, the performance drops even for large lattices.

Distributing the lattice among parallel processes can be implemented for grand canonical ensemble Monte Carlo simulations, with only slightly less accuracy in regions of second-order phase transitions. Thus we can apply this method to lattice models with the external magnetic field or the chemical potential, such as Hubbard $U - J$ model [5] described by Hamiltonian (1). Our work is the first step toward obtaining results for large lattices for this model.

Acknowledgments. This work was supported in part by MNiSzW within the project No. N519 579138. Numerical calculations were carried out in major part on the platforms of the Supercomputing and Networking Center in Poznań. The remaining part of simulations was performed on multicomputers at Faculty of Physics of Adam Mickiewicz University in Poznań.

References

1. Musiał, G., Dębski, L.: Lect. Notes in Comput. Sci. **2328**, 535 (2002)
2. Musiał, G.: Phys. Rev. B **69**, 024407 (2004)
3. Dębski, L., Musiał, G., Rogiers, J.: Lect. Notes in Comput. Sci. **3019**, 455 (2004)
4. Murawski, S., Kapcia, K., Pawłowski, G., Robaszkiewicz, S.: Acta Phys. Pol. A **121**(5–6), 1035 (2012)
5. Murawski, S., Kapcia, K., Pawłowski, G., Robaszkiewicz, S.: Acta Phys. Pol. **126**(4–A), A-110–A-113 (2014)
6. Binder, K.: Z. Phys. B Condens. Matter **43**(2), 119 (1981)
7. Pawłowski, G.: Eur. Phys. J. B **53**(4), 471 (2006)
8. Onsager, L.: Phys. Rev. **65**, 117 (1944)
9. Metropolis, N., Rosenbluth, A., Rosenbluth, M., Teller, A., Teller, E.: J. Chem. Phys. **21**, 1087 (1953)
10. Landau, D.P., Binder, K.: A Guide to Monte Carlo Simulations in Statistical Physics. Cambridge University Press, Cambridge (2000)
11. MPI Forum home page, http://www.mpi-forum.org/
12. Gropp, W., Lusk, E., Skjellum, A.: Using MPI - 2nd Edition: Portable Parallel Programming with the Message-Passing Interface. MIT Press, Cambridge (1999)
13. Van de Velde, E.F.: Concurrent Scientific Computing. Springer-Verlag, New York (1994)
14. Saarinen, J., Kaski, K., Viitanen, J.: Rev. Sci. Instrum. **20**, 2981 (1989)
15. Penna, T.J.P., de Oliveira, P.M.C.: J. Stat. Phys. **61**, 933 (1990)

16. Schleier, W., Besold, G., Heinz, K.: J. Stat. Phys. **66**, 1101 (1992)
17. Flanigan, M., Tamayo, P.: Phys. A **215**, 461 (1995)
18. Ren, R., Orkoulas, G.: J. Chem. Phys. **126**, 211102 (2007)
19. Uhlherr, A.: Comput. Phys. Commun. **155**, 31 (2003)
20. Preis, T., Virnau, P., Paul, W., Schneider, J.J.: J. Comput. Phys. **228**, 4468 (2009)
21. Block, B., Virnau, P., Preis, T.: Comput. Phys. Commun. **181**, 1549 (2010)
22. Adams, D.J.: J. Comput. Phys. **75**, 138 (1988)
23. Mehlig, B., Heermann, D.W., Forrest, B.M.: Phys. Rev. B **45**, 679 (1992)
24. Heffelfinger, G.S.: Comput. Phys. Commun. **128**, 219 (2000)
25. Diaconu, M., Puscasu, R., Stancu, A.: J. Optoelectron. Adv. Mater. **5**, 971 (2003)
26. Hackl, R., Matuttis, H.-G., Singer, J.M., Husslein, I., Morgenstern, Th: Phys. A **212**, 261 (1994)
27. Barkema, G.T., MacFarland, T.: Phys. Rev. E **50**, 1623 (1994)
28. Vanderbauwhede, W., Benkrid, K.: High-Performance Computing Using FPGAs. Springer, New York (2013)

The Second Workshop on Applied High Performance Numerical Algorithms in PDEs

Schwarz Preconditioner with Face Based Coarse Space for Multiscale Elliptic Problems in 3D

Leszek Marcinkowski[1]([⊠]) and Talal Rahman[2]

[1] Faculty of Mathematics, Informatics and Mechanics,
Institute of Applied Mathematics and Mechanics, University of Warsaw,
Banacha 2, 02-097 Warsaw, Poland
L.Marcinkowski@mimuw.edu.pl
[2] Department of Computing, Mathematics and Physics,
Bergen University College, Inndalsveien 28, 5063 Bergen, Norway
Talal.Rahman@hib.no

Abstract. We present a parallel preconditioner based on the domain decomposition for the finite element discretization of multiscale elliptic problems in 3D with highly heterogeneous coefficients. The proposed preconditioner is constructed using an abstract framework of the Additive Schwarz Method which is intrinsically parallel. The coarse space consists of multiscale finite element functions associated with the wire basket, and is enriched with functions based on solving carefully constructed generalized eigen value problem locally on each face. The convergence rate of the Preconditioned Conjugate Method with the proposed preconditioner is shown to be independent of the variations in the coefficients for sufficient number of eigenfunctions in the coarse space.

Keywords: Finite element method · Domain decomposition method · Additive Schwarz Method · Abstract coarse space

1 Introduction

In many applications, like in the porous media flow simulation where we model flow of water, gas and oil in reservoirs and aquifers, we need to numerically solve partial differential equations with highly heterogeneous coefficients representing for instance the permeability. It is known that high contrast in the coefficients causes many standard numerical methods to perform badly.

Domain decomposition methods are among the efficient solvers for systems of equations arising from the finite element discretizations of elliptic partial differential equations, cf. [22], and Additive Schwarz Methods (ASM) are among

L. Marcinkowski—This work was partially supported by Polish Scientific Grant 2011/01/B/ST1/01179 and Chinese Academy of Science Project: 2013FFGA0009 - GJHS20140901004635677.

T. Rahman—The author acknowledges the support of NRC through the DAADppp project 233989.

© Springer International Publishing Switzerland 2016
R. Wyrzykowski et al. (Eds.): PPAM 2015, Part II, LNCS 9574, pp. 345–354, 2016.
DOI: 10.1007/978-3-319-32152-3_32

the most popular domain decomposition methods, cf. e.g. [3,11,15,16] and references therein. In classical overlapping Additive Schwarz Methods the domain is divided into overlapping subdomains, where local subproblems are defined, and a coarse problem is defined globally for the scalability, cf. [22]. If subdomains are such that, in each subdomain, the variations in the coefficients are not too large, it is well known, that classical coarse spaces yield methods that are robust with respect to the variation, cf. e.g. [4,15,22]. In the recent years, the research has extended to highly heterogeneous coefficients, cf. e.g. [5–10,14,17–21,23,24]. In some of those works, the construction of coarse spaces have been based on enriching their coarse spaces with eigenfunctions of some generalized eigenvalue problems, cf. [1,6,7,9,14,21], resulting in methods that are robust with respect to any heterogeneity. This has been the source of our inspiration in this paper.

We propose a parallel additive Schwarz preconditioner for the finite element discretization of the self-adjoint elliptic second order problem in 3D with highly heterogeneous and highly varying coefficients. Preconditioned conjugate gradients method (cf. [12]) is used to solve the resulting preconditioned system. The preconditioner is based on the abstract Schwarz framework where the solution space is decomposed into subspaces associated with the overlapping subdomains, and a specially constructed multiscale coarse space associated with the wire basket of the decomposition, which is then enriched with functions based on solving generalized eigenvalue problems defined locally over each face. The present work is an extension to 3D of the recent work of [9] in 2D. The obtained bounds are independent of the geometries of the subdomains, and the heterogeneities in the coefficients.

The remainder of this paper is organized as follows, in Sect. 2 we present the finite element discretization. In Sect. 3 we present our coarse space introducing the multiscale finite element functions and the functions based on generalized eigenvalue problem on each face. Section 4 contains a description of the overlapping additive Schwarz preconditioner, and in Sect. 5 we briefly discuss the implementation issues.

2   Finite Element Discretization

The aim is to find an approximation to the solution of the following self-adjoint second order elliptic differential problem: find $u^* \in H_0^1(\Omega)$ such that

$$a(u^*, v) = f(v) \qquad \forall v \in H_0^1(\Omega), \tag{1}$$

where

$$a(u, v) = \int_\Omega \alpha(x)(\nabla u)^T \nabla v \, dx, \qquad f(v) = \int_\Omega fv \, dx. \tag{2}$$

Here Ω is a polygonal domain in the three dimensional space, $f \in L^2(\Omega)$ and α is a strictly positive and bounded function. Hence, we can always scale α by its minimal value; we further assume that $\alpha \geq 1$.

We introduce a quasi-uniform triangulation of the domain Ω, denoting it with $T_h(\Omega) = T_h = \{\tau\}$, which consists of tetrahedrons τ, and we let $h = \max_{\tau \in T_h} \operatorname{diam}(\tau)$ be the parameter of T_h, cf. e.g. [2] for more details.

Let V_h be the finite element space of continuous functions which are piecewise linear over the triangulation T_h and zero on the boundary $\partial\Omega$. The degrees of freedom are associated with the nodes or nodal points which are the vertices of tetrahedrons.

Note that on each element $\tau \in T_h$ the gradient $\nabla u_{|\tau}$ is a constant vector, hence for $u, v \in V^h$ we have $\int_\tau \alpha(x)\nabla u^T \nabla v\, dx = (\nabla u_{|\tau}^T \nabla v_{|\tau}) \int_\tau \alpha(x)\, dx$, hence we can assume that $\alpha(x)$ is piecewise constant over the elements of T_h.

Remark 1. We can also consider a more general case when our differential problem is defined with the following symmetric bilinear form

$$\int_\Omega (\nabla u)^T A(x) \nabla v\, dx,$$

where $A(x) \in (L^\infty(\Omega))^{3\times 3}$ is symmetric, and strictly positive definite over Ω in the following sense:

$$\exists\, C_1, C_0 > 0,\ \forall x \in \Omega,\ \forall \xi \in \mathbb{R}^3 \qquad C_0 \xi^T \xi \leq \xi^T A(x) \xi \leq C_1 \xi^T \xi.$$

We can scale A and assume that $C_0 = 1$ and the entries of A are piecewise constant functions over the elements of T_h. If the condition number of $A_{|\tau}$ in any $\tau \in T_h$ remain uniformly bounded, we remark that the result of this paper holds true also for this case.

The discrete FEM problem is formulated as follows: find $u_h \in V_h$ such that

$$a(u_h, v) = f(v) \qquad \forall v \in V_h. \tag{3}$$

The problem has a unique solution by the Lax-Milgram lemma and there are error estimates, see e.g. [2] and references therein. By formulating the discrete problem in the standard nodal basis $\{\phi_i\}_{x_i \in \Omega_h}$, we get the following system of algebraic equations

$$A_h u_h = f_h \tag{4}$$

where $A_h = (a(\phi_i, \phi_j))_{i,j}$, $f_h = (f_j)_{x_j \in \Omega_h}$ with $f_j = \int_\Omega f(x)\psi_i\, dx$, and $u_h = (u_i)_i$ with $u_i = u_h(x_i)$. Here $u_h = \sum_{x_i \in \Omega_h} u_i \phi_i$. The resulting system is symmetric and in general very ill-conditioned; any standard iterative method may perform badly due to the ill-conditioning of the system.

In this paper we present a method for solving such systems using the preconditioned conjugate method (cf. [12]) and propose an additive Schwarz preconditioner (cf. [22]). Let Ω be partitioned into a collection of disjoint open and connected substructures Ω_k, such that $\overline{\Omega} = \bigcup_{k=1}^N \overline{\Omega}_k$. We assume that the triangulation T_h is aligned with the subdomains Ω_k, that is any $\tau \in T_h$ is contained in one subdomain, hence, each subdomain Ω_k inherits the local triangulation $T_h(\Omega_k) = \{\tau \in T_h : \tau \subset \Omega_k\}$. Such a partition may be computed by a mesh partitioning software like e.g. METIS, cf. [13]. We make an additional assumption that the number of subdomains which share a vertex or an edge of an element of T_h is bounded by a constant. An important role is played by the interface $\Gamma = \bigcup_{k=1}^{N} \partial\Omega_k \backslash \partial\Omega$. The non-empty intersection of two subdomains $\partial\Omega_i \cap \partial\Omega_j$ not on $\partial\Omega$ is either a collection of 2D faces of elements of T_h, in which case we say that $\overline{\mathcal{F}}_{ij} = \partial\Omega_i \cap \partial\Omega_j$ is a generalized closed face, or it is a collection of closed edges of elements of T_h, in which case we say that $\overline{\mathcal{E}}_{ij} = \partial\Omega_i \cap \partial\Omega_j$ is a generalized closed edge, or it is a vertex of T_h. We define the wire basket of this partition as the sum of the closed edges of the elements of T_h, which are not on $\partial\Omega$ but are contained in more than two substructures, in other words those contained in any generalized edge, and we denote the wire basket by $\mathcal{W}$. We also define the local wire basket $\mathcal{W}_i = \mathcal{W} \cap \partial\Omega_i$ which will play a crucial role in our analysis. We define the sets of nodal points, $\Omega_h, \partial\Omega_h, \Omega_{k,h}, \mathcal{F}_{kl,h}, \mathcal{E}_{kl,h}, \mathcal{W}_{k,h}$ etc., as the sets of vertices of elements of T_h, which are in $\Omega, \partial\Omega, \Omega_k, \mathcal{F}_{kl}, \mathcal{E}_{kl}, \mathcal{W}_k$ etc., respectively.

3   Coarse Space

In our method, the key role is played by the global coarse space which is a space of discrete harmonic functions (cf. Sect. 3.1 below) and which consists of two space components: the multiscale coarse space component and the generalized face based eigenfunction space component.

3.1   Discrete Harmonic Extensions

We start by defining the discrete harmonic extensions. Local subspaces $V_{h,k}$ are defined as restrictions, of the space V_h to $\overline{\Omega}_k$, that is

$$V_{h,k} = \{u_{|\overline{\Omega}_k} : u \in V_h\} = \{v \in C(\overline{\Omega}_k) : v_{|\tau} \in \mathcal{P}_1(\tau), \tau \in T_h(\Omega_k), v_{|\overline{\Omega}_k \cap \partial\Omega} = 0\},$$

and we let

$$V_{h,k}^0 = V_{h,k} \cap H_0^1(\Omega_k).$$

Let the local discrete harmonic extension operator $\mathcal{H}_k : V_{h,k} \to V_{h,k}$ be defined as the unique solution to the following local problem:

$$\begin{cases} a_k(\mathcal{H}_k u, v) = 0 & \forall v \in V_{h,k}^0 \\ \mathcal{H}_k u = u & \text{on} \quad \partial\Omega_k. \end{cases} \tag{5}$$

where $a_k(u, v) = \int_{\Omega_k} \alpha(x)\nabla u^T \nabla v \, dx$. A function $u \in V_{h,k}$ is discrete harmonic in Ω_k if $u_{|\overline{\Omega}_k} = \mathcal{H}_k u \in V_{h,k}$. For $u \in V_h$, if all its restrictions to local subdomains are discrete harmonic then u is said to be piecewise discrete harmonic over the partition. Note that a discrete harmonic function in $V_{k,h}$ is uniquely defined by its values at the nodal points in $\partial\Omega_{k,h}$.

3.2   Multiscale Coarse Space Component

We define the multiscale component of the coarse space here. We need a few extra definitions. Let $V_h(\mathcal{F}_{kl})$ be the space of all traces of functions from V_h onto $\overline{\mathcal{F}}_{kl}$ and let $V_{h,0}(\mathcal{F}_{kl})$ be its subspace of functions taking zero values at the nodal points of $\mathcal{W}_h \cap \mathcal{F}_{kl,h}$, i.e. the nodal points on the boundary of $\mathcal{F}_{kl}$.

Note that as α is piecewise constant over T_h, it may have jumps across the 2D common faces of two neighboring elements (i.e. tetrahedrons) in T_h. For any face $f \subset \mathcal{F}_{kl}$ we define $\overline{\alpha}_f = \max\{\alpha_{|\tau_1}, \alpha_{|\tau_2}\}$ where $\tau_1 \in T_h(\Omega_k)$ and $\tau_2 \in T_h(\Omega_l)$ are two neighboring elements such that f is their common face.

With each face $\mathcal{F}_{kl}$, we associate a bilinear form $a_{\mathcal{F}_{kl},h} : V_h(\mathcal{F}_{kl}) \times V_h(\mathcal{F}_{kl}) \to \mathbb{R}$, which is defined as

$$a_{\mathcal{F}_{kl},h}(u,v) = \sum_{f \subset \mathcal{F}_{kl}} \int_f \overline{\alpha}_f \nabla u \nabla v \, ds,$$

where the sum is over the 2D faces of the elements of T_h forming the face $\mathcal{F}_{kl}$, and the integral is over each such 2D face. Note that $u \in V_h(\mathcal{F}_{kl})$ is continuous, and its restriction to such a face f is a linear polynomial.

Analogously, associated with the face $\mathcal{F}_{kl}$, we define a scaled discrete weighted L^2 inner product, a symmetric bilinear form $b_{\mathcal{F}_{kl},h} : V_h(\mathcal{F}_{kl}) \times V_h(\mathcal{F}_{kl}) \to \mathbb{R}$, as

$$b_{\mathcal{F}_{kl},h}(u,v) = \sum_{x \in \mathcal{F}_{kl,h}} \overline{\alpha}_x u(x) \, v(x),$$

where $\overline{\alpha}_x = \max_{x \in \partial\tau} \alpha_{|\tau}$, i.e. is equal to the maximal value of $\alpha_{|\tau}$ over all elements τ sharing the node x as a vertex.

Finally, we introduce the multiscale coarse space component $V_{ms} \subset V_h$ as the space of functions whose degrees of freedom are associated with the nodal points of $\mathcal{W}_h$. For any function $u \in V_{ms}$, on each face $\mathcal{F}_{kl}$, it is defined as the solution of the following generalized face problem,

$$\begin{cases} a_{\mathcal{F}_{kl},h}(u_{|\mathcal{F}_{kl}}, v) = 0 & \forall v \in V_{h,0}(\mathcal{F}_{kl}) \\ u_{|\mathcal{F}_{kl}} = u & \text{on} \quad \partial\mathcal{F}_{kl} = \mathcal{W} \cap \overline{\mathcal{F}}_{kl}. \end{cases} \tag{6}$$

and inside each subdomain, it is defined as the discrete harmonic extension in the sense of (5). So, once $u \in V_{ms}$ is known at the nodal points of $\mathcal{W}_h$, its values at the nodal points of each face can be computed by solving (6), and then its values at the nodal points of each subdomain can be computed by solving (5).

Proposition 1. *The problem (6) has a unique solution.*

Proof. Note that if the form $a_{\mathcal{F}_{kl},h}(v,v)$ is zero for any $v \in V_{h,0}(\mathcal{F}_{kl})$, then it means that v is constant on each 2D face $f \subset \mathcal{F}_{kl}$. The continuity of v yields that v is equal to a single constant over all 2D faces contained in $\mathcal{F}_{kl}$. Finally this constant is zero because v is zero at the nodal points on the wire basket (boundary of the face). This proves that the form is positive definite in $V_{h,0}(\mathcal{F}_{kl})$.

Let $\hat{u}$ be a function equal to u at the nodal points of $\partial \mathcal{F}_{kl}$ and equal to zero at all nodal points in the interior of the face $\mathcal{F}_{kl}$, i.e. not belonging to $\mathcal{W}_h$. Then $\tilde{u} := u_{\mathcal{F}_{kl}} - \hat{u}$ is in $V_{h,0}(\mathcal{F}_{kl})$ and we can rewrite (6) as: find $\tilde{u} \in V_{h,0}(\mathcal{F}_{kl})$ such that

$$a_{\mathcal{F}_{kl},h}(\tilde{u}, v) = -a_{\mathcal{F}_{kl},h}(\hat{u}, v) \qquad \forall v \in V_{h,0}(\mathcal{F}_{kl}),$$

which obviously has a unique solution due the positive definiteness of the bilinear form in $V_{h,0}(\mathcal{F}_{kl})$.

Finally $u \in V_{ms}$ is discrete harmonic in Ω_k and thus the values of u in $\Omega_{k,h}$ are uniquely defined by its values on the wire basket and on faces using (5). We note that the dimension of V_{ms} equals the number of all nodes in the set $\mathcal{W}_h$.

3.3   Generalized Face Based Eigenfunction Space Component

We introduce a face based generalized eigenvalue problem: find $(\lambda, \psi) \in \mathbb{R} \times V_{h,0}(\mathcal{F}_{kl})$ such that

$$a_{\mathcal{F}_{kl},h}(\psi, v) = \lambda b_{\mathcal{F}_{kl},h}(\psi, v) \qquad \forall v \in V_{h,0}(\mathcal{F}_{kl}) \tag{7}$$

Since both bilinear forms are symmetric and positive definite in $V_{h,0}(\mathcal{F}_{kl})$, there exist real and positive eigenvalues and their respective $b_{\mathcal{F}_{kl},h}$-orthogonal and normalized eigenvectors satisfying (7) such that

$$0 < \lambda_1^{\mathcal{F}_{kl}} \le \lambda_2^{\mathcal{F}_{kl}} \le \dots \le \lambda_M^{\mathcal{F}_{kl}},$$

and

$$b_{\mathcal{F}_{kl},h}(\psi_j^{\mathcal{F}_{kl}}, \psi_i^{\mathcal{F}_{kl}}) = 0 \quad j \ne i, \qquad b_{\mathcal{F}_{kl},h}(\psi_j^{\mathcal{F}_{kl}}, \psi_j^{\mathcal{F}_{kl}}) = 1.$$

Here M is the dimension of $V_{h,0}(\mathcal{F}_{kl})$.

For any $1 \le n \le M$ we can define a orthogonal projection: $\pi_n^{\mathcal{F}_{kl}} : V_{h,0}(\mathcal{F}_{kl}) \to span\{\psi_j^{\mathcal{F}_{kl}}\}_{j=1}^n \subset V_{h,0}(\mathcal{F}_{kl})$ as

$$\pi_n^{\mathcal{F}_{kl}} v = \sum_{j=1}^n b_{\mathcal{F}_{kl},h}(v, \psi_j^{\mathcal{F}_{kl}}) \psi_j^{\mathcal{F}_{kl}}. \tag{8}$$

By a simple algebraic argument (similar to those in [21] or [9]) we get the following lemma.

Lemma 1. *The operator* $\pi_n^{\mathcal{F}_{kl}}$ *is* $a_{\mathcal{F}_{kl},h}$-*orthogonal projection and moreover*

$$\|v - \pi_n^{\mathcal{F}_{kl}} v\|_{b,\mathcal{F}_{kl}}^2 \le \frac{1}{\lambda_{n+1}^{\mathcal{F}_{kl}}} \|v - \pi_n^{\mathcal{F}_{kl}} v\|_{a,\mathcal{F}_{kl}}^2 \quad \forall v \in V_{h,0}(\mathcal{F}_{kl}),$$

where $\|v\|_{a,\mathcal{F}_{kl}}^2 = a_{\mathcal{F}_{kl},h}(v, v)$ *and* $\|v\|_{b,\mathcal{F}_{kl}}^2 = b_{\mathcal{F}_{kl},h}(v, v)$.

We further assume that a nonnegative number $n(\mathcal{F}_{kl})$, not greater than the dimension of $V_{h,0}(\mathcal{F}_{kl})$, is known or given for each face $\mathcal{F}_{kl}$. Then for each eigenvector $\psi_j^{\mathcal{F}_{kl}}$, $1 \le j \le n(\mathcal{F}_{kl})$ we define $\Psi_j^{\mathcal{F}_{kl}} \in V_h$ which is equal to $\psi_j^{\mathcal{F}_{kl}}$ on the face $\mathcal{F}_{kl}$, zero on the remaining faces and everywhere on the wire basket $\mathcal{W}$, and finally discrete harmonic inside each subdomain in the sense of (5) defining uniquely its values at all interior nodes of the subdomain. We are now able to introduce the face based eigenfunction space component which is

$$V_{h,n}^{\mathcal{F}_{kl}} = span\{\Psi_j^{\mathcal{F}_{kl}}\}_{j=1}^{n(\mathcal{F}_{kl})}, \qquad \forall \mathcal{F}_{kl} \subset \Gamma.$$

Finally, our coarse space is defined as follows:

$$V_0 := V_{ms} + \sum_{\mathcal{F}_{kl} \subset \Gamma} V_{h,n}^{\mathcal{F}_{kl}}. \tag{9}$$

4   Additive Schwarz Method (ASM) Preconditioner

We define our preconditioner utilizing the abstract framework of ASM, i.e. we introduce a decomposition of the global space V_h into the sum of smaller subspaces of V_h, and define symmetric positive definite bilinear forms on the subspaces; cf. [22]. In our present work, we consider only the original bilinear form $a(u, v)$, i.e. (2), on each subspace.

The coarse space is defined in the previous section, cf. (9). The local subspace V_k associated with the subdomain Ω_k, is defined as the space of all functions $u \in V_h$ which take the value zero at all nodal points that lie outside $\overline{\Omega}_k$. It is easy to see that $V_h = \sum_{k=1}^{N} V_k$. Now, including the coarse space, we have the following decomposition:

$$V_h = V_0 + \sum_{k=1}^{N} V_k.$$

The additive Schwarz operator $T : V_h \rightarrow V_h$ is defined in terms of the projection like operators, $T_k, k = 0 \cdots N$, as follows, i.e. $T = T_0 + \sum_{k=1}^{N} T_k$, where the coarse space projection like operator, $T_0 : V_h \rightarrow V_0$, is defined as

$$a(T_0 u, v) = a(u, v) \qquad \forall v \in V_0,$$

and the local subspace projection operators, $T_k : V_h \rightarrow V_k$, are defined as

$$a(T_k u, v) = a(u, v) \qquad \forall v \in V_k, \qquad k = 1, \ldots, N.$$

Under the Schwarz framework, the problem (3) is then reformulated as the following equivalent preconditioned system,

$$T u_h = g, \tag{10}$$

where $g = g_0 + \sum_{k=1}^{N} g_k$ with $g_0 = T_0 u_h^*$, $g_k = T_k u_h^*$, $k = 1, \ldots, N$, and u_h^* the exact solution. Note that the right hand side vectors, $g_k, k = 0 \cdots, N$, can be calculated without explicitly knowing the exact solution.

4.1   An Estimate of the Condition Number

We present the main result of this paper, namely, an estimate of the condition number of the preconditioned system (3), which is given in the following theorem.

Theorem 1. *There exist positive constants c and C such that*

$$c(1 + \max_{\mathcal{F}_{kl}}(\lambda_{n+1}^{\mathcal{F}_{kl}})^{-1})^{-1}\, a(u,u) \leq a(Tu,u) \leq C\, a(u,u) \qquad \forall u \in V_h,$$

where $\lambda_{n+1}^{\mathcal{F}_{kl}}$ and $n = n(\mathcal{F}_{kl})$ are as defined in Sect. 3.3, and c, C are constants independent of α, h and the number of subdomains.

A Sketch of the Proof. The proof is based on the abstract Schwarz framework, where we need to verify the three key assumptions of the framework, see [22] for the framework. The first two assumptions, that is, the local stability and strengthened Schwarz-Cauchy inequalities, follow immediately from standard arguments. The last assumption, that is the assumption on the stable decomposition, is less trivial.

We propose the following decomposition of u, $u = u_0 + \sum_{i=1}^{N} u_i$, with $u_i \in V_i$ for $i = 0, \ldots, N$. For any $u \in V_h$ we let $u_0 \in V_0$ be defined as follows. Let $u_{ms} \in V_{ms}$ be equal to u at all nodes of $\mathcal{W}_h$. The restriction of $u - u_{ms}$ to a face $\mathcal{F}_{kl}$ is then a function in $V_{h,0}(\mathcal{F}_{kl})$. Let $u_{kl} \in V_{h,n}^{\mathcal{F}_{kl}}$ be equal to $\pi_n^{\mathcal{F}_{kl}}(u - u_{ms})$ (cf. (8)). Note that u_{kl} is zero at all wire basket nodes and discrete harmonic inside subdomains. Now, by letting $u_0 = u_{ms} + \sum_{\mathcal{F}_{kl} \subset \Gamma} u_{kl}$, and $u_i = I_h(\theta_i(u - u_0))$, where $\{\theta_i\}$ is the standard partition of unity with respect to the partition $\{\Omega_i\}$ and I_h the standard nodal interpolation operator.

Now using the above decomposition, and the estimates of Lemma 1 we can show that

$$a(u_0, u_0) + \sum_{k=1}^{N} a(u_k, u_k) \leq C(1 + \max_{\mathcal{F}_{kl}}(\lambda_{n+1}^{\mathcal{F}_{kl}})^{-1}) a(u, u),$$

where C is a constant independent of α, h and number of subdomains. The proof of the theorem then follows from the abstract Schwarz framework, cf. e.g. [22].

Remark 2. The idea is to collect eigenfunctions with the smallest eigenvalues (bad eigenmodes) into the coarse space, whereby removing their influence on the convergence. Normally, the bad modes are associated with the channels (regions with large coefficients) crossing the interface $\mathcal{F}_{kl}$. The number of eigenfunctions, $n(\mathcal{F}_{kl})$, required for the robustness, can either be preassigned from experience or chosen adaptively by setting a threshold and choosing those eigenfunctions whose eigenvalues are smaller than the threshold.

Remark 3. Although, an explicit dependence on the mesh parameters does not appear in the convergence estimate of Theorem 1, it is not difficult to tell that this dependence is somehow hidden in the eigenvalues of the face eigenvalue problems, and as soon as the influence of the bad eigenmodes have been removed, it will start to show, at least numerically. However, if we choose the threshold, cf. Remark 2, to be in the order of $\frac{h}{H}$, it is straightforward to see that the convergence will be in the order of $\frac{H}{h}$.

5   Implementation Issues

In this section, we briefly discuss the implementation of our ASM preconditioner. We propose to use the preconditioned conjugate gradient iteration (cf. e.g. [12]) for the system (10). Constructing the coarse space requires the solution of the generalized eigenvalue problem (7) on each subdomain face (interface), the first few eigenfunctions corresponding to the smallest eigenvalues are then included in the coarse space. Prescribing a threshold λ_0, and then computing the eigenpairs with eigenvalues smaller than λ_0, we can get an automatic way to enrich the coarse space. The simplest way would be to compute a fixed number of eigenpairs, e.g. $n = 5$ or so, this may however not guarantee robustness as the number of channels crossing a face may be much larger. In each step of PCG we compute a residual vector which requires solving the coarse problem and local subproblems, cf. [22]. All these problems are independent so they can be solved in parallel. The local subdomain problems are solved locally on their respective subdomains. The coarse problem is global, and although its dimension equals the number of nodes on the wire basket plus the number of local eigenfunctions, the coarse stiffness matrix is quite sparse. However, if we add too many of the eigenfunctions, the coarse space may become too large and the coarse problem too expensive, on the other hand, if we add too few eigenfunctions then the condition number may be too large and the convergence of the iterative scheme too slow.

References

1. Bjørstad, P.E., Koster, J., Krzyżanowski, P.: Domain decomposition solvers for large scale industrial finite element problems. In: Sørevik, T., Manne, F., Moe, R., Gebremedhin, A.H. (eds.) PARA 2000. LNCS, vol. 1947, pp. 373–383. Springer, Heidelberg (2001). http://dx.doi.org/10.1007/3-540-70734-4_44
2. Brenner, S.C., Scott, L.R.: The Mathematical Theory of Finite Element Methods. Texts in Applied Mathematics, vol. 15, 3rd edn. Springer, New York (2008). http://dx.doi.org/10.1007/978-0-387-75934-0
3. Brenner, S.C., Wang, K.: Two-level additive Schwarz preconditioners for C^0 interior penalty methods. Numer. Math. **102**(2), 231–255 (2005). http://dx.doi.org/10.1007/s00211-005-0641-2
4. Dohrmann, C.R., Widlund, O.B.: An overlapping Schwarz algorithm for almost incompressible elasticity. SIAM J. Numer. Anal. **47**(4), 2897–2923 (2009). http://dx.doi.org/10.1137/080724320
5. Efendiev, Y., Galvis, J., Lazarov, R., Margenov, S., Ren, J.: Robust two-level domain decomposition preconditioners for high-contrast anisotropic flows in multiscale media. Comput. Methods Appl. Math. **12**(4), 415–436 (2012). http://dx.doi.org/10.2478/cmam-2012-0031
6. Efendiev, Y., Galvis, J., Lazarov, R., Willems, J.: Robust domain decomposition preconditioners for abstract symmetric positive definite bilinear forms. ESAIM Math. Model. Numer. Anal. **46**(5), 1175–1199 (2012). http://dx.doi.org/10.1051/m2an/2011073

7. Efendiev, Y., Galvis, J., Vassilevski, P.S.: Spectral element agglomerate alge-braic multigrid methods for elliptic problems with high-contrast coefficients. In: Huang, Y., Kornhuber, R., Widlund, O., Xu, J. (eds.) Domain Decompo-sition Methods in Science and Engineering XIX. Lecture Notes in Computa-tional Science and Engineering, vol. 78, pp. 407–414. Springer, Heidelberg (2011). http://dx.doi.org/10.1007/978-3-642-11304-8_47

8. Galvis, J., Efendiev, Y.: Domain decomposition preconditioners for multiscale flows in high-contrast media. Multiscale Model. Simul. **8**(4), 1461–1483 (2010). http://dx.doi.org/10.1137/090751190

9. Gander, M.J., Loneland, A., Rahman, T.: Analysis of a new harmonically enriched multiscale coarse space for domain decomposition methods (2015, submitted)

10. Graham, I.G., Lechner, P.O., Scheichl, R.: Domain decomposition for multi-scale PDEs. Numer. Math. **106**(4), 589–626 (2007). http://dx.doi.org/10.1007/s00211-007-0074-1

11. Griebel, M., Oswald, P.: On the abstract theory of additive and multiplicative Schwarz algorithms. Numer. Math. **70**(2), 163–180 (1995)

12. Hackbusch, W.: Iterative Solution of Large Sparse Systems of Equations. Applied Mathematical Sciences, vol. 95. Springer, New York (1994). Translated and revised from the 1991 German original

13. Karypis, G., Kumar, V.: A fast and highly quality multilevel scheme for partition-ing irregular graphs. SIAM J. Sci. Comput. **20**(1), 359–392 (1999)

14. Klawonn, A., Radtke, P., Rheinbach, O.: FETI-DP methods with an adaptive coarse space. SIAM J. Numer. Anal. **53**(1), 297–320 (2015)

15. Mandel, J., Brezina, M.: Balancing domain decomposition for problems with large jumps in coefficients. Math. Comp. **65**(216), 1387–1401 (1996)

16. Marcinkowski, L.: A balancing Neumann-Neumann method for a mortar finite element discretization of a fourth order elliptic problem. J. Numer. Math. **18**(3), 219–234 (2010). http://dx.doi.org/10.1515/JNUM.2010.011

17. Pechstein, C.: Finite and Boundary Element Tearing and Interconnecting Solvers for Multiscale Problems. Lecture Notes in Computational Science and Engineering, vol. 90. Springer, Heidelberg (2013)

18. Pechstein, C., Scheichl, R.: Analysis of FETI methods for multiscale PDEs. Numer. Math. **111**(2), 293–333 (2008). http://dx.doi.org/10.1007/s00211-008-0186-2

19. Pechstein, C., Scheichl, R.: Scaling up through domain decomposition. Appl. Anal. **88**(10–11), 1589–1608 (2009). http://dx.doi.org/10.1080/00036810903157204

20. Scheichl, R., Vainikko, E.: Additive Schwarz with aggregation-based coarsening for elliptic problems with highly variable coefficients. Computing **80**(4), 319–343 (2007). http://dx.doi.org/10.1007/s00607-007-0237-z

21. Spillane, N., Dolean, V., Hauret, P., Nataf, F., Pechstein, C., Scheichl, R.: Abstract robust coarse spaces for systems of PDEs via generalized eigenproblems in the overlaps. Numer. Math. **126**(4), 741–770 (2014). http://dx.doi.org/10.1007/s00211-013-0576-y

22. Toselli, A., Widlund, O.: Domain Decomposition Methods–Algorithms and Theory. Springer Series in Computational Mathematics, vol. 34. Springer, Berlin (2005)

23. Van Lent, J., Scheichl, R., Graham, I.G.: Energy-minimizing coarse spaces for two-level Schwarz methods for multiscale PDEs. Numer. Linear Algebra Appl. **16**(10), 775–799 (2009). http://dx.doi.org/10.1002/nla.641

24. Vassilevski, P.S.: Multilevel Block Factorization Preconditioners: Matrix-Based Analysis and Algorithms for Solving Finite Element Equations. Springer, New York (2008)

A Compact Parallel Algorithm for Spherical Delaunay Triangulations

Florian Prill[✉] and Günther Zängl

Deutscher Wetterdienst, Offenbach, Germany
florian.prill@dwd.de

Abstract. We present a data-parallel algorithm for the construction of Delaunay triangulations on the sphere. Our method combines a variant of the classical Bowyer-Watson point insertion algorithm [2,14] with the recently published parallelization technique by Jacobsen et al. [7]. It resolves a breakdown situation of the latter approach and is suitable for practical implementation due to its compact formulation. Some complementary aspects are discussed such as the parallel workload, floating-point arithmetics and an application to interpolation of scattered data.

Keywords: Spherical delaunay triangulation · Parallel computing · Computational geometry · Interpolation

1 Introduction

Spherical geodesic grids are currently used in a number of models in weather and climate research [8,15], offering global quasi-uniform resolution and lending themselves to massive parallelism. The representation of geophysical data on these large meshes has recently attracted new interest in several classical problems in computational geometry such as interpolation and grid generation [4,7]. A natural building block for many of these algorithms are the Delaunay triangulation, which in two dimensions maximizes the minimum measure of angles of all the triangles in the triangulation, and the Voronoi diagram, of which the Delaunay triangulation is a dual graph.

Extensive literature exists on the construction of Delaunay tesselations where the different types of algorithms can be classified into incremental insertion, divide-and-conquer and gift-wrapping approaches [13]. The special case of spherical meshes was treated, e.g., in [11]. However, algorithms for data-parallel execution have been studied in much less detail. An exception is the recent publication by Jacobsen et al. [7], which simplifies the technically complex merge process of local triangulations by independently triangulating spherical caps. A drawback lies in the fact that the described approach theoretically suffers from breakdown situations when the spherical caps are chosen too small. Moreover, the approach in [7] is not a self-contained description in the sense that it delegates the local subproblems to planar triangulation methods.

© Springer International Publishing Switzerland 2016
R. Wyrzykowski et al. (Eds.): PPAM 2015, Part II, LNCS 9574, pp. 355–364, 2016.
DOI: 10.1007/978-3-319-32152-3_33

In this paper we combine the data-parallel approach [7] with a variant of the incremental Bowyer-Watson algorithm from [1,2,14] for the sphere. Combining the two algorithms offers several advantages: No stereographic projection is required for the local problems, which substantially simplifies the algorithm. Besides, the parallel construction does not rely on a proper *a priori* choice of constants and it does not fail due to insufficient subset radii. The result is an algorithm whose relative compactness may be attractive for practitioners although divide-and-conquer methods run faster in the worst-case limit.

This paper is organized as follows: The details of the local and the parallel components of the algorithm are described in Sect. 2. We briefly discuss complementary aspects such as parallel workload and floating-point arithmetics. In Sect. 3 numerical results are presented for large triangulations of several million points and several hundred parallel processes. We also apply the algorithm to interpolating functions of two variables on the sphere, which is of considerable importance in meteorological applications, cf. [4].

2   Parallel Construction of the Delaunay Tesselation

2.1   Data-Parallel Algorithm

For the domain decomposition, we follow the approach by Jacobsen et al. [7]: We are given a point set on the unit sphere $\{\mathbf{p}_i\}_{i=1,\ldots,N} \subset S^2$. When executed in parallel on q processes, the Delaunay triangulation is constructed in three steps:

Algorithm 1. Data-parallel algorithm

p.1. Cover the sphere by spherical caps with centers $\{\mathbf{q}_i\}_{i=1,\ldots,q}$, radii $\{\theta_i\}_i$.

p.2. Construct independent triangulations $T_1, \ldots, T_q$, see Sect. 2.2.

p.3. Synchronize and merge into a global tesselation T.

A triangle is *Delaunay* if its circumcircle encloses no other point from the triangulation point set. This has consequences for data-parallel execution, since simplices in the tesselation may not be Delaunay wrt. points that are not in the local subset. Jacobsen et al. [7] therefore formulate a *global Delaunay criterion*: If triangles with circumcenter $\mathbf{c}_i$ and radius r_i satisfy

$$\|\mathbf{c}_i - \mathbf{q}\| + r_i < \theta \tag{1}$$

then this is sufficient for the Delaunay property in the global point set. Here, $\| \cdot \|$ denotes the geodesic on the sphere. The local algorithm that is discussed below meets the criterion (1) by construction. The only remaining requirement for the choice of subsets in step p.1 is that the whole sphere must be covered.

Several partitioning strategies are possible for the covering of the sphere: Our approach is to construct an auxiliary triangulation of points $\mathbf{v}_1, \ldots, \mathbf{v}_q$, $q \ll N$, selecting $\{\theta_k\}_{k=1}^q$ as the largest radii of adjacent coarse triangle circumcircles, see Fig. 1a, for example. Then the merge step p.3 merely consists of removing duplicate triangles from the triangulation T based on their global vertex indices.

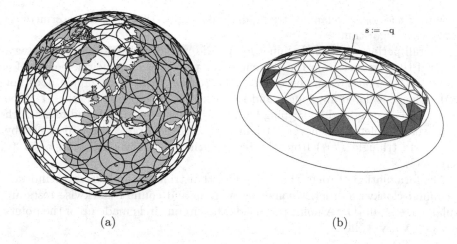

$$\mathbf{s} := -\mathbf{q}$$

(a)   (b)

Fig. 1. Covering by spherical caps, generated by $q = 200$ spiral points (see Sect. 3). Each cap gives rise to a local triangulation problem, executed by one parallel process. Figure 1b: state of Algorithm 2 after insertion of 10 % of a global point set. The shaded region denotes the set of cells that would presently violate the Delaunay condition (1), i.e. $\mathbf{t}[\Delta] > -1$. The insertion loop terminates after the spherical cap with radius $\theta = 0.8$ has been covered by the triangulation.

This operation is of time complexity $\mathcal{O}(\#T\, q\, \log(q))$ applying a heap based q-way merge. It should also be noted that for many applications the merging of local tesselations (step p.3) is not at all required, especially when the Delaunay triangulation serves as a building block for other parallel algorithms (see the data-fitting algorithm in Sect. 3.2 as an example).

2.2   Local Algorithm

Each independent subtask p.2 is solved by Algorithm 2 below, where the Delaunay construction by successive point insertion operates as follows: As points are inserted into the triangulation, triangles are deleted whose circumcenters enclose a new point $\mathbf{p}$, evacuating a polyhedral cavity. New triangles then fill this cavity, connecting $\mathbf{p}$ to its edges.

Compared to the classical Bowyer-Watson algorithm the main difference is that the points are processed beginning at the center $\mathbf{q}$ of the spherical cap, ordered along a search direction $\mathbf{s} := -\mathbf{q}$. Thus ordering the points by their distance to the cap center, the algorithm keeps track of the number of invalid triangles created by points inside the given cap radius, where by the term *invalid* we denote triangles that violate the global Delaunay condition (1). Even outside the given radius, the algorithm keeps inserting sufficiently many points into the triangulation until (at least) all triangles within the given radius satisfy the global criterion. Only if all these triangles have been resolved, the algorithm stops. In contrast to [7] the parallel method therefore does not fail due to an insufficient overlap of the local Delaunay triangulations. Figure 1b shows an intermediate state of the algorithm.

Overall performance of the inner loop is improved by the following strategy, cf. [1]: If the distance from the current point $\mathbf{p}_i$ to the circumcircle center of a triangle Δ is greater than the circumcircle radius, then this triangle does not need to be considered any longer and is marked as "complete" ($\mathtt{a}[\Delta] \leftarrow 1$). The inner loop is accelerated by keeping track of the smallest triangle index j_0 which is not yet complete. Whenever a triangle is removed from T, it is replaced by a complete triangle $T(j_1)$ from the back of the triangulation list, in order to increase j_0.

The formulation of Algorithm 2 assumes that triangles and edges are indexed in counter-clockwise order. Geometric in-circle and counter-clockwise tests are explained, e.g., in [11]: A point $\mathbf{p}$ is inside the circumcircle made up of the points $\Delta = \{\mathbf{v}_1, \mathbf{v}_2, \mathbf{v}_3\}$, iff

$$circum\_circle(\Delta, \mathbf{p}) : \qquad \det(\mathbf{v}_2 - \mathbf{p}, \mathbf{v}_3 - \mathbf{p}, \mathbf{v}_1 - \mathbf{p}) > 0.$$

A point $\mathbf{p}$ is located counter-clockwise relative to a directed arc $\mathbf{v}_1 \to \mathbf{v}_2$ when

$$ccw(\mathbf{v}_1, \mathbf{v}_2, \mathbf{p}) : \qquad \det(\mathbf{v}_1, \mathbf{v}_2, \mathbf{p}) \leq 0.$$

Robust evaluation of these expressions wrt. floating-point arithmetics requires additional considerations, see Sect. 2.3.

The algorithm avoids the common technique of enclosing the vertices in a large triangular bounding box ("super-triangle") but instead connects all boundary triangles to a "ghost point" [13]. An array o is used to store for each triangle the edge opposite to the ghost point, or $\mathtt{o}[\Delta] = -1$ for interior triangles. When a triangle Δ' is formed by attaching a new point to the edge i of another triangle Δ, then the following lookup table can be used to compute $\mathtt{o}[\Delta]$:

$$\mathtt{o}[\Delta'] \leftarrow Q_{i,\mathtt{o}[\Delta]+2} \quad \text{where} \quad Q := \begin{bmatrix} -1 & -1 & 1 & 2 \\ -1 & 2 & -1 & 1 \\ -1 & 1 & 2 & -1 \end{bmatrix}.$$

2.3   Further Discussion

Estimate for the Size of the Auxiliary Triangulation. The question arises if an algorithm that parallelizes with overlapping spherical caps is still efficient. The following geometrical argument (which also holds for [7]) may provide some insight: A global triangulation with n_p vertices contains $n_c = 2n_p - 4$ cells. Under the assumption of a uniform point density on the unit sphere, this yields the following estimate for the average mesh size:

$$\Delta x \approx (4\pi/n_c)^{\frac{1}{2}} = 2[\pi/(2n_p - 4)]^{\frac{1}{2}}.$$

Algorithm 2. Local algorithm

    input : point set $\{\mathbf{p}_i\}_{i=1,\ldots,N}$; spherical cap $\{\mathbf{s},\theta\}$
    output: triangulation T

1   sort the points $\{\mathbf{p}_i\}_{i=1,\ldots,N}$ such that $\{\mathbf{p}_i \cdot \mathbf{s}\}_i$ increases monotonically
2   **foreach** edge e of the initial triangle $\{\mathbf{p}_1, \mathbf{p}_2, \mathbf{p}_3\}$ **do**
3   append a triangle $\Delta := \{\mathbf{e}^1, \mathbf{e}^2, \mathbf{g}\}$ with a "ghost point" $\mathbf{g}$, set $o[\Delta] \leftarrow 0$

4   $j_0 \leftarrow 0$; $k \leftarrow 0$; $\mathbf{a}[\cdot] \leftarrow 0$; $\mathbf{t}[\cdot] \leftarrow 1$
5   **foreach** point $\mathbf{p}_i$, $i = 4,\ldots,N$ **do**
6   **if** $(\mathbf{p}_i \cdot \mathbf{s} > -\cos(\theta))$ **and** $(k = 0)$ **then exit** loop
7     $j \leftarrow j_0$; $j_0 \leftarrow -1$; $j_1 \leftarrow \max\{l = j,\ldots,\#T \mid \mathbf{a}[l] = 1\}$
8   clear e-list and o-list
9   **foreach** $j \leq \#T$ where $\mathbf{a}[j] = 0$ **do**
10   set $\mathbf{c}, r \leftarrow$ circumcenter, radius of $T(j)$
11   **if** $(-\mathbf{p}_i \cdot \mathbf{s} < t[j])$ **then** $t[j] \leftarrow -1$; $k \leftarrow k - 1$
12       $j_{\min,0} \leftarrow j_0$; **if** $(j_0 = -1)$ **then** $j_0 \leftarrow j$
13   **if** $(o[j] \neq -1)$ **then**
14         $inside \leftarrow ccw(E(o[j]; j), \mathbf{p}_i)$
15   **else**
16         $inside \leftarrow circum\_circle(T(j), \mathbf{p}_i)$

17   **if** $inside$ **then**
18   append $E(1,\ldots,3;j)$ to e-list and $Q(1,\ldots,3;o[j]+2)$ to o-list
19   **if** $(t[j] > -1)$ **then** $k \leftarrow k - 1$
20   **if** $(j_1 > j)$ **then**
21   replace $T(j)$ by $T(j_1)$
22           $j \leftarrow j - 1$; $j_1 \leftarrow \max\{l = j,\ldots,j_1 \mid \mathbf{a}[l] = 1\}$
23   **else**
24   replace $T(j)$ by $T(\#T)$
25   **else if** $(o[j] = -1)$ **then**
26   **if** $(\mathbf{p}_i - \mathbf{c}) \cdot \mathbf{s} > r$ **then** $\mathbf{a}[j] \leftarrow 1$; $j_0 \leftarrow j_{\min,0}$
27   remove duplicate edges from e-list
28   **foreach** edge e in e-list **do**
29   append triangle $\Delta := \{\mathbf{e}^1, \mathbf{e}^2, \mathbf{p}_i\}$ to T with center $\mathbf{c}$, radius r
30   set $o[\Delta] := $ o-list(e)
31   **if** $(\cos(r) > \mathbf{p}_i \cdot \mathbf{s})$ **and** $(\|\mathbf{c} + \mathbf{s}\| - r < \theta)$ **then**
32         $k \leftarrow k + 1$; $\mathbf{t}[\Delta] \leftarrow \cos(\|\mathbf{c} + \mathbf{s}\| + r)$

33 remove $j \in T$ where $(o[j] \neq -1)$ or $\|\mathbf{c}(j) + \mathbf{s}\| + r(j) > \|\mathbf{p}_i + \mathbf{s}\|$

From this we can estimate that the auxiliary triangulation with q points leads to spherical caps with radius $r_0 \approx 2[\pi/(2q - 4)]^{\frac{1}{2}}$. The number of points $n_{p,\text{cap}}$ contained in the spherical cap can, on the other hand, be approximated by:

$$n_{p,\text{cap}} = A_{\text{cap}}\frac{n_{\text{p}}}{4\pi} = 2\pi[1 - \cos(r_0)]\frac{n_{\text{p}}}{4\pi} = (n_p/2)[1 - \cos(r_0)]. \qquad (2)$$

Inserting r_0 into Eq. (2) and solving for q yields the following estimate:

If each subtask of the parallel triangulation algorithm shall process a workload of $n_{p,\text{cap}}$ points out of a given a point set with n_p points, then the size of the auxiliary triangulation and the number of subtasks should be approximately chosen as

$$q = \lceil 2 + 2\pi \, \text{acos}^{-2}(1 - 2n_{p,\text{cap}}/n_p) \rceil \,.$$

Equation (2) also allows for a basic estimate of the parallel efficiency E_q of the algorithm: We make the simplifying assumption that the runtime T_1 of Algorithm 2 is linearly bounded by n_p. By Eq. (2) we have:

$$E_q = \frac{T_1}{q\,T_q} = \frac{2}{q[1 - \cos(\sqrt{\frac{4\pi}{2q-4}})]} \,.$$

The theoretical parallel efficiency is then $\lim_{q\to\infty} E_q = 2/\pi \approx 0.637$. In fact, the best known planar algorithms are not linear but have $\mathcal{O}(n \log n)$ complexity and we may expect superlinear speedup for the parallel version which does less work than the corresponding serial algorithm. This holds true, however, only as long as the parallel communication, the pre-sorting, and the merge process are not taken into account.

Complexity of the Local Algorithm and Task Parallelization. The worst-case complexity of the serial algorithm is of the order $\mathcal{O}(n_{p,\text{cap}}^2)$. This estimate, however, does not take into account the short-cut of the inner loop, which skips triangles that have been marked by $a[\Delta] = 1$. In practice, Algorithm 2 runs fairly well for the average case.

Since the point insertion strategy requires all parallel threads to share the same view on the current state of the triangulation, the speed-up that can be expected from (shared memory) task parallelism is inherently limited. The inner loop (line 9 in Algorithm 2) however, that evacuates the polyhedral cavity, may be parallelized as follows:

1. In ll.20, do not remove triangles but "mark" them: $a[\Delta] \leftarrow 2$.
2. Loop l.9 may then be executed in a thread-parallel fashion.
3. At regular intervals, e.g., every 50000 marked triangles, revisit triangles with $a[\Delta] = 2$ (single-threaded) and remove them. If possible, move "complete" triangles from the back of the list as in the sequential variant of Algorithm 2.

The sorting and the merging stages can take advantage of multiple threads as well but have not been investigated here. Experimental results for the multithreaded version are shown in Sect. 3.

Robust Determinantal Tests. The influence of round-off errors in geometric algorithms was already pointed out in Lawson's original paper [9] for the evaluation of determinants and the case of planar triangulations. We achieve robust computation of determinants with a *floating-point filter*: According to [3] the absolute error of the floating-point approximation $\tilde{e}$ for the arithmetic expression

$$e := \det\left(\mathbf{v}^1, \mathbf{v}^2, \mathbf{v}^3\right), \quad \mathbf{v}^i \in \mathbb{R}^3, \tag{3}$$

can be bounded by

$$|\tilde{e} - e| \leq \tilde{e}_{\text{sup}}\, ind_e\, 2^{-p}. \tag{4}$$

Here, p is the mantissa length, $p = 52$ for the standard double precision data type, and by $\tilde{e}_{\text{sup}}$ we denote an upper bound for $|\tilde{e}|$ which can be estimated for points $\mathbf{v}^i \in \mathbb{R}^3$ on the sphere based on the approximation $|\tilde{v}_i^j| \approx |v_j^i| \leq 1$. The index ind_e denotes an operation-dependent integer value which can be computed a priori see [3], e.g., for rules for the different arithmetic expressions (9 multiplications and 5 additions/subtractions). In summary we have

$$\tilde{e}_{\text{sup}} \leq 6 \quad \text{and} \quad ind_e = 8.$$

Therefore, in view of Eq. (4), the following approach has proven to lead to a sufficiently robust method with a reasonably small overhead: At runtime all determinantal tests are evaluated in double precision. It is then assumed that the floating-point evaluation of the determinant (3) has the correct sign, if

$$|\widetilde{\det}\left(\mathbf{v}^1, \mathbf{v}^2, \mathbf{v}^3\right)| > C := 6 \cdot 8 \cdot 2^{-52}.$$

Only for cases where the determinant evaluates to smaller absolute values than $C_0 := 1.1 \cdot 10^{-14} > C$ we switch to quadruple (128 bit) precision.

3 Numerical Experiments

In this section we give experimental results for Algorithm 1. Two example point sets are used:

A. $n_p := 2,000,000$ random points on the unit sphere, and
B. $n_p := 2,949,120$ points of the icosahedral-bisection grid used by the ICON global atmospheric model [15].

3.1 Strong Scaling Tests

For our tests, the spherical caps are chosen at q generalized spiral points, a set of points approximately equally spaced on the sphere [10]. In spherical coordinates (θ, ϕ) these points are given by

$$\theta_k := \text{acos}(h_k)\,, \quad h_k := -1 + 2\,(k-1)/(q-1), \quad 1 \leq k \leq q,$$

$$\phi_k := \left[\phi_{k-1} + \frac{3.6}{\sqrt{q(1 - h_k^2)}}\right] (\text{mod } 2\pi), \qquad 2 \leq k < q, \quad \phi_1 := \phi_q := 0\,.$$

Table 1. Timings of scaling results, compared to the sequential STRIPACK algorithm [12]. Average in seconds over all processes; overall runtime including parallel communication is given in parentheses. The lower part of the table shows results for the hybrid variant, each run with 4 OpenMP threads.

	STRIPACK	Serial run	$q = 10$	40
Set A	22.6–2464.4	641.69	131.18 (150.84)	40.81 (56.29)
Set B	189.06	960.89	194.02 (203.88)	61.99 (70.56)
	100	200	400	800
Set A	14.40 (27.38)	7.36 (17.12)	4.31 (10.50)	2.26 (6.34)
Set B	20.77 (24.64)	10.29 (13.90)	6.21 (9.60)	3.20 (5.35)
	100×4	200×4	400×4	800×4
Set A	8.32 (17.49)	4.48 (11.40)	2.83 (7.59)	1.71 (4.82)
Set B	11.40 (15.75)	5.93 (9.63)	3.83 (7.37)	2.32 (4.98)

Wallclock timings for Algorithm 2 are given in Table 1. Here the setup of the spherical caps (auxiliary triangulation) is negligible for the runtime measurements. Comparison to results obtained with the well-known sequential STRIPACK algorithm [12] shows that with our implementation of Algorithm 1 at least $q = 10$ parallel processes are required to yield competitive wallclock timings. It can also be seen, however, that the timings of the STRIPACK package differ widely depending on the ordering of point set. With the rare exception of *identical* sorting keys $\{\mathbf{p}_i \cdot \mathbf{s}\}_i$ this is not the case for Algorithm 1. Results were obtained on a Cray XC40 with Intel Haswell-E5-2670 processors and a hybrid MPI/OpenMP implementation. Note that the multi-threaded experiments were conducted with enabled hyperthreading, s. t. only 2 physical cores were required per task.

Uniqueness of the Tesselation. If there exist coplanar points lying on a common empty circle, then the Delaunay triangulation is not unique. This situation is not uncommon for point sets generated for computational meshes such as point set B. When Algorithm 1 is computed in parallel, we avoid invalid tesselations by deliberately disturbing the perfectly symmetric point sets. After cyclically adding a constant $\epsilon \ll 1$ to the point coordinates and re-normalization onto the sphere the resulting triangulation proves to be unique in all considered tests.

3.2 Application to Barycentric Interpolation

In this section we briefly describe the application of the Delaunay algorithm to the following data fitting problem on the sphere: Given data sites $\{\mathbf{x}_i\}_{i=1}^n$ and real numbers $\{r_i\}_{i=1}^n$, find a function s defined on the unit sphere which interpolates the data in the sense that

$$s(\mathbf{x}_i) = r_i, \quad i = 1, \ldots, n.$$

For the planar case many bivariate interpolation methods exist that have an analog for the sphere and rely on an auxiliary triangulation, see, e.g., [6].

The barycentric interpolation scheme is one example for these methods. Let a triangulation of the data sites $\mathbf{x}_i$ be given. For practical applications it is sufficiently accurate to make a polyhedral approximation of the spherical triangles [4]. The weights $\{u_j\}_{j=1}^3$ for a point $\mathbf{p}$ wrt. the planar triangle $\mathbf{v}_1$, $\mathbf{v}_2$, $\mathbf{v}_3$ are computed from the relations

$$\mathbf{p} = \sum_{j=1}^3 u_j\mathbf{v}_j, \quad \sum_{j=1}^3 u_j = 1. \tag{5}$$

In summary, we get a continuous interpolating function $s(\mathbf{x}) = \sum_j u_j r_j$ from the following three steps:

1. Construct the Delaunay triangulation using Algorithm 1.
2. Locate the containing triangle $\{\mathbf{v}_j\}_{j=1}^3$ for every destination point.
3. Compute the barycentric weights using formula (5).

If no assumptions on the set of destination points are made, Step 2 deserves additional remarks: The containing triangles can be efficiently located by means of an octree data structure of the triangle bounding boxes [5]. Traversing the octree provides a short-list of triangles for which an *inside triangle test* must be performed. For reasons of robustness this should combine several tests: We chose to check the sides of the triangle with the dot product and, additionally, to test for valid barycentric coordinates $u_j \in [0, 1]$.

Note that for data-parallel execution the Delaunay algorithm requires only synchronization of the data site locations for Step 1, provided that the the the data sites and the destination points are both decomposed into the same (convex-shaped) domains. Weights on each partition are then computed only for the local partition of the global Delaunay triangulation. Therefore no communication needs to be established for the Delaunay triangulation itself.

References

1. Bourke, P.: Efficient triangulation algorithm suitable for terrain modelling or an algorithm for interpolating irregularly-spaced data with applications in terrain modelling. In: Pan Pacific Computer Conference, Beijing, China (1989)
2. Bowyer, A.: Computing Dirichlet tessellations. Comput. J. **24**(2), 162–166 (1981)
3. Burnikel, C., Funke, S., Seel, M.: Exact geometric computation using cascading. IJCGA **11**(3), 245–266 (2001)
4. Carfora, M.F.: Interpolation on spherical geodesic grids: a comparative study. J. Comput. Appl. Math. **210**(1–2), 99–105 (2007)
5. de Berg, M., Cheong, O., van Kreveld, M., Overmars, M.: Computational Geometry: Algorithms and Applications, 3rd edn. Springer, Heidelberg (2008)
6. Fasshauer, G.E., Schumaker, L.L.: Scattered data fitting on the sphere. In: Fasshauerand, G.E. (ed.) Mathematical Methods for Curves and Surfaces II, pp. 117–166. Vanderbilt University Press, Nashville (1998)

7. Jacobsen, D.W., Gunzburger, M., Ringler, T., Burkardt, J., Peterson, J.: Parallel algorithms for planar and spherical Delaunay construction with an application to centroidal Voronoi tessellations. Geosci. Model Dev. **6**(4), 1353–1365 (2013)

8. Ju, L., Ringler, T., Gunzburger, M.: Voronoi tessellations and their application to climate and global modeling. In: Lauritzen, P., Jablonowski, C., Taylor, M., Nair, R. (eds.) Numerical Techniques for Global Atmospheric Models. Lecture Notes in Computational Science and Engineering, vol. 80, pp. 313–342. Springer, Heidelberg (2011)

9. Lawson, C.L.: C^1 surface interpolation for scattered data on a sphere. Rocky Mt. J. Math. **14**(1), 177–202 (1984)

10. Rakhmanov, E.A., Saff, E., Zhou, Y.: Minimal discrete energy on the sphere. Math. Res. Lett **1**(6), 647–662 (1994)

11. Renka, R.J.: Interpolation of data on the surface of a sphere. ACM Trans. Math. Softw. **10**(4), 417–436 (1984)

12. Renka, R.J.: Algorithm 772: STRIPACK: Delaunay triangulation and voronoi diagram on the surface of a sphere. ACM Trans. Math. Softw. **23**(3), 416–434 (1997)

13. Shewchuk, J.R.: Lecture notes on Delaunay mesh generation. Department of Electrical Engineering and Computer Sciences, University of California at Berkeley (2012)

14. Watson, D.F.: Computing the n-dimensional Delaunay tessellation with application to Voronoi polytopes. Comput. J. **24**(2), 167–172 (1981)

15. Zängl, G., Reinert, D., Rípodas, P., Baldauf, M.: The ICON (ICOsahedral Nonhydrostatic) modelling framework of DWD and MPI-M: description of the nonhydrostatic dynamical core. Q. J. R. Meteorol. Soc. **141**(687), 563–579 (2015)

On Conforming Local Post-refinement
of Adjacent Tetrahedral and Hexahedral Meshes

Sergey Korotov$^{(\boxtimes)}$ and Talal Rahman

Department of Computing, Mathematics and Physics,
Bergen University College, Inndalsveien 28, 5020 Bergen, Norway
{sergey.korotov,talal.rahman}@hib.no

Abstract. In this note we propose a local post-refinement technique, which can be used to provide the overall conformity of tetrahedral and hexahedral meshes meeting at the planar interface, which presents a quite common situation in many simulations of real-life problems. The same technique can be also used for the case of two adjacent non-matching hexahedral meshes.

Keywords: Finite element method · Tetrahedral mesh · Hexahedral mesh · Mesh conformity · Pyramidal finite element

1   Introduction

The issue of nonconformity of the overall mesh through various interfaces, where (sub)-meshes of different types or with different geometric characteristics meet, often appears in real-life problems, e.g. in computational fluid dynamics, see [8–10] and references therein. In some situations one may need to provide the overall conformity of such hybrid meshes using some computationally non-expensive (local) post-refinements, for example for construction of reliable a posteriori error estimates controlling computational errors of various types (see e.g. [3] for some relevant discussion).

In this note we propose some local post-refinement technique that produces a conforming mesh over the whole domain and keeps most of the original subdomain meshes intact. As we handle element faces of triangular and quadrilateral shapes, introducing a new type of elements - in our case pyramids, as those having faces of both shapes, is unavoidable (cf. [7]). Our approach is different from those proposed in [7–10] for similar or close situations and can be considered as certain generalization of the idea from our earlier work [3].

In what follows, all triangles, tetrahedra, quadrilaterals, and hexahedra, are considered as closed sets. We will only deal with convex quadrilaterals and hexahedra. Construction of finite element approximations on hybrid meshes consisting of finite elements of different types is presented e.g. by Wieners in [11]. Relevant material on constructing and handling various pyramidal finite elements can be found in recent papers [1,5,6] by Křížek et al.

© Springer International Publishing Switzerland 2016
R. Wyrzykowski et al. (Eds.): PPAM 2015, Part II, LNCS 9574, pp. 365–370, 2016.
DOI: 10.1007/978-3-319-32152-3_34

2   Main Result

In Fig. 1, we illustrate some conforming post-refinement when triangular and quadrilateral meshes meet along some line in the solution domain but do not match each other. Using this technique, we can "eliminate" hanging nodes, possibly introducing some new (auxiliary but non-hanging) nodes. It is worth to notice that no other types of elements (besides triangles and quadrilaterals) ever appear in this two-dimensional situation.

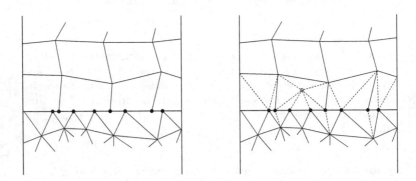

Fig. 1. Some conforming post-refinement of non-matching triangular and quadrilateral meshes. By bullets we mark the hanging nodes appearing in the common interface, the circle (in the right) denotes the auxiliary node introduced.

In what follows, we present the generalization of this idea for the three-dimensional case. Let the polyhedral domain $\Omega \subset \mathbb{R}^3$ be divided into two non-overlapping subdomains Ω^1 and Ω^2 sharing a rectangular interface $\mathcal{S}$ (more general situations will be discussed later on). Assume that in the subdomains Ω^1 and Ω^2 conforming (face-to-face) tetrahedral and hexahedral meshes $\mathcal{M}^1$ and $\mathcal{M}^2$, respectively, are independently generated. Obviously, $\mathcal{M}^1$ and $\mathcal{M}^2$ never meet face-to-face across $\mathcal{S}$. In Fig. 2 (left) we present an example of two non-matching planar partitions, induced by tetrahedral and hexahedral meshes on the interface $\overline{\mathcal{S}}$.

Now we show how to make a local post-refinement of $\mathcal{M}^1$ and $\mathcal{M}^2$ so that the resulting overall hybrid mesh (consisting of tetrahedra, hexahedra and pyramids - differently from the two-dimensional case, in 3D we unavoidably need some new elements – pyramids – having both, triangular and rectangular faces, to provide the overall conformity) over $\overline{\Omega} = \overline{\Omega}^1 \cup \overline{\Omega}^2$ is conforming. Let

$$\mathcal{T} = \{T_i, \ i = 1, \ldots, n\} \quad \text{and} \quad \mathcal{Q} = \{Q_j, \ j = 1, \ldots, m\}$$

denote the two partitions (into triangles and quadrilaterals) of $\overline{\mathcal{S}}$ induced by the tetrahedra and hexahedra from the meshes $\mathcal{M}^1$ and $\mathcal{M}^2$, respectively. It is clear that

$$\overline{\mathcal{S}} = \bigcup_{i=1,\ldots,n;\, j=1,\ldots,m} T_i \cap Q_j. \tag{1}$$

Now consider only those sets from the right-hand side of (1) for which

$$\text{meas}_2 \, (T_i \cap Q_j) > 0,$$

where the symbol meas_2 denotes the area of the planar domains. They are convex polygons, see Fig. 2 (right), and each of these polygons can be conformly refined into several triangles using polygon's nodal points only, see Fig. 3 (left). The resulting triangulation, called *coupling triangulation* and denoted by **T**, is conforming and it covers the whole interface $\overline{S}$, see Fig. 3 (right).

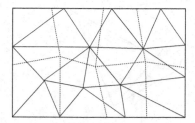 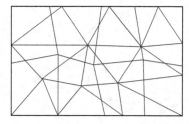

Fig. 2. Traces of triangular (bold lines) and quadrilateral (dotted lines) faces of mesh elements adjacent to the common interface $\overline{S}$ (left). An associated splitting of this interface into convex polygons (right).

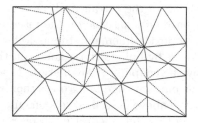 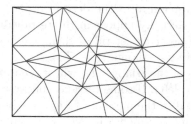

Fig. 3. Splitting of the convex polygons of the interface $\overline{S}$ into triangles (left). The coupling triangulation **T** (right).

Let $\mathbf{T} = \{t_k, \, k = 1, \ldots, \ell\}$, where t_k are the elements of the coupling triangulation. From the construction of the coupling triangulation, we observe that for each $k \in \{1, \ldots, \ell\}$, one has $\text{meas}_2 \, t_k > 0$ and there are indices $i_k \in \{1, \ldots, n\}$ and $j_k \in \{1, \ldots, m\}$ such that

$$t_k \subseteq T_{i_k} \quad \text{and} \quad t_k \subseteq Q_{j_k}. \tag{2}$$

Now we run over all triangles $t_k \in \mathbf{T}$ and consider the mesh $\mathcal{M}^1$. For each t_k we find an unique tetrahedron in $\mathcal{M}^1$, which has the triangle T_{i_k} defined in (2) as its face. Within this tetrahedron we form a subtetrahedron which is a

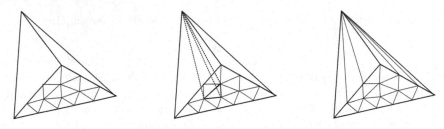

Fig. 4. An illustration on splitting the tetrahedron from $\mathcal{M}^1$ adjacent to $\overline{S}$.

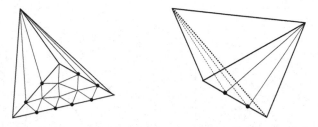

Fig. 5. Splitting of tetrahedra which share only an edge with $\overline{S}$ providing the conformity in the refined "tetrahedral part" Ω^1.

convex hull of the triangle t_k and the vertex of the taken tetrahedron opposite to t_k, see Fig. 4 (center). Finally, any tetrahedron from $\mathcal{M}^1$, having a face lying on $\overline{S}$, is refined as sketched in Fig. 4 (right). Still some hanging nodes and edges may remain after the described above refinement of the tetrahedral mesh $\mathcal{M}^1$. They actually appear in the tetrahedra that have only an edge on the interface $\overline{S}$. In order to get rid of them, we perform the next refinement step as follows. Consider a triangle $T_i \in \mathcal{T}$, $i \in \{1, \ldots, n\}$. Mark by bullets those nodes on the edges of the triangle T_i, which are induced by the coupling triangulation $\mathbf{T}$, see Fig. 5 (left). If some tetrahedron from $\mathcal{M}^1$ has only an edge lying on $\overline{S}$ then we refine it in the manner of Fig. 5 (right). After the above described refinements of relevant tetrahedra in $\mathcal{M}^1$ we produce a conforming tetrahedral mesh over $\overline{\Omega}^1$ the traces of which on $\overline{S}$ coincide with the defined above coupling triangulation $\mathbf{T}$, for more details see [3].

Further, we show how we proceed with the "hexahedral part" of the domain, i.e. with the mesh $\mathcal{M}^2$. For each $t_k \in \mathbf{T}$ we can find an unique hexahedron in $\mathcal{M}^2$, which has the quadrangle Q_{j_k} defined in (2) as its face, see Fig. 6 (left) where Q_{j_k} is covered by a corresponding part of the coupling triangulation including the triangle t_k itself. Let us now select some point inside the hexahedron (marked by the bullets in Fig. 6), and take it as an auxiliary node for the new mesh made in the following way. First, we decompose the hexahedron into 6 pyramids being convex hulls of this interior point and 6 faces of the hexahedron. The pyramid with the quadrilateral base Q_{j_k} is further split into tetrahedra, using the traces of the coupling triangulation, in the manner of Fig. 6 (right) (cf. Figure 4 with a similar refinement for tetrahedra). Third, the reminding quadrilateral faces of

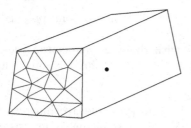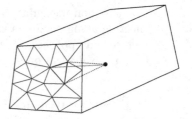

Fig. 6. A hexahedron from the mesh $\mathcal{M}^2$ and its face lying on $\overline{S}$ with the splitting induced by the coupling triangulation (left). Formation of a tetrahedron in this hexahedron using the triangle t_k and some auxiliary node (marked by bullet) inside (right).

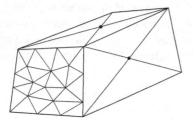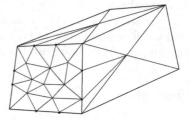

Fig. 7. On splitting of faces of a hexahedron from "hexahedral part" Ω^2.

the hexahedron (besides that one opposite to Q_{j_k}) we split by diagonals and introduce four more new nodes for the new mesh, which are the intersection points of these diagonals inside each face, see Fig. 7 (left). Fourth, on each of the edges of the quadrilateral Q_{j_k} we mark by bullets all the vertices stemming from the coupling triangulation and connect them to the intersection points inside corresponding faces of the hexahedron in the manner of Fig. 7 (right). Now, we proceed as follows: the splitting of the pyramid with the base Q_{j_k} was explained earlier, the pyramid with the base opposite to Q_{j_k} is not split, and each of the four remaining pyramids is split into tetrahedra which are the convex hulls of that interior point (inside the hexahedron) and the triangles constituting the splitting of the bases of the pyramids. We repeat this procedure for all hexahedra adjacent to the interface $\overline{S}$. It is obvious that the final refinement of the mesh $\mathcal{M}^2$ is conforming, also it conformly fits the earlier described refinement of the mesh $\mathcal{M}^1$. However, in addition to tetrahedra (and hexahedra) we are forced now to introduce different mesh elements – pyramids.

3   Final Remarks, Open Problems

The above presented technique can also be applied for more general situations:

– Two adjacent domains Ω^1 and Ω^2, meet at the planar interface, which is not a rectangle but any polygon (even nonconvex one).

- The case of two non-matching hexahedral meshes of adjacent polyhedral domains meeting along some planar interface.
- The case of several polyhedral domains with different associated meshes (tetrahedral or hexahedral ones) meeting at disjoint planar polygonal interfaces.

Usage of the coupling triangulation may lead, in principle, to thin tetrahedral elements. However, due to recent results in the finite element analysis, see e.g. [2,4], it does not bring difficulties in the context of finite element methods. Moreover, we can always avoid producing undesired tetrahedral shapes by merging/shifting some nodes associated with thin triangles in the coupling triangulation if needed.

An optimal selection of the interior points inside the hexahedra is an open problem. We should also notice that the above approach does not seem be easily modified to the case of general hexahedral meshes, i.e. those having some nonconvex elements. The above two issues will be addressed to in our next paper.

References

1. Chen, C.M., Křížek, M., Liu, L.: Numerical integration over pyramids. Adv. Appl. Math. Mech. **5**, 309–320 (2013)
2. Hannukainen, A., Korotov, S., Křížek, M.: The maximum angle condition is not necessary for convergence of the finite element method. Numer. Math. **120**, 79–88 (2012)
3. Juntunen, M., Korotov, S.: Conforming post-refinement of non-matching tetrahedral meshes. In: Proceedings of the Mascot, Madrid, Spain (2013, to appear)
4. Křížek, M.: On the maximum angle condition for linear tetrahedral elements. SIAM J. Numer. Anal. **29**, 513–520 (1992)
5. Liu, L., Davies, K.B., Yuan, K., Křížek, M.: On symmetric pyramidal finite elements. Dyn. Continuous Discrete Impulsive Syst. Ser. B Appl. Algorithms **11**, 213–227 (2004)
6. Liu, L., Davies, K.B., Křížek, M., Guang, L.: On higher order pyramidal finite elements. Adv. Appl. Math. Mech. **3**, 131–140 (2011)
7. Owen, S.J., Canann, S.A., Saigal, S.: Pyramidal elements for maintaining tetrahedra to hexahedra conformability. In: AMD, Trends in Unstructured Mesh Generation, vol. 220, pp. 1–7. ASME (1997)
8. Qin, N., Carnie, G., LeMoigne, A., Liu, X., Shahpar, S.: Buffer layer method for linking two non-matching multi-block structured grids. In: AIAA 2009–1361 (2009)
9. Song, S., Wan, M., Wang, S., Wang, D., Zou, Z.: Robust and quality boundary constrained tetrahedral mesh generation. Commun. Comput. Phys. **14**, 1304–1321 (2013)
10. Wang, Y., Qin, N., Carnie, G., Shahpar, S.: Zipper layer method for linking two dissimilar structured meshes. J. Comput. Phys. **225**, 130–148 (2013)
11. Wieners, C.: Conforming discretizations on tetrahedrons, pyramids, prisms and hexahedrons. University of Stuttgart, Bericht 97/5, pp. 1–9 (1997)

Fast Static Condensation for the Helmholtz Equation in a Spectral-Element Discretization

Immo Huismann[✉], Jörg Stiller, and Jochen Fröhlich

Institute of Fluid Mechanics, Center for Advancing Electronics Dresden (cfaed),
01062 Dresden, Germany
Immo.Huismann@tu-dresden.de

Abstract. Current research in computational fluid dynamics focuses on higher-order methods. These possess a more extensive coupling between degrees of freedom, resulting in a larger runtime per degree of freedom compared to low-order methods. This work tries to tackle this issue by combining the static condensation method with tensor-product and sum factorization, leading to a well-scaling solver for the HELMHOLTZ equation.

Keywords: Spectral element method · Static condensation · Elliptic equations · Substructuring

1 Introduction

The algorithmic landscape in computational fluid dynamics (CFD) has changed in the last few years. While industrial codes stick to the well researched low-order methods, high-order methods are gaining more and more interest in the scientific community [2]. These methods, i.e. the discontinuous GALERKIN and spectral element methods (SEM), combine the convergence properties of spectral techniques with the versatility of finite volumes or elements, fueling the movement towards higher polynomial orders to lower the error.

Yet the application of higher polynomial degrees incorporates a major drawback: The tighter coupling inside the elements leads to more work which, furthermore, scales super-linearly with the number of collocation points inside an element [6]. Moreover, incompressible CFD codes spend up to 90 % of the computation time in pressure solvers, and these scale, hence, badly as well.

This paper focuses on regaining and retaining linear complexity throughout the whole solution process of the pressure solver, from the operator execution, to the preconditioner to the number of iterations. To achieve this, the static condensation method [9], also known as SCHUR complement or substructuring, is applied. The approach is well known for SEM [1], but the implementations proposed in the literature scale super-linearly with the number of degrees of freedom. This paper combines this method with further factorization techniques, which were sketched briefly in [5], attaining linear complexity and paving parts of the road to higher polynomial degrees.

© Springer International Publishing Switzerland 2016
R. Wyrzykowski et al. (Eds.): PPAM 2015, Part II, LNCS 9574, pp. 371–380, 2016.
DOI: 10.1007/978-3-319-32152-3_35

2   Spectral-Element Discretization of the Helmholtz Equation

The problem considered is the so-called HELMHOLTZ equation [2]

$$\lambda u - \Delta u = f \tag{1}$$

in a domain Ω, where u is the function to solve for, f is the right-hand side, λ a non-negative constant and Δ the LAPLACE operator. The equation is very widely encountered in NAVIER-STOKES solvers, e.g., it is obtained when discretizing diffusion terms implicitly or when solving the pressure equation. A GALERKIN formulation transforms (1) to the weak form, in which the domain Ω is decomposed into n_e non-overlapping cuboidal elements Ω_e. In each of these, a nodal tensor-product basis is employed. For all three directions, GAUSS-LOBATTO-LEGENDRE (GLL) polynomials of degree p constitute the basis [2], as illustrated for the two-dimensional case in Fig. 2. The specific node choice allows GAUSS-LOBATTO quadrature for the integrals of the weak form for each direction in every element. The resulting mass matrix of the standard element $\mathbf{M}$ is diagonal, whereas the stiffness matrix of the standard element $\mathbf{L}$ is a full matrix.

Using an element-wise storage, the discretized HELMHOLTZ equation can be written as

$$\mathcal{A}\left(\{\mathbf{H}_e\mathbf{u}_e\}_e\right) = \mathcal{A}\left(\{\mathbf{F}_e\}_e\right) \tag{2}$$

where an array $\mathbf{u}_e$ denotes the coefficients of a variable u inside an element Ω_e. For cuboidal elements

$$\mathbf{H}_e = \lambda\mathbf{M}\otimes\mathbf{M}\otimes\mathbf{M} + \mathbf{L}\otimes\mathbf{M}\otimes\mathbf{M} + \mathbf{M}\otimes\mathbf{L}\otimes\mathbf{M} + \mathbf{M}\otimes\mathbf{M}\otimes\mathbf{L}. \tag{3}$$

For readability, the metric coefficients were omitted. The assembly operation, $\mathcal{A}$, also known as direct stiffness summation, joins contributions over element boundaries and, for the described basis, reduces to the summation of nodal contributions across adjoining element faces.

A tensor product $\mathbf{C}\otimes\mathbf{B}\otimes\mathbf{A}$ constitutes a matrix whose application corresponds to the consecutive applications of $\mathbf{A}$, $\mathbf{B}$, and $\mathbf{C}$ in the first, second, and third direction inside the element, respectively. For more details on the SEM the reader is referred to [2,6] and for tensor products and their properties to [8] due to restrictions of space.

Due to the tensor-product formulation of (3), the HELMHOLTZ operator can be evaluated in $\mathcal{O}\left(p^4\right)$ multiplications per element. While the number is far lower than the expected $\mathcal{O}\left(p^6\right)$, it still scales super-linearly with the number of degrees of freedom and obstructs the path to higher polynomial degrees. The following sections propose adequate methods to remedy this issue.

3   The Static Condensation Method

Equation (1) is of elliptic type, i.e. values in the interior of the domain solely depend upon the boundary values and the right-hand side. The GALERKIN formulation retains this property, but (2) works on all degrees of freedom, wasting compute power and slowing the iteration process.

Reducing the number of degrees of freedom leads to a lower amount of work for iterative solvers and has a positive effect on the condition of the system. One approach to this end is the static condensation method, also known as SCHUR complement method [9], the main idea of which is the eradication of unrequired degrees of freedom from the equation system.

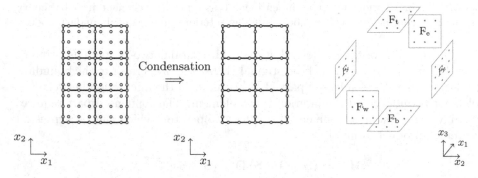

Fig. 1. Illustration of condensation. Left: Two-dimensional static condensation process eliminating degrees of freedom inside the elements. Right: Exploded view of the faces of one three-dimensional element, including compass notation for the faces and face collocation nodes. For an unobstructed view boundary edges and vertices are not displayed.

Applying the static condensation to the whole domain is feasible for simple geometries and leads to astonishingly fast solvers [7], but is not affordable for more complex domains. This paper focuses, hence, on the usage of the method inside each element, decreasing the number of degrees of freedom from $\mathcal{O}\left(p^3 n_e\right)$ to $\mathcal{O}\left(p^2 n_e\right)$ and fairing the condition number of the equation system. The first result is a lower number of degrees of freedom, as depicted in Fig. 1. The second one is a changed element HELMHOLTZ operator in (2), which now incorporates stronger coupling between element boundaries. Since only the change in the operator inside each element Ω_e needs to be discussed, not the interaction between elements, the element subscript is dropped.

The values in an element $\mathbf{u}$ can be categorized as either interior values $\mathbf{u}_\mathrm{I}$ denoted by subscript I, or boundary values $\mathbf{u}_\mathrm{B}$, denoted by subscript B, as illustrated in Fig. 2. When expanding the notation towards matrices, e.g. $\mathbf{H}_\mathrm{BI}$ maps from the interior to the boundary, the equation system reads as

$$\begin{pmatrix} \mathbf{H}_\mathrm{BB} & \mathbf{H}_\mathrm{IB} \\ \mathbf{H}_\mathrm{BI} & \mathbf{H}_\mathrm{II} \end{pmatrix} \begin{pmatrix} \mathbf{u}_\mathrm{B} \\ \mathbf{u}_\mathrm{I} \end{pmatrix} = \begin{pmatrix} \mathbf{F}_\mathrm{B} \\ \mathbf{F}_\mathrm{I} \end{pmatrix}, \tag{4}$$

which in turn leads to

$$\mathbf{u}_\mathrm{I} = \mathbf{H}_\mathrm{II}^{-1}\left(\mathbf{F}_\mathrm{I} - \mathbf{H}_\mathrm{IB}\mathbf{u}_\mathrm{B}\right) \tag{5}$$

$$\hat{\mathbf{H}}\mathbf{u}_\mathrm{B} := \big(\underbrace{\mathbf{H}_\mathrm{BB}}_{\hat{\mathbf{H}}^\mathrm{prim}} - \underbrace{\mathbf{H}_\mathrm{BI}\mathbf{H}_\mathrm{II}^{-1}\mathbf{H}_\mathrm{IB}}_{\hat{\mathbf{H}}^\mathrm{cond}}\big)\mathbf{u}_\mathrm{B} = \underbrace{\mathbf{F}_\mathrm{B} - \mathbf{H}_\mathrm{BI}\mathbf{H}_\mathrm{II}\mathbf{F}_\mathrm{I}}_{\hat{\mathbf{F}}}. \tag{6}$$

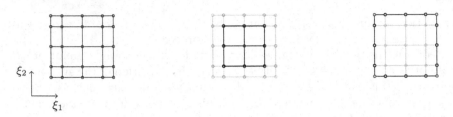

Fig. 2. Decomposition of a two-dimensional tensor-product element into boundary nodes and inner nodes. Left: all nodes, middle: interior nodes, right: boundary nodes.

The operator of the condensed system, $\hat{\mathbf{H}}$, contains two parts: The primary part $\hat{\mathbf{H}}^{\mathrm{prim}}$ is the restriction of the original HELMHOLTZ operator to the boundary nodes, whereas the condensed part $\hat{\mathbf{H}}^{\mathrm{cond}}$ consists of the interaction between the element boundary and the interior of the element. The condensed part requires the inverse of the inner element HELMHOLTZ operator, which can be expressed in tensor-product form [8]

$$\mathbf{H}_{\mathrm{II}}^{-1} = (\mathbf{S} \otimes \mathbf{S} \otimes \mathbf{S}) \, \mathbf{D}^{-1} \, (\mathbf{S} \otimes \mathbf{S} \otimes \mathbf{S})^{T} . \tag{7}$$

The diagonal matrix $\mathbf{D}$ is given, for the standard element, as

$$\mathbf{D} = \lambda \mathbf{I} \otimes \mathbf{I} \otimes \mathbf{I} + \mathbf{\Lambda} \otimes \mathbf{I} \otimes \mathbf{I} + \mathbf{I} \otimes \mathbf{\Lambda} \otimes \mathbf{I} + \mathbf{I} \otimes \mathbf{I} \otimes \mathbf{\Lambda}, \tag{8}$$

where $\mathbf{I}$ is the identity matrix. The transformation matrix $\mathbf{S}$ and the eigenvalue matrix $\mathbf{\Lambda}$ are defined by the generalized eigenvalue problem

$$\mathbf{S}^{T} \mathbf{L}_{\mathrm{II}} \mathbf{S} = \mathbf{\Lambda}, \quad \mathbf{S}^{T} \mathbf{M}_{\mathrm{II}} \mathbf{S} = \mathbf{I}. \tag{9}$$

4   Factorization of the Condensed Operator

The static condensation itself leads to fewer degrees of freedom. Yet this does not imply that the operator requires fewer operations. While the primary part of the HELMHOLTZ operator can be implemented in a way that requires $\mathcal{O}\left(p^{3}\right)$ multiplications, the condensed part, $\hat{\mathbf{H}}^{\mathrm{cond}}$, poses more problems. Due to the diagonal mass matrix, the HELMHOLTZ operator works along mesh lines in the element, coupling the element interior only with values on the faces and neither with the edges, nor with the vertices of the element. Hence, previous implementations mapped from faces to all faces, similar to Algorithm 1, requiring $\mathcal{O}\left(p^{4}\right)$ multiplications for application of the operator when using tensor products. Thus they preferred the usage of matrix-matrix multiplications due to a lower number of operations [1].

The usage of Algorithm 1 provides no gain in complexity compared to the full HELMHOLTZ operator, only in the condition number of the equation system. The goal of this section is, hence, two-fold: First, derive an operator-evaluation that scales linearly with the number of degrees of freedom, second, create an algorithm which is competitive to a matrix-product based implementation of

Algorithm 1. Evaluation of the condensed operator in a direct face-to-face variant.

1: **for** all $j \in \mathcal{I}$ **do**
2:    $\hat{\mathbf{v}}_{\mathrm{F}_j} \leftarrow \sum_{i \in \mathcal{I}} \hat{\mathbf{H}}^{\mathrm{cond}}_{\mathrm{F}_j \mathrm{F}_i} \hat{\mathbf{u}}_{\mathrm{F}_i}$

Algorithm 1. The derivation is presented for the east face in compass notation, as shown in Fig. 1, but the operators for the other faces can be deduced in the same fashion. A subscript i in F_i denotes a variable on face i, where the short-hands w, e, s, n, b, and t are employed for the faces west, east, south, north, bottom, and top, respectively. The set $\mathcal{I} = \{\mathrm{w}, \mathrm{e}, \mathrm{s}, \mathrm{n}, \mathrm{b}, \mathrm{t}\}$ serves as a short-hand when referring to all faces.

The condensed operator from face east to face east, $\hat{\mathbf{H}}^{\mathrm{cond}}_{\mathrm{F}_e \mathrm{F}_e}$, is

$$\hat{\mathbf{H}}^{\mathrm{cond}}_{\mathrm{F}_e \mathrm{F}_e} = \mathbf{H}_{\mathrm{F}_e \mathrm{I}} \mathbf{H}_{\mathrm{II}}^{-1} \mathbf{H}_{\mathrm{IF}_e}$$

where the operators from the inner element to the face can be deduced from (3) as

$$\mathbf{H}_{\mathrm{F}_e \mathrm{I}} = \mathbf{M}_{\mathrm{II}} \otimes \mathbf{M}_{\mathrm{II}} \otimes \mathbf{L}_{p\mathrm{I}} \tag{10}$$

$$\mathbf{H}_{\mathrm{IF}_e} = \left(\mathbf{M}_{\mathrm{II}} \otimes \mathbf{M}_{\mathrm{II}} \otimes \mathbf{L}_{p\mathrm{I}} \right)^{T} = \mathbf{H}_{\mathrm{F}_e \mathrm{I}}^{T}. \tag{11}$$

The matrices are of size $n_\mathrm{p} \times n_\mathrm{p}$, $n_\mathrm{p} \times 1$, and $1 \times n_\mathrm{p}$, where $n_\mathrm{p} = p - 1$. Thus, the operators in (10) and (11) can be applied in $3n_\mathrm{p}^3$ multiplications, if and only if the row and column matrices are applied first and last, respectively. Yet the inverse of the inner element HELMHOLTZ operator requires $\mathcal{O}\left(n_\mathrm{p}^4\right)$ multiplications due to the three-dimensional tensor products. When merging both tensor-product operations, the above expands to

$$\hat{\mathbf{H}}^{\mathrm{cond}}_{\mathrm{F}_e \mathrm{F}_e} = \underbrace{\left(\mathbf{M}_{\mathrm{II}} \mathbf{S} \otimes \mathbf{M}_{\mathrm{II}} \mathbf{S} \otimes \mathbf{L}_{p\mathrm{I}} \mathbf{S} \right)}_{\mathbf{H}_{\mathrm{F}_e \mathrm{E}}} \mathbf{D}^{-1} \underbrace{\left(\mathbf{M}_{\mathrm{II}} \mathbf{S} \otimes \mathbf{M}_{\mathrm{II}} \mathbf{S} \otimes \mathbf{L}_{p\mathrm{I}} \mathbf{S} \right)^{T}}_{\mathbf{H}_{\mathrm{EF}_e}}. \tag{12}$$

The left part, $\mathbf{H}_{\mathrm{F}_e \mathrm{E}}$, is a mapping from the inner element eigenspace, denoted by the subscript E, to face east, while the right part is the corresponding mapping from the face to the eigenspace. Both consist of a tensor-product, either with a reducing, or a prolonging matrix. Treating these matrices first, and last, respectively, leads to $3n_\mathrm{p}^3$ multiplications for the operator. For the diagonal matrix $\mathbf{D}$, n_p^3 multiplications are required. Thus, one face to face operator can be implemented in $7n_\mathrm{p}^3$ and, hence, Algorithm 1 in $6 \cdot 6 \cdot 7n_\mathrm{p}^3 = 252n_\mathrm{p}^3$.

Linear scaling with the degrees of freedom, $\mathcal{O}\left(p^3\right)$, was attained for the condensed operator, but the leading coefficient is forbiddingly high, rendering the technique quite irrelevant. But as all faces first map to the eigenspace, and then to the faces, further factorization is possible: The contributions from the different faces can be summed in the eigenspace, eliminating the need to map from each face to every other one, leading to Algorithm 2. Evaluating it with tensor products then only requires $37n_\mathrm{p}^3$ multiplications, compared to the $36n_\mathrm{p}^4$ of a matrix-based implementation of Algorithm 1.

Algorithm 2. Evaluation of the condensed part that accumulates contributions in the eigenspace and then maps back to the faces.

1: $\tilde{u} \leftarrow \sum_{i \in \mathcal{I}} \hat{H}_{EF_i} \hat{u}_{F_i}$
2: $\tilde{v} \leftarrow D^{-1} \tilde{u}$
3: **for all** $j \in \mathcal{I}$ **do**
4:     $\hat{v}_{F_j} \leftarrow \hat{H}_{F_j E} \tilde{v}$

Furthermore, the reconstruction of internal degrees of freedom and the computation of the right-hand side for the condensed system need to be treated. Using the same factorization techniques, both can be factorized to require $3n_p^4 + 19n_p^3$ multiplications. While still scaling super-linearly in the degrees of freedom, the above attains a gain of one power compared to previous implementations [3].

5   Comparison of Operator Implementations

The last chapter focused upon operation counts to compare the different algorithms. From this perspective a tensor-product based implementation of Algorithm 2 starts being faster than a matrix-product implementation of Algorithm 1 at a polynomial degree of $p = 3$. Yet, the efficiency of implementations will differ in practice. Tensor products require many low-dimensional matrix operations, whereas highly-optimized libraries, e.g. BLAS, empower the usage of large operators. Moreover, the "real-world" efficiency is influenced by the capabilities of the compiler and the hardware.

This section compares three different implementations of the condensed operator $\hat{H}$, all using different variants of the condensed part: A matrix-product implementation of Algorithm 1, short-handed MP1, a tensor-product implementation of Algorithm 2, called TP2, and a matrix-product implementation of Algorithm 2, named MP2.

For MP1, $36n_p^4 n_e$ multiplications are required per application, and components of the primary part can be incorporated into the matrices. In TP2, $37n_p^3 n_e$ multiplications are needed for the condensed part, whereas MP2 uses more than $12n_p^5 n_e$. This comparison will treat the latter two unfairly; both only require the eigenvalues of the inner element HELMHOLTZ operator in every element, and constant matrices, whereas MP1 needs the face-to-face matrices in every element. But storing these matrices is not feasible at the time being, e.g. for a polynomial degree of $p = 15$ they would occupy more than one gigabyte of RAM for just 128 elements. Thus, only meshes with constant element width are utilized, and only one set of matrices is stored, leading to far higher cache efficiency than normal for the method. To incorporate a more general method, the study includes MP2.

For a HELMHOLTZ parameter of $\lambda = \pi$, the operators were applied 100 times for different polynomial degrees ranging between 2 and 32, each with 512 elements. The program was compiled with the Intel Fortran compiler v. 2015 and run on one core of an Intel Xeon E5-2690. The Intel MKL provided DGEMM from BLAS.

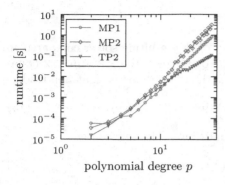 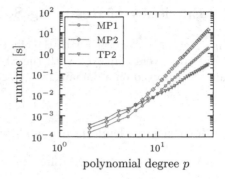

Fig. 3. Comparison of three implementations: matrix-product face-to-face variant (MP1), matrix-product face-to-eigenspace (MP2), and tensor-product based face-to-eigenspace variant (TP2). Left: operator setup times for 512 elements, right: operator runtimes for 512 elements.

Figure 3 depicts the runtimes for operator execution and setup. The setup times show the expected $\mathcal{O}\left(p^5\right)$ and $\mathcal{O}\left(p^3\right)$ slopes for large p and are in the same order as the operator execution or lower. Hence, they are negligible in an iterative scheme.

The execution times exhibit three main features: First, the matrix-based variant MP2 is never the fastest, by a factor ranging from 2 to 50. Second, MP1 is the fastest for low polynomial degrees, whereas from 10 upward TP2 takes over and gets up to 5 times faster for $p = 32$. Third, the expected asymptotic behaviour is seen for large polynomial degrees.

The new operator is not faster for all cases. As the tensor-product operations are far more numerous, this was to be expected. Yet, the linear scaling is achieved, enabling higher polynomial degrees. Additionally, the storage requirements were lowered to $\mathcal{O}\left(p^3 n_e\right)$, which allows for in-homogeneous meshes.

6 Factorization of the Preconditioner

For the case of the condensed equation, diagonal and block preconditioner were investigated in [1] for a fixed number of elements. The face-wise version proved to be rather efficient, lowering the condition number from $\mathcal{O}\left(p^2\right)$ to $\mathcal{O}\left(1\right)$, the relation between condition number and number of elements was not investigated in [1]. Yet the face-wise preconditioners need to be calculated and their storage requirements scale with $\mathcal{O}\left(p^4 n_e\right)$, as they were applied as matrices, destroying linearity and exceeding memory for large polynomial degrees. But a slightly altered preconditioner can be factorized: The condensed operator, introduced in (6), from a face to itself can be factorized via (9), extracting $\mathbf{M}_{\mathrm{II}}\mathbf{S}$ on the left and $\mathbf{S}^T\mathbf{M}_{\mathrm{II}}$ on the right. The remainder is diagonal and can be combined with the contributions from other elements. Hence, a block preconditioner can be constructed by a eigenvalue transformation via $\mathbf{S}$ and the inverse of the diagonal mentioned above. Its application costs $\mathcal{O}\left(p^3 n_e\right)$ multiplications and, thus, scales linearly with the number of degrees of freedom.

7   Efficiency of Resulting Solver

This section is concerned with the efficiency of the resulting solver compared to standard solvers employed with the SEM. To this end, the HELMHOLTZ equation (1) is solved on a domain $\Omega = [0,2]^3$ with the continuous right-hand side

$$f(x) = \left(\lambda + 3\pi^2\right) \sin\left(\pi x_1\right) \sin\left(\pi x_2\right) \sin\left(\pi x_3\right), \tag{13}$$

and homogeneous DIRICHLET boundary conditions. The solution is

$$u(x) = \sin\left(\pi x_1\right) \sin\left(\pi x_2\right) \sin\left(\pi x_3\right). \tag{14}$$

The HELMHOLTZ parameter is set to $\lambda = 0$, which leads to the harder to solve LAPLACE equation and is the parameter required for pressure solvers. The results for small HELMHOLTZ parameters provided no large deviations. Three solver implementations were compared, all based on the conjugate gradient method (CG) [4]: A diagonally preconditioned version for the full equation system (pfCG), a CG solver for the condensed one (cCG), and a block-preconditioned variant (bpcCG) of the latter. The last two utilize the tensor-product variant of Algorithm 2 to evaluate the element-wise condensed HELMHOLTZ operator.

For polynomial degrees ranging from 2 to 32 and 8 elements in each direction, i.e. $n_e = 512$, the solvers were run 11 times, only the last 10 runtimes were averaged. The iteration process stops after a reduction of the L2 residual by 12 orders of magnitude. Figure 4 visualizes the iteration count and the measured runtimes of the solvers. The most prominent feature is the higher slope of the iteration count of pfCG, which is due to the larger operation complexity and higher condition number. While the slope of both condensed solvers is lower, the preconditioned version outperforms the non-preconditioned one.

When comparing the runtimes of the solvers, the iteration count plays its part, but is far less noticeable than expected: For low polynomial degrees, the

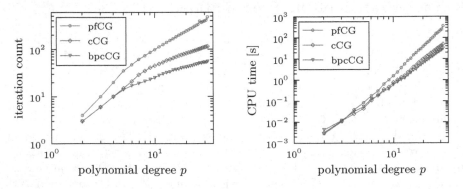

Fig. 4. Iteration count and CPU time to solution for three different solvers (pfCG: full system, preconditioned CG, cCG: CG solver for condensed system, and bpcCG: block-preconditioned CG for the condensed system) when varying the polynomial degrees p and keeping $n_e = 8^3$. Left: iteration count, right: time to solution.

runtimes of the condensed solvers are nearly indistinguishable and pfCG is only by a factor of two slower. But starting from $p = 8$ on, a noticeable speedup presents itself when using the condensed system and pfCG and bpcCG produce nearly the same runtimes. While the iteration count of bpcCG is far lower, the face-wise preconditioning requires additional $24n_{\mathrm{p}}^3 n_{\mathrm{e}}$ multiplications, penalizing the method. Hence, the new solver only leads to gains from $p = 12$ on, where it is faster than cCG by a factor of 1.5 to 2 and by a factor of up to 10 times faster than pfCG. As the test case is rather easy, even for unpreconditioned solvers, further work is required to evaluate the solvers, e.g. by using in-homogeneous grids, possibly leading to further gains for the new method.

8   Conclusions

The paper proposed a solver for the HELMHOLTZ equation on cuboidal elements that heavily relies on tensor-product evaluations. Starting from a tensor-product notation of the condensed HELMHOLTZ equation, an implementation with linear scaling was derived. Its efficiency was compared with different approaches and starting from a polynomial degree of 10, the proposed evaluation method provided a runtime benefit over optimized matrix-based approaches for homogeneous grids. Furthermore, the storage requirements now only scale linearly with the number of degrees of freedom, enabling the usage of higher polynomial degrees. Additionally, the preprocessing steps of the condensation process were sped up with the tensor-product evaluation, compared to older implementations [3].

Based on the new operator, a preconditioned conjugate gradient method was implemented for the condensed system and compared to a standard solver for the SEM. The combination of the new methods allows the usage of polynomial degrees ranging from normal FEM up to spectral level, without any change in algorithm. Moreover no assumptions were made on the grid topology, only the shape of the elements is restricted to cuboidal.

The current implementation gains only a marginal advantage through the preconditioning, the preconditioner is too cost-intensive for the linearly scaling operator evaluation. Thus, future work will focus on lowering the amount of operations for the preconditioner, to yield further performance gains with the algorithm.

Acknowledgment of Funds. This work is supported in part by the German Research Foundation (DFG) within the Cluster of Excellence 'Center for Advancing Electronics Dresden' (cfaed).

References

1. Couzy, W., Deville, M.: A fast Schur complement method for the spectral element discretization of the incompressible Navier-Stokes equations. J. Comput. Phys. **116**(1), 135–142 (1995)

2. Deville, M., Fischer, P., Mund, E.: High-Order Methods for Incompressible Fluid Flow. Cambridge University Press, Cambridge (2002)
3. Haupt, L., Stiller, J., Nagel, W.E.: A fast spectral element solver combining static condensation and multigrid techniques. J. Comput. Phys. **255**, 384–395 (2013)
4. Hestenes, M.R., Stiefel, E.: Methods of conjugate gradients for solving linear systems. J. Res. Natl. Bur. Stan. **49**(6), 409–436 (1952)
5. Huismann, I., Haupt, L., Stiller, J., Fröhlich, J.: Sum factorization of the static condensed helmholtz equation in a three-dimensional spectral element discretization. PAMM **14**(1), 969–970 (2014)
6. Karniadakis, G., Sherwin, S.: Spectral/hp Element Methods for CFD. Oxford University Press, New York (1999)
7. Kwan, Y.Y., Shen, J.: An efficient direct parallel spectral-element solver for separable elliptic problems. J. Comput. Phys. **225**(2), 1721–1735 (2007)
8. Lynch, R., Rice, J., Thomas, D.: Direct solution of partial difference equations by tensor product methods. Numerische Mathematik **6**(1), 185–199 (1964)
9. Wilson, E.L.: The static condensation algorithm. Int. J. Numer. Meth. Eng. **8**(1), 198–203 (1974)

An Iterative Regularization Algorithm for the TV-Stokes in Image Processing

Leszek Marcinkowski[1]([⊠]) and Talal Rahman[2]

[1] Faculty of Mathematics, Informatics and Mechanics,
Institute of Applied Mathematics and Mechanics,
University of Warsaw, Banacha 2, 02-097 Warszawa, Poland
L.Marcinkowski@mimuw.edu.pl
[2] Department of Computing, Mathematics and Physics,
Bergen University College, Inndalsveien 28, 5063 Bergen, Norway
Talal.Rahman@hib.no

Abstract. Image denoising is one of the fundamental problems in the image processing. In a PDE based approach for image processing, the simplest possible method for denoising is to solve the heat equation. However such a diffusion equation will destroy sharp edges in the image. An approach known for preserving the edges while denoising is called the classical Rudin-Osher-Fatemi (ROF) method based on the total variation (TV) regularization. Recently, an algorithm, also known as the TV-Stokes, based on two minimization steps involving the smoothing of the tangential field and then the reconstruction of the image has been proposed. The latter produces images without the blocky effect which we observe in the case of the ROF model. An iterative regularization method for the total variation based image restoration has recently been proposed giving significant improvement over the classical method in the quality of the restored image. In this paper we propose a similar algorithm for the TV-Stokes denoising algorithm.

Keywords: Iterative regularization · Total variation · TV-Stokes · Denoising

1 Introduction

Recovering an image from a noisy and blurry image is an inverse problem which is solved via variational methods, e.g. cf [1,7]. This requires the minimization of some energy functional.

By the Euler-Lagrange formulation it results into a set of nonlinear partial differential equations which are then solved using say, the gradient-descent iteration, see for instance [12] for the classical model of Rudin, Osher and Fatemi

L. Marcinkowski—This work was partially supported by Polish Scientific Grant N/N201/0069/33.
T. Rahman—The author acknowledges the support of NRC through DAADppp project 233989.

R. Wyrzykowski et al. (Eds.): PPAM 2015, Part II, LNCS 9574, pp. 381–390, 2016.
DOI: 10.1007/978-3-319-32152-3_36

(ROF model), which is based on the total variation (TV) regularization of the intensity (gray level), and [9,11] for an improved model (TV-Stokes model) which is based on the total variation regularization of the tangential field of the intensity. The drawback of such algorithms is that their convergence is very slow, particularly for large images. There exist now algorithms which are much faster, those based on the dual formulation of the underlying models, see for instance [2,3,8]

An iterative regularization algorithm for the ROF model has recently been proposed, cf. [10], giving significant improvement over the classical method in the quality of the restored image. The main purpose of this paper is to propose a similar algorithm for the TV-Stokes model, and its dual formulation for faster convergence. The paper is organized as follows: in Sect. 2 we present the iterative regularization algorithm for the ROF model, and in Sect. 3 we propose a similar algorithm for the TV-Stokes model. In Sect. 4 we describe Chambolle's iteration for the dual formulation of the TV-Stokes model, which we use for the numerical experiments of Sect. 5.

2   Iterative Regularization for the TV Denoising

Let the noisy image d_0 represented as scalar $L^2(\Omega)$ function be given. The classical denoising method is based on the minimization problem:

$$\min_d \int_\Omega |\nabla d|\, d\mathbf{x} \; + \; \frac{\lambda}{2} \int_\Omega (d_0 - d)^2 d\mathbf{x}, \tag{1}$$

where λ is a constant which is used to balance between the smoothing of the image and the fidelity to the input image. It is difficult to know how to choose λ. An equivalent formulation of (1) is the following constrained minimization problem, cf. e.g. [4]:

$$\min_{\|d_0-d\|_{L^2}^2=\sigma^2} \int_\Omega |\nabla d|\, d\mathbf{x}, \tag{2}$$

where σ is the noise level. One often has a reasonable estimate of the noise level. In the original paper [12], a gradient projection method was used to solve (2). The method is known for its good edge preserving capability. It suffers however from its blocky effect on the resulting image. Not just that, it looses quite easily the high frequency part of the image as well. The recently proposed iterative regularization method [10], an algorithm which is based on the original TV denoising algorithm, has proven to give a much better result than the constrained denoising algorithm of ROF.

Given d_0, λ, and $v_0 = 0$. For $k = 0, 1, 2, \ldots$, find the minimizer d_{k+1} of the following minimization problem,

$$\min_d \int_\Omega |\nabla d|\, d\mathbf{x} \; + \; \frac{\lambda}{2} \int_\Omega (d_0 + v_k - d)^2 d\mathbf{x}, \tag{3}$$

and update

$$v_{k+1} = v_k + d_0 - d_{k+1}. \tag{4}$$

Algorithm: TV Iterative Regularization

Given d_0 and λ ;

Initialize counter: $k = 0$;

Set: $v_0 = 0$;

while *not converged* **do**

> Initialize counter: $n = 0$;
>
> Set: $u^0 = v_k + d_0$;
>
> **while** *not converged* **do**
>
> > Calculate u^{n+1}:
> >
> > $$\frac{u^{n+1} - u^n}{\Delta t} = \nabla \cdot \left(\frac{\nabla u^n}{|\nabla u^n|} \right) + \lambda(u^0 - u^n) \qquad (5)$$
> >
> > Update counter: $n = n + 1$;
>
> **end**
>
> Set: $d_{k+1} = u^n$;
>
> Update:
>
> $$v_{k+1} = v_k + d_0 - d_{k+1} \qquad (6)$$
>
> Update counter: $k = k + 1$;

end

Algorithm 1. Iterative regularization for ROF denoising.

For stopping the iterative procedure, a reasonable criterion to use is the discrepancy principle, that is to stop the iteration the first time the residual $\|d_0 - d_k\|_{L^2}$ is of the same order as the noise level σ, cf. [10]. We know that the problem (3) has a unique solution. It is shown in [10] that d_k will converge to the original noisy image d_0 as we continue to iterate beyond the discrepancy point.

2.1 Discrete Algorithm

The algorithm consists of two loops. The first one, which will be the outer loop, we call it the k-loop. In each iteration of the k-loop, we need the minimizer of the classical ROF model, which we do by the descent technique, iterating over an artificial time step to steady state.

2.2 Discretization

For the time discretization we use an explicit scheme, where, in each time step, the nonlinear term is calculated using values from the previous time step and is therefore a known quantity. Each vertex of the rectangular grid corresponds to the position of a pixel or pixel center where the image intensity variable d is defined, cf. Fig. 1 (Right).

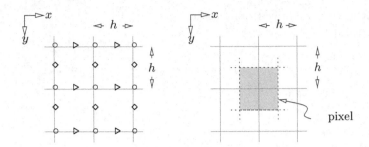

Fig. 1. Left: the computational grid with approximating points for the variables d, d_x, and d_y, represented by ○,▷, and ◇, respectively. Right: mapping the computational grid onto the pixels.

For the space discretization, we approximate the derivatives by finite differences using the standard forward/backward difference operators $D_x^\pm$ and $D_y^\pm$, and the centered difference operators C_x^h and C_y^h, respectively in the x and y directions, as $D_x^\pm f = \pm\frac{f(x\pm h,y)-f(x,y)}{h}$, $D_y^\pm f = \pm\frac{f(x,y\pm h)-f(x,y)}{h}$, $C_x^h f = \frac{f(x+h,y)-f(x-h,y)}{2h}$, and $C_y^h f = \frac{f(x,y+h)-f(x,y-h)}{2h}$ for any function f, where h correspond to the $h-$spacing. We introduce two average operators A_x and A_y as $A_x f = (f(x,y) + f(x+h,y))/2$ and $A_y f = (f(x,y) + f(x,y+h))/2$.

The discrete approximation of (5), thus, takes the following form:

$$\nabla \cdot \left(\frac{\nabla u^n}{|\nabla u^n|}\right) + \lambda(u^0 - u^n) \approx D_x^- \left(\frac{D_x^+ u^n}{T_1^n}\right) + D_y^- \left(\frac{D_y^+ u^n}{T_2^n}\right) + \lambda(u^0 - u^n), \quad (7)$$

where T_1^n is defined as $T_1^n = \sqrt{\left(D_x^+ u^n\right)^2 + \left(A_x(C_y^h u^n)\right)^2 + \epsilon}$, and T_2^n as $T_2^n = \sqrt{\left(D_y^+ u^n\right)^2 + \left(A_y(C_x^h u^n)\right)^2 + \epsilon}$. Here ϵ is a small number.

3   Iterative Regularization for the TV-Stokes

3.1   The TV-Stokes Denoising

Let the noisy image d_0 represented as scalar $L^2(\Omega)$ function be given. We compute $\tau_0 = \nabla^\perp d_0$. The algorithm is then defined in two steps, see [9,11]. In the first step, writing the tangent vector as $\tau = (v, u)$, we solve the following minimization problem:

$$\min_\tau \int_\Omega (|\nabla v| + |\nabla u|)\, d\mathbf{x} + \frac{\delta}{2}\int_\Omega |\tau - \tau_0|^2\, d\mathbf{x} \quad (8)$$

subject to $\nabla \cdot \tau = 0$, where δ is a constant which is used to balance between the smoothing of the tangent field and the fidelity to the input tangent field. Once we have the smoothed tangent field, we can get the corresponding normal field $\mathbf{n} = (u, -v)$. In the second step, we reconstruct our image by fitting it to the normal field through solving the following minimization problem:

$$\min_d \int_\Omega \left(|\nabla d| - \nabla d\frac{\mathbf{n}}{|\mathbf{n}|}\right) d\mathbf{x} + \frac{\lambda}{2}\int_\Omega (d_0 - d)^2 d\mathbf{x}. \quad (9)$$

As before, let $\|d - d_0\|_{L^2}^2 = \sigma^2$ be the estimated noise variance. This can be estimated using statistical methods. If the exact noise variance cannot be obtained, then an approximate value may be used. In which case, a larger value would result in over-smoothing and a smaller value would result in under-smoothing.

3.2 Iterative Regularization

Given $d_0, s_0 = 0, \delta$ and λ. For $k = 0, 1, 2, \ldots$, in the first step, we compute $\tau_0 = \nabla^\perp(d_0 + s_k)$, and we solve the following minimization problem.

$$\min_\tau \int_\Omega (|\nabla v| + |\nabla u|)\, dx + \frac{\delta}{2} \int_\Omega |\tau - \tau_0|^2\, dx, \tag{10}$$

subject to $\nabla \cdot \tau = 0$.

Once we have the smoothed tangent field, we get the corresponding normal field $\mathbf{n} = (u, -v)$. In the second step, we reconstruct our image by fitting it to the normal field through solving the following minimization problem.

$$\min_d \int_\Omega \left((|\nabla d| - \nabla d \frac{\mathbf{n}}{|\mathbf{n}|} \right) dx + \frac{\lambda}{2} \int_\Omega (d_0 + s_k - d)^2 dx, \tag{11}$$

and update $s_{k+1} = s_k + d_0 - d_{k+1}$. For stopping of the iterative procedure, we use the discrepancy principle, that is to stop the iteration the first time the residual $\|d_0 - d_k\|_{L^2}$ is of the same order as the noise level σ, cf. [10]. It is possible to show that d_k will converge to the original noisy image d_0 as we continue to iterate beyond the discrepancy point.

3.3 Discrete Algorithm

The algorithm consists of two loops, the outer loop being the k-loop as before. In each iteration of the k-loop, the two minimizing steps of the TV-Stokes algorithms is performed. The discrete algorithm is in Algorithm 2 below.

3.4 Discretization

For the time discretization, we use an explicit scheme, where, in each time step, the nonlinear term is calculated using values from the previous time step and is therefore a known quantity. As before, each vertex of the rectangular grid corresponds to the position of a pixel or pixel center where the image intensity variable d is defined, cf. Fig. 1 (Right).

For the space discretization again we use a staggered grid, cf. Fig. 1 (Left). We approximate the derivatives by finite differences using the standard forward/backward difference operators $D_x^\pm$ and $D_y^\pm$, and the centered difference operators C_x^h and C_y^h, respectively in the x and y directions, as described in Sect. 2.

The discrete approximation of (15)–(17) are as follows

$$\frac{v^{n+1} - v^n}{\Delta t} = D_x^- \left(\frac{D_x^+ v^n}{T_1^n(v)} \right) + D_y^- \left(\frac{D_y^+ v^n}{T_2^n(v)} \right) + \delta(v^0 - v^n) + D_x^- q^n \quad (12)$$

$$\frac{u^{n+1} - u^n}{\Delta t} = D_x^- \left(\frac{D_x^+ u^n}{T_1^n(u)} \right) + D_y^- \left(\frac{D_y^+ u^n}{T_2^n(u)} \right) + \delta(u^0 - u^n) + D_y^- q^n \quad (13)$$

$$\frac{q^{n+1} - q^n}{\Delta t} = D_x^+ v^n + D_y^+ u^n \quad (14)$$

where $T_1^n(u)$ is defined as $T_1^n(u) = \sqrt{\left(D_x^+ u^n\right)^2 + \left(A_x(C_y^h u^n)\right)^2 + \epsilon}$ and $T_2^n(u)$ as $T_2^n(u) = \sqrt{\left(D_y^+ u^n\right)^2 + \left(A_y(C_x^h u^n)\right)^2 + \epsilon}$. Analogously, we define $T_1^n(v)$ and $T_2^n(v)$ by replacing u with v.

Algorithm: TV-Stokes iterative regularization

Given d_0, δ and λ ;

Initialize counter: $k = 0$;

Set: $s_0 = 0$;

while *not converged* **do**

  Initialize counter: $n = 0$;

  Set: $w^0 = d_0 + s_k$ and $(v^0, u^0) = \nabla^\perp w^0$, $q^0 = 0$;

  while *not converged* **do**

    Calculate $\tau^{n+1} = (v^{n+1}, u^{n+1})$:

$$\frac{v^{n+1} - v^n}{\Delta t} = \nabla \cdot \left(\frac{\nabla v^n}{|\nabla v^n|} \right) - \delta \left(v^n - v^0 \right) + \frac{\partial q^n}{\partial x} \quad (15)$$

$$\frac{u^{n+1} - u^n}{\Delta t} = \nabla \cdot \left(\frac{\nabla u^n}{|\nabla u^n|} \right) - \delta \left(u^n - u^0 \right) + \frac{\partial q^n}{\partial y} \quad (16)$$

$$\frac{q^{n+1} - q^n}{\Delta t} = \frac{\partial v^n}{\partial x} + \frac{\partial u^n}{\partial y} \quad (17)$$

    Update counter: $n = n + 1$;

  end

  Set $\mathbf{n} = (u^{n+1}, -v^{n+1})$;

  while *not converged* **do**

    Calculate w^{n+1}:

$$\frac{w^{n+1} - w^n}{\Delta t} = \nabla \left(\frac{\nabla d}{|\nabla d|} - \frac{\mathbf{n}}{|\mathbf{n}|} \right) + \lambda(w^0 - w^n) \quad (18)$$

    Update counter: $n = n + 1$;

  end

  Set: $d_{k+1} = w^{n+1}$;

  Update:

$$s_{k+1} = s_k + d_0 - d_{k+1} \quad (19)$$

  Update counter: $k = k + 1$;

end

Algorithm 2. Iterative regularization for TV Stokes denoising.

The discrete approximation of (18) is defined as follows

$$\frac{w^{n+1} - w^n}{\Delta t} = D_x^- \left(\frac{D_x^+ w^n}{T_3^n} - n_1 \right) + D_y^- \left(\frac{D_y^+ w^n}{T_4^n} - n_2 \right) + \lambda(w^0 - w^n) \quad (20)$$

where T_3^n is defined as $T_3^n = \sqrt{\left(D_x^+ w^n\right)^2 + \left(A_x(C_y^h w^n)\right)^2 + \epsilon}$ and T_4^n as $T_4^n = \sqrt{\left(D_y^+ w^n\right)^2 + \left(A_y(C_x^h w^n)\right)^2 + \epsilon}$. and n_k, for $k = 1$ and 2, respectively as $n_1 = \frac{u}{\sqrt{u^2 + (A_x(A_y v))^2) + \epsilon}}$ and $n_2 = \frac{-v}{\sqrt{v^2 + (A_y(A_x u))^2) + \epsilon}}$.

4   Chambolle's Algorithm

In this section we present a dual approach for solving our TV Stokes iterative regularization, cf. [5,6,8]. We consider the image in $L^2(\Omega)$ be approximated on the regular mesh and be represented as $d \in \mathbb{R}^{N \times N}$. The derivative matrices, corresponding to the u and v, are then computed naturally from d using appropriate finite differences, which again constitute the pair of matrices corresponding to the tangential vector $\tau = (v, u)$.

4.1   First Step

In the first step, we consider the minimization problem (10):

$$\min_{\nabla \tau = 0} \int_\Omega (|\nabla v| + |\nabla u|)\, d\mathbf{x} + \frac{\delta}{2} \int_\Omega |\tau - \tau_0|^2 \, d\mathbf{x}. \quad (21)$$

Using a dual formulation of the TV norm we can write

$$\int_\Omega (|\nabla v| + |\nabla u|)\, d\mathbf{x} = \max_{\mathbf{G}} \int_\Omega \langle \tau, \nabla \cdot \mathbf{G} \rangle \, dx,$$

where $\langle \mathbf{x}, \mathbf{y} \rangle = x_1 y_1 + x_2 y_2$ for $\mathbf{x}, \mathbf{y} \in \mathbb{R}^2$, and $\mathbf{G} = (\mathbf{g}_1, \mathbf{g}_2)^T$ is the dual variable such that $\mathbf{g}_i \in C_c^1(\Omega)^2$ and $|\mathbf{g}_i|_\infty \leq 1$. Using this, (10) can be reformulated as

$$\min_{\nabla \tau = 0} \max_{\mathbf{G}} \int_\Omega \langle \tau, \nabla \cdot \mathbf{G} \rangle dx + \frac{\delta}{2} \int_\Omega |\tau - \tau_0|^2 \, d\mathbf{x}. \quad (22)$$

Here $\nabla \cdot \mathbf{G} = (\nabla \cdot \mathbf{g}_1, \nabla \cdot \mathbf{g}_2)^T$. We define the orthogonal projection Π_Y onto $Y = \{\tau : \nabla \cdot \tau = 0\}$ as

$$\Pi_Y \begin{bmatrix} \tau_1 \\ \tau_2 \end{bmatrix} = \begin{bmatrix} \tau_1 \\ \tau_2 \end{bmatrix} - \nabla \triangle^\dagger \nabla \cdot \begin{bmatrix} \tau_1 \\ \tau_2 \end{bmatrix}. \quad (23)$$

We note that $\nabla \cdot \tau = 0$ is equivalent to $\Pi_Y \tau = \tau$; using this, and exchanging min and max, we get

$$\max_{\mathbf{G}} \min_{\tau} \int_\Omega \langle \tau, \Pi_Y \nabla \cdot \mathbf{G} \rangle dx + \frac{\delta}{2} \int_\Omega |\tau - \tau_0|^2 \, d\mathbf{x}. \quad (24)$$

Minimizing with respect to τ we get

$$\tau = \tau_0 - \frac{1}{\delta}\Pi_Y \nabla \cdot \mathbf{G}. \tag{25}$$

Substituting it back, we obtain the dual problem:

$$\min_{\mathbf{G}} \int_\Omega |\Pi_Y \nabla \cdot \mathbf{G} - \delta\tau_0|^2 \, dx. \tag{26}$$

This problem can be solved using Chambolle's fixed point iteration (cf. [3]):

$$\mathbf{G}^{n+1} = \frac{\mathbf{G}^n + \Delta t \nabla [\Pi_Y \nabla \cdot \mathbf{G} - \delta\tau_0]}{1 + \Delta t \nabla [\Pi_Y \nabla \cdot \mathbf{G} - \delta\tau_0]} \tag{27}$$

Fig. 2. Denoising of Lena image, with noise level ≈ 8, $\delta = .16$ and $\mu = 0.20$.

In practice we compute an approximation of Π_Y using the following discrete gradient, discrete divergence and discrete Laplace operator. For $d \in \mathbb{R}^{N\times N}$ representing an image on a 2D grid let

$$\nabla^h d = (dD^T, Dd)^T, \quad \nabla^h \cdot (p_1, p_2) = -p_1 D - D^T p_2, \tag{28}$$

where D is differentiation matrix. Then $\triangle^h = -dDD^T - D^T Dd$ and the discrete projection becomes: $\Pi_Y^h = I - \nabla^h(\triangle^h)^\dagger \nabla^h$. Because we know SVD of $\triangle^h$, thus the action of $(\triangle^h)^\dagger$ can be computed using discrete cosine and sine matrices with the aid of the Fast Fourier Transform requiring only $O(N^2 \log_2(N))$ operations.

4.2   Second Step

In the second step, we have an unconstrained minimization problem (11). Using the dual formulation of the TV norm, the problem can be reformulated as

$$\min_{d} \max_{\mathbf{g}\in C_c^1(\Omega)^2:|\mathbf{g}|_\infty \leq 1} \int_\Omega d\, \nabla \cdot \left(\mathbf{g} + \frac{\mathbf{n}}{|\mathbf{n}|}\right) dx + \frac{\lambda}{2}\int_\Omega (d_0 + s_k - d)^2 dx. \tag{29}$$

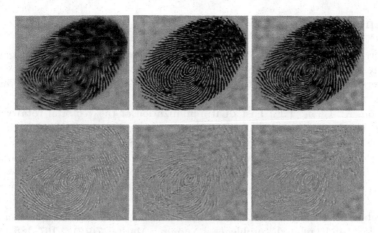

Fig. 3. Denoising of fingerprint image, with noise level ≈ 6.4, $\delta = .16$ and $\mu = 0.20$.

Exchanging the min and max, and minimizing with respect to d, we get

$$d = d_0 + s_k - \frac{1}{\lambda}\nabla \cdot \left(\mathbf{g} + \frac{\mathbf{n}}{|\mathbf{n}|}\right). \tag{30}$$

Substituting it back, we obtain the dual problem:

$$\min_{\mathbf{g}} \int_{\Omega} \left|\lambda(d_0 + s_k) - \nabla \cdot \left(\mathbf{g} + \frac{\mathbf{n}}{|\mathbf{n}|}\right)\right|^2 d\mathbf{x}. \tag{31}$$

Using Chambolle's fixed point iteration we get

$$\mathbf{g}^{n+1} = \frac{\mathbf{g}^n + \Delta t \nabla[\nabla \cdot \left(\mathbf{g}^n + \frac{\mathbf{n}}{|\mathbf{n}|}\right) - \lambda(d_0 + s_k)]}{1 + \Delta t \nabla[\nabla \cdot \left(\mathbf{g}^n + \frac{\mathbf{n}}{|\mathbf{n}|}\right) - \lambda(d_0 + s_k)]}. \tag{32}$$

5   Numerical Results

The algorithm has been applied to the Lena and the fingerprint image, and the results are shown in Figs. 2 and 3, respectively, showing three iterations of the iterative regularization algorithm, with denoised images in the first row and their corresponding difference images (difference between the noisy image and the denoised image) in the second row. The preliminary results shown in the figures proves that he proposed algorithm works well.

Acknowledgements. We would like to thank Bin Wu for the numerical experiments.

References

1. Aubert, G., Kornprobst, P.: Mathematical Problems in Image Processing: Partial Differential Equations and the Calculus of Variations, With a foreword by Olivier Faugeras. Applied Mathematical Sciences, vol. 147, 2nd edn. Springer, New York (2006)
2. Bresson, X., Chan, T.F.: Fast dual minimization of the vectorial total variation norm and applications to color image processing. Inverse Probl. Imaging **2**(4), 455–484 (2008). http://dx.doi.org/10.3934/ipi.2008.2.455
3. Chambolle, A.: An algorithm for total variation minimization and applications. J. Math. Imaging Vision **20**(1–2), 89–97 (2004). http://dx.doi.org/10.1023/B:JMIV.0000011320.81911.38, special issue on mathematics and image analysis
4. Chambolle, A., Lions, P.L.: Image recovery via total variation minimization and related problems. Numer. Math. **76**(2), 167–188 (1997). http://dx.doi.org/10.1007/s002110050258
5. Chan, T.F., Golub, G.H., Mulet, P.: A nonlinear primal-dual method for total variation-based image restoration. In: Berger, M.-O., Deriche, R., Herlin, I., Jaffré, J., Morel, J.-M. (eds.) ICAOS '96. Lecture Notes in Control and Information Sciences, vol. 219, pp. 241–252. Springer, London (1996). http://dx.doi.org/10.1007/3-540-76076-8_137
6. Chan, T.F., Golub, G.H., Mulet, P.: A nonlinear primal-dual method for total variation-based image restoration. SIAM J. Sci. Comput. **20**(6), 1964–1977 (1999). http://dx.doi.org/10.1137/S1064827596299767
7. Chan, T.F., Shen, J.: Image Processing and Analysis, Variational, PDE, Wavelet, and Stochastic Methods. Society for Industrial and Applied Mathematics (SIAM), Philadelphia (2005). http://dx.doi.org/10.1137/1.9780898717877
8. Elo, C.A., Malyshev, A., Rahman, T.: A dual formulation of TV-Stokes algorithm for image denoising. In: Tai, X.C., Mørken, K., Lysaker, M., Lie, K.A. (eds.) SSVM 2009. LNCS, vol. 5567, pp. 307–318. Springer, Heidelberg (2009)
9. Litvinov, W.G., Rahman, T., Tai, X.C.: A modified TV-Stokes model for image processing. SIAM J. Sci. Comput. **33**(4), 1574–1597 (2011). http://dx.doi.org/10.1137/080727506
10. Osher, S., Burger, M., Goldfarb, D., Xu, J., Yin, W.: An iterative regularization method for total variation-based image restoration. Multiscale Model. Simul. **4**(2), 460–489 (2005). (electronic) http://dx.doi.org/10.1137/040605412
11. Rahman, T., Tai, X.-C., Osher, S.J.: A TV-Stokes denoising algorithm. In: Sgallari, F., Murli, A., Paragios, N. (eds.) SSVM 2007. LNCS, vol. 4485, pp. 473–483. Springer, Heidelberg (2007)
12. Rudin, L., Osher, S., Fatemi, E.: Nonlinear total variation based noise removal algorithms. Phys. D **60**, 259–268 (1992)

Discretization of the Drift-Diffusion Equations with the Composite Discontinuous Galerkin Method

Konrad Sakowski[1,2]([⊠]), Leszek Marcinkowski[2], Pawel Strak[1],
Pawel Kempisty[1], and Stanislaw Krukowski[1,3]

[1] Institute of High Pressure Physics, Polish Academy of Sciences,
ul. Sokolowska 29/37, 01-142 Warsaw, Poland
konrad@unipress.waw.pl

[2] Faculty of Mathematics, University of Warsaw, Banacha 2, 02-097 Warsaw, Poland

[3] Interdisciplinary Centre for Mathematical and Computational Modelling,
Warsaw University, ul. Pawinskiego 5a, 02-106 Warsaw, Poland

Abstract. We present three variants of discretization of the stationary van Roosbroeck equations. They are the Composite Discontinuous Galerkin Methods, in standard symmetric/non-symmetric version, and the Weakly Over-Penalized Symmetric Interior Penalty method.

Numerical simulations of gallium nitride semiconductor devices are presented. Results of these simulations serve as a base to perform the convergence analysis of the presented methods. Errors of approximations obtained with these methods are compared with each other.

Keywords: Drift-diffusion · van Roosbroeck equations · Composite Discontinuous Galerkin Method · Weakly Over-Penalized Symmetric Interior Penalty Method

1 Introduction

Modelling of the semiconductor heterostructures is important part of the design process of new devices. Numerical simulations reduce costs of development by replacing time-inefficient and expensive physical experiments, which must be performed with sufficient care. Moreover a scientist does not have an insight into the physics of the experiment to the extend available in the computer modelling. On the other hand, numerical simulation allows to examine precisely every modelled aspect of a device, to instantly analyze values which cannot be measured physically, or where measurement is expensive and time-consuming.

In modelling of luminescent semiconductor devices, simulations with the drift-diffusion model [1,2] are quite efficient. The drift-diffusion model consists of three nonlinear elliptic differential equations. Unfortunately coefficients of these equations are exponentially nonlinear in the unknown variables and they strongly vary from one physical layer into another. To resolve this problem, we use the

© Springer International Publishing Switzerland 2016
R. Wyrzykowski et al. (Eds.): PPAM 2015, Part II, LNCS 9574, pp. 391–400, 2016.
DOI: 10.1007/978-3-319-32152-3_37

Discontinuous Galerkin Method (DGM)[3,4], and since the boundaries of the layers are known, we decided to apply Composite DGM [5]. Therefore we allow discontinuities only at the interfaces between layers, and inside the layers we use the standard conforming continuous Finite Element Method (FEM)[6]. Also, in two- or three-dimensions, these method allows to use independent meshes inside different layers. It is possible to make fine grid on more important parts of a device and coarse grid on the rest.

2   Drift-Diffusion Model

The differential problem is to find functions $\psi, F_n, F_p : \Omega \to \mathbb{R}$, where Ω is an interval in $\mathbb{R}$, polygon in $\mathbb{R}^2$ or polyhedron in $\mathbb{R}^3$, such that

$$\nabla \cdot \Big(\varepsilon_0 \varepsilon(x) \nabla \psi(x)\Big) = -qC(x, \psi, n, p),$$
$$\nabla \cdot \big(\mu_n(x) n(x, \psi, F_n) \nabla F_n(x)\big) = qR(x, \psi, n, p), \qquad (1)$$
$$\nabla \cdot \big(\mu_p(x) p(x, \psi, F_p) \nabla F_p(x)\big) = -qR(x, \psi, n, p),$$

with Dirichlet boundary conditions on Σ_D and homogeneous Neumann boundary conditions on $\Sigma_N := \partial\Omega \backslash \Sigma_D$. In one dimension, we assume $\Sigma_N = \emptyset$. We define

$$n(x) := N_c(x) \exp\left(\frac{F_n(x) - E_c(x) + q\psi(x)}{kT}\right),$$
$$p(x) := N_v(x) \exp\left(\frac{E_v(x) - F_p(x) - q\psi(x)}{kT}\right), \qquad (2)$$

The symbols' meaning is as follows: ψ is the electrostatic potential, F_n and F_p are the quasi-Fermi levels, n and p are the electron concentration and hole concentration. Operator C is the electrostatic charge and R is the recombination rate, and we assume that these operators do not involve differentiation of unknown functions. Other symbols stand either for material parameters or for physical constants, and we assume they are piecewise constant. We omit the description of physical details of the drift-diffusion system. We refer the reader to [1,7,8].

Therefore the drift-diffusion system in the formulation (1) is a system of nonlinear elliptic differential equations. This system is written in an unscaled form, with the potential ψ in volts and quasi-Fermi levels F_n, F_p in joules.

3   Discretizations

3.1   Discrete Problem Definition

Discrete Space. Semiconductor luminescent devices are divided into physical layers, which vary in material composition, doping level and physical properties. We take into account this natural partition and we assume that we have $\mathcal{E}$ such that $\Omega = \bigcup_{E \in \mathcal{E}} \overline{E}$ and $E_1 \cap E_2 = \emptyset$ for any $E_1, E_2 \in \mathcal{E}$. Also, in computational practice, we often artificially split one physical layer to improve accuracy of numerical solutions. We assume that $\mathcal{E}$ is a triangulation [9] and we treat it as a coarse grid for Composite Discontinuous Galerkin Method. Then we define so-called broken Sobolev space $X^s := \{v \in L_2(\Omega) : \forall E \in \mathcal{E} \quad v|_E \in H^s(E)\}$. We will write $v_E := v|_E$.

Assume that for any $E \in \mathcal{E}$ we have triangulations $\mathcal{T}_E := \mathcal{T}_{E,h_E}(E)$ of E, where $h_E := \max\{\text{diam}(\tau) : \tau \in \mathcal{T}_E\}$. We define $\mathcal{T}_h := \bigcup_{E \in \mathcal{E}} \mathcal{T}_E$, where $h := \max\{h_E : E \in \mathcal{E}\}$. Then on every E we define a discrete space $X_{h_E}(E)$ of piecewise linear functions on the triangulation $\mathcal{T}_E$:

$$X_{h_E} := X_{h_E}(E) := \left\{ \mathbf{p}_E \in \mathcal{C}(\overline{E}) : \forall \tau \in \mathcal{T}_E \quad \mathbf{p}_E\big|_\tau \in \mathbb{P}_1(\tau) \right\} \tag{3}$$

We assume that $\{\mathcal{T}_h\}_h$ is a regular family of triangulations with all finite elements being affine-equivalent to a single reference finite element, i.e. it satisfies assumptions (H1), (H2) of [9].

Finally we define a discrete space X_h as $X_h = \prod_{E \in \mathcal{E}} X_{h_E}(E)$. Let e be an edge of some $E \in \mathcal{E}$. Since $\mathcal{E}$ is a triangulation, then either $e = \partial E_1 \cap \partial E_2$ for some $E_1, E_2 \in \mathcal{E}$ or $e \subset \partial E \cap \partial \Omega$ for some $E \in \mathcal{E}$. Therefore we define operators $[\cdot] := [\cdot]_e : X^1 \to L_2(e)$, $\{\cdot\} := \{\cdot\}_e : X^1 \to L_2(e)$ as

$$[\mathbf{p}] := \begin{cases} \mathbf{p}_{E_1} - \mathbf{p}_{E_2} & \text{if } e = \partial E_1 \cap \partial E_2, \\ \mathbf{p}_E & \text{if } e = \partial E \cap \partial \Omega, \end{cases}$$

$$\{\mathbf{p}\} := \begin{cases} \frac{1}{2}(\mathbf{p}_{E_1} + \mathbf{p}_{E_2}) & \text{if } e = \partial E_1 \cap \partial E_2, \\ \mathbf{p}_E & \text{if } e = \partial E \cap \partial \Omega, \end{cases} \tag{4}$$

where ν_E is a normal vector to E.

By Γ we denote a set of all internal and boundary edges of $\mathcal{E}$. Then Γ is a sum of disjoint sets Γ_D, Γ_N and Γ_I, where

$$\Gamma_D := \{e \in \Gamma : e \subset \Sigma_D\}, \qquad \Gamma_N := \{e \in \Gamma : e \subset \Sigma_N\},$$
$$\Gamma_I := \{e \in \Gamma : e \subset \text{int}(\Omega)\}, \qquad \Gamma_{DI} := \Gamma_D \cup \Gamma_I. \tag{5}$$

Therefore Γ_D (resp. Γ_N) contains edges lying on the boundary, where Dirichlet (resp. Neumann) boundary conditions are imposed and Γ_I comprises all internal edges.

3.2   Composite Discontinuous Galerkin Method

We start with a general elliptic problem, which will be further used to construct a Composite Discontinuous Galerkin discretizations. We propose three kind of discrete problems: first one is based on Weakly Over-Penalized Symmetric Interior Penalty (WOPSIP) method [10], while the other formulations rely on standard symmetric and non-symmetric Discontinuous Galerkin Methods [3]. In each case we use the composite formulation [5], i.e. inside every $E \in \mathcal{E}$ we use the Finite Element Method on the triangulation $\mathcal{T}_E$.

Differential Problem. Let $\Omega \subset \mathbb{R}^d$, $d \in \{1, 2, 3\}$ be an interval or a rectangle and let $\partial\Omega = \overline{\Sigma_D \cup \Sigma_N}$, where $\Sigma_D \cap \Sigma_N = \emptyset$. For $d = 1$ we assume $\Sigma_N = \emptyset$. Let $\hat{z} \in H^1(\Omega) \cap L_\infty(\Omega)$ stands for a Dirichlet boundary condition on Σ_D. Let $a \in L_\infty(\Omega)$, $0 < a_0 < a$ for some $a_0 \in \mathbb{R}$.

Our problem is to find $z \in \hat{z} + H^1_{0,\Sigma_D}(\Omega)$ such that

$$a(z, \vartheta) = f(\vartheta) \quad \forall \vartheta \in H^1_{0,\Sigma_D}(\Omega), \qquad \left.\frac{\partial z}{\partial \nu}\right|_{\Sigma_N} = 0, \qquad (6)$$

where

$$a(z, \vartheta) := \int_\Omega a(x)\nabla z(x) \cdot \nabla\vartheta(x)dx, \quad f(\vartheta) := \int_\Omega f(x)\vartheta(x)dx.$$

In this general setting, the problem is linear.

Discrete Problems. All the problems to be presented are of the following form. Find $p_h \in X_h$ such that for every $\varphi_h \in X_h$ it satisfies

$$a_h(p_h, \varphi_h) = f_h(\varphi_h),$$
$$a_h(p_h, \varphi_h) := A(p_h, \varphi_h) + sC(p_h, \varphi_h) + tC(\varphi_h, p_h) + G(p_h, \varphi_h)$$
$$f_h(\varphi_h) := B(\varphi_h) + H(\varphi_h),$$

where

$$A(p_h, \varphi_h) := \sum_{E \in \mathcal{E}} \int_E a(x)\nabla p_h \cdot \nabla\varphi_h dx, \quad B(\varphi_h) := \sum_{E \in \mathcal{E}} \int_E f(x)\varphi_h dx,$$

$$C(p_h, \varphi_h) := -\sum_{e \in \Gamma_{DI}} \int_e \{a\nabla p_h \cdot \nu\}[\varphi_h]ds,$$

$$G(p_h, \varphi_h) := \sum_{e \in \Gamma_{DI}} \eta_e \int_e [p_h][\varphi_h]ds, \qquad H(\varphi_h) := \sum_{e \in \Gamma_D} \eta_e \int_e [\hat{z}][\varphi_h]ds,$$

$$\eta_e := \begin{cases} 2\sigma_e h_E^{-q} & e \in \Gamma_D, e \subset E \in \mathcal{E}, \\ \sigma_e\left(h_{E_1}^{-q} + h_{E_2}^{-q}\right) & e \in \Gamma_I, e \subset E_1 \cap E_2, E_1 \in \mathcal{E}, E_2 \in \mathcal{E}. \end{cases} \qquad (7)$$

Note that these operators depend on parameters of the elliptic problem $a, f, \hat{z}$, and on penalty coefficients η_e. η_e is a penalty coefficient for an interface e, and it depends on the triangulation parameters and penalty parameters $\sigma_e > 0$. Also for $d = 1$ integrals over e simplify to $\int_e f(x)ds = f(e)$.

The discrete problems which we would like to discuss vary by the definition of operator a_h. Depending on the problem, parameters $s, t \in \{0, 1\}$, $q \in \{1, 2\}$ will be chosen accordingly.

Composite Weakly Over-Penalized Symmetric Interior Penalty Method (CWOPSIP). We would like to apply the Weakly Over-Penalized Symmetric Interior Penalty method (WOPSIP, [10]) to our generalized problem (6). We would like to allow discontinuities only on the edges of $E \in \mathcal{E}$.

In this case we choose $s = t = 0$, so a_h is very simple and it is symmetric. However it comes at cost of high penalty parameter ($q = 2$).

Composite Non-symmetric Standard Discontinuous Galerkin Method (CNDGM). The problem is formed by application of the Green theorem to the differential problem on the subdomains $E \in \mathcal{E}$ and then by addition of the penalty operators. Thus we take $s = 1, t = 0, q = 1$. The penalty parameters are now inversely proportional to the first power of the triangulation parameters, in contrast to WOPSIP method. Unfortunately this problem is not symmetric.

Composite Symmetric Standard Discontinuous Galerkin Method (CSDGM). This problem results from symmetrization of the previous problem (see [5]) by simply taking $s = t = q = 1$.

3.3   Comments on Theoretical Results

We are also working on the theoretical analysis of the solution existence and error estimates of the discretizations of the drift-diffusion system. Our analysis of CSDGM and CWOPSIP is undergoing and our results obtained so far inspired us to perform the numerical simulations described in this paper.

These results refer to the equilibrium case for $d \in \{1, 2\}$, and also the formulation of the problem is simplified (see for example [11,12]). By the equilibrium case we mean a situation where no bias is applied to the device, i.e. a device is disconnected from the power source [1]. Then, by physical arguments, the quasi-Fermi levels are constant and in fact the only unknown function is the potential. To prove the error estimates for the discretization proposed by us, we use similar approach to presented in [13] for the Navier-Stokes equation. This method, however, cannot be utilized for our problem in general, as it implies a differential solution to be unique. This is generally not the case for the drift-diffusion system in non-equilibrium case.

Therefore our early results indicate that in the equilibrium case for $d = 1$ the error estimate for the CWOPSIP method is $\|\psi - \psi_h\|_{H^1(\Omega)} = O(h)$. Unfortunately due to high penalty term, this estimate does not hold for $d = 2$ for the general grids. We only have $\|\psi - \psi_h\|_{H^1(\Omega)} = O(h^{1/2})$. However, if we use the CSDGM, with normal penalty term, then we regain the original estimate, i.e. $\|\psi - \psi_h\|_{H^1(\Omega)} = O(h)$.

Detailed proof of these results will be published elsewhere.

4   Numerical Experiments

In numerical experiments, we would like to check what are the convergence rates of the discretizations presented in Sect. 3 achieved in practice. In simulations we

Table 1. Schemata of devices used in the simulations.

PN junction				
Layer	Material	Donor doping	Acceptor doping	Length
n-type	GaN	2×10^{18} cm^{-3}	0	100 nm
p-type	GaN	0	2×10^{19} cm^{-3}	100 nm

Blue laser				
Layer	Material	Donor doping	Acceptor doping	Length
n-base	GaN	3×10^{18} cm^{-3}	0	1000 nm
n-cladding	Al$_{0.1}$Ga$_{0.9}$ N	3×10^{18} cm^{-3}	0	500 nm
n-waveguide	GaN	3×10^{18} cm^{-3}	0	100 nm
quantum well	In$_{0.2}$Ga$_{0.8}$ N	0	0	4 nm
p-EBL	Al$_{0.2}$Ga$_{0.8}$N	0	2×10^{19} cm^{-3}	20 nm
p-cladding	Al$_{0.1}$Ga$_{0.9}$N	0	1×10^{19} cm^{-3}	500 nm

use one-dimensional drift-diffusion system (1). Details of the algorithm used for solution of nonlinear discrete equations is described in [14], and practical applications of the presented methods may be found in [15].

We distinguish between two cases. In the equilibrium case, which is an object of our theoretical study (see Sect. 3.3), we are generally interested only in error of ψ, as other unknowns are constant by the physical reasons. On the other hand, in the non-equilibrium case, there is some voltage applied to the contacts of a device and all unknowns must be computed.

Unfortunately, there is no explicit solution given for any non-trivial devices, therefore as a reference in computing errors we use CSDGM solutions with $K = 1024$, where K is a number of nodes per $E \in \mathcal{E}$. For every function f_K taken into account, we compute the relative errors defined as

$$\text{error}_{L_2(\Omega)} := \frac{\|f_K - f_{\text{ref}}\|_{L_2(\Omega)}}{\|f_{\text{ref}}\|_{L_2(\Omega)}}, \quad \text{error}_{H^1(\Omega)} := \frac{\|f_K - f_{\text{ref}}\|_{H^1(\Omega)}}{\|f_{\text{ref}}\|_{H^1(\Omega)}}, \quad (8)$$

where f_{ref} is a numerical solution computed on a fine grid, as mentioned before. Errors are presented in function of K. Note that $h = c/K$ for some constant c. Also we made the penalty parameters to be dependent on the average value of elliptic equations' coefficients (compare [5]). Otherwise we would have to choose carefully the penalty parameter for every $e \in \Gamma_{DI}$, what is impractical for complex devices.

4.1   Results

We start from a simple device, a pn junction. It consists of two physical layers (Table 1). We additionally divide these layers to introduce additional narrow layers near the interface of the n-type, p-type and contacts of the device to improve the convergence. Then in every layer we setup K equidistant nodes. Simulation is in one-dimension.

We start with the equilibrium case (Table 2), where we present relative errors of CNDGM, CSDGM and CWOPSIP numerical solutions for the potential ψ.

Table 2. Relative errors of the potential ψ. Simulation were performed for the pn junction in the equilibrium state.

K	CNDGM $L_2(\Omega)$		$H^1(\Omega)$		CSDGM $L_2(\Omega)$		$H^1(\Omega)$		CWOPSIP $L_2(\Omega)$		$H^1(\Omega)$	
Function: ψ												
2	8.7e-02		4.6e-01		8.7e-02		4.6e-01		8.7e-02		4.6e-01	
4	1.5e-02	5.8	2.3e-01	2.0	1.5e-02	5.8	2.3e-01	2.0	1.5e-02	5.8	2.3e-01	2.0
8	4.0e-03	3.8	1.2e-01	1.9	4.0e-03	3.8	1.2e-01	1.9	4.0e-03	3.8	1.2e-01	1.9
16	1.2e-03	3.3	5.8e-02	2.0	1.2e-03	3.3	5.8e-02	2.0	1.2e-03	3.3	5.8e-02	2.0
32	2.9e-04	4.1	2.9e-02	2.0	3.0e-04	4.1	2.9e-02	2.0	3.0e-04	4.1	2.9e-02	2.0
64	7.3e-05	4.0	1.4e-02	2.0	7.4e-05	4.0	1.4e-02	2.0	7.4e-05	4.0	1.4e-02	2.0
128	1.8e-05	4.0	7.2e-03	2.0	1.8e-05	4.0	7.2e-03	2.0	1.8e-05	4.0	7.2e-03	2.0
256	4.4e-06	4.1	3.5e-03	2.0	4.4e-06	4.1	3.5e-03	2.0	4.4e-06	4.2	3.5e-03	2.0

Table 3. Relative error of ψ, F_n and F_p. Simulation were performed for the pn junction under 1 V bias.

K	CNDGM $L_2(\Omega)$		$H^1(\Omega)$		CSDGM $L_2(\Omega)$		$H^1(\Omega)$		CWOPSIP $L_2(\Omega)$		$H^1(\Omega)$	
Function: ψ												
2	7.8e-02		4.8e-01		7.8e-02		4.8e-01		7.9e-02		4.8e-01	
4	3.0e-02	2.6	2.7e-01	1.8	3.0e-02	2.6	2.7e-01	1.8	3.0e-02	2.6	2.7e-01	1.8
8	7.1e-03	4.3	1.3e-01	2.0	7.1e-03	4.2	1.3e-01	2.0	7.1e-03	4.2	1.3e-01	2.0
16	1.5e-03	4.9	6.7e-02	2.0	1.5e-03	4.9	6.7e-02	2.0	1.5e-03	4.9	6.7e-02	2.0
32	3.7e-04	4.0	3.4e-02	2.0	3.7e-04	4.0	3.4e-02	2.0	3.7e-04	4.0	3.4e-02	2.0
64	9.2e-05	4.0	1.7e-02	2.0	9.2e-05	4.0	1.7e-02	2.0	9.2e-05	4.0	1.7e-02	2.0
128	2.3e-05	4.0	8.3e-03	2.0	2.3e-05	4.0	8.3e-03	2.0	2.3e-05	4.0	8.3e-03	2.0
256	5.5e-06	4.1	4.1e-03	2.0	5.5e-06	4.1	4.1e-03	2.0	5.5e-06	4.1	4.1e-03	2.0
Function: F_n												
2	3.0e-03		1.0		3.0e-03		1.0		1.3e-02		1.0	
4	2.2e-03	1.4	1.0	1.0	2.2e-03	1.4	1.0	1.0	9.8e-03	1.4	1.0	1.0
8	1.5e-03	1.4	1.0	1.0	1.5e-03	1.4	1.0	1.0	7.1e-03	1.4	1.0	1.0
16	1.1e-03	1.4	9.8e-01	1.0	1.1e-03	1.4	9.8e-01	1.0	5.1e-03	1.4	1.0	1.0
32	7.5e-04	1.4	9.6e-01	1.0	7.5e-04	1.4	9.6e-01	1.0	3.6e-03	1.4	1.1	1.0
64	5.0e-04	1.5	9.3e-01	1.0	5.1e-04	1.5	9.3e-01	1.0	2.6e-03	1.4	1.1	0.9
128	3.2e-04	1.6	8.7e-01	1.1	3.3e-04	1.6	8.7e-01	1.1	1.8e-03	1.4	1.3	0.9
256	1.9e-04	1.7	7.6e-01	1.1	1.9e-04	1.7	7.6e-01	1.1	1.3e-03	1.4	1.7	0.8
Function: F_p												
2	2.0e-03		1.0		2.0e-03		1.0		1.0e-02		1.0	
4	1.6e-03	1.3	1.0	1.0	1.6e-03	1.3	1.0	1.0	7.6e-03	1.3	1.0	1.0
8	1.2e-03	1.3	1.0	1.0	1.2e-03	1.3	1.0	1.0	5.6e-03	1.4	1.0	1.0
16	8.5e-04	1.4	9.8e-01	1.0	8.5e-04	1.4	9.8e-01	1.0	4.1e-03	1.4	1.0	1.0
32	5.9e-04	1.4	9.6e-01	1.0	6.0e-04	1.4	9.6e-01	1.0	2.9e-03	1.4	1.1	1.0
64	4.0e-04	1.5	9.3e-01	1.0	4.0e-04	1.5	9.3e-01	1.0	2.1e-03	1.4	1.1	0.9
128	2.6e-04	1.5	8.7e-01	1.1	2.6e-04	1.5	8.7e-01	1.1	1.5e-03	1.4	1.3	0.9
256	1.5e-04	1.7	7.6e-01	1.1	1.5e-04	1.7	7.6e-01	1.1	1.0e-03	1.4	1.7	0.8

Table 4. Relative error of the carrier concentrations n and p. Simulation were performed for the pn junction under 1 V bias.

	CNDGM				CSDGM				CWOPSIP			
K	$L_2(\Omega)$		$H^1(\Omega)$		$L_2(\Omega)$		$H^1(\Omega)$		$L_2(\Omega)$		$H^1(\Omega)$	
Function: n												
2	2.0e-01		8.3e-01		2.0e-01		8.3e-01		2.0e-01		8.3e-01	
4	1.5e-01	1.3	7.9e-01	1.0	1.5e-01	1.3	7.9e-01	1.1	1.5e-01	1.3	7.9e-01	1.1
8	6.7e-02	2.3	5.7e-01	1.4	6.7e-02	2.3	5.7e-01	1.4	6.7e-02	2.3	5.7e-01	1.4
16	2.0e-02	3.4	3.1e-01	1.9	2.0e-02	3.4	3.1e-01	1.9	2.0e-02	3.4	3.1e-01	1.9
32	5.0e-03	3.9	1.5e-01	2.1	5.0e-03	3.9	1.5e-01	2.1	5.0e-03	3.9	1.5e-01	2.1
64	1.3e-03	3.9	7.3e-02	2.0	1.3e-03	3.9	7.3e-02	2.0	1.3e-03	3.9	7.3e-02	2.0
128	3.2e-04	4.0	3.7e-02	2.0	3.2e-04	4.0	3.7e-02	2.0	3.2e-04	4.0	3.7e-02	2.0
256	7.6e-05	4.2	1.8e-02	2.0	7.6e-05	4.2	1.8e-02	2.0	7.6e-05	4.2	1.8e-02	2.0
Function: p												
2	9.6e-02		7.4e-01		9.7e-02		7.4e-01		9.8e-02		7.5e-01	
4	4.9e-02	2.0	5.8e-01	1.3	4.9e-02	2.0	5.8e-01	1.3	4.9e-02	2.0	5.8e-01	1.3
8	1.8e-02	2.7	3.2e-01	1.8	1.8e-02	2.7	3.2e-01	1.8	1.8e-02	2.7	3.2e-01	1.8
16	5.1e-03	3.5	1.5e-01	2.2	5.2e-03	3.5	1.5e-01	2.2	5.2e-03	3.5	1.5e-01	2.2
32	1.4e-03	3.7	6.8e-02	2.2	1.4e-03	3.7	6.8e-02	2.2	1.4e-03	3.7	6.8e-02	2.2
64	3.6e-04	3.9	3.3e-02	2.1	3.6e-04	3.9	3.3e-02	2.1	3.6e-04	3.9	3.3e-02	2.1
128	8.9e-05	4.0	1.6e-02	2.0	8.9e-05	4.0	1.6e-02	2.0	8.9e-05	4.0	1.6e-02	2.0
256	2.1e-05	4.2	7.9e-03	2.1	2.1e-05	4.2	7.9e-03	2.1	2.1e-05	4.2	7.9e-03	2.1

Table 5. Relative error of the potential ψ, n and p. Simulation were performed for the laser under 2 V bias.

	CNDGM				CSDGM				CWOPSIP			
K	$L_2(\Omega)$		$H^1(\Omega)$		$L_2(\Omega)$		$H^1(\Omega)$		$L_2(\Omega)$		$H^1(\Omega)$	
Function: ψ												
16	1.4e-03		1.1e-01		1.2e-03		1.1e-01		1.4e-03		1.1e-01	
32	4.5e-04	3.1	5.8e-02	1.9	4.5e-04	2.7	5.8e-02	1.9	4.5e-04	3.1	5.8e-02	1.9
64	1.4e-04	3.2	2.9e-02	2.0	1.4e-04	3.2	2.9e-02	2.0	1.4e-04	3.2	2.9e-02	2.0
128	3.8e-05	3.6	1.4e-02	2.0	3.9e-05	3.6	1.4e-02	2.0	3.9e-05	3.6	1.4e-02	2.0
256	9.4e-06	4.1	7.0e-03	2.1	9.6e-06	4.1	7.0e-03	2.1	9.6e-06	4.1	7.0e-03	2.1
Function: n												
16	6.6e-02		6.0e-01		6.8e-02		6.1e-01		6.6e-02		6.0e-01	
32	2.9e-02	2.2	4.4e-01	1.4	2.9e-02	2.3	4.4e-01	1.4	2.9e-02	2.2	4.4e-01	1.4
64	1.0e-02	2.9	2.7e-01	1.6	1.0e-02	2.9	2.7e-01	1.6	1.0e-02	2.9	2.7e-01	1.6
128	2.8e-03	3.5	1.4e-01	1.9	2.8e-03	3.5	1.4e-01	1.9	2.8e-03	3.5	1.4e-01	1.9
256	7.3e-04	3.9	7.2e-02	2.0	7.3e-04	3.9	7.2e-02	2.0	7.3e-04	3.9	7.2e-02	2.0
Function: p												
16	1.2e-03		1.8e-01		1.1e-03		1.8e-01		1.2e-03		1.8e-01	
32	3.0e-04	3.9	8.8e-02	2.0	3.1e-04	3.7	8.8e-02	2.0	3.0e-04	3.9	8.8e-02	2.0
64	7.6e-05	4.0	4.4e-02	2.0	7.8e-05	4.0	4.4e-02	2.0	7.7e-05	4.0	4.4e-02	2.0
128	1.9e-05	4.0	2.2e-02	2.0	1.9e-05	4.0	2.2e-02	2.0	1.9e-05	4.0	2.2e-02	2.0
256	4.6e-06	4.1	1.1e-02	2.0	4.7e-06	4.1	1.1e-02	2.0	4.6e-06	4.1	1.1e-02	2.0

These results indicate clearly that errors of all these methods converge linearly to zero in $H^1(\Omega)$ norm as $h \to 0$. For $L_2(\Omega)$ norm, the errors drop quadratically in h, as it would be expected for plain finite element method. Also note that for given K errors are similar for all the presented discretization method.

Then we pass to non-equilibrium simulations for $1\,V$ bias (Table 3). For the potential ψ, the conclusion is as in equilibrium case. On the other hand, for the quasi-Fermi levels the situation is much worse. For CNDGM and CSDGM discretizations we observe sublinear convergence on both norms, and the $H^1(\Omega)$ convergence is much slower. For CWOPSIP, we observe $L_2(\Omega)$ convergence only. Having in mind that we do not have exact solution, it is hard to determine whether there is any $H^1(\Omega)$ convergence at all in any case.

However, in Table 4 we also included convergence results for derived functions n, p. We did so because the drift-diffusion equations may be expressed with a few sets of functions [16]. In particular, we can use ψ, n, p instead of ψ, F_n, F_p. An argument in favor of the latter choice is that it simplifies the equations. Also the carrier concentrations n, p may vary by dozens orders of magnitude over computational domains, while quasi-Fermi functions stay on the same level. Having in mind definition (2), we may think of F_n (resp. $-F_p$) as a logarithm of n (resp. p). That said, we observe that for the carrier concentrations n, p the situation is much more promising. Convergence is linear in $\|\cdot\|_{H^1(\Omega)}$ and quadratic in $\|\cdot\|_{L_2(\Omega)}$. Errors are similar for all the methods taken into account. This observation also explains, how could ψ convergence be as good as in equilibrium case while F_n, F_p error is much worse, as generally the drift-diffusion equations' coefficients and right hand sides are directly dependent on n, p, not on F_n, F_p.

In second approach we proceed to more complex device - blue InGaN laser. The structure used in this simulation (Table 1) is simplified a little in comparison with the real laser structure, but it resembles its essential features: a GaN base, AlGaN claddings, an InGaN quantum well and an electron blocking layer.

The results are presented in Table 5. Generally they agree with the conclusions drawn before, i.e. quadratic $L_2(\Omega)$ convergence and linear $H^1(\Omega)$ convergence of ψ, n, p, but it can be seen later (from K above 64). Errors of all methods are similar for a given K.

5   Conclusions

We presented three methods of discretization of the drift-diffusion equations with Composite Discontinuous Galerkin Method [5]. These methods are based on the standard symmetric and non-symmetric Discontinuous Galerkin Method [3], and on Weakly Over-Penalized Interior Penalty Method [10].

We demonstrate results of one-dimensional numerical simulations of gallium nitride semiconductor devices with the presented methods. These results indicate that L_2 convergence of the electrostatic potential and carrier concentrations is quadratic and H^1 convergence linear. We also observed sublinear L_2 convergence for the quasi-Fermi levels, while the H^1 convergence is very slow, if it exists at all. Errors of all three methods are on similar levels.

Acknowledgements. The research was funded by Polish National Science Center on the basis of the decision DEC-2011/03/D/ST3/02071.

References

1. Selberherr, S.: Analysis and Simulation of Semiconductor Devices. Springer, Wien (1984)
2. Markowich, P.A., Ringhofer, C.A., Schmeiser, C.: Semiconductor Equations. Springer, Wien (1990)
3. Riviere, B.: Discontinuous Galerkin Methods for Solving Elliptic and Parabolic Equations: Theory and Implementation. Society for Industrial and Applied Mathematics, Philadelphia (2008)
4. Arnold, D.N., Brezzi, F., Cockburn, B., Marini, L.D.: Unified analysis of discontinuous Galerkin methods for elliptic problems. SIAM J. Numer. Anal. **39**(5), 1749–1779 (2001)
5. Dryja, M.: On discontinuous Galerkin methods for elliptic problems with discontinuous coefficients. Comput. Methods Appl. Math. **3**(1), 76–85 (2003)
6. Brenner, S., Scott, R.: The Mathematical Theory of Finite Element Methods. Springer, New York (2008)
7. Sze, S., Ng, K.: Physics of Semiconductor Devices. Wiley-Interscience, Berlin (2006)
8. Jerome, J.W.: The approximation problem for Drift-Diffusion systems. SIAM Rev. **37**(4), 552–572 (2012)
9. Ciarlet, P.G.: The Finite Element Method for Elliptic Problems. North-Holland Publishing Company, Amsterdam (1978)
10. Brenner, S.C., Owens, L., Sung, L.Y.: A weakly over-penalized symmetric interior penalty method. Electron. Trans. Numer. Anal. **30**, 107–127 (2008)
11. Jerome, J.W.: Analysis of Charge Transport. Springer, Berlin (1996)
12. Jerome, J.W.: Consistency of semiconductor modeling: an existence/stability analysis for the stationary van Roosbroeck system. SIAM J. Appl. Math. **45**(4), 565–590 (1985)
13. Girault, V., Riviere, B., Wheeler, M.F.: A discontinuous Galerkin method with nonoverlapping domain decomposition for the stokes and Navier-Stokes problems. Math. Comput. **74**(249), 53–84 (2005)
14. Sakowski, K., Marcinkowski, L., Krukowski, S.: Modification of the Newton's method for the simulations of gallium nitride semiconductor devices. In: Wyrzykowski, R., Dongarra, J., Karczewski, K., Waśniewski, J. (eds.) PPAM 2013, Part II. LNCS, vol. 8385, pp. 551–560. Springer, Heidelberg (2014)
15. Sakowski, K., Marcinkowski, L., Krukowski, S., Grzanka, S., Litwin-Staszewska, E.: Simulation of trap-assisted tunneling effect on characteristics of gallium nitride diodes. J. Appl. Phys. **111**(12), 123115 (2012)
16. Polak, S.J., den Heijer, C., Schilders, W.H.A., Markowich, P.: Semiconductor device modelling from the numerical point of view. J. Numer. Methods Eng. **24**, 763–838 (1987)

Additive Nonoverlapping Schwarz for h-p Composite Discontinuous Galerkin

Piotr Krzyżanowski[(✉)]

University of Warsaw, Warsaw, Poland
p.krzyzanowski@mimuw.edu.pl

Abstract. A second order elliptic problem with piecewise constant coefficient in 2-D or 3-D is considered. The problem is discretized by a composite *h-p* finite element method, using continuous functions in subregions where the coefficient is constant and applying discontinuous Galerkin interior penalty method to couple them. The resulting discrete problem is solved by a two-level nonoverlapping additive Schwarz method. Condition number estimate of the preconditioned system, depending on the relative sizes of the underlying grids and on the relative degrees of finite elements used on the fine and coarse grids, is provided. In particular, the rate of convergence of the method is independent of the jumps of the coefficient.

Keywords: Nonoverlapping additive Schwarz method · Discontinuous Galerkin *h-p* discretization · Interior penalty method · Discontinuous coefficient

1 Introduction

In this paper we consider a second order elliptic equation

$$-\mathrm{div}(\varrho \nabla u) = f,$$

with homogeneous Dirichlet boundary condition, where the diffusion coefficient ϱ is a discontinuous, piecewise constant function. The problem is discretized by a composite continuous–discontinuous Galerkin (cG–dG) finite element method, using a continuous *h-p* discretization in regions where ϱ is constant. A weighted interior penalty method is then used to glue the solution across the interfaces on which large jump of ϱ may occur. Our goal in this paper is to analyze a two-level nonoverlapping additive Schwarz method (ASM) preconditioner, see e.g. [7], which introduces in a natural way a coarse grain parallelism and improves the convergence rate of an iterative solver. Such problem has already been considered by Dryja in [3] for linear finite elements in 2D, where a multilevel ASM was designed and analyzed. Here, we generalize the approach of [3] to the case of *h-p* composite cG–dG discretization.

This research has been partially supported by the Polish National Science Centre grant 2011/01/B/ST1/01179.

R. Wyrzykowski et al. (Eds.): PPAM 2015, Part II, LNCS 9574, pp. 401–410, 2016.
DOI: 10.1007/978-3-319-32152-3_38

For the h-p, but dG–only discretization of the problem, with globally constant coefficient, $\varrho \equiv 1$, Antonietti and Houston [1] have proved that the condition number of the system preconditioned with two–level nonoverlapping ASM is at most $O(p^2 H/h)$ where H is the coarse mesh size. However, numerical experiments therein have revealed that if the coarse space contained piecewise polynomial functions of order p, the observed condition number behaved rather like $O(p H/h)$. In this paper we prove this conjecture for the cG–dG h-p discretization, showing a more general result, that if the coarse space contains piecewise polynomial functions of order $q \leq p$, the condition number is $O(p^2/\max\{q,1\}) \cdot O(H/h)$. To author's knowledge, this is the first theoretical result of this kind. In addition, the bound holds independently of the jumps of the diffusion coefficient ϱ, provided the jumps of ϱ are aligned with the coarse grid.

The paper is organized as follows. In Sect. 2, differential problem and its cG–dG discretization are formulated. In Sect. 3, a nonoverlapping two-level ASM for solving the discrete problem is designed and analyzed. We conclude with final remarks in Sect. 5.

For nonnegative scalars x, y, we shall write $x \lesssim y$ if there exists a positive constant C, independent of: x, y, the fine and coarse mesh parameters h, H, the orders of the finite element spaces p, q, and of jumps of the diffusion coefficient ϱ as well, such that $x \leq Cy$. If both $x \lesssim y$ and $y \lesssim x$, we shall write $x \simeq y$.

The norm of a function f from the Sobolev space $H^k(S)$ will be denoted by $\|f\|_{k,S}$, while the seminorm of f will be denoted by $|f|_{k,S}$.

2   Differential Problem and Its cG–dG h-p Discretization

Let Ω be a bounded open polyhedral domain in R^d, $d \in \{2,3\}$, with Lipschitz boundary $\partial\Omega$. We consider the following variational problem for prescribed $f \in L^2(\Omega)$ and $\varrho \in L^\infty(\Omega)$:

Find $u^* \in H_0^1(\Omega)$ such that

$$a(u^*, v) = (f, v)_\Omega, \qquad \forall v \in H_0^1(\Omega), \tag{1}$$

where

$$a(u, v) = \int_\Omega \varrho\, \nabla u \cdot \nabla v\, dx, \qquad (f, v)_\Omega = \int_\Omega fv\, dx.$$

We assume that Ω can be partitioned into M nonoverlapping polyhedral subregions $D_1, \ldots, D_M$, $\bar{\Omega} = \bigcup_{m=1}^M \bar{D}_m$, with the property that ϱ restricted to any of these subregions is some positive constant. Thus, we allow for jumps of the coefficient only across boundaries of these subregions. Problems of this kind arise in many engineering applications, e.g. in modelling composite construction materials or layered electronic devices.

In what follows we will analyze a preconditioner for a system of algebraic equations arising from a discretization of (1) with composite cG–dG h-p finite element method. The corresponding finite element spaces and the discrete problem are introduced in the following subsection.

2.1   Finite Element Spaces and cG–dG *h-p* Discretization

Our discretization will follow the approach of [3], with necessary modifications to incorporate higher order finite elements. Let us first introduce the coarse mesh $\mathcal{T}_H$ — an affine, shape-regular triangulation of Ω into N disjoint simplicial subdomains (triangles in 2D, tetrahedrons in 3D), $\mathcal{T}_H = \{\Omega_1, \ldots, \Omega_N\}$. We do not assume that $\mathcal{T}_H$ is matching, but we do require that the number of neighbors is bounded for every element of $\mathcal{T}_H$ by an absolute constant $\mathcal{N}$. We also assume that $\mathcal{T}_H$ resolves the subregions $\{D_m\}$ in the sense that each D_m, $m = 1, \ldots, M$, is a sum of certain elements from $\mathcal{T}_H$. In this way, the jumps of ϱ can only occur at the boundaries of coarse mesh subdomains; inside Ω_i the coefficient is constant:

$$\varrho_{|\Omega_i} = \varrho_i > 0, \qquad i = 1, \ldots, N.$$

In accordance with the above notation, if necessary, for any function φ defined on Ω, we shall write φ_i to denote its restriction (not necessarily constant) to Ω_i:

$$\varphi_i := \varphi_{|\Omega_i}.$$

For $\Omega_i \in \mathcal{T}_H$ we define $H_i = \operatorname{diam}(\Omega_i)$ and further we set $H = (H_1, \ldots, H_N)$.

By Γ^0 we denote the interface of the coarse mesh, that is, the set of all common faces (edges in 2D) of elements from $\mathcal{T}_H$, so that $e \in \Gamma^0$ if and only if $e = \partial\Omega_i \cap \partial\Omega_j$ is of positive measure for some $i \neq j$. An analogous set of faces on $\partial\Omega$ will be denoted by Γ^∂. Finally, we define $\Gamma = \Gamma^0 \cup \Gamma^\partial$, the skeleton of the coarse mesh.

Next, in each Ω_i, $i = 1, \ldots, N$, let us introduce an affine, shape regular, quasi-uniform and matching simplicial triangulation $\mathcal{T}_{h_i}(\Omega_i)$, where h_i is the mesh parameter, i.e. $h_i = \max\{\operatorname{diam}(K) : K \in \mathcal{T}_{hi}(\Omega_i)\}$. We will refer to $\mathcal{T}_{h_i}(\Omega_i)$ as the local triangulation of subdomain Ω_i. With $p \geq 1$ and $i = 1, \ldots, N$, we define the corresponding local (continuous) finite element spaces as

$$V_{h_i}^p(\Omega_i) = \{v \in C(\Omega_i) : v_{|K} \in \mathcal{P}^p(K) \quad \forall K \in \mathcal{T}_{h_i}(\Omega_i)\},$$

where $\mathcal{P}^p$ is the space of polynomials of degree at most p, see Fig. 1.

Let $h = (h_1, \ldots, h_N)$ collect the parameters of local meshes. We define the global fine mesh on Ω,

$$\mathcal{T}_h = \{K \in \mathcal{T}_{h_i}(\Omega_i) : i = 1, \ldots, N\}$$

and assume for simplicity that the mesh is shape regular and quasiuniform. Finally we define the global finite element space over $\mathcal{T}_h$, in which we will approximate the solution of (1), as

$$V_h^p = \{v \in L^2(\Omega) : v_{|\Omega_i} \in V_{h_i}^p(\Omega_i)\}. \tag{2}$$

It consists of piecewise polynomial functions, which are continuous inside subdomains Ω_i, but may be discontinuous across the coarse mesh skeleton Γ.

We discretize (1) by the composite cG–dG method, using the symmetric weighted interior penalty discontinuous Galerkin method to enforce weak continuity across Γ, see for example [3,5]:

Find $u_h^* \in V_h^p$ such that

$$\mathcal{A}_h(u_h^*, v_h) = (f, v_h)_\Omega, \qquad \forall v_h \in V_h^p, \tag{3}$$

where

$$\mathcal{A}_h(u, v) \equiv A_h(u, v) - S_h(u, v) - S_h(v, u)$$

and

$$A_h(u, v) \equiv \sum_{i=1}^{N} (\varrho \nabla u, \nabla v)_{\Omega_i} + \sum_{e \in \Gamma} \langle \gamma [u], [v] \rangle_e, \tag{4}$$

$$S_h(u, v) \equiv \sum_{e \in \Gamma} \langle \{\varrho \nabla u\}, [v] \rangle_e. \tag{5}$$

Here we use standard notation

$$(u, v)_{\Omega_i} = \int_{\Omega_i} u\, v\, dx \quad \text{and} \quad \langle u, v \rangle_e = \int_e u\, v\, d\sigma.$$

In (4)–(5), for $e \in \Gamma^0$ such that $e = \partial\Omega_i \cap \partial\Omega_j$, we define (recalling that by convention $u_i = u_{|\Omega_i}$):

$$\{\varrho \nabla u\} = \overline{\varrho}\,(\nabla u_i + \nabla u_j), \qquad [u] = u_i\, n_i + u_j\, n_j,$$

where n_i is the unit normal vector pointing outward Ω_i, and (cf. [3])

$$\overline{\varrho} = \frac{\varrho_i \varrho_j}{\varrho_i + \varrho_j}.$$

Note that also $\{\varrho \nabla u\} = \omega_i \varrho_i \nabla u_i + \omega_j \varrho_j \nabla u_j$, where $\omega_i = \varrho_j/(\varrho_i + \varrho_j)$, as in [5]; in particular, we have $\omega_i + \omega_j = 1$ and $\omega_i \varrho_i = \omega_j \varrho_j = \dfrac{\varrho_i \varrho_j}{\varrho_i + \varrho_j} = \overline{\varrho}$. Moreover, we set (cf. [6])

$$\gamma_{|e} = \delta \cdot \frac{\overline{\varrho}\, p^2}{\underline{h}},$$

where $\underline{h} = \min\{h_i, h_j\}$, and δ is a positive penalty constant.

On $e \in \Gamma^\partial$ which is a face of Ω_i, we set $\{\varrho \nabla u\} = \varrho_i \nabla u_i$, $[u] = u_i\, n_i$ and $\gamma = \delta\, \varrho_i p^2 / h_i$.

For sufficiently large δ the discrete problem (3) is well-defined, according to the following lemma:

Lemma 1. *There exists positive δ_0 independent of h, H, p and ϱ such that if $\delta \geq \delta_0$, the bilinear form $\mathcal{A}_h(\cdot, \cdot)$ is symmetric, positive definite, and there holds*

$$\mathcal{A}_h(u, u) \simeq A_h(u, u) \quad \forall u \in V_h^p, \tag{6}$$

that is, $\mathcal{A}_h(\cdot, \cdot)$ is uniformly spectrally equivalent to $A_h(\cdot, \cdot)$.

Proof. The proof follows the lines of [3,6] and thus is omitted.

In what follows we shall take as granted that δ is a fixed constant such that $\delta \geq \delta_0$.

3   Nonoverlapping Two Level Additive Schwarz Method

The condition number of the discrete problem (3) can be prohibitively large, affected by the degree of the polynomials used, the fine mesh size, and by the magnitude of jumps in ϱ. Thus, for an iterative solution of (3), some preconditioning is necessary. In this section we consider a nonoverlapping ASM proposed in [1] for a fully discontinuous Galerkin discretization of (1), i.e. when the finite element functions are allowed inside Ω_i. In this paper, we assume a subspace consisting of functions discontinuous only across Γ.

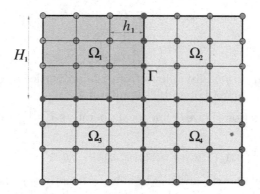

Fig. 1. Example decomposition into four subdomains $\Omega_1, \ldots, \Omega_4$ in which the diffusion coefficient ϱ is constant. Continuous finite elements of order p are used inside subdomain Ω_i, but discontinuities are allowed across Γ.

Let us introduce a decomposition of V_h^p:

$$V_h^p = V_0 + \sum_{i=1}^{N} V_i, \tag{7}$$

where for $i = 1, \ldots, N$ the local spaces are

$$V_i = \{v \in V_h^p : v_{|\Omega_j} = 0 \text{ for all } j \neq i\}, \tag{8}$$

so that V_i is a zero–extension of functions from $V_h^p(\Omega_i)$. Note that V_h^p is a direct sum of the local spaces.

The coarse space is

$$V_0 = \{v \in V_h^p : v_{|\Omega_i} \in \mathcal{P}^q(\Omega_i) \text{ for all } i = 1, \ldots, N\}$$

with $0 \leq q \leq p$, so that $V_0 \subset V_h^p$. Observe that functions from V_0 are in general discontinuous across Γ.

The coarse space considered here is richer than the usual choice of subdomainwise constant functions, e.g. [3]. In order to exploit the enlarged number of degrees of freedom in the convergence analysis, we need the following approximation lemma which we state without proof due to space restrictions:

Lemma 2. *For any* $u \in H^1(\Omega_i)$, $i = 1, \ldots, N$, *there exists* $u_i^{(0)} \in \mathcal{P}^q(\Omega_i)$ *such that*

$$|u - u_i^{(0)}|_{0,\Omega_i} \lesssim \frac{H_i}{\tilde{q}} |u|_{1,\Omega_i}, \tag{9}$$

$$|u - u_i^{(0)}|_{1,\Omega_i} \lesssim |u|_{1,\Omega_i}, \tag{10}$$

$$|u - u_i^{(0)}|_{0,\partial\Omega_i} \lesssim \left(\frac{H_i}{\tilde{q}}\right)^{1/2} |u|_{1,\Omega_i}, \tag{11}$$

where $\tilde{q} = \max\{q, 1\}$.

Proof. For $q = 0$, we choose as usual [7, Corollary A.15]

$$u_i^{(0)} = \bar{u}_i := \frac{1}{|\Omega_i|} \int_{\Omega_i} u \, dx.$$

The proof for positive q easily follows from the properties of the Babuška–Suri interpolant, cf. [2, Lemma 4.5] and [6, Lemma 4.4].

Using decomposition (7) we define local operators $T_i : V_h^p \to V_i$, $i = 1, \ldots, N$, by "inexact" solvers

$$\mathcal{A}_h(T_i u, v) = \mathcal{A}_h(u, v) \qquad \forall v \in V_i,$$

so that on each subdomain one has to solve a (relatively) small system of linear equations for degrees of freedom restricted only to $V_{h_i}^p(\Omega_i)$ (see Sect. 4 for details); for $j \neq i$ we set $(T_i u)_{|\Omega_j} = 0$.

The coarse solve operator is $T_0 : V_h^p \to V_0$ defined analogously as

$$\mathcal{A}_h(T_0 u, v_0) = \mathcal{A}_h(u, v_0) \qquad \forall v_0 \in V_0. \tag{12}$$

Finally, the preconditioned operator is

$$T = T_0 + \sum_{i=1}^{N} T_i. \tag{13}$$

Our main result here is the following theorem, which bounds the condition number of T in terms of p, q, H and h.

Theorem 1. *The preconditioned operator T defined in (13) is symmetric with respect to $\mathcal{A}_h(\cdot, \cdot)$ and satisfies*

$$\beta^{-1} \mathcal{A}_h(u, u) \lesssim \mathcal{A}_h(Tu, u) \lesssim \mathcal{A}_h(u, u) \qquad \forall u \in V_h^p,$$

where

$$\beta = \frac{p^2}{\max\{1, q\}} \max_{i=1,\ldots,N} \frac{H_i}{h_i}.$$

Therefore, the condition number of the preconditioned operator T is $O(\beta)$, independently of the jumps in ϱ.

Proof. Here we sketch the proof of the Theorem, following the abstract theory of additive Schwarz methods, cf. [7, Theorem 2.7], and prove three key properties for $T_0, T_1, \ldots, T_N$:

Strengthened Cauchy–Schwarz Inequality. It is straightforward to verify that if $u \in V_i$ and $v \in V_j$ with $i, j \in \{1, \ldots, N\}$ then $\mathcal{A}_h(u, v) = 0$, if only Ω_i and Ω_j are not neighbors, i.e. they do not share a face in $\mathcal{E}_H^0$. Thus, we conclude that then

$$\mathcal{A}_h(u, v) \leq \epsilon_{ij} \mathcal{A}_h(u, u)^{1/2} \mathcal{A}_h(v, v)^{1/2}$$

and the spectral radius of $\epsilon = (\epsilon_{ij})_{i,j=1}^N$ is bounded by $\mathcal{N}$ which has been assumed an absolute constant, see Sect. 2.1.

Local Stability. For all $i = 0, 1, \ldots, N$

$$\mathcal{A}_h(u, u) \leq \omega\, \mathcal{A}_h(u, u) \qquad \forall u \in V_i,$$

with absolute constant ω. This is an obvious consequence of Lemma 1. Since subdomainwise constant functions form a subspace in V_0, and both bilinear forms coincide on this subspace, we additionally conclude that $\omega \geq 1$.

Stable Decomposition. We have to prove that there exist: a decomposition of $u \in V_h^p$,

$$u = \sum_{i=0}^{N} u^{(i)} \qquad \text{with } u^{(i)} \in V_i \tag{14}$$

and C_0 independent of h, H, p, q, ϱ, such that

$$\sum_{i=0}^{N} \mathcal{A}_h(u^{(i)}, u^{(i)}) \leq C_0^2 \mathcal{A}_h(u, u) \qquad \forall u \in V_h^p. \tag{15}$$

In order to construct the stable decomposition of $u \in V_h^p$, we first define $u^{(0)} \in V_0$ such that on each $\Omega_i \in \mathcal{T}_H$, $i = 1, \ldots, N$,

$$u^{(0)}\big|_{\Omega_i} = u_i^{(0)},$$

where $u_i^{(0)}$ is as specified in Lemma 2.

Next, for $i = 1, \ldots, N$, we define $u^{(i)} \in V_i$, as a zero-extension of the restriction of $u - u^{(0)}$ to Ω_i:

$$u^{(i)} = \begin{cases} u_i - u_i^{(0)} & \text{on } \Omega_i, \\ 0 & \text{elsewhere.} \end{cases}$$

where we recall that by convention $u_i \equiv u|_{\Omega_i}$.

Then for $i = 1, \ldots, N$ we have

$$\mathcal{A}_h(u^{(i)}, u^{(i)}) = \varrho_i |\nabla u^{(i)}|_{0,\Omega_i}^2 + \sum_{e \subset \partial \Omega_i} \gamma_e |[u^{(i)}]|_{0,e}^2.$$

We have by (10)

$$\varrho_i |\nabla u^{(i)}|_{0,\Omega_i}^2 = \varrho_i |\nabla(u - u_i^{(0)})|_{0,\Omega_i}^2 \lesssim \varrho_i |\nabla u|_{0,\Omega_i}^2.$$

For the second term, using (11), we conclude that

$$\sum_{e \subset \partial \Omega_i} \gamma_e |[u^{(i)}]|_{0,e}^2 \lesssim \varrho_i \frac{p^2}{h_i} \frac{H_i}{\tilde{q}} |\nabla u|_{0,\Omega_i}^2$$

and in consequence

$$A_h(u^{(i)}, u^{(i)}) \lesssim \varrho_i \left(1 + \frac{p^2}{h_i} \frac{H_i}{\tilde{q}}\right) |\nabla u|_{0,\Omega_i}^2.$$

Summing over all subdomains and making use of Lemma 1 we arrive at

$$\sum_{i=1}^{N} A_h(u^{(i)}, u^{(i)}) \lesssim \frac{p^2}{\tilde{q}} \max_{i=1,\dots,N} \frac{H_i}{h_i} A_h(u, u).$$

It remains to obtain a bound for

$$A_h(u^{(0)}, u^{(0)}) = \sum_{i=1}^{N} \varrho_i |\nabla u_i^{(0)}|_{0,\Omega_i}^2 + \sum_{e \in \Gamma} \gamma |[u^{(0)}]|_{0,e}^2.$$

Observe that by (10) we have $|\nabla u_i^{(0)}|_{0,\Omega_i} \leq |\nabla(u - u_i^{(0)})|_{0,\Omega_i} + |\nabla u|_{0,\Omega_i} \lesssim |\nabla u|_{0,\Omega_i}$, so in consequence

$$\sum_{i=1}^{N} \varrho_i |\nabla u_i^{(0)}|_{0,\Omega_i}^2 \lesssim \sum_{i=1}^{N} \varrho_i |\nabla u|_{0,\Omega_i}^2 \lesssim A_h(u, u).$$

Next, for any $e \in \Gamma^0$ such that $e = \partial \Omega_i \cap \partial \Omega_j$ we have

$$|[u - u^{(0)}]|_{0,e}^2 \lesssim |(u - u^{(0)})_i|_{0,e}^2 + |(u - u^{(0)})_j|_{0,e}^2 + |[u]|_{0,e}^2,$$

so, applying triangle inequality,

$$\gamma_{|e} |[u - u^{(0)}]|_{0,e}^2 \lesssim \delta \frac{\varrho_i p^2}{h_i} |u_i - u_i^{(0)}|_{0,e}^2 + \delta \frac{\varrho_j p_j^2}{h_j} |u_j - u_j^{(0)}|_{0,e}^2 + \gamma_{|e} |[u]|_{0,e}^2.$$

Since similar bound can also be obtained for $e \in \Gamma^\partial$, we conclude that

$$\sum_{e \in \Gamma} \gamma |[u^{(0)}]|_{0,e}^2 \lesssim \sum_{e \in \Gamma} \gamma |[u]|_{0,e}^2 + \sum_{i=1}^{N} \varrho_i \frac{p^2}{h_i} |u_i - u_i^{(0)}|_{0,\partial \Omega_i}^2.$$

Now, the first term is not greater than $A_h(u, u)$ and the last term has already been estimated by $\beta A_h(u, u)$. Thus we conclude that

$$A_h(u^{(0)}, u^{(0)}) \lesssim \beta A_h(u, u),$$

which completes the proof.

Remark 1. *The approach described above can be generalized in several directions, e.g. allowing nonconforming meshes, varying FE degree and fully discontinuous elements inside subdomains D_i. Moreover, the recently introduced method, cf. [4], with a potential for increased amount of parallelism, can also be adapted to the discretization considered here. The generalized theory will appear in a forthcoming paper.*

4   Practical Remarks and Parallel Implementation

According to Theorem 1, Krylov subspace based iterative methods, such as, for example, the Conjugate Gradient method, will converge to the solution u of the system

$$Tu = f \tag{16}$$

at the rate proportional to the square root of β. In particular, Theorem 1 guarantees that the convergence rate bound will remain constant when H_i and h_i are scaled by the same factor — a situation commonly considered when assessing the weak scalability of an iterative solver. The same holds true when both p and q are increased by the same factor.

Let us also remind the reader that the transformed equation (16) is just the original discrete equation preconditioned in a specific way with a matrix–free preconditioner [7], so that the Preconditioned Conjugate Gradient method can easily be implemented. The action of the preconditioner, in turn, corresponds to the solution of the local problems (8) which can straightforwardly be applied in parallel; the coarse problem (12) is also solved independently of the local problems.

5   Conclusions

A preconditioner based on a nonoverlapping additive Schwarz method for composite continuous–discontinuous Galerkin h-p discretization of a second order elliptic PDE with discontinuous diffusion coefficient has been analyzed.

It has been shown that if the coarse space is imposed on a mesh with parameter H and consists of subdomainwise polynomials of q–th order, then the condition number is bounded by

$$O(p^2 / \max\{q, 1\}) \cdot O(H/h)$$

independently of the jumps of the diffusion coefficient ϱ.

This result provides an explicit estimate of the condition number on the polynomial degree used to define the coarse space. In particular, choosing $q = p$ for all subdomains, the condition number grows only linearly with p; on another extreme, if $q = 0$, the dependence is quadratic, which can be quite disappointing for high order approximations.

In addition, the condition estimate also retains the linear dependence on H/h and the independence of the jumps of ϱ — already known for nonoverlapping ASM developed for low order discontinuous Galerkin approximations.

Acknowledgement. The author would like to thank Max Dryja for comments on an early draft of the paper. This research has been partially supported by the Polish National Science Centre grant 2011/01/B/ST1/01179.

References

1. Antonietti, P.F., Houston, P.: A class of domain decomposition preconditioners for hp-discontinuous Galerkin finite element methods. J. Sci. Comput. **46**(1), 124–149 (2011)
2. Babuška, I., Suri, M.: The h-p version of the finite element method with quasi-uniform meshes. RAIRO Modél. Math. Anal. Numér. **21**(2), 199–238 (1987)
3. Dryja, M.: On discontinuous Galerkin methods for elliptic problems with discontinuous coefficients. Comput. Methods Appl. Math. **3**(1), 76–85 (2003). (electronic)
4. Dryja, M., Krzyżanowski, P.: A massively parallel nonoverlapping additive Schwarz method for discontinuous Galerkin discretization of elliptic problems. Numer. Math. **132**, 347–367 (2015)
5. Ern, A., Stephansen, A.F., Zunino, P.: A discontinuous Galerkin method with weighted averages for advection-diffusion equations with locally small and anisotropic diffusivity. IMA J. Numer. Anal. **29**(2), 235–256 (2009)
6. Süli, E., Schwab, C., Houston, P.: hp-DGFEM for partial differential equations with nonnegative characteristic form. In: Cockburn, B., Karniadakis, G.E., Shu, C.-W. (eds.) Discontinuous Galerkin Methods. Lecture Notes in Computer Science and Engineering, vol. 11, pp. 221–230. Springer, Heidelberg (2000)
7. Toselli, A., Widlund, O.: Domain Decomposition Methods-Algorithms and Theory. Springer Series in Computational Mathematics, vol. 34. Springer, Heidelberg (2005)

Minisymposium on High Performance Computing Interval Methods

Up-to-date Interval Arithmetic: From Closed Intervals to Connected Sets of Real Numbers

Ulrich Kulisch[✉]

Institut für Angewandte und Numerische Mathematik,
Karlsruher Institut für Technologie, 76128 Karlsruhe, Germany
Ulrich.Kulisch@kit.edu

Abstract. This paper unifies the representations of different kinds of computer arithmetic. It is motivated by the book *The End of Error* by John Gustafson [5]. Here interval arithmetic just deals with connected sets of real numbers. These can be closed, open, half-open, bounded or unbounded.

In an earlier paper [19] the author showed that different kinds of computer arithmetic like floating-point arithmetic, conventional interval arithmetic for closed real intervals and arithmetic for interval vectors and interval matrices can all be derived from an abstract axiomatic definition of computer arithmetic and are just special realizations of it. A computer operation is defined via a monotone mapping of an arithmetic operation in a complete lattice onto a complete sublattice.

This paper shows that the newly defined unum and ubound arithmetic [5] can be deduced from the same abstract mathematical model. To a great deal unum and ubound arithmetic can be seen as an extension of arithmetic for closed real intervals to open and half-open real intervals, just to connected sets of real numbers. Deriving computer executable formulas for ubound arithmetic on the base of pure floating-point numbers (without the IEEE 754 exceptions) leads to a closed calculus that is totally free of exceptions, i.e., any arithmetic operation of the set $+, -, \cdot, /$, and the dot product for ubounds together with a number of elementary functions always delivers a ubound as result. This wonderful property is suited moving correct and rigorous machine computation more into the centre of scientific computing.

Keywords: Computer arithmetic · Interval arithmetic · Axiomatic definition · Unum and ubound arithmetic · Arithmetic for connected sets of real numbers · Exact dot product · Exception-free computer arithmetic

1   Introduction

The first section briefly reviews the development of arithmetic for scientific computing from a mathematical point of view from the early days of floating-point arithmetic to conventional interval arithmetic until the latest step of unum and ubound arithmetic.

© Springer International Publishing Switzerland 2016
R. Wyrzykowski et al. (Eds.): PPAM 2015, Part II, LNCS 9574, pp. 413–434, 2016.
DOI: 10.1007/978-3-319-32152-3_39

1.1 Early Floating-Point Arithmetic

Early computers designed and built by Konrad Zuse, the Z3 (1941) and the Z4 (1945), are among the first computers that used the binary number system and floating-point for number representation [4,26]. Both machines carried out the four basic arithmetic operations of addition, subtraction, multiplication, division, and the square root by hardware. In the Z4 floating-point numbers were represented by 32 bits. They were used in a way very similar to what today is IEEE 754 single precision arithmetic. The technology of those days was poor (electromechanical relays, electron tubes). It was complex and expensive. To avoid frequent interrupts special representations and corresponding wirings were available to handle the three special values: 0, ∞, and *indefinite* (for $0/0$, $\infty \cdot 0$, $\infty - \infty$, ∞/∞, and others).

These early computers were able to execute about 100 flops (floating-point operations per second). For comparison: With a mechanic desk calculator or a modern pocket calculator a trained person can execute about 1000 arithmetic operations (somewhat reliably) per day. The computer could do this in 10 s. This was a gigantic increase in computing speed by a factor of about 10^4.

Over the years the computer technology was drastically improved. This permitted an increase of the word size and of speed. Already in 1965 computers were on the market (CDC 6600) that performed 10^5 flops. At these speeds a conventional error analysis of numerical algorithms, that estimates the error of each single arithmetic operation, becomes questionable. Examples can be given which illustrate that computers after very few operations sometimes deliver a completely absurd result [30]. For example it can be easily shown that for a certain system of two linear equations with two unknowns even today's computers deliver a result of which possibly not a single digit is correct. Such results strongly suggest to use the computer more for computing close two-sided bounds on the solution rather than, as now, approximations with unknown accuracy.

1.2 The Standard for Floating-Point Arithmetic IEEE 754

Continuous progress in computer technology allowed extra features such as additional word sizes and differences in the coding and numbers of special cases. To stabilize the situation a standard for floating-point arithmetic was developed and internationally adopted in 1985. It is known as the IEEE 754 floating-point arithmetic standard. Until today the most used floating-point format is double precision. It corresponds to about 16 decimal digits. A revision of the standard IEEE 754, published in 2008, added another word size of 128 bits.

During a floating-point computation exceptional events like underflow, overflow or division by zero may occur. For such events the IEEE 754 standard reserves some bit patterns to represent special quantities. It specifies special representations for $-\infty$, $+\infty$, -0, $+0$, and for NaN (not a number). Normally, an overflow or division by zero would cause a computation to be interrupted. There are, however, examples for which it makes sense for a computation to continue. In IEEE 754 arithmetic the general strategy upon an exceptional event is to deliver a result and continue the computation. This requires the result of operations on or resulting in special values to be defined. Examples are: $4/0 = \infty$, $-4/0 = -\infty$, $0/0 = \texttt{NaN}$, $\infty - \infty = \texttt{NaN}$, $0 \cdot \infty = \texttt{NaN}$, $\infty/\infty = \texttt{NaN}$, $1/(-\infty) = -0$, $-3/(+\infty) = -0$, $\log 0 = -\infty$, $\log x = \texttt{NaN}$ when $x < 0$, $4 - \infty = -\infty$. When a $\texttt{NaN}$ participates in a floating-point operation, the result is always a $\texttt{NaN}$. The purpose of these special operations and results is to allow programmers to postpone some tests and decisions to a later time in the program when it is more convenient.

The standard for floating-point arithmetic IEEE 754 has been widely accepted and has been used in almost every processor developed since 1985. This has greatly improved the portability of floating-point programs. IEEE 754 floating-point arithmetic has been used successfully in the past. Many computer users are familiar with all details of IEEE 754 arithmetic including all its exceptions like *underflow*, *overflow*, $-\infty$, $+\infty$, $\texttt{NaN}$, $-0, +0$, and so on. Seventy years of extensive use of floating-point arithmetic with all its exceptions makes users believe that this is the only reasonable way of using the computer for scientific computing. IEEE 754 is quasi taken as an axiom of computing.

By the time the original standard IEEE 754 was developed, early microprocessors were on the market. They were made with a few thousand transistors, and ran at 1 or 2 MHz. Arithmetic was provided by an 8-bit adder. Dramatic advances in computer technology, in memory size, and in speed have been made since 1985. Arithmetic speed has gone from megaflops (10^6 flops), to gigaflops (10^9 flops), to teraflops (10^{12} flops), to petaflops (10^{15} flops), and it is already approaching the exaflops (10^{18} flops) range. This even is a greater increase of computing speed since 1985 than the one from a hand calculator to the first electronic computers! A qualitative difference goes with it. At the time of the megaflops computer a conventional error analysis was recommended. Today the PC is a gigaflops computer. For the teraflops or petaflops computer conventional error analysis is no longer practical.

Computing indeed has already reached astronomical dimensions! With increasing speed, problems that are dealt with become larger and larger. Extending pure floating-point arithmetic by operations for elements that are not real numbers and perform trillions of operations with them appears questionable. What seemed to be reasonable for slow speed computers needs not to be so for computers that perform trillions of operations in a second. A compiler could detect exceptional events and ask the user to treat them as for any other error message.

The capability of a computer should not just be judged by the number of operations it can perform in a certain amount of time without asking whether the computed result is correct. It should also be asked how fast a computer can compute correctly to 3, 5, 10 or 15 decimal places for certain problems. If the question were asked that way, it would very soon lead to better computers. Mathematical methods that give an answer to this question are available for many problems. Computers, however, are at present not designed in a way that allows these methods to be used effectively. Computer arithmetic must move strongly towards more reliability in computing. Instead of the computer being merely a fast calculating tool it must be developed into a scientific instrument of mathematics.

1.3   Conventional Interval Arithmetic

Issues just mentioned were one of the reasons why interval arithmetic has been invented. Conventional interval arithmetic just deals with **bounded and closed real intervals.** Formulas for the basic arithmetic operations for these intervals are easily derived. Interval arithmetic became popular after the book [24] by R.E. Moore was published in 1966. It was soon further exploited by other well known books by G. Alefeld and J. Herzberger [1,2] or by E. Hansen [6,7] for instance, and others. Interval mathematics using conventional interval arithmetic has been developed to a high standard over the last few decades. It provides methods which deliver results with guarantees.

Since the 1970-ies until lately [12,13,27,33] attempts were undertaken to extend the arithmetic for closed and bounded real intervals to unbounded intervals. However, inconsistencies to deal with $-\infty$ and $+\infty$ have occurred again and again. If the real numbers $\mathbb{R}$ are extended by $-\infty$ and $+\infty$ then unusual and unsatisfactory operations are to be dealt with like $\infty - \infty$, $0 \cdot \infty$, or ∞/∞.

1.4   The Proposed Standard for Interval Arithmetic IEEE P1788

In April 2008 the author of this article published a book [20] in which the problems with the infinities and other exceptions are definitely eliminated. Here interval arithmetic just deals with sets of real numbers. Since $-\infty$ and $+\infty$ are not real numbers, they cannot be elements of a real interval. They only can be bounds of a real interval. Formulas for the arithmetic operations for bounded and closed real intervals are well established in conventional interval arithmetic. It is shown in the book that these formulas can be extended to closed and unbounded real intervals by a continuity principle. For a bound $-\infty$ or $+\infty$ in an interval operand the bounds for the resulting interval can be obtained from the formulas for bounded real intervals by applying well established rules of real analysis for computing with $-\infty$ and $+\infty$. It is also shown in the book that obscure operations like $\infty - \infty$ or ∞/∞ do not occur in the formulas for the operations for unbounded real intervals. *This new approach to arithmetic for bounded and unbounded closed real intervals leads to an algebraically closed calculus which is free of exceptions. It remains free of exceptions if the operations are mapped on a floating-point screen by the monotone, upwardly directed rounding,* for definition see Definition 3. Intervals bring the continuum on the computer. An interval between two floating-point bounds represents the continuous set of real numbers between these bounds.

A few months after publication of the book [20] the IEEE Computer Society founded a committee IEEE P1788 for developing a standard for interval arithmetic in August 2008. A motion, presented by the author, to include arithmetic for unbounded real intervals where $-\infty$ and $+\infty$ may be bounds but not elements of unbounded real intervals has been accepted by IEEE P1788.

With little hardware expenditure interval arithmetic can be made as fast as simple floating-point arithmetic. The lower and the upper bound of an arithmetic operation easily can be computed simultaneously. With more suitable processors, rigorous methods based on interval arithmetic could be comparable in speed to today's approximate methods. As computers speed up, interval arithmetic becomes a principal and necessary tool for controlling the precision of a computation as well as the accuracy of the computed result.

Floating-point arithmetic and interval arithmetic are distinct calculi. Floating-point arithmetic as specified by IEEE 745 is full of complicated constructs, data and events like rounding to nearest, overflow, underflow, $+\infty$, $-\infty$, $+0$, -0 as numbers, or operations like $\infty - \infty$, ∞/∞, $0 \cdot \infty$. In contrast to this reasonably defined interval arithmetic leads to an exception-free calculus. It is thus only reasonable to keep the two calculi strictly separate. Mentioning IEEE 754 arithmetic in IEEE 1788 already confronts the reader with all its complicated constructs.

1.5 Advanced Computer Arithmetic

The book [20] deals with computer arithmetic in a more general sense than usual. It shows how the arithmetic and mathematical capability of the digital computer can be enhanced in a quite natural way. This is motivated by the desire and the need to improve the accuracy of numerical computing and to control the quality of computed results.

Advanced computer arithmetic extends the accuracy requirements for the elementary floating-point operations as defined by the arithmetic standard IEEE 754 to the customary product spaces of computation: the complex numbers, the real and complex intervals, the real and complex vectors and matrices, and the real and complex interval vectors and interval matrices. All computer approximations of arithmetic operations in these spaces should deliver a result that differs from the correct result by at most one rounding. For all these product spaces this accuracy requirement leads to operations which are distinctly different from those traditionally available on computers. This expanded set of arithmetic operations is taken as a definition of what is called **advanced computer arithmetic** in [20]. Programming environments that provide advanced computer arithmetic have been available since 1980 [9,10,12,22,33,34].

Advanced computer arithmetic is then used to develop algorithms for computing highly accurate and guaranteed bounds for a number of standard problems of numerical analysis like systems of linear equations, evaluation of polynomials or other arithmetic expressions, numerical integration, optimization problems, and many others [12,13]. These can be taken as higher order arithmetic operations. Essential for achieving these results is an exact dot product.

In vector and matrix spaces[1] the dot product of two vectors is a fundamental arithmetic operation. It is fascinating that this basic operation is also a mean to increase the speed of computing besides of the accuracy of the computed result. Actually the simplest and fastest way for computing a dot product of two floating-point vectors is to compute it exactly. Here the products are just shifted and added into a wide fixed-point register on the arithmetic unit. By pipelining, the exact dot product can be computed in the time the processor needs to read the data, i.e., it comes with utmost speed. This high speed is obtained by totally avoiding slow intermediate access to the main memory of the computer.

Any method that computes a dot product correctly rounded to the nearest floating-point number also has to consider the values of the summands. This results in a more complicated method with the outcome that it is necessarily slower than a conventional computation of the dot product in floating-point arithmetic. Experience with a prototype development in 1994 [3,17] shows that a hardware implementation of the **exact dot product** can be expected to be three to four times faster than the latter and it is faster by more than one magnitude than any method for computing a correctly rounded dot product. The main difference, however, is accuracy. There are many applications where a correctly rounded or otherwise precise dot product does not suffice to solve the problem. For details see [20, 28, 29, 31].

The hardware needed for the exact dot product is comparable to that for a fast multiplier by an adder tree, accepted years ago and now standard technology in every modern processor. The exact dot product brings the same speedup for accumulations at comparable costs.

In 2009 the author prepared a motion that requires inclusion of the exact dot product as essential ingredient for obtaining high accuracy in interval computations into the standard IEEE 1788. The motion was accepted. But in 2013, however, the motion was weakened by the committee to now just recommending an exact dot product. In practice a recommendation guarantees nonstandard behavior for different computing systems.

Advanced computer arithmetic certainly is a much more useful extension to pure floating-point arithmetic than all the exceptions provided by IEEE 754. All forms of speculation need to be removed from computing.

1.6   Unum and Ubound Arithmetic

While about 70 scientists from all over the world have been working on a standard for interval arithmetic for more than 6 years since August 2008, all of a sudden like out of nothing John Gustafson publishes a book: *The End of Error* [5]. Reading this book became a big surprise. It is a sound piece of work and it is hard to believe that a single person could develop so many nice ideas and put them together into a sketch of what might become the future of computing. Reading the book is fascinating. The situation very much reminds me to a text by Friedrich Schiller in his work *Demetrius*. It says:

---

[1] For real, complex, interval, and complex interval data.

Was ist die Mehrheit? Die Mehrheit ist der Unsinn,

Verstand ist stets bei wen'gen nur gewesen.

For almost 60 years interval arithmetic was defined for the set $\mathbb{IR}$ of closed and bounded real intervals. *The End of Error* expands this to the set $\mathbb{JR}$ of just connected sets of real numbers. These can be closed, open, half-open, bounded, or unbounded. The book shows that arithmetic for this expanded set is closed under addition, subtraction, multiplication, division, also square root, powers, logarithm, exponential, and many other elementary functions needed for technical computing, i.e., arithmetic operations for intervals of $\mathbb{JR}$ always lead to intervals of $\mathbb{JR}$ again. The calculus is free of exceptions. It remains free of exceptions if the bounds are restricted to a floating-point screen, for proof see Sect. 2. John Gustafson shows in his book that this new extension of conventional interval arithmetic opens new areas of applications and allows getting better results.

2   Axiomatic Definition of Computer Arithmetic

Frequently mathematics is seen as the science of structures. Analysis carries three kinds of structures: an algebraic structure, an order structure, and a topological or metric structure. These are coupled by certain compatibility properties, as for instance: $a \leq b \Rightarrow a + c \leq b + c$.

It is well known that floating-point numbers and floating-point arithmetic do not obey the algebraic rules of the real numbers $\mathbb{R}$. However, the rounding is a monotone function. So the changes to the order structure are minimal. **This is the reason why the order structure plays a key role for an axiomatic definition of computer arithmetic.**

We begin by listing a few well-known concepts and properties of ordered sets.

Definition 1. *A relation $\leq$ in a set M is called an* order relation, *and* $\{M, \leq\}$ *is called an* ordered set[2] *if for all $a, b, c \in M$ the following properties hold:*

(O1) $a \leq a$,   (reflexivity)
(O2) $a \leq b \ \wedge \ b \leq c \ \Rightarrow \ a \leq c$,   (transitivity)
(O3) $a \leq b \ \wedge \ b \leq a \ \Rightarrow \ a = b$,   (antisymmetry)

An ordered set M is called linearly *or* totally ordered *if in addition*

(O4) $a \leq b \vee b \leq a$ *for all $a, b \in M$.*   (linearly ordered)

An ordered set M is called

(O5) *a* lattice *if for any two elements $a, b \in M$, the $\inf\{a, b\}$ and the $\sup\{a, b\}$ exist.*   (lattice)

(O6) *It is called* conditional completely ordered *if for every bounded subset $S \subseteq M$, the $\inf S$ and the $\sup S$ exist.*

---

[2] Occasionally called a partially ordered set.

(O7) *An ordered set M is called* completely ordered *or a* complete lattice *if for every subset $S \subseteq M$, the* inf S *and the* sup S *exist.*   (complete lattice)

With these concepts the real numbers $\{\mathbb{R}, \leq\}$ are a conditional complete linearly ordered field.

In the definition of a complete lattice, the case $S = M$ is included. Therefore, inf M and sup M exist. Since they are elements of M, every complete lattice has a least and a greatest element.

If a subset $S \subseteq M$ of a complete lattice $\{M, \leq\}$ is also a complete lattice, $\{S, \leq\}$ is called a *complete sublattice* of $\{M, \leq\}$ if the two lattice operations inf and sup in both sets lead to the same result, i.e., if

$$\text{for all} \quad A \subseteq S, \quad \inf_M A = \inf_S A \quad \text{and} \quad \sup_M A = \sup_S A.$$

Definition 2. *A subset S of a complete lattice $\{M, \leq\}$ is called a* screen *of M, if every element $a \in M$ has upper and lower bounds in S and the set of all upper bounds of $a \in M$ has a least and the set of all lower bounds a greatest element in S. If a minus operator exists in M, a screen is called* symmetric, *if for all $a \in S$ also $-a \in S$.*

As a consequence of this definition a complete lattice and a screen have the same least and greatest element. It can be shown that a screen is a complete sublattice of $\{M, \leq\}$ with the same least and greatest element, [20].

Definition 3. *A mapping $\square : M \to S$ of a complete lattice $\{M, \leq\}$ onto a screen S is called a* rounding *if (R1) and (R2) hold:*

(R1) *for all* $a \in S$, $\square\, a = a$.   (projection)
(R2) $a \leq b \Rightarrow \square\, a \leq \square\, b$.   (monotone)

A rounding is called downwardly directed *resp.* upwardly directed *if for all $a \in M$*

(R3) $\square\, a \leq a$ *resp.* $a \leq \square\, a$.   (directed)

If a minus operator is defined in M, a rounding is called antisymmetric *if*

(R4) $\square\, (-a) = -\, \square\, a$,   *for all $a \in M$.*   (antisymmetric)

The monotone downwardly resp. upwardly directed roundings of a complete lattice onto a screen are unique. For the proof see [20].

Definition 4. *Let $\{M, \leq\}$ be a complete lattice and $\circ : M \times M \to M$ a binary arithmetic operation in M. If S is a screen of M, then a rounding $\square : M \to S$ can be used to approximate the operation $\circ$ in S by*

(RG) $a \,\boxdot\, b := \square\, (a \circ b)$, *for $a, b \in S$.*

If a minus operator is defined in M and S is a symmetric screen of M, then a mapping $\square : M \to S$ with the properties (R1,2,4) and (RG) is called a semimorphism[3].

Semimorphisms with antisymmetric roundings are particularly suited for transferring properties of the structure in M to the subset S. It can be shown [20] that semimorphisms leave a number of reasonable properties of ordered algebraic structures (ordered field, ordered vector space) invariant.

If an element $x \in M$ is bounded by $a \le x \le b$ with $a, b \in S$, then by (R1) and (R2) the rounded image $\square\, x$ is bounded by the same elements: $a \le \square\, x \le b$, i.e., $\square\, x$ is either the least upper (supremum) or the greatest lower (infimum) bound of x in S. Similarly, if for $x, y \in S$ the result of an operation $x \circ y$ is bounded by $a \le x \circ y \le b$ with $a, b \in S$, then by (R1), (R2), and (RG) also $a \le x \,\square\, y \le b$, i.e., $x \,\square\, y$ is either the least upper or the greatest lower bound of $x \circ y$ in S. If the rounding is upwardly or downwardly directed the result is the least upper or the greatest lower bound respectively.

In an earlier paper [19] the author applies the abstract formalism developed here to the most frequent models, floating-point arithmetic and arithmetic for closed real intervals. Essential properties and explicit formulas for the operations in these models can directly be derived from the abstract setting given in this section. We refrain from repeating this here and refer the reader to this earlier paper. Abstract settings of computer arithmetic for higher dimensional spaces like complex numbers, vectors and matrices for real, complex, and interval data can be developed following similar schemes. We briefly sketch this in Sect. 4. For more details see [20] and the literature cited there.

3   Unum and Ubound Arithmetic

In his recently published book *The End of Error* [5] John Gustafson develops a computing environment for real numbers and for sets of real numbers which is superior to conventional floating-point and interval arithmetic. A new number format, the *unum*[4], can more efficiently be used on computers with respect to many desirable properties like power consumption, storage requirements, bandwidth, parallelism concerns, and even speed. It gets mathematical rigor that even conventional interval arithmetic is not able to attain.

By obvious reasons John Gustafson's book strives for being upward compatible with IEEE 754 floating-point arithmetic and with traditional interval arithmetic. From the mathematical point of view, however, there is no need for doing this. Here we show that the new computing environment perfectly fits into an abstract mathematical approach to computer arithmetic as sketched in Sect. 2. Like conventional closed real intervals also unums and ubounds just deal

---

[3] The properties (R1,2,4) and (RG) of a semimorphism can be shown to be necessary conditions for a homomorphism between ordered algebraic structures. For more details see [20].

[4] Stands for **universal number**.

with sets of real numbers. $-\infty$ and $+\infty$ are not real numbers. They are just used as bounds to describe sets of real numbers. They are, however, themselves not elements of these sets. There is absolutely no need for introducing entities like $-0, +0, NaN$ (not a number) or NaI (not an interval) in this new computing environment. Focusing on the mathematical core of the new computing scheme leads to several additional simplifications.

A *unum* (Fig. 1) is a bit string of variable length that has six subfields: the *sign bit s, exponent, fraction, uncertainty bit u (ubit), exponent size,* and *fraction size*. The first three subfields describe a floating-point number. If the ubit is 0, the number is exact. If it is 1, it is inexact. An inexact unum can be interpreted as the set of all real numbers in the open interval between the floating-point part of the unum and the floating-point number one bit further from zero. The last two subfields, the exponent size and the fraction size are used to automatically shrink or enlarge the number of bits used for the representation of the exponent and the fraction part of the unum depending on results of operations. This automatic scaling adapts the word size to the needs of the computation. The set of all unums is denoted by $\mathbb{U}$. By the definition of unums $-\infty$ and $+\infty$ are elements of $\mathbb{U}$.

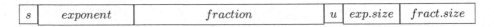

s	exponent	fraction	u	exp.size	fract.size

Fig. 1. The universal number format *unum*.

A *ubound* is a single unum or a pair of unums that represent a mathematical interval of the real line. Closed endpoints are represented by exact unums (ubit = 0), and open endpoints are represented by inexact unums (ubit = 1). So the ubit in a unbound's bound describes the kind of bracket that is used in the representation of the ubound. It is closed, if the ubit is 0 and it is open, if the ubit is 1. We denote the set of all ubounds by $\mathbb{JU}$. Later we shall occasionally denote an element $a \in \mathbb{JU}$ by $a = \langle a_1, a_2 \rangle$ where a_1, a_2 are floating-point numbers and each one of the angle brackets $\langle$ and $\rangle$ can be open or closed.

The ubit after the floating-point part of a unum can be 0 or 1. So the set of unums $\mathbb{U}$ is a superset of the set of floating-point numbers, $\mathbb{U} \supset \mathbb{F}$. Nevertheless the unums are a linearly ordered set $\{\mathbb{U}, \leq\}$. For positive floating-point numbers the unum with ubit 0 is less than the unum with ubit 1 and for negative floating-point numbers the unum with ubit 0 is greater than the unum with ubit 1. With the following notations $\overline{\mathbb{R}} := \mathbb{R} \cup \{-\infty, +\infty\}$ and $\overline{\mathbb{F}} := \mathbb{F} \cup \{-\infty, +\infty\}$ the ordered set $\{\overline{\mathbb{F}}, \leq\}$ is a screen of $\{\overline{\mathbb{R}}, \leq\}$. It is now easy to see that as a bit string the ordered set of unums $\{\mathbb{U}, \leq\}$ is also a screen of $\{\overline{\mathbb{R}}, \leq\}$. It is a larger, i.e., a finer screen than $\{\overline{\mathbb{F}}, \leq\}$.[5]

The directed roundings ∇ resp. $\triangle$ can now be extended as mappings from the extended set of real numbers $\overline{\mathbb{R}}$ onto the set of unums $\mathbb{U}$, $\nabla : \overline{\mathbb{R}} \to \mathbb{U}$ and

---

[5] This makes it plausible that unum arithmetic can lead to better results than floating-point arithmetic.

$\triangle : \overline{\mathbb{R}} \to U$. It is easy to see that they are again related by the property:

$$\nabla (-a) = - \triangle a \quad \text{and} \quad \triangle (-a) = - \nabla a. \tag{1}$$

These roundings ∇ and $\triangle$ can most frequently be used to map intervals or sets of real numbers onto ubounds. Here ∇ delivers the lower bound and $\triangle$ the upper bound. This allows to express the ubit of the unum by the bracket of the ubound. Exact unums are expressed by closed endpoints, by square brackets. A closed endpoint is an element of the ubound. Inexact unums are expressed by open endpoints, by round brackets. An open endpoint is just a bound but not an element of the ubound.

We illustrate these roundings by simple examples. We use the decimal number system, a fraction part of three digits, and a space before the ubit. The following results are possible:

$\nabla (0.543216) = 0.543\ 1 = (0.543, \qquad \triangle (0.543216) = 0.543\ 1 = 0.544),$
$\nabla (0.543) = 0.543\ 0 = [0.543, \qquad \triangle (0.543) = 0.543\ 0 = 0.543],$
$\nabla (-0.543216) = -0.543\ 1 = (-0.544, \quad \triangle (-0.543216) = -0.543\ 1 = -0.543),$
$\nabla (-0.543) = -0.543\ 0 = [-0.543, \qquad \triangle (-0.543) = -0.543\ 0 = -0.543].$

Let now $\mathbb{JR}$ denote the set of bounded or unbounded real intervals where each bound can be open or closed. So $\mathbb{JR}^6$ denotes the set of open or closed or half-open intervals of real numbers. Besides of the empty set every interval of $\mathbb{JR}$ can be expressed by round and/or square brackets. If the bracket adjacent to a bound is round, the bound is not an element of the interval; if it is square the bound is an element of the interval.

With set inclusion as an order relation the ordered set $\{\mathbb{JR}, \subseteq\}$ is a complete lattice. The infimum of two or more elements of $\{\mathbb{JR}, \subseteq\}$ is the intersection and the supremum is the convex hull. The subset of $\mathbb{JR}$ where all bounds are unums of U is denoted by $\mathbb{JU}$. Then $\{\mathbb{JU}, \subseteq\}$ is a screen of $\{\mathbb{JR}, \subseteq\}$. In both sets $\mathbb{JR}$ and $\mathbb{JU}$ the infimum of two or more elements of $\mathbb{JR}$ and $\mathbb{JU}$ is the intersection and the supremum is the interval (convex) hull. The least element of both sets $\mathbb{JR}$ and $\mathbb{JU}$ is the empty set $\varnothing$ and the greatest element is the set $\mathbb{R} = (-\infty, +\infty)$. Elements of $\mathbb{JR}$ and $\mathbb{JU}$ are denoted by bold letters.

Definition 5. *For elements $a, b \in \mathbb{JR}$ we define arithmetic operations $\circ \in \{+, -, \cdot, /\}$ as set operations*

$$a \circ b := \{a \circ b \mid a \in a \wedge b \in b\}. \tag{2}$$

Here for division we assume that $0 \notin b$.

Explicit formulas for the operations $a \circ b, \circ \in \{+, -, \cdot, /\}$ can be obtained in great similarity to the operations in $\overline{\mathbb{IR}}$. For derivation see [20]. However, each bound of the resulting interval in $\mathbb{JR}$ can now be open or closed.

It is a well established result that under Definition (2) $\mathbb{JR}$ is a closed calculus, i.e., the result $a \circ b$ is again an element of $\mathbb{JR}$. For details see [5].

---

[6] We do not introduce a separate symbol for the subset of bounded such intervals here as in the case of real intervals.

Remark 1: A bound of the result $\boldsymbol{a} \circ \boldsymbol{b}$ in (2) is closed if and only if the ubit of the adjacent number is zero, i.e., the number is an exact unum. This can only happen if both operands for computing the bound come from closed interval bounds. In case of an inexact unum in any of the operands the bound is open.

Let us now denote an interval $\boldsymbol{a} \in \mathbb{JR}$ by $\boldsymbol{a} = \langle a_1, a_2 \rangle$, where each one of the angle brackets $\langle$ and $\rangle$ can be open or closed. Then we obtain by (2) immediately

$$-\boldsymbol{a} := (-1) \cdot \boldsymbol{a} = (-1) \cdot \{x \mid a_1 \leq x \leq a_2\} = \{x \mid -a_2 \leq x \leq -a_1\}$$
$$= \langle -a_2, -a_1 \rangle \in \mathbb{JR}. \tag{3}$$

$$- \langle a_1, a_2 \rangle = \langle -a_2, -a_1 \rangle. \tag{4}$$

[7]More precisely: If the lower bound of the interval $\boldsymbol{a}$ is open (resp. closed) then the upper bound of $-\boldsymbol{a}$ is open (resp. closed), and if the upper bound of $\boldsymbol{a}$ is open (resp. closed) then the lower bound in $-\boldsymbol{a}$ is open (resp. closed).

With (4) subtraction can be reduced to addition by $\boldsymbol{a} - \boldsymbol{b} = \boldsymbol{a} + (-\boldsymbol{b})$.

If in (4) $\boldsymbol{a} \in \mathbb{JU}$, then also $-\boldsymbol{a} \in \mathbb{JU}$, i.e., $\mathbb{JU}$ is a symmetric screen of $\mathbb{JR}$.

Between the complete lattice $\{\mathbb{JR}, \subseteq\}$ and its screen $\{\mathbb{JU}, \subseteq\}$ the monotone upwardly directed rounding $\diamondsuit : \mathbb{JR} \to \mathbb{JU}$ is uniquely defined by the following properties:

(R1) $\diamondsuit \, \boldsymbol{a} = \boldsymbol{a}$, for all $\boldsymbol{a} \in \mathbb{JU}$.   (projection)

(R2) $\boldsymbol{a} \subseteq \boldsymbol{b} \Rightarrow \diamondsuit \, \boldsymbol{a} \subseteq \diamondsuit \, \boldsymbol{b}$, for $\boldsymbol{a}, \boldsymbol{b} \in \mathbb{JR}$.   (monotone)

(R3) $\boldsymbol{a} \subseteq \diamondsuit \, \boldsymbol{a}$, for all $\boldsymbol{a} \in \mathbb{JR}$.   (upwardly directed)

For $\boldsymbol{a} = \langle a_1, a_2 \rangle \in \mathbb{JR}$ the result of the monotone upwardly directed rounding $\diamondsuit$ can be expressed by

$$\diamondsuit \, \boldsymbol{a} = \langle \nabla a_1, \triangle a_2 \rangle. \tag{5}$$

where again each one of the angle brackets $\langle$ and $\rangle$ can be open or closed.

Similarly to the case of closed real intervals of $\overline{\mathbb{IR}}$ we now define an order relation $\leq$ for intervals of $\mathbb{JR}$. For intervals $\boldsymbol{a} = \langle a_1, a_2 \rangle$, $\boldsymbol{b} = \langle b_1, b_2 \rangle \in \mathbb{JR}$, the relation $\leq$ is defined by $\boldsymbol{a} \leq \boldsymbol{b} :\Leftrightarrow \langle a_1 \leq \langle b_1 \wedge a_2 \rangle \leq b_2 \rangle$. So we have for instance: $[1, 2) \leq (1, 2]$, or $[-2, -1) \leq (-2, -1]$.

For the $\leq$ relation for intervals compatibility properties hold between the algebraic structure and the order structure in great similarity to the real numbers. For instance:

(OD1) $\boldsymbol{a} \leq \boldsymbol{b} \Rightarrow \boldsymbol{a} + \boldsymbol{c} \leq \boldsymbol{b} + \boldsymbol{c}$, for all $\boldsymbol{c}$.

(OD2) $\boldsymbol{a} \leq \boldsymbol{b} \Rightarrow -\boldsymbol{b} \leq -\boldsymbol{a}$.

(OD3) $0 \leq \boldsymbol{a} \leq \boldsymbol{b} \wedge \boldsymbol{c} \geq 0 \Rightarrow \boldsymbol{a} \cdot \boldsymbol{c} \leq \boldsymbol{b} \cdot \boldsymbol{c}$.

(OD4) $0 < \boldsymbol{a} \leq \boldsymbol{b} \wedge \boldsymbol{c} > 0 \Rightarrow 0 < \boldsymbol{a}/\boldsymbol{c} \leq \boldsymbol{b}/\boldsymbol{c} \wedge \boldsymbol{c}/\boldsymbol{a} \geq \boldsymbol{c}/\boldsymbol{b} > 0$.

With respect to set inclusion as an order relation arithmetic operations in $\{\mathbb{JR}, \subseteq\}$ are inclusion isotone by (2), i.e., $\boldsymbol{a} \subseteq \boldsymbol{b} \Rightarrow \boldsymbol{a} \circ \boldsymbol{c} \subseteq \boldsymbol{b} \circ \boldsymbol{c}$ or equivalently

---

[7] An integral number a in a ubound expression is interpreted as ubound $[a, a]$.

(OD5) $a \subseteq b \wedge c \subseteq d \Rightarrow a \circ c \subseteq b \circ d$, for all $\circ \in \{+, -, \cdot, /\}, 0 \notin b, d$ for $\circ = /$. (inclusion isotone)

Setting $c, d = -1$ in (OD5) delivers immediately $a \subseteq b \Rightarrow -a \subseteq -b$ which differs significantly from (OD2).

Using (1), (4), and (5) it is easy to see that the monotone upwardly directed rounding $\Diamond : \mathbb{JR} \to \mathbb{JU}$ is antisymmetric, i.e.,

(R4) $\Diamond(-a) = -\Diamond a$, for all $a \in \mathbb{JR}$. \hfill (antisymmetric).

Definition 6. *With the upwardly directed rounding* $\Diamond : \mathbb{JR} \to \mathbb{JU}$ *binary arithmetic operations in* $\mathbb{JU}$ *are defined by semimorphism:*

(RG) $a \diamondsuit b := \Diamond(a \circ b)$, *for all* $a, b \in \mathbb{JU}$ *and all* $\circ \in \{+, -, \cdot, /\}$.

Here for division we assume that a/b *is defined.*

If a ubound $a \in \mathbb{JU}$ is an upper bound of a ubound $x \in \mathbb{JR}$, i.e., $x \subseteq a$, then by (R1), (R2), and (R3) also $x \subseteq \Diamond x \subseteq a$. This means $\Diamond x$ is the least upper bound, the supremum of x in $\mathbb{JU}$. Similarly if for $x, y \in \mathbb{JU}$, $x \circ y \subseteq a$ with $a \in \mathbb{JU}$, then by (R1), (R2), (R3), and (RG) also $x \circ y \subseteq x \diamondsuit y \subseteq a$, i.e., $x \diamondsuit y$ is the least upper bound, the supremum of $x \circ y$ in $\mathbb{JU}$. Occasionally the supremum $x \diamondsuit y \in \mathbb{JU}$ of the result $x \circ y \in \mathbb{JR}$ is called the tightest enclosure of $x \circ y$.

Arithmetic operations in $\mathbb{JU}$ are inclusion isotone, i.e.,

(OD5) $a \subseteq b \wedge c \subseteq d \Rightarrow a \diamondsuit c \subseteq b \diamondsuit d$, for $\circ \in \{+, -, \cdot, /\}, 0 \notin b, d$ for $\circ = /$. (inclusion isotone)

This is a consequence of the inclusion isotony of the arithmetic operations in $\mathbb{JR}$, of (R2) and of (RG).

Since the arithmetic operations $x \circ y$ in $\mathbb{JR}$ are defined as set operations by (2) the operations $x \diamondsuit y$ for ubounds of $\mathbb{JU}$ defined by (RG) are not directly executable. The step from the definition of arithmetic by set operations to computer executable operations still requires some effort. We discuss this question in the next section. For details see also [5, 20].

3.1   Executable Ubound Arithmetic

We now consider the question how executable formulas for ubound arithmetic can be obtained. Let $a = \langle a_1, a_2 \rangle, b = \langle b_1, b_2 \rangle \in \mathbb{JR}$. Arithmetic in $\mathbb{JR}$ is defined by

$$a \circ b := \{a \circ b \mid a \in a \wedge b \in b\}, \tag{6}$$

for all $\circ \in \{+, -, \cdot, /\}, 0 \notin b$ in case of division. The function $a \circ b$ is continuous with respect to both variables. The set $a \circ b$ is the range of the function $a \circ b$ over the product set $a \times b$ with or without the boundaries depending on the open-closedness of a and b. Since a and b are intervals of $\mathbb{JR}$ the set $a \times b$ is a simply connected subset of $\overline{\mathbb{R}}^2$, ($\overline{\mathbb{R}} := \mathbb{R} \cup \{-\infty, +\infty\}$). In such a region the range $a \circ b$ of the function $a \circ b$ is also simply connected. Therefore

$$a \circ b = \langle \inf(a \circ b), \sup(a \circ b) \rangle, \tag{7}$$

i.e., for $a, b \in \mathbb{IR}$, $0 \notin b$ in case of division, $a \circ b$ is again an interval of $\mathbb{IR}$.

The angle brackets on the right hand side of (7) depend on the open-closed endpoints of the intervals a and b. The elements $-\infty$ and $+\infty$ can occur as bounds of real intervals. But they are themselves not elements of these intervals.

Neither the set definition (6) of the arithmetic operations $a \circ b$, $\circ \in \{+, -, \cdot, /\}$, nor the form (7) can be executed on the computer. So we have to derive more explicit formulas.

We demonstrate this in case of addition. By (OD1) we obtain $a_1 \leq a$ and $b_1 \leq b \Rightarrow a_1 + b_1 \leq \inf(a+b)$. On the other hand $\inf(a+b) \leq a_1 + b_1$. From both inequalities we obtain by (O3): $\inf(a + b) = a_1 + b_1$. Analogously one obtains $\sup(a + b) = a_2 + b_2$. Thus

$$a + b = \langle \inf(a + b), \sup(a + b) \rangle = \langle a_1 + b_1, a_2 + b_2 \rangle.$$

Similarly by making use of (OD1,2,3,4) for intervals of $\mathbb{IR}$ and the simple sign rules $-(a \cdot b) = (-a) \cdot b = a \cdot (-b), -(a/b) = (-a)/b = a/(-b)$ explicit formulas for all interval operations can be derived, [20].

Actually the infimum and supremum in (7) is taken for operations with the bounds. For bounded intervals $a = \langle a_1, a_2 \rangle$ and $b = \langle b_1, b_2 \rangle \in \mathbb{IR}$ the following formula holds for all operations with $0 \notin b$ in case of division:

$$a \circ b = \langle \min_{i,j=1,2}(a_i \circ b_j), \max_{i,j=1,2}(a_i \circ b_j) \rangle \text{ for } \circ \in \{+, -, \cdot, /\}. \tag{8}$$

Now we get by (RG) for intervals of $\mathbb{IU}$

$$a \diamondsuit b := \diamondsuit(a \circ b) = \langle \nabla \min_{i,j=1,2}(a_i \circ b_j), \triangle \max_{i,j=1,2}(a_i \circ b_j) \rangle$$

and by the monotonicity of the roundings ∇ and $\triangle$:

$$a \diamondsuit b = \langle \min_{i,j=1,2}(a_i \triangledown b_j), \max_{i,j=1,2}(a_i \triangle b_j) \rangle.$$

For bounded and nonempty intervals $a = \langle a_1, a_2 \rangle$ and $b = \langle b_1, b_2 \rangle$ of $\mathbb{IU}$ the unary operation $-a$ and the binary operations addition, subtraction, multiplication, and division are shown in the following tables. For details see [20]. Therein the operator symbols for intervals are denoted by $+, -, \cdot, /$.

Minus operator $-a = \langle -a_2, -a_1 \rangle.$
Addition $\langle a_1, a_2 \rangle + \langle b_1, b_2 \rangle = \langle a_1 \triangledown b_1, a_2 \triangle b_2 \rangle.$
Subtraction $\langle a_1, a_2 \rangle - \langle b_1, b_2 \rangle = \langle a_1 \triangledown b_2, a_2 \triangle b_1 \rangle.$

In real analysis division by zero is not defined. In interval arithmetic, however, the interval in the denominator of a quotient may contain zero. We consider this case also.

Multiplication	$\langle b_1, b_2 \rangle$	$\langle b_1, b_2 \rangle$	$\langle b_1, b_2 \rangle$
$\langle a_1, a_2 \rangle \cdot \langle b_1, b_2 \rangle$	$b_2 \leq 0$	$b_1 < 0 < b_2$	$b_1 \geq 0$
$\langle a_1, a_2 \rangle, a_2 \leq 0$	$\langle a_2 \triangledown b_2, a_1 \triangle b_1 \rangle$	$\langle a_1 \triangledown b_2, a_1 \triangle b_1 \rangle$	$\langle a_1 \triangledown b_2, a_2 \triangle b_1 \rangle$
$a_1 < 0 < a_2$	$\langle a_2 \triangledown b_1, a_1 \triangle b_1 \rangle$	$\langle min(a_1 \triangledown b_2, a_2 \triangledown b_1), max(a_1 \triangle b_1, a_2 \triangle b_2) \rangle$	$\langle a_1 \triangledown b_2, a_2 \triangle b_2 \rangle$
$\langle a_1, a_2 \rangle, a_1 \geq 0$	$\langle a_2 \triangledown b_1, a_1 \triangle b_2 \rangle$	$\langle a_2 \triangledown b_1, a_2 \triangle b_2 \rangle$	$\langle a_1 \triangledown b_1, a_2 \triangle b_2 \rangle$

Division, $0 \notin \boldsymbol{b}$	$\langle b_1, b_2 \rangle$	$\langle b_1, b_2 \rangle$
$\langle a_1, a_2 \rangle / \langle b_1, b_2 \rangle$	$b_2 < 0$	$b_1 > 0$
$\langle a_1, a_2 \rangle, a_2 \leq 0$	$\langle a_2 \triangledown b_1, a_1 \triangle b_2 \rangle$	$\langle a_1 \triangledown b_1, a_2 \triangle b_2 \rangle$
$\langle a_1, a_2 \rangle, a_1 < 0 < a_2$	$\langle a_2 \triangledown b_2, a_1 \triangle b_2 \rangle$	$\langle a_1 \triangledown b_1, a_2 \triangle b_1 \rangle$
$\langle a_1, a_2 \rangle, 0 \leq a_1$	$\langle a_2 \triangledown b_2, a_1 \triangle b_1 \rangle$	$\langle a_1 \triangledown b_2, a_2 \triangle b_1 \rangle$

The general rule for computing the set $\boldsymbol{a}/\boldsymbol{b}$ with $0 \in \boldsymbol{b}$ is to remove its zero from the interval $\boldsymbol{b}$ and perform the division with the remaining set.[8] Whenever zero in $\boldsymbol{b}$ is an endpoint of $\boldsymbol{b}$, the result of the division can be obtained directly from the above table for division with $0 \notin \boldsymbol{b}$ by the limit process $b_1 \to 0$ or $b_2 \to 0$ respectively. The results are shown in the table for division with $0 \in \boldsymbol{b}$. Here, the round brackets stress that the bounds $-\infty$ and $+\infty$ are not elements of the interval.

Division, $0 \in \boldsymbol{b}$	$\boldsymbol{b} =$	$\langle b_1, b_2 \rangle$	$\langle b_1, b_2 \rangle$
$\langle a_1, a_2 \rangle / \langle b_1, b_2 \rangle$	$\langle 0, 0 \rangle$	$b_1 < b_2 = 0$	$0 = b_1 < b_2$
$\langle a_1, a_2 \rangle = \langle 0, 0 \rangle$	$\varnothing$	$\langle 0, 0 \rangle$	$\langle 0, 0 \rangle$
$\langle a_1, a_2 \rangle, a_1 < 0, a_2 \leq 0$	$\varnothing$	$\langle a_2 \triangledown b_1, +\infty)$	$(-\infty, a_2 \triangle b_2)$
$\langle a_1, a_2 \rangle, a_1 < 0 < a_2$	$\varnothing$	$(-\infty, +\infty)$	$(-\infty, +\infty)$
$\langle a_1, a_2 \rangle, 0 \leq a_1, 0 < a_2$	$\varnothing$	$(-\infty, a_1 \triangle b_1)$	$(a_1 \triangledown b_2, +\infty)$

In the case that zero is an interior point of the denominator, two different versions to solve the problem can be offered. One could be to return the entire set of real numbers $(-\infty, +\infty)$. The other one would be to split the interval $\langle b_1, b_2 \rangle$ into the two distinct sets $\langle b_1, 0 \rangle$ and $(0, b_2 \rangle$. Division by these two sets leads to two distinct unbounded real intervals. The results of the two divisions are already shown in the table for division by $0 \in \boldsymbol{b}$. The computer could return the two results as an improper interval where the left hand bound is greater than the right hand bound together with an appropriate information for the user. This second version for division by an interval that contains zero as an interior point has been used to develop the extended interval Newton method which allows computing all zeros of a function in a given interval. For details see [20].

---

[8] This is in full accordance with function evaluation: When evaluating a function over a set, points outside its domain are simply ignored.

Four kinds of *unbounded intervals* come from division by an interval of $\mathbb{JU}$ that contains zero:

$$\varnothing, \quad (-\infty, a\rangle, \quad \langle b, +\infty), \quad \text{and} \quad (-\infty, +\infty). \tag{9}$$

Arithmetic for bounded intervals can easily be extended to these new elements. The first rule is that any operation with the empty set $\varnothing$ returns the empty set as result. By continuity reasons the rules for bounded real intervals can also be executed if a bound becomes $-\infty$ or $+\infty$. Doing this, only rules for computing with $-\infty$ or $+\infty$ are needed which are well established in real analysis. Obscure operations like $\infty - \infty$ or ∞/∞ do not occur. For proof see [20].

Intervals of $\mathbb{JU}$ are connected sets of real numbers. $-\infty$ and $+\infty$ are not elements of these intervals. So multiplication of any such interval by 0 can only have 0 as the result. This very naturally leads to the following rules:

$$(-\infty, a\rangle \cdot 0 = \langle b, +\infty) \cdot 0 = (-\infty, +\infty) \cdot 0 = 0. \tag{10}$$

For intervals of $\mathbb{JU}$ we can now state:

Arithmetic for closed, open, and half-open, bounded or unbounded real intervals of $\mathbb{JU}$ is free of exceptions, i.e., arithmetic operations for intervals of $\mathbb{JU}$ always lead to intervals of $\mathbb{JU}$ again.

This is in sharp contrast to other models of interval arithmetic which consider $-\infty$ and $+\infty$ as elements of unbounded real intervals. In such models obscure arithmetic operations like $\infty - \infty, \infty/\infty, 0 \cdot \infty$ occur which require introduction of unnatural superficial objects like *NaI* (Not an Interval).

High speed by support of hardware and programming languages is vital for all kinds of interval arithmetic to be more widely accepted by the scientific computing community. Right now no commercial processor provides interval arithmetic or unum and ubound arithmetic by hardware. In the author's book *Computer Arithmetic and Validity – Theory, Implementation, and Applications*, second edition 2013 [20] considerable emphasis is put on speeding up interval arithmetic. The book shows that interval arithmetic for diverse spaces can efficiently be provided on the computer if two features are made available by fast hardware:

I. Fast and direct hardware support for double precision interval arithmetic and
II. a fast and exact multiply and accumulate operation or, an exact dot product (EDP).

Realization of I. and II. is discussed at detail in the book [20]. It is shown that I. and II. can be obtained at very little hardware cost. With I. interval arithmetic would be as fast as simple floating-point arithmetic. The simplest and fastest way for computing a dot product is to compute it exactly. To make II. conveniently available a new data format *complete* is used together with a few very restricted arithmetic operations. By pipelining the EDP can be computed in the time the processor needs to read the data, i.e., it comes with utmost speed. I. and II. would boost both the speed of a computation and the accuracy of the result. Fast hardware for I. and II. must be supported by future processors. Computing the dot product exactly even can be considerably faster than computing it conventionally in double or extended precision floating-point arithmetic.

Modern processor architecture is coming very close to what is requested here. See [8], and in particular pp. 1–1 to 1–3 and 2–5 to 2–6. These processors provide register space of 16 K bits. Only about 4 K bits suffice for a *complete register* which allows computing a dot product exactly at extreme speed for the double precision format. We now discuss a frequent application of this.

4   A Sketch of Arithmetic for Matrices with Ubound Components

The axioms for computer arithmetic shown in Sect. 2 also can be applied to define computer arithmetic in higher dimensional spaces like complex numbers, vectors and matrices for real, complex, interval and ubound data, for instance. Here we briefly sketch how arithmetic for matrices with interval and ubound components could be embedded into the axiomatic definition of computer arithmetic outlined in Sect. 2.

Let $\{\overline{\mathbb{R}}, +, \cdot, \leq\}$ be the completely ordered set of real numbers and $\{\mathbb{U}, \leq\}$ the symmetric screen of unums. In the ordered set of $n \times n$ matrices $\{M_n\mathbb{R}, +, \cdot, \leq\}$ we consider intervals $\mathbb{J}M_n\mathbb{R}$ and $\mathbb{J}M_n\mathbb{U}$ where all bounds can be open or closed. Let $\mathbb{P}M_n\mathbb{R}$ denote the power set[9] of $M_n\mathbb{R}$. Then $\mathbb{P}M_n\mathbb{R} \supset \mathbb{J}M_n\mathbb{R} \supset \mathbb{J}M_n\mathbb{U}$. $\mathbb{J}M_n\mathbb{R}$ is an upper[10] screen of $\mathbb{P}M_n\mathbb{R}$ and $\mathbb{J}M_n\mathbb{U}$ is a screen of $\mathbb{J}M_n\mathbb{R}$. We consider the monotone upwardly directed roundings $\square : \mathbb{P}M_n\mathbb{R} \to \mathbb{J}M_n\mathbb{R}$ and $\diamondsuit : \mathbb{J}M_n\mathbb{R} \to \mathbb{J}M_n\mathbb{U}$. They are uniquely defined.

For matrices $\boldsymbol{A}, \boldsymbol{B} \in \mathbb{J}M_n\mathbb{R}$ the set definition of arithmetic operations

$$\boldsymbol{A} \circ \boldsymbol{B} := \{A \circ B \mid A \in \boldsymbol{A} \wedge B \in \boldsymbol{B}\}, \circ \in \{+, \cdot\} \tag{11}$$

does not lead to an interval again. The result is a more general set. It is an element of the power set of matrices. To obtain an interval again the upwardly directed rounding from the power set onto the set of intervals of $\square : \mathbb{P}M_n\mathbb{R} \to \mathbb{J}M_n\mathbb{R}$ has to be applied. With it arithmetic operations for intervals $\boldsymbol{A}, \boldsymbol{B} \in \mathbb{J}M_n\mathbb{R}$ are defined by

(RG) $\boldsymbol{A} \;\boxdot\; \boldsymbol{B} := \square\, (\boldsymbol{A} \circ \boldsymbol{B}), \circ \in \{+, -, \cdot\}.$

As in the case of conventional intervals subtraction can be expressed by negation and addition.

The set $\mathbb{J}M_n\mathbb{U}$ of intervals of computer representable matrices is a screen of $\mathbb{J}M_n\mathbb{R}$. To obtain arithmetic for intervals $\boldsymbol{A}, \boldsymbol{B} \in \mathbb{J}M_n\mathbb{U}$ once more the monotone upwardly directed rounding, now denoted by $\diamondsuit : \mathbb{J}M_n\mathbb{R} \to \mathbb{J}M_n\mathbb{U}$ is applied:

(RG) $\boldsymbol{A} \;\lozenge\; \boldsymbol{B} := \diamondsuit\, (\boldsymbol{A} \;\boxdot\; \boldsymbol{B}), \circ \in \{+, \cdot\}.$

---

[9] The power set of a set M is the set of all subsets of M.
[10] For definition see [20].

This leads to the best possible operations in the interval spaces $JM_n\mathbb{R}$ and $JM_n\mathbb{U}$.

Because of the set definition of the arithmetic operations, however, these best possible operations are not directly executable on a computer. Therefore, we are now going to express them in terms of computer executable formulas. For details see [20].

To do this, we consider the set of $n \times n$ matrices $M_n J\mathbb{R}$. The elements of this set have components that are intervals of $J\mathbb{R}$. With the operations and the order relation $\leq$ of the latter, we define operations $\boxplus$, $\boxdot$, and an order relation $\leq$ in $M_n J\mathbb{R}$ by employing the conventional definition of the operations for matrices. With $\boldsymbol{A} = (\boldsymbol{a}_{ij})$, $\boldsymbol{B} = (\boldsymbol{b}_{ij}) \in M_n J\mathbb{R}$ let be

$$\boldsymbol{A} \boxplus \boldsymbol{B} := (\boldsymbol{a}_{ij} + \boldsymbol{b}_{ij}) \wedge \boldsymbol{A} \boxdot \boldsymbol{B} := \left(\sum_{\nu=1}^{n} \boldsymbol{a}_{i\nu} \cdot \boldsymbol{b}_{\nu j} \right) \wedge \boldsymbol{A} \leq \boldsymbol{B} :\Leftrightarrow \boldsymbol{a}_{ij} \leq \boldsymbol{b}_{ij}, \text{i,j}$$

$$= 1(1)\text{n}.$$

Here $+, \cdot$ are the operations in $J\mathbb{R}$ as defined in (2) and $\sum$ denotes the repeated summation in $J\mathbb{R}$.

Remark 2. The bounds of the components of the product matrix $\boldsymbol{A} \boxdot \boldsymbol{B}$ will be open in the majority of cases. This is a simple consequence of Remark 1. In a bit weaker form this also holds for the addition $\boldsymbol{A} \boxplus \boldsymbol{B}$.

We now define a mapping

$$\chi : M_n J\mathbb{R} \to J M_n \mathbb{R}$$

which for matrices $\boldsymbol{A} = (\boldsymbol{a}_{ij}) \in M_n J\mathbb{R}$ with $\boldsymbol{a}_{ij} = \langle a_{ij}^{(1)}, a_{ij}^{(2)} \rangle \in J\mathbb{R},$[11] $i, j = 1(1)n$, has the property

$$\chi \boldsymbol{A} = \chi(\boldsymbol{a}_{ij}) = \chi(\langle a_{ij}^{(1)}, a_{ij}^{(2)} \rangle) := \langle (a_{ij}^{(1)}), (a_{ij}^{(2)}) \rangle. \tag{12}$$

Obviously χ is a one-to-one mapping of $M_n J\mathbb{R}$ onto $J M_n \mathbb{R}$ and an order isomorphism with respect to $\leq$. It can be shown that χ is also an algebraic isomorphism for the operations addition and multiplication, i.e.,

$$\chi \boldsymbol{A} \;\boxcircle\; \chi \boldsymbol{B} = \chi(\boldsymbol{A} \;\boxcircle\; \boldsymbol{B}), \circ \in \{+, \cdot\}.$$

For the proof in case of closed intervals $\boldsymbol{a}_{ij}, \boldsymbol{b}_{ij} \in I\mathbb{R}$ see [20].

Whenever two structures are isomorphic, corresponding elements can be identified with each other. This allows us to define an inclusion relation even for elements $\boldsymbol{A} = (\boldsymbol{a}_{ij})$, $\boldsymbol{B} = (\boldsymbol{b}_{ij}) \in M_n J\mathbb{R}$ by

$$\boldsymbol{A} \subseteq \boldsymbol{B} :\Leftrightarrow \boldsymbol{a}_{ij} \subseteq \boldsymbol{b}_{ij}, \text{ for all i,j=1(1)n.}$$

and

$$(a_{ij}) \in \boldsymbol{A} = (A_{ij}) :\Leftrightarrow a_{ij} \in A_{ij}, \text{ for all } i, j = 1(1)n.$$

---

[11] The angle brackets $\langle$ and $\rangle$ here denote the interval bounds. Each one of them can be open or closed.

This convenient definition allows for the interpretation that a matrix $\boldsymbol{A} = (\boldsymbol{a}_{ij}) \in M_n \mathbb{JR}$ also represents a set of matrices as demonstrated by the following identity:

$$\boldsymbol{A} = (A_{ij}) \equiv \{(a_{ij}) \mid a_{ij} \in A_{ij}, \text{ i,j=1(1)n}\}.$$

[12]Both matrices contain the same elements.

With the monotone upwardly directed rounding $\Diamond : \mathbb{JR} \to \mathbb{JU}$ a rounding $\Diamond : M_n \mathbb{JR} \to M_n \mathbb{JU}$ and operations in $M_n \mathbb{JU}$ can now be defined by

$$\Diamond \boldsymbol{A} := (\Diamond \boldsymbol{a}_{ij}),$$

$$\boldsymbol{A} \diamondsuit \boldsymbol{B} := \Diamond (\boldsymbol{A} \boxdot \boldsymbol{B}), \circ \in \{+, \cdot\}.$$

Now it can be shown (for the proof in case of closed intervals see [20]) that the mapping χ establishes an isomorphism

$$\chi \boldsymbol{A} \diamondsuit \chi \boldsymbol{B} = \chi (\boldsymbol{A} \diamondsuit \boldsymbol{B}), \circ \in \{+, \cdot\},$$

i.e., the structures $\{M_n \mathbb{JU}, \diamondsuit, \diamondsuit, \leq, \subseteq\}$ and $\{\mathbb{J}M_n \mathbb{U}, \diamondsuit, \diamondsuit, \leq, \subseteq\}$ can be identified with each other.

This isomorphism reduces the optimal, best possible but not computer executable operations in $\mathbb{J}M_n \mathbb{U}$, to the operations in $M_n \mathbb{JU}$. We analyze these operations more closely.

For matrices $\boldsymbol{A} = (\boldsymbol{a}_{ij})$, $\boldsymbol{B} = (\boldsymbol{b}_{ij}) \in M_n \mathbb{JU}$, $\boldsymbol{a}_{ij}, \boldsymbol{b}_{ij} \in \mathbb{JU}$ arithmetic operations are defined by

$$\boldsymbol{A} \diamondsuit \boldsymbol{B} := \Diamond (\boldsymbol{A} \boxplus \boldsymbol{B}) \quad \wedge \quad \boldsymbol{A} \diamondsuit \boldsymbol{B} := \Diamond (\boldsymbol{A} \boxdot \boldsymbol{B})$$

with the rounding $\Diamond \boldsymbol{A} := (\Diamond \boldsymbol{a}_{ij})$. This leads to the following formulas for the operations in $M_n \mathbb{JU}$:

$$\boldsymbol{A} \diamondsuit \boldsymbol{B} = (\Diamond (\boldsymbol{a}_{ij} + \boldsymbol{b}_{ij})) = (\boldsymbol{a}_{ij} \diamondsuit \boldsymbol{b}_{ij}), \tag{13}$$

$$\boldsymbol{A} \diamondsuit \boldsymbol{B} = \Diamond (\boldsymbol{A} \boxdot \boldsymbol{B}) = \left(\Diamond \sum_{\nu=1}^{n} (\boldsymbol{a}_{i\nu} \cdot \boldsymbol{b}_{\nu j}) \right). \tag{14}$$

These operations are executable on a computer. The componentwise addition in (13) can be performed by means of the addition in $\mathbb{JU}$. The multiplications in (14) are to be executed using the multiplication in $\mathbb{JR}$. Then the lower bounds and the upper bounds are to be added in $\mathbb{R}$. Finally the rounding $\Diamond : \mathbb{JR} \to \mathbb{JU}$ has to be executed.

With $\boldsymbol{a}_{ij} = \langle a_{ij}^1, a_{ij}^2 \rangle$, $\boldsymbol{b}_{ij} = \langle b_{ij}^1, b_{ij}^2 \rangle \in \mathbb{JU}$, (14) can be written in a more explicit form:

$$\boldsymbol{A} \diamondsuit \boldsymbol{B} = \left(\langle \nabla \sum_{\nu=1}^{n} \min_{r,s=1,2} (a_{i\nu}^r b_{\nu j}^s), \ \triangle \sum_{\nu=1}^{n} \max_{r,s=1,2} (a_{i\nu}^r b_{\nu j}^s) \rangle \right). \tag{15}$$

---

[12] The round brackets here denote the matrix braces.

Here the products $a_{i\nu}^r b_{\nu j}^s$ are elements of $\mathbb{R}$ (and in general not of $\mathbb{U}$). **The summands (products of double length) are to be correctly accumulated in $\mathbb{R}$ by the exact scalar product.** Finally the sum of products is rounded only once by $\triangledown$ resp. $\triangle$ from $\mathbb{R}$ onto $\mathbb{U}$. The angle brackets in (15) denote the interval bounds. Each one of them can be open or closed. The large round brackets denote the matrix braces. In the vast majority of cases the angle brackets will be open. Only in the very rare case that a sum before rounding is an exact unum the angle bracket is closed.

5   Short Term Progress

Compared with conventional interval arithmetic *The End of Error* [5] means a huge step ahead. For being more energy efficient and other reasons it controls the word size of the interval bounds in dependence of intermediate results and keeps it as small as possible. To avoid mathematical shortcomings it extends the basic set from closed real intervals to connected sets of real numbers. All this are laudable and most natural goals. The entire step, however, may be too big to get realized on computers that can be bought on the market in the near future.

So it may be reasonable to look for a smaller step which might have a more realistic chance. As such the introduction of the ubit into the floating-point bounds at the cost of shrinking the excessive exponent sizes of the IEEE 754 floating-point formats by one bit would already be a great step ahead. It would allow an extension of conventional interval arithmetic to closed, open, half-open, bounded, and unbounded sets of real numbers. By the way it would reduce the register memory for computing the dot product exactly in case of the double precision format, for instance, from excessive 4000 to only about 2000 bit. As side effect the exact dot product brings speed and associativity for addition.

Acknowledgement. The author owes thanks to Goetz Alefeld and Gerd Bohlender as well as to two unknown referees for useful comments on the paper. He gratefully acknowledges e-mail exchange with John Gustafson on the contents of the paper. Gerd Bohlender presented the paper at PPAM 2015.

References

1. Alefeld, G., Herzberger, J.: Einführung in die Intervallrechnung, Informatik 12. Bibliographisches Institut, Mannheim Wien Zürich (1974)
2. Alefeld, G., Herzberger, J.: Introduction to Interval Computations. Academic Press, New York (1983)
3. Baumhof, C., A new VLSI vector arithmetic coprocessor for the PC. In: Knowles, S., McAllister, W.H. (eds.) Proceedings of 12th Symposium on Computer Arithmetic ARITH, Bath, England, 19–21 July 1995, pp. 210–215. IEEE Computer Society Press, Piscataway (1995)
4. De Beauclair, W.: Rechnen mit Maschinen. Vieweg, Braunschweig (1968)
5. Gustafson, J.L.: The End of Error. CRC Press, Taylor and Francis Group, A Chapman and Hall Book, Boca Raton (2015)

6. Hansen, E.R.: Topics in Interval Analysis. Clarendon Press, Oxford (1969)
7. Hansen, E.R.: Global Optimization Using Interval Analysis. Marcel Dekker Inc., New York (1992)
8. INTEL: Intel Architecture Instruction Set Extensions Progamming Reference, 319433-017, December 2013. http://software.intel.com/en-us/file/319433-017pdf
9. Klatte, R., Kulisch, U., Neaga, M., Ratz, D., Ullrich, C.: PASCAL-XSC: Sprachbeschreibung mit Beispielen. Springer, Heidelberg (1991). http://www2.math.uni-wuppertal.de/xsc/, http://www.xsc.de/
10. Klatte, R., Kulisch, U., Neaga, M., Ratz, D., Ullrich, C.: PASCAL-XSC - Language Reference with Examples. Springer, Heidelberg (1992). http://www2.math.uni-wuppertal.de/xsc/, http://www.xsc.de/. Russian translation MIR, Moscow, 1995, third edition 2006. http://www2.math.uni-wuppertal.de/xsc/, http://www.xsc.de/
11. Hammer, R., Hocks, M., Kulisch, U., Ratz, D.: Numerical Toolbox for Verified Computing I: Basic Numerical Problems (PASCAL-XSC). Springer, Heidelberg (1993). Russian translation MIR, Moskow (2005)
12. Klatte, R., Kulisch, U., Lawo, C., Rauch, M., Wiethoff, A.: C-XSC: A C++ Class Library for ExtendedScientific Computing. Springer, Heidelberg (1993). http://www2.math.uni-wuppertal.de/xsc/, http://www.xsc.de/
13. Hammer, R., Hocks, M., Kulisch, U., Ratz, D.: C++ Toolbox for Verified Computing: Basic Numerical Problems. Springer, Heidelberg (1995)
14. Kulisch, U.: An axiomatic approach to rounded computations. TS Report No. 1020, Mathematics Research Center, University of Wisconsin, Madison, Wisconsin (1969) and Numer. Math. **19** , 1–17 (1971)
15. Kulisch, U.: Implementation and Formalization of Floating-Point Arithmetics, IBM T. J. Watson-Research Center, Report Nr. RC, pp. 1–50 (1973). Invited talk at the Caratheodory Symposium, Athens, September 1973. In: The Greek Mathematical Society, C. Caratheodory Symposium, pp. 328–369 (1973). In: Computing **14**(4608), 323–348 (1975)
16. Kulisch, U.: Grundlagen des Numerischen Rechnens - Mathematische Begründung der Rechnerarithmetik. Bibliographisches Institut, Mannheim Wien Zürich (1976). ISBN 3-411-01517-9
17. Kulisch, U., Teufel, T., Hoefflinger, B.: Genauer und trotzdem schneller: Ein neuer Coprozessor für hochgenaue Matrix-und Vektoroperationen. Titelgeschichte. Elektronik **26**, 52–56 (1994)
18. Kulisch, U.W.: Complete interval arithmetic and its implementation on the computer. In: Cuyt, A., Krämer, W., Luther, W., Markstein, P. (eds.) Numerical Validation in Current Hardware Architectures. LNCS, vol. 5492, pp. 7–26. Springer, Heidelberg (2009)
19. Kulisch, U.: An axiomatic approach to computer arithmetic with an appendix on interval hardware. In: Wyrzykowski, R., Dongarra, J., Karczewski, K., Waśniewski, J. (eds.) PPAM 2011, Part II. LNCS, vol. 7204, pp. 484–495. Springer, Heidelberg (2012)
20. Kulisch, U.: Computer Arithmetic and Validity: Theory, Implementation, and Applications. de Gruyter, Berlin (2008)
21. Kulisch, U., Snyder, V.: The exact dot product. Prepared for and sent to IEEE P1788 (2009, to be published in TOMS)
22. Kulisch, U. (ed.): PASCAL-XSC: A PASCAL Extension for Scientific Computation, Information Manual and Floppy Disks. B. G. Teubner, Stuttgart (1987)
23. Kulisch, U.: Mathematics and speed for interval arithmetic - a complement to IEEE P1788. Prepared for and sent to IEEE P1788 (2014, to be published in TOMS)

24. Moore, R.E.: Interval Analysis. Prentice Hall Inc., Englewood Cliffs (1966)
25. Pryce, J.D. (ed.): P1788, IEEE Standard for Interval Arithmetic. http://grouper. ieee.org/groups/1788/email/pdfOWdtH2mOd9.pdf
26. Rojas, R., Rechenmaschinen, K.Z.: sechzig Jahre Computergeschichte. In: Spektrum der Wissenschaft, pp. 54–62. Spektrum Verlag, Heidelberg (1997)
27. Sun Microsystems, Interval Arithmetic Programming Reference, Fortran 95. Sun Microsystems Inc., Palo Alto (2000)
28. Oishi, S., Tanabe, K., Ogita, T., Rump, S.M.: Convergence of Rump's method for inverting arbitrarily ill-conditioned matrices. J. Comput. Appl. Math. **205**, 533–544 (2007)
29. Rump, S.M.: Kleine Fehlerschranken bei Matrixproblemen. Dissertation, Universität Karlsruhe (1980)
30. Rump, S.M.: How reliable are results of computers? Jahrbuch berblicke Mathematik (1983)
31. Rump, S.M.: Solving algebraic problems with high accuracy. In: Kulisch, U., Miranker, W.L. (eds.) Proceedings of the Symposium on New Approach to Scientific Computation, IBM Research Center, Yorktown Heights, NY, pp. 51–120. Academic Press, New York (1983)
32. IBM: IBM System/370 RPQ. High Accuracy Arithmetic, SA 22–7093-0, IBM Deutschland GmbH (Department 3282, Böblingen) (1984)
33. IBM: IBM High-Accuracy Arithmetic Subroutine Library (ACRITH). IBM Deutschland GmbH (Department 3282, Böblingen) (1983), third edition (1986). 1. General Information Manual, GC 33-6163-02. 2. Program Description and User's Guide, SC 33-6164-02. 3. Reference Summary, GX 33-9009-02
34. IBM: ACRITH–XSC: IBM High Accuracy Arithmetic– Extended Scientific Computation. Version 1, Release 1. IBM Deutschland GmbH (Department 3282, Böblingen) (1990). 1. General Information, GC33-6461-01. 2. Reference, SC33-6462-00.3. Sample Programs, SC33-6463-00. 4. How To Use, SC33-6464-00. 5. Syntax Diagrams, SC33-6466-00

Optimizing Cloud Use Under Interval Uncertainty

Vladik Kreinovich[(✉)] and Esthela Gallardo

Department of Computer Science, University of Texas at El Paso,
El Paso, TX 79968, USA
vladik@utep.edu, egallardo5@miners.utep.edu

Abstract. One of the main advantages of cloud computing is that it helps the users to save money: instead of buying a lot of computers to cover all their computations, the user can rent the computation time on the cloud to cover the rare peak spikes of computer need. From this viewpoint, it is important to find the optimal division between in-house and in-the-cloud computations. In this paper, we solve this optimization problem, both in the idealized case when we know the complete information about the costs and the user's need, and in a more realistic situation, when we only know interval bounds on the corresponding quantities.

Keywords: Cloud computing · Interval uncertainty

1 Formulation of the Problem

What is Cloud Computing. According to the official definition provided by the US National Institute of Standards and Technology (NIST), "Cloud computing is a model for enabling ubiquitous, convenient, on-demand network access to a shared pool of configurable computing resources (e.g., networks, servers, storage, applications, and services) that can be rapidly provisioned and released with minimal management effort or service provider interaction" [20]. There are many other definitions which concentrate on different aspects of cloud computing; see, e.g., [8,17,23,28].

One of the important aspects of cloud computing is that instead of performing all the computations on his/her own computer, a user can sometimes rent computing time from a computer-time-rental company.

How Much Computation Time Should We Rent? Renting is usually more expensive than buying and maintaining one's own computer, so if the user needs the same amount of computations day after day, cloud computing is not a good financial option. However, if a peak need for computing occurs rarely, it is often cheaper to rent the corresponding computation time than to buy a lot of computing power and idle it most of the time.

Once the user knows his/her computational requirements, the proper question is: should we use the cloud at all? if yes, how much computing power should

© Springer International Publishing Switzerland 2016
R. Wyrzykowski et al. (Eds.): PPAM 2015, Part II, LNCS 9574, pp. 435–444, 2016.
DOI: 10.1007/978-3-319-32152-3_40

we buy for in-house computations and how much computation time should we rent from the cloud company? how much will it cost?

Finally, if a cloud company offers a multi-year deal with fixed rates, should we take it or should we buy computation time on a year-by-year basis?

Why This Is Important. Surprisingly, while the main purpose of cloud computing is to save user's money, most cloud users are computer folks with little knowledge of economics. As a result, often, they make wrong financial decisions about the cloud use; see, e.g., [29]. It is important to come up with proper recommendations for using cloud computing.

What We Do in This Paper. In this paper, we provide the desired financial recommendations, first under the idealized assumption that we have a complete information, and then, in a more realistic situation of interval uncertainty.

Comment. It is worth mentioning that in this paper, we only consider the financial aspects of cloud computing, i.e., the idea that we rent computing time. In this analysis, we do not take into account "minimal management efforts" aspects of cloud computing – e.g., the fact that the system automatically takes care of allocating resources. Because we only use the financial aspects of cloud computing as renting computing time, our recommendations are applicable not only to cloud computing per se, but to any situation when a user can buy computing time – e.g., to renting computing time on mainframe computers.

2   How Much Computations to Perform In-House and How Much In Cloud: Case of Complete Information

Main Idea. The overall computation costs can be decomposed into fixed costs (buying computer(s)) and variable costs (maintaining computers). When we use a cloud, there is no fixed cost (since we do not need to buy a computer), but the variable cost is much higher. This is the main idea behind our computations.

Case of Complete Information: Description. Let us first consider the idealized case when we have complete information about our needs and about all the costs.

This means, first, that we know the cost of keeping a certain level of computational ability in-house. Let us pick some time quantum (e.g., day or hour). Then, the overall cost (fixed + variable) of buying and maintaining the corresponding computers is proportional to these computer's computational ability – i.e., the number of computing operations (e.g., Teraflops) that these computers can perform in this time unit. Let c_0 denote the cost per unit of computations. Then, if we buy computers with computational ability x_0, we pay $c_0 \cdot x_0$ for these computers.

This also means that we know the (variable) cost of computing in the cloud. Let us denote this cost by c_1. So, if one day, we need to perform x computations in the cloud, we have to pay the amount $c_1 \cdot x$.

As we have mentioned, computing in the cloud is usually more expensive than computing in-house. Part of this extra cost is the cost of moving data, another part is the overhead to support the computing staff, marketing staff, etc. As a result, $c_1 > c_0$.

Complete knowledge also means that we know the user's needs. This means that for each possible computation need x, we know the probability that one of the days, we will need to perform exactly x computations. These probabilities can be estimated by analyzing the previous needs: if we needed x computations in 10 % of the days, this means that the probability of needing x computations is exactly 10 %.

The probability distribution is usually described either by a cumulative distribution function (cdf) $F(x) = \text{Prob}(X \leq x)$, or by a probability density function (pdf) $\rho(x)$ for which the probability to be within an interval $[\underline{x}, \overline{x}]$ is equal to the integral $\int_{\underline{x}}^{\overline{x}} \rho(x)\, dx$, and the overall probability is 1: $\int_0^\infty \rho(x)\, dx = 1$.

The relationship between pdf and cdf is straightforward:

- $F(x)$ is the integral of pdf: $F(x) = \int_0^x \rho(t)\, dt$;
- vice versa, the pdf is the derivative of the cdf: $\rho(x) = \dfrac{dF}{dx}$.

What is the Cost of Buying x_0 Computational Abilities and Doing All Other Computations in the Cloud? We want to select the amount x_0 of computing power to buy, so that everything in excess of x_0 will be sent to the cloud. We want to select this amount so that the expected overall cost of computations is the smallest possible.

So, to find the corresponding value x_0, let us compute how much it will cost the user to buy x_0 equipment and to rent all other computation time. We already know that the cost of buying and maintaining an equipment with capacity x_0 is equal to $c_0 \cdot x_0$.

The expected cost of using the cloud can be obtained by adding the costs multiplied by the corresponding probabilities. We need computations in the cloud when $x > x_0$, For each such value x, we need to rent the amount $x - x_0$ in the cloud. The cost of such renting is $c_1 \cdot (x - x_0)$. The probability of needing exactly x computations is proportional to $\rho(x)$. To be more precise, the probability that we need between x and $x + \Delta x$ computations is equal to $\rho(x) \cdot \Delta x$; thus, the expected cost of using the cloud is therefore equal to the sum of the products $(c_1 \cdot (x - x_0)) \cdot (\rho(x) \cdot \Delta x)$, i.e., to the value $\sum c_1 \cdot (x - x_0) \cdot \rho(x) \cdot \Delta x$. In the limit, when $\Delta x \to 0$, this sum tends to the integral $\int_{x_0}^\infty c_1 \cdot (x - x_0) \cdot \rho(x)\, dx$. Thus, the overall cost is equal to the sum of the in-house and in-the-cloud costs:

$$C(x_0) = c_0 \cdot x_0 + c_1 \cdot \int_{x_0}^\infty (x - x_0) \cdot \rho(x)\, dx. \tag{1}$$

Let Us Use This Cost Expression to Find the Optimal Value x_0. We want to find the value x_0 for which the cost expression (1) attains its smallest possible value. To find this minimizing value, we need to differentiate the expression (1) with respect to x_0 and equate the corresponding derivative to 0.

To make this differentiation easier, let us transform the expression (1) by using integration by parts $\int_a^b u\, dv = u \cdot v|_a^b - \int_a^b v\, du$. Here, $\rho(x) = \dfrac{d(F(x) - 1)}{dx}$, so we can take $u = x - x_0$ and $v = F(x) - 1$. The product $u \cdot v = (x - x_0) \cdot (F(x) - 1)$ is equal to 0 on both endpoints $x = x_0$ and $x = \infty$, so we get

$$C(x_0) = c_0 \cdot x_0 - c_1 \cdot \int_{x_0}^{\infty} (F(x) - 1)\, dx.$$

Since $F(x) \leq 1$, it is convenient to swap the signs and get the expression

$$C(x_0) = c_0 \cdot x_0 + c_1 \cdot \int_{x_0}^{\infty} (1 - F(x))\, dx. \tag{2}$$

The derivative of this sum is equal to the sum of the derivatives. The derivative of the second term can be obtained from the fact that the derivative of the integral is equal to the integrated function. Thus, the equation $\dfrac{dC(x_0)}{dx_0} = 0$ becomes $c_0 - c_1 \cdot (1 - F(x_0)) = 0$, i.e., equivalently,

$$F(x_0) = 1 - \frac{c_0}{c_1}. \tag{3}$$

This formula can be simplified even further if we take into account that for each $p \in [0, 1]$, the value x for which $F(x) = p$ is known as the p-th *quantile*. For example, for $p = 0.5$, we have the *median*, for $p = 0.25$ and $p = 0.75$, we have *quartiles*, for $p = 0.1, 0.2, \ldots, 0.9$ we have *deciles*, etc.

So, we arrive at the following conclusion.

How Many Computations to Perform In-House: Optimal Solution. If we know the costs c_0 and c_1 per computation in house and in the cloud, and we also know the probability distribution $F(x)$ describing the user's needs, then the optimal amount x_0 of computational power to buy is determined by the formula (3), i.e., x_0 is a quantile corresponding to $p = 1 - \dfrac{c_0}{c_1}$.

Once we know the optimal value x_0, we can then compute the corresponding cost by using the formula (2).

Discussion. In the extreme case when $c_1 = c_0$, there is no sense to buy anything at all: we can perform all the computations in the cloud. As the cloud costs c_1 increases, the threshold x_0 increases, so when c_1 is very high, it does not make sense to use the cloud at all.

Example. The user's need is usually described by the *power law* distribution, in which, for some threshold t, we have:

- $1 - F(x) = 1$ for $x \leq t$ and then
- $1 - F(x) = \left(\dfrac{x}{t}\right)^{-\alpha}$ for some $\alpha > 0$.

Power law is ubiquitous in many financial situations, see, e.g., [1–4, 9, 10, 14–16, 18, 19, 21, 22, 24–27].

In this case, the formula (3) takes the form

$$\left(\frac{x_0}{t}\right)^{-\alpha} = \frac{c_0}{c_1}.$$

By raising both sides by the power $-1/\alpha$ and multiplying both sides by the threshold t, we conclude that

$$x_0 = t \cdot \left(\frac{c_0}{c_1}\right)^{-1/\alpha} = t \cdot \left(\frac{c_1}{c_0}\right)^{1/\alpha}. \tag{4}$$

Substituting this expression into the formula (2), we can compute the expected cost. This cost consists of two parts: $c_0 \cdot x_0$ and the integral; we will denote the integral part by I. Let us compute both parts and then add them up. Here,

$$c_0 \cdot x_0 = c_0 \cdot t \cdot \left(\frac{c_1}{c_0}\right)^{1/\alpha} = t \cdot c_0^{1-1/\alpha} \cdot c_1^{1/\alpha} \tag{5}$$

Since $1 - F(x) = t^\alpha \cdot x^{-\alpha}$, the integral I takes the form

$$I = \int_{x_0}^{\infty} (1 - F(x)) \, dx = c_1 \cdot t^\alpha \cdot \int_{x_0}^{\infty} x^{-\alpha} \, dx = c_1 \cdot t^\alpha \cdot \frac{x_0^{1-\alpha}}{\alpha - 1}.$$

Substituting the value (4) into this formula, we get

$$I = c_1 \cdot t^\alpha \cdot t^{1-\alpha} \cdot \left(\frac{c_1}{c_0}\right)^{(1-\alpha)/\alpha} \cdot \frac{1}{\alpha - 1},$$

i.e., to

$$I = t \cdot c_0^{1-1/\alpha} \cdot c_1^{1/\alpha} \cdot \frac{1}{\alpha - 1}. \tag{6}$$

By comparing (6) and (4), we can see that $I = c_0 \cdot x_0 \cdot \dfrac{1}{\alpha - 1}$, thus

$$C(x_0) = c_0 \cdot x_0 + I = c_0 \cdot x_0 \cdot \left(1 + \frac{1}{\alpha - 1}\right) = c_0 \cdot x_0 \cdot \frac{\alpha}{\alpha - 1}.$$

Dividing both the numerator and the denominator of this fraction by α, we get the final formula for the cost:

$$C(x_0) = c_0 \cdot x_0 \cdot \frac{1}{1 - \dfrac{1}{\alpha}}. \tag{7}$$

Discussion. The difference between the overall cost (7) and the in-house cost $c_0 \cdot x_0$ is the expected cost of using the cloud.

The larger α, the faster the probabilities of the need for computing power x decrease with x, and thus, the smaller should be the expected cost of using the cloud. And indeed, when α increases, the factor in (7) tends to 1, meaning that the cost of in-the-cloud computations tends to 0.

3   How Much Computations to Perform In-House and How Much In Cloud: Case of Interval Uncertainty

Formulation of the Problem. In the previous section, we considered the idealized case when we know the exact costs c_0 and c_1 and the exact probabilities $F(x)$. In practice, we rarely know the exact costs and probabilities. At best, we know the bounds on these quantities, i.e.:

- we know the interval $[\underline{c}_0, \overline{c}_0]$ of possible values of in-house cost c_0;
- we know the interval $[\underline{c}_1, \overline{c}_1]$ of possible values of the in-the cloud cost c_1; and
- for each computation amount x, we know the interval $[\underline{F}(x), \overline{F}(x)]$ of possible values of the cdf $F(x)$; these bounds are also known as a *p-box*; see, e.g., [5–7].

How to Select x_0 in Case of Interval Uncertainty: Analysis of the Problem. For any selection of the value x_0, different values $c_0 \in [\underline{c}_0, \overline{c}_0]$ and $c_1 \in [\underline{c}_1, \overline{c}_1]$, and for different functions $F(x) \in [\underline{F}(x), \overline{F}(x)]$, the formula (2) leads to different values of the cost $C(x_0)$.

We do not know the probabilities of different values c_i or different functions $F(x)$, all we know is the bounds. In this case, the only information that we have about the cost $C(x_0)$ corresponding to a selection x_0 is that this cost belongs to the interval $[\underline{C}(x_0), \overline{C}(x_0)]$, where:

- the value $\underline{C}(x_0)$ is the smallest possible value of the cost, and
- $\overline{C}(x_0)$ is the largest possible value of the cost.

In such case of interval uncertainty, natural requirements leads to the following decision making procedure [11–13]:

- we select a parameter $\alpha_H \in [0, 1]$ that describes the user's degree of optimism-pessimism, and
- we select the alternative x_0 for which the combination

$$\alpha_H \cdot \underline{C}(x_0) + (1 - \alpha_H) \cdot \overline{C}(x_0)$$

is the smallest possible.

Here:

- the value $\alpha_H = 1$ (corresponding to full optimism) means that we only consider the best-case (optimistic) scenarios;
- the value $\alpha_H = 0$ (corresponding to full pessimism) means that we only consider the worst-case (pessimistic) scenarios;
- values α_H between 0 and 1 means that we take both best-case and worst-case scenarios into account.

For the formula (2), it is easy to find the smallest and the largest value of $C(x_0)$: from the formula (2), we get

$$\underline{C}(x_0) = \underline{c}_0 \cdot x_0 + \underline{c}_1 \cdot \int_{x_0}^{\infty} (1 - \overline{F}(x)) \, dx. \tag{8}$$

and

$$\overline{C}(x_0) = \overline{c}_0 \cdot x_0 + \overline{c}_1 \cdot \int_{x_0}^{\infty} (1 - \underline{F}(x)) \, dx. \tag{9}$$

Thus, the above procedure means that we need to optimize the function

$$C_H(x_0) = c_{0,H} \cdot x_0 + c_{1,H} \cdot \int_{x_0}^{\infty} (1 - F_H(x)) \, dx, \tag{10}$$

where we denoted

$$c_{0,H} = \alpha_H \cdot \underline{c}_0 + (1 - \alpha_H) \cdot \overline{c}_0; \tag{11}$$

$$c_{1,H} = \alpha_H \cdot \underline{c}_1 + (1 - \alpha_H) \cdot \overline{c}_1; \tag{12}$$

$$F_H(x) = \alpha_H \cdot \overline{F}(x) + (1 - \alpha_H) \cdot \underline{F}(x). \tag{13}$$

Differentiating the expression (10) with respect to x_0 and equating the derivative to 0, we conclude that $\dot{c}_{0,H} = c_{1,H} \cdot (1 - F_H(x_0))$, i.e., that

$$F_H(x_0) = 1 - \frac{c_{0,H}}{c_{1,H}}. \tag{14}$$

Resulting Recommendation. To find the optimal value x_0:

- we should first find the parameter α_H corresponding to the user's optimism-pessimism level;
- then, we compute the values $c_{0,H} = \alpha_H \cdot \underline{c}_0 + (1 - \alpha_H) \cdot \overline{c}_0$, $c_{1,H} = \alpha_H \cdot \underline{c}_1 + (1 - \alpha_H) \cdot \overline{c}_1$, and the function $F_H(x) = \alpha_H \cdot \overline{F}(x) + (1 - \alpha_H) \cdot \underline{F}(x)$;
- after that, we find the value x_0 for which $F_H(x_0) = 1 - \frac{c_{0,H}}{c_{1,H}}$.

Once we find the optimal value x_0, we can use the formulas (8) and (9) to find the range of possible values of costs.

4   Auxiliary Question: When Is It Beneficial to Sign A Multi-Year Contract?

Formulation of the Problem. Let us assume that we have an average yearly amount X of computations to perform in the cloud, and we expect the same amount for the few following years. For this year's computations, the cloud company offers us the rate of c_1 per computation; for a T-year contract, the price will be $c_T < c_1$. Shall we sign a contract?

Additional Information that We Need to Make a Decision. To decide which is more beneficial, we need to take into account two things:

- first, computers improve year after year, so the computing cost steadily decreases; let $v < 1$ be a yearly decrease in cost; this means that next year, computing in the cloud will cost $v \cdot c_1$ per computation, the year after that $v^2 \cdot c_1$, etc.;

- we also need to take into account that paying a certain amount a next year is less painful that paying the same amount a this year, since we could invest a, get interest, pay a next year, and keep the interest; from this viewpoint, paying a certain amount a next year is equivalent to paying $a \cdot q$ this year, where the discounting parameter $q < 1$ depends on the current interest rate.

Analysis of the Problem. In the case of year-by-year payments:

- we pay the amount $c_1 \cdot X$ this year,
- we pay the amount $v \cdot c_1 \cdot X$ next year,
- we pay the amount $v^2 \cdot c_1 \cdot X$ the year after,
- ..., and
- we pay the amount $v^{T-1} \cdot c_1 \cdot X$ in the last (T-th) year.

By using discounting, we find out that:

- paying $v \cdot c_1 \cdot X$ next year is equivalent to paying $q \cdot v \cdot c_1 \cdot X$ this year;
- paying $v^2 \cdot c_1 \cdot X$ in Year 3 is equivalent to paying $q^2 \cdot v^2 \cdot c_1 \cdot X$ this year;
- ..., and
- paying $v^{T-1} \cdot c_1 \cdot X$ in Year T is equivalent to paying $q^{T-1} \cdot v^{T-1} \cdot c_1 \cdot X$ this year.

Thus, year-by-year payments are equivalent to paying the following amount right away:

$$c_1 \cdot X + v \cdot q \cdot c_1 \cdot X + v^2 \cdot q^2 \cdot c_1 \cdot X + \ldots + v^{T-1} \cdot q^{T-1} \cdot c_1 \cdot X$$
$$= c_1 \cdot X \cdot (1 + q \cdot v + q^2 \cdot v^2 + \ldots + q^{T-1} \cdot v^{T-1}).$$

By using the formula for the sum of the geometric progression, we conclude that this cost is equal to

$$c_1 \cdot X \cdot \frac{1 - (q \cdot v)^T}{1 - q \cdot v}.$$

Alternatively, if we sign a contract, then we pay the same amount $c_T \cdot X$ every year. By using discounting, we find out that:

- paying $c_T \cdot X$ next year is equivalent to paying $q \cdot c_T \cdot X$ this year;
- paying $c_T \cdot X$ in Year 3 is equivalent to paying $q^2 \cdot c_T \cdot X$ this year;
- ..., and
- paying $c_T \cdot X$ in Year T is equivalent to paying $q^{T-1} \cdot c_T \cdot X$ this year.

Thus, these payments are equivalent to paying the following amount right away:

$$c_T \cdot X + q \cdot c_T \cdot X + q^2 \cdot c_T \cdot X + \ldots + q^{T-1} \cdot c_T \cdot X$$
$$= c_T \cdot X \cdot (1 + q + q^2 + \ldots + q^{T-1}).$$

By using the formula for the sum of the geometric progression, we conclude that this cost is equal to

$$c_T \cdot X \cdot \frac{1 - q^T}{1 - q}.$$

By comparing these two numbers, and dividing both sides of the resulting inequality by the common factor X, we arrive at the following conclusion.

When It Is Beneficial to Sign a Multi-Year Contract: Recommendation. It is beneficial to sign a multi-year contract if

$$c_T \cdot \frac{1 - q^T}{1 - q} \leq c_1 \cdot \frac{1 - (q \cdot v)^T}{1 - q \cdot v}.$$

Acknowledgments. This work was supported in part by the National Science Foundation grants HRD-0734825 and HRD-1242122 (Cyber-ShARE Center of Excellence) and DUE-0926721.

The authors are thankful to the anonymous referees for valuable suggestions.

References

1. Beirlant, J., Goegevuer, Y., Teugels, J., Segers, J.: Statistics of Extremes: Theory and Applications. Wiley, Chichester (2004)
2. Chakrabarti, B.K., Chakraborti, A., Chatterjee, A.: Econophysics and Sociophysics: Trends and Perspectives. Wiley, Berlin (2006)
3. Chatterjee, A., Yarlagadda, S., Chakrabarti, B.K.: Econophysics of Wealth Distributions. Springer, Milan (2005)
4. Farmer, J.D., Lux, T.: Applications of statistical physics in economics and finance. Spec. Issue J. Econ. Dyn. Control. 32(1), 1–320 (2008)
5. Ferson, S.: Risk Assessment with Uncertainty Numbers: RiskCalc. CRC Press, Boca Raton (2002)
6. Ferson, S., Kreinovich, V., Ginzburg, L., Myers, D.S., Sentz, K.: Constructing Probability Boxes and DempsterShafer Structures, Sandia National Laboratories, Albuquerque, New Mexico, Report SAND2002-4015 (2013)
7. Ferson, S., Kreinovich, V., Oberkampf, W., Ginzburg, L.: Experimental Uncertainty Estimation and Statistics for Data Having Interval Uncertainty, Sandia National Laboratories, Report SAND-0939 (2007)
8. Furht, B., Escalanate, A.: Handbook of Cloud Computing. Springer, New York (2010)
9. Gabaix, X., Parameswaran, G., Vasiliki, P., Stanley, H.E.: Understanding the cubic and half-cubic laws of financial fluctuations. Physica A **324**, 1–5 (2003)
10. Gomez, C.P., Shmoys, D.B.: Approximations and randomization to boost CSP techniques. Ann. Oper. Res. **130**, 117–141 (2004)
11. Hurwicz, L.: Optimality criteria for decision making under ignorance. Cowles Commission Discussion Paper, Statistics. No. 370 (1951)
12. Kreinovich, V.: Decision making under interval uncertainty (and beyond). In: Guo, P., Pedrycz, W. (eds.) Human-Centric Decision-Making Models for Social Sciences, pp. 163–193. Springer, Heidelberg (2014)
13. Luce, R.D., Raiffa, R.: Games and Decisions: Introduction and Critical Survey. Dover, New York (1989)
14. Mandelbrot, B.: The Fractal Geometry of Nature. Freeman, San Francisco (1983)
15. Mandelbrot, B., Hudson, R.L.: The (Mis)behavior of Markets: A Fractal View of Financial Turbulence. Basic Books, New York (2006)

16. Mantegna, R.N., Stanley, H.E.: An Introduction to Econophysics: Correlations and Complexity in Finance. Cambridge University Press, Cambridge (1999)
17. Marinescu, D.: Cloud Computing: Theory and Practice. Morgan Kaufmann, Waltham (2013)
18. Markovich, N. (ed.): Nonparametric Analysis of Univariate Heavy-Tailed Data: Research and Practice. Wiley, Chichester (2007)
19. McCauley, J.: Dynamics of Markets: Econophysics and Finance. Cambridge University Press, Cambridge (2004)
20. Mell, P., Grance, T.: The NIST Definition of Cloud Computing, pp. 145–800. US National Institute of Standards and Technology Special Publication, Gaithersburg (2011)
21. Rachev, S.T., Mittnik, S.: Stable Paretian Models in Finance. Wiley, New York (2000)
22. Resnick, S.I.: Heavy-Tail Phenomena: Probabilistic and Statistical Modeling. Springer, New York (2007)
23. Rhoton, J.: Cloud Computing Explained. Recursive Press, London (2010)
24. Roehner, B.: Patterns of Speculation - A Study in Observational Econophysics. Cambridge University Press, Cambridge (2002)
25. Stanley, H.E., Amaral, L.A.N., Gopikrishnan, P., Plerou, V.: Scale invariance and universality of economic fluctuations. Physica A **283**, 31–41 (2000)
26. Stoyanov, S.V., Racheva-Iotova, B., Rachev, S.T., Fabozzi, F.J.: Stochastic models for risk estimation in volatile markets: a survey. Ann. Oper. Res. **176**, 293–309 (2010)
27. Vasiliki, P., Stanley, H.E.: Stock return distributions: tests of scaling and universality from three distinct stock markets. Phys. Rev. E: Stat., Nonlin, Soft Matter Phys. **77**(3), 2 (2008). Publ. 037101
28. Velte, A.T., Velte, T.J., Elsenpeter, R.: Cloud Computing: A Practical Approach. McGraw-Hill, New York (2010)
29. Whittmann, A.: Does the cloud keep pace with Moore's law? Information Week, No. 1327, 12 March 2012, p. 42 (2012)

The TOPSIS Method in the Interval Type-2 Fuzzy Setting

Ludmila Dymova[1], Pavel Sevastjanov[1($\boxtimes$)], and Anna Tikhonenko[2]

[1] Institute of Computer and Information Science, Czestochowa University
of Technology, Dabrowskiego 73, 42-200 Czestochowa, Poland
sevast@icis.pcz.p
[2] Faculty of Mathematics and Natural Sciences, Cardinal Stefan Wyszyski
University in Warsaw, Warsaw, Poland
http://www.icis.pcz.pl/
http://www.uksw.edu.pl/

Abstract. The technique for establishing order preference by similarity
to the ideal solution ($TOPSIS$) now is probably one of most popular
method for Multiple Criteria Decision Making ($MCDM$). The method
was primarily developed for dealing with real-valued data.

Nevertheless, in practice often it is hard to present precisely exact
ratings of alternatives with respect to local criteria and as a result these
ratings are seen as fuzzy values. A number of papers have been devoted
to fuzzy extension of the $TOPSIS$ method in the literature, but only a
few works provided the type-2 fuzzy extensions, whereas such extensions
seem to be very useful for solution of many real-world problem, e.g., Multiple Criteria Group Decision Making problem. Since the proposed type-
2 fuzzy extensions of the $TOPSIS$ method have some limitations and
drawbacks, in this paper we propose an interval type-2 fuzzy extension
of the $TOPSIS$ method realised with the use of α-cuts representation of
the interval type-2 fuzzy values ($IT2FV$). This extension is free of the
limitations and drawbacks of the known methods. The proposed method
is realised for the cases of perfectly normal and normal $IT2FV$s.

Keywords: Interval type-2 fuzzy extension · TOPSIS · α-cuts

1 Introduction

A technique for establishing order performance by similarity to the ideal solution
($TOPSIS$) was first developed by Hwang and Yoon [18] for solving ($MCDM$)
problems.

Currently the $TOPSIS$ method is very popular, but we cannot say that it is
the best one for the solution of all $MCDM$ problems. In our opinion, the search
for some unique best method for the solution of $MCDM$ problems is senseless.
In practice, we usually try to choose among a great number of existing methods
such that seems to be maximally appropriate to the specificity of the problem

© Springer International Publishing Switzerland 2016
R. Wyrzykowski et al. (Eds.): PPAM 2015, Part II, LNCS 9574, pp. 445–454, 2016.
DOI: 10.1007/978-3-319-32152-3_41

and preferences of decision makers. Generally, the choice of the method for the solution of $MCDM$ problem is a context dependent problem [30].

In [15], we have shown that the classical $TOPSIS$ method may be considered as a modified weighted sum of local criteria. Although the weighted sum is the most popular approach to the solution of $MCDM$ problems in some cases it cannot be used. An important property of weighted sum aggregation is that the small values of some local criteria may be counterbalanced by large values of other ones in the final assessment. For example, a high percent of goods of low quality in most cases cannot be counterbalanced by low production costs, just as the low professional qualifications of medical staff usually cannot be compensated for by the high quality of diagnostic equipment and so on. In some fields, e.g., in ecological modelling, the weighted sum is not used for aggregation [27]. The reason behind this is that in practice there are cases when if any local criterion is totally dissatisfied then the considered alternative should be rejected from the consideration completely. Since this compensative property of weighted sum aggregation is in many applications undesirable, a decision maker may prefer to use, e.g., weighted geometric aggregation and a more cautious decision maker may prefer aggregation based on the "principle of maximal pessimism" [15].

Nevertheless, the general idea of $TOPSIS$ method, i.e., establishing the order of preference by similarity to ideal solution seems to be very attractive and fruitful. Therefore, in [15] we introduced other types of local criteria aggregation in the $TOPSIS$ method and developed a method for the generalisation of different aggregation modes. In classical methods for $MCDM$, the ratings and weights of criteria are known precisely. In the classical $TOPSIS$ method, the ratings of alternatives and the weights of criteria are presented by real values, too. The classical $TOPSIS$ method has been successfully used in various fields [9,24]. A comprehensive survey of $TOPSIS$ method applications is presented in [3]. However, sometimes it is difficult to determine precisely the real values of the rating of alternatives with respect to local criteria, and as a result, these ratings are presented as fuzzy values. Some papers have been devoted to fuzzy extensions of the $TOPSIS$ method in the literature, but these extensions are not complete since the ideal solutions are usually presented as real values (not as fuzzy values) or as fuzzy values which are not attainable in the decision matrix [1,4,5,16,22,31]. In most of papers [1,2,6,8,11–13,19,23,28,29], a defuzzification of elements of fuzzy decision matrix is used, which leads inevitable to the loss of important information and may provide wrong results. On the other hand, when we deal with the problem of multiple criteria group decision making, the fuzzy rating of alternatives may be presented by different experts in different ways. The use of type-2 fuzzy sets and operations on type-2 fuzzy values makes it possible to avoid this problem. Such approach was used by Chen and Lee [10] for the solution of group $MCDM$ problem using trapezoidal interval type-2 fuzzy values in the framework of $TOPSIS$ method. Unfortunately, in this work, for evaluation of trapezoidal interval type-2 fuzzy values the heuristic expression provided real-valued evaluation of rating of alternatives was used. A technique based on generalised interval -valued trapezoidal fuzzy value in the framework of $TOPSIS$ method was proposed in [25]. The drawback of this work is that only reference points (real values) of interval-valued trapezoidal fuzzy values were used for the evaluation of positive and negative ideal solutions.

Therefore, in this paper we propose a new approach to the interval type-2 fuzzy extension of $TOPSIS$ method using the α-cuts representation of the interval type-2 fuzzy values ($IT2FV$). This extension is free of the limitations and drawbacks of the known methods. Here we restrict ourselves to the consideration only triangular perfectly normal and normal $IT2FV$s.

The rest of paper is set out as follows: In Sect. 2, we recall some basic definitions needed for the subsequent analysis. Section 3 is devoted to the interval type-2 fuzzy extension of the $TOPSIS$ method using the α-cuts. Section 4 contains some concluding remarks.

2   Preliminaries

2.1   The Basics of the $TOPSIS$ Method

Suppose a $MCDM$ problem is based on m alternatives A_1, A_2,...,A_m and n local criteria C_1, C_2,...,C_n. Each alternative is evaluated with respect to the n criteria. All the ratings are assigned to alternatives and presented in the decision matrix $D[x_{ij}]_{m \times n}$, where x_{ij} is the rating of alternative A_i with respect to the criterion C_j. Let $W = [w_1, w_2, ..., w_n]$ be the vector of local criteria weights satisfying $\sum_{j=1}^{n} w_j = 1$.

The $TOPSIS$ method consists of the following steps [18]:

1. Normalize the decision matrix:

$$r_{ij} = \frac{x_{ij}}{\sqrt{\sum_{k=1}^{m} x_{kj}^2}}, \ i = 1, ..., m; \ j = 1, ..., n. \tag{1}$$

   Multiply the columns of normalized decision matrix by the associated weights:

$$v_{ij} = w_j \times r_{ij}, \ i = 1, ..., m; \ j = 1, ..., n. \tag{2}$$

2. Determine the positive ideal and negative ideal solutions, respectively, as follows:

$$A^+ = \{v_1^+, v_2^+, ..., v_n^+\}$$
$$= \{(\max_i v_{ij} \,|j \in K_b) \ (\min_i v_{ij} \,|j \in K_c)\}, \tag{3}$$

$$A^- = \{v_1^-, v_2^-, ..., v_n^-\}$$
$$= \{(\min_i v_{ij} \,|j \in K_b) \ (\max_i v_{ij} \,|j \in K_c)\}, \tag{4}$$

   where K_b is a set of benefit criteria and K_c is a set of cost criteria.
3. Obtain the distances of the existing alternatives from the positive ideal and negative ideal solutions: two Euclidean distances for each alternatives are, respectively, calculated as follows:

$$S_i^+ = \sqrt{\sum_{j=1}^{n} (v_{ij} - v_j^+)^2}, \ i = 1, ..., m,$$
$$S_i^- = \sqrt{\sum_{j=1}^{n} (v_{ij} - v_j^-)^2}, \ i = 1, ..., m. \tag{5}$$

4. Calculate the relative closeness to the ideal alternatives:

$$RC_i = \frac{S_i^-}{S_i^+ + S_i^-}, \quad i = 1, 2, ..., m, \quad 0 \leq RC_i \leq 1. \tag{6}$$

5. Rank the alternatives according to their relative closeness to the ideal alternatives: the bigger RC_i, the better alternative A_i.

2.2   Type-2 Fuzzy Sets and Their α-cuts Representation

Here we present the basic terminology used in this paper.

A type-2 fuzzy set $(T2FS)$ $\tilde{A}$ is defined as follows:

Definition 1 [20,21].

$$\tilde{A} = \int_{\forall x \in X} \int_{\forall u \in J_x \subseteq [0,1]} \mu_{\tilde{A}}(x, u)/(x, u), \tag{7}$$

where $\int \int$ denotes the union over all admissible values of x and u, and $\mu_{\tilde{A}}(x, u)$ is a type-2 membership function.

A $T2FS$ is three dimensional $(3D)$. The Vertical Slice (VS) is the two dimensional $(2D)$ plane in the u and $\mu_{\tilde{A}}(x, u)$ axes for a single value of $x = x'$, then VS is defined by the equation

$$VS(x') = \mu_{\tilde{A}}(x', u) \equiv \mu_{\tilde{A}}(x') = \int_{u \in J_{x'}} f_{x'}(u)/u, \tag{8}$$

where $f_{x'}(u) \in [0,1]$ is called the secondary grade and J_x represents the domain of the secondary membership function called the secondary domain. Of course, the VS is a type-1 fuzzy set $(T1FS)$ in $[0,1]$. The Vertical Slice Representation (VSR) of $T2FS$ is represented by the union of all the vertical slices.

The Footprint Of Uncertainty (FOU) is derived from the union of all primary memberships:

$$FOU(\tilde{A}) = \int_{x \in X} J_x. \tag{9}$$

According to [20], the FOU is represented by the lower and upper membership functions:

$$FOU(\tilde{A}) = \int_{x \in X} [\underline{\mu_{\tilde{A}}(x)}, \overline{\mu_{\tilde{A}}(x)}]. \tag{10}$$

Interval type-2 fuzzy set $(IT2FS)$ is defined to be a $T2FS$ where all its secondary grades are equal to 1 $(\forall f_x(u) = 1)$. A $IT2FS$ can be completely determined using its FOU given in equation (10). The α-cuts of $IT2FS$ $\tilde{A}$ are defined in [17] as follows: $\tilde{A}_{\tilde{\alpha}} = \{(x, u) \, | \, f_x(u) \geq \tilde{\alpha}\}$.

But as we are dealing with $IT2FS$ this definition may be substantially simplified. Since $IT2FS$ can be completely determined using its FOU, the α-cuts of $IT2FS$ may be represented by the α-cuts of its $FOU(\tilde{A})$.

Definition 2. The α-cuts of $IT2FS$ $\hat{A}$ are presented as follows:

$$\hat{A}_\alpha = \left\{ x \, \Big| \, \underline{\mu_{\hat{A}}(x)} \geq \alpha, \overline{\mu_{\tilde{A}}(x)} \geq \alpha \right\}.$$

Definition 3 [17] (Perfectly Normal *IT2FS*). A *IT2FS* $\hat{A}$, is said to be perfectly normal if $\sup \overline{\mu}_{\hat{A}}(x) = \sup \underline{\mu}_{\hat{A}}(x) = 1$.

Since α-cuts of type 1 fuzzy sets are based on the corresponding intervals, the α-cuts of *IT2FS* may be presented by corresponding intervals with interval-valued bounds as follows:

$$\hat{A}_{\alpha} = \left[\left[\overline{x}_{\alpha}^{L}, \underline{x}_{\alpha}^{L} \right], \left[\underline{x}_{\alpha}^{U}, \overline{x}_{\alpha}^{U} \right] \right]. \tag{11}$$

Then the perfectly normal interval type-2 fuzzy value (*IT2FV*) may be presented as follows:

$$\hat{A} = \bigcup_{\alpha} \alpha \left[\left[\overline{x}_{\alpha}^{L}, \underline{x}_{\alpha}^{L} \right], \left[\underline{x}_{\alpha}^{U}, \overline{x}_{\alpha}^{U} \right] \right]. \tag{12}$$

Definition 4 [17] (Normal *IT2FS*). A *IT2FS* $\hat{A}$, is said to be normal if $\sup \overline{\mu}_{\hat{A}} (x) = 1$ and $\sup \underline{\mu}_{\hat{A}}(x) = h < 1$, where h is the lower membership function height. The normal $\widetilde{IT}2FV$ may be presented as follows:

$$\hat{A} = \begin{cases} \bigcup_{\alpha} \alpha [[\overline{x}_{\alpha}^{L}, \underline{x}_{\alpha}^{L}], [\underline{x}_{\alpha}^{U}, \overline{x}_{\alpha}^{U}]], \alpha \leq h, \\ \bigcup_{\alpha} \alpha [\overline{x}_{\alpha}^{L}, \overline{x}_{\alpha}^{U}], \alpha > h, \end{cases} \tag{13}$$

An important problem in the implementation of *TOPSIS* method in the interval type-2 fuzzy setting is the comparison of *IT2FV*s. Since our approach is based on the α-cut representation of *IT2FV* the problem reduces to the interval comparison. Based on the comparison of the most popular methods for interval comparison, it was shown in [26] that to obtain a measure of distance between intervals which additionally indicates which interval is greater/lesser, the following value which represents the distance between the centers of compared intervals A and B may be successfully used.

$$\Delta_{A-B} = \left(\frac{1}{2}(a^{L} + a^{U}) - \frac{1}{2}(b^{U} + b^{L}) \right). \tag{14}$$

3   An Extension of the TOPSIS Method Under Interval Type-2 Fuzzy Uncertainty

3.1   The Case of Perfectly Normal Triangular Interval Type-2 Fuzzy Values

Let $A_1, A_2, ..., A_m$ be alternatives, $C_1, C_2, ..., C_n$ be local criteria and $w_1, w_2, ..., w_n$ be real-valued weights of local criteria such that $\sum_{i=1}^{n} w_i = 1$. Let $D \left[\hat{X}_{ij} \right]_{n \times m}$ be the decision matrix, where $\hat{X}_{ij}$ is the perfectly normal interval type-2 fuzzy value representing the rating of alternative A_i with respect to the criterion C_j.

Let $\overline{x}_{ij}^{L}$, $\underline{x}_{ij}^{L}$, x_{ij}^{M}, $\underline{x}_{ij}^{U}$, $\overline{x}_{ij}^{U}$ be the reference points of perfectly normal *IT2FV* $\hat{X}_{ij}$ It can be represented as follows

$$\hat{X}_{ij} = \left\{ \left[\overline{x}_{ij}^{L}, \underline{x}_{ij}^{L} \right], x_{ij}^{M}, \left[\underline{x}_{ij}^{U}, \overline{x}_{ij}^{U} \right] \right\}. \tag{15}$$

The first step of the method is the normalization of the decision matrix. In [7], the following method for normalization of the decision matrix with ratings presented by type 1 fuzzy values was proposed:

$$r_{ij} = \left(\frac{x_{ij}^L}{x_j^+}, \frac{x_{ij}^M}{x_j^+}, \frac{x_{ij}^U}{x_j^+} \right), \ i = 1, 2, ...m, \ j \in K_b, \tag{16}$$

where $x_j^+ = \max_j(x_{ij}^U), \ j \in K_b,$

$$r_{ij} = \left(\frac{x_j^-}{x_{ij}^U}, \frac{x_j^-}{x_{ij}^M}, \frac{x_j^-}{x_{ij}^L} \right), \ i = 1, 2, ...m, \ j \in K_c, \tag{17}$$

where $x_j^- = \min_j(x_{ij}^L), \ j \in K_c$.

In (16) and (17), $x_{ij}^L, x_{ij}^M, x_{ij}^U$ are the reference points of triangular fuzzy values.

We can see that normalizations of benefit and cost fuzzy ratings of alternatives are made in different ways. As a result these normalizations preserve the property that supports of normalized fuzzy numbers belong to the interval $[0,1]$.

This method can be extended to use it for the normalization of decision matrices with perfectly normal interval type-2 fuzzy ratings as follows:

$$\hat{r}_{ij} = \left(\left[\frac{\underline{x}_{ij}^L}{x_j^+}, \frac{\overline{x}_{ij}^L}{x_j^+} \right], \frac{x_{ij}^M}{x_j^+}, \left[\frac{\underline{x}_{ij}^U}{x_j^+}, \frac{\overline{x}_{ij}^U}{x_j^+} \right] \right), \ i = 1, 2, ...m, \ j \in K_b, \tag{18}$$

where $x_j^+ = \max_j(\overline{x}_{ij}^U), \ j \in K_b,$

$$\hat{r}_{ij} = \left(\left[\frac{x_j^-}{\overline{x}_{ij}^U}, \frac{x_j^-}{\underline{x}_{ij}^U} \right], \frac{x_j^-}{x_{ij}^M}, \left[\frac{x_j^-}{\underline{x}_{ij}^L}, \frac{x_j^-}{\overline{x}_{ij}^L} \right] \right), \ i = 1, 2, ...m, \ j \in K_c, \tag{19}$$

where $x_j^- = \min_j(\overline{x}_{ij}^L), \ j \in K_c$.

Using the normalization procedure based on the expressions (18) and (19), from the normalized decision matrix is obtained. The next step is the obtaining of positive ideal A^+ and the negative ideal A^- solutions, respectively. Then in our case, the expressions (3) and (4) are extended as follows:

$$
\begin{aligned}
A^+ &= \{\hat{\nu}_1^+, \hat{\nu}_2^+, ..., \hat{\nu}_n^+\} \\
&= \{ ([\overline{\nu}_1^{+L}, \underline{\nu}_1^{+L}], \nu_1^{+M}, [\underline{\nu}_1^{+U}, \overline{\nu}_1^{+U}]), ..., ([\overline{\nu}_n^{+L}, \underline{\nu}_n^{+L}], \nu_n^{+M}, [\underline{\nu}_n^{+U}, \overline{\nu}_n^{+U}]) \} \\
&= \{ \max_i \{\hat{r}_{ij}\} \, | j \subset K_b, \{\min_i \{\hat{r}_{ij}\} \, | j \subset K_c \}
\end{aligned}
\tag{20}
$$

$$
\begin{aligned}
A^- &= \{\hat{\nu}_1^-, \hat{\nu}_2^-, ..., \hat{\nu}_n^-\} \\
&= \{ ([\overline{\nu}_1^{-L}, \underline{\nu}_1^{-L}], \nu_1^{-M}, [\underline{\nu}_1^{-U}, \overline{\nu}_1^{-U}]), ..., ([\overline{\nu}_n^{-L}, \underline{\nu}_n^{-L}], \nu_n^{-M}, [\underline{\nu}_n^{-U}, \overline{\nu}_n^{-U}]) \} \\
&= \{ \min_i \{\hat{r}_{ij}\} \, | j \subset K_b, \{\max_i \{\hat{r}_{ij}\} \, | j \subset K_c \}
\end{aligned}
\tag{21}
$$

In (20) and (21), $\hat{r}_{ij} = ([\overline{r}_{ij}^L, \underline{r}_{ij}^L], r_{ij}^M, [\underline{r}_{ij}^U, \overline{r}_{ij}^U])$.

We can see that to obtain the positive ideal A^+ and negative ideal A^- solutions the operation of comparison of $\hat{r}_{ij}$ is needed (see $\min_i \{\hat{r}_{ij}\}$ and $\max_i \{\hat{r}_{ij}\}$ in (20) and (21)).

To avoid the use of the heuristic approaches to the perfectly normal interval type-2 fuzzy values comparison, we shall represent $\hat{r}_{ij}$ by the sets of corresponding α-cuts. Then after comparison of $\hat{r}_{ij}$ on α-cuts with the use of approach to the interval comparison presented by the expression (14) we shall aggregate the results of comparison into the final real-valued estimation of difference between normal interval type-2 fuzzy values $\Delta(\hat{r}_{ij} - \hat{r}_{oj})$, the sign of which indicates what the compared normal interval type-2 fuzzy value is greater/lesser and the $abs(\Delta(\hat{r}_{ij} - \hat{r}_{oj}))$ represents the difference between them.

Therefore, first we should develop the method for comparison of interval type-2 fuzzy value on the α-cuts.

At the first step we extend the conventional method for interval subtraction. It is easy to see that α-cut of $\hat{r}_{ij}$ may be presented by corresponding intervals with interval-valued bounds (see (13)) as follows:

$$\hat{r}_{ij\alpha} = \left[\left[\overline{r}^L_{ij\alpha}, r^L_{ij\alpha} \right], \left[\underline{r}^U_{ij\alpha}, \overline{r}^U_{ij\alpha} \right] \right]. \tag{22}$$

Using the formal extension of conventional operation of interval subtraction to the case of perfectly normal interval type-2 fuzzy values we can present this extended operation as follows:

$$\begin{aligned} r_{ij\alpha} - r_{oj\alpha} &= \left[\left[\overline{r}^L_{ij\alpha}, r^L_{ij\alpha} \right] - \left[\underline{r}^U_{oj\alpha}, \overline{r}^U_{oj\alpha} \right], \left[\underline{r}^U_{ij\alpha}, \overline{r}^U_{ij\alpha} \right] - \left[\overline{r}^L_{oj\alpha}, r^L_{oj\alpha} \right] \right] \\ &= \left[\left[\overline{r}^L_{ij\alpha} - \overline{r}^U_{oj\alpha}, r^L_{ij\alpha} - \underline{r}^U_{oj\alpha} \right], \left[\underline{r}^U_{ij\alpha} - r^L_{oj\alpha}, \overline{r}^U_{ij\alpha} - \overline{r}^L_{oj\alpha} \right] \right] \\ &== \left[\left[r_{ij\alpha} - r_{oj\alpha} \right]^L, \left[r_{ij\alpha} - r_{oj\alpha} \right]^U \right]. \end{aligned} \tag{23}$$

Then in the spirit of the method for interval comparison presented by the expression (14) we provide several averaging procedures (24, 25) obtaining finally the real valued estimation of difference between compared perfectly normal interval type-2 fuzzy values on the α-cuts $\Delta(r_{ij\alpha} - r_{oj\alpha})$ (see (26)).

$$\begin{aligned} \left[\left[\Delta(r_{ij\alpha} - r_{oj\alpha}) \right] \right] &= \tfrac{1}{2} \left[\left[r_{ij\alpha} - r_{oj\alpha} \right]^L + \left[r_{ij\alpha} - r_{oj\alpha} \right]^U \right] \\ &= \tfrac{1}{2} \left[\left[\overline{r}^L_{ij\alpha} + \underline{r}^U_{ij\alpha}, -\overline{r}^U_{oj\alpha} - r^L_{oj\alpha} \right], \left[r^L_{ij\alpha} + \overline{r}^U_{ij\alpha}, -\underline{r}^U_{oj\alpha} - \overline{r}^L_{oj\alpha} \right] \right] \\ &= \left[\left[\Delta^L(r_{ij\alpha} - r_{oj\alpha}) \right], \left[\Delta^U(r_{ij\alpha} - r_{oj\alpha}) \right] \right]. \end{aligned} \tag{24}$$

$$\begin{aligned} \left[\Delta(r_{ij\alpha} - r_{oj\alpha}) \right] &= \tfrac{1}{2} \left[\left[\Delta^L(r_{ij\alpha} - r_{oj\alpha}) \right] + \left[\Delta^U(r_{ij\alpha} - r_{oj\alpha}) \right] \right] \\ &= \tfrac{1}{4} \left[\overline{r}^L_{ij\alpha} + \underline{r}^U_{ij\alpha} + r^L_{ij\alpha} + \overline{r}^U_{ij\alpha}, - \left(\overline{r}^U_{oj\alpha} + r^L_{oj\alpha} + \underline{r}^U_{oj\alpha} + \overline{r}^L_{oj\alpha} \right) \right]. \end{aligned} \tag{25}$$

$$\Delta(r_{ij\alpha} - r_{oj\alpha}) = \tfrac{1}{8} \left(\overline{r}^L_{ij\alpha} + \underline{r}^U_{ij\alpha} + r^L_{ij\alpha} + \overline{r}^U_{ij\alpha} - \left(\overline{r}^U_{oj\alpha} + r^L_{oj\alpha} + \underline{r}^U_{oj\alpha} + \overline{r}^L_{oj\alpha} \right) \right). \tag{26}$$

To get the final real valued estimation of difference between compared perfectly normal interval type-2 fuzzy values, we aggregate $\Delta(r_{ij\alpha} - r_{oj\alpha})$ as follows:

$$\Delta(\hat{r}_{ij} - \hat{r}_{oj}) = \frac{\sum\limits_{\alpha} \alpha \Delta(r_{ij\alpha} - r_{oj\alpha})}{\sum\limits_{\alpha} \alpha} \tag{27}$$

The last expression indicates that the contribution of α-cut to the overall estimation of $\Delta(\hat{r}_{ij} - \hat{r}_{oj})$ increases along with the rise of its number.

The second step is the obtaining the distances of the existing alternatives from the positive ideal and negative ideal solutions.

Obviously, from the definitions (20) and (21) we have $\Delta(\hat{\nu}_j^+ - \hat{r}_{ij}) \geq 0$, $j \in K_b$, $\Delta(\hat{r}_{ij} - \hat{\nu}_j^+) \geq 0$, $j \in K_c$, $\Delta(\hat{r}_{ij} - \hat{\nu}_j^-) \geq 0$, $j \in K_c$ and $\Delta(\hat{\nu}_j^- - \hat{r}_{ij}) \geq 0$, $j \in K_b$, $i = 1, 2, ..., m$. Therefore the corresponding distances may be presented as follows (see more detailed analysis in [15]).

$$S_i^+ = \sum_{j \in K_b} w_j \Delta(\hat{\nu}_j^+ - \hat{r}_{ij}) + \sum_{j \in K_c} w_j \Delta(\hat{r}_{ij} - \hat{\nu}_j^+), \, i = 1, 2, ..., m, \tag{28}$$

$$S_i^- = \sum_{j \in K_c} w_j \Delta(\hat{r}_{ij} - \hat{\nu}_j^-) + \sum_{j \in K_b} w_j \Delta(\hat{\nu}_j^- - \hat{r}_{ij}), \, i = 1, 2, ..., m. \tag{29}$$

3.2   The Case of Normal Interval Type-2 Fuzzy Values

In the case of normal $IT2FVs$ for $\alpha > h$ we deal with usual intervals on α-cuts (see expression 13). Therefore, the differences between compared normal $IT2FVs$ on the α-cuts such that $\alpha > h$ my be presented as the difference between usual intervals represented by the expression (14). Hence the expression (26) takes the form:

If $\alpha > h$ then

$$\Delta \left(r_{ij_\alpha} - r_{oj_\alpha} \right) = \tfrac{1}{2} \left(\left(r_{ij_\alpha}^L + r_{ij_\alpha}^U \right) - \left(r_{oj_\alpha}^L + r_{oj_\alpha}^U \right) \right). \tag{30}$$

If $\alpha \leq h$ then

$$\Delta(r_{ij_\alpha} - r_{oj_\alpha}) = \tfrac{1}{8} \left(\overline{r}_{ij_\alpha}^L + \underline{r}_{ij_\alpha}^U + \underline{r}_{ij_\alpha}^L + \overline{r}_{ij_\alpha}^U - \left(\overline{r}_{oj_\alpha}^U + \underline{r}_{oj_\alpha}^L + \underline{r}_{oj_\alpha}^U + \overline{r}_{oj_\alpha}^L \right) \right). \tag{31}$$

The other needed for calculation expressions (20),(21),(27),(28),(29) are not changed.

We have not found in the literature an example of the solution of $MCDM$ problem with the use of $TOPSIS$ method under interval type-2 fuzzy uncertainty, when the ratings of alternatives in the decision matrix are presented by the triangular normal and perfectly normal $IT2FVs$. Therefore we cannot compare our results with those obtained using other approaches.

Nevertheless, according to our experience [14] the use of approach based of the direct interval extension of $TOPSIS$ method with the operation of interval comparison based on the operation of interval subtraction makes it possible to avoid the problems concerned with the real-valued representations of intervals by their lower or upper bounds which may lead to the wrong results.

It the current paper, we extend this operation to the case of intervals with interval-valued bounds representing $IT2FVs$ on the α-cuts.

As it was showed in [14], the use of α-cut representation of fuzzy values allows us to avoid many problems of intermediate defuzzyfication of ratings and fuzzy values comparison based on the representation intervals on α-cuts by their real-valued representations. Therefore, in this paper we extend this approach to the case of the use of $TOPSIS$ method under interval type-2 fuzzy uncertainty.

4   Conclusion

The extension of the $TOPSIS$ method under interval type-2 fuzzy uncertainty is proposed. This extension is provided by the use of α-cut representation of triangular interval type-2 fuzzy values $IT2FVs$. To implement this approach, a method for comparison of intervals with interval-valued bounds representing $IT2FVs$ on the α-cuts is proposed. This method is based on the extended operation of interval subtraction.

As illustrative examples, the numerical methods for the solution of $MCDM$ problems when the decision matrices are presented by the triangular normal and perfectly normal $IT2FVs$ are considered.

Acknowledgements. The research has been supported by the grant financed by National Science Centre (Poland) on the basis of decision number DEC-2013/11/B/ST6/00960.

References

1. Anisseh, M., Piri, F., Shahraki, M.R., Agamohamadi, F.: Fuzzy extension of TOPSIS model for group decision making under multiple criteria. Artif. Intell. Rev. **38**, 325–338 (2012)
2. Bao, Q., Ruan, D., Shena, Y., Hermans, E., Janssens, D.: Improved hierarchical fuzzy TOPSIS for road safety performance evaluation. Knowl. Based Syst. **32**, 84–90 (2012)
3. Behzadian, M., Otaghsara, S.K., Yazdani, M., Ignatius, J.: A state-of the-art survey of TOPSIS application. Expert Syst. Appl. **39**, 13051–13069 (2012)
4. Bäyäközkan, G., Çifçi, G.: A combined fuzzy AHP and fuzzy TOPSIS based strategic analysis of electronic service quality in healthcare industry. Expert Syst. Appl. **39**, 2341–2354 (2012)
5. Bäyäközkan, G., Çifçi, G.: A novel hybrid MCDM approach based on fuzzy DEMATEL, fuzzy ANP and fuzzy TOPSIS to evaluate green suppliers. Expert Syst. Appl. **39**, 3000–3011 (2012)
6. Chamodrakas, I., Martakos, D.: A utility-based fuzzy TOPSIS method for energy efficient network selection in heterogeneous wireless networks. Appl. Soft Comput. **12**, 1929–1938 (2012)
7. Chen, C.-T.: A fuzzy approach to select the location of distribution center. Fuzzy Sets Syst. **118**, 65–73 (2001)
8. Chen, M.F., Tzeng, G.H.: Combining grey relation and TOPSIS concepts for selecting an expatriate host country. Math. Comput. Model. **40**, 1473–1490 (2004)
9. Chen, Y.-J.: Structured methodology for supplier selection and evaluation in a supply chain. Inf. Sci. **181**, 1651–1670 (2011)
10. Chen, S.-M., Lee, L.-W.: Fuzzy multiple attributes group decision-making based on the interval type-2 TOPSIS method. Expert Syst. Appl. **37**, 2790–2798 (2010)
11. Chu, T.C.: Facility location selection using fuzzy TOPSIS under group decisions. Int. J. Uncertainty Fuzziness Knowl. Based Syst. **10**, 687–701 (2002a)
12. Chu, T.C.: Selecting plant location via a fuzzy TOPSIS approach. Int. J. Adv. Manuf. Technol. **20**, 859–864 (2002b)

13. Chu, T.C., Lin, Y.C.: A fuzzy TOPSIS method for robot selection. Int. J. Adv. Manuf. Technol. **21**, 284–290 (2003)
14. Dymova, L., Sevastjanov, P., Tikhonenko, A.: A direct interval extension of TOP-SIS method. Expert Syst. Appl. **40**, 4841–4847 (2013)
15. Dymova, L., Sevastjanov, P., Tikhonenko, A.: An approach to generalization of fuzzy TOPSIS method. Inf. Sci. **238**, 149–162 (2013)
16. Fan, Z.-P., Feng, B.: A multiple attributes decision making method using individual and collaborative attribute data in a fuzzy environment. Inf. Sci. **179**, 3603–3618 (2009)
17. Hamrawi, H., Coupland, S.: Type-2 fuzzy arithmetic using alpha-planes. In: IFSA-EUSFLAT Conference, pp. 606–612, Lisbon, Portugal (2009)
18. Hwang, C.L., Yoon, K.: Multiple Attribute Decision Making Methods and Applications. Springer, Heidelberg (1981)
19. Iç, Y.T.: Development of a credit limit allocation model for banks using an integrated fuzzy TOPSIS and linear programming. Expert Syst. Appl. **39**, 5309–5316 (2012)
20. Mendel, J.M.: Uncertain Rule-Based Fuzzy Logic Systems: Introduction and New Directions. Prentice Hall, Upper Saddle River (2001)
21. Mendel, J.M., John, R.I.: Type-2 fuzzy sets made simple. IEEE Trans. Fuzzy Syst. **10**, 117–127 (2002)
22. Mokhtarian, M.N., Hadi-Vencheh, A.: A new fuzzy TOPSIS method based on left and right scores: an application for determining an industrial zone for dairy products factory. Appl. Soft Comput. **12**, 2496–2505 (2012)
23. Paksoy, T., Pehlivan, N.Y., Kahraman, C.: Organizational strategy development in distribution channel management using fuzzy AHP and hierarchical fuzzy TOPSIS. Expert Syst. Appl. **39**, 2822–2841 (2012)
24. Peng, Y., Wang, G., Wang, H.: User preferences based software defect detection algorithms selection using MCDM. Inf. Sci. **191**, 3–13 (2012)
25. Rashid, T., Beg, I., Husnine, S.M.: Robot selection by using generalized interval-valued fuzzy numbers with TOPSIS. Appl. Soft Comput. **21**, 462–468 (2014)
26. Sevastjanov, P., Tikhonenko, A.: Direct interval extension of TOPSIS method. In: Wyrzykowski, R., Dongarra, J., Karczewski, K., Waśniewski, J. (eds.) PPAM 2011, Part II. LNCS, vol. 7204, pp. 504–512. Springer, Heidelberg (2012)
27. Silvert, W.: Ecological impact classification with fuzzy sets. Ecol. Model. **96**, 1–10 (1997)
28. Tsaur, S.H., Chang, T.Y., Yen, C.H.: The evaluation of airline service quality by fuzzy MCDM. Tourism Manage. **23**, 107–115 (2002)
29. Vahdani, B., Mousavi, S.M., Tavakkoli-Moghaddam, R.: Group decision making based on novel fuzzy modified TOPSIS method. Appl. Math. Model. **35**(9), 4257–4269 (2011)
30. Zimmermann, H.J., Zysno, P.: Latest connectives in human decision making. Fuzzy Sets Syst. **4**, 37–51 (1980)
31. Xu, Z., Chen, J.: An interactive method for fuzzy multiple attribute group decision making. Inf. Sci. **177**, 248–263 (2007)

A Study on Vectorisation and Paralellisation of the Monotonicity Approach

Iwona Skalna[⊠] and Jerzy Duda

AGH University of Science and Technology, Krakow, Poland
skalna@agh.edu.pl, jduda@zarz.agh.edu.pl

Abstract. Solving parametric interval linear systems is one of the fundamental problems of interval computations. When the solution of a parametric linear system is a monotone function of interval parameters, then an interval hull of the parametric solution set can be computed by solving at most $2n$ real systems. If only some of the elements of the solution are monotone functions of parameters, then a good quality interval enclosure of the solution set can be obtained. The monotonicity approach, however, suffers from poor performance when dealing with large scale problems. Therefore, in this paper an attempt is made to improve the efficiency of the monotonicity approach. Techniques such as vectorisation and parallelisation are used for this purpose. The proposed approach is verified using some illustrative examples from structural mechanics.

Keywords: Parametric interval linear systems · Monotonicity approach · Vectorisation · Parallelisation

1 Introduction

Solving systems of linear equations is essential in modern engineering. Highly complex physical systems, which would require extremely complex formulae to describe, are approximated with high accuracy by a very large set of linear equations. In order to get reliable results, uncertainty, which is inevitable in real life problems, should be taken into account in any computations. Therefore, solving linear systems is one of the fundamental problems of interval computations, wherein "to solve a system" usually means to enclose a *parametric solution set* by an interval vector as tightly as possible.

The assumption that the coefficients of a linear system vary independently within given ranges is rarely satisfied in practice. That is why recently a big effort was made to develop methods that are able to solve the so-called *parametric interval linear systems*, i.e., linear systems with elements being functions of parameters that are allowed to vary within given intervals. Until now, several such methods have been developed, one can mention approximate methods described in [4,5,11,17,18].

The tightest is the resulting interval vector, the better. The narrowest possible interval enclosure is called the *interval hull solution* (or simply the hull).

© Springer International Publishing Switzerland 2016
R. Wyrzykowski et al. (Eds.): PPAM 2015, Part II, LNCS 9574, pp. 455–463, 2016.
DOI: 10.1007/978-3-319-32152-3_42

When the parametric solution is monotone with respect to all the parameters, the hull can be computed by solving at most $2n$ real systems. The monotonicity approach was investigated, e.g., in [6,12,15,20]. Generally, checking monotonicity is a very complex task and for large scale problems the monotonicity based methods are inefficient. Therefore, in this paper an attempt is made to reduce the computational time of the monotonicity approach. Vectorisation and parallelisation techniques are used for this purpose. It is worth to add that the monotonicity approach is extensively used in the interval global optimisation. Thus, the improvement of the efficiency of the monotonicity approach will significantly influence the efficiency of the interval global optimisation and monotonicity based methods.

The paper is organised as follows. The Sect. 2 contains preliminaries on solving parametric interval linear systems with two disjoint sets of parameters. In the Sect. 3, the monotonicity approach for computing interval hull solution is outlined. Section 4 presents general concepts of vectorisation and parallelisation. This is followed by a description of the monotonicity approach. Next, some illustrative examples of truss structures and the results of computational experiments are presented. The paper ends with concluding remarks.

2   Preliminaries

Italic font will be used for real quantities, while bold italic font will denote their interval counterparts. Let $\mathbb{IR}$ denote a set of real compact intervals $\boldsymbol{x} = [\underline{x}, \overline{x}] = \{x \in \mathbb{R} \mid \underline{x} \leqslant x \leqslant \overline{x}\}$. For two intervals $a, b \in \mathbb{IR}$, $a \geqslant b$, $a \leqslant b$ and $a = b$ will mean that, resp., $\underline{a} \geqslant \overline{b}, \overline{a} \geqslant \underline{b}$, and $\underline{a} = \underline{b} \wedge \overline{a} = \overline{b}$. $\mathbb{IR}^n$ will denote interval vectors and $\mathbb{IR}^{n \times n}$ will denote square interval matrices [10]. The midpoint $\check{x} = (\underline{x} + \overline{x})/2$ and the radius $r(\boldsymbol{x}) = (\overline{x} - \underline{x})/2$ are applied to interval vectors and matrices componentwise.

Definition 1. *A parametric linear system*

$$A(p)x = b(p) \tag{1}$$

is a linear system with elements that are real valued functions of a K-dimensional vector of parameters $p = (p_1, \ldots, p_K) \in \Re^K$, i.e., for each $i, j = 1, \ldots, n$,

$$\begin{aligned} A_{ij} : \Re^K \ni (p_1, \ldots, p_K) &\to A_{ij}(p_1, \ldots, p_K) \in \Re, \\ b_i : \Re^K \ni (p_1, \ldots, p_K) &\to b_i(p_1, \ldots, p_K) \in \Re. \end{aligned} \tag{2}$$

Functions describing the elements of a parametric linear system can be generally divided into affine-linear and nonlinear. However, nonlinear dependencies can be easily reduced to affine-linear using affine arithmetic [1]. Therefore, the following consideration are limited to the affine-linear case.

Remark: Obviously, the transformation from nonlinear to affine-linear dependencies causes some loss of information [1], nevertheless, the approach based on affine arithmetic is worth considering as it is simple and quite efficient.

Definition 2. *A parametric interval linear system with affine dependencies is given by*

$$A(p)x = b(p), \tag{3}$$

where $A(p) = A^{(0)} + \sum_{k=1}^{K} A^{(k)} p_k$, $b(p) = b^{(0)} + \sum_{k=1}^{K} b^{(k)} p_k$, $A^{(i)} \in \Re^{n \times n}$, *and* $b^{(i)} \in \Re^n$ *(i = 1, ..., K).*

If the involved parameters are subject to uncertainty, which means that they allowed to vary within given intervals (the interval-based model of uncertainty is adopted in this paper), then a *parametric interval linear system* is obtained.

Definition 3. *A parametric interval linear system is an infinite set (family) of parametric real linear systems*

$$\{A(p)x = b(p) \mid p \in \boldsymbol{p}\}. \tag{4}$$

The family (4) is usually written in a compact form as

$$A(\boldsymbol{p})x(\boldsymbol{p}) = b(\boldsymbol{p}). \tag{5}$$

Definition 4. *A parametric (united) solution set of the system (5) is a set of solutions to all systems from the family (4), i.e.,*

$$S(\boldsymbol{p}) = \{ x \mid \exists p \in \boldsymbol{p}, \ A(p)x = b(p) \}. \tag{6}$$

In order that the solution set be bounded, the parametric matrix $A(\boldsymbol{p})$ must be regular, i.e., $A(p)$ must be non-singular for each $p \in \boldsymbol{p}$.

In general case, the problem of computing the solution set (6) as well as its hull are NP-hard. Therefore, usually an outer interval enclosure, i.e., the vector $x^{\text{out}} \supset S(\boldsymbol{p})$, is computed instead. However, when the solution of a parametric system is monotone with respect to all parameters, then the hull of the solution set can be computed with polynomial cost in n and K. If the solution is monotone with respect to some of the parameters, then a good quality outer solution can be computed with polynomial cost in n and K.

3 Monotonicity Approach

For the sake of completeness of the paper, a brief reminder of the monotonicity approach is presented below.

Let $E^K = \{e \in \mathbb{R}^K \mid e_k \in \{-1, 0, 1\}, k = 1, \ldots, K\}$. For $p \in \mathbb{IR}^K$, $e \in E^K$, $p_k^e = \underline{p}_k$ if $e_k = -1$, $p_k^e = \check{p}_k$ if $e_k = 0$, and $p_k^e = \bar{p}_k$ if $e_k = 1$.

Theorem 1. *Let $A(\boldsymbol{p})$ be regular and let the functions $x_i(p) = \{A^{-1}(p) \cdot b(p)\}_i$ be monotone on an interval box $\boldsymbol{p} \in \mathbb{IR}^K$, with respect to each parameter p_k (k = 1, ..., K). Then, for each $i = 1, \ldots, n$,*

$$\{\Box S(\boldsymbol{p})\}_i = \left[\left\{ A\left(p^{-e^i}\right)^{-1} b\left(p^{-e^i}\right) \right\}_i, \left\{ A\left(p^{e^i}\right)^{-1} b\left(p^{e^i}\right) \right\}_i \right], \tag{7}$$

where $e_k^i = \text{sign} \frac{\partial x_i}{\partial p_k}$, $k = 1, \ldots, K$.

Now consider the family of parametric linear equations (4) and assume that $A_{ij}(p)$ and $b_i(p)$ $(i, j = 1, \ldots, n)$ are continuously differentiable in $\boldsymbol{p}$. If x is a solution to the system $A(p)x = b(p)$, then $x = A(p)^{-1}b(p)$, which means that x is a function of p. Thus, the global monotonicity properties of the solution with respect to each parameter p_k can be verified by checking the sign of derivatives $\frac{\partial x}{\partial p_k}(p)$ on the domain $\boldsymbol{p}$. The differentiation of the Eq. (1) with respect to p_k $(k = 1, \ldots, K)$ yields

$$\left\{ A(p) \frac{\partial x}{\partial p_k}(p) = \frac{\partial b}{\partial p_k}(p) - \frac{\partial A(p)}{\partial p_k} x(p) \; \middle| \; p \in \boldsymbol{p} \right\}. \tag{8}$$

Since $A_{ij}(p)$, $b_j(p)$ are affine linear functions of p, thus $\frac{\partial A_{ij}}{\partial p_k}$, $\frac{\partial b_i}{\partial p_k}$ are constant on $\boldsymbol{p}$. Hence, the approximation of $\frac{\partial x}{\partial p_k}(p)$ can be obtained by solving the following K parametric linear systems

$$A(\boldsymbol{p}) \frac{\partial x}{\partial p_k} = b'(\boldsymbol{x}^*) \tag{9}$$

where $b'(\boldsymbol{x}^*) = \{ b^{(k)} - A^{(k)} x^* \mid x^* \in \boldsymbol{x}^* \}$ and $\boldsymbol{x}^*$ is some initial solution to the system (5).

For a fixed i $(1 \leqslant i \leqslant n)$, let $\boldsymbol{D}_{ki}$ denotes the interval estimate of $\{ \frac{\partial x_i}{\partial p_k}(p) \mid p \in \boldsymbol{p} \}$ obtained by solving the Eq. (9). Now assume that each $\boldsymbol{D}_{ki}$ $(k = 1, \ldots, K)$ has a constant sign or equals 0. Then, in order to calculate the hull of $\{S(\boldsymbol{p})\}_i$, the elements of the vector e^i must be determined as follows: $e_k^i = 1$ if $\boldsymbol{D}_{ki} \geqslant 0$, $e_k^i = 0$ if $\boldsymbol{D}_{ki} \equiv 0$, and $e_k^i = -1$ if $\boldsymbol{D}_{ki} \leqslant 0$. If the sign of some of the partial derivatives was not determined definitely, then a new vector of parameters is constructed by substituting the respective endpoints for interval parameters. The process of determining the sign of derivatives restarts and continues until no further improvement is obtained. The algorithm of the method is presented below. Parts of code in Algorithm 1 which are candidates for parallelisation and vectorisation are indicated by comments.

4   Parallelisation and Vectorisation

Parallelisation is the process of converting sequential code into a multi-threaded one in order to use available processors simultaneously. The parallelisation process often also includes *vectorisation*, because contemporary central processor units are able to perform operations on multiple data in a single instruction. This ability is called SIMD (single instruction multiple data). It allows to convert an algorithm from a scalar implementation, in which a single instruction can deal with one pair of operands at a time, to a vector process, where a single instruction can refer to a vector of operands (series of adjacent values). Vectorisation can be carried out either automatically by contemporary C++ compilers or forced by a programmer usually by using an appropriate pragma. Vectorisation not always brings performance improvement due to additional data movement and pipeline synchronisation. Thus, the vectorisation can be profitable for loops that

---

Algorithm 1. Monotonicity approach

$x_0 \supseteq \Box\{x \mid A(p)x = b(p) \text{ for some } p \in \boldsymbol{p}\}$
// potential candidate for parallelisation
for $k = 1$ to K **do**
    // potential candidate for vectorisation
    $\boldsymbol{D}_k \supseteq \Box\{y \mid \exists p \in \boldsymbol{p} \; A(p)y = \partial b/\partial p_k - \partial A(p)/\partial p_k x_0\}$
end for
// potential candidate for parallelisation
for $i = 1$ to n **do**
  for $k = 1$ to K **do**
      // Assign a value to e_k^i based on $\boldsymbol{D}_{ki}$
  end for
  // potential candidates for vectorisation
  $\boldsymbol{x}_i^{\min} \supseteq \Box\{x \mid A(p^{-e})x = b(p^{-e})\}_i$
  $\boldsymbol{x}_i^{\max} \supseteq \Box\{x \mid A(p^e)x = b(p^e)\}_i$
  $\boldsymbol{x}_i^{\text{out}} = [\underline{\boldsymbol{x}}_i^{\min}, \overline{\boldsymbol{x}}_i^{\max}]$
end for

---

run for a suitable number of iterations. Such loops however, must meet certain constraints including continuous access of memory, no data dependency and only single exit from the loop.

The newest Intel and AMD processors implements Advanced Vector Extensions (AVX) instruction set that operates on 256 bit SIMD registers. For double precision floating point numbers this allows to perform basic mathematical operations on 4 numbers at once. Example of addition with SIMD is shown in Fig. 1. In the experiments presented in this paper, the newest Intel C++ compiler (16.0) is used. This compiler efficiently analyse the code and indicates which loops are worth to be vectorised. It can also be forced to vectorise other loops by using #pragma simd. Both these mechanisms are used in the experiments to improve the efficiency of the monotonicity approach.

While vectorisation plays only a supporting role in parallelisation process, the main benefit can be achieved by transforming the code so that it is able to utilise many threads simultaneously. This is realised in either task parallelism model (the so called fork-join parallelism) or single program multiple data (SPMD) model. For a single multi-core processor the first model is usually implemented as parallel constructs nest in a straightforward manner [21].

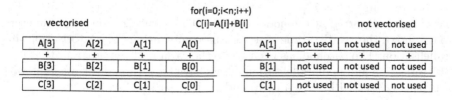

Fig. 1. Loops vectorisation

Similarly to the vectorisation, parallelisation can be done automatically by a C++ compiler or can be guided by a user. When using an Intel compiler three methods can be used: Threading Building Blocks (TBB) or Cilk Plus (originally developed in MIT) and auto parallelisation with OpenMP. The two first methods use work-stealing strategy, in which each processor maintains its own local queue and when the local queue is empty, the worker randomly steals work from victim worker queues, while in OpenMP a master thread forks a specified number of slave threads and divides a task among them. In our experiments all three types of parallelisation are used.

5   Numerical Examples

To check the performance of the monotonicity approach some illustrative examples of structural mechanical systems are considered. The obtained results are compared with the results given by a method (it should be added that there are several such methods [2,4,12,17], however a great majority of them yields very similar results, especially for the problem considered here) for computing outer interval solution of parametric interval linear systems.

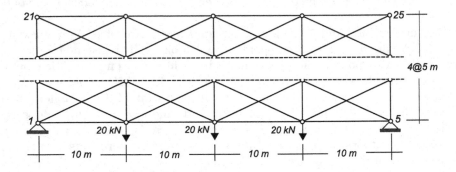

Fig. 2. Example 1: 5-bay 4-floor plane truss structure

Example 1 (*5-bay 4-floor plane truss structure*). For the plane truss structure shown in Fig. 2 the displacements of the nodes are computed. The truss is subjected to downward forces $P_2 = P_3 = P_4 = 20[\text{kN}]$ as depicted in the figure; Young's modulus $Y = 2.0 \times 10^{11}[\text{Pa}]$, cross-section area $C = 0.0001[\text{m}^2]$, length of horizontal bars is $L = 10[\text{m}]$, and the lenght of vertical bars is $H = 5[\text{m}]$. The truss is fully supported at the nodes 1 and 5. This gives 72 interval parameters. Table 1 shows the relative times for various combinations of vectorisation and parallelisation. The baseline has been set to the variant of the program that used no vectorisation and no parallelisation. Tests have been run on the machine with Intel Xeon 1220v2 CPU with 4 cores and no hyper-threading ability.

The times presented in the table show the benefits that can be achieved by the vectorisation of the monotonicity algorithm, which are 10 % on average.

Table 1. Comparison of the computation times for Example 1

	Bridge_1	Bridge_4	Bridge_5	Bridge_6
no vectorisation; no parallelisation	76.09	76.38	77.06	77.27
forced vectorisation; no parallelisation	98.04	97.4	98.17	98.04
auto vectorisation; no parallelisation	69.61	69.79	70.09	70.18
auto vectorisation; auto parallelisation	69.14	71.76	21.26	21.67
no vectorisation; Cilk parallelisation	22.68	22.71	22.80	23.17
auto vectorisation; Cilk parallelisation	20.72	20.59	19.23	19.36
auto vectorisation; TBB parallelisation	18.68	19.93	19.03	19.06

Automatic parallelisation does not improve the processing times and in one case it consumes even more time. This is due to the fact, that compiler cannot be sure that the processing data are fully independent. When Cilk and TBB methods have been used, the improvement is significant and when combined with vectorisation they can improve the processing times up to four times.

Example 2 (Baltimore bridge built in 1870). Consider the plane truss structure shown in Fig. 3 subjected to downward forces of $P_1 = 80[kN]$ at node 11, $P_2 = 120[kN]$ at node 12 and P_1 at node 15; Young's modulus $Y = 2.1 \times 10^{11}$ $[Pa]$, cross-section area C = 0.004$[m^2]$, and length $L = 1[m]$. Assume that the stiffness of 23 bars is uncertain by $\pm 5\%$. This gives 23 interval parameters.

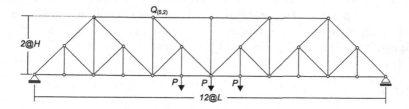

Fig. 3. Example 2: Baltimore bridge

The comparison of the performance of different variants of code vectorisation and paralellisation is presented in Table 2. Again the tests have been run on Intel Xeon 1220v3 processor with 4 cores. For the more complex problem the benefit in processing times is higher and equals up to 15 %. When we combine either Cilk or TBB parallelisation method with vectorisation the overall improvement reaches more than four times. The difference between both two fork-joint models is unconvincing, so each of them can be successfully applied. The overall computational experiments prove, that the more complex problem is the more is the benefit from using parallelisation and vectorisation.

Table 2. Comparison of the computation times for Example 2

	Bridge_1	Bridge_4	Bridge_5	Bridge_6
no vectorisation; no parallelisation	150.08	150.29	149.52	149.86
forced vectorisation; no parallelisation	203.61	204.03	203.25	203.39
auto vectorisation; no parallelisation	131.73	131.32	130.59	130.55
auto vectorisation; auto parallelisation	131.21	130.11	129.61	129.56
no vectorisation; Cilk parallelisation	41.67	41.57	41.31	40.85
auto vectorisation; Cilk parallelisation	36.73	35.89	35.75	36.06
auto vectorisation; TBB parallelisation	36.09	35.57	35.39	35.76

6   Conclusions

Checking the sign of the derivatives is a clue to test the global monotonicity of the solution of parametric linear systems. The global monotonicity enables calculating the interval hull solution easily by solving at most $2n$ real systems. The main deficiency of the monotonicity approach is its poor performance. As shown by the performed experiments, the performance of the monotonicity approach can be improved by using techniques such as vectorisation and parallelisation, which are available for contemporary C++ and Fortran compilers like Visual C++, Gnu cpp or Intel compiler in Parallel Studio XE.

The presented methodology can be applied to any problem which requires solving linear systems with input data dependent on uncertain parameters. The monotonicity approach is also a crucial acceleration techniques for interval global optimisation applied for the problem of solving parametric interval linear systems. The improved version of monotonicity approach can significantly decrease the computational time of the interval global optimisation.

References

1. de Figueiredo, L.H., Stolfi, J.: Self-validated numerical methods and applications. In: Brazilian Mathematics Colloquium Monographs, IMPA/CNPq, Rio de Janeiro, Brazil (1997)
2. Hladík, M.: Enclosures for the solution set of parametric interval linear systems. Int. J. Appl. Math. Comput. Sci. **22**(3), 561–574 (2012)
3. Jansson, C.: Interval linear systems with symmetric matrices, skew-symmetric matrices and dependencies in the right hand side. Computing **46**(3), 265–274 (1991)
4. Kolev, L.: A method for outer interval solution of linear parametric systems. Reliable Comput. **10**, 227–239 (2004)
5. Kolev, L.: Outer solution of linear systems whose elements are affine functions of interval parameters. Reliable Comput. **6**, 493–501 (2002)
6. Kolev, L.: Solving linear systems whose elements are non-linear functions of intervals. Numer. Algorithms **37**, 213–224 (2004)

7. Muhanna, R.L., Erdolen, A.: Geometric uncertainty in truss systems: an interval approach. In: Rafi, L., Muhanna, R.L.M., (eds.): Proceedings of the NSF Workshop on Reliable Engineering Computing: Modeling Errors and Uncertainty in Engineering Computations, Savannah, Georgia, USA, 22–24 February 2006, pp. 239–247 (2006)

8. Muhanna, R., Kreinovich, V., Solin, P., Cheesa, J., Araiza, R., Xiang, G.: Interval finite element method: new directions. In: Rafi, L., Muhannah, R.L.M., (eds.) Proceedings of the NSF Workshop on Reliable Engineering Computing (REC), Svannah, Georgia USA, 22–24 February 2006, pp. 229–244 (2006)

9. Neumaier, A.: Worst case bounds in the presence of correlated uncertainty. In: Rafi, L., Muhannah, R.L.M., (eds.): Proceedings of the NSFWorkshop on Reliable Engineering Computing (REC), Savannah, Georgia USA, 22–24 February 2006, pp. 113–114 (2006)

10. Neumaier, A.: Interval Methods for Systems of Equations. Encyclopedia of Mathematics and Its Applications. Cambridge University Press, Cambridge (1990)

11. Popova, E.D.: On the solution of parametrised linear systems. In: Kraemer, W., J.W.v.G., (eds.) Scientific Computing, Validated Numerics, Interval Methods, Kluwer Academic Publishers, pp. 127–138 (2001)

12. Popova, E., Lankov, R., Bonev, Z.: Bounding the response of mechanical structures with uncertainties in all the parameters. In: Rafi, L., Muhannah, R.L.M., (eds.) Proceedings of the NSF Workshop on Reliable Engineering Computing (REC), Svannah, Georgia USA, 22–24 February 2006, pp. 245–265 (2006)

13. Pownuk, A.: Calculations of displacement in elastic and elastic-plastic structures with interval parameters. In: 33rd Solid Mechanics Conference, Zakopane, Poland, pp. 160–161, September 2000

14. Rao, S., Berke, L.: Analysis of uncertain structural systems using interval analysis. AIAA J. $35(4)$, 727–735 (1997)

15. Rohn, J.: A method for handling dependent data in interval linear systems. Technical report 911, Academy of Sciences of the Czech Republic, Czech Republic (2004)

16. Rump, S.: Verification methods for dense and sparse systems of equations. In: Herzberger, J. (ed.) Topics in Validated Computations. Studies in Computational Mathematics, vol. 5, pp. 63–135. Elsevier, Amsterdam (1994)

17. Skalna, I.: A method for outer interval solution of parametrized systems of linear interval equations. Reliable Comput. $12(2)$, 107–120 (2006)

18. Skalna, I.: Methods for solving systems of linear equations of structure mechanics with interval parameters. Comput. Assist. Mech. Eng. Sci. $10(3)$, 281–293 (2003)

19. Skalna, I.: Evolutionary optimization method for approximating the solution set hull of parametric linear systems. In: Boyanov, T., Dimova, S., Georgiev, K., Nikolov, G. (eds.) NMA 2006. LNCS, vol. 4310, pp. 361–368. Springer, Heidelberg (2007)

20. Skalna, I.: On checking the monotonicity of parametric interval solution of linear structural systems. In: Wyrzykowski, R., Dongarra, J., Karczewski, K., Wasniewski, J. (eds.) PPAM 2007. LNCS, vol. 4967, pp. 1400–1409. Springer, Heidelberg (2008)

21. Cytron, R., Lipkis, J., Schonberg, E.: A compiler-assisted approach to SPMD execution. In: Proceedings of the 1990 ACM/IEEE Conference on Supercomputing (Supercomputing 1990), CA, USA, pp. 398–406. IEEE Computer Society Press, Los Alamitos (1990)

Preliminary Experiments with an Interval Model-Predictive-Control Solver

Bartłomiej Jacek Kubica[1,2(✉)]

[1] Institute of Control and Computation Engineering,
Warsaw University of Technology, Warsaw, Poland
bartlomiej_kubica@sggw.pl
[2] Department of Applied Informatics, Faculty of Applied Informatics
and Mathematics, Warsaw University of Life Sciences,
Nowoursynowska 159, 02-776 Warsaw, Poland

Abstract. Model-Predictive Control (MPC) is a popular advanced control technique, known for its robustness and simplicity in taking control constraints into account. In recent years, the interest grows in applying interval methods to compute MPC. The paper applies interval methods in a simple case. Numerical results for a benchmark problem are presented.

Keywords: Interval computations · Model-predictive control · Dual mode · Zonotopes · Multithreaded programming · TBB

1 Introduction

Model-Predictive Control (MPC) can be applied for several systems. It has important advantages with respect to classical PID controllers. They include robustness and possibility of taking constraints on the control variables into account. Also, it is natural to use it for nonlinear systems, while PID requires linear ones – at least locally.

As it reduces the problem of finding the control to an optimization problem, interval methods can be applied. This fact has been noticed by MPC experts, who developed interval algorithms to compute the control, satisfying certain conditions on the predicted behavior; cf. [5,16,17]. Yet the problem is hard and it seems to high knowledge on interval algorithms is required to provide well-tuned algorithms. General-purpose interval solvers do not seem to be adequate to features of this specific problem. Hence, we would like to encourage other interval researchers to contribute to this important topic. Some interesting examples of such contributions are [6,21,22].

2 Model-Predictive Control

We want to compute the control $u_k = \left(u_k(1), \ldots, u_k(n_u)\right)^T \in \mathbb{R}^{n_u}$ at state k. We might not know the current state $x_k = \left(x_k(1), \ldots, x_k(n_x)\right)^T$ of the system precisely, but we have some information on it (e.g., the interval $\mathbf{x}_k \subseteq \mathbb{R}^{n_x}$ guaranteed to contain x_k).

© Springer International Publishing Switzerland 2016
R. Wyrzykowski et al. (Eds.): PPAM 2015, Part II, LNCS 9574, pp. 464–473, 2016.
DOI: 10.1007/978-3-319-32152-3_43

For any sequence of controls $u_k, u_{k+1}, \ldots, u_{k+N}$ we can compute predictions of the states: $x_{k+1|k}, x_{k+2|k}, \ldots, x_{k+N|k}$. These predictions are made, based on the model (hence the name of the approach) of the system:

$$x_{k+1} = f(x_k, u_k, w_k), \qquad (1)$$

where w_k is the disturbance. Function $f(\cdot)$ can, in general, be nonlinear.

In the simplest case, there are no disturbances, $x_{k+1} = f(x_k, u_k)$ and – for a known control policy – we can compute precise predictions of all future states.

If we have a quality measure $Q(u, x)$, we can optimize it:

$$\min_u Q(u, x) \qquad (2)$$

s.t.

$$x_{l+1} = f(x_l, u_l), \; l = k, k+1, \ldots, l+N,$$

thus reducing a control problem to an optimization problem. In the above formulae, we denote the whole sequences of controls and states by: $u = (u_k, \ldots, u_{k+N})^T$, $x = (x_k, \ldots, x_{k+N})^T$.

What is the prediction horizon N? It depends on the problem what control horizon we have. In general, the control horizon can be finite or infinite. The prediction horizon is finite, usually, and its appropriate choice is crucial for both correctness and convergence of the control algorithm. Details on how to choose it can be found, e.g., in [22]; in our study we treat it as given.

When do we solve the above optimization problem and what control do we, actually, apply? There are various possibilities (see, e.g., [21]), but usually, we solve the optimization problem at each stage k, apply the *first* control u_k and, in the next stage, we re-compute further control variables.

Please note, the control is (or – can be) suboptimal, because:

– a finite horizon N is considered in each stage, while the process will (probably) continue after it,
– the disturbances might affect the result.

2.1 Dual Mode MPC

PID regulators can be applied to linear systems – or at least linearized ones. In a small vicinity of the linearization point (the state and reference control value), PID regulators (and many other more or less traditional approaches to control) perform reasonably well. The most important goal for MPC is to drive the system into the proper region, where less sophisticated algorithms are sufficient.

Such an approach, i.e., switching between MPC and a less advanced algorithm, is called *dual mode MPC* and is the most common situation (not the only one, though – e.g., [6]) to apply interval methods; cf. [5,16]. We adopt this approach, also.

In this case, the main feature of MPC, in which we are interested, is *robustness*.

When our goal is to change the system state robustly, the disturbances can be treated in the *min-max* manner:

$$\min_{u} \max_{w} Q(x, u, w).$$

What we want is making the system move to the given state set *for any* possible disturbances values.

This approach can be extended by using interval representations of random variables (see, e.g., [11,14]). This results in robust optimization of the objective's expected value:

$$\min_{u} \mathbb{E}_{w} Q(x, u, w).$$

This interesting approach will not be considered in the paper, but can be a fruitful subject of future research.

3   Implementation

We want to solve the equality-constrained optimization problem (2). There are many instances of the interval branch-and-bound method that can be applied to it; cf. [7,9,23].

Before we describe the overall algorithm, let us consider the constraints under solution.

We adopt the notation of [10], where intervals and interval vectors are denoted by boldface italics (e.g., **x**, **y**, etc.) while inclusion functions, evaluated in interval arithmetic – by sanserif letters (e.g., f, g, etc.).

3.1   Solving Constraints

The problem has several non-typical properties. The most important of them is the structure of constraints. Actually, they give explicit formulae for some variables and it is very easy to process them using hull-consistency [4]. Also, this is equivalent to narrowing using the interval Newton operator:

$$\mathbf{x}_{l+1}(i) = \operatorname{mid} \mathbf{x}_{l+1}(i) - \frac{f(\mathbf{x}_l, \mathbf{u}_l) - \operatorname{mid} \mathbf{x}_{l+1}(i)}{\partial\left(\frac{f(\mathbf{x}_l, \mathbf{u}_l) - x_{l+1}(i)}{\partial x_{l+1}(i)}\right)} = \tag{3}$$

$$= \operatorname{mid} \mathbf{x}_{l+1}(i) - \frac{f(\mathbf{x}_l, \mathbf{u}_l) - \operatorname{mid} \mathbf{x}_{l+1}(i)}{-1} =$$

$$= f(\mathbf{x}_l, \mathbf{u}_l)$$

This means, we can use the formulae:

$$\mathbf{x}_{k+1}^{new} = f(\mathbf{x}_k, \mathbf{u}_k), \qquad \mathbf{x}_{k+1} = \mathbf{x}_{k+1} \cap \mathbf{x}_{k+1}^{new}, \tag{4}$$

for narrowing.

Also, verifying $\mathbf{x}_{k+1}^{new} \subseteq \mathbf{x}_{k+1}$ proves that there is a (unique) solution – as in other forms of the interval Newton operator (cf. [7,9,12,23]).

If we want to solve a constraint with respect to $x_k(i)$, $i = 1, \ldots, n_x$ or to $u_k(i)$, $i = 1, \ldots, n_u$, the situation is a bit more difficult. Unlike [18] we do not assume that constraints can be reformulated to obtain explicit formulae for these variables, so we have to use a Newton operator in a more traditional form. We choose the Gauss-Seidel operator:

$$\mathbf{x}_k^{new}(i) = \check{x}_k(i) - \Big(\mathsf{f}\big(\check{u}_k(i), \check{x}_k(i)\big) - \sum_{j=1,j\neq i}^{n_x} \mathbf{A}_{ij}^x \cdot \big(\mathbf{x}_k(j) - \check{x}_k(j)\big) +$$

$$- \sum_{j=1}^{n_u} \mathbf{A}_{ij}^u \cdot \big(\mathbf{u}_k(j) - \check{u}_k(j)\big) - \mathbf{x}_{k+1}\Big)/\mathbf{A}_{ii}^x,$$

$$\mathbf{u}_k^{new}(i) = \check{u}_k(i) - \Big(\mathsf{f}\big(\check{u}_k(i), \check{x}_k(i)\big) - \sum_{j=1}^{n_x} \mathbf{A}_{ij}^x \cdot \big(\mathbf{x}_k(j) - \check{x}_k(j)\big) +$$

$$- \sum_{j=1,j\neq i}^{n_u} \mathbf{A}_{ij}^u \cdot \big(\mathbf{u}_k(j) - \check{u}_k(j)\big) - \mathbf{x}_{k+1}\Big)/\mathbf{A}_{ii}^u.$$

In the above formulae, $\mathbf{A}^x$ and A^u are submatrices of the Jacobi matrix of $\mathsf{f}(\mathbf{u}_k, \mathbf{x}_k)$ and quantities with a check denote midpoints (for brevity).

These narrowing operators are applied for all $i = 1, \ldots, n_x$ and for all $i = 1, \ldots, n_u$ – to narrow all $\mathbf{x}_k$'s and $\mathbf{u}_k$'s (it is an underdetermined equation, cf. [12,13]).

Forward and backward iteration. In the beginning we narrow forward all states starting from the initial state (to obtain initial bounds on x_k for $k = 1, \ldots, N$. Then, we perform backward narrowing for states $k = N - 1, N - 2, \ldots, 1$ to narrow bounds for x_k's and u_k's.

Both forward and backward narrowing are performed after each bisection – always starting at k, for which one of the control variable domains had been bisected. As in [18], when no improvement is obtained for some $\mathbf{x}_k$, we break narrowing as it will improve nothing further, without another bisection.

3.2   Branch-and-Bound Method

We have two kinds of variables: u and x. Boxes contain both of them – the pair $(\mathbf{u}, \mathbf{x})$ – but bisection is performed over u's, only. Ranges on x's are computed according to the transfer Eq. (4).

Our implementation is going to be parallel – the concurrency is obtained using the task-based approach of TBB [2]. A convenient version of the b&b algorithm is the one presented in [19]. It can be expressed by the pseudocode, presented in Algorithm 1.

Algorithm 1. Branch-and-bound method – Lyudvin's version

Require: $\mathbf{u}^{(0)}, \mathbf{x}^{(0)}, \mathbf{f}, Q, \varepsilon$
1: $L = \{(\mathbf{u}^{(0)}, \mathbf{x}^{(0)})\}$
2: $\mathbf{q} = Q(\mathbf{u}^{(0)}, \mathbf{x}^{(0)})$
3: $\mathbf{q}_{opt} = \overline{q}$ {upper bound on the global minimum}
4: $\gamma = \underline{q} + \frac{\text{wid} \, \mathbf{q}}{10^5}$ {threshold value}
5: **repeat**
6:     **for all** $(\mathbf{u}, \mathbf{x}) \in L$ **do**
7:         narrow $\mathbf{x}$ and $\mathbf{u}$
8:         $\mathbf{q} = Q(\mathbf{u}, \mathbf{x})$
9:         **if** $(\underline{q} < threshold)$ **then**
10:             put $(\mathbf{u}, \mathbf{x})$ to L_{cache}
11:         **else**
12:             put $(\mathbf{u}, \mathbf{x})$ to L_{other}
13:         **end if**
14:     **end for**
15:     **while** $(L_{cache}! = \emptyset)$ **do**
16:         take $(\mathbf{u}, \mathbf{x})$ from L_{cache}
17:         process $(\mathbf{u}, \mathbf{x})$ using narrowing tests
18:         update q_{opt}, if possible
19:         **if** (the box has been discarded) **then**
20:             **continue**
21:         **else if** $(\text{wid} \, \mathbf{u} < \varepsilon)$ **then**
22:             put $(\mathbf{u}, \mathbf{x})$ to L_{sol}
23:         **else**
24:             bisect $\mathbf{u}$ and put subboxes to L_{cache} or L_{other}, according to the objective
                values
25:         **end if**
26:     **end while**
27:     $\gamma = \gamma \cdot 1.1$ {update the threshold value}
28:     $L = L_{other}$
29: **until** $L_{other} == \emptyset$
30: discard all $(\mathbf{u}, \mathbf{x}) \in L_{sol}$ such that $Q(\mathbf{u}, \mathbf{x}) > q_{opt}$
31: **return** L_{sol}

3.3   Processing a Box

In the b&b schema, it is important to choose proper tools to process a single box. Some of them have already been presented in Subsect. 3.1. Other techniques include:

– discarding boxes for which the objective exceeds the value q_{opt}; see Algorithm 1,
– using gradients to prune the boxes, cutting off regions with too high objective value – in a similar manner as [20].

In this context, an interesting is to use so-called *total derivatives*.

3.4   Total Derivatives

Q, as defined in (2), is q function of several variables u_k and x_k for $k = 1, \ldots, N$. Please note, these variables are not independent: each state x_k depends on earlier states and controls.

The gradient of a function is the vector of partial derivatives. But in our case an interesting information is carried by the vector of all total derivatives with respect to u_k, i.e.:

$$\frac{dQ}{du_k} = \frac{\partial Q}{\partial u_k} + \sum_{i=k+1}^{N} \frac{\partial Q}{\partial x_i} \cdot \frac{dQ}{dx_i}. \tag{5}$$

Thanks to using total derivatives, we can perform a form of the monotonicity test (see, e.g., [9]), which would not be applicable, otherwise, as the problem is equality constrained.

Unfortunately, computing total derivatives is a relatively time-consuming operation, that would not always be worthwhile. Comparison of algorithm versions using total derivatives or not, will be presented in Sect. 4.

3.5   Wrapping Effect

Wrapping effect is encountered in virtually all problems in which interval computations are chained – computing sequences, solving differential equations, difference equations, etc. Some papers, e.g., [5,17], suggest using the arithmetic of zonotopes [15] to prevent the exponential increase of errors. Others, e.g., [18], suggest using Taylor arithmetic.

The author tried to use zonotopes, but did not obtain satisfying results. Even if the computations were a bit more precise, the overhead of computing all the derivatives was very large. Computing $f(\cdot)$ in simple interval arithmetic performed better.

This might be caused by some features of our implementation, e.g., we do not use the *cascade reduction*, described in [15]. This is going to be subject to future research.

Also, we tested switching between using and not using zonotopes, according to the box width. If the maximal diameter is below a given threshold value, we use the approximation based on zonotopes; for larger boxes – the traditional interval arithmetic is used.

3.6   Filtering

In this paper we assume that we can observe the state of the system precisely. However, in many practical problems the state is not measured with a significant error. In this case, we have to estimate the current state value x_k from current (and possibly previous) system outputs y_k.

Solving the equations system $g(\mathbf{u}_k, \mathbf{x}_k) = \mathbf{y}_k$ using interval methods is straightforward (cf. [8]). Also, zonotopes can be applied directly (see, e.g., [5]).

4   Computational Experiments

Numerical experiments have been performed on a computer with 4 cores (allowing hyper-threading), i.e., an Intel Core i7-3632QM with 2.2 GHz clock. The machine ran under control of a 64-bit Manjaro 0.8.8 GNU/Linux operating system with the GCC 4.8.2, glibc 2.18 and the Linux kernel 3.10.22-1-MANJARO.

The solver has been written in C++ and compiled using the GCC compiler (with the option enabling C++11 standard). The C-XSC library (version 2.5.3) [1] was used for interval computations. The parallelization was done with TBB 4.2, update 2 [2]. OpenBLAS 0.2.8 [3] was linked for BLAS operations.

The parallelization of four threads has been considered.

4.1   Example – The Pendulum Problem

This example is taken from [18]. We have a horizontally movable carriage on which a pendulum in mounted. We want to set the acceleration $u \in \mathbb{R}$ of the carriage to control position q and velocity $\dot{q}$ of the pendulum. Specifically, we want to set it to a zero position and keep it there.

The state is $x = (q, \dot{q})^T \in \mathbb{R}^2$. The initial state is assumed in [18] to be $(-\pi, 0)^T$

The system can be described by the following differential equations:

$$\dot{x}(1) = x(2), \tag{6}$$
$$\dot{x}(2) = K_{\sin} \cdot \sin x(1) - K_{\cos} \cdot u \cdot \cos x(2).$$

This results in the following transition equations:

$$x_{k+1}(1) = x_k(1) + \delta \cdot x_k(2), \tag{7}$$
$$x_{k+1}(2) = x_k(2) + \delta \cdot \Big(K_{\sin} \cdot \sin x(1) - K_{\cos} \cdot u \cdot \cos x(2) \Big).$$

The objective to minimize is:

$$Q(u, x) = \sum_{k=1}^{N} \Big(u_k^2 + 0.5 x_k(1)^2 + 0.5 x_k(2)^2 \Big).$$

Values of all parameters and bounds are given in tables in [18].

4.2   Numerical Results

We consider nine versions of the b&b algorithm. Description of each of them consists of two parts:

– does it use zonotopes: "no zono","zono" or "zono-switch" – switching between not using them for boxes with width exceeding a given threshold (equal to 1.0),
– does it use total derivatives: "no total", "total" or tot-switch" – switching between not using them for boxes with width exceeding a given threshold (also, equal to 1.0).

Table 1. Results the example with no additional constraints on states

Alg. version	f evals	∇f evals	obj.evals	∇ obj.evals	∇ total	bisecs	sol.boxes	time
no zono, no total	62272197	5236810	10755619	5691059	—	2855715	661842	148.9 s
no zono, total	471898	1673420	142083	0	36263	22468	12	7.9 s
no zono, tot-switch	1591850	773594	409509	119795	14087	77126	12	5.8 s
zono, no total	56008017	56008062	9813837	5115933	—	2572451	587292	304.3 s
zono, total	623772	2696607	224399	0	46063	30951	8	12.1 s
zono, tot-switch	1781893	2370898	638748	132553	13089	87305	8	13.0 s
zono-switch, no total	55822335	55147023	9435497	5104823	—	2562597	587292	296.9 s
zono-switch, total	464108	1920419	140145	0	35737	22205	8	8.8 s
zono-switch, tot-switch	1590710	1545488	409112	119795	14003	77084	8	9.2 s

The following notation is used in the Table 1:

- alg. version: "no zono, no total", "no zono, total", "no zono, tot-switch", "zono, no total", "zono, total", "zono, tot-switch", "zono-switch, no total", "zono-switch, total" or "zono-switch, tot-switch" (cf. the above paragraph),
- f evals – number of f's evaluations,
- ∇f evals – number of f's gradients evaluations,
- obj.evals – number of Q's evaluations,
- ∇ obj.evals – number of Q's gradients evaluations,
- ∇ total – number of Q's total derivative's vectors evaluations,
- bisecs –number of bisections,
- sol.boxes –number of resulting boxes in the list L_{sol},
- time – execution time.

5   Conclusions

We presented an attempt to apply interval methods in solving MPC problems. Results obtained by other researcher are compared briefly and commented.

It occurred very worthwhile to use total derivatives – both in terms of computation time and accuracy. To the best knowledge of the author, using total derivatives was an original idea, not considered in other papers. It is worth noting that total derivatives work best for relatively small boxes (for too large ones, their interval enclosures are too wide). Hence, switching between using or not using them occurred to be worthwhile, also. It outperforms the version using total derivatives always at least for the case of not using zonotopes. Improving the heuristic for this switching is going to be one of the subjects of future research.

As for using zonotopes, our results differ from these obtained by others. E.g., using the Kühn method, based on zonotopes, did not occur worthwhile in our experiments; cf. [5] or [17]. Reasons of that remain to be determined.

As for switching between using zonotopes and not using them, they give intermediate results. The version not using zonotopes at all is always the most efficient – even, if a bit less accurate (higher number of solution boxes resulting from the algorithm).

There are still wide possibilities of tuning the algorithm. They are going to be studied in our future research.

Acknowledgments. The author is grateful to Adam Woźniak for inspiration, interesting discussions, support and all the invaluable help.

References

1. C++ eXtended Scientific Computing library (2014). http://www.xsc.de
2. Intel Threading Building Blocks (2014). http://www.threadingbuildingblocks.org
3. OpenBLAS library (2014). http://xianyi.github.com/OpenBLAS/
4. Benhamou, F., Goualard, F., Granvilliers, L., Puget, J.F.: Revising hull and box consistency. In: International Conference on Logic Programming, pp. 230–244. The MIT Press (1999)
5. Bravo, J.M., Alamo, T., Camacho, E.F.: Robust MPC of constrained discrete-time nonlinear systems based on approximated reachable sets. Automatica **42**(10), 1745–1751 (2006)
6. Dombrovskii, V.V., Chausova, E.V.: Model predictive control for linear systems with interval and stochastic uncertainties. Reliable Comput. **19**(4), 351–360 (2014)
7. Hansen, E., Walster, W.: Global Optimization Using Interval Analysis. Marcel Dekker, New York (2004)
8. Jaulin, L., Kieffer, M., Didrit, O., Walter, E.: Applied Interval Analysis. Springer, London (2001)
9. Kearfott, R.B.: Rigorous Global Search: Continuous Problems. Kluwer, Dordrecht (1996)
10. Kearfott, R.B., Nakao, M.T., Neumaier, A., Rump, S.M., Shary, S.P., van Hentenryck, P.: Standardized notation in interval analysis. Vychislennyie tiehnologii (Computational technologies) **15**(1), 7–13 (2010)
11. Kubica, B.J.: Estimating utility functions of network users - an algorithm using interval computations. Ann. Univ. Timisoara **40**, 121–134 (2002)
12. Kubica, B.J.: Interval methods for solving underdetermined nonlinear equations systems. Reliable Comput. **15**, 207–217 (2011)
13. Kubica, B.J.: Presentation of a highly tuned multithreaded interval solver for underdetermined and well-determined nonlinear systems. Numer. Algorithms **70**, 1–35 (2015). http://dx.doi.org/10.1007/s11075-015-9980-y
14. Kubica, B.J., Malinowski, K.: Interval random variables and their application in queueing systems with long-tailed service times. In: Lawry, J., Miranda, E., Miranda, A., Li, S., Gil, M.A., aw Grzegorzewski, P., Hyrniewicz, O. (eds.) Soft Methods for Integrated Uncertainty Modelling. Advances in Soft Computing, vol. 37, pp. 393–403. Springer, Heidelberg (2006)
15. Kühn, W.: Rigorously computed orbits of dynamical systems without the wrapping effect. Computing **61**(1), 47–67 (1998)
16. Limon, D., Alamo, T., Bravo, J., Camacho, E., Ramirez, D., de la Peña, D.M., Alvarado, I., Arahal, M.: Interval arithmetic in robust nonlinear MPC. In: Assessment and Future Directions of Nonlinear Model Predictive Control, pp. 317–326. Springer (2007)

17. Limon, D., Bravo, J., Alamo, T., Camacho, E.: Robust MPC of constrained nonlinear systems based on interval arithmetic. IEE Proc. Control Theory Appl. **152**(3), 325–332 (2005)
18. Lyre, F., Poignet, P.: Nonlinear model predictive control via interval analysis. In: 44th IEEE Conference on Decision and Control, 2005 and 2005 European Control Conference, CDC-ECC 2005, pp. 3771–3776. IEEE (2005)
19. Lyudvin, D.Y., Shary, S.P.: Testing implementations of pps-methods for interval linear systems. Reliable Comput. **19**(2), 176–196 (2013). SCAN 2012 Proceedings
20. Martínez, J.A., Casado, L.G., García, I., Sergeyev, Y.D., Tóth, B.: On an efficient use of gradient information for accelerating interval global optimization algorithms. Numer. Algorithms **37**(1–4), 61–69 (2004)
21. Rauh, A., Hofer, E.P.: Interval methods for optimal control. In: Variational Analysis and Aerospace Engineering, pp. 397–418. Springer (2009)
22. Rauh, A., Senkel, L., Kersten, J., Aschemann, H.: Interval methods for sensitivity-based model-predictive control of solid oxide fuel cell systems. Reliable Comput. **19**(4), 361–384 (2014)
23. Shary, S.P.: Finite-dimensional Interval Analysis. XYZ (2013), electronic book (in Russian). http://www.nsc.ru/interval/Library/InteBooks/SharyBook.pdf. Accessed 15 May 2014

Interval Nine-Point Finite Difference Method for Solving the Laplace Equation with the Dirichlet Boundary Conditions

Malgorzata A. Jankowska[✉]

Institute of Applied Mechanics, Poznan University of Technology,
Jana Pawla II 24, 60-965 Poznan, Poland
malgorzata.jankowska@put.poznan.pl

Abstract. An interval version of the conventional nine-point finite difference method for solving the two-dimensional Laplace equation with the Dirichlet boundary conditions is proposed. This interval scheme is interesting due to the fact that the local truncation error of the conventional method is of the high (fourth) order, but it becomes of the sixth order for square mesh. In the theoretical approach presented, this error is bounded by some interval values and we can prove that the exact solution belongs to the interval solutions obtained.

Keywords: Interval nine-point finite difference method · Interval arithmetic · Laplace equation

1 Introduction

Interval methods together with interval arithmetic are eagerly developed since the early papers by Sunaga [17] and Moore [11,12] appeared. Such methods represent quite a new approach to numerical computations. It is due to the fact that interval solutions obtained with such methods include the exact solution of the problem. An approach that we can choose to construct an interval method for solving the initial-boundary value problems is based on conventional finite difference methods (FDMs). As examples, we can mention an interval FDM of Crank-Nicolson type for solving the heat conduction equation with the Dirichlet and mixed boundary conditions [5,6,9], an interval backward FDM for solving the diffusion equation [4], an interval FDM for solving the wave equation [18,19], an interval central FDM for solving the Poisson equation [3] and some interval FDM applied for numerical modeling of skin tissue heating [10]. Verified solutions of the 3D Laplace equation are also presented in [8]. Finally, an approach of Nakao et al. [13]. They introduce the constructive a priori error estimates for a full discrete approximation of the heat equation.

A conventional nine-point FDM was considered e.g. by Kantorovich and Krylov [7], Orszag and Israeli [14], Rosser [15], Anderson et al. [1], Boisvert [2] because of the high order of the local truncation error. This error is $O\left(h^4 + k^4\right)$

© Springer International Publishing Switzerland 2016
R. Wyrzykowski et al. (Eds.): PPAM 2015, Part II, LNCS 9574, pp. 474–484, 2016.
DOI: 10.1007/978-3-319-32152-3_44

(compare also with [16]), but it becomes $O\left(h^6\right)$ for square mesh (see e.g. [7,15]). An interval FDM for solving the Laplace equation with the Dirichlet boundary conditions proposed in the paper is based on the conventional nine-point FD scheme. In the theoretical approach presented, the local truncation error of the conventional method is bounded by some interval values. In such a form it represents a scheme for computation of guaranteed results and we can prove that the exact solution belongs to the interval solutions obtained.

2   Dirichlet Problem and Conventional Nine-Point Finite Difference Method

We consider the Laplace equation with the Dirichlet boundary conditions

$$\nabla^2 u\left(x,t\right) \equiv \frac{\partial^2 u}{\partial x^2}\left(x,y\right) + \frac{\partial^2 u}{\partial y^2}\left(x,y\right) = 0, \quad (x,y) \in \Omega, \tag{1}$$

$$u\left(x,y\right) = \varphi\left(x,y\right), \quad (x,y) \in \partial\Omega. \tag{2}$$

Subsequently, we take a rectangular domain Ω, with the boundary $\partial\Omega$, such that $\Omega = \{(x,y) : 0 \leq x \leq a,\ 0 \leq y \leq b\}$. We have $u(0,y) = \varphi_1(y)$, $u(x,0) = \varphi_2(x)$, $u(a,y) = \varphi_3(y)$, $u(x,b) = \varphi_4(x)$, and we also assume that $\varphi_1(0) = \varphi_2(0)$, $\varphi_2(a) = \varphi_3(0)$, $\varphi_3(b) = \varphi_4(a)$, $\varphi_4(0) = \varphi_1(b)$.

First, we choose two integers n and m. Then, we find the mesh constants h and k such that $h = a/n$ and $k = b/m$. Hence, we get the grid points (x_i, y_j) where $x_i = ih$ for $i = 0, 1, \ldots, n$ and $y_j = jk$ for $j = 0, 1, \ldots, m$.

Now we derive the nine-point finite difference scheme together with an appropriate local truncation error. First, we expand u in the Taylor series about (x_i, y_j) and evaluate it at the points (x_{i+1}, y_j) and (x_{i-1}, y_j). We have

$$u\left(x_{i+1}, y_j\right) = u\left(x_i, y_j\right) + \frac{\partial u}{\partial x}\left(x_i, y_j\right)h + \frac{1}{2}\frac{\partial^2 u}{\partial x^2}\left(x_i, y_j\right)h^2 + \frac{1}{6}\frac{\partial^3 u}{\partial x^3}\left(x_i, y_j\right)h^3$$
$$+ \frac{1}{24}\frac{\partial^4 u}{\partial x^4}\left(x_i, y_j\right)h^4 + \frac{1}{120}\frac{\partial^5 u}{\partial x^5}\left(x_i, y_j\right)h^5 + \frac{1}{720}\frac{\partial^6 u}{\partial x^6}\left(\xi_i^+, y_j\right)h^6, \tag{3}$$

$$u\left(x_{i-1}, y_j\right) = u\left(x_i, y_j\right) - \frac{\partial u}{\partial x}\left(x_i, y_j\right)h + \frac{1}{2}\frac{\partial^2 u}{\partial x^2}\left(x_i, y_j\right)h^2 - \frac{1}{6}\frac{\partial^3 u}{\partial x^3}\left(x_i, y_j\right)h^3$$
$$+ \frac{1}{24}\frac{\partial^4 u}{\partial x^4}\left(x_i, y_j\right)h^4 - \frac{1}{120}\frac{\partial^5 u}{\partial x^5}\left(x_i, y_j\right)h^5 + \frac{1}{720}\frac{\partial^6 u}{\partial x^6}\left(\xi_i^-, y_j\right)h^6, \tag{4}$$

where $\xi_i^+ \in (x_i, x_{i+1})$, $\xi_i^- \in (x_{i-1}, x_i)$. We add the formulas (3)–(4) and we get

$$u\left(x_{i+1}, y_j\right) + u\left(x_{i-1}, y_j\right) = 2u\left(x_i, y_j\right) + \frac{\partial^2 u}{\partial x^2}\left(x_i, y_j\right)h^2 + \frac{1}{12}\frac{\partial^4 u}{\partial x^4}\left(x_i, y_j\right)h^4$$
$$+ \frac{1}{720}h^6\left[\frac{\partial^6 u}{\partial x^6}\left(\xi_i^+, y_j\right) + \frac{\partial^6 u}{\partial x^6}\left(\xi_i^-, y_j\right)\right]. \tag{5}$$

As a consequence of the Darboux's theorem we have

$$\frac{\partial^6 u}{\partial x^6}\left(\xi_i, y_j\right) = \frac{1}{2}\left[\frac{\partial^6 u}{\partial x^6}\left(\xi_i^+, y_j\right) + \frac{\partial^6 u}{\partial x^6}\left(\xi_i^-, y_j\right)\right], \tag{6}$$

for some $\xi_i \in (x_{i-1}, x_{i+1})$. Subsequently, we use the notation

$$\delta_x^2 u(x_i, y_j) = u(x_{i+1}, y_j) - 2u(x_i, y_j) + u(x_{i-1}, y_j),$$
$$\delta_y^2 u(x_i, y_j) = u(x_i, y_{j+1}) - 2u(x_i, y_j) + u(x_i, y_{j-1}).$$

Then, with (6) the formula (5) can we written as

$$\left[1 + \frac{1}{12}h^2 \frac{\partial^2}{\partial x^2}\right] \frac{\partial^2 u}{\partial x^2}(x_i, y_j) = \frac{\delta_x^2 u(x_i, y_j)}{h^2} - \frac{1}{360}h^4 \frac{\partial^6 u}{\partial x^6}(\xi_i, y_j). \tag{7}$$

Note that applying the above procedure with respect to y_j we have

$$\left[1 + \frac{1}{12}k^2 \frac{\partial^2}{\partial y^2}\right] \frac{\partial^2 u}{\partial y^2}(x_i, y_j) = \frac{\delta_y^2 u(x_i, y_j)}{k^2} - \frac{1}{360}k^4 \frac{\partial^6 u}{\partial y^6}(x_i, \eta_j), \tag{8}$$

for some $\eta_j \in (y_{j-1}, y_{j+1})$.

Now we substitute $\partial^2 u/\partial x^2 (x_i, y_j)$ and $\partial^2 u/\partial y^2 (x_i, y_j)$, obtained from (7)–(8), to the Laplace Eq. (1) expressed at the grid points (x_i, y_j). We have

$$\frac{\delta_x^2 u(x_i, y_j)}{h^2} + \frac{\delta_y^2 u(x_i, y_j)}{k^2}$$
$$+ \frac{1}{12}\frac{k^2}{h^2}\left[\frac{\partial^2 u}{\partial y^2}(x_{i+1}, y_j) - 2\frac{\partial^2 u}{\partial y^2}(x_i, y_j) + \frac{\partial^2 u}{\partial y^2}(x_{i-1}, y_j)\right]$$
$$+ \frac{1}{12}\frac{h^2}{k^2}\left[\frac{\partial^2 u}{\partial x^2}(x_i, y_{j+1}) - 2\frac{\partial^2 u}{\partial x^2}(x_i, y_j) + \frac{\partial^2 u}{\partial x^2}(x_i, y_{j-1})\right] \tag{9}$$
$$= \left[1 + \frac{1}{12}k^2 \frac{\partial^2}{\partial y^2}\right]\frac{1}{360}h^4 \frac{\partial^6 u}{\partial x^6}(\xi_i, y_j) + \left[1 + \frac{1}{12}h^2 \frac{\partial^2}{\partial x^2}\right]\frac{1}{360}k^4 \frac{\partial^6 u}{\partial y^6}(x_i, \eta_j).$$

Then, we express the second derivatives of u in (9) as follows

$$\frac{\partial^2 u}{\partial y^2}(x_{i+1}, y_j) = \frac{1}{k^2}\delta_y^2 u(x_{i+1}, y_j) - \frac{k^2}{12}\frac{\partial^4 u}{\partial y^4}\left(x_i, \overline{\eta}_j^{(1)}\right), \tag{10}$$

$$\frac{\partial^2 u}{\partial y^2}(x_i, y_j) = \frac{1}{k^2}\delta_y^2 u(x_i, y_j) - \frac{k^2}{12}\frac{\partial^4 u}{\partial y^4}\left(x_i, \overline{\eta}_j^{(2)}\right), \tag{11}$$

$$\frac{\partial^2 u}{\partial y^2}(x_{i-1}, y_j) = \frac{1}{k^2}\delta_y^2 u(x_{i-1}, y_j) - \frac{k^2}{12}\frac{\partial^4 u}{\partial y^4}\left(x_{i-1}, \overline{\eta}_j^{(3)}\right), \tag{12}$$

$$\frac{\partial^2 u}{\partial x^2}(x_i, y_{j+1}) = \frac{1}{h^2}\delta_x^2 u(x_i, y_{j+1}) - \frac{h^2}{12}\frac{\partial^4 u}{\partial x^4}\left(\overline{\xi}_i^{(1)}, y_{j+1}\right), \tag{13}$$

$$\frac{\partial^2 u}{\partial x^2}(x_i, y_j) = \frac{1}{h^2}\delta_x^2 u(x_i, y_j) - \frac{h^2}{12}\frac{\partial^4 u}{\partial x^4}\left(\overline{\xi}_i^{(2)}, y_j\right), \tag{14}$$

$$\frac{\partial^2 u}{\partial x^2}(x_i, y_{j-1}) = \frac{1}{h^2}\delta_x^2 u(x_i, y_{j-1}) - \frac{h^2}{12}\frac{\partial^4 u}{\partial x^4}\left(\overline{\xi}_i^{(3)}, y_{j-1}\right), \tag{15}$$

where $\overline{\xi}_i^{(1)}, \overline{\xi}_i^{(2)}, \overline{\xi}_i^{(3)} \in (x_{i-1}, x_{i+1})$ and $\overline{\eta}_j^{(1)}, \overline{\eta}_j^{(2)}, \overline{\eta}_j^{(3)} \in (y_{j-1}, y_{j+1})$. Substituting (10)–(15) to (9), we obtain

$$\alpha\left[u(x_{i+1}, y_j) + u(x_{i-1}, y_j)\right] + \beta\left[u(x_i, y_{j+1}) + u(x_i, y_{j-1})\right] + u(x_{i+1}, y_{j+1})$$
$$+ u(x_{i+1}, y_{j-1}) + u(x_{i-1}, y_{j+1}) + u(x_{i-1}, y_{j-1}) - 20u(x_i, y_j) = \widehat{R}_{i,j}, \tag{16}$$

where

$$\widehat{R}_{i,j} = c_1 \left[\frac{\partial^6 u}{\partial x^6}(\xi_i, y_j) + \frac{1}{12}k^2 \frac{\partial^8 u}{\partial y^2 \partial x^6}(\xi_i, y_j) \right]$$

$$+ c_2 \left[\frac{\partial^6 u}{\partial y^6}(x_i, \eta_j) + \frac{1}{12}h^2 \frac{\partial^8 u}{\partial x^2 \partial y^6}(x_i, \eta_j) \right] \qquad (17)$$

$$+ c_3 \left[\frac{\partial^4 u}{\partial y^4}\left(x_{i+1}, \overline{\eta}_j^{(1)}\right) - 2\frac{\partial^4 u}{\partial y^4}\left(x_i, \overline{\eta}_j^{(2)}\right) + \frac{\partial^4 u}{\partial y^4}\left(x_{i-1}, \overline{\eta}_j^{(3)}\right) \right]$$

$$+ c_4 \left[\frac{\partial^4 u}{\partial x^4}\left(\overline{\xi}_i^{(1)}, y_{j+1}\right) - 2\frac{\partial^4 u}{\partial x^4}\left(\overline{\xi}_i^{(2)}, y_j\right) + \frac{\partial^4 u}{\partial x^4}\left(\overline{\xi}_i^{(3)}, y_{j-1}\right) \right],$$

$$\alpha = 2\frac{5k^2 - h^2}{h^2 + k^2}, \quad \beta = 2\frac{5h^2 - k^2}{h^2 + k^2},$$

$$c_1 = \frac{h^6 k^2}{30(h^2 + k^2)}, \quad c_2 = \frac{k^6 h^2}{30(h^2 + k^2)}, \quad c_3 = \frac{k^6}{12(h^2 + k^2)}, \quad c_4 = \frac{h^6}{12(h^2 + k^2)}.$$

From the boundary conditions we have

$$u(x_0, y_j) = \varphi_1(y_j), \quad u(x_n, y_j) = \varphi_3(y_j), \quad j = 0, 1, \ldots, m,$$
$$u(x_i, y_0) = \varphi_2(x_i), \quad u(x_i, y_m) = \varphi_4(x_i), \quad i = 1, 2, \ldots, n-1. \qquad (18)$$

Now we transform the exact formula (16) with (17)–(18) into the appropriate separate forms in accordance with the position in the grid. Such approach is reasonable in the case of an explicit formulation of the interval couterpart of the conventional nine-point finite difference method. We have

$$u(x_2, y_2) + \alpha u(x_2, y_1) + \beta u(x_1, y_2) - 20u(x_1, y_1)$$
$$= -u(x_2, y_0) - u(x_0, y_2) - u(x_0, y_0) - \alpha u(x_0, y_1) - \beta u(x_1, y_0) + \widehat{R}_{1,1}, \qquad (19)$$

$$u(x_2, y_{j+1}) + u(x_2, y_{j-1}) + \alpha u(x_2, y_j) + \beta[u(x_1, y_{j+1}) + u(x_1, y_{j-1})]$$
$$- 20u(x_1, y_j) = -u(x_0, y_{j+1}) - u(x_0, y_{j-1}) - \alpha u(x_0, y_j) + \widehat{R}_{1,j},$$
$$j = 2, 3, \ldots, m-2, \qquad (20)$$

$$u(x_2, y_{m-2}) + \alpha u(x_2, y_{m-1}) + \beta u(x_1, y_{m-2}) - 20u(x_1, y_{m-1}) = -u(x_2, y_m)$$
$$- u(x_0, y_m) - u(x_0, y_{m-2}) - \alpha u(x_0, y_{m-1}) - \beta u(x_1, y_m) + \widehat{R}_{1,m-1}, \qquad (21)$$

$$u(x_{i+1}, y_2) + u(x_{i-1}, y_2) + \alpha[u(x_{i+1}, y_1) + u(x_{i-1}, y_1)] + \beta u(x_i, y_2)$$
$$- 20u(x_i, y_1) = -u(x_{i+1}, y_0) - u(x_{i-1}, y_0) - \beta u(x_i, y_0) + \widehat{R}_{i,1},$$
$$i = 2, 3, \ldots, n-2, \qquad (22)$$

$$u(x_{i+1}, y_{j+1}) + u(x_{i+1}, y_{j-1}) + u(x_{i-1}, y_{j+1}) + u(x_{i-1}, y_{j-1}) + \alpha[u(x_{i+1}, y_j)$$
$$+ u(x_{i-1}, y_j)] + \beta[u(x_i, y_{j+1}) + u(x_i, y_{j-1})] - 20u(x_i, y_j) = \widehat{R}_{i,j},$$
$$i = 2, 3, \ldots, n-2, \quad j = 2, 3, \ldots, m-2, \qquad (23)$$

$$u\left(x_{i+1}, y_{m-2}\right) + u\left(x_{i-1}, y_{m-2}\right) + \alpha\left[u\left(x_{i+1}, y_{m-1}\right) + u\left(x_{i-1}, y_{m-1}\right)\right] + \beta u\left(x_i, y_{m-2}\right)$$
$$- 20u\left(x_i, y_{m-1}\right) = -u\left(x_{i+1}, y_m\right) - u\left(x_{i-1}, y_m\right) - \beta u\left(x_i, y_m\right) + \widehat{R}_{i,m-1},$$
$$i = 2, 3, \ldots, n-2, \tag{24}$$

$$u\left(x_{n-2}, y_2\right) + \alpha u\left(x_{n-2}, y_1\right) + \beta u\left(x_{n-1}, y_2\right) - 20u\left(x_{n-1}, y_1\right) = -u\left(x_n, y_2\right)$$
$$- u\left(x_n, y_0\right) - u\left(x_{n-2}, y_0\right) - \alpha u\left(x_n, y_1\right) - \beta u\left(x_{n-1}, y_0\right) + \widehat{R}_{n-1,1}, \tag{25}$$

$$u\left(x_{n-2}, y_{j+1}\right) + u\left(x_{n-2}, y_{j-1}\right) + \alpha u\left(x_{n-2}, y_j\right) + \beta\left[u\left(x_{n-1}, y_{j+1}\right) + u\left(x_{n-1}, y_{j-1}\right)\right]$$
$$- 20u\left(x_{n-1}, y_j\right) = -u\left(x_n, y_{j+1}\right) - u\left(x_n, y_{j-1}\right) - \alpha u\left(x_n, y_j\right) + \widehat{R}_{n-1,j},$$
$$j = 2, 3, \ldots, m-2, \tag{26}$$

$$u\left(x_{n-2}, y_{m-2}\right) + \alpha u\left(x_{n-2}, y_{m-1}\right) + \beta u\left(x_{n-1}, y_{m-2}\right) - 20u\left(x_{n-1}, y_{m-1}\right) = -u\left(x_n, y_m\right)$$
$$- u\left(x_n, y_{m-2}\right) - u\left(x_{n-2}, y_m\right) - \alpha u\left(x_n, y_{m-1}\right) - \beta u\left(x_{n-1}, y_m\right) + \widehat{R}_{n-1,m-1}. \tag{27}$$

The formulas (19)–(27) can be given in the following matrix form

$$Cu = \widehat{E}_C + \widehat{E}_L, \tag{28}$$

where

$$u_i = \left[u\left(x_i, y_1\right), u\left(x_i, y_2\right), \ldots, u\left(x_i, y_{m-1}\right)\right]^{\mathrm{T}}, \quad i = 1, 2, \ldots, n-1,$$

$$C = \begin{bmatrix} D & G & 0 & \ldots & 0 & 0 & 0 \\ G & D & G & \ldots & 0 & 0 & 0 \\ 0 & G & D & \ldots & 0 & 0 & 0 \\ \vdots & \vdots & \vdots & \ddots & \vdots & \vdots & \vdots \\ 0 & 0 & 0 & \ldots & D & G & 0 \\ 0 & 0 & 0 & \ldots & G & D & G \\ 0 & 0 & 0 & \ldots & 0 & G & D \end{bmatrix}, u = \begin{bmatrix} u_1 \\ u_2 \\ u_3 \\ \vdots \\ u_{n-3} \\ u_{n-2} \\ u_{n-1} \end{bmatrix}, \widehat{E}_C = \begin{bmatrix} e_{C\,1} \\ e_{C\,2} \\ e_{C\,3} \\ \vdots \\ e_{C\,n-3} \\ e_{C\,n-2} \\ e_{C\,n-1} \end{bmatrix}, \widehat{E}_L = \begin{bmatrix} e_{L\,1} \\ e_{L\,2} \\ e_{L\,3} \\ \vdots \\ e_{L\,n-3} \\ e_{L\,n-2} \\ e_{L\,n-1} \end{bmatrix}, \tag{29}$$

$$D = \begin{bmatrix} -20 & \beta & 0 & \ldots & 0 & 0 & 0 \\ \beta & -20 & \beta & \ldots & 0 & 0 & 0 \\ 0 & \beta & -20 & \ldots & 0 & 0 & 0 \\ \vdots & \vdots & \vdots & \ddots & \vdots & \vdots & \vdots \\ 0 & 0 & 0 & \ldots & -20 & \beta & 0 \\ 0 & 0 & 0 & \ldots & \beta & -20 & \beta \\ 0 & 0 & 0 & \ldots & 0 & \beta & -20 \end{bmatrix}, G = \begin{bmatrix} \alpha & 1 & 0 & \ldots & 0 & 0 & 0 \\ 1 & \alpha & 1 & \ldots & 0 & 0 & 0 \\ 0 & 1 & \alpha & \ldots & 0 & 0 & 0 \\ \vdots & \vdots & \vdots & \ddots & \vdots & \vdots & \vdots \\ 0 & 0 & 0 & \ldots & \alpha & 1 & 0 \\ 0 & 0 & 0 & \ldots & 1 & \alpha & 1 \\ 0 & 0 & 0 & \ldots & 0 & 1 & \alpha \end{bmatrix}, \tag{30}$$

and

$$e_{C\,1,1} = -u\left(x_2, y_0\right) - u\left(x_0, y_2\right) - u\left(x_0, y_0\right) - \alpha u\left(x_0, y_1\right) - \beta u\left(x_1, y_0\right),$$
$$e_{C\,1,j} = -u\left(x_0, y_{j+1}\right) - u\left(x_0, y_{j-1}\right) - \alpha u\left(x_0, y_j\right), \quad j = 2, 3, \ldots, m-2,$$
$$e_{C\,1,m-1} = -u\left(x_2, y_m\right) - u\left(x_0, y_m\right) - u\left(x_0, y_{m-2}\right) - \alpha u\left(x_0, y_{m-1}\right) \tag{31}$$
$$- \beta u\left(x_1, y_m\right),$$

$$e_{C\,i,1} = -u\left(x_{i+1}, y_0\right) - u\left(x_{i-1}, y_0\right) - \beta u\left(x_i, y_0\right), \quad i = 2, 3, \ldots, n-2$$

$$e_{C\,i,j} = 0, \quad i = 2, 3, \ldots, n-2, \quad j = 2, 3, \ldots, m-2, \tag{32}$$

$$e_{C\,i,m-1} = -u\left(x_{i+1}, y_m\right) - u\left(x_{i-1}, y_m\right) - \beta u\left(x_i, y_m\right), \quad i = 2, 3, \ldots, n-2,$$

$$e_{C\,n-1,1} = -u\left(x_n, y_2\right) - u\left(x_n, y_0\right) - u\left(x_{n-2}, y_0\right) - \alpha u\left(x_n, y_1\right) - \beta u\left(x_{n-1}, y_0\right),$$

$$e_{C\,n-1,j} = -u\left(x_n, y_{j+1}\right) - u\left(x_n, y_{j-1}\right) - \alpha u\left(x_n, y_j\right), \quad j = 2, 3, \ldots, m-2,$$

$$e_{C\,n-1,m-1} = -u\left(x_n, y_m\right) - u\left(x_n, y_{m-2}\right) - u\left(x_{n-2}, y_m\right) - \alpha u\left(x_n, y_{m-1}\right) \tag{33}$$
$$- \beta u\left(x_{n-1}, y_m\right),$$

$$e_{L\,i} = \left[\widehat{R}_{i,1}, \widehat{R}_{i,2}, \ldots, \widehat{R}_{i,m-1}\right]^{\mathrm{T}}, \quad i = 1, 2, \ldots, n-1. \tag{34}$$

Note that $\dim C = (n-1) \times (n-1)$, $\dim u = \dim \widehat{E}_C = \dim \widehat{E}_L = (n-1) \times 1$, $\dim D = \dim G = (m-1) \times (m-1)$.

Remark 1. Let $u_{i,j}$ approximate $u\left(x_i, t_j\right)$. If we also ignore the error terms given in components of $\widehat{E}_L$, then from (19)–(27) (or (28) with (29)–(34)) with (17), we get the conventional nine-point finite difference method with the local truncation error of order $O\left(h^4 + k^4\right)$.

3   Interval Nine-Point Finite Difference Method

Now we make the following assumptions about values in midpoints of the derivatives included in the local truncation error $\widehat{R}_{i,j}$ of the conventional method. For the interval approach applied to (16) (or (19)–(27)) with (17) we assume that there exist the intervals $M_{i,j}^{(X)}$, $M_{i,j}^{(Y)}$, $S_{i,j}^{(X)}$, $S_{i,j}^{(Y)}$, $Q_{i,j}^{(X)}$, $Q_{i,j}^{(Y)}$, such that the following relations hold

– for $i = 1, 2, \ldots, n-1$, $j = 1, 2, \ldots, m-1$, $\xi_i \in \left(x_{i-1}, x_{i+1}\right)$, $\eta_j \in \left(y_{j-1}, y_{j+1}\right)$,

$$\frac{\partial^6 u}{\partial x^6}\left(\xi_i, y_j\right) \in M_{i,j}^{(X)} = \left[\underline{M}_{i,j}^{(X)}, \overline{M}_{i,j}^{(X)}\right], \quad \frac{\partial^6 u}{\partial y^6}\left(x_i, \eta_j\right) \in M_{i,j}^{(Y)} = \left[\underline{M}_{i,j}^{(Y)}, \overline{M}_{i,j}^{(Y)}\right], \tag{35}$$

$$\frac{\partial^8 u}{\partial y^2 \partial x^6}\left(\xi_i, y_j\right) \in S_{i,j}^{(X)} = \left[\underline{S}_{i,j}^{(X)}, \overline{S}_{i,j}^{(X)}\right], \quad \frac{\partial^8 u}{\partial x^2 \partial y^6}\left(x_i, \eta_j\right) \in S_{i,j}^{(Y)} = \left[\underline{S}_{i,j}^{(Y)}, \overline{S}_{i,j}^{(Y)}\right], \tag{36}$$

– for $i = 1, 2, \ldots, n-1$, $\overline{\xi}_i^{(1)}, \overline{\xi}_i^{(2)}, \overline{\xi}_i^{(3)} \in \left(x_{i-1}, x_{i+1}\right)$,

$$\frac{\partial^4 u}{\partial x^4}\left(\overline{\xi}_i^{(1)}, y_{j+1}\right) \in Q_{i,j}^{(X)} = \left[\underline{Q}_{i,j}^{(X)}, \overline{Q}_{i,j}^{(X)}\right], \quad j = 0, 1, \ldots, m-2,$$

$$\frac{\partial^4 u}{\partial x^4}\left(\overline{\xi}_i^{(2)}, y_j\right) \in Q_{i,j}^{(X)} = \left[\underline{Q}_{i,j}^{(X)}, \overline{Q}_{i,j}^{(X)}\right], \quad j = 1, 2, \ldots, m-1, \tag{37}$$

$$\frac{\partial^4 u}{\partial x^4}\left(\overline{\xi}_i^{(3)}, y_{j-1}\right) \in Q_{i,j}^{(X)} = \left[\underline{Q}_{i,j}^{(X)}, \overline{Q}_{i,j}^{(X)}\right], \quad j = 2, 3, \ldots, m,$$

– for $j = 0, 1, \ldots, m - 1$, $\overline{\eta}_j^{(1)}, \overline{\eta}_j^{(2)}, \overline{\eta}_j^{(3)} \in (y_{j-1}, y_{j+1})$,

$$\frac{\partial^4 u}{\partial y^4}\left(x_{i+1}, \overline{\eta}_j^{(1)}\right) \in Q_{i,j}^{(Y)} = \left[\underline{Q}_{i,j}^{(Y)}, \overline{Q}_{i,j}^{(Y)}\right], \quad i = 0, 1, \ldots, n - 2,$$

$$\frac{\partial^4 u}{\partial y^4}\left(x_i, \overline{\eta}_j^{(2)}\right) \in Q_{i,j}^{(Y)} = \left[\underline{Q}_{i,j}^{(Y)}, \overline{Q}_{i,j}^{(Y)}\right], \quad i = 1, 2, \ldots, n - 1, \quad (38)$$

$$\frac{\partial^4 u}{\partial y^4}\left(x_{i-1}, \overline{\eta}_j^{(3)}\right) \in Q_{i,j}^{(Y)} = \left[\underline{Q}_{i,j}^{(Y)}, \overline{Q}_{i,j}^{(Y)}\right], \quad i = 2, 3, \ldots, n.$$

Now we denote by X_i, $i = 0, 1, \ldots, n$, Y_j, $j = 0, 1, \ldots, m$ the intervals such that $x_i \in X_i$ and $y_j \in Y_j$. Furthermore, $\Phi_1 = \Phi_1(Y)$, $\Phi_2 = \Phi_2(X)$, $\Phi_3 = \Phi_3(Y)$, $\Phi_4 = \Phi_4(X)$ denote interval extensions of the functions $\varphi_1 = \varphi_1(y)$, $\varphi_2 = \varphi_2(x)$, $\varphi_3 = \varphi_3(y)$, $\varphi_4 = \varphi_4(x)$, respectively. If we substitute (35)–(38) to (19)–(27) with (17), then we get an interval nine-point finite difference scheme of the form

$$U_{2,2} + \alpha U_{2,1} + \beta U_{1,2} - 20U_{1,1} = -U_{2,0} - U_{0,2} - U_{0,0} - \alpha U_{0,1} - \beta U_{1,0} + R_{1,1}, \quad (39)$$

$$U_{2,j+1} + U_{2,j-1} + \alpha U_{2,j} + \beta\left[U_{1,j+1} + U_{1,j-1}\right] - 20U_{1,j} = -U_{0,j+1} - U_{0,j-1}$$
$$- \alpha U_{0,j} + R_{1,j},$$
$$j = 2, 3, \ldots, m - 2, \quad (40)$$

$$U_{2,m-2} + \alpha U_{2,m-1} + \beta U_{1,m-2} - 20U_{1,m-1} = -U_{2,m} - U_{0,m} - U_{0,m-2}$$
$$- \alpha U_{0,m-1} - \beta U_{1,m} + R_{1,m-1}, \quad (41)$$

$$U_{i+1,2} + U_{i-1,2} + \alpha\left[U_{i+1,1} + U_{i-1,1}\right] + \beta U_{i,2} - 20U_{i,1} = -U_{i+1,0} - U_{i-1,0}$$
$$- \beta U_{i,0} + R_{i,1},$$
$$i = 2, 3, \ldots, n - 2, \quad (42)$$

$$U_{i+1,j+1} + U_{i+1,j-1} + U_{i-1,j+1} + U_{i-1,j-1} + \alpha\left[U_{i+1,j} + U_{i-1,j}\right]$$
$$+ \beta\left[U_{i,j+1} + U_{i,j-1}\right] - 20U_{i,j} = R_{i,j},$$
$$i = 2, 3, \ldots, n - 2, \quad j = 2, 3, \ldots, m - 2, \quad (43)$$

$$U_{i+1,m-2} + U_{i-1,m-2} + \alpha\left[U_{i+1,m-1} + U_{i-1,m-1}\right] + \beta U_{i,m-2} - 20U_{i,m-1}$$
$$= -U_{i+1,m} - U_{i-1,m} - \beta U_{i,m} + R_{i,m-1},$$
$$i = 2, 3, \ldots, n - 2, \quad (44)$$

$$U_{n-2,2} + \alpha U_{n-2,1} + \beta U_{n-1,2} - 20U_{n-1,1} = -U_{n,2} - U_{n,0} - U_{n-2,0} - \alpha U_{n,1}$$
$$- \beta U_{n-1,0} + R_{n-1,1}, \quad (45)$$

$$U_{n-2,j+1} + U_{n-2,j-1} + \alpha U_{n-2,j} + \beta \left[U_{n-1,j+1} + U_{n-1,j-1} \right]$$
$$- 20 U_{n-1,j} = -U_{n,j+1} - U_{n,j-1} - \alpha U_{n,j} + R_{n-1,j},$$
$$j = 2,3,\ldots,m-2, \tag{46}$$

$$U_{n-2,m-2} + \alpha U_{n-2,m-1} + \beta U_{n-1,m-2} - 20 U_{n-1,m-1} = -U_{n,m} - U_{n,m-2}$$
$$- U_{n-2,m} - \alpha U_{n,m-1} - \beta U_{n-1,m} + R_{n-1,m-1}, \tag{47}$$

where

$$R_{i,j} = c_1 \left[M_{i,j}^{(X)} + \frac{1}{12} k^2 S_{i,j}^{(X)} \right] + c_2 \left[M_{i,j}^{(Y)} + \frac{1}{12} h^2 S_{i,j}^{(Y)} \right]$$
$$+ c_3 \left[Q_{i+1,j}^{(Y)} - 2 Q_{i,j}^{(Y)} + Q_{i-1,j}^{(Y)} \right] + c_4 \left[Q_{i,j+1}^{(X)} - 2 Q_{i,j}^{(X)} + Q_{i,j-1}^{(X)} \right] \tag{48}$$

and

$$U_{0,j} = \Phi_1 \left(Y_j \right), \quad U_{n,j} = \Phi_3 \left(Y_j \right), \quad j = 0,1,\ldots,m,$$
$$U_{i,0} = \Phi_2 \left(X_i \right), \quad U_{i,m} = \Phi_4 \left(X_i \right), \quad i = 1,2,\ldots,n-1. \tag{49}$$

The interval method (39)–(47) with (48) can be also represented in the following matrix form

$$CU = E_C + E_L, \tag{50}$$

where

$$U = \begin{bmatrix} U_1 \\ U_2 \\ U_3 \\ \vdots \\ U_{n-3} \\ U_{n-2} \\ U_{n-1} \end{bmatrix}, \quad E_C = \begin{bmatrix} E_{C\,1} \\ E_{C\,2} \\ E_{C\,3} \\ \vdots \\ E_{C\,n-3} \\ E_{C\,n-2} \\ E_{C\,n-1} \end{bmatrix}, \quad E_L = \begin{bmatrix} E_{L\,1} \\ E_{L\,2} \\ E_{L\,3} \\ \vdots \\ E_{L\,n-3} \\ E_{L\,n-2} \\ E_{L\,n-1} \end{bmatrix}, \tag{51}$$

$$U_i = [U_{i,1}, U_{i,2}, \ldots, U_{i,m-1}]^{\mathrm{T}}, \quad i = 1,2,\ldots,n-1, \tag{52}$$

and

$$E_{C\,1,1} = -U_{2,0} - U_{0,2} - U_{0,0} - \alpha U_{0,1} - \beta U_{1,0},$$
$$E_{C\,1,j} = -U_{0,j+1} - U_{0,j-1} - \alpha U_{0,j}, \quad j = 2,3,\ldots,m-2, \tag{53}$$
$$E_{C\,1,m-1} = -U_{2,m} - U_{0,m} - U_{0,m-2} - \alpha U_{0,m-1} - \beta U_{1,m},$$

$$E_{C\,i,1} = -U_{i+1,0} - U_{i-1,0} - \beta U_{i,0}, \quad i = 2,3,\ldots,n-2,$$
$$E_{C\,i,j} = 0, \quad i = 2,3,\ldots,n-2, \quad j = 2,3,\ldots,m-2, \tag{54}$$
$$E_{C\,i,m-1} = -U_{i+1,m} - U_{i-1,m} - \beta U_{i,m} \quad i = 2,3,\ldots,n-2,$$

$$E_{C\ n-1,1} = -U_{n,2} - U_{n,0} - U_{n-2,0} - \alpha U_{n,1} - \beta U_{n-1,0},$$
$$E_{C\ n-1,j} = -U_{n,j+1} - U_{n,j-1} - \alpha U_{n,j}, \quad j = 2, 3, \ldots, m-2, \qquad (55)$$
$$E_{C\ n-1,m-1} = -U_{n,m} - U_{n,m-2} - U_{n-2,m} - \alpha U_{n,m-1} - \beta U_{n-1,m},$$

$$E_{L\ i} = [R_{i,1}, R_{i,2}, \ldots, R_{i,m-1}]^{\mathrm{T}}, \quad i = 1, 2, \ldots, n-1. \qquad (56)$$

Note that $\dim U = \dim E_C = \dim E_L = (n-1) \times 1$.

The last thing that should be considered is the method for solving the interval linear system of Eq. (50). Following [12] let us consider a finite system of linear algebraic equations of the form $Ax = b$, where A is an n-by-n matrix and b is an n-dimensional vector (the coefficients of A and b are real or interval values). Then, from [12], we have the following theorem.

Theorem 1. *If we can carry out all the steps of a direct method for solving $Ax = b$ in the interval arithmetic (if no attempted division by an interval containing zero occurs, nor any overflow or underflow), then the system has a unique solution for every real matrix in A and every real vector in b, and the solution is contained in the resulting interval vector X.*

Taking into account Theorem 1, we can formulate the subsequent theorem that concerns the interval solution obtained with the interval nine-point finite difference method considered. The interval solution is such that the exact solution belongs to it.

Theorem 2. *Let us assume that the local truncation error of the nine-point finite difference scheme can be bounded by the appropriate intervals. Moreover, let $\Phi_1 = \Phi_1(Y)$, $\Phi_2 = \Phi_2(X)$, $\Phi_3 = \Phi_3(Y)$, $\Phi_4 = \Phi_4(X)$ denote interval extensions of the functions $\varphi_1 = \varphi_1(y)$, $\varphi_2 = \varphi_2(x)$, $\varphi_3 = \varphi_3(y)$, $\varphi_4 = \varphi_4(x)$ given in the boundary conditions (2) formulated for the Laplace Eq. (1). If $u(0, y_j) \in U_{0,j}$, $u(a, y_j) \in U_{n,j}$, $j = 0, 1, \ldots, m$, $u(x_i, 0) \in U_{i,0}$, $u(x_i, b) \in U_{i,m}$, $i = 0, 1, \ldots, n$ and the linear system of Eq. (50) corresponding to the interval nine-point finite difference method (39)–(47) can be solved with some direct method in the interval arithmetic (if no attempted division by an interval containing zero occurs, nor any overflow or underflow), then for the interval solutions considered it can be shown that $u(x_i, t_j) \in U_{i,j}$, $i = 1, 2, \ldots, n-1$, $j = 1, 2, \ldots, m-1$.*

4 Conclusions

The main contribution of the paper is an explicit formulation of the nine-point finite difference scheme together with the local truncation error terms that are ignored in a conventional approach. Based on it we managed to propose some interval nine-point finite difference scheme. Taking into account the matrix representation (50), the interval method considered allows us to compute a guaranteed result. Further research would concern an algorithm for an approximation of the endpoints of the error terms with some possibly high order finite differences and a computer implementation of the interval method in the interval conventional or multiple-precision floating-point arithmetic.

Acknowledgments. The paper was supported by the Poznan University of Technology (Poland) through Grant No. 02/21/DSPB/3463.

References

1. Anderson, D.A., Tannehill, J.C., Pletcher, R.H.: Computational Fluid Mechanics and Heat Transfer. Hemisphere Publishing, New York (1984)
2. Boisvert, R.F.: Families of high order accurate discretizations of some elliptic problems. SIAM J. Sci. Stat. Comput. **2**(3), 268–284 (1981)
3. Hoffmann, T., Marciniak, A., Szyszka, B.: Interval versions of central difference method for solving the Poisson equation in proper and directed interval arithmetic. Found. Comput. Decis. Sci. **38**(3), 193–206 (2013)
4. Jankowska, M.A.: An interval backward finite difference method for solving the diffusion equation with the position dependent diffusion coefficient. In: Wyrzykowski, R., Dongarra, J., Karczewski, K., Waśniewski, J. (eds.) PPAM 2011, Part II. LNCS, vol. 7204, pp. 447–456. Springer, Heidelberg (2012)
5. Jankowska, M.A.: An interval finite difference method of Crank-Nicolson type for solving the one-dimensional heat conduction equation with mixed boundary conditions. In: Jónasson, K. (ed.) PARA 2010, Part II. LNCS, vol. 7134, pp. 157–167. Springer, Heidelberg (2012)
6. Jankowska, M.A., Sypniewska-Kaminska, G.: Interval finite difference method for solving the one-dimensional heat conduction problem with heat sources. In: Manninen, P., Öster, P. (eds.) PARA. LNCS, vol. 7782, pp. 473–488. Springer, Heidelberg (2013)
7. Kantorovich, L., Krylov, V.: Approximate Methods of Higher Analysis. Interscience Publishers, New York (1958)
8. Manikonda, S., Berz, M., Makino, K.: High-order verified solutions of the 3D Laplace equation. WSEAS Trans. Comput. **4**(11), 1604–1610 (2005)
9. Marciniak, A.: An interval version of the Crank-Nicolson method – the first approach. In: Jónasson, K. (ed.) PARA 2010, Part II. LNCS, vol. 7134, pp. 120–126. Springer, Heidelberg (2012)
10. Mochnacki, B., Piasecka-Belkhayat, A.: Numerical modeling of skin tissue heating using the interval finite difference method. MCB Mol. Cell. Biomech. **10**(3), 233–244 (2013)
11. Moore, R.E.: Interval Analysis. Prentice-Hall, Englewood Cliffs (1966)
12. Moore, R.E., Kearfott, R.B., Cloud, M.J.: Introduction to Interval Analysis. SIAM, Philadelphia (2009)
13. Nakao, M., Kimura, T., Kinoshita, T.: Constructive a priori error estimates for a full discrete approximation of the heat equation. SIAM J. Numer. Anal. **51**(3), 1525–1541 (2013)
14. Orszag, S., Israeli, M.: Numerical simulation of viscous incompressible flows. Annu. Rev. Fluid Mech. **6**, 281–318 (1974)
15. Rosser, J.: Nine-point difference solutions for Poisson's equation. Comput. Math. Appl. **1**(3–4), 351–360 (1975)
16. Singer, I., Turkel, E.: High-order finite difference methods for the Helmholtz equation. Comput. Meth. Appl. Mech. Eng. **163**(1–4), 343–358 (1998)
17. Sunaga, T.: Theory of interval algebra and its application to numerical analysis. Res. Assoc. Appl. Geom. (RAAG) Mem. **2**, 29–46 (1958)

18. Szyszka, B.: The central difference interval method for solving the wave equation. In: Wyrzykowski, R., Dongarra, J., Karczewski, K., Waśniewski, J. (eds.) PPAM 2011, Part II. LNCS, vol. 7204, pp. 523–532. Springer, Heidelberg (2012)
19. Szyszka, B.: An interval version of Cauchy's problem for the wave equation. In: AIP Conference Proceedings, vol. 1648 (2015)

Workshop on Complex Collective Systems

How Do People Search: A Modelling Perspective

Isabella von Sivers$^{(\boxtimes)}$, Michael J. Seitz, and Gerta Köster

Munich University of Applied Sciences, Lothstr. 64, 80335 München, Germany
{isabella.von_sivers,m.seitz,gerta.koester}@hm.edu

Abstract. The simulation of pedestrian movement is an important tool to ensure safety whenever many people have to be evacuated or pass through an environment. Although there are many simulation models for pedestrian dynamics, crucial aspects of human behaviour are still being neglected. One of those behaviours is the search strategy humans use to find someone or something within a building. We present three possible search strategies for pedestrian simulation. Two are often used as default implementations: random search and the optimal solution. The third more plausibly agrees with findings from psychology, neuroscience and related fields: a nearest room heuristic. We compare and evaluate the strategies, present simulation results for two concrete scenarios, and give a recommendation for computer models of human search behaviour.

Keywords: Pedestrian dynamics · Search strategies · Cognitive heuristics · Travelling salesman problem

1 Introduction

Whenever many people come together, it is important to consider how they move, navigate or, more generally, behave to ensure their safety and comfort. Studying these dynamics has become a wide and active research field [7]. The resulting complex systems must be formally analysed, implemented and validated, that is, tested with controlled experiments or field observations.

One approach to study the collective dynamics of pedestrian behaviour is through microscopic computer simulation of pedestrian and crowd motion (see [9,34] for reviews). Cellular automata (e.g. [4]) and social force models (e.g. [15]) are two major approaches. Phenomena, such as congestions and lane formation, were reproduced with these models. More recent alternatives are the Optimal Steps Model [27,30] and the Gradient Navigation Model [8]. While most of the models focus on the locomotion and route-choice of pedestrians, modern concepts from psychology and neuroscience are rarely considered.

In emergencies, but also in daily life, pedestrians often seek a specific spatial goal: the exit door, their missing family members, or a certain office. This leads us to the question: How do people search? Chu and Law [5] introduced general search behaviour for pedestrians in an evacuation simulation but did not give information on how to model the concrete search strategy.

© Springer International Publishing Switzerland 2016
R. Wyrzykowski et al. (Eds.): PPAM 2015, Part II, LNCS 9574, pp. 487–496, 2016.
DOI: 10.1007/978-3-319-32152-3_45

In psychology, biology, and neuroscience researchers investigate the mechanics and strategies that animals and humans use to perceive, navigate and search their surroundings [2,10,16,23]. However, at this point, human search strategies are not sufficiently well understood to easily carry results over to computer algorithms.

Our goal is to analyse three fundamentally different search strategies through specific but relevant search scenarios so that we can give first recommendations on when to use which strategy. Two of the strategies, the random search and the shortest path, are well defined mathematical algorithms and therefore are convenient for implementation. But are these good choices for the simulation of human behaviour? The third strategy is based on heuristic decision making in the spirit of [13]. For this heuristic, each agent searches the closest room that has not been searched yet [3,33].

For the evaluation of the strategies, we draw on findings from biology, psychology, and neuroscience that relate to our research questions. Due to the vast amount and breadth of related literature, our discussion must remain incomplete. However, we hope to start a discussion on search strategies for simulation models of pedestrian behaviour.

The contribution is structured as follows: First we describe the search problem formally and present the scenarios we use for our investigation. With this background, we discuss the three search strategies and their plausibility. Then, we show simulation results produced with the three strategies. Finally, we discuss the impact of our study, its limitations, and future directions.

2   Search Geometry and Simulation Model

We limit our study to search tasks in a building well known to the person. This has two main reasons. First, in everyday situations, pedestrians are likely to be in a well known building, like their office building, their favourite shopping mall, or their family home. Second, search in an unknown building would add additional complexity and touch on additional fields, such as cognitive maps, path integration, and visual landmarks [6].

In our study, virtual pedestrians – called *agents* – start their search in one room of a building and try to find a target of which it does not know the location. They do know, however, the locations of the entrances to all rooms. We assume that, as soon as an agent enters a room, they realize whether or not the target is in the room, that is, whether or not to stop the search.

We formally describe the building through a graph. The building consists of M rooms, which are connected through N doors. On both sides of each door, we place a vertex. Thus, we have $2N$ vertices in total. Each vertex is assigned to the room in which it is located. Two vertices are connected by an edge if they are assigned to the same room or if they are located at the two sides of the same door. Hence, there are at least $2N - 1$ and at most $N(N + 1)/2$ edges in the graph. The weight on an edge is the Euclidean distance between the two vertices. See Fig. 1 for an illustration of such a graph.

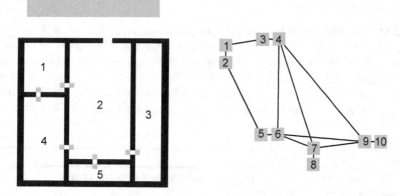

Fig. 1. Example of a building and the corresponding graph. Left: map of the building with arbitrarily numbered rooms (blue numbers), and vertices next to each door (yellow rectangles). Right: graph representation (Color figure online).

This representation of a building is derived from the idea of visibility graphs [1,18,22]. Alternative representations are, for instance, graphs with only one vertex at each door, and graphs with rooms as vertices, which are dual to the description with rooms as edges. For an overview of graph representations of buildings see [11]. Other possible weights on the edges are heuristic assessments of distances or the actual travelling time necessary to walk from one vertex to another. However, whether any of these graphs accurately describes the representation of a building in the human mind remains an open question.

In this study, we use the Optimal Steps Model [27,28] with the personal space approach [30] for agent locomotion. Once a vertex is chosen as intermediate goal, agents follow the shortest path to the target while avoiding close proximity to other agents or walls. For more details on the microscopic pedestrian simulation model see [27,28,30]. The search strategies themselves are independent of the locomotion model.

3   Comparison of Search Strategies

We present three search strategies for two buildings in which the agents know all rooms. A complete search would lead through all rooms for all three strategies. Agents search rooms only once, assuming that they know which rooms they have visited already. However, rooms *can be* entered several times. This may be the case if a room has to be crossed to reach another room. The search path describes the route the agent chooses to walk through the building until the target is found. The length of the search path is the distance the agent covers to find the target or to search all rooms. Each search algorithms compiles a list of targets for the pedestrians and changes the target if required during the simulation. Hence, the algorithms are executed at the beginning of the simulation and at each time step of the locomotion model.

3.1   Random Search

The agent randomly selects the next room among the rooms it has not yet visited. No search direction or room in the building is prioritized in the search sequence. The algorithmic formulation of this strategy is very simple and fast. It has complexity $\mathcal{O}(M)$, where M is the number of rooms. It is easy to implement an requires only a list of room numbers and a list of the vertices. See Algorithm 1. However, it is also an inefficient search strategy in terms of path lengths.

Algorithm 1. Random search.

---

1   initialise lists with vertices and room numbers based on the graph
    representation of the building;
2   shuffle list of room numbers;
3   **while** *search target not found* **do**
4 | target vertex is closest vertex from the first room of the room number list;
5 | go to the target vertex;
6 |_   delete first number from the list of room numbers;

In simple situations, humans are likely to find the shortest path connecting all search targets, provided they have a complete understanding of the geometry. This was shown in several experiments on path finding with multiple targets on a picture or in a room [3,33]. Although the experimental results are limited to small environments where all targets could be seen by the participants, it is highly doubtful that pedestrians randomly search a well-known building. Still, there may be cases, e.g. involving extreme stress or great geometrical complexity, where human behaviour is correctly represented by a random search.

3.2   Optimal Search

For the optimal search, we additionally assume that agents know all distances in the building. Formally, this means that they know the underlying graph of the building with all weights on the edges. Furthermore, we assume that the agents are capable of calculating the optimal way through the building, that is, they minimise the length of the search path under the constraint that they visit every room at least once. This optimisation problem is a modification of the generalised travelling salesman problem [31] with possible repetitions of the sets [19] and without return [21]. As a modification of a travelling salesman problem, it is known to be NP-hard [21,25]. The algorithm for one agent is shown in Algorithm 2. For simplicity, we use a brute force backtracking algorithm [24] to solve the implicit travelling salesman problem with a complexity of at least $\mathcal{O}((N - 1)!)$, where N is the number of vertices. Other more, efficient implementations (e.g. [17,25]) could be used but are not the focus of our study.

It has been shown that mammals have a detailed cognitive representation of their neighbourhood [14]. Following these findings, we argue that humans

---

Algorithm 2. Optimal search algorithm. To compute the optimal order, we use a brute force backtracking algorithm [24].

---

1 initialise lists with vertices and room numbers based on the graph representation of the building;
2 compute optimal vertex order (vertices may be represented in this list more than once);
3 **while** *search target not found* **do**
4 ⎜   target vertex is first vertex of the vertex list;
5 ⎜   go to the target vertex;
6 ⎿   delete first vertex from the vertex list;

---

also have a representation of their surroundings, at least in well-known limited areas. This information may be enough to gauge rough distances but may still be insufficient for a complete optimal search. Furthermore, humans are often unable to optimise their decisions because they lack information, or because the the problem is too difficult to solve [12,13]. Hence, we consider it questionable that humans actually optimise their search path through a complete building before starting the search, unless the buildings is very small and well-known.

3.3 Nearest Room Heuristic

As last strategy we propose that an agent searches the closest room next it has not yet visited. This is inspired by the ability of humans to process and to remember the layout of a building [14] and to plan their search path ahead. The algorithm is formulated in Algorithm 3. It has complexity $\mathcal{O}(M \cdot 2N)$ where M is the room number and N is the number of doors. Thus it needs more computation time than the random search but is much faster than the optimal search.

Algorithm 3. Nearest room heuristic.

---

1 initialise lists with vertices and room numbers based on the graph representation of the building;
2 **while** *search target not found* **do**
3 ⎜   compute distances for all vertices in the vertex list to current position;
4 ⎜   sort list (short to long distance);
5 ⎜   target vertex is first vertex of the list;
6 ⎜   go to the target vertex;
7 ⎿   delete all vertices from the vertex list that are assigned to the current room;

This strategy has already been proposed as a heuristic humans might use to solve travelling salesman problems [3,33]. Furthermore, it follows the psychological paradigm that humans employ fast and simple heuristics with limited information from the environment to solve problems [12,13,32]. Hence, we argue that the nearest room heuristic is a plausible starting point to introduce search behaviour that is based on psychological findings into pedestrian simulation.

3.4   Simulation Experiments

In this section, we look at two concrete simulation scenarios where agents search for a missing person: a very simple building with five rooms and a slightly more complex geometry with 200 agents seeking safety. We study the individual search trajectories.

In Fig. 2, one agent starts its search in the middle room on the bottom. The missing person is not in the building. Therefore, all rooms must be visited before the search stops. With the random search, edges are used multiple times and the whole search path is rather unstructured. Furthermore, for every new run of the simulation, the sequence of visited rooms changes. Hence, the search time and the length of the search path varies considerably. The optimal search provides the fastest and shortest complete search route. With the nearest room heuristic, the agents follows a path that is relatively close in length to the optimal path but is not the same. In several other simple simulation experiments, we found that the trajectory of the nearest room heuristic is often identical to the one of the optimal search. However, we have not yet substantiated this observation with a formal investigation.

Clearly, the success of a search strategy depends on the building's topography, the initial location of the searching person, and the location of the search target. If we had put the missing person in the left room on the bottom in Fig. 2, then the nearest room heuristic would yield the shortest search path. If the target was located in the upper left room, the random search would have been the best choice.

In the second scenario, we study the evacuation of a building with 200 agents, including 6 agents that represent parents missing one child each. We assume that the children remain at their starting positions – a reaction typical for frightened

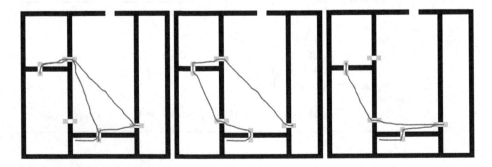

Fig. 2. Trajectories (blue lines) of an agent searching a building. The agent starts in the middle room on the bottom. Left: random search. Middle: nearest room heuristic. Right: optimal search (Color figure online).

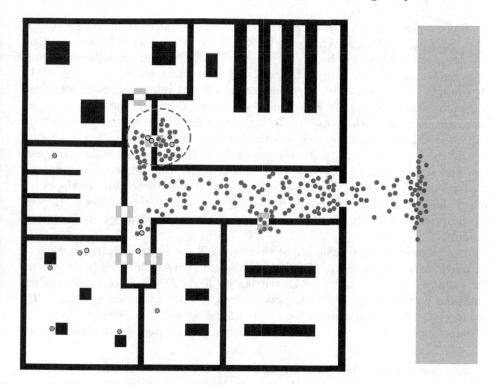

Fig. 3. Screenshot of the evacuation of a building with 200 persons including 6 parents (light blue circles with solid black border) and their 6 children (light green circles with dotted black border) to the safe area at the right. The congestion at the door of the upper right room is caused by two parents who try to enter the room against the flow to search for their children (Color figure online).

individuals [20]. The parents search with the nearest room heuristic. They evacuate when reunited with their children. The other pedestrians evacuate individually.

Figure 3 shows a screen shot after a few seconds of simulation. In the lower left room, the first parent has reunited with his or her child. At the door of the upper right room, a congestion has formed. It is caused by two searching parents trying to enter the room against the flow. This different behaviour of parents – returning into a building against the evacuation route – can also be observed in real evacuations [26,29].

4   Discussion

In this paper, we compared three possible search strategies for pedestrian simulation: the random search, the nearest room heuristic, and the optimal search. We formally described the strategies as algorithms, used them for simulation, and discussed their plausibility as models of human behaviour.

Since humans are able to process and to remember at least the rough layout of a building, random search seems limited to situations where these abilities are severely inhibited by the circumstances. Findings from psychology suggest that humans rather use simple heuristics to solve their problems. Thus, optimal search, which demands enormous computational resources, does not seem adequate either, except for very simple scenarios. However, the nearest room heuristic often yields the same results suggesting that heuristic decision making can be as efficient as mathematical optimisation. Furthermore the nearest room heuristics is in accordance with the psychological paradigm that humans use fast and simple heuristics to solve problems. In simulations experiments, agents that adopted this strategy found a satisfying solution – often close to the optimum. The search route was structured and the paths seemed realistic. Hence, we propose the nearest room heuristic as default search strategy in computer models.

Nonetheless, there is little doubt that humans use more complex strategies. In particular, humans use additional information if available. In an evacuation, for example, parents may have knowledge on likely locations of their missing children. Furthermore, social interactions may provide additional guidance. To us, incorporating more information processing – including uncertainty – seems a promising next step in research on search strategies for pedestrian simulations.

Acknowledgement. This work was funded by the German Federal Ministry of Education and Research through the projects MEPKA on mathematical characteristics of pedestrian stream models (grant number 17PNT028) and MultikOSi on assistance systems for urban events – multi criteria integration for openness and safety (grant number 13N12824). The authors also acknowledge the support by the Faculty Graduate Center CeDoSIA of TUM Graduate School at Technische Universität München, Germany.

References

1. Arikan, O., Chenney, S., Forsyth, D.: Efficient multi-agent path planning. In: Magnenat-Thalmann, N., Thalmann, D. (eds.) Computer Animation and Simulation 2001. Eurographics, pp. 151–162. Springer, Vienna (2001)
2. Barnett-Cowan, M., Bülthoff, H.H.: Human path navigation in a three-dimensional world. Behav. Brain Sci. **36**, 544–545 (2013)
3. Blaser, R., Wilber, J.: A comparison of human performance in figural and navigational versions of the traveling salesman problem. Psychol. Res. **77**(6), 761–772 (2013)
4. Burstedde, C., Klauck, K., Schadschneider, A., Zittartz, J.: Simulation of pedestrian dynamics using a two-dimensional cellular automaton. Phys. A: Stat. Mech. Appl. **295**, 507–525 (2001)
5. Chu, M., Law, K.: Computational framework incorporating human behaviors for egress simulations. J. Comput. Civ. Eng. **27**(6), 699–707 (2013)
6. Collett, T.S., Graham, P.: Animal navigation: path integration, visual landmarks and cognitive maps. Curr. Biol. **14**(12), R475–R477 (2004)

7. Daamen, W., Duives, D.C., Hoogendoorn, S.P. (eds.): The Conference in Pedestrian and Evacuation Dynamics (PED 2014), Transportation Research Procedia, vol. 2, pp. 1–818. Elsevier, Delft, The Netherlands (2014). www.sciencedirect.com/science/journal/23521465/2/
8. Dietrich, F., Köster, G.: Gradient navigation model for pedestrian dynamics. Phys. Rev. E **89**(6), 062801 (2014)
9. Duives, D.C., Daamen, W., Hoogendoorn, S.P.: State-of-the-art crowd motion simulation models. Transp. Res. Part C: Emerg. Technol. **37**, 193–209 (2013)
10. Etienne, A.S., Maurer, R., Berlie, J., Reverdin, B., Rowe, T., Georgakopoulos, J., Seguinot, V.: Navigation through vector addition. Nat. **396**(6707), 161–164 (1998)
11. Franz, G., Mallot, H., Wiener, J.: Graph-based models of space in architecture and cognitive science - a comparative analysis. In: Proceedings of the 17th International Conference on Systems Research, Informatics and Cybernetics, vol. 3038 (2005)
12. Gigerenzer, G.: Why heuristics work. Perspect. Psychol. Sci. **3**(1), 20–29 (2008)
13. Gigerenzer, G., Todd, P.M., A.B.C. Research Group: Simple Heuristics That Make Us Smart. Oxford University Press, Oxford (1999)
14. Hafting, T., Fyhn, M., Molden, S., Moser, M.B., Moser, E.I.: Microstructure of a spatial map in the entorhinal cortex. Nat. **436**(7052), 801–806 (2005)
15. Helbing, D., Molnár, P.: Social force model for pedestrian dynamics. Phys. Rev. E **51**(5), 4282–4286 (1995)
16. Hutchinson, J.M.C., Gigerenzer, G.: Simple heuristics and rules of thumb: where psychologists and behavioural biologists might meet. Behav. Process. **69**(2), 97–124 (2005). The Proceedings of the Meeting of the Society for the Quantitative Analyses of Behavior (SQAB 2004)
17. Karapetyan, D., Gutin, G.: Efficient local search algorithms for known and new neighborhoods for the generalized traveling salesman problem. Eur. J. Oper. Res. **219**(2), 234–251 (2012). http://www.sciencedirect.com/science/article/pii/S0377221712000288
18. Kneidl, A., Borrmann, A., Hartmann, D.: Generation and use of sparse navigation graphs for microscopic pedestrian simulation models. Adv. Eng. Inform. **26**(4), 669–680 (2012)
19. Laporte, G., Nobert, Y.: Generalized travelling salesman problem through n sets of nodes: an integer programming approach. INFOR J. **21**(1), 61–75 (1983)
20. Leach, J.: Why people freeze in an emergency: temporal and cognitive constraints on survival responses. Aviat. Space Environ. Med. **75**(6), 539–542 (2004). http://www.ingentaconnect.com/content/asma/asem/2004/00000075/00000006/art00011
21. Lenstra, J.K., Kan, A.: On general routing problems. Netw. **6**(3), 273–280 (1976)
22. Lozano-Pérez, T., Wesley, M.A.: An algorithm for planning collision-free paths among polyhedral obstacles. Commun. ACM **22**(10), 560–570 (1979)
23. Moser, E.I., Kropff, E., Moser, M.B.: Place cells, grid cells, and the brain's spatial representation system. Annu. Rev. Neurosci. **31**(1), 69–89 (2008)
24. Näher, S.: The travelling salesman problem. In: Vöcking, B., Alt, H., Dietzfelbinger, M., Reischuk, R., Scheideler, C., Vollmer, H., Wagner, D. (eds.) Algorithms Unplugged, pp. 383–391. Springer, Berlin Heidelberg (2011)
25. Pop, P.C., Matei, O., Sabo, C.: A new approach for solving the generalized traveling salesman problem. In: Blesa, M.J., Blum, C., Raidl, G., Roli, A., Sampels, M. (eds.) HM 2010. LNCS, vol. 6373, pp. 62–72. Springer, Heidelberg (2010)
26. Proulx, G.: Evacuation from a single family house. In: Proceedings of the 4th International Symposium on Human Behaviour in Fire. Robinson College, Cambridge, UK, pp. 255–266 (2009)

27. Seitz, M.J., Köster, G.: Natural discretization of pedestrian movement in continuous space. Phys. Rev. E **86**(4), 046108 (2012)
28. Seitz, M.J., Köster, G.: How update schemes influence crowd simulations. J. Stat. Mech.: Theor. Exp. **7**, P07002 (2014)
29. Sime, J.D.: Affiliative behaviour during escape to building exits. J. Environ. Psychol. **3**(1), 21–41 (1983)
30. von Sivers, I., Köster, G.: Dynamic stride length adaptation according to utility and personal space. Transp. Res. Part B: Methodol. **74**, 104–117 (2015)
31. Srivastava, S., Kumar, S., Garg, R., Sen, P.: Generalized traveling salesman problem through n sets of nodes. CORS J. **7**, 97–101 (1969)
32. Tversky, A., Kahneman, D.: Judgment under uncertainty: Heuristics and biases. Sci. **185**, 1124–1131 (1974)
33. Wiener, J., Ehbauer, N., Mallot, H.: Planning paths to multiple targets: Memory involvement and planning heuristics in spatial problem solving. Psychol. Res. **73**(5), 644–658 (2009)
34. Zheng, X., Zhong, T., Liu, M.: Modeling crowd evacuation of a building based on seven methodological approaches. Build. Environ. **44**(3), 437–445 (2009)

A Sandpile Cellular Automata-Based Approach to Dynamic Job Scheduling in Cloud Environment

Jakub Gasior[1]([⊠]) and Franciszek Seredynski[2]

[1] Systems Research Institute, Polish Academy of Sciences, Warsaw, Poland
j.gasior@ibspan.waw.pl
[2] Cardinal Stefan Wyszynski University, Warsaw, Poland

Abstract. The paper presents a general framework studying issues of effective load balancing and scheduling in highly parallel and distributed environments such as currently built Cloud computing systems. We propose a novel approach based on the concept of the Sandpile cellular automaton: a decentralized multi-agent system working in a critical state at the edge of chaos. Our goal is providing fairness between concurrent job submissions by minimizing slowdown of individual applications and dynamically rescheduling them to the best suited resources.

Keywords: Sandpile cellular automata · Self-organization · Cloud computing · Load-balancing

1 Introduction

In this paper, we consider the aspect of effective load balancing in cloud computing (CC) systems, i.e., the process of distributing the workload among various computing nodes to improve both resource utilization and job response time. We formulate a model defined as follows: given a set of virtual resources in the Cloud $(M_1, M_2, ..., M_n)$, a number of Cloud clients $(U_1, U_2, ..., U_k)$ and a random set of parallel applications (also jobs or tasks) run by the clients $(J_1, J_2, ..., J_i)$, find such an allocation of jobs to the resources to maximize the system throughput.

We are interested in parallel and distributed algorithms working in environments with only limited, local information. We propose a fully decentralized and adaptive load balancing scheme to solve the studied problem. The working hypothesis of our approach relies on the *Self-Organized Criticality* theory (SOC) described by Bak in [1]. SOC describes a property of complex systems that consists of a critical state formed by self-organization at the border of order and chaos. To that aim, we extend the *Sandpile* cellular automaton, one of the first systems where SOC properties were observed. In our model each CC node in the system is characterized as a cell of the proposed automaton.

The remainder of this paper is organized as follows. In Sect. 2, we present the works related to the distributed scheduling and load balancing in the Grid and

© Springer International Publishing Switzerland 2016
R. Wyrzykowski et al. (Eds.): PPAM 2015, Part II, LNCS 9574, pp. 497–506, 2016.
DOI: 10.1007/978-3-319-32152-3_46

Cloud computing systems. In Sect. 3, we describe the proposed Cloud system model. Section 4 presents the proposed solution of dynamic load balancing and scheduling scheme based on the *Sandpile* model. The experimental evaluation of the proposed approach is given in Sect. 5. We end the paper in Sect. 6 with some conclusions and indications for future work.

2   State of the Art

Distributed scheduling has been a widely studied subject in the context of real-time systems, where users define time constraints for their jobs and applications. In [5] authors proposed a dynamic load balancing algorithm based mainly on the *Peer-to-Peer (P2P)* technology aiming at illustrating the application potentials of gossiping protocols. Authors describe this problem analogously to averaging a set of distributed numbers using decentralized aggregation and membership management.

In [3] authors proposed a solution based on building a consensus over heterogeneous networks, i.e. networks whose nodes have different speeds. However, the study assumes that all jobs are available in the system from the start, which simplifies it into a static optimization problem instead of a dynamic one, considered here.

In [4] authors proposed a decentralized load balancing approach designed for computing jobs in computational P2P systems consisting of nodes which know only their nearest neighbors defined by a one hop communication path. Another decentralized scheduler for Bag-of-Tasks applications ensuring a fair and efficient use of the resources by providing a similar share of the platform to every application was presented in [2].

Finally, a self-organizing model for non-clairvoyant load-balancing in large-scale decentralized systems was proposed in [6]. Authors employed the Sandpile model with two different interconnection topologies, based on a ring and a small-world graph and using a gossiping-based version of the agent system. Instead of propagating a real avalanche, the gossiping protocol forwards the avalanche virtually until a new state of equilibrium is found. The proposed solution was found to reduce the overhead of intermediate migrations and increase the overall throughput of the system.

3   Cloud Model

3.1   System and User Model

System consists of a set of geographically distributed Cloud nodes $M_1, M_2, ..., M_m$, which are connected to each other via a wide area network. Each node M_i is described by a parameter m_i, which denotes the number of identical processors P_i and its computational power s_i, characterized by a number of operations per unit of time it is capable of performing [7].

Users $(U_1, U_2, ..., U_n)$ submit their jobs to the system, expecting their completion before a required deadline. Job (denoted as J_k^j) is jth job produced (and owned) by user U_k. Each job has varied parameters defined as a tuple $< r_k^j, size_k^j, t_k^j, d_k^j >$, specifying its release date $r_k^j > 0$; its size $1 \leq size_k^j \leq m_m$, that is referred to as its processor requirements or *degree of parallelism*; its execution requirements t_k^j defined by a number of operations and deadline d_k^j.

3.2   Problem Formulation

A typical way of assessing system's performance is measuring the completion time of submitted jobs. However, in a system with non-zero release dates, the completion time of the job is not appropriate to evaluate the actual performance. Consequently, when scheduling multiple jobs, as far as fairness is concerned, the most suited metric seems to be the maximum *Slowdown* (also known as *Stretch*). The *Slowdown* of a job J_k^j is defined as the ratio of its response time under the concurrent scheduling policy $(C_k^j - r_k^j)$ over its response time in a dedicated mode, i.e., when it is the only application executed on the whole platform:

$$\varsigma_k^{i,j} = \frac{C_k^j - r_k^j}{p_k^{i,j}}. \tag{1}$$

The objective is to allocate a batch of local jobs to the available Cloud nodes M_i and minimize the global system *Slowdown*, ς_{max} thereby enforcing a fair trade-off between all submitted applications. We consider minimization of the time ς_{max}^i on each Cloud node M_i over the system in such a way that the global *Slowdown* is defined as: $\varsigma_{max} = avg_i\{\varsigma_{max}^i\}$. The resulting optimization problem can be formulated as follows:

$$Minimize\left(\varsigma_{max}\right). \tag{2}$$

We assume that there is no centralized control and the assignment of jobs to available resources within the Cloud is governed exclusively by specialized agents assigned to individual Cloud nodes. Agents interact with each other under a set of rules specified by a *Sandpile* CA. Its specific mode of operation will be explained in detail in the subsequent sections.

4   Dynamic Load Balancing and Scheduling Based on Sandpile Model

4.1   Designing an Efficient Scheduler

To construct our Sandpile model-based scheduler, we employ a two - dimensional grid, that is discrete and discontinuous concerning space and time. Let us use two indices (x, y) to number the cellular automaton's cells. Each cell represents one of the distributed Cloud nodes $M_1, M_2, ..., M_m$, as defined in the previous Section.

Just like in a classic *Sandpile* automaton, every agent A_i is monitoring its assigned resource denoted as M_i, waiting for a new event to alter the status of the local queue (height of the grain slope). It can be either retrieving a job from the client, new job being pushed into the queue from one of its neighbors or a job failure enforcing recalculation of the local schedule. As soon as such an event occurs, agent A_i estimates the potential changes in the local workload. The following step is determined by a set of transition rules. If the resource is in a state of equilibrium, the agent waits for further developments. Otherwise, the transition rule is triggered and the agent initiates an avalanche, sending surplus workload to the neighboring nodes.

The agent has the same objective as the whole platform: minimizing the *Slowdown* metrics among all jobs in his local job queue. We rely on the relation between the *Slowdown* and *Completion Time* of the job J_k^j executed on machine M_i defined previously in Eq. 1. The first step in realizing this goal is construction of an accurate availability summary which describes node's capacity to process new jobs submitted by clients. The availability of a node can be characterized by the size (duration) of a free *Time Slot* that can be allocated to the arriving jobs.

Estimation of this parameter is based on the model proposed in [2] and provided in the example depicted in Fig. 1 visualizing five jobs scheduled for execution on exemplary CC node with two processing elements. Variables below the Gantt chart $(d_1^1, ..., d_1^5)$ represent deadline values of each job in the local queue. Let us further assume that a new job J_{new} will be released to the system at time r_{new} with execution time p_{new} and deadline d_{new}.

In the example of Fig. 1(b), deadline of the new job J_{new}, d_{new} lies between deadlines d_3 and d_4, respectively. Thus, the earliest starting time for the job J_{new} is after jobs J_1, J_2, and J_3 (that is all jobs with $d_j < d_{new}$) have been completed. Similarly, the latest completion time for the job J_{new} must ensure that jobs J_4 and J_5 (that is all jobs with $d_j > d_{new}$) will not miss their deadlines, and also that deadline d_{new} is not exceeded. In practice, it enforces rescheduling of jobs J_4 and J_5 to the end of the local job queue as visualized in Fig. 1(b).

We proceed by reconstruction of the local schedule at the time of arrival of the new job J_{new}, i.e., r_{new}. We begin by sorting the local job queue according to the *Earliest Deadline First* scheduling policy and compute the latest starting time of each job J_j such that no job J_{j+1} with $j + 1 \geq j$ misses its specified deadline. Let us further assume that a variable C_k denotes the moment at which $k - 1$ job in the queue is expected to finish its execution. It can be calculated by adding the remaining execution time of the $k - 1$ jobs to the starting time r_{new}, as follows:

$$C_k = r_{new} + \sum_{j=1}^{k-1} p_j. \qquad (3)$$

Then, assuming that the new job J_{new} would be at position k in the local queue, we can calculate the size of the available *Time Slot* τ_i that can potentially be devoted to job J_{new} between the moment at which the previous job is going to finish (C_k), and the deadline of the new job (d_{new}) or the last moment at which next job must start (x_{k+1}), whichever comes first:

$$\tau = min(d_{new}, x_{k+1}) - C_k. \tag{4}$$

As visible in Fig. 1(c), the above procedure is performed individually for each processor in the considered cloud node. By combining the data from separate processors in each node we are able to provide a complete availability overview necessary for dynamic rescheduling process performed by the proposed *Sandpile* CA-based scheduler described in detail in the following Section.

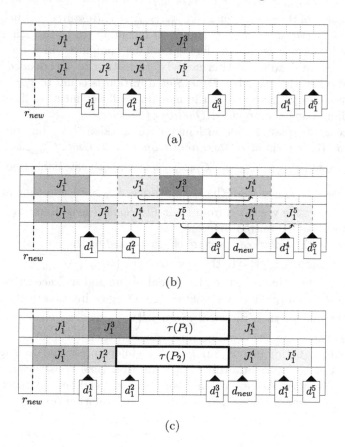

Fig. 1. Estimation of the available *Time Slot* on node M_i: Jobs J_1, J_2 and J_3 are processed as soon as possible, while jobs J_4 and J_5 are processed as late as possible. The size of the available free *Time Slot* τ may differ between processors.

4.2 The Sandpile Scheduler Dynamics

Our proposed solution is a local neighborhood diffusion approach which employs overlapping domains to achieve system-wide load balancing and fairness between individual Cloud nodes. We assume that the state of node M_i at the time t is represented by a variable $S_i(t)$, while neighborhood $V_i(t)$ defines the set of its neighbors N_i in the communication network. In our case, we consider the *Moore* neighborhood (formed by $z = 8$ cells surrounding a given cell) with periodic boundary conditions.

A set of possible states of our *Sandpile* scheduler consists of three modes: whenever a resource becomes overburdened in comparison with its neighbors, its local *Maximum Completion Time*, C^i_{max} exceeds the local average by a specified *threshold* value $\left(C^i_{max} - \overline{C^i_{max}} > C_{threshold}\right)$. For simplicity's sake we employ in our work a value of threshold equal to $C_{threshold} = 1$. Resources in such state are considered as *Overloaded*. They will send all incoming traffic to their neighbors, as well as any surplus workload that cannot be completed in a required time frame (i.e., before deadline). Alternatively, transition to *Overloaded* state can be triggered when the available *Time Slot* duration is equal to $\tau_i = 0$. In such cases, at least one job assigned to the machine will not meet its deadline, thus negatively impacting the provided *Quality of Service*.

Resource that are not overburdened with workload can be considered as *Underloaded*. Their estimated *Maximum Completion Time*, C^i_{max} is lower than the local average $\left(\overline{C^i_{max}} - C^i_{max} > C_{threshold}\right)$, while the available free *Time Slot* duration is usually longer than the local average $\left(\tau_i > \overline{\tau}\right)$ and they are capable of accepting excessive workload from their *Overloaded* neighbors, as well as any incoming workload submitted by users.

Resources in the *Balanced* state are characterized by the estimated *Maximum Completion Time*, C^i_{max} close to the local average $\left(|C^i_{max} - \overline{C^i_{max}}| \leq C_{threshold}\right)$. They will run jobs, which exist in their local queue and will accept new jobs as well. Such a state can be alternatively triggered by estimation of the free *Time Slots* slightly smaller than the local average $\left(0 \ll \tau_i < \overline{\tau}\right)$ ensuring that they are capable of meeting the deadline constraints of their local jobs.

Every agent (node) in our *Sandpile* scheduler (presented in Algorithm 1) gathers up-to-date information in each transition time constructing complete availability summary. Agents inform their nearest neighbors of their workload levels (and free *Time Slots*, τ_i) and update this information throughout program execution.

The rescheduling process is invoked whenever a new job J_{new} arrives to node M_i at time r_{new} (Algorithm 1: Line 8). It could be sent by the Cloud user, one of the neighboring nodes or rescheduled due to earlier job failure. After estimating the free *Time Slot* duration and *Completion Time* of the new job queue, state of the node is updated according to previously defined transition rules (Algorithm 1: Line 11).

If the arrival of the job triggers the *Overloaded* transition rule (i.e., causes workload imbalance), the excessive jobs will be sent to one of the available neighbors. Scheduling agent will find a set of neighboring nodes which are suitable to execute the job J_{new}. Nodes are first sorted in a descending order according to the available *Time Slots* and the job J_{new} is sent to a machine with the longest available free *Time Slot* (Algorithm 1: Lines 12–18).

Algorithm 1. Pseudo-code of the *CA-Stretch* Scheduling Algorithm

---

1: **Input:** Q: Node's M_i local job queue
2: **Input:** r_{new}: Release time of the job J_{new}
3: **Input:** p_{new}: Processing time of the job J_{new} on machine M_i
4: **Input:** S_i: Current state of the machine M_i
5: **Input:** V_i: Neighborhood of the machine M_i

6: *Initialize Iteration Counter, $T \leftarrow 0$*
7: **for all** $A_i \in A$ **do** {In parallel}
8:     *Trigger(J_{new})*
9:     *Calculate Completion Time of Local Job Queue, C_{max}^i.*
10:    *Calculate Free Time Slot Duration, τ_i.*
11:    *Update Cell State, $S_i(t_n) = f(S_i(t_{n-1}), V_i(t_{n-1}))$*

12:    **if** S_i == *Overloaded* **then**
13:        **while** S_i == *Overloaded* **do**
14:            **for all** $M_i \in V_i$ **do**
15:                *Sort machines in V_i by non-decreasing Time Slots, $\tau_1 \leq \tau_2 \leq ... \leq \tau_n$.*
16:                *Send job J_{new} to the machine with the longest Time Slot, τ_i.*
17:            **end for**
18:        **end while**
19:    **else**
20:        **if** $\tau \geq p_{new}$ **then**
21:            *Add job J_{new} to machine's M_i queue.*
22:            *Schedule job J_{new} in the free Time Slot, τ_i.*
23:        **else**
24:            **for all** $M_i \in V_i$ **do**
25:                *Sort machines in V_i by non-decreasing Time Slots, $\tau_1 \leq \tau_2 \leq ... \leq \tau_n$.*
26:                *Send job J_{new} to the machine with the longest τ_i.*
27:            **end for**
28:        **end if**
29:    **end if**
30:    *Update Iteration Counter, $T \leftarrow T + 1$*
31: **end for**

In a case of nodes in *Underloaded* and *Balanced* states, the job J_{new} will be added to their local queue. As long as the schedule can be accommodated before the required deadline, a job will be allocated for execution in the available *Time Slots* (Algorithm 1: Lines 21–22). Alternatively, an excessive job will be sent to one of the neighboring nodes (Algorithm 1: Lines 24–27). Because this process may trigger new events in the adjacent nodes, the avalanches (migrations of jobs) will iteratively continue throughout the entire system until a global state of equilibrium is achieved.

5 Experimental Analysis and Performance Evaluation

5.1 Simulation Testbed

To study the performance of the proposed dynamic load balancing and rescheduling algorithm, we have conducted several simulation experiments under three Cloud system scales: a small-scale Cloud system composed of $m = 16$ CC nodes, a medium-scale Cloud system composed of $m = 64$ CC nodes and a large-scale Cloud system composed of $m = 144$ CC nodes.

The number of processing elements in each node is generated by a Gaussian probability distribution function with mean 6 and variance 1. The computational capacity of the processing elements is similarly generated with mean 4 (instruction per time unit) and variance 1. The nominal bandwidth of the network connecting every two nodes is assumed to be generated with mean 2 (instructions transferred per time unit) and variance 1. The number of users is fixed at 8, 32, and 72 for the small-scale, medium-scale, and large-scale systems, respectively.

The execution requirements of submitted jobs are normally distributed with mean 10 (instructions) and variance of 3 (instructions). Each job is composed of a number of threads, where the number of threads is randomly and uniformly selected from the following set (1, 2, 3, 4). For each client, the generation rate of the new jobs is Poisson distributed with rate (mean) of 5 (time units). The efficiency of the analyzed job scheduling methods is measured in terms of:

- **Average Turnaround Time:** denote the total number of simulated jobs as N, the completion time of a single job J_k^j as C_k^j, the arrival time as r_k^j, and the *Average Turnaround Time* is defined as: $\frac{\sum_{j=1}^{N}(C_k^j - r_k^j)}{N}$;
- **Slowdown:** the ratio of the response time under the concurrent scheduling policy over its response time in dedicated mode.

5.2 Simulation Results

An efficient scheduler has to be scalable and capable of dealing with heterogeneity; good performance should not be only restricted to established architectures but also be invariant to scale and type of architecture. We have considered multiple scenarios of heterogeneous architectures and compared the performance of the proposed job scheduling algorithm (denoted further as *CA-Stretch*) with that of several well known scheduling approaches, such as *First Come First Served* (FCFS), *Shortest Job First* (SJF), *Longest Job First* (LJF) and *Shortest Remaining Time First* (SRTF). We conducted several simulations with workload size equal to 5000 jobs, scheduled within each Cloud system scale.

Figure 2(a) shows the average performance results of analyzed algorithms under different system sizes. As can be seen, the proposed job scheduling algorithm significantly outperforms other scheduling methods especially in large scale CC systems. It has been also found that *CA-Stretch* scheduling scheme has shown the best performance while producing the shortest *Average Turnaround Time* and *Slowdown* results as depicted in Fig. 2(b) and (c), respectively.

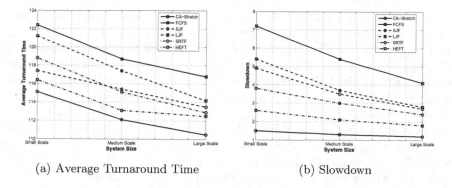

(a) Average Turnaround Time   (b) Slowdown

Fig. 2. Performance results of conducted experiments with multiple scheduling heuristics for a total of $n = 5000$ jobs scheduled within $m = 16, 64$ and 144 CC nodes: (a) Average Turnaround Time, (b) Slowdown.

This could be due to the fact that *CA-Stretch* is a job scheduling technique balancing the arriving workload locally on individual Cloud nodes, in which the workload that is assigned to each resource varies over time according to its available computational capacity and availability of the free *Time Slots* in a local schedule. As the job submission rate changes over time our solution is capable of adaptively rescheduling the arriving workload to the most appropriate resources. This avoids situations where some nodes become too busy or too idle. This reduces the *Average Turnaround Times* of the jobs, and - in the effect - decreases the overall system *Slowdown*, providing a fair trade-off between all submitted jobs.

It can be also stated that the proposed scheduling procedure distributes the submitted workload among the Cloud nodes considerably more evenly than other compared techniques in all analyzed system scales, thus offering greater flexibility and scalability. This is due to the fact that *CA-Stretch* takes into consideration the amount and duration of the available *Time Slots* on each Cloud node to evenly distribute the workload within the system, while the SOC properties of the *Sandpile* model guarantee equalization of workload in all overlapping local neighborhoods, and in the result, the whole simulated system platform.

6   Conclusions

We have proposed a novel parallel and distributed algorithm based on the *Sandpile* cellular automata model for dynamic load balancing and rescheduling in the Cloud environment. In our solution, computing resources are under the control of agents which can interact locally within an established neighborhood. In such a system, jobs arrive to resources and accumulate similarly to grains of sand. When a given agent detects any inequalities, it sends excessive jobs to his neighbors, often starting an avalanche, which may be propagated throughout the entire system until a new state of equilibrium is met.

We presented the rules of local interactions among agents providing a global behavior of the system. We also addressed common issues of resource heterogeneity and communication overheads by taking several parameters into account such as processing power of computing nodes and their communication latency. Proposed decentralized scheduling approach is particularly convenient for large-scale distributed environments. The objective of our scheduler is to ensure fairness among applications, by minimizing the *Slowdown* of all submitted jobs. Our solution inherits the benefits of both static and dynamics scheduling strategies. The proposed algorithm is robust and scalable due to implemented on-demand rescheduling mechanisms, which have a great impact on enhancing its performance over other scheduling strategies in dynamic environments.

References

1. Bak, P., Tang, C., Wiesenfeld, K.: Self-organized criticality: an explanation of the 1/f noise. Phys. Rev. Lett. **59**, 381–384 (1987)
2. Celaya, J., Marchal, L.: A fair decentralized scheduler for bag-of-tasks applications on desktop grids. In: 2010 10th IEEE/ACM International Conference on Cluster, Cloud and Grid Computing (CCGrid), pp. 538–541, May 2010
3. Franceschelli, M., Giua, A., Seatzu, C.: Load balancing over heterogeneous networks with gossip-based algorithms. In: Proceedings of the Conference on American Control Conference, ACC 2009, pp. 1987–1993. IEEE Press, Piscataway (2009)
4. Hu, J., Klefstad, R.: Decentralized load balancing on unstructured peer-2-peer computing grids. In: NCA, pp. 247–250. IEEE Computer Society (2006)
5. Jelasity, M., Montresor, A., Babaoglu, O.: A modular paradigm for building self-organizing peer-to-peer applications. In: Di Marzo Serugendo, G., Karageorgos, A., Rana, O.F., Zambonelli, F. (eds.) ESOA 2003. LNCS (LNAI), vol. 2977, pp. 265–282. Springer, Heidelberg (2004)
6. Laredo, J., Bouvry, P., Guinand, F., Dorronsoro, B., Fernandes, C.: The sandpile scheduler. Cluster Comput. **17**, 1–14 (2014)
7. Tchernykh, A., Schwiegelshohn, U., Yahyapour, R., Kuzjurin, N.: On-line hierarchical job scheduling on grids with admissible allocation. J. Sched. **13**(5), 545–552 (2010)

Conflict Solution According to "Aggressiveness" of Agents in Floor-Field-Based Model

Pavel Hrabák[1]([✉]) and Marek Bukáček[2]

[1] Institute of Information Theory and Automation, Czech Academy of Sciences,
Pod Vodarenskou vezi 4, 182 08 Prague, Czech Republic
hrabak@utia.cas.cz
[2] Faculty of Nuclear Sciences and Physical Engineering,
Czech Technical University in Prague, Trojanova 13, 120 00 Prague, Czech Republic
bukacma2@fjfi.cvut.cz

Abstract. This contribution introduces an element of "aggressiveness" into the Floor-Field based model with adaptive time-span. The aggressiveness is understood as an ability to win conflicts and push through the crowd. From experiments it is observed that this ability is not directly correlated with the desired velocity in the free flow regime. The influence of the aggressiveness is studied by means of the dependence of the travel time on the occupancy of a room. A simulation study shows that the conflict solution based on the aggressiveness parameter can mimic the observations from the experiment.

Keywords: Floor-Field model · Conflict solution · Aggressiveness

1 Introduction

This article focuses on a microscopic study of a simulation tool for pedestrian flow. The object of the study is a simulation of one rather small room with one exit and one multiple entrance, which may be considered as one segment of a large network. The behaviour of pedestrians in such environment has been studied by our group by means of variety experiments [4,6] from the view of the boundary induced phase transition (this has been studied theoretically for Floor-Field model in [9]). Observing data from these experiments we have found out that each participant has different ability to push through the crowd. Therefore, this article is motivated by the aim to mimic such behaviour by simple cellular model, which may be applied in simulations of apparently heterogeneous scenarios as [13,16].

The original model is based on the Floor-Field Model [7,12,14] with *adaptive time-span* [5] and *principle of bonds* [10]. The adaptive time span enables to model heterogeneous stepping velocity of pedestrians; the principle of bonds helps to mimic collective behaviour of pedestrians in lines. It is worth noting that there is a variety of modifications of the Floor-Field model capturing different aspects of pedestrian flow and evacuation dynamics. Quite comprehensive summary can be found in [15].

© Springer International Publishing Switzerland 2016
R. Wyrzykowski et al. (Eds.): PPAM 2015, Part II, LNCS 9574, pp. 507–516, 2016.
DOI: 10.1007/978-3-319-32152-3_47

In this article we focus on the solution of conflicts, which accompany all cellular models with parallel update, i.e., when more agents decide to enter the same site/cell. In such case, one of the agents can be chosen at random to win the conflict, the randomness can be executed proportionally to the hopping probability of conflicting agents [7]. The unresolved conflicts play an important role in models of pedestrian evacuation. The aim to attempt the same cell may lead to the blocking of the motion. This is captured by the friction parameter μ denoting the probability that none of the agent wins the conflict. An improvement is given by the friction function [17], which raises the friction according to the number of conflicting agents.

In our approach we introduce an additional property determining the agent's ability to win conflicts, which may be understood as agent's aggressiveness. This characteristics has been inspired by the analyses of repeated passings of pedestrians through a room under various conditions from free flow to high congestion. As will be shown below, this characteristic significantly affects the time spent by individual agents in the room, which is referred to as the *travel time*. Similar heterogeneity in agents behaviour has been used in [11], where the "aggressiveness" has been represented by the willingness to overtake.

2   Experiment

The introduction of the aggressiveness as an additional model parameter is motivated by the microscopic analyses of the experimental data from the experiment "passing-through" introduced in [6]. The set-up of the experiment is shown in Fig. 1. Participants of the experiment were entering a rectangular room in order to pass through and leave the room via the exit placed at the wall opposite to the entrance.

The inflow rate of pedestrians has been controlled in order to study the dependence of the phase (free flow or congested regime) on the inflow rate α. In order to keep stable flow through the room, pedestrians were passing the room repeatedly during all runs of the experiment.

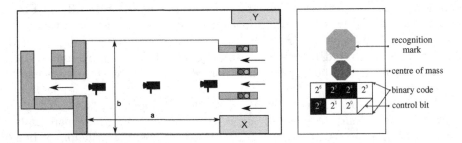

Fig. 1. Taken from [6]. Left: experimental setting of the experiment, a = 7.2 m, b = 4.4 m. After the passage through the exit, participant returned to the area Y waiting for another entry. Right: sketch of pedestrian's hat used for automatic image recognition.

Each participant has been equipped by a hat with unique binary code. The automatic image recognition enables us not only to restore the pedestrians trajectories but more over to assign all trajectories to individual participants. This enables the study of individual properties of the pedestrians under a bride scale of conditions, since for each participant there are 20 to 40 records of their passings.

One of the investigated quantities is the travel-time $TT = T_{out} - T_{in}$ denoting the length of the time interval a pedestrian spent in the room between the entrance at T_{in} and the egress at T_{out}. To capture the pedestrians behaviour under variety of conditions, the travel time is investigated with respect to the average number of pedestrians in the room N_{mean} defined as

$$N_{mean} = \frac{1}{T_{out} - T_{in}} \int_{T_{in}}^{T_{out}} N(t)dt, \tag{1}$$

where $N(t)$ stands for the number of pedestrians in the room at time t. Figure 2 shows the scatter plot of all pairs (N_{mean}, TT) gathered over all runs of experiment and all participants.

The reaction of participants to the occupancy of the room significantly differs. There are two basic characteristics that can be extracted: the mean travel time in the free-flow regime (0–7 pedestrians) and the slope of the travel-time dependence on the number of pedestrians in the congested regime

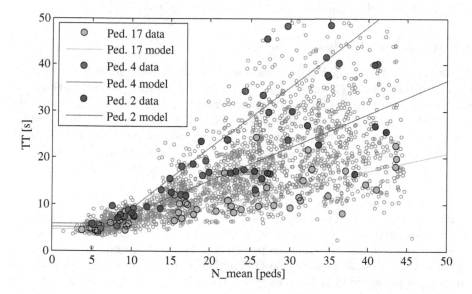

Fig. 2. Scatter plot of the travel time TT with respect to the occupancy N_{mean} extracted from the experiment. Three participants are highlighted. Their travel time is approximated by the piecewise linear model. We can see that Ped. 2 has lower desired velocity in free regime but higher ability to push through the crowd in comparison to Ped. 4.

(10–45 pedestrians). The former is given by the desired velocity, the latter reflects the ability to push through the crowd, referred to as the aggressiveness. This observation corresponds to the piece-wise linear model for each pedestrian

$$TT = \frac{S}{v_0(i)} + \mathbf{1}_{\{N>7\}}(N - 7) \cdot \text{slope}(i) + \text{noise} \qquad (2)$$

where $S = 7.2\,\text{m}$, $v_0(i)$ is the free-flow velocity of the pedestrian i, slope(i) is the unique coefficient of the linear model for pedestrian i. The breakpoint $N = 7$ depends from the room geometry. The weighted mean of the R^2 value of the model (2) is 0.688.

Detailed description of the experiment and its analyses has been presented at the conference TGF 15 and will be published in the proceedings [3]. Videos capturing the exhibition of the aggressive behaviour are available at http://gams.fjfi.cvut.cz/peds.

3   Model Definition

The model adapts the principle of the known Floor-Field cellular model. The playground of the model is represented by the rectangular two-dimensional lattice $\mathbb{L} \subset \mathbb{Z}^2$ consisting of cells $x = (x_1, x_2)$. Every cell may be either occupied by one agent or empty. Agents are moving along the lattice by hopping from their current cell $x \in \mathbb{L}$ to a neighbouring cell $y \in \mathcal{N}(x) \subset \mathbb{L}$, where the neighbourhood $\mathcal{N}(x)$ is Moore neighbourhood, i.e., $\mathcal{N}(x) = \{y \in \mathbb{L};\ \max_{j=1,2} |x_j - y_j| \le 1\}$.

3.1   Choice of the New Target Cell

Agents choose their target cells y from $\mathcal{N}(x)$ stochastically according to probabilistic distribution $P(y \mid x;\ \text{state of } \mathcal{N}(x))$, which reflects the "attractiveness" of the cell y to the agent. The "attractiveness" is expressed by means of the static field S storing the distances of the cells to the exit cell $E = (0,0)$, which is the common target for all agents. For the purposes of this article, the euclidean distance has been used, i.e., $S(y) = \sqrt{|y_1|^2 + |y_2|^2}$. Then it is considered $P(y \mid x) \propto \exp\{-k_S S(y)\}$, for $y \in \mathcal{N}(x)$. Here $k_S \in [0, +\infty)$ denotes the parameter of sensitivity to the field S.

The probabilistic choice of the target cell is further influenced by the occupancy of neighbouring cells and by the diagonality of the motion. An occupied cell is considered to be less attractive, nevertheless, it is meaningful to allow the choice of an occupied cell while the principle of bonds is present (explanation of the principle of bonds follows below). Furthermore, the movement in diagonal direction is penalized in order to suppress the zig-zag motion in free flow regime and support the symmetry of the motion with respect to the lattice orientation.

Technically this is implemented as follows. Let $O_x(y)$ be the identifier of agents occupying the cell y from the point of view of the agent sitting in cell x, i.e. $O_x(x) = 0$ and for $y \ne x$ $O_x(y) = 1$ if y is occupied and $O_x(y) = 0$ if y is empty. Then $P(y \mid x) \propto (1 - k_O O_x(y))$, where $k_O \in [0, 1]$ is again the parameter of sensitivity to the occupancy ($k_O = 1$ means that occupied cell will never be chosen). Similarly can be treated the diagonal motion defining the diagonal movement identifier as $D_x(y) = 1$ if $(x_1 - y_1) \cdot (x_2 - y_2) \neq 0$ and $D_x(y) = 0$ otherwise. Sensitivity parameter to the diagonal movement is denoted by $k_D \in [0, 1]$ ($k_D = 1$ implies that diagonal direction is never chosen).

The probabilistic choice of the new target cell can be than written in the final form

$$P(y \mid x) = \frac{\exp\left\{ -k_S S(y) \right\} \left(1 - k_O O_x(y)\right) \left(1 - k_D D_x(y)\right)}{\sum_{z \in \mathcal{N}(x)} \exp\left\{ -k_S S(z) \right\} \left(1 - k_O O_x(z)\right) \left(1 - k_D D_x(z)\right)}. \tag{3}$$

It is worth noting that the site x belongs to the neighbourhood $\mathcal{N}(x)$, therefore the Eq. (3) applies to $P(x \mid x)$ as well.

3.2 Updating Scheme

The used updating scheme combines the advantages of fully-parallel update approach, which leads to necessary conflicts, and the asynchronous clocked scheme [8] enabling the agents to move at different rates.

Each agent carries as his property the *own period* denoted as τ, which represents his desired duration between two steps, i.e., the agent desires to be updated at times $t = k\tau$, $k \in \mathbb{Z}$. Such principle enables to model different velocities of agents, but undesirably suppresses the number of conflicts between agents with different τ. To prevent this, we suggest to divide the time-line into isochronous intervals of the length $h > 0$. During each algorithm step $k \in \mathbb{Z}$ such agents are updated, whose desired time of the next actualization lies in the interval $[kh, (k+1)h)$. A wise choice of the interval length h in dependence on the distribution of τ leads to the restoration of conflicts in the model. It is worth noting that we use the concept of adaptive time-span, i.e., the time of the desired actualization is recalculated after each update of the agent, since it can be influenced by the essence of the motion, e.g., diagonal motion leads to a time-penalization, since it is $\sqrt{2}$ times longer. For more detail see e.g. [5]. This is an advantage over the probabilistic approach introduced in [1].

3.3 Principle of Bonds

The principle of bonds is closely related to the possibility of choosing an occupied cell. An agent who chooses an occupied cell builds a bond to the agent sitting in the chosen cell. This bond lasts until the motion of the blocking agent or until the next activation of the bonded agent. The idea is that the bonded agents attempt to enter their chosen cell immediately after it becomes empty.

3.4 Aggressiveness and Solution of Conflicts

The partially synchronous updating scheme of agents leads to the kind of conflicts that two ore more agents are trying to enter the same cell. This occurs when

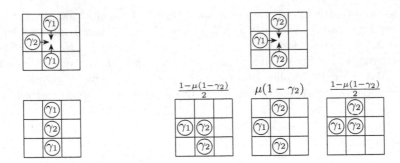

Fig. 3. Conflict solution for $\gamma_1 < \gamma_2$. Left: more aggressive wins the conflict over two less aggressive. Right: the conflict of two more aggressive can resolve by the blocking of the movement.

more agents choose as their target cell the same cell, or when more agents are bonded to the same agent, which becomes empty. The mechanism of the conflict solution is the same in both cases. Each agent carries an information about his ability to "win" conflicts which is here referred to as aggressiveness and denoted by letter $\gamma \in [0,1]$. The conflict is always won by agents with highest γ.

If there are two or more agents with the highest γ, the friction parameter μ plays a role. In this article we assume that the higher is the aggressiveness γ, the less should be the probability that none of the agents wins the conflict. Therefore, the conflict is not solved with probability $\mu(1-\gamma)$ (none of the agents move). With complement probability $1 - \mu(1-\gamma)$ the conflict resolves to the motion of one of the agents. This agent is chosen randomly with equal probability from all agents involved in the conflict having the highest γ. The mechanism of the friction can be easily modified. An example of conflict solution is depicted in Fig. 3.

4   Impact of the Aggressiveness Element

The effect of the aggressiveness has been studied by means of the simulation. Results stressed in this article come from the simulations with parameters given by Table 1. The values of τ and γ are distributed among agents uniformly and independently on each other.

The simulation set-up has been designed according to the experiment, i.e., the room of the size $7.2\,\mathrm{m} \times 4.4\,\mathrm{m}$ has been modelled by the rectangular

Table 1. Values of parameters used for simulation.

k_S	k_O	k_D	h	μ	τ	γ
3.5	1	0.7	0.1 s	0.5	$\{.25, .4\}$	$\{0, 1\}$

lattice 18 sites long and 11 sites wide. The size of one cell therefore corresponds to $0.4\,\mathrm{m} \times 0.4\,\mathrm{m}$. The exit is placed in the middle of the shorter wall, the open boundary is modelled by a multiple entrance on the opposite wall. New agents are entering the lattice stochastically with the mean inflow rate α [pedestrians/second]. The inflow rate is a controlled parameter. For more detailed description of the simulation we refer the reader to [2].

It has been shown that such system evinces the boundary induced phase transition from the free flow (low number of agents in the lattice) to the congestion regime (high number of pedestrians in the lattice) via the transient phase (number of pedestrians fluctuating between the low and high value). Therefore, wise choice of different inflow rates α covering the all three phases, enables us to study the dependence of the travel time TT on the average number of agents in the lattice N_{mean}. When simulating with parameters from Table 1, the correct choice of inflow rate is $\alpha \in [1, 3]$.

Figure 4 shows the dependence of the travel time $TT = T_{\mathrm{out}} - T_{\mathrm{in}}$ on the average number of agents in the lattice N_{mean} calculated according to (1). Measured data consisting of pairs (N_{mean}, TT) are aggregated over simulations for inflow rate values $\alpha \in \{1, 1.5, 1.8, 2.0, 2.3, 2.7, 3.0\}$; for each inflow α twenty runs of the simulation were performed. Each run simulates $1000\,\mathrm{s}$ of the introduced scenario starting with empty room. Agents were distributed into four groups according to their own period τ and aggressiveness γ, namely "fast aggressive" ($\tau = 0.25$, $\gamma = 1$), "fast calm" ($\tau = 0.25$, $\gamma = 0$), "slow aggressive" ($\tau = 0.4$, $\gamma = 1$), and "slow calm" ($\tau = 0.4$, $\gamma = 0$).

In the graph of the Fig. 4 we can see the average travel time for each group calculated with respect to the occupancy of the room. It is evident that for low occupancy up to 10 agents in the room the mean travel time for each group levels at a value corresponding to the free flow velocity given by the own updating period. For the occupancy above 20 agents in the lattice, the linear growth of the mean travel time with respect to N_{mean} is obvious. Furthermore, the average travel time for fast-calm corresponds to the travel time of slow aggressive. The Fig. 4 shows two auxiliary graphs presenting the dependence of TT on N_{mean} for systems with homogeneity in γ (left) or in τ (right). From the graphs we can conclude that the heterogeneity in aggressiveness γ reproduces the desired variance in the slope of the graph without the non-realistic high variance in free flow generated by the heterogeneity of own updating frequency.

The influence is even more evident from the graph in Fig. 5 representing a plot of all travel time entries with respect to the time of the exiting T_{out}. Right graph shows the box-plots of the travel time for four groups measured after $500\,\mathrm{s}$ from the initiation, i.e., in the steady state of the system. We can see that in this view, the aggressiveness plays more important role than the desired velocity of agents.

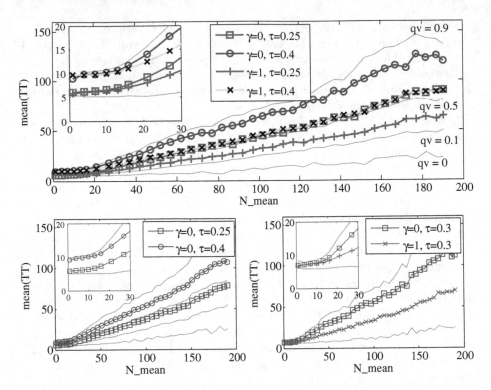

Fig. 4. Dependence of the mean travel time TT on the average occupancy N_{mean} for each group of agents. Gray lines represent the quantiles of the travel time regardless to the groups. Top: heterogeneity in both, γ and τ. Bottom left: heterogeneity in τ. Bottom right: heterogeneity in γ.

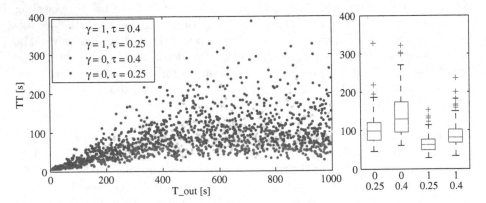

Fig. 5. Left: development of travel time TT in time for one run of the simulation. The value TT is plotted against the time of the exit T_{out} to ensure that values corresponding to the same time stem from similar conditions near the exit. Inflow rate $\alpha = 3\,\text{ped/s}$. The agent group is indicated by the color. Right: box-plots of the travel time for entries with $T_{\text{in}} > 500\,\text{s}$ (i.e. in the steady state) (Color figure online).

5   Conclusions and Future Work

The article introduced a parameter of aggressiveness as an additional characteristics of agents in the Floor-Field model with adaptive time. This parameter is understood as an ability to win conflicts. Therefore the heterogeneity of agents is given by their desired velocity (determined by the own period τ) and their ability to win conflicts referred to as the aggressiveness.

The simulation study shows that the aggressiveness has significant influence in the regime with high occupation of the room, i.e., in the dense crowd, and on the other hand has no effect in the free flow, as desired. The linear dependence of the travel time on the number of pedestrians in the agents neighbourhood seems to be a good tool how to measure the ability of agents/pedestrians to push through the crowd. The independence of this ability on the desired velocity of agents is very important to mimic the aspect that some "fast" pedestrians can be significantly slowed down by the crowd while some "slow" pedestrians can push through the crowd more effectively.

We believe that such feature can be very useful in the simulation of the evacuation or egress of large complexes as e.g. football stadiums, where the less aggressive pedestrians (parents with children, fragile women) can be slowed down and leave the facility significantly later than the average. The model reproduces this aspect even in the case of the homogeneity in own period τ.

In the future we plan to study this aspect in more detail. Mainly we would like to focus on the joint distribution of the desired velocity τ and the aggressiveness γ among the population and study its impact by means of the proposed model.

Acknowledgements. This work was supported by the Czech Science Foundation under grants GA13-13502S (P. Hrabák) and GA15-15049S (M. Bukáček). Further support was provided by the CTU grant SGS15/214/OHK4/3T/14.

References

1. Bandini, S., Crociani, L., Vizzari, G.: Heterogeneous pedestrian walking speed in discrete simulation models. In: Chraibi, M., Boltes, M., Schadschneider, A., Seyfried, A. (eds.) Traffic and Granular Flow '13, pp. 273–279. Springer International Publishing, Cham (2015)
2. Bukáček, M., Hrabák, P.: Case study of phase transition in cellular models of pedestrian flow. In: Wąs, J., Sirakoulis, G.C., Bandini, S. (eds.) ACRI 2014. LNCS, vol. 8751, pp. 508–517. Springer, Heidelberg (2014)
3. Bukáček, M., Hrabák, P., Krbálek, M.: Individual microscopic results of bottleneck experiments. In: Traffic and Granular Flow '15. Springer International Publishing (2015, to appear). arXiv:1603.02019 [physics.soc-ph]
4. Bukáček, M., Hrabák, P., Krbálek, M.: Experimental analysis of two-dimensional pedestrian flow in front of the bottleneck. In: Chraibi, M., Boltes, M., Schadschneider, A., Seyfried, A. (eds.) Traffic and Granular Flow '13, pp. 93–101. Springer International Publishing, Cham (2015)

5. Bukáček, M., Hrabák, P., Krbálek, M.: Cellular model of pedestrian dynamics with adaptive time span. In: Wyrzykowski, R., Dongarra, J., Karczewski, K., Waśniewski, J. (eds.) PPAM 2013, Part II. LNCS, vol. 8385, pp. 669–678. Springer, Heidelberg (2014)

6. Bukáček, M., Hrabák, P., Krbálek, M.: Experimental study of phase transition in pedestrian flow. In: Daamen, W., Duives, D.C., Hoogendoorn, S.P. (eds.) Pedestrian and Evacuation Dynamics 2014, Transportation Research Procedia, vol. 2, pp. 105–113. Elsevier Science B.V. (2014)

7. Burstedde, C., Klauck, K., Schadschneider, A., Zittartz, J.: Simulation of pedestrian dynamics using a two-dimensional cellular automaton. Phys. A **295**(3–4), 507–525 (2001)

8. Cornforth, D., Green, D.G., Newth, D.: Ordered asynchronous processes in multi-agent systems. Phys. D **204**(1–2), 70–82 (2005)

9. Ezaki, T., Yanagisawa, D., Nishinari, K.: Analysis on a single segment of evacuation network. J. Cell. Automata **8**(5–6), 347–359 (2013)

10. Hrabák, P., Bukáček, M., Krbálek, M.: Cellular model of room evacuation based on occupancy and movement prediction: comparison with experimental study. J. Cell. Automata **8**(5–6), 383–393 (2013)

11. Ji, X., Zhou, X., Ran, B.: A cell-based study on pedestrian acceleration and over-taking in a transfer station corridor. Phys. A **392**(8), 1828–1839 (2013)

12. Kirchner, A., Schadschneider, A.: Simulation of evacuation processes using a bionics-inspired cellular automaton model for pedestrian dynamics. Phys. A **312**(1–2), 260–276 (2002)

13. Kłeczek, P., Wąs, J.: Simulation of pedestrians behavior in a shopping mall. In: Wąs, J., Sirakoulis, G.C., Bandini, S. (eds.) ACRI 2014. LNCS, vol. 8751, pp. 650–659. Springer, Heidelberg (2014)

14. Kretz, T.: Pedestrian traffic, simulation and experiments. Ph.D. thesis, Universität Duisburg-Essen, Germany (2007)

15. Schadschneider, A., Chowdhury, D., Nishinari, K.: Stochastic Transport in Complex Systems: From Molecules to Vehicles. Elsevier Science B.V., Amsterdam (2010)

16. Spartalis, E., Georgoudas, I.G., Sirakoulis, G.C.: CA crowd modeling for a retirement house evacuation with guidance. In: Wąs, J., Sirakoulis, G.C., Bandini, S. (eds.) ACRI 2014. LNCS, vol. 8751, pp. 481–491. Springer, Heidelberg (2014)

17. Yanagisawa, D., Kimura, A., Tomoeda, A., Nishi, R., Suma, Y., Ohtsuka, K., Nishinari, K.: Introduction of frictional and turning function for pedestrian outflow with an obstacle. Phys. Rev. E **80**, 036110 (2009)

Computer Simulation of Traffic Flow Based on Cellular Automata and Multi-agent System

Magda Chmielewska, Mateusz Kotlarz, and Jarosław Wąs[✉]

Department of Applied Computer Science, Faculty of Electrical Engineering,
Automatics, IT and Biomedical Engineering,
AGH University of Science and Technology,
Al. Mickiewicza 30, 30-059 Kraków, Poland
jarek@agh.edu.pl

Abstract. The paper describes a microscopic simulation of traffic flow phenomenon based on Cellular Automata and Multi-agent System. The simulation enables the study of the complexity of the traffic system and can provide current information about road capacity. A car is represented as a set of several neighboring cells, as an extension of Nagel-Shreckenberg model devoted for urban traffic simulation. The car is represented as an agent whose decisions are based on the actual situation on the road including neighboring cars decisions. The model also contains traffic lights and mechanisms such as right-of-way, route planning or lane changing which help to simulate more complex behavior of vehicles.

Keywords: Simulation · Traffic flow · Cellular Automata · Agent-based model

1 Introduction

Nowadays computer simulations are often chosen to study plenty of phenomenons. It is a safe way to expand our knowledge without conducting many expensive experiments. We can simulate phenomenons which are too difficult or simply cannot be recreated in the field. One of these phenomenons is traffic flow. Many scientists have become interested in methods, that would give the best traffic model and provide similar results in real life. We are going to mention two models that we have found especially interesting.

Nagel-Schreckenberg model [6] is a theoretical model of freeway traffic based on Cellular Automata where the car is represented as one cell and contains information only about velocity. In every iteration the car moves forward the number of cells that is equal to its velocity. The model is too simple to simulate all of the traffic occurring in the city, because it doesn't simulate any road regulations or the behavior of cars on crossroads. Another interesting model is proposed by Rolf Hoffmann [5]. It is based on Global Cellular Automata (GCA) with access algorithms to model traffic. This model uses dynamic links between

© Springer International Publishing Switzerland 2016
R. Wyrzykowski et al. (Eds.): PPAM 2015, Part II, LNCS 9574, pp. 517–527, 2016.
DOI: 10.1007/978-3-319-32152-3_48

potential global neighbors and gives us the possibility to change our neighbors state, something that is not allowed in classical CA.

In this paper, the authors present the model of traffic flow based on Cellular Automata and Multi-Agent System [1,3,7,10] containing simulation of cars and traffic lights. The result of this work is an application based on modificated Nagel-Schreckenberg model - extended by us to be used in urban environment.

2   Proposed Model

The main goal was to create traffic model with simple update rules, which takes into account all urban circumstances such as traffic lights, changing lanes etc. We propose discrete, non-deterministic, rule-based model, which extends well-known Nagel-Schreckenberg model [6].

The Nagel-Schreckenberg model was originally designed for freeway traffic. Hence it has some disadvantages in case of urban traffic simulation. The first one is unrealistic acceleration and deceleration of vehicles. In the NaSch model the system updates every second ($dt = 1\,\mathrm{s}$). This means that acceleration or deceleration rate of each vehicle equals $7.5\,\mathrm{m/s^2}$. This value is too high to represent urban traffic where drivers accelerate and brake more smooth in comparison to the highways (Table 1).

Table 1. Acceleration and deceleration rates (based on [2,9]).

Source	Typical values ($\frac{m}{s^2}$)
ITE (1982)	Maximal acceleration: 1.5–3.6
	Maximal deceleration: 1.5–2.4
	Normal deceleration: 0.9–1.5
Gipps (1981)	Normal acceleration: 0.9–1.5
	Maximal breaking: 3.0
Firstbus (private communication)	acceleration (buses): 1.2–1.6
Intelligent Driver Model (IDM)	Acceleration: 1.0
	Comfortable deceleration: 1.5

In order to adapt the model to city conditions, it was necessary to propose several modifications.

2.1   Road Representation

We propose to divide the road network into smaller cells (1 m). Due to such a spatial discretization each vehicle has to occupy more that one cell at a time-step (according to its length). This solution is inspired by the work [4]. Each car is divided into smaller parts: a *head* and a *tail* (other pieces). The most important part is the *head* which determines the position of the car on the road. This approach enables to distinguish several types of vehicles (for instance cars,

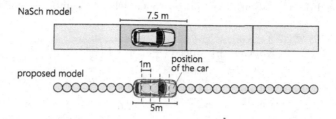

Fig. 1. A comparison between NaSch model and modified model.

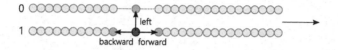

Fig. 2. Connections between cells at the road segment.

trucks, buses or HGVs). Each vehicle occupies as many cells as it is long (rounded up to the nearest number). For instance, if the car is 4.5 m long, it occupies 5 cells (Fig. 1).

The whole road network is composed of single roads. Each road consists of single sections which are divided into single cells. Road segments need to be connected in order to enable traffic flow (Fig. 2).

At single road segment each cell has a connection with its neighbors, i.e. the next cell, the previous one, and the cells on the left and right (if they exist). In this approach the road network is similar to the graph where cells correspond to the graph nodes (also called vertices) and connections between cells correspond to graph edges.

2.2 Modification of NaSch Rules

In the proposed model the system updates every 500 ms ($dt = 0.5$ s). The possible speeds of vehicles are shown in Table 2. In order to obtain the acceleration rates of 1–2 m/s^2 (for urban traffic) we introduced decimal fractions to represent speed and acceleration of each vehicle.

We propose following modification of update rules:

1. **Randomization** (with the probability $p = 0.15$):

$$behaviour_{t+1} = \begin{cases} max(behaviour_{t+1} - 2, -2) & \text{if the vehicle is entering the road,} \\ behaviour_{t+1} & \text{otherwise.} \end{cases}$$

$$(1)$$

2. **Accelerating or breaking:**

$$v_{t+1} = \begin{cases} max(v_t - dec, 0) & \text{if } behavior < 0, \\ v_t & \text{if } behavior = 0, \\ min(v_t + acc, v_{max}) & \text{if } behavior > 0, \end{cases}$$

$$(2)$$

Table 2. Speed discretization in the proposed model (n – number of cells, dt – time step).

Speed [$\frac{n}{dt}$]	Speed [$\frac{m}{s}$]	Speed [$\frac{km}{h}$]
10	20	72
9	18	64.8
8	16	57.6
7	14	50.4
6	12	43.2
5	10	36
4	8	28.8
3	6	21.6
2	4	14.4
1	2	7.2

where acc – acceleration which depends on a type of vehicle, dec – deceleration which depends on the road situation.

3. **Vehicle movement:**

$$x_{t+1} \rightarrow x_t + v_{floor},\qquad(3)$$

where v_{floor} is a speed rounded down.

We introduced variable *behavior* which represents the behavior of each vehicle. Values less than 0 represent braking, 0 means that vehicle should keep its speed, 1 means that it can accelerate (if the maximum velocity is not reached). Randomization probability was by default set to 0.15.

2.3   Modeling of Car Behaviour

Each vehicle treats other drivers on the road like obstacles. There are two kinds of obstacles:

– static – obstacles which don't move (velocity always equals 0),
– dynamic – obstacles which can move (velocity ≥ 0).

Visibility of the obstacle depends on its type. There are obstacles which are visible for every vehicle (e.g. cells occupied by other vehicles or traffic lights). The other obstacles can be ignored by the approaching vehicle under certain conditions (e.g. extreme points which are visible only for vehicles that have been assigned to the specific route).

Behavior of each vehicle depends on the obstacles which are within its sight:

$$s(n) = db_{normal}(v_n) + v_n,\qquad(4)$$

where db_{normal} – normal braking distance (from v_n to 0), v_n – speed of vehicle n.

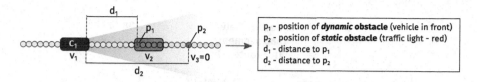

Fig. 3. The example of two obstacle types – dynamic (p_1) and static (p_2).

At this distance the car is looking for obstacles of both types. If some obstacle is found, the *real distance* is calculated. In case of dynamic obstacle, the real distance is the distance plus its emergency stopping distance:

$$d_{real} = d + db_{em}(v_{obstacle}), \tag{5}$$

where d – distance to obstacle, db_{em} – emergency braking distance (from $v_{obstacle}$ to 0), $v_{obstacle}$ – velocity of obstacle (equals 0 if obstacle is static) (Fig. 3).

Finally, the behavior of each vehicle is based on its speed and the minimum of calculated real distances to obstacles that were found (if both types of obstacles have been found):

$$d_{real} = min(d_{real_1}, d_{real_2}), \tag{6}$$

where: d_{real_1}, d_{real_2} – real distances to found obstacles.

2.4 Changing Lane and Route Assignment

There are two motivations to lane change maneuver:

- profitable – when it enables to keep higher speed and pass some slower cars ahead,
- necessary – in order to reach some point of the road network (and it's not possible from the current lane of the road) (Fig. 4).

Regardless of driver's motivation the safety criterion has to be satisfied before the maneuver is allowed:

$$\begin{cases} b(v_b) + d_1 - len > b(v), \\ b(v) + d_2 > b(v_f), \end{cases} \tag{7}$$

where $b(v_b)$, $b(v)$, $b(v_b)$ – emergency braking distance of vehicles c_b, c and c_f, d_1 – distance between c_b and c, d_2 – distance between c and c_f, len – length of vehicle c.

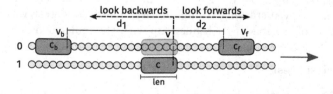

Fig. 4. The example of change lane maneuver.

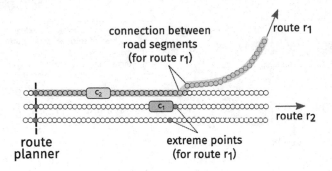

Fig. 5. The example of route assignment. Vehicles c_1 and c_2 are following route r_1. Vehicle c_1 has to change lane in order to follow assigned route.

Necessary lane changing is connected with route planning. Routes are assigned to vehicles approaching some crossroads. The point of such assignment is so-called route planner (see Fig. 5). For each route there are also some extreme points, which define the last possible road cells where the lane has to be changed. If lane change maneuver is not allowed (due to safety criterion which is not satisfied) the car has to stop before the extreme point and wait until the road is clean.

2.5   Right-of-way

In an urban area there are a lot of intersections without traffic lights where drivers must yield right-of-way. In proposed model such points are called priority points. There are two kinds of priority points:

– crossing points – places where vehicles enter the main road,
– entering points – places where vehicles only cut some other road lane.

This cells are blocked when entering the road is not allowed due to approaching vehicles. A priority point is connected with one or more checkpoints which are situated on the main road at the collision points (see Fig. 6).

3   Simulation Results

In order to check proposed model two tests have been made. The first one was carried out to obtain the relation between main traffic characteristics (velocity, density and flow) by increasing traffic density. The second test was comparing the arrival time of the bus at each bus stop with the real schedule.

3.1   The First Test

In the application a three-lane road was created in a shape of the loop. In one point of the road a car generator was placed. On the opposite site of the

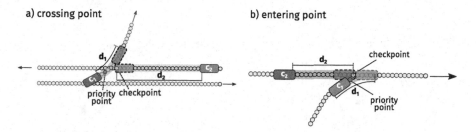

a) crossing point   b) entering point

Fig. 6. The examples of both types of priority points (a – crossing point, b – entering point). Checkpoints are situated at the cross point of the roads.

loop the measurement point was placed. Measurement point has three counters: c_f (flow counter), v_c (velocity counter) and d_c (density counter) and it takes three parameters: dt_m (measurement interval) and d_m (measurement distance).

Three traffic flow characteristics (flow, velocity and density) were calculated in every measurement interval (formulas based on [8]):

$$J_n = \cdot \frac{60}{dt_m \cdot n_{lanes}} [veh/h/lane] \tag{8}$$

$$v_n = \frac{c_v}{c_f} [m/dt] \tag{9}$$

$$\rho_n = \frac{c_d}{dt_m \cdot 60 \cdot ips \cdot n_{lanes}} \cdot \frac{1000}{d_m} [veh/km/lane] \tag{10}$$

where dt_m – measurement interval, d_m – measurement distance, n_{lanes} – number of lanes, ips – iterations per second (Fig. 7).

The relation between traffic characteristics shows that the maximum flow (occurred by density about 45 vehicles per hour) is about 1800 vehicles per hour (at the single lane) and that the maximum density that can be reached is about 170 vehicles per kilometer (Fig. 8).

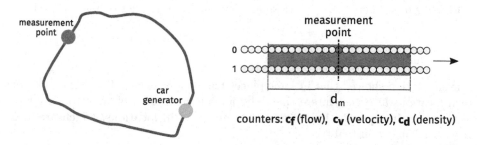

Fig. 7. The road designed for test purposes (left) and the example showing the placement of measurement point (right).

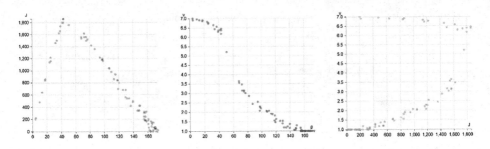

Fig. 8. Diagrams of relation between (from left): $J(\rho)$, $v(\rho)$, $v(J)$.

3.2 The Bus Test

The second part of model validation was comparing the time at each bus stop with the real schedule. All bus stops are shown in the Fig. 9. The whole route of the bus normally takes 20 min.

To compare the times, 10 single tests had been made. The results are shown in the Fig. 10. The minimum difference between schedule and simulation departure equals only 1 s (at the bus stop number five), and the maximum reached 70 s (at the first bus stop).

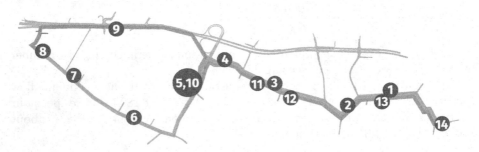

Fig. 9. All bus stops (line 168) in simulation area (in the order they are visited).

3.3 Traffic Jams in Simulation

The simulation showed that in some places traffic jams occur. Three examples of such locations are most interesting. The jams happen on yield-controlled inter-sections (without traffic lights) and they occurred in locations, which are also very problematic in reality.

It was useful and important observation, which shows that measured traffic data was correct and initial state of the simulation leads to the same traffic problems as in real life (Fig. 11).

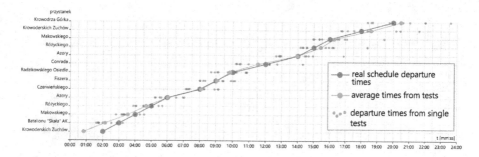

Fig. 10. Graph showing times of departure at each bus stop.

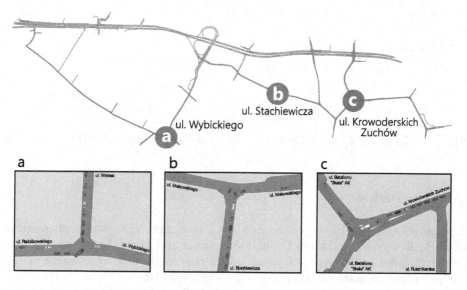

Fig. 11. Traffic jams occurring during simulation.

3.4   Simulation Performance

Although the application is divided into two separate threads (first thread is responsible for calculation and the second for visualization part) calculating of vehicle positions in next iteration is not parallel. After some modifications it could be implemented in parallel and for sure it would improve application performance.

The essential part of model testing was to measure the time of calculations. In order to obtain the maximum possible number of vehicles during the test we generated, at every generation point, as many vehicles, as possible (the number of vehicles reached over 1900). Obtained times (at each iteration of simulation) are plotted on a graph (Fig. 12). Time of calculation depends on simulation area – chosen area was the northwestern part of Cracow (Poland) with area containing roads with length of 33 km (considering each lane separately).

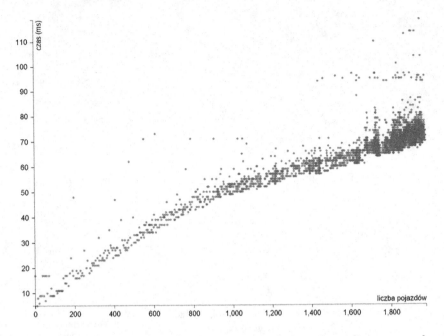

Fig. 12. The graph showing times of model calculations depending on number of vehicles.

4   Conclusions

The purpose of this paper was to develop a new method to study phenomena of traffic flow which combines Cellular Automata and Multi-Agent System. The model includes more elements which affect the traffic flow and represent the car as an autonomic unit whose decisions can affect other traffic participants decisions. This model is also enhanced with traffic lights, right-of-way and lane changing rules and route planning. The model can be further extended and the application can be modified to enable parallel calculations.

One of the most valuable advantages of the application is that it can be used as a powerful tool which can help to test different traffic scenarios on the roads (before they are even built). The biggest advantage of this solution is reduction of time, costs and resources.

References

1. Bonabeau, E.: Agent-based modeling: methods and techniques for simulating human systems. Proc. Natl. Acad. Sci. U. S. A. **99**(3), 7280–7287 (2002)
2. Bonsall, P., Liu, R., Young, W.: Modelling safety-related driving behaviour-impact of parameter values. Transp. Res. Part A Policy Pract. **39**(5), 425–444 (2005). http://www.sciencedirect.com/science/article/pii/S0965856405000261

3. Davidsson, P.: Multi agent based simulation: beyond social simulation. In: Moss, S., Davidsson, P. (eds.) MABS 2000. LNCS (LNAI), vol. 1979, pp. 97–107. Springer, Heidelberg (2001)
4. Hartman, D.: head leading algorithm for urban traffic modeling. In: 16th International European Simulation Symposium and Exhibition ESS, pp. 297–302 (2004)
5. Hoffmann, R.: GCA-w algorithms for traffic simuation. Acta Phys. Pol. B Proc. Suppl. **4**, 183–200 (2011)
6. Nagel, K., Schreckenberg, M.: A cellular automaton model for freeway traffic. J. Phys. I Fr. **2**(12), 2221–2229 (1992)
7. Płaczek, B.: A traffic model based on fuzzy cellular automata. J. Cell. Automata **8**, 261–282 (2013). ISSN 1557–5969
8. Schadschneider, A., Chowdhury, D., Nishinari, K.: Stochastic Transport in Complex Systems: From Molecules to Vehicles. Elsevier, Amsterdam (2011)
9. Treiber, M., Kesting, A.: Traffic Flow Dynamics. Springer, Heidelberg (2013)
10. Wąs, J., Bieliński, R., Gajewski, B., Orzechowski, P.: Problematyka modelowania ruchu miejskiego z wykorzystaniem automatów komórkowych. AUTOMATYKA **13**(3), 1207–1217 (2009)

A Stochastic Optimal Velocity Model for Pedestrian Flow

Antoine Tordeux[1($\boxtimes$)] and Andreas Schadschneider[2]

[1] Computer Simulation for Fire Safety and Pedestrian Traffic, Bergische Universität Wuppertal, Wuppertal, Germany
a.tordeux@fz-juelich.de
[2] Institut für Theoretische Physik, Universität zu Köln, Köln, Germany

Abstract. We propose a microscopic stochastic model to describe 1D pedestrian trajectories obtained in laboratory experiments. The model is based on optimal velocity (OV) functions and an additive noise determined by the inertial Ornstein-Uhlenbeck process. After statistical estimation of the OV function and noise parameters, we explore the model by simulation. The results show that the stochastic approach gives a good description of the characteristic relation between speed and spacing (fundamental diagram) and its variability. Moreover, it can reproduce the observed stop-and-go waves, bimodal speed distributions, and nonzero speed or spacing autocorrelations.

Keywords: Unidirectional pedestrian streams · Stochastic optimal velocity model · Statistical estimation of the parameters · Ornstein-Uhlenbeck process

1 Introduction

The analysis and modeling of pedestrian dynamics has attracted a lot of attention during the last decades [11,29]. Empirically, data have been obtained from experiments in laboratory conditions [7,15] with software to automatically extract the trajectories from video recordings [9]. These investigations allowed to establish many features of pedestrian dynamics [30], e.g. the unimodal shape of the fundamental flow-density diagram or the presence of stop-and-go waves as characteristics of unidirectional pedestrian streams [30,31]. Interestingly, these phenomena do not only hold for pedestrians but are also observed for vehicle or bike motion in 1D showing a certain universality in streams composed of human agents and related self-driven flows [39].

Numerous models have been developed to understand and analyze the characteristics of self-driven flows [6,11,29]. The unimodal shape of the fundamental diagram is already found in simple models like the Asymmetric Simple Exclusion Process (ASEP) [22] where it is related to the exclusion principle. More generally it is well explained microscopically by phenomenological monotone relations between the agent speed and distance spacing with the neighbor (usually called *optimal velocity* (OV), see [3]). The relation reflects the tendency to

© Springer International Publishing Switzerland 2016
R. Wyrzykowski et al. (Eds.): PPAM 2015, Part II, LNCS 9574, pp. 528–538, 2016.
DOI: 10.1007/978-3-319-32152-3_49

respect safety spacings to avoid collision due to unexpected movements of the neighbors. It is observed with both pedestrians [2] and drivers [4].

Nonlinear traffic waves and instability were the topics of the pioneering papers in the 1950's and early 1960's [10]. Microscopic continuous models defined by systems of differential equations were initially used [27]. The inertial optimal velocity models based on the OV function and defined by systems of ordinary or delayed equations are ones the most investigated traffic models [3,23]. Traffic waves are analyzed through instability of uniform solutions [25] or mapping to macroscopic soliton equations [20]. Generally speaking, it seems that the introduction of delays and deterministic inertial mechanisms generates instability of the uniform solution and the emergence and stable propagation of stop-and-go waves.

Many microscopic models describing nonuniform dynamics are stochastic [17]. A noise is added to differential systems of continuous models for both pedestrian [13,21] and road vehicle [33,34]. Yet, the stochastic aspect does not seem to be preponderant in the dynamics, especially in the formation of stop-and-go waves. In continuous pedestrian models, the noise is used for ambiguous situations (e.g. conflicts) in which two or more behavioral alternatives are equivalent [13] or to model heterogeneous pedestrian behaviors [28]. Few studies have shown that the noise plays a major role (see [12] for bidirectional streams and the formation of lanes). For road traffic models, probabilistic distributions of the parameters are also used to model heterogeneous driving styles [26], and stochastic noises are introduced to model perception errors [34] or to switch from a stationary state to an other [33]. The use of white noises or time-correlated ones does not impact the global dynamics of the second order models [34].

In this paper, we show that the introduction of a specific additive noise in a first order model can impact the dynamics and generate stop-and-go phenomena without requirement of deterministic instabilities. The noise is relaxed at the second order through a Langevin equation. After calibration, we observe by simulation that the model is able to give a good description of pedestrian dynamics and notably the stop-and-go waves. The paper is organized as following. The stochastic OV model is defined in Sect. 2. The description model calibration is presented in Sect. 3. The simulation of the model and comparison to the real data are done in Sect. 4. Conclusions are proposed in Sect. 5.

2   Stochastic Optimal Velocity Model

Initially, the optimal velocity model is a second-order model for which the speed is relaxed to an optimal speed depending on the spacing (headway) [3]. The relaxation is determined by an OV function $V : \Delta x \mapsto V(\Delta x)$. Nowadays, any approach based on the OV function is called *OV model* or *extended OV model*. The minimal OV model is [27]

$$\mathrm{d}x_n(t) = V(\Delta x_n(t))\,\mathrm{d}t, \qquad (1)$$

with $x_n(t)$ the position of agent n at time t and $\Delta x_n(t) = x_{n+1}(t) - x_n(t)$ the distance spacing, $x_{n+1}(t)$ being the position of the first predecessor $n + 1$.

The uniform solutions are stable in this model if the optimal speed function is increasing which is a natural assumption. The minimalist OV model is too simple to reasonably describe wave phenomena. More realistic dynamics are obtained if an inertia is introduced through reaction (or relaxation) time parameters such as in the *ordinary second order OV model* [3]

$$\begin{cases} dx_n(t) = v_n(t)\, dt, \\ dv_n(t) = \frac{1}{b}[V(\Delta x_n(t)) - v_n(t)]\, dt, \end{cases} \tag{2}$$

with $v_n(t)$ the agent speed and $b > 0$ the relaxation time parameter. The OV function calibrates the fundamental diagram while stop-and-go waves can be obtained if the reaction times are sufficiently high for that the stability condition fails.

In the literature, stochastic OV models are classically related to discrete models of interacting particle systems [18,19]. Here, we propose to use stochastic OV models by adding a stochastic noise to the continuous minimalist model (1). The noise is centered and stationary, with finite variance. It models other random factors affecting the speed besides the spacing. We denote $W(t)$ the Wiener process such that $W(t, s) - W(t)$ is normally distributed with mean zero, variance s, and independent to $W(t)$ for all t and s. In order to introduce a non-vanishing noise autocorrelation, we use the model

$$\begin{cases} dx_n(t) = V(\Delta x_n(t))\, dt + \varepsilon_n(t)\, dt, \\ d\varepsilon_n(t) = -\frac{1}{b}\varepsilon_n(t)\, dt + a\, dW_n(t), \end{cases} \tag{3}$$

with a the amplitude of the noise and $b > 0$ the relaxation time parameter. The noise $\varepsilon_n(t)$ is the solution of a Langevin equation. It is a standard stochastic process called the Ornstein-Uhlenbeck process, for which the autocorrelation tends to zero exponentially. The noise randomly oscillates around zero making positive and negative corrections to the optimal speed at random instants with independent increments. This behavior is consistent with action-point traffic models and observations that drivers react at discrete random times [32,35,36,38]. The model (3) is close to the deterministic second order OV model (2). Yet with the stochastic approach, the inertia only affects the noise. The uniform solutions are linearly stable in the model (3) in the deterministic case where $a = 0$ as soon as $V(\cdot)$ is strictly increasing. However, the trajectories obtained from the model with the additive noise describe nonuniform solutions with stop-and-go waves (see Fig. 1, the simulation details are given in Sect. 4). Yet, oppositely to the unstable deterministic approaches, there are no generic problems of collision and backward motion (see for instance [8,37]).

3   Calibration of the Parameters

The data we use to calibrate and evaluate[1] the models are pedestrian trajectories in a ring over laboratory conditions [1]. The experiments in a ring with length of

---

[1] There is no split of the data; both calibration and evaluation steps are done with the global data sample.

27 m and width of 0.7 m. Several experiments were carried out with different level of densities (the pedestrians numbers go from 14 to 70 with 11 tested density levels) and uniform initial distribution. The trajectories are measured on two segments with length of 4 m using the software PeTrack [5] with a time resolution of 0.04 s (frame-rate 25 fps). The variables used for the model calibration are the distance spacing and speed

$$\Delta x(t) = x_1(t) - x(t) \quad \text{and} \quad v_{\delta t}(t) = \tfrac{1}{\delta t}\big(x(t + \delta t/2) - x(t - \delta t/2)\big). \quad (4)$$

with x_1 the position of the predecessor. The spacings are measured instantaneously while the speeds have to be averaged over time intervals of length $\delta t = 0.8$ s to avoid effects of the pedestrian step frequency that is close to 0.7 s [24].

The OV function models a phenomenological relation between the speed and the spacing. Two main states are classically distinguished: (1) the free state, when the spacing is large and the speed is equal to the maximal desired speed and (2) the congested (or interactive) state, when the spacing is small and the speed depends on the spacing. Both road traffic and pedestrian observations show clear correlations between speed and spacing in congested regimes. This suggests that the spacing is proportional to the speed to keep a constant safety time gap to react to unexpected behaviors of the predecessor [2,4]. Therefore, we assume that the OV function is piecewise linear

$$V_p(\Delta x) = \min\big\{v_0, \max\{0, (\Delta x - \ell)/T\}\big\}, \qquad p = (v_0, T, \ell), \quad (5)$$

with v_0 the desired (or maximal) speed, T the time gap and ℓ the longitudinal length of the pedestrian. We propose to estimate these parameters microscopically by using $K = 5251$ pseudo-independent measures from the sample of trajectories (by waiting 5 s between each observation). We denote the observations by $(\Delta x_k, v_k)$, $k = 1, \ldots, K$, where the speed v_k is averaged over $\delta t = 0.8$ s. For a given pedestrian k, the residuals $R_k(p)$ of the model are

$$R_k(p) = V_p(\Delta x_k) - v_k. \quad (6)$$

As in [16], the parameters are estimated by minimizing the empirical variance of the residuals

$$\tilde{p} = \arg\min_p \textstyle\sum_k R_k^2(p). \quad (7)$$

This estimation by least squares maximizes the likelihood under the assumption that the residuals are independent and normal, and has in general good properties if the noise repartition is compact. The observations, the estimations of the parameters and the histogram of the residuals are given in Fig. 2. The $R^2 = 0.78$ of the estimation (the proportion of the variance explained by the model) reveals a good fit of the model. Moreover the distribution of the residuals is relatively compact. Note that the fit can be slightly improved by using sigmoid OV functions with 4 parameters ($R^2 = 0.80$).

The empirical estimation of the variance of the residuals maximizing the likelihood is $\tilde{\sigma}_R^2 = \tfrac{1}{K} \sum_k R_k^2(\tilde{p})$. The stationary variance and δt-autocorrelation

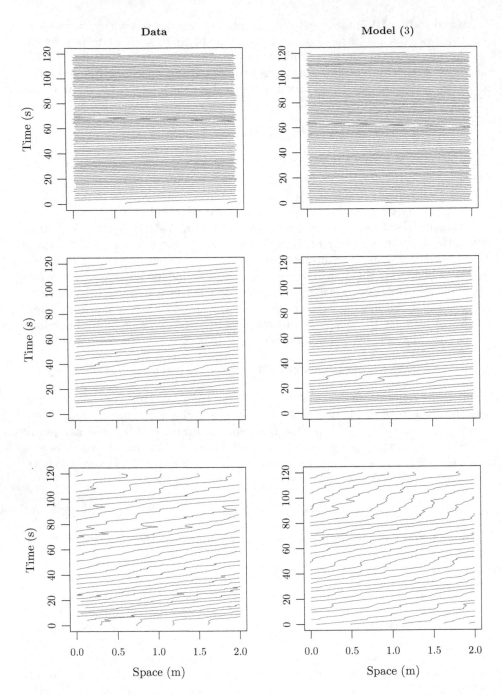

Fig. 1. Trajectories on a segment of length 2 m. From top to bottom: 25 (free state), 45 (slightly congested state) and 62 pedestrians (congested state) on the ring of length 27 m. From left to right: Real data and the calibrated stochastic model (3).

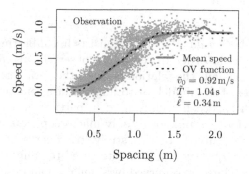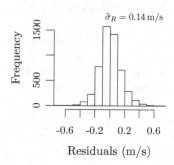

Fig. 2. Statistical estimation of the parameters (left panel) and histogram of the residuals (right panel). $\tilde{\sigma}_R$ is the empirical standard deviation of the residuals. $\delta t = 0.8\,\mathrm{s}$ in the calculus of the speed. Global sample of observations.

of the Ornstein-Uhlenbeck process are $var(\varepsilon) = a^2 b/2$ and $\tilde{c}_{\delta t} = e^{-\delta t/b}$. These relations allow to obtain the estimators for b and a

$$\tilde{b} = -\delta t / \log(\tilde{c}_{\delta t}) \qquad \text{and} \qquad \tilde{a} = \tilde{\sigma}_R \sqrt{2/\tilde{b}}. \tag{8}$$

The estimations for all the data are $\tilde{a} \approx 0.09\,\mathrm{ms}^{-3/2}$ and $\tilde{b} \approx 4.38\,\mathrm{s}$. Note that the value of the relaxation time b is close to $5\,\mathrm{s}$ that is approximately 10 times larger than the value $\tau \approx 0.5\,\mathrm{s}$ generally used with force-based pedestrian models based on a relaxation process (see for instance [13]). Estimations by class of spacing show clear relations between the noise parameters and this variable. The results are shown in Fig. 3. We can see for $\tilde{b}$ particular uni-modal shapes in the congested phase where $\Delta x \le \ell + v_0 T \approx 1.3\,\mathrm{m}$. For the free phase where $\Delta x \ge \ell + v_0$, the values are relatively constant. The shape of the parameter $\tilde{a}$ is more irregular. It will be assumed constant on $\Delta x \le 0.95$ and $\Delta x \ge 0.95\,\mathrm{m}$ in the simulations.

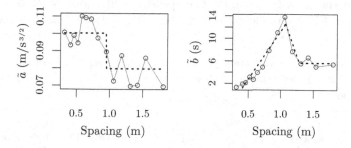

Fig. 3. Statistical estimation of the noise parameters by class of spacing. The dotted lines are the linear approximations used in the simulations.

4   Simulation Results

In the analysis of complex systems, the *top-down* method consists in calibrating the parameters of a microscopic model in order to reproduce observed macroscopic behaviors. It requires knowledge about the relation between the parameter values and the macroscopic properties of the model, or to implement a sensitivity analysis. The top-down approach has been mainly used in particle physics where the microscopic particle behaviors are unknown (only macroscopic quantities such as the temperature are measured). In this study, the microscopic performances (i.e. the trajectories) are observed and directly used to calibrate the parameters. The macroscopic behaviors are observed by simulation and used to validate the calibrated models. This *bottom-up* method allows to control both local and global dynamics.

We evaluate the model (3) by comparing simulation results to the real data. A similar setup as in the real experiments is reproduced for the simulations (from 14 to 70 pedestrians in a ring of length 27 m). The models are simulated by using explicit Euler-Maruyama schemes [14]. The discretisation of the relaxed noise model (3) is

$$\begin{cases} x_n(t+dt) = x_n(t) + dt\, V_{\tilde{p}}(\Delta x_n(t)) + dt\, \varepsilon_n(t), \\ \varepsilon_n(t+dt) = (1 - dt/\tilde{b})\, \varepsilon_n(t) + \sqrt{dt}\, \tilde{a}\, \xi_n(t), \end{cases} \tag{9}$$

with $(\xi_n(t), n, t)$ independent normal random variables. The time step dt is set to 1e-3 s.

The stochastic model is firstly evaluated by looking at the mean, standard deviation and correlation of the speed and spacing for the global sample of observations (see Table 1). The trajectories for 25, 45 and 62 pedestrians are presented in Fig. 1. Some stop-and-go waves propagate when the density increases as in real data (see Fig. 1, middle and bottom panels). Yet we do not observe the collision and backward motion problems frequently related with the unstable deterministic approaches [8,37]. The autocorrelations of the speed and spacing also give good fits to the data (see Fig. 4). The speed distributions by class of spacing are plotted in Fig. 5. We clearly observe bimodal distributions for intermediate

Table 1. Mean, standard deviation (in m and m/s) and correlation for the spacing Δx and speed $v_{\delta t}$ of a pedestrian and his/her predecessor (Δx^1 and $v_{\delta t}^1$) for global sample of observations. $\delta t = 0.8$ s.

	Δx		$v_{\delta t}$		Δx^1		$v_{\delta t}^1$	
	Data	Mod. (3)	Data	Mod. (3)	Data	Mod. (3)	Data	Mod. (3)
Mean	0.68	0.67	0.32	0.32	0.68	0.67	0.32	0.31
Std-dev	0.33	0.34	0.30	0.30	0.33	0.35	0.30	0.30
Corr. Δx	1	1	0.87	0.87	0.79	0.76	0.87	0.87
$v_{\delta t}$	0.87	0.87	1	1	0.85	0.84	0.97	0.97

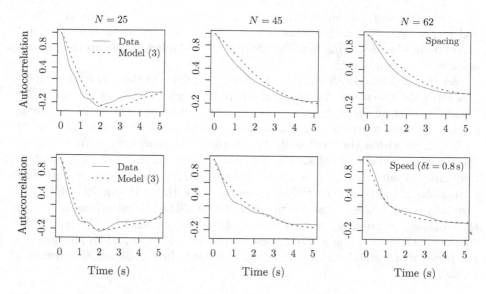

Fig. 4. Autocorrelation function for the spacing (top panels) and the speed (bottom panels) for $N = 25$ (free state, left panels), $N = 45$ (slightly congested state, middle panels) and $N = 62$ (congested state, right panels).

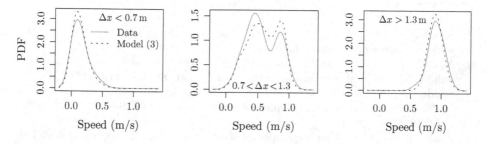

Fig. 5. Probability density function of the speed $v_{\delta t}$ with $\delta t = 0.8\,\mathrm{s}$ by class of spacing Δx. Global sample of observations.

spacings within the data and model (3) (see Fig. 5, middle panel). This result is consistent with stable propagation of the stop-and-go waves.

5   Conclusion

A first order pedestrian model based on Optimal Velocity functions and additive stochastic noise is proposed and calibrated using real pedestrian data on a ring. The model gives realistic descriptions of pedestrian trajectories in one dimension. Mean values and correlations of the speed and spacing are relatively well fitted through piecewise linear OV function with three parameters. Stop-and-go phenomena at congested density levels, bimodal speed distributions, and nonzero speed and spacing autocorrelations are obtained thanks to the relaxed noise at the second order.

As the classical deterministic OV models, inertia mechanisms are used to generate collective waves. Yet the inertia here is stochastic, without deterministic instability of the uniform solution. Also, and oppositely to classical deterministic models, there is no requirement of using nonlinear dynamics to obtain (nonlinear) traffic waves within the stochastic OV approach. Moreover, we do not observe the generic problems of collision and motion backward that are unfortunately frequently obtained with the unstable deterministic approaches. The statistical estimation of the relaxation time is close to 5 s for the noisy model, while it is generally around 0.5 s for the deterministic Ansatz. The relaxation mechanism of the stochastic approach is clearly not that of the classical models. This makes the stochastic OV model a new way to describe accurately stop-and-go phenomena. For pedestrian dynamics in two dimensions, it has to be completed by a direction model.

Acknowledgment. We thank the Deutsche Forschungsgemeinschaft (DFG) for support under the grants Scha 636/9-1 and SE 1789/4-1.

References

1. Forschungszentrum   Jülich   and   Wuppertal   University. www.asim.uni-wuppertal.de/datenbank
2. Asja, J., Appert-Rolland, C., Lemercier, S., Pettré, J.: Properties of pedestrians walking in line: fundamental diagrams. Phys. Rev. E **85**, 036111 (2012)
3. Bando, M., Hasebe, K., Nakayama, A., Shibata, A., Sugiyama, Y.: Dynamical model of traffic congestion and numerical simulation. Phys. Rev. E **51**(2), 1035–1042 (1995)
4. Banks, J.H.: Average time gaps in congested freeway flow. Transport. Res. A-Pol. **37**, 539–554 (2003)
5. Boltes, M.: Software `PeTrack`, FZ Jülich. www.ped.fz-juelich.de/petrack
6. Chowdhury, D., Santen, L., Schadschneider, A.: Statistical physics of vehicular traffic and some related systems. Phys. Rep. **329**(4–6), 199–329 (2000)
7. Daamen, W., Hoogendoorn, S.: Controlled experiments to derive walking behaviour. EJTIR **3**(1), 39–59 (2003)
8. Davis, L.C.: Modifications of the optimal velocity traffic model to include delay due to driver reaction time. Phyica A **319**, 557–567 (2003)
9. Dollar, P., Wojek, C., Schiele, B., Perona, P.: Pedestrian detection: an evaluation of the state of the art. IEEE Trans. Pattern Anal. Mach. Intell. **34**(4), 743–761 (2012)
10. Gazis, D.C.: The origins of traffic theory. Oper. Res. **50**(1), 69–77 (2002)
11. Helbing, D.: Traffic and related self-driven many-particle systems. Rev. Mod. Phys. **73**, 1067–1141 (2001)
12. Helbing, D., Farkas, I., Vicsek, T.: Freezing by heating in a driven mesoscopic system. Phys. Rev. Lett. **84**, 1240–1243 (2000)

13. Helbing, D., Molnár, P.: Social force model for pedestrian dynamics. Phys. Rev. E **51**, 4282–4286 (1995)
14. Higham, D.: An algorithmic introduction to numerical simulation of stochastic differential equations. SIAM Rev. **43**(3), 525–546 (2001)
15. Holl, S., Seyfried, A.: Hermes - an evacuation assistant for mass events. Innovatives Supercomput. Dtschl. inSiDE **7**(1), 60–61 (2009)
16. Hoogendoorn, S.P., Daamen, W., Landman, R.: Microscopic calibration and validation of pedestrian models - cross-comparison of models using experimental data. In: Waldau, N., Gattermann, P., Knoflacher, H., Schreckenberg, M. (eds.) Pedestrian and Evacuation Dynamics 2005, pp. 329–340. Springer, Heidelberg (2007)
17. Jost, D., Nagel, K.: Probabilistic traffic flow breakdown in stochastic car following models. In: Hoogendoorn, S.P., Luding, S., Bovy, P.H., Schreckenberg, M., Wolf, D.E. (eds.) Traffic and Granular Flow 03, pp. 87–103. Springer, Heidelberg (2005)
18. Kanai, M., Nishinari, K., Tokihiro, T.: Stochastic optimal velocity model and its long-lived metastability. Phys. Rev. E **72**, 035102 (2005)
19. Kanai, M., Nishinari, K., Tokihiro, T.: Analytical study on the criticality of the stochastic optimal velocity model. J. Phys. A **39**(12), 2921 (2006)
20. Komatsu, T.S., Sasa, S.-I.: Kink soliton characterizing traffic congestion. Phys. Rev. E **52**(5), 5574–5582 (1995)
21. Kosiński, R., Grabowski, A.: Langevin equations for modeling evacuation processes. Acta. Phys. Pol. B Proc. Suppl. **3**(2), 365–376 (2010)
22. Liggett, M.: Interacting Particle Systems. Classics in Mathematics. Springer Science and Business Media, Heidelberg (2004)
23. Newell, G.F.: Nonlinear effects in the dynamics of car-following. Oper. Res. **9**(2), 209–229 (1961)
24. Olivier, A.-H., Kulpa, R., Pettre, J., Cretual, A.: A step-by-step modeling, analysis and annotation of locomotion. Comput. Animat. Virtual Worlds **22**, 421–433 (2011)
25. Orosz, G., Wilson, R.E., Stepan, G.: Traffic jams: dynamics and control. Proc. R. Soc. A **368**(1957), 4455–4479 (2010)
26. Ossen, S., Hoogendoorn, S.P., Gorte, B.G.: Inter-driver differences in car-following: a vehicle trajectory based study. Transport. Res. Rec. **121–129**, 2008 (1965)
27. Pipes, L.A.: An operational analysis of traffic dynamics. J. Appl. Phys. **24**(3), 274–281 (1953)
28. Portz, A., Seyfried, A.: Analyzing stop-and-go waves by experiment and modeling. In: Peacock, R., Kuligowski, E., Averill, J. (eds.) Pedestrian and Evacuation Dynamics, pp. 577–586. Springer, New York (2010)
29. Schadschneider, A., Chowdhury, D., Nishinari, K.: Stochastic Transport in Complex Systems. From Molecules to Vehicles. Elsevier Science Publishing Co Inc., Amsterdam (2010)
30. Schadschneider, A., Klüpfel, H., Kretz, T., Rogsch, C., Seyfried, A.: Fundamentals of pedestrian and evacuation dynamics. In: Bazzan, A., Klügl, F. (eds.) Multi-Agent Systems for Traffic and Transportation Engineering, pp. 124–154. IGI Global, Hershey (2009)
31. Seyfried, A., Portz, A., Schadschneider, A.: Phase coexistence in congested states of pedestrian dynamics. In: Bandini, S., Manzoni, S., Umeo, H., Vizzari, G. (eds.) ACRI 2010. LNCS, vol. 6350, pp. 496–505. Springer, Heidelberg (2010)
32. Todosiev, E.: The action point model of the driver-vehicle system. Ph.D. thesis, Ohio State University (1963)
33. Tomer, E., Safonov, L., Havlin, S.: Presence of many stable nonhomogeneous states in an inertial car-following model. Phys. Rev. Lett. **84**(2), 382–385 (2000)

34. Treiber, M., Kesting, A., Helbing, D.: Delays, inaccuracies and anticipation in microscopic traffic models. Physica A **360**(1), 71–88 (2006)
35. Wagner, P.: How human drivers control their vehicle. EPJ B **52**(3), 427–431 (2006)
36. Wagner, P., Lubashevsky, I.: Empirical basis for car-following theory development. Technical report, German Aerospace Center, Germany (2006)
37. Wilson, R.E., Berg, P., Hooper, S., Lunt, G.: Many-neighbour interaction and non-locality in traffic models. Eur. J. Phys. B **39**(3), 397–408 (2004)
38. Zgonnikov, A., Lubashevsky, I., Kanemoto, S., Miyazawa, T., Suzuki, T.: To react or not to react? Intrinsic stochasticity of human control in virtual stick balancing. J. R. Soc. Interface **11**, 2014063 (2014)
39. Zhang, J., Mehner, W., Holl, S., Boltes, M., Andresen, E., Schadschneider, A., Seyfried, A.: Universal flow-density relation of single-file bicycle, pedestrian and car motion. Phys. Lett. A **378**(44), 3274–3277 (2014)

On the Evacuation Module SigmaEva Based on a Discrete-Continuous Pedestrian Dynamics Model

Ekaterina Kirik[1,2](✉), Andrey Malyshev[1], and Maria Senashova[1]

[1] Institute of Computational Modelling SB RAS,
Akademgorodok 50/44, Krasnoyarsk 660036, Russia
kirik@icm.krasn.ru
[2] Siberian Federal University, Krasnoyarsk, Russia

Abstract. The discrete-continuous model is a novel contribution to mathematical modeling of pedestrian dynamics. This model is of individual type; people (particles) move in a continuous space, – in this sense the model is continuous. But the number of directions for the particles to move is limited, – in this sense the model is discrete. The model is realized in the computer evacuation module SigmaEva. This article is focused on a presenting of the model and computational aspects and the model is discussed in respect with discrete and continuous models.

Keywords: Pedestrian dynamics model · Discrete-continuous approach · Evacuation modeling · Fundamental diagram

1 Introduction

This study is aimed at the novel approach of mathematical simulation of pedestrian movement – a discrete-continuous pedestrian dynamics model. The model is of the individual type meaning that trajectories of every person are simulated.

An idea of the discrete-continuous approach is to omit solution of the differential equations system (as it is done in social-force models [2,3] and to extract speed for every person from empirical data taking into account the local densities [8]. We use an analytical expression of speed versus density by Kholshevnikov and Samoshin [5], one can use another speed-density dependencies. A probability approach is used to find direction for each pedestrian in the next step (as in stochastic cellular automata models, i.e. [1,12,15]). A procedure to calculate probabilities to move in each direction is adopted from the previously presented stochastic cellular automata floor field model [7]. Directed movement is given by using the static floor field (as in cellular automata approach) that shows a distance from every point of the space to the nearest exit. So person at each time step is allowed to move in a continuous space, – in this sense the model is continuous, but the number of directions for the particles to move is limited and predetermined by a user, – in this sense the model is discrete. Such

© Springer International Publishing Switzerland 2016
R. Wyrzykowski et al. (Eds.): PPAM 2015, Part II, LNCS 9574, pp. 539–549, 2016.
DOI: 10.1007/978-3-319-32152-3_50

approach seems to be fruitful and useful from mathematical and computational view points and practical applications.

The model is realized in the computer evacuation module SigmaEva. In the article a mathematics of the model and the most important computational issues are presented. The model is discussed in respect with social-force (continuous) models and floor field stochastic cellular automata (discrete) models.

The article is organized as follows. In the next section, the mathematical statement of the pedestrian movement modeling problem is presented. It is followed by the decision section (a movement equation, a direction choice, a speed formula are presented), discussion section.

2   Statement of the Problem

A continuous modeling space including an infrastructure (obstacles) $\Omega \in R^2$ are known. People (particles) may move to a free space only[1].

Every particle is considered as a hard flat round disk with diameter d_i, initial positions of particles are given by coordinates of disks centers $x_i(0) = (x_i^1(0), x_i^2(0))$, $i = \overline{1, N}$, N is number of particles (it is supposed that these are coordinates of mass center projections).

Every particle is assigned with the free movement speed[2] v_i^0, square of projection s_i. It is supposed that while moving people do not exceed a maximal speed (a free movement speed) $v_i(t) \leq v_i^0$, and each person controls speed taking into account the local density $v_i(t) = f(\rho)$.

Each time step t every particle may move in one of predetermined directions $e_i(t) = \{e^\alpha(t), \alpha = \overline{1, q}\}$, the model parameter q is the number of directions (for example, a set of directions uniformly distributed around a circle is considered here). Particles that cross target line leave the modeling space.

To orient in the space particles use the static floor field S. Let the nearest exit is assumed as a target point of each pedestrian.

The goal is to model an individual people movement to the target point taking into account interactions with the environment (other particles, obstacles, modelling border (walls)).

3   Mathematical Model

3.1   Preliminary Step – Static Field S

To model directed movement a "map" which stores the information on the shortest distance to the target point is used. The unit of this distance is meter, $[m]$. Such map is saved in so called static floor field that was originally introduced for discrete cellular automata model [1,12,15]. It does not change with time and is independent of the presence of the particles.

---

[1] There is unified coordinate system, and all data are given in this system.
[2] We assume that the free movement speed is random normal distributed magnitude with some mathematical expectation and dispersion [5,6].

We use field S which increases radially from the exit and it is zero in the exit(s) line(s) [7] that differs from original approach. To calculate field S modeling space Ω is covered by a discrete orthogonal grid with cell some cm in size (the size of the cell may vary from 10 to 80 cm in depends on sizes of the modelling area), and Dijkstras algorithm with 16-nodes pattern is used, for instance. Distance to the exit from arbitrary point in Ω is given by bidirectional interpolation among nearest nodes.

3.2 Movement Equation

A person movement equation is derived from a finite-difference presentation of the velocity $v(t)e(t) \approx \frac{x(t)-x(t-\Delta t)}{\Delta t}$. This expression allows to present new position of the particle. For each instant t, the coordinates of i-th particle is given by the formula:

$$x_i(t) = x_i(t - \Delta t) + v_i(t)e_i(t)\Delta t, i = \overline{1, N} \tag{1}$$

where $x_i(t - \Delta t)$ denotes the particle's position at time $t - \Delta t$; $v_i(t)$ is speed measures in m/s; $e_i(t)$ is the unit direction vector. Time shift $\Delta t = 0.25$, [s], is assumed to be fixed.

Unknown values in (1) for each time step for each particle are speed $v_i(t)$ and direction $e_i(t)$. As it was said above we omit to describe forces that act on person, solve system of differential equations and, as a result, get velocity vector $v_i(t)$. We propose to get speed from experimental data (fundamental diagram). In this case in contrast with force-based models we have an opportunity to divide task of finding the velocity vector to two parts. At first, one need to choose the direction for particle to move. Then speed, taking into account the local density in the direction, is calculated using some speed-density dependance (Fig. 1).

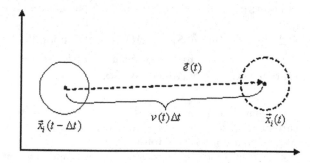

Fig. 1. Movement scheme, the previous and the next position of the particle.

3.3 Choosing of the Direction to Move

In this discrete-continuous model we took inspiration from our previously presented stochastic CA FF model [7]. We consider desired direction as a random discrete variable with a probability distribution. All predetermined directions are possible values which are assigned with some probabilities. The desired direction is chosen according to this probability distribution. Probabilities vary dynamically and issued from the following facts. Pedestrians keep themselves at a certain distance from other people and obstacles. The tighter the people flow and the more in a hurry a pedestrian, the smaller this distance. During movement, people follow at least two strategies: the shortest path and the shortest time.

Thus personal probabilities to move in each direction each time step have contributions: (a) the main driven force (given by destination point), (b) interaction with other pedestrians, (c) interaction with an infrastructure (non movable obstacles). The highest probability is given to direction that has most preferable conditions for movement considering other particles and obstacles and strategy of the people movement (the shortest path and/or the shortest time)[3].

The following procedure is applied for every particle to calculate transition probability for each of predetermined directions and then to choose a direction.

Let particle i has current coordinate $x(t - \Delta t)$. The probability to move from this position to direction $e^\alpha(t) = \left(\cos \frac{2\pi}{q} \alpha, \sin \frac{2\pi}{q} \alpha \right)$, $\alpha = \overline{1, q}$ is the following:

$$p_\alpha^i(t) = \frac{\tilde{p}_\alpha^i(t)}{Norm} = \frac{\exp\left[k_S^i \triangle S_\alpha - k_P^i \rho(r_\alpha^*) - k_W^i (1 - \frac{r_\alpha^*}{r}) 1(\triangle S_\alpha) \right] W(r_\alpha^* - \frac{d_i}{2})}{Norm};$$

(2)

$Norm = \sum\limits_{\alpha=1}^{q} \tilde{p}_\alpha^i(t)$. Visibility radius r $(r \geq \max\{\frac{d_i}{2}\})$, $[m]$, is model parameter representing the maximum distance at which people and obstacles influence on the probability value in the given direction. Obstacles can reduce visibility radius r to r_α^* (see Fig. 2). People density $0 \leq \rho(r_\alpha^*) \leq 1$ is estimated in the visibility area, see paragraph below. Function $1(\cdot)$ is Heaviside unit step function. There are model parameters: $k_S^i > 0$ is field S-sensitive parameter; $k_W^i > 0$ is wall-sensitive parameter; $k_P^i > 0$ is density-sensitive parameter. Information on parameters is below.

$\Delta S_\alpha = S(t - \Delta t) - S_\alpha$, where $S(t - \Delta t)$ is static floor field in the point $x_i(t - \Delta t)$, S_α is static floor field in the point $x = x_i(t - \Delta t) + 0, 1e_i^{\alpha(t)}$. The movement to the target point is controlled by ΔS_α[4].

---

[3] In contrast with original floor field models [1,12,15] to take in to account other people we use current local density in the direction instead of dynamical field D which store "historical" data of the flow intensivity.

[4] The reason why we use field S radially increasing from exit and gradient ΔS_α instead of original issues is the following. Originally pure values of field S (which radially *decreases* from exit) are used in the probability formula in floor field models, e.g. [1,4,10–12,15]. We propose to use only a value of gradient ΔS_α. From a mathematical view point, it yields the same result [7], but computationally this trick has a great advantage. The values of field S may be too high (it depends on the modelling space Ω size); in this case, $\exp(k_S S_\alpha)$ can appear uncomputable. This is a significant limitation of the models. At the same time, $0 \leq |\Delta S_\alpha| < 1$, and a value of $\exp(k_S S_\alpha)$ is computable.

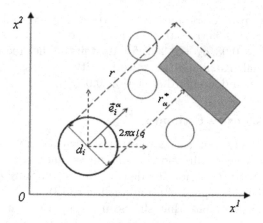

Fig. 2. Visibility area of particle i in direction α. Given visibility radius r is reduced to the r_α^* due to the presence of the obstacle.

Function $W\left(r_\alpha^* - \frac{d_i}{2}\right) = \begin{cases} 1, r_\alpha^* - \frac{d_i}{2} > w; \\ 0, r_\alpha^* - \frac{d_i}{2} \leq w \end{cases}$ controls approaching to obstacles[5]. Model parameter $0 \leq w \leq 1$, [m], is coefficient of inadherence to obstacles.

If $Norm = 0$ than particle does not leave present position[6].

If $Norm \neq 0$ than desired direction $e_i(t)$ is considered as discrete random value with distribution that is given by transition probabilities calculated. Exact direction $e_i(t) = e_i^{\hat{\alpha}}(t) = \left(\cos \frac{2\pi}{q}\hat{\alpha}, \sin \frac{2\pi}{q}\hat{\alpha}\right)$ is determined according to standard procedure for discrete random quantities.

3.4 Speed Calculation

Person's speed is density dependent [5,6,13,14]. We assume that only conditions in front of the person influence on speed. It is motivated by a front line effect which is well pronounced while flow moves in open boundary conditions. Front line people move with free movement velocity while middle part is waiting a free space that is necessary to make a step. It leads to a spreading of the flow. If not to take into account such effect simulation will be slower then real process. Thus only density $\rho_i(\hat{\alpha})$ in direction chosen $e_i(t) = e_i^{\hat{\alpha}}(t)$ is required to determine speed. According [5,6] current speed is

$$v_i(t) = v_i^{\hat{\alpha}}(t) = \begin{cases} v_i^0(1 - a_l \ln \frac{\rho_i(\hat{\alpha})}{\rho^0}, \rho_i(\hat{\alpha}) > \rho^0; \\ v_i^0, \qquad\qquad \rho_i(\hat{\alpha}) \leq \rho^0. \end{cases} \tag{3}$$

where ρ^0 is limit people density until which people may move with free movement speed (it means that the local density does not influence on people's speed);

---

[5] Note, function $W(\cdot)$ "works" with nonmovable obstacles only.

[6] Actually this situation is impossible. Only function $W(\cdot)$ may give (mathematic) zero to probability. If $Norm = 0$, a particle is surrounded by obstacles.

parameter a_l shows people adaptation to current density while moving on different way types ($\rho^0 = 0.06, a_1 = 0.295$ is for horizontal way; $\rho^0 = 0.1, a_2 = 0.4$, for down stairs; $\rho^0 = 0.08, a_3 = 0.305$, for upstairs). The free movement speed v_i^0 is random normal distributed magnitude with, for instance, mathematical expectation 1.66 m/s and standard deviation 0.083 m/s.

3.5 The Local Density Estimate

One more important question is how to estimate density in the visibility area. It follows from (2) that every direction should be assigned with the local density each time step for every particle. The density in the choosing direction $\rho_i(\hat{\alpha})$ is necessary to calculate the speed $v_i^{\hat{\alpha}}(t)$ according to (3).

We use analytical estimate that shows an occupation rate of the visibility area (see Fig. 2), and thus the density estimate is dimensionless and varies in the interval $[0, 1)$. Analytical estimate supposes to calculate total square S_2 of other particles that intersect the visibility area with square $S_1 = d_i \cdot r_\alpha^*$. There were developed an algorithm to identify positions of each disc intersecting visibility area, and the geometric formulas to calculate the intersection area. Density is $\rho(r_\alpha^*) = \frac{S_2}{S_1}, \in [0, 1), [m^2/m^2]$. Some words about the quality of this density estimate are given in the following subsection. Next idea is to use a weighted density function to take into account distance between considered particle and particles in the visibility area.

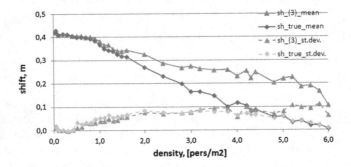

Fig. 3. Average calculated shift ("sh_(3)_mean"), average true shift ("sh_true_mean"), corresponding standard deviations ("sh_(3)_st.dev", "sh_true_st.dev")

The rectangular shape of the visibility area (i.e., instead of conus) is motivated by the fact that a particle may move just along the direction vector, and only particles that are in the visibility area influence on the possibility of the movement.

3.6 New Position, Collisions

For every particle i, $i = \overline{1, N}$ to define new position one substitute in (1) the direction $e_i^{\hat{\alpha}}(t)$ and the speed $v_i^{\hat{\alpha}}(t)$. If the particle moves from the position

$x_i(t - \Delta t)$ to a new position $x_i(t) = x_i(t - \Delta t) + v_i^{\hat{\alpha}}(t)e_i^{\hat{\alpha}}(t)\Delta t$ without collisions (intersections) with other particles a new coordinate for the particle is fixed[7].

To imagine what a difference is between average calculated shift $shift = \sum_{i=1}^{N} v_i^{\hat{\alpha}}(t)\Delta t$ and average true shift for different densities see Fig. 3 (curves "sh_(3)_mean" and "sh_true_mean" correspondingly). Standard deviations are given by curves "sh_(3)_st.dev." and "sh_true_st.dev." in Fig. 3. Data were obtained during a simulation experiment in a corridor $2 \times 50\,\mathrm{m}^2$ under periodic boundary conditions [8].

When new positions of all particles are calculated then a correction of possible positions is applied. It takes place if there are particles l and m intending to "the same" position: $|x_l(t) - x_m(t)| \leq \frac{d_l}{2} + \frac{d_m}{2} - \Delta d$. In this case conflict resolution procedure is applied[8].

4   Discussion

The model was implemented in the SigmaEva computer simulation module. This section deals with computational aspects of the model and program realization.

4.1   Model Parameters k_S^i, k_P^i, k_W^i

Probability formula (2) contains non-dimensional model parameters: k_S^i, k_P^i, k_W^i.

Parameter $k_S^i > 0$ is field S-sensitive parameter which can be interpreted as knowledge of the shortest way to the exit (or a wish to move to the exit). The equality $k_S^i = 0$ means that the pedestrian ignores the information from field S and move randomly. The higher k_S^i, the better directed the movement.

Parameter $k_W^i \geq 0$ is wall-sensitive parameter which determines the effect of walls and obstacles. We assume that people avoid obstacles only moving towards the destination point. When people make detours (in this case $\Delta S_{\hat{\alpha}} < 0$) approaching the obstacles is not excluded.

Parameter $k_P^i > 0$ is the density-sensitive parameter and determines the effect of the people density. The higher parameter k_P^i, the more pronounced the shortest time strategy for the particle.

Note that probabilities are density adaptive; the low people density lowers the effect of density-sensitive term, and role of the shortest path strategy increases automatically. But this automatic property is not enough. Ideally a time-spatial adaptation for parameters k_P^i and k_W^i is required. All parameters may be assigned individually or unified for all involved particles.

For wide range of evacuation tasks that intend directed movement under normal conditions the set numerous simulation experiment showed that $k_S^i = 40$, $k_P^i = 6$, $k_W^i = 2$ are appropriate.

---

[7] Note (!) that positions of other particles are taken in to account for time $t - \Delta t$. As in cellular automata models parallel update is used here.

[8] Only here we operate with coordinates obtained for time t. As in floor field cellular automata models movement of all involved particles is denied with probability μ. One of candidates moves to the desired cell with probability $1 - \mu$. A person allowed to move is chosen randomly.

4.2   Model Parameters q

The number of predetermined directions $q > 0$ is model parameter. High value of q seems to be natural. Probability-based approaches give directed movement when one direction has got considerably higher probability then the others. In discrete cellular automata models with four cell neighborhood it is pronounced very sharp. So the higher q, the less pronounced this fact. In this case some number of directions have got more or less the same conditions in front of them. It leads to more or less the same probabilities for such directions. For instance, the vectors of probabilities (2) for some particle for t and $t + \Delta t$ are[9]

α	1	2	3	4	5	6	7	8	9	10	11	12	13	14	15	16
t	0	0	0	0	0.001	0.002	0.005	0.028	0.128	0.175	0.34	0.196	0.124	0	0	0
$t + \Delta t$	0	0	0	0	0	0	0	0.025	0.114	0.184	0.30	0.114	0.101	0.108	0.04	0.015

It results in very well pronounced swaying movement because step by step there is now one direction with probability considerably exceeding other directions. To exclude it we use a computational trick that artificially scale probabilities relatively each other:

$$\hat{p}_\alpha^i(t) = \frac{(\tilde{p}_\alpha^i(t))^6}{\sum\limits_{\alpha=1}^{q} (\tilde{p}_\alpha^i(t))^6}. \tag{4}$$

The corresponding scaled vectors of probabilities (4) for the same particle for t and $t + \Delta t$ are

α	1	2	3	4	5	6	7	8	9	10	11	12	13	14	15	16
t	0	0	0	0	0	0	0	0	0.003	0.018	0.942	0.035	0.002	0	0	0
$t + \Delta t$	0	0	0	0	0	0	0	0	0.003	0.050	0.941	0.003	0.001	0.002	0	0

One can see that one direction with expressed probability value for both steps. This computational trick smoothes movement of the particle. Our experience after numerous simulation experiments is that $q = 16$ gives satisfying dynamics.

4.3   Discrete-Continuous Approach with Respect to Discrete and Continuous Approaches

Usually real-life phenomena are continuous. At the same time computer calculations are discrete by nature; and discrete models are more convenient for computer simulation and save computational time.

---

[9] Zero probabilities means very low values which are about $10^{-4} - 10^{-16}$ here.

The last one encouraged developing of discrete approach to model pedestrian dynamics. Discrete models are discrete by space and time. Such models operate with spaces sampled into cells of some size (often 40 cm × 40 cm in size), which are either empty or occupied by the only person (exclusive principle). Usually each particle can move to one of its four (the von Neumann neighborhood) or eight (Moore neighborhood) adjacent cells or stay in the present cell at each discrete time step $t \rightarrow t + 1$; i.e., $v_{max} = 1[step]$. There are models that allow $v_{max} > 1[step]$. Discrete models describe people dynamics (including interaction with environment) in a rule-based way. Stochasticity of the people movement is captured by calculating probabilities to move in every direction. It gives relative mathematical simplicity to models, and it is the pros. And the cons are problems concerning adjusting real building sizes to discrete model space. Sometimes in a focus of (fire) engineering investigation is a changing of a door size on 5–15 cm. Discrete models can not pick up such point. But it is very important for the main application of pedestrian evacuation models that are fire safety problems.

Continuous models imply that people move in the continuous space. Generally there are no restrictions on movement except the nearest people and obstacles. From practical applications continuity of the space is the pros; and cons of traditional continuous approach are problems with movement equation. In Newtonian mechanics to define trajectory of an object (person) one need to know current position and velocity vector (or its time derivative that is acceleration). Acceleration vector is given usually by forces that act on object (the second Newton's law) $m\boldsymbol{a} = \boldsymbol{F}$. The main goal (and problem) in such models is to give analytical expressions to describe forces $\boldsymbol{F}$ and solve a system of N differential equations. The forces $\boldsymbol{F}$ are not physic here. Forces of an interaction with other people, obstacles and main driving force are matter of consideration in this approach. A numerical solution of such system is time consuming in itself, moreover stability of the numerical methods and a solution needs the relaxation time (time step) to be very small $\Delta t \approx 0.01\,\mathrm{s}$.

The discrete-continuous model omits the solution of the differential equations system and "extracts" velocity scalar for each particle from empirical data taking into account the local densities. We use speed-density dependence from (3); one can use the other fundamental diagram. Set of possible directions is predetermined. Desired direction for every particle each time step is considered as a random variable with some probability distribution. This distribution is time-spatially dependent and gives the highest probability to the direction that has the most preferable conditions for movement considering other particles and obstacles and strategy of the people movement (the shortest path and/or the shortest time). The directed movement is given by using the static floor field that shows a distance from each point of the modelling space to the exit. So person at each time step is allowed to move in a continuous space (in this sense model is continuous), but number of directions where particles may move is limited and predetermined by a user (in this sense model is discrete). Time step $\Delta t = 0.25\,\mathrm{s}$, that is approximate duration of one human step.

Such combination of discrete and continuous approaches seems to be fruitful and useful from both mathematical statement and practical applications. Of course model is "computationally heavier" in comparing with pure discrete models but this is a price for flexibility and applicability. Comparing with force-based models this model is "faster" considerably: $\Delta t = 0.25\,$s versus $\Delta t \approx 0.01\,$s.

Correctness of the model was checked by agreement of a simulated velocity-density dependence (fundamental diagram) with the empirical data [8,9]. Validation with fundamental diagrams in periodic and open boundary conditions shows good dynamical properties: maintaining velocity according to local "directed" density, an initial density and the free movement speed maintain approximately till $0,5[pers/m]$, flow diffusion realizes if it is possible, model full flow rises with increasing bottleneck width. A comparison with experimental data says that model results are within an existing conception of the speed-density dependence of people flow. We continue to develop and investigate the model, improve and fast computational scheme by using high performance computing.

References

1. Burstedde, V., Klauck, K., Schadschneider, A., Zittartz, J.: Simulation of pedestrian dynamics using a 2-dimensional cellular automaton. Physica A **295**, 507–525 (2001)
2. Chraibi, M., Seyfried, A., Schadschneider, A.: Generalized centrifugal-force model for pedestrian dynamics. Phys. Rev. E **82**, 046111 (2010)
3. Helbing, D., Farkas, I., Vicsek, T.: Simulating dynamical features of escape panic. Nature **407**, 487–490 (2001)
4. Henein, C.M., White, T.: Macroscopic effects of microscopic forces between agents in crowd models. Physica A **373**, 694–718 (2007)
5. Kholshevnikov, V.: Forecast of human behavior during fire evacuation. In: Proceedings of the International Conference on Emergency Evacuation of People from Buildings - EMEVAC, pp. 139–153. Belstudio, Warsaw (2011)
6. Kholshevnikov, V., Samoshin, D.: Evacuation and human behavior in fire. Academy of State Fire Service, EMERCOM of Russia, Moscow (2009)
7. Kirik, E., Yurgel'yan, T., Krouglov, D.: On realizing the shortest time strategy in a CA FF pedestrian dynamics model. Cybern. Syst. **42**(1), 1–15 (2011)
8. Kirik, E., Malyshev, A., Popel, E.: Fundamental diagram as a model input - direct movement equation of pedestrian dynamics. In: Weidmann, U., Kirsch, U., Schreckenberg, M. (eds.) Pedestrian and Evacuation Dynamics 2012, pp. 691–702. Springer, Cham (2013)
9. Kirik, E., Malyshev, A.: On validation of SigmaEva pedestrian evacuation computer simulation module with bottleneck flow. J. Comput. Sci. **5**, 847–850 (2014)
10. Kretz, T., Schreckenberg, M.: The F.A.S.T.-Model. arxiv:cs.MA/0804.1893v1
11. Yanagisawa, D., Nishinari, K.: Mean-field theory for pedestrian outflow through an exit. Phys. Rev. E **76**, 061117 (2007)
12. Nishinari, K., Kirchner, A., Namazi, A., Schadschneider, A.: Extended floor field CA model for evacuation dynamics. IEICE Trans. Inf. Syst. E **87–D**, 726–732 (2004)

13. Predtechenskii, V.M., Milinskii, A.I.: Planing for foot traffic flow in buildings. American Publishing, New Dehli (1978). Translation of Proektirovanie Zhdanii s Uchetom organizatsii Dvizheniya Lyudskikh potokov, Stroiizdat Publishers, Moscow (1969)
14. Schadschneider, A., Klingsch, W., Kluepfel, H., Kretz, T., Rogsch, C., Seyfried, A.: Evacuation dynamics: empirical results, modeling and applications. In: Meyers, R.A. (ed.) Encyclopedia of Complexity and System Science, pp. 3142–3192. Springer, New York (2009)
15. Schadschneider, A., Seyfried, A.: Validation of CA models of pedestrian dynamics with fundamental diagrams. Cybern. Syst. **40**(5), 367–389 (2009)

Towards Effective GPU Implementation of Social Distances Model for Mass Evacuation

Adrian Kłusek[1]([⊠]), Paweł Topa[1], and Jarosław Wąs[2]

[1] Institute of Computer Science, AGH University of Science and Technology,
al. Mickiewicza 30, 30-059 Kraków, Poland
klusek@student.agh.edu.pl
[2] Institute of Applied Computer Science, AGH University of Science and Technology,
al. Mickiewicza 30, 30-059 Kraków, Poland
{topa,jarek}@agh.edu.pl

Abstract. The Social Distances (SD) model for massive evacuation is based on a Cellular Automata and agent-based representation of pedestrians. When parallel processors are used, this approach creates a high performance simulation. In this paper, we present a new algorithm for Social Distance that is highly optimized for GPU computations. The original algorithms were redesigned in order to efficiently exploit the power of graphics processors. The performance of the SD model executed on a GPU is several times greater than the performance of the same algorithm executed on a normal CPU. It is now possible to simulate at least 10^6 pedestrians in real time.

Keywords: Pedestrian dynamics · GPGPU · Cellular Automata

1 Introduction

In microscopic crowd dynamics models, we take into consideration the behavior and dynamics of a particular entity (an agent). One can distinguish two basic groups of models. On the one hand is the force-based Social Force Model [9], which is the most popular microscopic (continuous) model. On the other hand there are the Cellular Automata–based models, which use the idea of static and dynamic floors fields [3,4,7,8,10,19,20].

Among the discrete models of crowd dynamics, the application of potential fields [4,8,10,11,20] is especially appreciated. This group of models uses a set of fields which modifies the transition function of an applied cellular automaton. Usually a static potential field with defined POIs of pedestrians (Points of Interest defined as short term and long term aims/attractors of particular pedestrians) is used together with a dynamic potential field generated by moving pedestrians (as analogy to chemo-taxis). Recently, other types of floor fields were proposed, for instance proxemic floor field [7].

In this paper we pay special attention to a GPU implementation of CA-based crowd simulations. The usefulness of GPUs for Cellular Automata modeling has

© Springer International Publishing Switzerland 2016
R. Wyrzykowski et al. (Eds.): PPAM 2015, Part II, LNCS 9574, pp. 550–559, 2016.
DOI: 10.1007/978-3-319-32152-3_51

been demonstrated in many papers. Rybacki et al. [15] use GPUs for implementing a few well-known CA rules (e.g. Game of Life, Parity). The article concludes that the usefulness of GPU-based Cellular Automata algorithms strongly depends on the models that are to be simulated. Bilotta et al. [1] has ported the MAGFLOW model [16] (lava flow simulation) to the CUDA environment. The algorithms were carefully optimized for GPU architecture, and they benefited from the usage of shared memory. Also, Topa [18] investigated the possibilities of constructing an efficient GPU-aware algorithms for a CA-based model for water flow. Blecic et al. [2] used Nvidia Kepler architecture to implement a model of urban dynamics. The authors also implemented an additional version that uses shared memory, but it gives only a small increase in performance due to the hardware managed cache. D'Ambrosio [5] applied configurations with multiple GPUs to provide a fast analysis of scenarios of potential wildfires. This team also applied GPU computations for risk assessment of areas threatened by volcano eruption [6].

Models of crowd dynamics based on the Cellular Automata paradigm also perform with high efficiency when implemented for a GPU. Miao et al. [12] demonstrated how a GPU can be used for modeling pedestrian evacuations. The application of a GPU for modeling crowd dynamics using two classical models, Social Force and Social Distances, was studied by Wąs et al. [13,14]. They proposed a naive implementation in which the computational domain was simply distributed among the GPU threads. The algorithms were not optimized for the specific GPU architecture but also made it possible to speed up calculation up to 2–4 times.

In this paper, we present a new implementation of the Social Distances model for mass evacuation (described in detail in [21]), optimized for GPU computation. The general architecture is still based on our previous implementations [13,14], however the algorithms were completely redesigned to achieve a high level of parallelism. In the next section, the foundations of both methods are briefly presented. Next, we show how the algorithms were redesign and re-implemented for GPU architecture. Sample simulation results and performance evaluation are presented and discussed at the end.

2   Implementation of Social Distances for CPU

We have decided to implement a new CPU-based version of the Social Distances model for mass evacuation. It will be used to evaluate the performance of a new GPU-based version. In this section, we briefly present the assumptions we made in rewriting the algorithms.

Efficient implementations of these models exist, and they were studied and optimized many times so far. In the current version we have decided to pay special attention to instruction flows. Three parts of the model algorithm generate a large number of divergent execution paths:

1. Finding visible areas.
2. Calculating the pedestrians' compressibility factor.

3. Calculating the cost function.

In general, the branch instructions were replaced with multidimensional arrays of pre-calculated data.

2.1 Visibility Areas

Before each pedestrian is moved to a new location, its potential directions of movement are considered. The modeled area is represented by a two-dimensional grid. Each cell corresponds to a square 25×25 cm. In a single cell there can be only one entity, either an obstacle, a POI or a person. In the algorithm, the area is represented by a 2-dimensional table of characters, with values corresponding to entities. If cell is occupied by a person, a value in the table indicates its orientation (see Fig. 1).

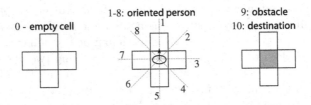

Fig. 1. A single cell can be empty (0) or occupied by a person (1–8), obstacle (9) or POI (10).

The algorithm uses another table with the same dimensions as the main table that stores information about the POIs. In this case, each cell contains information about the minimum number of steps to be carried from the site to the nearest target. This table is updated only when POIs change their location. What is more, there is the possibility to locate a different number of POIs and indicate specific target to specific group of people.

In order to determine the visible fields around the pedestrian, the program uses a two-dimensional (9×5 elements) array of structures (table moves):

```
struct Move{
    int dx;
    int dy;
    int position;
};
```

Fields dx and dy contain vectors of translation between the central cell and its neighbors within radius 1. Field position contains information about which cell in its vicinity will be occupied after the movement. In table moves, the row with index 0 is unused in order to avoid additional operations on indexes.

For the situation shown in Fig. 2, a person in time step $t1$ sees fields that are marked with a dot. In the main table, the cell that corresponds to this person has value 1 (compare Fig. 1). It is used as an index for the table moves. Each row from table moves contains fields that are visible for this person (fields with dots). In this case, the corresponding row contains the following values:

moves [1][] = { {−1, 0, 3}, {−1, 1, 0}, {0, 1, 1},
  { 1, 1, 2}, { 1, 0, 5} }

Fig. 2. The example of the transition a person from one place to another.

For each potential destination cell, we calculate a cost function. The direction with the lowest value from the cost function is chosen. A new location is calculated in the following way:

new.x = old.x + moves[1][selected_direction].dx;
new.y = old.y + moves[1][selected_direction].dy;

The value of the position field is used as an index for another table int directions[9] = {8, 1, 2, 7, 1, 3, 6, 5, 4};, which indicates a pedestrian's orientation after the movement:

occupancy[new.x][new.y]
  = directions[moves[1][selected_direction].position];

2.2   Compressibility Factor

The compressibility factor is used to determine whether the new position is comfortable enough. Figure 3 illustrates how the mutual orientation of two neighboring persons influences this value.

Once again, the computationally-intensive code that discovers pedestrians' mutual configuration is replaced by tables with pre-calculated data. The algorithm uses a 4-dimensional table:

float forces [2][2][9][11];

The two first dimensions are indexed with a vector indicating the mutual location of two neighboring pedestrians. In order to simplify the table, only absolute values of vector coordinates are used. The third dimension is indexed according to the orientation of the pedestrian, for which we calculate the movement. The last dimension is indexed by a value taken from the occupancy table (0 - free, 1–8 pedestrian with a given orientation, 9 - obstacle and 10 - POI).

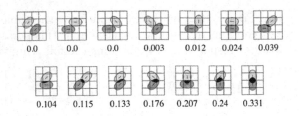

Fig. 3. Reciprocal orientations of two pedestrians represented by grey ellipses and the calculated ratios of cross-sections marked in black (compressibility parameter) and ellipse size for cell size dg = 0.25 m.

In Fig. 2 where two people are standing side by side, the compressibility factor is calculated as follows:

1. The absolute value of the translation vector of people is taken. The result is a vector [1,1].
2. The orientation of the currently considered pedestrian is 1.
3. The orientation of its neighbor is 8.
4. The compressibility factor can be read from `forces[1][1][1][8]` and is equal to 0.0. This is the third configuration from Fig. 3.

When the compressibility factor for potential location is lower than the threshold parameter set for the given simulation, the pedestrian moves to a new location.

2.3   Cost Function

Another area of optimization is the code that calculates the cost function (see Eq. 1). Its aim is bringing about movement towards POIs (Points Of Interests) by avoiding obstacles and walls and establishing patterns of following other pedestrians using the idea of a dynamic floor field. The value of the function (Eq. 1) depends on the actual state of the world. We take into account values of: static floor field S_{ij}, dynamic floor field D_{ij}, density in the vicinity of the cell f_{ij}, distance between cell f_{ij} and the POI, avoiding walls parameter W and inertia parameter I.

$$cost(f_{ij}) = S_{ij} + (dens(f_{ij}) + \alpha\,dist(s, f_{ij})) * W * I$$
$$dens(f_{ij}) = e^{\delta D_{ij}} \tag{1}$$

If the value of the destination cell corresponds to a free area or a POI, its value is $dens(f_{ij}) = e^0$. In the other cases, the value is e^δ (delta can be set as a parameter form the simulation). Table `float ePowTheta[11]` contains pre-calculated values of this expressions. It is indexed by a value stored in a destination cell.

3   Migrating Social Distances Model from CPU to GPU

In order to efficiently exploit the architecture and resources of a GPU, some implementational assumptions were made. We can group them into these two sections:

1. Specific memory allocation and usage,
2. Thread scheduling.

3.1   Specific Memory Allocation and Usage

In order to speedup memory transfers from the host to the GPU, we use the **cudaMallocHost** function to allocate required memory. It also guarantees that allocated memory is not pageable. Data is copied from the host to the GPU only once at the beginning of computation, and later the program uses only GPU memory. To avoid copying the data between the CPU and GPU during the simulation, we use graphics interoperability to improve their storing. The array with people is located in the VBO(Vertex Buffer Object) on the GPU, which is mapped to CUDA pointers during the finding of new locations and is rendered in the next step without any copying.

The lookup tables that are widely used in this algorithm are stored in the constant memory. This memory is cached, and access to it significantly reduces the required bandwidth.

Texture memory is used to store the world map. Consecutive warps maintain pedestrians which are located near to each other. Reading by texture fetch function we have opportunity to achieve better bandwidth.

3.2   Thread Scheduling

In a single step of the simulation, a pedestrian is able to move only to the neighboring cell. Due to this fact, we can construct an algorithm that does not use any synchronization between threads. To do that, the occupancy table is logically partitioned into 9-element containers (see Fig. 4). Cells in each container are indexed with values of 0–8. For all people we compute the information about their type of place in the current time step. In the next step, we divide the whole process of moving people into 9 sub-steps.

```
for(subStep=0; subStep < 9; subStep++){
    run_threads_for_persons_in_cells_of_type(subStep);
}
```

During the computation, a single thread calculates a new position for one pedestrian. Pedestrians are processed in an order defined by the index inside the container. First, all pedestrians that occupy cells with index 0 within all containers are processed. Next, all pedestrian that occupy cells with index 1 within all containers and so on. Pedestrians who have already been moved in a previous sub-step will not be moved. This is done to avoid moving the same person more than once in one simulation step.

4 Results

The algorithm described in the previous sections was implemented in C++ with the CUDA programming environment. A version for computation only only a CPU was also implemented.

Figure 5 shows a snapshot from a simulation intended to resemble pilgrims evacuating after The Holy Mass celebrated by Pope John Paul II in 2002 in Błonia Park in Cracow. This event brought together approximately 2.5 million pilgrims, spread over an area of 48 hectares. In fact, Cracow's Błonia Park is often used to host large gatherings.

The input for the application was a map of Błonia Park converted into a bitmap (obstacles are marked with black). During simulation, green dots represent pedestrian and red areas are their goals. One pixel of the bitmap represents a square the size of 25×25 cm.

In order to examine the performance of the implementations, a series of tests was carried out. The Intel i7 architecture processor was examined, as well as four Nvidia GPUs, each with a different Compute Capability: GeForce GT240M (Tesla architecture with CC 1.2, 48 cores), Quadro FX 4800 (Tesla architecture with CC 1.3, 192 cores), GeForce GTX 480 (Fermi architecture with CC 2.0, 480 cores) GeForce GT 755M (Kepler architecture with CC 2.1, 384 cores).

The results show (see Figs. 6 and 7) that Intel i7 is able to outperform only a very old GPU, the GT240M, which was made for laptops. The best results were achieved by a GeForce GTX480, which is a regular graphics card. Our most advanced GPU, the GeForce GT755M, is also a laptop version, which means that it has reduced capabilities compared to the desktop version.

The performance tests checked the time execution configuration with various initial number of pedestrians. Moreover, we checked how initial pedestrian distribution influenced the results. In the tests with a high initial density (see Fig. 6), the population of pedestrians was tightly distributed in neighboring cells. In configuration with low density (see Fig. 7), pedestrian are distributed every fourth cell. The main goal of such scenarios is to compare a case where most pedestrians are blocked and more conditions have to be check, with case where pedestrians have free space in their vicinity. It is worth to notice that the initial pedestri-

0	1	2	0	1	2	0	1	2
3	4	5	3	4	5	3	4	5
6	7	8	6	7	8	6	7	8
0	1	2	0	1	2	0	1	2
3	4	5	3	4	5	3	4	5
6	7	8	6	7	8	6	7	8
0	1	2	0	1	2	0	1	2
3	4	5	3	4	5	3	4	5
6	7	8	6	7	8	6	7	8

Fig. 4. Occupancy is partitioned into 9-element logical containers. Cells in each container are indexed with values of 0–8.

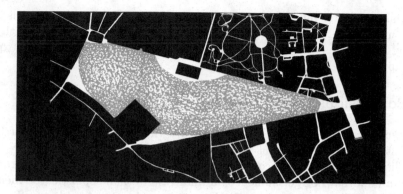

Fig. 5. Screenshot of the results from the simulation of crowd evacuation from Błonia Park in Cracow. See text for details.

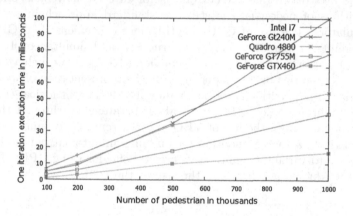

Fig. 6. Performance of Social Distances model with optimized algorithms executed on various GPUs. Configuration with high initial density of pedestrians.

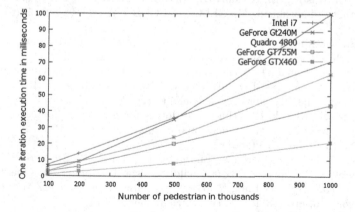

Fig. 7. Performance of Social Distances model with optimized algorithms executed on various GPUs. Configuration with low initial density of pedestrians

ans' density slightly influences the results. Performance for configurations with higher initial density is only slightly better than in a configuration with low density. This means that configurations with low density were unable to exploit all computational resources.

5   Conclusions

The results presented in this paper illustrate that the optimal usage of GPU architecture requires algorithms that are designed for massive parallel processing. We believe that algorithms constructed with the very specific architecture of GPUs in mind have bigger influence on final results than sophisticated memory management. The most modern GPUs have a cache that is able to efficiently speedup transactions to/from global memory [17].

It should be stressed that the current use of CPUs in movement and behavioral algorithms of pedestrians is still the standard approach in professional crowd simulators. The use of CPUs is still more convenient than GPUs, due to the possibility of handling complex instructions and simultaneously creating multi-variant patterns of pedestrians' behaviors. However, we believe that the improvements proposed in this paper, namely the representation of space in the form of arrays, the partitioning of occupancy into 9-element logical containers, the special coding of objects, the removal of if-statements through the use of multidimensional arrays, the use of VBO and the removal of unnecessary data copying constitute a good foundation for building an automatized methodology for creating multi-variant simulations of crowd behavior using GPUs. From our point of view, due to its efficiency, the use of the GPU is well worth further study.

References

1. Bilotta, G., Rustico, E., Hérault, A., Vicari, A., Russo, G., Del Negro, C., Gallo, G.: Porting and optimizing MAGFLOW on CUDA. Ann. Geophys. **54**(5), 662–670 (2011)
2. Blecic, I., Cecchini, A., Trunfio, G.: Cellular automata simulation of urban dynamics through GPGPU. J. Supercomputing **65**(2), 614–629 (2013)
3. Bukáček, M., Hrabák, P.: Case study of phase transition in cellular models of pedestrian flow. In: Wąs, J., Sirakoulis, G.C., Bandini, S. (eds.) ACRI 2014. LNCS, vol. 8751, pp. 508–517. Springer, Heidelberg (2014)
4. Burstedde, C., Klauck, K., Schadschneider, A., Zittartz, J.: Simulation of pedestrian dynamics using a two-dimensional cellular automaton. Phys. A Stat. Mech. Appl. **295**(3–4), 507–525 (2001)
5. D'Ambrosio, D., Di Gregorio, S., Filippone, G., Rongo, R., Spataro, W., Trunfio, G.: A multi-GPU approach to fast wildfire hazard mapping. In: Obaidat, M.S., Filipe, J., Kacprzyk, J., Pina, N. (eds.) Simulation and Modeling Methodologies. AISC, vol. 256, pp. 183–195. Springer, Heidelberg (2014)
6. D'Ambrosio, D., Filippone, G., Marocco, D., Rongo, R., Spataro, W.: Efficient application of GPGPU for lava flow hazard mapping. J. Supercomputing **65**(2), 630–644 (2013)

7. Ezaki, T., Yanagisawa, D., Ohtsuka, K., Nishinari, K.: Simulation of space acquisition process of pedestrians using proxemic floor field model. Phys. A Stat. Mech. Appl. **391**(1–2), 291–299 (2012)
8. Gwizdałła, T.M.: Some properties of the floor field cellular automata evacuation model. Phys. A Stat. Mech. Appl. **419**, 718–728 (2015)
9. Helbing, D., Molnár, P.: Social force model for pedestrian dynamics. Phys. Rev. E **51**, 4282–4286 (1995)
10. Hrabák, P., Bukáček, M., Krbálek, M.: Cellular model of room evacuation based on occupancy and movement prediction: comparison with experimental study. J. Cell. Automata **8**(5–6), 383–393 (2013)
11. Kirik, E., Malyshev, A.: On validation of SigmaEva pedestrian evacuation computer simulation module with bottleneck flow. J. Comp. Sci. **5**(5), 847–850 (2014)
12. Miao, Q., Lv, Y., Zhu, F.: A cellular automata based evacuation model on GPU platform. In: 15th International IEEE Conference on Intelligent Transportation Systems, pp. 764–768 (2012)
13. Mróz, H., Wąs, J.: Discrete vs. continuous approach in crowd dynamics modeling using GPU computing. Cybern. Syst. **45**(1), 25–38 (2014)
14. Mróz, H., Wąs, J., Topa, P.: The use of GPGPU in continuous and discrete models of crowd dynamics. In: Wyrzykowski, R., Dongarra, J., Karczewski, K., Waśniewski, J. (eds.) PPAM 2013, Part II. LNCS, vol. 8385, pp. 679–688. Springer, Heidelberg (2014)
15. Rybacki, S., Himmelspach, J., Uhrmacher, A.M.: Experiments with single core, multi-core, and GPU based computation of cellular automata. In: 2009 First International Conference on Advances in System Simulation, pp. 62–67 (2009)
16. Spataro, W., Lupiano, W., Lupiano, V., Avolio, M., D'Ambrosio, D., Trunfio, G.: High detailed lava flows hazard maps by a cellular automata approach. In: Pina, N., Kacprzyk, J., Filipe, J. (eds.) Simulation and Modeling Methodologies, Technologies and Applications. Advances in Intelligent Systems and Computing, vol. 197, pp. 85–99. Springer, Heidelberg (2013)
17. Topa, P.: Cellular automata model tuned for efficient computation on GPU with global memory cache. In: 22nd Euromicro International Conference on Parallel, Distributed, and Network-Based Processing, PDP 2014, Torino, Italy, 12–14 February 2014, pp. 380–383 (2014)
18. Topa, P., Mlocek, P.: GPGPU implementation of cellular automata model of water flow. In: Wyrzykowski, R., Dongarra, J., Karczewski, K., Waśniewski, J. (eds.) PPAM 2011, Part I. LNCS, vol. 7203, pp. 630–639. Springer, Heidelberg (2012)
19. Vihas, C., Georgoudas, I.G., Sirakoulis, G.C.: Cellular automata incorporating follow-the-leader principles to model crowd dynamics. J. Cell. Automata **8**(5–6), 333–346 (2013)
20. Wei, X., Song, W., Lv, W., Liu, X., Fu, L.: Defining static floor field of evacuation model in large exit scenario. Sim. Modell. Pract. Theo. **40**, 122–131 (2014)
21. Wąs, J., Lubaś, R.: Towards realistic and effective agent-based models of crowd dynamics. Neurocomputing **146**, 199–209 (2014)

GPU and FPGA Parallelization
of Fuzzy Cellular Automata for the Simulation
# of Wildfire Spreading

Vasileios G. Ntinas[1], Byron E. Moutafis[1], Giuseppe A. Trunfio[2],
and Georgios Ch. Sirakoulis[1(✉)]

[1] Department of Electrical and Computer Engineering, School of Engineering,
Democritus University of Thrace, University Campus, 67100 Xanthi, Greece
{vntinas,vyromout,gsirak}@ee.duth.gr
[2] DADU, University of Sassari, Piazza Duomo, 6, 07041 Alghero, SS, Italy
trunfio@uniss.it

Abstract. This paper presents a Fuzzy Cellular Automata (FCA) model with the aim to cope with the computational complexity and data uncertainties that characterize the simulation of wildfire spreading on real landscapes. Moreover, parallel implementations of the proposed FCA model, on both GPU and FPGA, are discussed and investigated. According to the results, the parallel models exhibit significant speedups over the corresponding sequential algorithm. As a possible application, the proposed model could be embedded on a portable electronic system for real-time prediction of fire spread scenarios.

Keywords: Forest fire spreading · Cellular Automata · Fuzzy theory · GPU implementation · Hardware · Parallelization

1 Introduction

Wildfires are a frequent source of environmental disasters, which affect the flora, the fauna and endanger human lives. Prediction of wildfires spread is crucial to the society, so research has been focused on mathematical and practical models of this phenomenon in order to support the prevention of human losses, environmental hazards and economic disasters. Starting from the surface fire equations introduced by Rothermel [1], several models have been proposed to tackle the difficulties of wildfire spreading [2,3]. However, to achieve realistic simulations, many phenomena and environmental parameters (e.g. wind speed, terrain slope, fuel bed, humidity) have to be concerned in these models.

Consequently, they are often computationally complex, which results in decreasing performance of the corresponding software implementations. According to the literature, such a complexity can be tackled by the Cellular Automata (CA) approach, which are considered a fine alternative to Differential Equations for many physical systems and processes, especially regarding environmental modeling [2–5]. Furthermore, the inherent parallelism of CA results in improved

© Springer International Publishing Switzerland 2016
R. Wyrzykowski et al. (Eds.): PPAM 2015, Part II, LNCS 9574, pp. 560–569, 2016.
DOI: 10.1007/978-3-319-32152-3_52

performance, either when implemented on Graphics Processing Units (GPUs) [5] or in hardware, like the Field Programmable Gate Array (FPGA) [6].

Nevertheless, in many cases, depending on the specific environmental parameters, the problem cannot be defined exactly because of its complicated dynamics and the vagueness of environmental conditions. To overcome these difficulties, an enhanced CA approach, combining CA with fuzzy logic [7], could be of certain interest. However, the application of fuzziness in CA rules and states cannot be considered an efficient one, unless the CA parallelism is exploited to reduce the computational needs of such a hybrid model.

The purpose of this paper is to propose an efficient, robust and dynamic Fuzzy Cellular Automata (FCA) model able to simulate wildfire spreading, whose results would be smoother and closer to the real wildfires spreading behavior. The proposed FCA model will be able to advance its performance and accuracy by applying corresponding fuzzy rules depending on the environmental parameters taken into consideration for the area under study. Moreover, the model's performance would allow real-time prediction by parallelization on GPU and FPGA, thus, fire departments could exploit its features and utilize the simulation results during wildfire's evolution. The computational investigation shows that the resulting simulation times in both parallel implementations, i.e. GPUs and FPGA, allow for the FCA model to be considered as a promising real time decision system able to manage efficiently fast optimization of risk-mitigation interventions and fire-fighting activities.

2   Fuzzy Cellular Automaton Model Description

CA are models of physical systems, where space and time are discrete and interactions are local [8]. A CA is characterized by five properties [9]: (i) the number of spatial dimensions (in our case two); (ii) the width of each side of the lattice; (iii) the width of the neighborhood of each cell, including the cells itself; (iv) the states of the CA cells; (v) the CA transition function F, which computes the state of each cell at the time step $(t + 1)$ as a function of the state of its neighboring cells at the time t. CA have sufficient expressive dynamics to represent phenomena of arbitrary complexity [10] and, at the same time, they can be simulated exactly by digital computers, because of their intrinsic discreteness, i.e. the topology of the simulated system is reproduced in the simulating device. Furthermore, they can easily handle complicated boundary and initial conditions, inhomogeneities and anisotropies [11,12].

The FCA model adopted in this study aims to cope with the computational complexity and data uncertainties that characterize the simulation of wildfire spreading. As outlined below, the proposed model can be parameterized in order to simulate the phenomenon more accurately, without important reduction in performance. The model parameters that influence the simulated fire behavior are associated with many environmental characteristics, such as topography of the area (slope and orientation toward the sun), wind (speed and direction), weather conditions (humidity, temperature and precipitation), fuel and oxygen density. In real-life situations, input data on the above characteristics is often imprecise and vague, which causes unreliability in the results of fire spread simulations. However, the fuzziness of the proposed FCA model allows the usage of unclear, misleading or erroneously-defined data inputs.

More specifically, the proposed model is based on a 2D FCA with Moore neighborhood, which has been selected to provide a fair variety of spreading directions. The states of each CA cell are associated with a linguistic variable of *fire intensity*, whose possible values should be selected appropriately, so that the model would match with the real forest fire spreading phenomenon. Herewith, the following five states are proposed:

$$S_{i,j}^t = \{No\ Fire, Low\ Fire, Medium\ Fire, High\ Fire, Burnt\}$$

Moreover, there is an extra state in which the cell is *non-flammable*, so that the cells that are initialized in this state do not take place in the computations. In addition, as shown in Fig. 1, we use triangular-shaped membership functions to state the evolution of CA, which represent a fair compromise between accuracy and simplicity.

The CA transition function (TF) computes the next state of each cell, considering as inputs the states of the eight Moore neighbors N, NW, W, SW, S, SE, E and NE and the current state of the cell. First, the TF uses the following linguistic rules for each neighboring cell:

1. *If* neighbor is Low Fire *then* cell state is Low Fire
2. *If* neighbor is Medium Fire *then* cell state is Medium Fire
3. *If* neighbor is High Fire *then* cell state is High Fire
4. *If* neighbor is Burnt *then* cell state is Burnt

Note that the weights for diagonal neighbors are smaller that those for side neighbors because of the difference in distance. Subsequently, the TF modifies its internal state as follows:

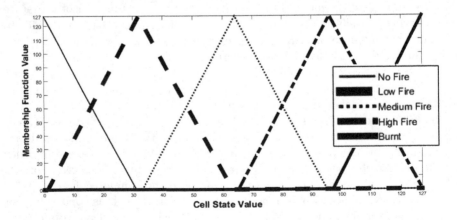

Fig. 1. Membership function representing the CA cell states.

1. *If* cell state is No Fire *then* cell state is No Fire
2. *If* cell state is Low Fire *then* cell state is Medium Fire
3. *If* cell state is Medium Fire *then* cell state is High Fire
4. *If* cell state is High Fire *then* cell state is Burnt
5. *If* cell state is Burnt *then* cell state is Burnt

It is worth noting that also the latter rules are affected by weights, which can be calibrated to account for the specific aspects of the real problem. As it has been mentioned, the model can be enriched with more fuzzy rules to represent more abrupt changes of the fire spreading in real-life situation. As a simple example, and for readability reasons, we only refer to the enhancement of the model with North wind and terrain acclivity from East to West (check also Table 1). In particular, to apply the North wind effects, the following rules have been added to the TF:

For the South, the South-West and the South-East neighbors:

1. *If* neighbor is Low Fire *then* cell state is No Fire
2. *If* neighbor is Medium Fire *then* cell state is No Fire
3. *If* neighbor is High Fire *then* cell state is No Fire

For the North, the North-West and the North-East neighbors:

1. *If* neighbor is Low Fire *then* cell state is Medium Fire
2. *If* neighbor is Medium Fire *then* cell state is High Fire
3. *If* neighbor is High Fire *then* cell state is Burnt

Also, to apply the terrain acclivity effects to the model the following rules have been added to the TF:

For the West, the South-West and the North-West neighbors:

1. *If* neighbor is Low Fire *then* cell state is No Fire
2. *If* neighbor is Medium Fire *then* cell state is No Fire
3. *If* neighbor is High Fire *then* cell state is No Fire

For the East, the North-East and the South-East neighbors:

1. *If* neighbor is Low Fire *then* cell state is Medium Fire
2. *If* neighbor is Medium Fire *then* cell state is High Fire
3. *If* neighbor is High Fire *then* cell state is Burnt

The weights of the above rules can be chosen according to the wind intensity and the terrain slope (e.g., as a result of a calibration process). As last step of the TF, the defuzzification phase gives the next state value of each CA cell according to the Fuzzy Mean method:

$$\psi = \sum_{i=1}^{N} w_i M_i c_i \Big/ \sum_{i=1}^{N} w_i M_i \,, \tag{1}$$

where ψ is the next state value of the cell, w_i is the weight of the ith rule, M_i is the value of the membership function, c_i is the centroid of the output states membership function for the i_{th} rule and N equals to the number of rules.

Table 1. Weights increments corresponding to the presented example with North wind and East-West terrain acclivity, where w is the weight in absence of wind and slope (i.e. $w = 2$ or $w = 1$ for adjacent and diagonal neighbors respectively).

	E–NE–SE Rules					W–NW–SW Rules				
	Next State					Next State				
	No Fire	Low	Medium	High	Burnt	No Fire	Low	Medium	High	Burnt
No Fire										
Low			$0.35 \times w$			$0.07 \times w$				
Medium				$0.35 \times w$		$0.07 \times w$				
High					$0.35 \times w$	$0.07 \times w$				
Burnt										

2.1 GPU Implementation

The FCA model can be easily implemented both in hardware and GPU to cope with the high computational requirements of real time applications. The GPU implementation was written using the C-language Compute Unified Device Architecture (CUDA). In CUDA, the GPU activation is obtained by writing device functions called kernels. When a kernel is issued by the CPU, a number of threads execute its code in parallel on different data. In the developed GPU-FCA, each thread executes the TF described above for a different cell of the CA.

Before the beginning of each simulation, the current CA is initialized through a CPU-to-GPU copy operation (i.e., from host to device global memory). Also, at the end of each CA step a device-to-device memory copy operation is used to re-initialize the *current* CA values with the *next* values (both stored in the GPU global memory as arrays). When, the CA state is required by the CPU (e.g., for graphical output purposes), a GPU-to-CPU copy is carried out. Instead of using the GPU shared memory to cache the CA states, in this study the parallel TF exploited the CUDA option for in-creasing the L1 cache size, which is a reasonable option for recent GPUs.

A frequent issue raised by the GPGPU parallelization of CA models is that only a small fraction of the cells perform actual computation at each step. Hence, launching one thread for each of the automaton cells would result in a certain amount of dissipation of the GPU computational power. For this reason, in the developed GPU-FCA the grid of threads is dynamically computed during the simulation in order to keep low the number of computationally irrelevant threads. More in details, at each CA step the procedure involves the computation of the smallest common rectangular bounding box (CRBB) that includes any cells on the current fire fronts plus their neighboring cells. Then, all the kernels required by the CA step are mapped on such CRBB. Note that using recent GPU devices, the CRBB computation in the GPU can be efficiently carried out at each step using the *atomicMin* and *atomicMax* CUDA primitives in the same kernel implementing the transition function [5]. Clearly, the efficiency of the CRBB approach depends on the actual size of the burned area with respect to the whole automaton. However, more efficient implementations (e.g. based on a dynamic list of the burning cells [5]) will be investigated in the future.

Given that most wildfires have small sizes, the ideal situations for exploiting the GPU with the FCA model are those requiring the simulation of a number of fires (e.g. from hundreds to many thousands) on the same landscape. This is for example required in the case of parameter calibration processes [13,15,16] simulations with data assimilation [14] or computation of risk maps [5]. The developed GPU-FCA can be easily extended to cope with the relevant problem of simulating in parallel a large number of fires as done in [5].

2.2 Hardware Implementation

Modern computers offer sufficient processing power to handle most of the analysis that several complex environmental phenomena require. Though, the application of a general-purpose computer in some cases may not be desirable, or it is even impossible, due to high power consumption and significant size. Portable, embedded general-purpose processors may however be unable to handle more complex computational tasks. Another method to achieve the speed-up execution of the proposed FCA model in such embedded systems, is to use the potential of available FPGA-devices. They enable parallel processing of data using custom digital structures. Besides, CA circuit design reduces to the design of a single, relatively simple cell and the layout is uniform. The whole mask for a large CA array (the cells with their internal connections as well as the interconnection between cells) can be generated by a repetitive procedure so no silicon area is wasted on long interconnection lines and because of the locality of processing, the length of critical paths is minimal and independent of the number of cells [6]. The hardware implementation of our FCA was written in Very High Speed Integrated Circuits Description Language (VHDL), which exploits the advantages of the hardware offering small execution time compared with the corresponding software implementation. Moreover, it takes advantage of CA inherent parallelism resulting in a minimum algorithmic complexity equal to $O(n^2)$, which hardly depends on the size of the FCA grid. On the other hand, the CA grids maximum size depends strongly on the hardware resources that a targeted FPGA can provide. Every CA cell of the grid is assumed as an independent component, so that the computation of the next state of the model is fully parallel and can be made in constant time. Therefore, the execution time is expressed as *time steps* × *cycles per cell* × *clock period*.

In the proposed FCA-FPGA, to tackle with the lack of floating point operations, the following adjustments have been applied:

1. The range of the membership functions and its values have been set to 0–127, i.e. seeking for a 7-bit unsigned number expression.
2. The triangular membership functions have integer slope, so that the membership function values could be calculated by bit shifting instead of floating point multiplication. In the same way, the weights have been selected such that there would be needed only bit shifting for the calculation.

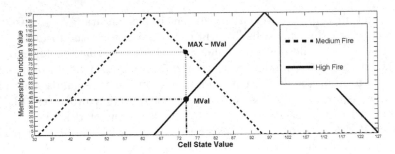

Fig. 2. The equivalent membership functions considered for hardware implementation of FCA model.

3. Finally, the division required for the defuzzification has been implemented by subtracting the denominator from the numerator and incrementing a counter until the result is negative. The quotient is the final value of the counter.

The architecture of the hardware implementation consists of a 2D grid with connected cell components taking into account the Moore neighborhood. In order to reduce significantly the hardware resources needed for membership function's calculation, the following process is proposed. First of all, the triangular membership functions symmetry has been utilized in order to compute the value only from the left side of the triangle, which has positive slope, so the usage of negative numbers for calculation has been avoided. Additionally, the maximum number of overlapping membership functions is two, so that in each iteration only two of them can be active. These active membership functions values are complementary to the maximum value. Taking advantage of this, only one value is needed to be calculated and then its complementary value is obtained by subtraction from the maximum value, as it is shown in Fig. 2.

Finally, the procedure of the next state computation is based on a FSMD (Finite State Machine Datapath), which is analytically presented in Fig. 3. The FSMD has been used to achieve re-usage of hardware, hence the computation of each neighbors' rules are made in the same datapath but with different input values. In every time step of the simulation, the next state calculation in each cell has been implemented by computing the rules for its own value and then the rules for neighbors' values. Nevertheless, this process takes place in a more serial fashion and is followed by the calculation of the Fuzzy Mean of Eq. (1).

3   Results

A preliminary assessment of the aforementioned FCA implementations was carried out by simulating 150 CA steps on a 500×500 grid of cells. Two test cases have been devised: the first on a flat terrain without wind and the second under the wind and slope conditions described in Sect. 2. In addition, to increase the computational complexity, each test case had two ignition points and the landscape included two non-flammable patches. The weights of the FCA rules, which

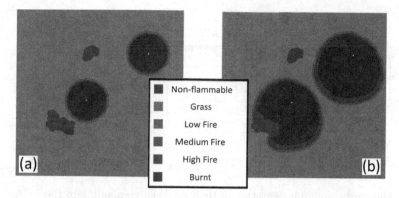

Fig. 3. Simulation results for (a) homogeneous propagation (b) propagation with north wind and terrain acclivity from East to West. The white spots are the ignition points.

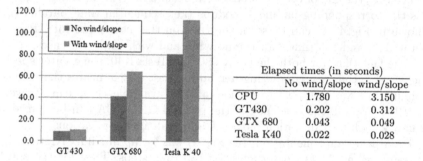

Elapsed times (in seconds)

	No wind/slope	wind/slope
CPU	1.780	3.150
GT430	0.202	0.312
GTX 680	0.043	0.049
Tesla K40	0.022	0.028

Fig. 4. The achieved speedups and elapsed times with GPU-FCA on different GPU devices.

can not be listed here for space reasons, have been chosen according to reasonable wind intensity and the slope of the terrain. The corresponding simulation results are presented in Fig. 3.

The software tests were performed on an Intel® Core^{TM} i7-4510U CPU @ 2.00 GHz, with 8 GB RAM, running Windows 7 64-bit. Besides the parallelized FCAs, also a sequential implementation was developed, using Ms Visual Studio 2013, and compiled with speed optimization turned on (i.e., /O2). For such sequential runs, the simulation times were 1.78 s in no wind/slope propagation and 3.15 s for fire propagation with wind and slope, respectively.

To carry out a preliminary assessment of the advantages provided by the GPU parallelization, the GPU-FCA model was executed on different GPUs: a consumer-level nVidia Geforce GT 430 (96 cores, 268.8 GFLOPs), a nVidia Geforce GTX 680 (1536 cores, 3090.4 GFLOPS), and a powerful Tesla K 40 accelerator (2880 cores, 4290 GFLOPs). On each GPU device, the computing time of the two experiments, with and without wind and slope, was measured. This makes sense because the number of cells included into the MRBB during the simulations was significantly different in the two test cases (i.e. 89232 vs.

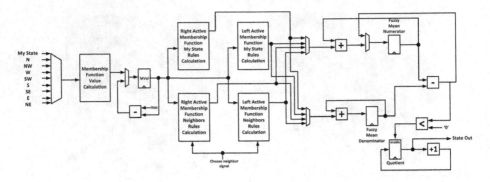

Fig. 5. The hardware implementation of the FSMD corresponding to the FCA model.

151904 cells at the end of the 150 steps). The results in terms of elapsed times as well as the corresponding parallel speedups (i.e., sequential time/parallel time) are shown in Fig. 4. As can be seen, the runs on the most powerful GPU were accelerated 81 and 113 times, for the no-wind and with-wind test cases, respectively. The fact that, for both the GTX 680 and Tesla K40, the greater speedup was achieved in the most computing-demanding test case indicates that the simpler simulation was not able to exploit all the available computing power of such GPUs. Overall, the efficiency of the parallel GPU-FCA can be considered satisfactory both for calibration purposes and real-time applications.

The hardware implementation (see Fig. 5) was set up in Xilinx Kintex Ultra-Scale resulting in 27,000 clock cycles for 150 time steps. This, for the specific target FPGA device, results in 0.135 ms or 200 MHz, thus the speedup is considered really great. Moreover, the specified period or in correspondence the time needed for the FCA to end its iterations remains steady no matter the increment of the specified CA grid as far as this can be implemented in the specific device. It is clear the resulting FPGA can be considered as a basic component of a portable electronic system able to provide real time information concerning the forest fire propagation on the part under test of the examined area. In doing so, GPS (Global Position System) tracking and GIS (Geographical Information System) provide the resulted FPGA processor with real data about possible roads and routes and feedbacks for the area under study. This could also be considered as part of a complete portable electronic system, possibly equipped also with Wi-Fi transceiver transmitter for communication reasons as well as with proper sensors, i.e. temperature, humidity and wind speed/direction sensors.

4   Conclusions

In this paper a FCA model, implemented in GPUs and FPGA, was proposed to simulate wildfire spreading. The simulation results indicate the efficiency of the proposed approach and the resulting speedup compared to the sequential execution of the code. As a future work, many more simulations in different environmental conditions will be carried out and test cases on real landscapes will be studied. Moreover, a methodology for advancing the fuzzy CA rules will be introduced.

References

1. Rothermel, R.C.: A mathematical model for predicting fire spread in wildland fuels. Technical report INT-115, USDA, Forest Service, Intermountain Forest and Range Experiment Station, Ogden, UT (1972)
2. Karafyllidis, I., Thanailakis, A.: A model for predicting forest fire spreading using cellular automata. Ecol. Model. **99**, 87–97 (1997)
3. Trunfio, G.A., D'Ambrosio, D., Rongo, R., Spataro, W., Di Gregorio, S.: A new algorithm for simulating wildfire spread through cellular automata. ACM Trans. Model. Comput. Simul. **22**, 1–26 (2011)
4. Avolio, M.V., Di Gregorio, S., Trunfio, G.A.: A randomized approach to improve the accuracy of wildfire simulations using cellular automata. J. Cell. Automata **9**(3–4), 209–223 (2014)
5. Di Gregorio, S., Filippone, G., Spataro, W., Trunfio, G.A.: Accelerating wildfire susceptibility mapping through GPGPU. J. Parallel Distrib. Comput. **73**(8), 1183–1194 (2013)
6. Progias, P., Sirakoulis, G.C.: An FPGA processor for modelling wildfire spreading. Math. Comput. Model. **57**, 1436–1452 (2013)
7. Mraz, M., Zimic, N., Lapanja, I., Bajec, I.: Fuzzy cellular automata: from theory toapplications. In: 12th IEEE International Conference on Tools with Artificial Intelligence, pp. 320–323 (2000)
8. von Neumann, J.: Theory of Self Reproducing Automata. University of Illinois Press, Urbana (1966)
9. Kalogeiton, V.S., Papadopoulos, D.P., Georgilas, I.P., Sirakoulis, G.C., Adamatzky, A.I.: Cellular automaton model of crowd evacuation inspired by slime mould. Int. J. Gen. Syst. **43**(4), 354–391 (2015)
10. Saravakos, P., Sirakoulis, G.C.: Modeling behavioral traits of employees in a workplace with cellular automata. In: Wyrzykowski, R., Dongarra, J., Karczewski, K., Waśniewski, J. (eds.) PPAM 2013, Part II. LNCS, vol. 8385, pp. 689–698. Springer, Heidelberg (2014)
11. Sirakoulis, G., Adamatzky, A.: Robots and Lattice Automata. Springer, Heidelberg (2015)
12. Was, J., Sirakoulis, G.C., Bandini, S.: Cellular Automata, Proceedings of 11th International Conference on Cellular Automata for Research and Industry, ACRI 2014, vol. 8751. Springer, Heidelberg (2014)
13. Artés, T., Cencerrado, A., Corts, A., Margalef, T.: Enhancing computational efficiency on forest fire forecasting by time-aware Genetic Algorithms. J. Supercomput. **71**(5), 1869–1881 (2015)
14. Xue, H., Gu, F., Hu, X.: Data assimilation using sequential Monte Carlo methods in wildfire spread simulation. ACM Trans. Model. Comput. Simul. **22**(4), 23 (2012)
15. Topa, P.: Cellular automata model tuned for efficient computation on GPU with global memory cache. In: PDP 2014 Proceedings, pp. 380–383 (2014)
16. Was, J., Mrz, H., Topa, P.: GPGPU computing for microscopic simulations of crowd dynamics, Computing and Informatics (2014, in press)

eVolutus: A New Platform
for Evolutionary Experiments

Paweł Topa[1,3(✉)], Maciej Komosinski[2], Jarosław Tyszka[3], Agnieszka Mensfelt[2],
Sebastian Rokitta[4], Aleksander Byrski[1], and Maciej Bassara[1]

[1] Department of Computer Science, AGH University of Science and Technology,
Al. Mickiewicza 30, 30-059 Kraków, Poland
topa@agh.edu.pl
[2] Institute of Computing Science, Poznan University of Technology,
Piotrowo 2, 60-965 Poznan, Poland
[3] Research Centre in Cracow, ING PAN Institute of Geological Sciences,
Polish Academy of Sciences, Senacka 1, 31-002 Kraków, Poland
[4] Helmholtz-Centre for Polar and Marine Research, Alfred Wegener Institute,
Am Handelshafen 12, 27570 Bremerhaven, Germany

Abstract. eVolutus is a new software platform designed for modeling
evolutionary and population dynamics of living organisms. Single-celled
eukaryotes, foraminifera, are selected as model organisms that have occupied the marine realm for at least 500 Ma and left an extraordinary fossil
record preserved in microscopic shells. This makes them ideal objects
for testing general evolutionary hypotheses based on studying multiscale genotypic, phenotypic, ecologic and macroevolutionary patterns.
Our platform provides a highly configurable environment for conducting
evolutionary experiments at various spatiotemporal scales.

Keywords: Artificial life · Multi-agent systems · Foraminifera

1 Introduction

Recent advances in the area of evolutionary computation and genetic algorithms
(GAs) are mostly motivated by performance of these algorithms in the context
of optimization. However, to simulate evolutionary processes within a realistic
paleobiologic and paleoecologic "deep time" context, an *in-silico* artificial life
approach [12,13] is employed here to reconstruct ecological and evolutionary
patterns of foraminifera by implementation of realistic principles known from
living organisms.

We simulate foraminifers with their iterative ontogenetic and morphogenetic
growth stages, adequately controlled by their semi-genetic codes and continuously interacting with their microhabitat. Mutations of traits can be either
introduced with controlled randomness, or user-defined, allowing the operator
to define and simulate various scenarios. The crucial concept of fitness function
is given in an indirect way as a complex function calculated based on current
(local) environmental conditions and the instantaneous state of the individual.

© Springer International Publishing Switzerland 2016
R. Wyrzykowski et al. (Eds.): PPAM 2015, Part II, LNCS 9574, pp. 570–580, 2016.
DOI: 10.1007/978-3-319-32152-3_53

We use foraminifera, single-celled eukaryotes that occupy marine benthic and pelagic zones throughout the world. Foraminifera have an extraordinary fossil record since Cambrian (500 million years ago). This makes them an ideal model often used for testing general evolutionary hypotheses [22,24,25].

We have previously introduced a new generation of morphogenetic models that can successfully predict basic architecture patterns of foraminiferal shells following the moving reference system [17,21,26]. The original model has now been extended by new components, i.e., size of the first chamber and thickness of the shell wall. They are introduced in order to produce more realistic shell morphology, as well as to achieve proper behavior in the microhabitat.

As a platform for modeling microhabitats and evolutionary processes that affect generation of foraminifers we use Framsticks [14] and AgE EMAS (Evolutionary Multi-Agent System) [4] environments. Both platforms have the ability to simulate biological evolution and offer complementary features.

eVolutus provides an environment that can be used by experimenters not skilled in computer programming, and tutorials for defining and tuning experiments by using simple script languages.

1.1 Foraminifera

Foraminifera are single-celled eukaryotes that mainly occupy benthic and pelagic habitats. Benthic foraminifera live either on the sea floor around the water/sediment interface, or within the top 10 cm of soft, usually fluidal, sediment. Planktonic foraminifera live in the open ocean, floating in the photic and subphotic zone of the water column [3].

Most foraminifers produce multichambered shells covering their soft cytoplasmic bodies. Foraminifera grow by successive construction of new chambers. We observe an enormous variety of shell shapes and chambers, however for many species spheroidal chambers may be a close approximation.

Communication between the cell and the external environment is provided through an aperture, a hole located in each chamber, connecting it to the successive one. Foraminifera extend 'granuloreticulopodia' i.e., granular network pseudopodia through these apertures in order to monitor the microhabitat, gather food, attach, and move, as well as to transmit signals within the cell [3,23].

Foraminifera mostly feed on single-celled algae and their detrital remains, which makes them strongly dependent on their availability in time and space. The most common temporal variability is reflected in seasonality of temperature and nutrient availability. These factors have a direct impact on distribution, life history strategies, reproduction, and population dynamics of foraminifera [10,23].

Foraminiferal life spans range from a few weeks in some planktonic foraminifera up to a few years in larger benthic foraminifera [8,10]. A typical life cycle of benthic foraminifera is characterized by an alternation of two methods of reproduction: sexual (in haploid generation) and asexual (in diploid generation) [8]. A haploid generation is equipped with one set of chromosomes, while diploid generation has two sets of chromosomes. This complex life cycles helps foraminifera in adjusting to variable (e.g. seasonal) conditions and allows them to create diverse and flexible life history strategies [23]. Planktonic foraminifera have only diploid generations and use a sexual method of reproduction.

1.2 AgE: A Lightweight Agent-Based Computing Framework

The AgE platform has been successfully implemented in different versions (Java, Python, .NET, Erlang) and utilized to implement systems with different applications, e.g. computing [4], decision support [20], simulation [7] or data integration [5]. We focus on a specific architecture dedicated to easily implement distributed version of AgE (actual implementation was carried out using Java technology and Hazelcast framework).

Lightweight agents execute pseudo-concurrently, by repeatedly executing certain step callback function. From their point of view, lightweight agents effectively execute in parallel. The platform emulates this parallelism by introducing several constraints on message propagation and change visibility.

1.3 Framsticks: Artificial Life Simulator

The Framsticks simulator [19], since its initial releases in 1996, has been used as a computing engine in a number of diverse applications [14]. They were mostly concerned with modeling complex adaptive systems, multiple autonomous agents and the process of evolution. This includes comparison of genetic encodings [18], estimating symmetry of evolved and designed agents [11], employing similarity measure to organize evolved constructs [15,16], optimization of fuzzy controllers that can be understood by a human [9], experiments with synthetic neuroethology, modeling foraminiferal genetics, morphology, and evolution [17], and simulating plastic neural nets and their evolution. This environment was also used to study various biological phenomena [14]: communication, emergence, flocking, evolution of restraint, predator–prey coevolution, semiosis, speciation, and user-driven (interactive, aesthetic) evolution.

Framsticks is implemented in C++, and it has its own virtual machine and a high-level scripting language (FramScript) that allows users to develop their own experiments, visualizations, GUI, neurons, and macros. The simulator runs on all major desktop and mobile platforms. It supports multi-threading and has a network server so that computations can be distributed. The C++ SDK and network clients are open source. These features make Framsticks a suitable platform to implement eVolutus.

2   eVolutus Architecture

A general architecture of the eVolutus platform is presented in Fig. 1. Habitats can be modeled using either the Framsticks platform or the AgE platform. In both cases, these tools have to be equipped with components that describe shell development (morphology), processes of food acquisition, energy management and reproduction (physiology).

Our goal is a platform that provides the experimenter a high level of configurability. Framsticks, since it has been developed, has ability to define experiments and set up simulations with its own scripting language called *FramScript*. AgE is more generic framework and such the feature is implemented in eVolutus "from scratch".

With a "virtual fossilization" module, the simulation results are sampled in controllable spatiotemporal resolution and stored for post-processing and analysis, mimicking the accumulation of shells of dead foraminifera in sedimentary rocks. In real geological investigations such fossils are extracted and only their morphology and chemistry can be analyzed. Virtual fossilization provides all the information including genetic information and environmental conditions. It allows tracking genotype changes over consecutive generations and the determination of influences of environmental conditions. This module also allows for controlling spatial and temporal range of collected results.

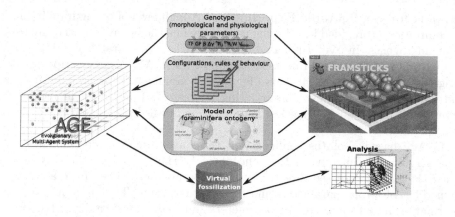

Fig. 1. The general architecture of the eVolutus simulation environment.

Both types of foraminiferal habitats can be represented using these frameworks. Model of planktonic habitat should cover the volume of an open ocean to depth of typical range of occurrence of foraminifera. For modeling the benthic habitat, only a very thin top layer of sediments should be modeled.

Agents in the model of planktonic habitat cannot move actively. They are moved using the random walk algorithm. The experimenter is able to define vectors of forces which represent simplified ocean currents. On the other hand, benthic foraminifera have the ability to move actively. They slowly travel through their habitat gathering nutrients.

The Genotype and the Model of Foraminifera Ontogenesis and Reproduction. The crucial part of our modeling platform is the model of foraminifera ontogenesis [17,21,26]. This relatively simple algorithm generates morphology of foraminiferal shell by using geometrical operations parameterized by up to 6 values. The model of foraminifera ontogenesis has been extended to new components and parameters, i.e., W as the thickness of chamber wall and R_1 – the radius of the first chamber. When virtual foraminifera are modeled in their virtual habitat, additional parameters–genes that controls their physiology and behaviors are necessary.

The crucial physiological component of the model of foraminifera evolution is reproduction. At this moment, three reproduction methods are recognized and implemented: sexual and two types of asexual modes [6]. Planktonic foraminifera use only the sexual method, while benthic foraminifera use all methods. Regardless of the type of reproduction, the ancestor agents are removed from the simulation and, if necessary, their genotypes and other parameters are saved for further analysis.

Another component of foraminifera physiology is food intake and energy management. Foraminifers gather food located in proximity neighborhood, store it as energy, and constantly use to grow and sustain all other life processes.

Support for Configurable Experiments. A high level of configurability and customization is provided by allowing users to define the behavior of the environment and agents. In eVolutus based on Framsticks, configurability is inherited in a natural way from the Framsticks simulator itself. A dedicated scripting language called *FramScript* [14] allows for high flexibility in defining experiments and setting up simulations. This scripting language is tailored for artificial life experiments and as such, it has less features than full-fledged languages like Java or Python.

In AgE-based eVolutus configuration tools are designed and implemented from scratch. We introduced the concept of "kernels", i.e., a relatively short functions that are created directly by the experimenter[1]. They have a strictly defined list of formal parameters and the returned value. They may just return information about environment parameters but they can contain instructions that calculate desired value. Currently the kernels are coded using JavaScript language and in this form they are directly invoked from the Java code via Java Scripting API [1].

3   Demonstration of Platform Capabilities and Sample Results

Planktonic Foraminifera: Population Dynamics and Evolving Genes. eVolutus which uses AgE as an environment for modeling evolution of agents is implemented in Java language. Currently our implementation performs only sequential processing, however parallel implementation can be easily introduced. Module of virtual fossilization uses MongoDB as a database engine. The reason why we choose this DBMS (Data Base Management System) is fact the collected data have a relatively simple schema of relations between collected records.

---

[1] The name "kernel" used in CUDA or OpenCL GPU programming means a short function executed by many GPU processing units in parallel.

Tools that will be used to view and analyze the stored data are designed and implemented at this moment.

The habitat is represented as the set of cells that can be organized as a grid in the basic version. The state of each cell is described by several parameters that reflect environmental condition, i.e., nutrients availability, insolation, temperature, salinity, ocean currents etc.

In order to save computational resources each cell can be occupied by many agents and we do not track their individual locations. Agents located in the same cell use the same resources and experience the same environmental conditions. We also assume that during sexual reproduction, only agents in the same block can exchange their genotypes.

Figure 2 presents a dynamics of planktonic foraminifera population. Parameters for this simulation were adjusted to achieve the dynamics of the population that follows Lotka-Volterra model [2]. The simulation covered 1200 days of real time. The habitat space has dimensions of $1000\,\text{m} \times 1000\,\text{m} \times 100\,\text{m}$. Food is delivered at a constant rate to each cell. Agents consume food and use it to maintain vital functions. They energy needs are proportional to the volume of shell. Agents developed shell and reproduce when they reach maturation age. Reproduction is sexual – agents produce gametes which are randomly paired.

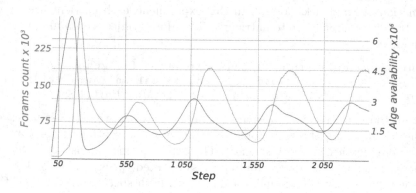

Fig. 2. Dynamics of foraminiferal population resemble classical predator-prey models such as the Lotka-Volterra model.

The simulation started with the randomly distributed population of the size of 5000 individuals. A small number of consumers (predators) allowed for rapid increase of amount of food. This led to the equally rapid growth of population— more than 200 000 individuals appeared in the habitat. They rapidly consumed the available food what in consequence reduced population. In further steps of simulation we observe characteristic pattern of shifted in time oscillation of food (prey) and population of foraminifers (predaters).

Dynamics of Haploid and Diploid Generations in Benthic Foraminifera.
The goal of this study was to test the implementation (software) by investigating the population dynamics of the benthic foraminifera taking into account haploid

and diploid generations. Foraminifers were placed on a virtual sea floor. Nutrients were appearing in random locations with a constant rate. Each foraminifer was moving toward the nearest nutrient item. Foraminifers were able to reproduce when they had accumulated a sufficient amount of energy. In the reproduction process, haploid and diploid generations were alternating.

Foraminifers were able to gather food located within their reticulopodia range. They were also able to sense nutrients located outside of the reticulopodia range, but within the defined chemotactic sensing range. In the following experiments there were two species of Foraminifera, differing in morphology and in behavioral strategy. In case of a shortage of nutrients, uniserial specimens switched to dormancy (hibernation) in order to save energy, while coiled specimens were exploring the environment, i.e., moving and searching for food.

These experiments have been implemented in Framsticks [19]. The implementation differs from the AgE implementation – here space is continuous, and it is not divided into "cells" in a grid. Both ploidy stages of foraminifera and nutrients are simulated as agents that are located in 3D. Time is measured in simulation steps. In each step, each agent can make a small movement. The lifespan of each agent is determined by its energy level.

In terms of technical implementation, the information about nutrient locations is transferred by a signal, which is a versatile way to send any kind of information in Framsticks [14,19]. In this experiment, each nutrient used a signal to broadcast a reference to nutrient's "MechPart" object containing nutrient coordinates:

```
var nutrient=Populations[1].add(ExpParams.foodgen);
nutrient.signals.add("food"); //add signal and name it "food"
nutrient.signals[0].value = nutrient.getMechPart(0);
```

In all the experiments the environmental conditions were the same except for the feeding rate, the reticulopodia range and the sensing range (but the ranges were identical for both species). The experiments demonstrated that the competition for food tends to eliminate less adapted species. This is shown in Fig. 3 which presents the virtual sea floor with both species present, and then

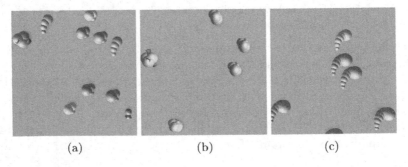

(a)   (b)   (c)

Fig. 3. A part of the sea floor surface with two populations of foraminifera species for feeding rate = 0.1 and different reticulopodia and sensing ranges. (a) In the beginning of the experiment (ranges = 3,6), (b) after the extinction of the uniserial species (ranges = 3,6), and (c) after the extinction of coiled species (ranges = 5,8).

Table 1. The number of simulation runs in which a given species (coiled:uniserial) survived for each combination of parameter values.

| | | Reticulopodia and sensing ranges [mm] | |
		3,6	5,8
Feeding rate	0.05	5:0	3:2
	0.1	2:3	0:5

after the extinction of one of the species. For each combination of parameter values, the simulation was run 5 times. Table 1 presents the survival ratio for both species.

Figure 4a presents the population dynamics of the experiment in which the hibernating species (uniserial) went extinct (feeding rate = 0.1, ranges = 3.6). The coiled population revealed cyclic abundance patterns that follow nutrient fluxes. The diploid individuals dominated coiled Foraminifera population due to a larger number of smaller offspring that move and actively seek for food. These results do not follow real-world observations of benthic foraminifers where haploids outbalance diploids. Diploids in the real world seem to have a much lower survival rate at a juvenile stage and they tend to grow to larger sizes that also explains their lower abundance [3,10]. Furthermore, benthic foraminifers often show more complex reproduction strategies that favor multiple cycles of asexual cloning [8,10] that, in consequence, affect the proportion of different generations. Implementing these new reproduction mechanisms and calibrating survival rates will increase the compatibility of this model with nature.

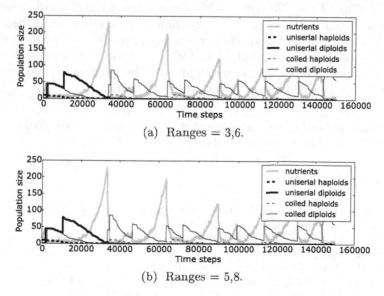

(a) Ranges = 3,6.

(b) Ranges = 5,8.

Fig. 4. The size of the uniserial haploids, uniserial diploids, coiled haploids, coiled diploids and nutrient populations in two experiments differing in the reticulopodia and sensing ranges. The feeding rate was 0.1.

Figure 4b presents the population dynamics of the experiment in which randomly moving species (coiled) went extinct (feeding rate = 0.1, ranges = 5,8). The diploid individuals were dominating again. For this set of parameter values, a better strategy was not to lose energy on movement.

4   Conclusions and Further Work

The eVolutus software platform presented here is under active development, and the initial tests of the habitat model proved that the proposed approach is efficient. We are able to easily configure the environment so it can produce results that follow well-known non-linear dynamics of ecological models [2].

Our further work will focus on testing and validating algorithms for reproduction and gene exchange. In the following steps, physiological properties like growth and motility will be implemented to tie individual behavior more to environmental parameters like food quality and quantity distributed in time and space. Another area of extensive works concerns large-scale simulations. Our goal is to extend simulations to the population size of 10^9–10^{12} individuals through geological time intervals from 10^3 to 10^8 years.

The AgE and Framsticks frameworks have shown complementary features. Both platforms support distributed computing and large scale experiments, however, for our purposes, additional mechanisms have to be implemented. This includes the need to map a regular grid of the habitat space into the available set of computational nodes, and providing methods of communication between nodes that process neighboring blocks.

Applicability of the eVolutus platform takes into account a broad range of population dynamic and evolutionary experiments that might be furtherassociated with other modelling methods (e.g. ecosystem models). The model cannot cover the overall complexity of physiological and ecological networks, nevertheless, it shows potential for further development. This approach, when tested on foraminifera, may also be extended for other organisms.

Acknowledgments. The research presented in the paper received partial support from Polish National Science Center (DEC-2013/09/B/ST10/01734).

References

1. The Java Scripting API. https://docs.oracle.com/javase/8/docs/technotes/guides/scripting/prog_guide/api.html
2. Berryman, A.: The origins and evolution of predator-prey theory. Ecology **73**(5), 1530–1535 (1992)
3. Brasier, M.: Microfossils. Springer, Heidelberg (1980)
4. Byrski, A., Dreżewski, R., Siwik, L., Kisiel-Dorohinicki, M.: Evolutionary multi-agent systems. Knowl. Eng. Rev. **30**(2), 171–186 (2015)
5. Dajda, J., Debski, R., Byrski, A., Kisiel-Dorohinicki, M.: Component-based architecture for systems, services and data integration in support for criminal analysis. J. Telecommun. Inf. Technol. **1**, 67–73 (2012)

6. Dettmering, C., Röttger, R., Hohenegger, J., Schmaljohann, R.: The trimorphic life cycle in foraminifera: observations from cultures allow new evaluation. Eur. J. Protistology **34**(4), 363–368 (1998)

7. Faber, L., Pietak, K., Byrski, A., Kisiel-Dorohinicki, M.: Agent-based simulation in AgE framework. In: Byrski, A., Oplatková, Z., Carvalho, M., Dorohnicki, M.K. (eds.) Advances in Intelligent Modelling and Simulation. SCI, vol. 416, pp. 55–83. Springer, Heidelberg (2012)

8. Goldstein, S.: Foraminifera: A Biological Overview. Kluwer Academic Pub., Dordrecht (1999)

9. Hapke, M., Komosinski, M.: Evolutionary design of interpretable fuzzy controllers. Found. Comput. Decis. Sci. **33**(4), 351–367 (2008)

10. Hohenegger, J.: Large Foraminifera: Greenhouse Constructions and Gardeners in the Oceanic Microcosm. Kagoshima University Museum, Kagoshima (2011)

11. Jaskowski, W., Komosinski, M.: The numerical measure of symmetry for 3D stick creatures. Artif. Life J. **14**(4), 425–443 (2008)

12. Komosinski, M.: The world of framsticks: simulation, evolution, interaction. In: Heudin, J.-C. (ed.) VW 2000. LNCS (LNAI), vol. 1834, pp. 214–224. Springer, Heidelberg (2000)

13. Komosinski, M., Adamatzky, A. (eds.): Artificial Life Models in Software, 2nd edn. Springer, London (2009)

14. Komosinski, M., Ulatowski, S.: Framsticks: creating and understanding complexity of life. In: Komosinski, M., Adamatzky, A. (eds.) Artificial Life Models in Software, chap. 5, pp. 107–148. Springer, London (2009)

15. Komosinski, M., Koczyk, G., Kubiak, M.: On estimating similarity of artificial and real organisms. Theory Biosci. **120**(3–4), 271–286 (2001)

16. Komosinski, M., Kubiak, M.: Quantitative measure of structural and geometric similarity of 3D morphologies. Complexity **16**(6), 40–52 (2011)

17. Komosinski, M., Mensfelt, A., Topa, P., Tyszka, J.: Application of a morphological similarity measure to the analysis of shell morphogenesis in foraminifera. In: Gruca, A., Brachman, A., Brachman, S., Czachórski, T. (eds.) ICMMI 2015. AISC, vol. 391, pp. 215–224. Springer, Heidelberg (2015)

18. Komosinski, M., Ulatowski, S.: Genetic mappings in artificial genomes. Theory Biosci. **123**(2), 125–137 (2004)

19. Komosinski, M., Ulatowski, S.: Framsticks web site (2015). http://www.framsticks.com

20. Krzywicki, D., Faber, L., Byrski, A., Kisiel-Dorohinicki, M.: Computing agents for decision support systems. Future Gener. Comput. Syst. **37**, 390–400 (2014)

21. Labaj, P., Topa, P., Tyszka, J., Alda, W.: 2D and 3D numerical models of the growth of foraminiferal shells. In: Sloot, P.M.A., Abramson, D., Bogdanov, A.V., Gorbachev, Y.E., Dongarra, J., Zomaya, A.Y. (eds.) ICCS 2003, Part I. LNCS, vol. 2657, pp. 669–678. Springer, Heidelberg (2003)

22. Lazarus, D., Hilbrecht, H., Spencer-Cervato, C., Therstein, H.: Sympatric speciation and phyletic change in Globorotalia truncatuloides. Paleobiology **21**(1), 28–51 (1995)

23. Murray, J.: Ecology and Palaeoecology of Benthic Foraminifera. Longman Scientific and Technical, Harlow, Wiley, New York (1991)

24. Pearson, P., Shackleton, N., Hall, M.: Stable isotopic evidence for the sympatric divergence of Globigerinoides trilobus and Orbulina universa (planktonic foraminifera). J. Geol. Soc. **154**, 295–302 (1997)

25. Strotz, L.C., Allen, A.P.: Assessing the role of cladogenesis in macroevolution by integrating fossil and molecular evidence. Proc. Nat. Acad. Sci. **110**(8), 2904–2909 (2013)
26. Tyszka, J., Topa, P.: A new approach to modeling of foraminiferal shells. Paleobiology **31**(30), 526–541 (2005)

Special Session on Algorithms, Methodologies and Frameworks for HPC in Geosciences and Weather Prediction

Accelerating Extreme-Scale Numerical Weather Prediction

Willem Deconinck[1]([✉]), Mats Hamrud[1], Christian Kühnlein[1],
George Mozdzynski[1], Piotr K. Smolarkiewicz[1], Joanna Szmelter[2],
and Nils P. Wedi[1]

[1] European Centre for Medium-Range Weather Forecasts,
ECMWF, Reading RG2 9AX, UK
{willem.deconinck,mats.hamrud,christian.kuehnlein,george.mozdzynski,
piotr.smolarkiewicz,nils.wedi}@ecmwf.int
[2] Wolfson School, Loughborough University, LE11 3TU Loughborough, UK
j.szmelter@lboro.ac.uk

Abstract. Numerical Weather Prediction (NWP) and climate simulations have been intimately connected with progress in supercomputing since the first numerical forecast was made about 65 years ago. The biggest challenge to state-of-the-art computational NWP arises today from its own software productivity shortfall. The application software at the heart of most NWP services is ill-equipped to efficiently adapt to the rapidly evolving heterogeneous hardware provided by the supercomputing industry. If this challenge is not addressed it will have dramatic negative consequences for weather and climate prediction and associated services. This article introduces Atlas, a flexible data structure framework developed at the European Centre for Medium-Range Weather Forecasts (ECMWF) to facilitate a variety of numerical discretisation schemes on heterogeneous architectures, as a necessary step towards affordable exascale high-performance simulations of weather and climate. A newly developed hybrid MPI-OpenMP finite volume module built upon Atlas serves as a first demonstration of the parallel performance that can be achieved using Atlas' initial capabilities.

Keywords: Numerical weather prediction · Climate simulation · High performance computing · Exascale

1 Introduction

Numerical Weather Prediction (NWP) and climate simulations have been intimately connected with progress in supercomputing since the first numerical forecast was made about 65 years ago. In the early days, computing power was gained through taking advantage of vector instructions on a large single-processor computer. With little modification to existing code bases, shared memory parallelization was introduced when more and more processors were added to one computer. When multi-node architectures became the norm, however, the required effort to

© Springer International Publishing Switzerland 2016
R. Wyrzykowski et al. (Eds.): PPAM 2015, Part II, LNCS 9574, pp. 583–593, 2016.
DOI: 10.1007/978-3-319-32152-3_54

port existing code to make use of distributed memory was significant, and often meant a rewrite or redesign of large parts of the code. Over a decade later, many NWP codes have grown in size to millions of lines making use of hybrid parallelization with distributed and shared memory on CPU architectures. As the traditional approaches to boosting CPU performance ran out of room [1], more nodes are being added to computing clusters, putting more and more strain on distributed memory parallclism, as well as ever increasing the energy bill. Hardware vendors are introducing alternatives to computing on traditional CPU architectures in order to keep increasing FLOP rates at reduced energy costs; for example offloading computations to GPU's or Intel MIC's. These alternatives are of less general purpose, and significant effort is required to adapt the existing code base to optimally use the hardware. The application software at the heart of most NWP services is ill-equipped to efficiently adapt to these rapidly evolving heterogeneous hardware. It is furthermore not yet clear which hardware, or combination of hardware will offer the best performance at minimal cost.

To address the challenge of extreme-scale, energy-efficient high-performance computing, the European Centre for Medium-Range Weather Forecasts (ECMWF) is leading the ESCAPE project, a European funded project involving regional NWP centres, Universities, HPC centres and hardware vendors, including an enterprise to explore the use of optical accelerators. The aim is to combine interdisciplinary expertise for defining and co-designing the necessary steps towards affordable, exascale high-performance simulations of weather and climate. Key project elements are to develop and isolate basic algorithmic ideas by breaking down NWP legacy codes into Weather & Climate Dwarfs in analogy to the Berkeley Dwarfs [2], and to classify these canonical building blocks with energy-aware performance metrics. The Berkeley Dwarfs identify archetypical computational and communication patterns, defining the priority areas for community research in numerical methods to enhance parallel computing and communication. The Berkeley dwarfs relevant to NWP contain algorithms responsible for: spectral methods, sparse linear algebra, dense linear algebra, unstructured meshes, structured grids, dynamic programming, and graphical models. Weather & Climate Dwarfs share these motifs but go further, by providing practical solutions to support co-design development and on which sustainable NWP services can be built. The distinct motivation to define Weather & Climate Dwarfs is to focus on maximizing computing performance with minimal energy consumption, i.e. the cost to solution, enveloped by the very strict requirement that NWP models need to run at speeds 200–300 times faster than real time. The ideas developed to facilitate extreme-scaling for the Weather & Climate Dwarfs, could then be implemented back in NWP models.

The Integrated Forecasting System (IFS) is one of such NWP models developed at the ECMWF. It relies on spectral transforms and corresponding computations in spectral space to evaluate horizontal derivatives. Spectral transforms on the sphere involve discrete spherical-harmonics transformations between physical (grid-point) space and spectral (spherical-harmonics) space. Extreme scaling performance of spectral transforms is crucial for its applicability in IFS on future hardware. Spectral transforms are identified as one of the Weather & Climate Dwarfs. Spectral transforms require data-rich global communication, and it is ultimately the communication overhead and not the computational burden that will limit the applicability of the spectral transform method at extreme scale. This is illustrated in Fig. 1 for 5 km and 2.5 km IFS simulations, where the communication cost amounts to 75 % of the transform cost on TITAN, the number 2 HPC system in the top 500 super computing list as in May 2015. To put these resolutions in perspective with the ECMWF operational requirements of solving 240 forecast days per day (or 10 days in 1 h), the 2.5 km resolution simulation is estimated to require 270,000 cores of a Cray XC-30 HPC, whereas the 5 km resolution simulation requires 25,000 cores to achieve this goal.

The cost of computing derivatives with the spectral transform method with the current state-of-the-art spectral transforms is still relatively high and hence the use of derivatives has been minimized carefully [3]. As part of the ESCAPE project, the spectral transform algorithms will be heavily scrutinized and improved upon by e.g. overlapping communications with computations. Hardware vendors will work together with scientists, and explore the use of bespoke accelerators based on optical computations.

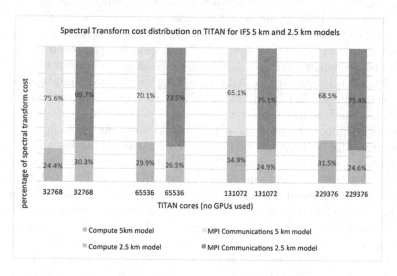

Fig. 1. Spectral transform cost distribution on TITAN for IFS 5 km and 2.5 km models. The total spectral transform cost on TITAN was in the range 29 to 51 % of the total wall time.

Notwithstanding the outcome of further applicability of the spectral transform method, alternative grid-point discretisation methods are desired to compute derivatives locally with compact grid-point stencils. Computations using compact grid-point stencils require only local — nearest neighbor — communication, which inherently scales much better and may better represent the locality of a physical process. Such discretisation methods are typically based on meshes composed of elements such as triangles, quadrilaterals, or lines. Examples include the finite volume method (FV), the finite element method (FE), or higher-order (>2) methods such as the Discontinuous Galerkin method (DG) and the Spectral Difference method (SD).

Atlas, a flexible parallel data structure framework that can handle a number of different structured grids and unstructured meshes, is being developed at the ECMWF, both to support research on alternative grid-point discretisation methods, as well as serving as the basis for leveraging the anticipated integration efforts resulting from the ESCAPE project. This Atlas framework will be introduced in detail in Sect. 2.

2   Atlas – A Flexible, Scalable, and Sustainable Model Infrastructure

2.1   Motivation

In order for NWP applications to optimally exploit future computer hardware emerging over the next 20–30 years, a flexible and dynamic data structure framework, named *Atlas* is being developed to serve as a foundation for a wide variety of applications, ranging from the use within European NWP models and for the development of alternative dynamical core modules [4], to development of applications responsible for pre- and post-processing of exponentially growing output data. It is imperative for Atlas to remain flexible and maintainable since the development of NWP models typically takes a decade, and NWP models typically last much longer than that.

The *Atlas* framework provides parallel distributed, flexible, object-oriented data structures for both structured grids and unstructured meshes on the sphere. It separates concerns of mathematical model formulation and numerical solutions from the cumbersome management of unstructured meshes, distributed memory parallelism and input/output of data. It is recognized that handling flexible object-oriented structures and carefully controlled memory-management, as would be required with the expected deepening of memory hierarchies in future hardware, is not easily achieved with the Fortran language, which is currently widely used in most NWP models. Hence, the language of choice for *Atlas* is C++, a highly performant language providing excellent object-orientation support, and building upon C's memory management proficiency. A Fortran2003 interface exports all of *Atlas*' functionality to Fortran applications, hence seamlessly introducing modular object-oriented concepts to legacy NWP models. Moreover, many other C++ based applications can directly benefit from *Atlas*.

2.2   Design

The application (e.g. the NWP model) starts by instructing *Atlas* to construct a *Grid* object, which provides a description of all longitude-latitude nodes in the model domain. The *Grid* object may not require significant memory in case it describes a structured grid in which simple formulas provide the longitude and latitude of every node on the sphere. In contrast, a *Mesh* object makes no assumption on any structure of the grid, and actually stores the horizontal coordinates of every node. Optionally, elements such as triangles, quadrilaterals and lines can be created by providing element-to-node connectivity tables. As the coordinates and the connectivity tables can have a significant memory footprint for large meshes, the *Mesh* is a distributed object, meaning that the mesh is subdivided in partitions which reside in memory on different parallel processes. With the *Grid* and the provision of a distribution scheme, *Atlas* generates a distributed *Mesh*. Specialized *FunctionSpace* objects can be created on demand, using the *Mesh*. A *FunctionSpace* describes in which manner *Fields*, objects that hold the memory of a full scalar, vector or tensor field, are discretized on the mesh. A straightforward *FunctionSpace* is the one where fields are discretized in the nodes of the grid. Other *FunctionSpace* objects could describe fields discretized in cell-centres of triangles and quadrilaterals, or in edge-centres of these elements. Yet another type of *FunctionSpace* can describe spectral fields in terms of spherical harmonics. *Fields* are stored contiguously in memory as a one-dimensional array and can be mapped to an arbitrary indexing mechanism to cater for the specific memory layout of NWP models, or a different memory layout that proves beneficial on emerging computer hardware. *Fields* are addressed by user-defined names and can be associated to *Metadata* objects, which store simple information like the units of the field or a time stamp. It is this flexibility and object-oriented design that leads to more maintainable and sustainable future-proof code. Figure 2 illustrates the object-oriented design and the relevant classes.

2.3   Massively Parallel Distribution

The MPI parallelisation relies on the distribution of the computational mesh with its unstructured horizontal index. The *Atlas* framework is responsible for generating a distributed mesh, and provides communication patterns using MPI to exchange information between the different partitions of the mesh. To minimise the cost of sending and receiving data, the distribution of the mesh is based on an equal regions domain decomposition algorithm optimal for a quasi-uniform node distribution on surface of the sphere [5,6]. The equal regions domain decomposition divides the sphere into bands oriented in zonal direction, and subdivides each band in a number of regions so that globally each region has the same number of nodes. Notably, the bands covering the poles are not subdivided, forming two polar caps. Figure 3 shows the partitioning of a spherical mesh with ~6.6 million nodes, quasi-uniformly distributed with grid spacing ~9 km, in 1600 partitions, the anticipated number of MPI tasks for this model resolution. The vertical direction is structured and not parallelized.

Because global communication across the entire supercomputing cluster are foreseen to become prohibitively expensive (see Fig. 1), numerical algorithms may be required to limit communication to – even physically located – nearest

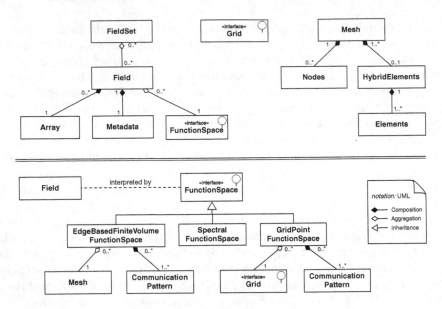

Fig. 2. Atlas data structure design.

Fig. 3. Equal regions domain decomposition (1600 partitions) for a fine mesh with ∼6.6 million nodes (∼9 km grid spacing).

neighbors. Such communication typically happens through thin halos (of 1 element wide) surrounding every mesh partition, hence creating an overlap with neighboring partitions. *Atlas* provides routines that expand the mesh partitions with these halos and provides communication patterns to update field values in the halos with field values from neighboring partitions. Figure 4 shows the equal regions domain decomposition into 32 partitions for a coarse mesh with ∼3500 nodes, quasi uniformly distributed with grid spacing ∼400 km, projected on a longitude-latitude domain. Figure 4 also illustrates the creation of internal and periodic halos. With periodic halos the periodicity in a longitude-latitude domain is treated exactly like any other internal boundary between different partitions. Notably, the partitions involving the poles are periodic with themselves.

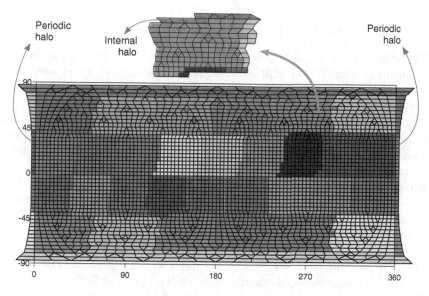

Fig. 4. Equal regions domain decomposition (32 partitions) for a coarse mesh with ~3500 nodes (~400 km grid spacing), projected on a longitude-latitude domain. Each partition is surrounded by a thin halo. Periodicity is provided through periodic halos.

3   Application in a Hybrid Finite Volume Module

As mentioned in Sect. 1, the NWP and climate models using the spectral transform method require global communication of large amounts of data. Realising the performance limitations of spectral transform methods at scale, a three-dimensional hybrid finite difference - finite volume module is developed [4], which only relies on nearest neighbor communication patterns as illustrated in Fig. 4. In the horizontal direction an unstructured median-dual edge-based finite volume method is used [7], while the vertical direction is discretized with a structured finite-difference method preferred with the vertically shallow nature of the atmosphere compared to the radius of the earth. Although capable to use arbitrary unstructured meshes in horizontal direction, bespoke meshes (such as in Fig. 4) based on reduced Gaussian grids [8,9] are generated, i.e. the location of its nodes coincides with those of the reduced Gaussian grid, which still facilitates the use of the spectral transform method. The generated mesh defines control volumes for a finite volume method while sharing the same data points as used in the spectral transform method. This approach allows a hybrid and evolutionary avenue to reach extreme scale, in which the finite volume module can initially be coupled to add extra functionality to the spectral transform model IFS, and could replace the spectral method when it appears to be advantageous in terms of performance, energy-efficiency, and forecast skill.

3.1   Memory Layout for Hybrid Parallelization

Additional performance can be gained using shared-memory parallelization techniques such as OpenMP, on multi-core architectures in which cores are distributed over nodes sharing the same memory. With recent hardware developments such as the Intel MIC's, the number of cores in one processor socket is increasing. The use of shared memory parallelisation when possible is often preferred to reduce the load with the MPI communications. The memory layout of fields plays an important role in optimal performance, especially with shared memory parallelization techniques.

In memory, a field is stored as a one-dimensional array that can be reinterpreted as a multi-dimensional array. The memory layout of a full 3D field can hence be reinterpreted with the horizontal unstructured index as the slowest moving index, followed by the structured vertical index, and the fastest index being the number of variables a field contains (e.g. scalar = 1, vector = 3, tensor = 9). For a scalar field, the "variable" index can be ignored. The memory layout for a three-dimensional scalar field is sketched in Fig. 5.

The advantage of this memory layout is twofold. First, it makes the vertical columns contiguous in memory, so that a halo-exchange involves contiguous chunks of memory and makes the packing and unpacking of send/receive buffers more efficient. Second, it favours the outer loop to be over the columns and

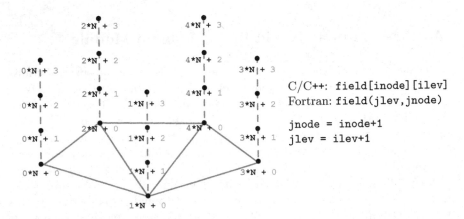

Fig. 5. Memory layout of a scalar field in 3D domain: solid lines show the underlying unstructured mesh; dashed lines mark the structured vertical columns with N denoting the number of vertical levels; and dots, numbered with memory offsets from the first index of the 1D array, represent the field values.

the inner loop to be over the levels within each column, with corresponding indices jnode and jlev, Fig. 5. Due to the unstructured nature of the horizontal jnode index, indirect addressing is required to access neighboring column data. The horizontal index for the outer loop then reduces the cost of this lookup by reusing the node specific computations for the entire column in the inner loop, giving the compiler the opportunity to optimise the inner loop in the vertical direction further with vector instructions, provided that computations for each vertical level are independent of each other. By using shared-memory parallelization with OpenMP over the outer horizontal index, further need for distributing the mesh is avoided.

3.2   Three-Dimensional Compressible Non-hydrostatic Dynamics

To illustrate the current status of the hybrid finite volume module, a baroclinic instability test case [10] is simulated to day 8 with results shown in Fig. 6. The simulation is three-dimensional and uses compressible non-hydrostatic dynamics. For a full description of the methodology, the reader is referred to [4].

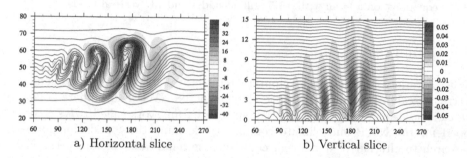

a) Horizontal slice   b) Vertical slice

Fig. 6. Simulation result using hybrid finite volume module for baroclinic instability test case [10] after 8 days. (a) Horizontal slice at surface showing meridional velocity (m/s). (b) Vertical slice at latitude ~53° N (X-axis in degrees, Y-axis in km) with shading showing vertical velocity (m/s) and solid contour lines showing isentropes (5 K interval).

The hybrid finite volume module is expected to perform well at extreme-scale due to local nearest neighbor communication patterns. Preliminary strong scaling results are reported in Fig. 7. These preliminary strong scaling results show acceptable promise with a parallel efficiency of 82 % at 40,000 CPU cores on ECMWF's Cray XC-30 HPC. It is expected that further improvement is possible by overlapping communication and computation, and by more carefully grouping communication buffers together.

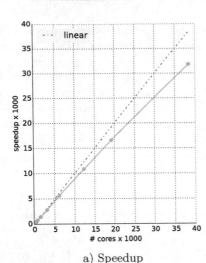
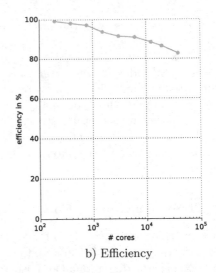

a) Speedup   b) Efficiency

Fig. 7. Preliminary strong scaling results for the hybrid finite-volume module on ECMWF's Cray XC-30 HPC. The results are created using a setup for solving hydrostatic equations, on a mesh with ∼5.5 million nodes and 137 vertical levels.

4 Remarks

With a variety of emerging computer architectures to tackle the challenge of extreme scale computing, a drastic change in the NWP and climate software stack is needed. A new model and datastructure framework called *Atlas* is developed for the NWP and climate community, forming the foundation for new research to alternative dynamical cores on one hand, and for an evolutionary modernisation of legacy NWP and climate codes on the other hand. During the ESCAPE project, this framework will be further developed and serve as a common infrastructure for leveraging the anticipated integration effort of the developments by the various partners involved. At the ECMWF, *Atlas* is already being succesfully used in a hybrid finite volume module. This finite volume module already shows promising scaling results, relying on *Atlas* for distributed memory parallelization and data management.

Acknowledgments. This work was supported in part by funding received from the European Research Council under the European Union's Seventh Framework Programme (FP7/2012/ERC Grant agreement no. 320375), and in part by the CRESTA project that has received funding from the European Community's Seventh Framework Programme (ICT-2011.9.13) under Grant Agreement no. 287703.

References

1. Sutter, H.: The free lunch is over: a fundamental turn toward concurrency in software. Dr. Dobb's J. **30**, 202–210 (2005)
2. Asanovic, K., Bodik, R., Catanzaro, B.C., Gebis, J.J., Keutzer, K., Patterson, D.A., Plishker, W.L., Shalf, J., Williams, S.W., Yelick, K.A.: The Landscape of Parallel Computing Research: A View from Berkeley. Technical Report, UC Berkeley (2006)
3. Wedi, N.P., Hamrud, M., Mozdzynski, G.: A fast spherical harmonics transform for global NWP and climate models. Mon. Weather Rev. **141**, 3450–3461 (2013)
4. Smolarkiewicz, P.K., Deconinck, W., Hamrud, M., Kühnlein, C., Mozdzynski, G., Szmelter, J., Wedi, N.P.: A hybrid all-scale finite-volume module for stratified flows on a rotating sphere. J. Comput. Phys. (2016, submitted)
5. Leopardi, P.: A partitioning of the unit sphere of equal area and small diameter. Electron. Trans. Numer. Anal. **25**, 309–327 (2006)
6. Mozdzynski, G.: A new partitioning approach for ECMWF's integrated forecasting system (IFS). In: Proceedings of the Twelfth ECMWF Workshop: Use of High Performance Computing in Meteorology, 30 October - 3 November 2006, Reading, UK, pp. 148–166, pp. 259–273. World Scientific (2007)
7. Szmelter, J., Smolarkiewicz, P.K.: An unstructured mesh discretisation in geospherical framework. J. Comput. Phys. **229**, 4980–4995 (2010)
8. Hortal, M., Simmons, A.J.: Use of reduced Gaussian grids in spectral models. Mon. Weather Rev. **119**, 1057–1074 (1991)
9. Wedi, N.P.: Increasing horizontal resolution in numerical weather prediction and climate simulations: illusion or panacea? Phil. Trans. R Soc. A **372**, 20130289 (2013). doi:10.1098/rsta.2013.0289
10. Jablonowski, C., Williamson, D.L.: A baroclinic instability test case for atmospheric model dynamical cores. Q. J. Roy. Meteorol. Soc. **132**, 2943–2975 (2006)

Scaling the GCR Solver Using a High-Level Stencil Framework on Multi- and Many-Core Architectures

Milosz Ciznicki[✉], Michal Kulczewski, Piotr Kopta, and Krzysztof Kurowski

Poznań Supercomputing and Networking Center, Poznań, Poland
{miloszc,kulka,pkopta,krzysztof.kurowski}@man.poznan.pl

Abstract. The recent advent of novel multi- and many-core architectures forces application programmers to deal with hardware-specific implementation details and to be familiar with software optimization techniques to benefit from new high-performance computing machines. An extra care must be taken for communication-intensive algorithms, which may be a bottleneck for forthcoming era of exascale computing. This paper aims to present a high level stencil framework implemented for the EULAG model that efficiently utilizes heterogeneous clusters. Only an efficient usage of both CPUs and GPUs with the flexible data decomposition method can lead to the maximum performance that scales communication-intensive elliptic solver with preconditioner.

Keywords: Stencils · CPU-GPU · SMP · NUMA · Elliptic solver

1 Introduction

The recent advent of novel multi- and many-core architectures, such as GPU and hybrid models, offer notable advantages over traditional supercomputers [1]. However, application programmers have to deal with hardware-specific implementation details and must be familiar with software optimization techniques to benefit from new high-performance computing machines. It is therefore of great importance to develop expertise in methods and algorithms for porting and adapting the existing and prospective modeling software to these new, yet already established machines.

Elliptic solvers of an elastic models are usually based on standard iterative algorithms for solving linear systems, e.g. CG, GMRES or GCR. Numerous reports on porting them to modern architectures are available [2]. However, in an anelastic solver for geophysical flows fast-acting physical processes may enter the elliptic problem implicitly. For simulating physical experiments with a high degree of anisotropy, additional preconditioning is necessary to improve matrix conditioning. Such preconditioner for anisotropic geometries often relies on the direct inversion using the Thomas algorithm. A comprehensive study on implementations of tridiagonal solvers on GPU found that it is possible to implement solvers which perform exceptionally well in the range of grid nodes [3].

© Springer International Publishing Switzerland 2016
R. Wyrzykowski et al. (Eds.): PPAM 2015, Part II, LNCS 9574, pp. 594–606, 2016.
DOI: 10.1007/978-3-319-32152-3_55

EULAG [4], an elastic model for simulating low Mach number flows under gravity, developed in the National Center of Atmospheric Research, is widely used in an international community and has a rich portfolio of applications. It features non-oscillatory forward-in-time (NFT) numerics, which are original and unique. It also employs preconditioned, nonsymmetric, generalized conjugate-residual type "Krylov" scheme [5] to solve an elliptic boundary value problem - reported to be among the most effective methods for solving difficult elliptic problems.

Our research is to provide novel methods to adapt scientific code to novel hardware architectures, taking EULAG as an example. As a result, a stencil framework is proposed that flexibly distributes data between CPUs and GPUs. In this paper we focus on efficient communication methods with efficient load balancing to scale the elliptic solver along with the preconditioner.

2   Related Works

A number of previous works have been focused on accelerating stencil computations on GPUs [6–11]. The works presented in [6, 7] used auto-tuning techniques to efficiently parallelize stencils on multi-core CPU and on many-core GPU. Other approaches employ multiple GPUs in solving stencil computations. They treat each GPU as an accelerator associated with separate MPI process where CPU acts as a management entity that does only the communication. Authors in [12] employed compiler based approach for automatic parallelisation of a code written in a domain specific language into CUDA. The work in [8] distributes stencil computations across multiple GPUs with explicit attention to the PCI express configuration.

Our work is more related to approaches that utilise multiple CPUs and GPUs simultaneously. Previous systems distributed stencil computations with simple decomposition methods with uniform partition where each processor and accelerator receives subdomains of the same size. For example, work in [9] utilises a high-level problem description to parallelise the code on the CPU and GPU clusters by combining OpenMP and CUDA. On the other hand, authors in [13] provide a framework that allows programmers to partition the data contiguously between CPU and GPU within single node. Unlike our work, their approach does not allow to carefully load balance the domain decomposition between heterogeneous architectures.

3   Description of the GCR Solver

The body of the elliptic solver code consists of five major routines (Fig. 1). The main routine advances the solution iteratively by calling other major computational routines. The routines *prforc* and *divrhs* initialize the solver. The former routine evaluates the first guess of the updated velocity, by combining the explicit part of the solution and the estimation of the generalized pressure gradient, while imposing an appropriate boundary condition. The latter routine evaluates the density weighted divergence of the velocity, and thus the initial residual error of the elliptic problem for pressure is computed. Among the most computationally intensive routine of the GCR solver is *laplc* that iteratively evaluates the generalized Laplacian operator (a combination of divergence and gradient) acting on residual errors. Another important part of the solver is the *precon* routine that accelerates the convergence of the variational scheme. By performing the direct matrix inversion in the vertical dimension of the grid, it is especially useful for large-scale simulations on thin spherical shells with grids characterized by a large anisotropy. The routine *precon* employs the sequential Thomas algorithm to solve tridiagonal systems of equations with the right hand side consisting of the horizontal divergence of the generalized horizontal gradient. This gradient is evaluated by *nablaCnablaxy*, which also belongs to the most computationally intensive routines of the GCR solver. With regard to the data access pattern, the computational loops within the elliptic solver can be simply divided into three categories: (i) reductions, (ii) implicit methods of the Thomas algorithm, and (iii) explicit methods of the stencils. In our work we focus on the explicit methods.

```
prforc();divrhs();precon();reduction();laplc()
for it=1..solver_iterations {
    reduction()
    if(exit) quit_for_loops;
    precon();laplc()
```

Fig. 1. The body of the elliptic solver code

4   High-Level Stencil Framework

This section describes the design goals of the proposed framework. The major objectives during development are described as follows.

The GCR solver code was ported from the FORTRAN 77 language. The framework is based on a domain-specific language (DSL) that expresses the stencil computations with similar to the Fortran semantics to ease the transition for the end users. The usage of the standard C++ language allows to avoid the non-standard language expression, non-standard programming models and requires no external libraries. The flexible architecture of the framework is suited for the computations on different configurations of the clusters that contain diverse number of CPUs and GPUs within the computational node. The framework is able to efficiently spread computations on the CPU-only clusters including the SMP machines, on the GPU-only clusters with fat nodes containing one GPU per single CPU core as well as on the hybrid clusters with powerful CPUs and GPUs. It provides the seamless subdomain decomposition to the blocks to efficiently utilize processors' memory hierarchy. Additionally, it allows the end user to manually tune the blocking size for the future processor architectures.

The framework contains the communication library with the unified interface that allows for the efficient intra-node and inter-node communication. Depending on the decomposition of a computational domain between processors, on a position of the actual processor and on the periodicity of the boundaries the library transparently chooses the most efficient communication method. The automatic parallelization for multi-CPU, multi-GPU and hybrid resources allows the users to write stencil functions that are translated to selected architectures. The usage of the C++ templates and provision of as much as possible of static information for the compiler improves the optimization during compilation. With the new generations of compilers the code will scale for the future architectures.

4.1 Programming Model

This section describes the implementation of the framework. Firstly, the structure of the framework is outlined with the initialization of the necessary resources. Than, the methodology of the writing and running user stencils with an example is given. Lastly, the domain decomposition method with the communication model is illustrated.

Framework Structure. The idea behind the parallelization of the computation between the processors is based on the data decomposition where each process updates the fixed part of the global domain called a subdomain. Since, the stencil computations require the neighbor points to update a point, the boundaries of the subdomains have to be communicated between processors. The communicated boundaries are saved in a designated buffer called a halo region. In order to efficiently utilize the data locality the OpenMP and MPI models are employed for the intra-node and inter-node communication. Each CPU and GPU have assigned separate MPI processes that are pinned to the selected cores. The GPU parallelization is done using CUDA whereas the CPU parallelization employs the OpenMP model.

Initialization. During the initialization of the framework user has to create the computational subdomain for each MPI process by using the `Geometry` and `Communicator` classes.

```
Geometry *geom = Geometry::init<DomainSize<NP,MP,LP>,
  HaloSize<HLEFT,HRIGHT,HBOT,HTOP,HGND,HSKY>, ProcessorGridSize<NPX,NPY,NPZ>>();
Communicator *comm = Communicator::init(geom);
```

The `Geometry` class initializes the 3D domain decomposition by using the `DomainSize`, `HaloSize` and `ProcessorGridSize` classes. The `ProcessorsGridSize` class specifies the number of the subdomains in each dimension. The decomposed subdomains may have different sizes with a restriction that each pair of the neighboring processors sharing boundary in the single dimension have the same boundary size in that dimension. The three values `NP`, `MP` and `LP` in `DomainSize` describe the size of the subdomain whereas the

HaloSize class characterizes the size of the halo region on each side. Furthermore, the Geometry class creates $n - 1$ OpenMP threads for CPU with n cores where the MPI processor of GPU is pinned to the $n-th$ core only if GPU is used. The Communicator class creates the specific communicator depending on the processor's architecture. To hide the communication time with the computation on GPU the communicator utilizes the CUDA streams to concurrently exchange the boundary data during the computation of the subdomain. The communicator based on the position of the processor within the global domain handles communication in a specific way. The processors inside the domain always communicate the data while the processors on the boundaries communicate the data for the periodic boundaries only and do not communicate them for the non-periodic boundaries. Additionally, for the decomposed domain with the single processor in a given dimension the data is exchanged using only the processor's local memory.

Stencils. The task of computing the stencils can be essentially divided into two parts. First, the stencil with an access pattern updating the domain point has to be defined. Second, the range within the computational domain on which stencil will be executed has to be provided. To enable this in the framework, the user defines the stencil functions and executes them through a kernel.

Writing Stencils. In the framework the stencils are defined as the C++ functors called the stencil functions. The 3D Laplacian function is defined as follows:

```
struct LaplcStencil {
 DEFINE_DO(const T *__restrict__ in_p,T *__restrict__ out_p,const T &cCoeff){
  IN3D(out_p, 0, 0, 0) = cCoeff * (
   IN3D(in_p, -1, 0, 0, CACHED) + IN3D(in_p,  1, 0, 0, CACHED) +
   IN3D(in_p,  0,-1, 0, CACHED) + IN3D(in_p,  0, 1, 0, CACHED) +
   IN3D(in_p,  0, 0,-1, CACHED) + IN3D(in_p,  0, 0, 1, CACHED)); }};
```

The DEFINE_DO macro allows to quickly define the functor. The stencil access pattern on the 3D domain is described by using the IN3D macro. There also exists IN1D and IN2D macros that allows to operate on the 1D and 2D domains. The functor through template parameters passes information about the domain dimensions and index parameters i, j, k to the macros. The first parameter of the IN3D macro takes a pointer to an array, the parameters from the second to the fourth take the positions of the domain point related to currently updated point. For example, IN3D(p, -1, 2, 0) returns indices of $(i-1, j+2, k)$. The last parameter of the IN3D macro is optional and gives a hint to the framework that the following point should be cached in the shared memory of GPU to improve efficiency. The decision which points which array should be cached is based on how many points of a given array is accessed. Typically, the array with the largest number of accessed points should it's values have cached to reduce the number of main memory accesses. However, if only single point is accessed per array the CACHED macro should not be used as it would degrade performance. Due to the small size of the shared memory the points of the single array can be

cached at time. The function parameters of stencil functions must begin with the
pointer to the array that is cached. In case of 3D Laplacian example the pointer
to the in_p is first. The details of the algorithm that does stencil computations
can be found in our previous work [14].

Running Stencils. In order to apply the stencil function the framework provides
the `kernel_conf` function that is used to invoke the 3D Laplacian on the 3D
computation domain as follows:

```
kernel_conf<type_t, DomainSize<NP,MP,LP>,
  StencilSize<SLEFT,SRIGHT,SBOT,STOP,SGND,SSKY>,
  HaloSize<HLEFT,HRIGHT,HBOT,HTOP,HGND,HSKY>,
  ComputeRegion<0+SLFET,NP-1-SRIGHT,0+SBOT,MP-1-STOP,0+SGND,LP-1-SSKY>,
  LaplcStencil, updateInner, Cache<TRUE,CSIZE>>(in_p, out_p, cCoeff);
```

The `kernel_conf` function is initialized with the type of the floating-point
calculations `type_t` such as float or double. Similarly to the `Geometry` class the
user provides `DomainSize` and `HaloSize`. The `StencilSize` describes the number of accessed neighboring points on each side of the stencil function. In our
example the 3D Laplacian provides `StencilSize<1,1,1,1,1,1>`. To restrict
the computation region of the stencil function the `ComputeRegion` class with
the range parameters is used. The stencil function will be applied from the
`SLEFT` index to the index `NP-1-SRIGHT` where NP is size of the subdomain in
the i direction. The `updateInner/updateLeft/`... enum allows to parallelism
the update of the boundary points with the inner points of the subdomain by
the utilization of the GPU streams. In our 3D Laplacian stencil example the
inner points use `updateInner` whereas the boundaries updates are modeled with
`updateLeft/updateRight`... so that the separate kernel calls are utilized for
each side. The `Cache` class drives the usage of the cache. The `TRUE` and `FALSE`
macros switch on and off the caching of the neighboring points of the stencil function, respectively. The cache size is automatically determined by the framework.
However, if needed the optional parameter `CSIZE` allows to manually control the
cache size in bytes. The order of the parameters of `kernel_conf` must be the
same as in the stencil function. The `kernel_conf` function executes the stencil
functions using OpenMP for CPU while for GPU the CUDA kernel functions
are called.

Domain Decomposition. There are number of the different decomposition
strategy options available. Typically systems use MPI-all parallelization scheme
with an uniform partition where each individual core maps to the MPI process
with no utilization of the shared memory on a compute node. This scheme is
very simple to implement and straightforward to run as it requires no knowledge
about the NUMA topology of the physical node. On the other hand, the number
of MPI messages required to exchange is a multiple of the number of cores thus
the communication overhead is substantial. Another choice is a strategy which
assigns single MPI process to the whole node. It decomposes the obtained sub-
block for a specified number of processors and accelerators and minimizes the

number of MPI processes thus the communication overhead, see the methodology described in [15]. The drawback of this method that the inner part of the sub-block is only decomposed in one dimension hence it is not flexible in balancing the load between the accelerators and processors. Additional strategy performs an uniform decomposition where pair of CPU and GPU are mapped to single MPI process. In this case the boundaries of the subdomain are updated and communicated by CPU whereas the inner points are handled by GPU. In this scenario CPU serves as a management entity and does not execute any computations thus it inefficiently uses CPUs. Our framework utilizes single MPI process per each processor with a flexible and efficient decomposition strategy to make best use of the hybrid architectures. The scheme can partition the domain in all three dimensions to non-uniform sub-blocks for the arbitrary number of the processors and accelerators. This scheme enables to compute on different cluster configurations that contain diverse number of CPUs and GPUs within the computational node. The framework is able to efficiently decompose the domain on the CPU-only clusters including the SMP machines, on the GPU-only clusters with fat nodes containing one GPU per single CPU core as well as on the hybrid clusters with powerful CPUs and GPUs. The partition mechanism is employed once before the compilation of the code for the target architecture thus the obtained decomposition is static during computations. This static decomposition allows the compiler to optimize the code for the stencil loops by utilization various techniques such as loop unrolling and vectorization. Once the subdomain for each processor is obtained, it is further decomposed to the optimal size for the cache blocking thus receiving the optimal size for the processor. The details of the subdomain decomposition is described in our previous work [14].

Communication and Computation Overlap. The framework utilizes the MPI communication to exchange messages between processors. The halo data transfer between the accelerators is conducted through CPU. CPU acts as a bridge that receives the data from the GPU then packs it to the MPI message and send to the host CPU of the target accelerator. The host CPU unpacks data and transfer it through PCI express to the target GPU. Please notice that the framework currently does not support GPUDirect RDMA to directly exchange data between GPUs located on different nodes without using CPUs as we did not have access to the cluster that supports it. The framework provides the aforementioned `Communicator` class to transfer data between processors and accelerators. The example usage of the class is showed below:

```
Communicator *comm = Communicator::init(geom);
comm->update(p_in,sizeLeft,shiftLeft,sizeRight,shiftRight,updateLeft);
// or
comm->update_beg(p_in,sizeLeft,shiftLeft,sizeRight,shiftRight,updateLeft);
// kernel execution
comm->update_end(p_in,sizeLeft,shiftLeft,sizeRight,shiftRight,updateLeft);
```

The class provides two methods of sending MPI messages: synchronous and asynchronous. The update method is used to apply synchronous communication whereas for the asynchronous communication type the pair of methods `update_beg` and `update_end` are exploited. The kernel execution is surrounded by asynchronous MPI calls to overlap communication with the computation. The `Communication` class is able to flexible exchange boundaries on each side of the domain. The `update` method requires six parameters where `p_in` is pointer to the 3D array containing data. The following pair of parameters defines the size and the shift of the boundary data received from the left processor whereas the next pair determines the size and the shift of the boundary data send to the right processor. The last parameter of the `update` method specifies the side of the update in this case the left update. For example, the domain is decomposed to the two processors in the i direction. There are defined two stencil functions on the domain. The computation range of the first stencil called the boundary stencil is constrained to left boundary of the domain while the second stencil call the inner stencil updates the inner points of the domain. The boundary stencil models the boundary condition and requires single point from the left shifted by one position, it is the `i-2` index. The inner stencil demands single point from the left with the `i-1` index. To fulfill the stencils requirements for the left and right processors the `update` method is called as follows:

```
// left processor call
comm->update(p_in,1,-1,1,0,updateLeft);
// right processor call
comm->update(p_in,1,0,1,-1,updateLeft);
```

To efficiently scale the code on large number of processors and accelerators the framework utilizes the overlapping of the communication with the computation. The idea of the overlapping is based on the separation of the computation of the boundary regions and the inner region. The separate kernel calls are employed for each side and an inside of the subdomain. The communication of the boundaries with the computation of the inside of the subdomain is overlapped and what is more the boundary kernels are computed in parallel to more efficiently utilize the accelerator resources. Figure 2 shows the flow of the overlapping method based on GPU of the 3D Laplacian stencil. To concurrently update the boundaries with the inside of the subdomain the seven streams are utilised. In case of more sophisticated stencils up to 27 streams are used. The kernel with the zero-copy memory is used to copy boundary data from GPU to the host CPU. The kernel with the zero-copy memory allows us on the fly change the 3D layout of the boundary data to linear ordered without using an intermediate buffer. The linear ordered data is directly passed to the MPI send function. The order of copying the boundaries is as follows first, the left and right boundaries are copied, next the bottom and top boundaries, and finally the ground and sky boundaries. The order is specified by the time needed to send the data through PCI express. The kernel updating inner of the subdomain is executed concurrently with the copy kernels. After the first boundary is copied to CPU it is send through MPI to the proper processor, depicted as Communication on the Figure. As soon as the halo region is received it is send back to GPU. Finally, the stencil updating the boundary is executed. This overlapping methodology allows us to concurrently execute five events: copy from and to GPU, compute the inner subdomain, MPI communication and compute the boundaries. Depending on the decomposition of the computational domain between processors, on the position of the actual processor and on the periodicity of the boundaries the library transparently chooses the most efficient communication method. The processors inside the domain always communicate the data while the processors on the boundaries communicate the data for the periodic boundaries only and do not communicate them for the non-periodic boundaries. Additionally, for the decomposition with the single processor in a given dimension the data is exchanged using only the processor's local memory.

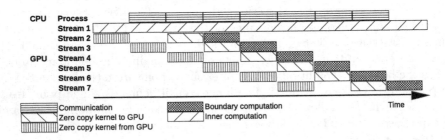

Fig. 2. Flow of the overlapping method based on GPU.

MPI Scheduler for NUMA. With the novel multi- and many-core architectures it is of great importance to place properly all process and threads on the underlying hardware. In [16] we proposed a method for mapping application topology to cluster. We used the MPI ping-pong benchmark to measure the sustained latency and bandwidth between all nodes to calculate the cost matrix. Next, the minimum path is calculated and eventually, the application topology is mapped to the hardware, using the Hilbert curve to calculate the spatial locality. We have extended this functionality especially to NUMA architectures, taking into account proper placement of OpenMP threads across system. The developer can rely on Intel or GNU facilities for proper thread placement, KMP and GOMP environments variables respectively. It allows users to give a hint to system on which cores OpenMP threads should be placed, as well as not to move threads between cores during the run. To pin MPI processes accordingly, one have to rely on Intel or OpenMPI facilities, which are not always available. To automate processes and threads placement on underlying hardware, we use the HWLOC, a portable hardware locality software package that provides a portable abstraction of the hierarchical topology of modern architectures. Using aforementioned HWLOC, we calculate distances between each pair of NUMA-nodes (which may by whole node in traditional cluster, socket or a processor in more SMP-like environment). Next, we find the minimum path as previously, and using the Hilbert curve we place each MPI process and its OpenMP threads on cores, where for most cases one MPI process per NUMA node is the most efficient allocation.

5   Experimental Evaluation

In this section, the strong and weak scaling results are presented using the GCR solver on the Piz-Daint cluster and on the Chimera SMP machine. The Piz-Daint cluster contains 5272 nodes equipped with an Intel Xeon E5-2670 CPU with 32 GB of RAM and Nvidia Tesla K20X GPU. The Chimera SMP machine has 2048 CPU cores with Intel Xeon E7-8837 CPUs clocked at 2.66 GHz with 16 TB of RAM.

The implementation of the GCR solver is validated using a standard benchmark test case for incompressible flow solvers. We simulate decaying turbulence of a homogeneous incompressible fluid. Here, only the simplified setup proposed by Taylor and Green is considered. The details of the problem can be found in our previous work [14].

5.1   Weak Scaling

The GCR solver is tested using five different versions of the code. The Fortran CPU MPI-all version is the original code developed in FORTRAN 77. The rest variants including C++ CPU MPI+OpenMP, C++ OpenMP, C++ GPU MPI+CUDA and C++ CPU-GPU MPI+OpenMP+CUDA are implemented with our framework. In order to evaluate the performance of all codes the number of floating-point operations are counted by calculating their occurrence in the source code. The MPI ranks in the MPI-all code are pinned to individual cores whereas for the MPI+OpenMP version single MPI rank is used for each CPU. The work is distributed across cores by using the OpenMP threads. In case of the GPU code single MPI rank is used. For the heterogeneous CPU-GPU case two MPI ranks are pinned to single CPU. First MPI rank executes seven OpenMP threads where second MPI rank is pinned to last CPU core and handles the execution of GPU. For ccNUMA architecture, OpenMP threads are distributed across the all allocated cores. For the MPI+OpenMP version, MPI ranks are placed on separate sockets (aka. NUMA nodes), with the corresponding set of OpenMP threads. Figure 3 shows the weak scaling results. The all codes almost reach the perfect linear scaling up to 512 nodes. Using 8 CPUs on the 488^3 domain size the C++ MPI+OpenMP version is 1.4× faster than the Fortran MPI-all code. Moving the code to GPUs for the same domain leads to 6× speedup comparing to the original Fortran code. The heterogeneous CPU-GPU code further improves speedup to 7× by distributing subdomains to CPUs.

5.2   Strong Scaling

Figures 3 and 4 show the strong scaling results for a fixed grid size 244^3 with the varying number of processors and accelerators for two-dimensional domain decomposition along the y and z directions. The results are presented for the best domain decompositions for all code variants. The CPU codes scale up to more than 100 nodes, however the run with GPU saturates at 128 GPUs count. In order to efficiently use the GPU resources it requires a minimal number of

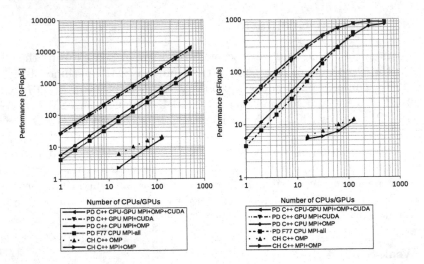

Fig. 3. Left: weak scaling for the 244^3 domain size per CPU/GPU. Right: strong scaling for the total 244^3 domain size. Used machines: PD - Piz-Daint supercomputer, CH - Chimera SMP.

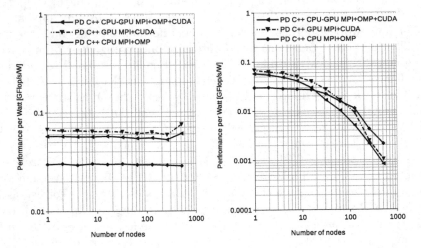

Fig. 4. Left: weak scaling for the 244^3 domain size per CPU/GPU. Right: strong scaling for the total 244^3 domain size. Used machines: PD - Piz-Daint supercomputer, CH - Chimera SMP.

domain points to saturate the memory bandwidth. The similar saturation can be observed with the CPU-GPU code.

6   Conclusions

In this work, the stencil framework is presented that utilizes a domain specific language to simplify the development of a stencil computations on heterogeneous architectures. The framework is written with C++ templates that provide portable code with no need for the additional dependencies. The C++ templates with the static domain decomposition allows the compiler to efficiently optimize the prepared code. The flexible domain decomposition scheme with the subdomain partition to fit the memory hierarchy of the target architecture supports load balancing the work between an arbitrary number of CPUs and GPUs. The resulting code with the communication overlap method achieves high scalability and a 7× speedup against the Fortran MPI-all code. The framework can be used with a good outcome on both the SMP machines and the heterogeneous CPU-GPU clusters.

The results from the evaluation tests showed that the heterogeneous cluster configurations promise relatively large energy savings. For future work, we want to develop the scheduling methods to dynamically allocate stencil tasks to various unit blocks to optimize the energy efficiency.

Acknowledgements. This work is supported by the Polish National Center of Science under Grant No. UMO-2011/03/B/ST6/03500. This research was supported in part by PL-Grid Infrastructure. This work was supported by a grant from the Swiss National Supercomputing Centre (CSCS) under project ID d25.

References

1. Kurzak, J., Bader, D., Dongarra, J.: Scientific Computing with Multicore and Accelerators. Computer and Information Science Series. Chapmann & Hall/CRC, Boca Raton (2010)
2. Georgescu, S., Okuda, H.: Conjugate gradients on multiple GPUs. Int J. Numer. Meth. Fluids **64**, 1254–1273 (2010)
3. Zhang, Y., Cohen, J.M., Owens, J.D.: Fast tridiagonal solvers on GPU. In: Newsletter ACM SIGPLAN Notices - PPoPP, vol. 45, p. 5 (2010)
4. Prusa, J.M., Smolarkiewicz, P.K., Wyszogrodzki, A.: Eulag a computational model for multiscale flows. Comput. Fluids **37**, 1193–1207 (2008)
5. Smolarkiewicz, P.K., Margolin, L.G.: Variational methods for elliptic problems in fluid models. In: Proceedings of ECMWF Workshop on Developments in Numerical Methods for Very High Resolution Global Models, vol. 7, pp. 137–159 (2000)
6. Kamil, S., Chan, C., Oliker, L., Shalf, J., Williams, S.: An auto-tuning framework for parallel multicore stencil computations. In: IEEE International Symposium on Parallel and Distributed Processing (IPDPS 2010), pp. 1–12. IEEE (2010)
7. Christen, M., Schenk, O., Burkhart, H.: Patus: a code generation and autotuning framework for parallel iterative stencil computations on modern microarchitectures. In: IEEE International Parallel and Distributed Processing Symposium (IPDPS 2011), pp. 676–687. IEEE (2011)
8. Lutz, T., Fensch, C., Cole, M.: PARTANS: an autotuning framework for stencil computation on multi-GPU systems. ACM Trans. Archit. Code Optim. (TACO) **9**(4), 59 (2013)
9. Blazewicz, M., Hinder, I., Koppelman, D.M., Brandt, S.R., Ciznicki, M., Kierzynka, M., Löffler, F., Schnetter, E., Tao, J.: From physics model to results: an optimizing framework for cross-architecture code generation. Sci. Program. **21**(1–2), 1–16 (2013)

10. Szustak, L., Rojek, K., Olas, T., Kuczynski, L., Halbiniak, K., Gepner, P.: Adaptation of MPDATA heterogeneous stencil computation to Intel Xeon Phi coprocessor. Sci. Program. (2015)
11. Wyrzykowski, R., Szustak, L., Rojek, K.: Parallelization of 2D MPDATA EULAG algorithm on hybrid architectures with GPU accelerators. Parallel Comput. **40**, 425–447 (2014)
12. Maruyama, N., Nomura, T., Sato, K., Matsuoka, S.: Physis: an implicitly parallel programming model for stencil computations on large-scale GPU-accelerated supercomputers. In: International Conference for High Performance Computing, Networking, Storage and Analysis (SC 2011), pp. 1–12. IEEE (2011)
13. Pereira, A.D., Ramos, L., Góes, L.F.: PSkel: a stencil programming framework for CPU-GPU systems. In: Practice and Experience, Concurrency and Computation (2015)
14. Rojek, K.A., Ciznicki, M., Rosa, B., Kopta, P., Kulczewski, M., Kurowski, K., Piotrowski, Z.P., Szustak, L., Wojcik, D.K., Wyrzykowski, R.: Adaptation of fluid model EULAG to graphics processing unit architecture. In: Practice and Experience, Concurrency and Computation (2014)
15. Xue, W., Yang, C., Fu, H., Wang, X., Xu, Y., Gan, L., Lu, Y., Zhu, X.: Enabling and scaling a global shallow-water atmospheric model on tianhe-2. In: IEEE 28th International Parallel and Distributed Processing Symposium, pp. 745–754. IEEE (2014)
16. Ciznicki, M., Kopta, P., Kulczewski, M., Kurowski, K., Gepner, P.: Elliptic solver performance evaluation on modern hardware architectures. In: Wyrzykowski, R., Dongarra, J., Karczewski, K., Waśniewski, J. (eds.) PPAM 2013, Part I. LNCS, vol. 8384, pp. 155–165. Springer, Heidelberg (2014)

Parallel ADI Preconditioners for All-Scale Atmospheric Models

Zbigniew P. Piotrowski[1]([✉]), Bartlomiej Matejczyk[2], Leszek Marcinkowski[3], and Piotr K. Smolarkiewicz[4]

[1] Institute of Meteorology and Water Management - National Research Institute, Podlesna 61, 01-673 Warsaw, Poland
zbigniew.piotrowski@imgw.pl

[2] Johann Radon Institute for Computational and Applied Mathematics, Linz, Austria
matejczyk.bartlomiej@gmail.com

[3] Faculty of Mathematics, Informatics and Mechanics, University of Warsaw, Warsaw, Poland
lmarcin@mimuw.edu.pl

[4] European Centre for Medium-Range Weather Forecasts, Reading, UK
piotr.smolarkiewicz@ecmwf.int

Abstract. Effective preconditioning lies at the heart of multiscale flow simulation, including a broad range of geoscientific applications that rely on semi-implicit integrations of the governing PDEs. For such problems, conditioning of the resulting sparse linear operator directly responds to the squared ratio of largest and smallest spatial scales represented in the model. For thin-spherical-shell geometry of the Earth atmosphere the condition number is enormous, upon which implicit preconditioning is imperative to eliminate the stiffness resulting from relatively fine vertical resolution. Furthermore, the anisotropy due to the meridians convergence in standard latitude-longitude discretizations becomes equally detrimental as the horizontal resolution increases to capture nonhydrostatic dynamics. Herein, we discuss a class of effective preconditioners based on the parallel ADI approach. The approach has been implemented in the established high-performance all-scale model EULAG with flexible computational domain distribution, including a 3D processor array. The efficacy of the approach is demonstrated in the context of an archetypal simulation of global weather.

Keywords: Deflation preconditioners · ADI · EULAG

1 Introduction

Modern numerical solvers, integrating complex and multiscale problems at the frontiers of geo- and astro-physics, constitute indispensable virtual laboratory for investigating inherently irreplicable phenomena, where traditional experimental approach becomes either impractical, infeasible or even entirely inapplicable. Many of these solvers target a particular range of scales and are essential for

© Springer International Publishing Switzerland 2016
R. Wyrzykowski et al. (Eds.): PPAM 2015, Part II, LNCS 9574, pp. 607–618, 2016.
DOI: 10.1007/978-3-319-32152-3_56

industrial or specialized scientific applications, such as Earth System Models for climate research, numerical forecasts suites from global- to meso-scale weather, down to small-scale CFD type models for cloud turbulence and wind engineering. In turn, multiscale solvers capable of integrating PDEs at scales from micro to stellar advance fundamental research in geo and astrophysics, contributing to understanding fundamental physical laws governing nonlinear phenomena. With the continuous advancement in supercomputing and the Big Data, spatial and temporal resolution increases and specialized applications are capable of capturing increasingly broader range of scales. This motivates further development and application of robust and accurate numerical techniques, capable of exploring finer scales admitted by improved resolution.

A notable example of such multiscale, multiphysics solver is an established model EULAG [12, 19],[1] recently extended to include a consistent soundproof-compressible formulation of PDEs governing atmospheric dynamics [20]. Highly efficient parallel formulation of EULAG with a 3D MPI domain distribution has been reported in [10].

Explicit integration of the fully compressible atmospheric equations is impractical due to the inherent stability limits related to rapid propagation of acoustic modes. Various numerical techniques, often relying on the operator splitting, are used to mitigate this problem. Effectively, they employ specialized implicit integration schemes applied to selected terms in the flow PDEs. Adopted mathematical simplifications may require solution filtering, possibly leading to numerical artefacts [9]. In turn, fully implicit techniques like Newton-Krylov solvers may be prohibitively expensive. Among the variety of existing methods, the recently reported consistent formulation of the EULAG numerics appear especially prospective as it allows for efficient multiscale integrations using robust forward-in-time integrators and preconditioned nonsymmetric Krylov-subspace solvers.

In realistic multiscale applications, the elliptic Poisson and Helmholtz boundary value problems (corresponding to soundproof and compressible formulation of the governing PDEs) are nonsymmetric, semi-definite, poorly conditioned and complicated. This may result from the anisotropy of grid resolution, planetary rotation, ambient large scale gradients and stratification, the use of curvilinear coordinates, or the imposition of partial-slip conditions along irregular lower boundaries. Despite enormous advancements in numerical methods for sparse linear systems, the effective solution of large linear problems depends (at least for the class of problems discussed) not necessarily on the choice of the best available iterative method, but rather on the artful preconditioning.

In global models, anisotropy of the Earth's atmosphere results in the condition number $\kappa \sim O(10^{10})$, or larger. Because the asymptotic convergence rate of conjugate gradient type methods is proportional to $\sqrt{\kappa}$, the use of directly invertible preconditioner that removes the stiffness associated with a relatively fine vertical resolution is imperative [17]. To date, the Generalized Conjugate Residual (GCR) Krylov solver of EULAG was supported by a deflation technique, in

---

[1] For comprehensive list of EULAG publications see model webpage at http://www2.mmm.ucar.edu/eulag/.

which splitting the preconditioner into implicit and explicit counterparts in the vertical and horizontal, respectively, lead to subsequent direct inversion in the vertical of the explicit horizontal part [22]. Variants with spectral decomposition in the horizontal were also considered [18,23].

ADI methods per se date back to the nineteen fifties [2], and their use as preconditioners for Krylov solvers in atmospheric models was advocated nearly two decades ago [15,22]. However, with the transition to massively parallel computing, parallel implementation of tridiagonal algorithms underlying the ADI approach has been dismissed (e.g., in favour of the deflation technique as specified above), especially that vertical direction in atmospheric codes remained standardly serial — for the sake of radiation and precipitation processes, inherently sequential in nature and difficult to parallelize. Notwithstanding, more recently it become clear that at the exascale computing mere parallelization in the horizontal may be insufficient. The recent development of the 3D parallelization in EULAG brought parallel tridiagonal algorithms and paved the way for high-performance ADI preconditioners. Further acceleration of variational Krylov solvers is important, in order to minimize their global reductions anticipated at the exascale computing. In this context ADI preconditioners appear promising and worthy exploring. This seems to be reflected in their ceaseless popularity in the current literature.

2   Model Framework

2.1   Analytic Formulation

The consistent formulation of EULAG's governing equations casts (and solves) the governing PDEs in generalized time-dependent curvilinear coordinates $(\bar{t}, \overline{\mathbf{x}}) \equiv (t, F(t, \mathbf{x}))$, where the coordinates $(t, \mathbf{x})$ of the physical space are orthogonal and stationary, but not necessarily Cartesian. In particular, global simulations employ the standard anholonomic latitude-longitude (lat-lon) spherical framework (Sect. 7.2 in [3]) for the physical space [12,19], in which components of the physical velocity vector are aligned at every point of the spherical shell with axes of a local Cartesian frame tangent to the lower surface of the shell; cf. Fig. 7.7 in [3]. For simplicity of the presentation, here we dismiss the time-dependency of the model coordinates; so, $(\bar{t}, \overline{\mathbf{x}}) \equiv (t, F(\mathbf{x}))$ in the formulae that follow. The governing equations for the physical velocity $\mathbf{u} = (u, v, w)$ and the potential temperature θ can compactly be written as

$$\frac{d\mathbf{u}}{d\bar{t}} = -\Theta\,\widetilde{\mathbf{G}}\overline{\nabla}\varphi - \mathbf{g}\Upsilon_B \frac{\theta'}{\theta_b} - \mathbf{f} \times (\mathbf{u} - \Upsilon_C \mathbf{u}_e) \ + \mathcal{M}'(\mathbf{u}, \mathbf{u}) + \overline{\mathcal{D}}_{\mathbf{v}}, \qquad (1)$$

$$\frac{d\theta'}{d\bar{t}} = -\overline{\mathbf{u}}^* \cdot \overline{\nabla}\theta_e + \overline{\mathcal{D}}_\theta, \qquad (2)$$

$$\frac{d\varrho}{d\bar{t}} = -\frac{\varrho}{\mathcal{G}}\overline{\nabla} \cdot \overline{\mathcal{G}}\overline{\mathbf{u}}^*. \qquad (3)$$

Here, the generalized density and pressure variables ϱ and φ are defined, respectively, for the [anelastic, compressible] PDEs as

$$\varrho := [\rho_b(z),\ \rho(\mathbf{x},t)]\ ,\quad \varphi := [c_p\theta_b\pi',\ c_p\theta_0\pi'], \tag{4}$$

where ρ and π denote the fluid density and the Exner pressure. The dimensionless coefficients are defined as

$$\Theta := \left[1,\ \frac{\theta(\mathbf{x},t)}{\theta_0}\right],\quad \Upsilon_B := \left[1,\ \frac{\theta_b(z)}{\theta_e(\mathbf{x})}\right],\quad \Upsilon_C := \left[1,\ \frac{\theta(\mathbf{x},t))}{\theta_e(\mathbf{x})}\right]. \tag{5}$$

On the LHS of (1) and (2), the total derivative $d/d\bar{t} = \partial/\partial\bar{t} + \overline{\mathbf{u}}^* \cdot \overline{\nabla}$, where $\overline{\mathbf{u}}^* = d\overline{\mathbf{x}}/d\bar{t}$ is the contravariant velocity in the computational space. The nabla operator $\overline{\nabla} \equiv \partial/\partial\overline{\mathbf{x}}$ represents the vector of partial derivatives corresponding to elementary finite differences in the model code. In the momentum Eq. (1), $\widetilde{\mathbf{G}}$ symbolizes a renormalized Jacobian matrix of the metric coefficients $\propto (\partial\overline{\mathbf{x}}/\partial\mathbf{x})$. The terms $\mathcal{M}(.,.)$ in (1) denote metric forces (viz. the generic Christoffel terms in the physical space), whereas $\mathbf{f}$ is the Coriolis acceleration, and $\mathcal{D}$ symbolizes the dissipation/diffusion operator. Subscript e refers to an ambient state, a particular solution to (1)–(3), while subscript b marks the anelastic base state [6]. The contravariant and physical velocities are related via

$$\overline{\mathbf{u}}^* = \widetilde{\mathbf{G}}^T\mathbf{u}, \tag{6}$$

and $\overline{\mathcal{G}}^2$ is the determinant of the metric tensor that defines the fundamental metric in the computational space.

2.2 Highlights of Numerical Approximations

Here we provide an outline of the nonoscillatory forward-in-time (NFT) numerical approximation strategy of EULAG. For comprehensive discussion and algorithmical details the interested reader is referred to [20] and references therein. The numerics of EULAG are unique, in that they rely on the Lagrangian/Eulerian congruence of governing equations integrated with the trapezoidal rule along flow trajectory in the 4D time-space continuum. In technical terms, however, the hydrodynamical solver of EULAG is reminiscent of the projection approach [1], where evaluation of an explicit part of the velocity $\mathbf{u}_{exp}$ is followed by the implicit completion of the solution, tantamount to the formulation and solution to the discrete boundary value problem (BVP) for pressure.[2] In particular, Eqs. (1)–(3) are cast in the conservation-law form of generalized transport equation for specific variable ψ (e.g. velocity component or potential temperature):

$$\frac{\partial \mathcal{G}\varrho\psi}{\partial\bar{t}} + \overline{\nabla} \cdot (\mathcal{G}\varrho\overline{\mathbf{u}}^*\psi) = \mathcal{G}\varrho R, \tag{7}$$

---

[2] This second step can be iterated when nonlinear terms resulting from, e.g., metric forces are present; cf. [19] for illustrative examples.

are integrated in time and space using MPDATA approach,[3] that provides a robust integrator for the entire system of the governing PDEs, including consistent formulation of a discrete BVP. The latter constraints the complete solution such as to assure that the updated solution satisfies the discrete mass continuity Eq. (3). In particular for the anelastic system, (3) takes an incompressible-like form

$$\frac{1}{\rho^*}\overline{\nabla} \cdot \rho^*\overline{\mathbf{u}}^* = 0,\tag{8}$$

where $\rho^* = \overline{\mathcal{G}}\rho_b$. therefore implying the Poisson BVP for φ

$$-\frac{\Delta t}{\rho^*}\frac{\partial}{\partial \overline{x}^j}\left[\rho^*\mathcal{E}\left(\widetilde{\mathcal{V}}^j - \widetilde{\mathcal{C}}^{jk}\frac{\partial \varphi}{\partial \overline{x}^k}\right)\right] = 0.\tag{9}$$

The problem in (9) can be though of as $\mathcal{L}(\varphi) - \mathcal{R} = 0$, wherein $\mathcal{E}$ and $\widetilde{\mathcal{C}}^{jk}$ denote 10 fields of known coefficients that generally vary in time and space; $\mathcal{E}\widetilde{\mathcal{V}}^j$ is the jth component of the explicit part of contravariant velocity solution; and repeating k indices imply the summation over the components of $\overline{\nabla}\varphi$; see [11] for the exposition. The multiplicative factor $-\Delta t/\rho^*$ assures the formal negative semi-definiteness of $\mathcal{L}$ and expresses the residual error $r = \mathcal{L}(\varphi) - \mathcal{R} \neq 0$ as the divergence of a local Courant number on the grid. Notably, for the compressible solver, the resulting Helmholtz BVP is composed of three Poisson-like operators and the term proportional to φ [20]. In either case, the resulting BVP is solved (subject to appropriate boundary conditions) using the GCR approach [4,21] — a robust preconditioned nonsymmetric Krylov-subspace solver akin to GMRES [14]. Given the updated pressure, and hence the updated contravariant velocity, the updated physical velocity components are constructed from (6).

A key element of our GCR machinery is the (left) operator preconditioning providing an estimate of the solution error $q = \varphi - \varphi_{exact}$ as

$$q = \mathcal{P}^{-1}(r),\tag{10}$$

where the preconditioner $\mathcal{P} \approx \mathcal{L}$ but is easier to invert than the $\mathcal{L}$. In EULAG, $\mathcal{P}$ closely matches $\mathcal{L}$ by only neglecting the cross derivative terms with coefficients $\widetilde{\mathcal{C}}^{jk}\big|_{k\neq j}$. In the remainder of the paper we elaborate on technical aspects of (10).

3   Implicit Inversion Preconditioning

3.1   Principles

Standard EULAG preconditioning relies on the direct inversion of the vertical component of the implicit operator, while evaluating horizontal part of the operator explicitly using a stationary Richardson iteration

$$\frac{q^{\mu+1} - q^\mu}{\Delta \tau} = P^z q^{\mu+1} + P^h q^\mu - r^\nu,\tag{11}$$

---

[3] MPDATA (for multidimensional positive definite advection transport algorithm) is a class of nonoscillatory forward-in-time flow solvers, widely documented in the literature; for a recent overview see [20] and references therein.

where P^z and P^h symbolize vertical and horizontal parts of $\mathcal{P} = P^z + P^h$; $\Delta\tau$ is a fixed pseudo-time step (selected such as to assure convergence of the iterative process; μ numbers preconditioner's iterations (usually a few to several), and r^ν is the residual error of the Krylov solver's νth iteration. Gathering all $q^{\mu+1}$ terms on the LHS, leads to the tridiagonal problem

$$(\Delta\tau^{-1}\mathcal{I} - P^z)q^{\mu+1} = \Delta\tau^{-1}q^\mu + P^h q^\mu - r^\nu := \widetilde{r}, \tag{12}$$

readily invertible with the Thomas algorithm concisely symbolized as

$$q^{\mu+1} = (\Delta\tau^{-1}\mathcal{I} - P^z)^{-1}\widetilde{r} \tag{13}$$

The developed ADI preconditioners enable extending the standard preconditioner (11) to admit implicitness in P^h and, thus, accelerate the convergence by increasing $\Delta\tau$. In particular, a two-dimensional ADI design leaves only single explicit direction. When operating on the global lat-lon grid, the natural choice is to treat the longitudinal direction implicitly, as the meridians converging towards poles can introduce considerable anisotropy in the coefficients of the horizontal operator P^h. While numerous formulations of a 2D ADI are possible, a particularly simple algorithm was provided by Peaceman and Rachford [8],

$$\begin{aligned}
\frac{q^{\mu+\frac{1}{2}} - q^\mu}{\Delta\tau_2} &= P^x q^{\mu+\frac{1}{2}} + P^z q^\mu &&+ P^y q^\mu - r^\nu, \\
\frac{q^{\mu+1} - q^{\mu+\frac{1}{2}}}{\Delta\tau_2} &= P^x q^{\mu+\frac{1}{2}} + P^z q^{\mu+1} &&+ P^y q^\mu - r^\nu, \tag{14}
\end{aligned}$$

where P^x and P^y symbolize, respectively, the longitude and latitude counterparts of $P^h = P^x + P^y$. To extend (14) to all three directions, we adopt the unconditionally stable 3D ADI Douglas [2] algorithm

$$\begin{aligned}
\frac{q^{\mu+\frac{1}{3}} - q^\mu}{\Delta\tau_3} &= P^x \frac{(q^{\mu+\frac{1}{3}} + q^\mu)}{2} + P^y q^\mu &&+ P^z q^\mu &&- r^\nu, \\
\frac{q^{\mu+\frac{2}{3}} - q^\mu}{\Delta\tau_3} &= P^x \frac{(q^{\mu+\frac{1}{3}} + q^\mu)}{2} + P^y \frac{(q^{\mu+\frac{2}{3}} + q^\mu)}{2} &&+ P^z q^\mu &&- r^\nu, \tag{15} \\
\frac{q^{\mu+1} - q^\mu}{\Delta\tau_3} &= P^x \frac{(q^{\mu+\frac{1}{3}} + q^\mu)}{2} + P^y \frac{(q^{\mu+\frac{2}{3}} + q^\mu)}{2} &&+ P^z \frac{(q^{\mu+1} + q^\mu)}{2} &&- r^\nu.
\end{aligned}$$

Grouping all implicit terms of (14) and (15) on the LHS, as in (12), shows that all three preconditioners share a common template algorithm

$$q^{\mu*} = (\Delta\widetilde{\tau}_i^{-1}\mathcal{I} - P^I)^{-1}\widetilde{r}^* \tag{16}$$

for the tridiagonal solver in $I = x, y, z$ directions at some intermediate step $\mu*$. $\widetilde{r}^*$ denotes the corresponding explicit elements of (14) and (15).[4] This substantiates the earlier assertion that the parallel implementation of the tri-diagonal inversion in one direction paves the way for the family of ADI preconditioners.

---

[4] Note that $\widetilde{\tau}_i = \tau_i$ for (11)–(14) but $\widetilde{\tau}_i = \tau_i/2$ in the (15).

3.2   Parallel Implementation

At the highest level, the implementation of algorithms (11)–(15) in EULAG follows the mathematical notation. Notably, as the first guess $q^{\mu=0} = 0$, the initlaisation of the algorithms needs bespoke script to avoid unnecessary memory references. In particular, since $\forall^I \, P^I q^{\mu=0} = 0$ in the first preconditioner iteration, it is only necessary to evaluate half of the explicit members of (11)–(15). Furthermore, the algorithms (11)–(15) can be judiciously mixed, whereupon we evaluate all members of the preconditioner family as they all can be useful in different classes of applications.

The BVP problems in EULAG rarely can be solved in single GCR iteration. For compressible equations the problem matrix is constant within given timestep. For anelastic equations in stationary coordinates it remains constant throughout the entire integration time. Inspection of the standard tridiagonal algorithm suggests that it is possible to precompute a good part of its forward step. Consequently, all variants of preconditioners in EULAG begin with a suitable initialization executed either once per simulation or time step, depending on the application at hand. This requires relatively large additional storage that is, however, rarely an issue, on the modern CPU clusters.[5]

The parallel tridiagonal inversion, if implemented in the form of naive parallel recurrence, suffers from the adverse load balance characteristics, i.e. only one core in the column may be active at the time. However, it is fairly easy to mitigate this problem if using two cores in vertical. With the given horizontal subdomain, at half of the gridpoints the forward tridiagonal sweep starts from one boundary while the remaining columns are evaluated from the forward sweep starting from the opposite boundary. Importantly, this technique does not noticeably alter the physical results. If the use of larger number of cores in the direction of tridiagonal inversion is needed and the cost of computations per core is significant, it is possible to use Pipelined Thomas Algorithm aiming at quickest possible fill of pipeline of computing cores in given direction. This is done by splitting the processor subdomain into chunks and communicating the results to the next core as soon as recurrence (acting in direction normal to the chunk surface) is executed; see Sect. 2 in [10] and references therein. While this strategy seems to be efficient for low speed computing cores, such as employed by IBM Bluegene/L architecture, we found the strategy in the spirit of recursive doubling [24] to be more effective on modern CPU clusters. The Thomas algorithm as employed for the A-grid discretization of EULAG for nonperiodic boundary conditions has the general form

$$e_k = A_k / (B_k - C_k e_{k-2}) \tag{17}$$

$$f_k = D_k(e_k) f_{k-2} + Q_k(e_k, rhs_{ADI}) \tag{18}$$

$$p_k = e_k p_{k+2} + f_k \tag{19}$$

---

[5] For the majority of geophysical applications, the main performance bottleneck is the memory bandwidth and access time, and not the main memory size. The latter is usually much larger than needed, as large number of timesteps and good scalability lead to routine use of hundreds of computing cores and tens of supercomputer nodes.

where k is indexes either x,y or z direction, A, B, C are known 3D matrices, D is a function of e_k and Q is a function of instantaneous right hand side rhs_{ADI} at a given step of an algorithm at hand from the suite (11)–(15). Because e_k can be precomputed, this allows to express last element of the recurrence by the first or the second element of the recurrence, for the last k index allowed being odd and even, respectively. For example, for the recurrence split in four-element segments

$$a_k = b_k a_{k-2} + c_k = b_k b_{k-2} a_{k-4} + b_k c_{k-2} + c_k. \tag{20}$$

The auxiliary product $b_k b_{k-2}$ and the remaining $b_k c_{k-2} + c_k$ term can be pre-computed at additional cost, but in fully parallel way. This reduces the cost of purely sequential part of the algorithm (per "chunk surface" gridpoint in computational subdomain) to single multiply/add operation along with the three memory loads and single store and we will later refer to it as LFF (Last From First) operation. For the whole computational domain the parallel tridiagonal algorithm employs two possible forms, on the sequence of $nproc$ computational cores denoted as subscripts in a given direction

$$LFF_1 \rightarrow SEND_{1 \ to \ 2} \rightarrow LFF_2 \rightarrow \dots \rightarrow SEND_{nproc-1 \ to \ nproc} \rightarrow LFF_{nproc}$$

or a variant employing one-dimensional MPI Gather/Scatter operations

$$MPIgather_{input \ for \ LFF} \rightarrow \forall_{1, nproc} LFF \rightarrow MPIscatter_{LFF \ results}.$$

Ultimately, the choice of optimal parallel inversion strategy depends on the problem and the computing architecture at hand.

4   Computational Efficiency and Performance

The suite of the preconditioners has been benchmarked for the baroclinic instability experiment (BAROC) [5], an archetype of the global weather. Details of BAROC implementation in EULAG are presented in [13] together with the grid convergence study and the comparison against the hydrostatic solutions in [5]. Further studies of BAROC included simulations with anelastic, pseudo-incompressible and fully compressible equations [20] using several explicit and implicit formulations of either flux-form Eulerian of semi-Lagrangian EULAG integrators.

The model setup assumes the analytically prescribed ambient state consisting of two mid-latitude zonal jets symmetric about the equator that are in unstable thermal wind equilibrium with the corresponding meridional distribution of the potential temperature. Initial velocity field is locally perturbed at the northern hemisphere, leading to the development of the instability manifested with the fastest growing eastward propagating Rossby mode of wavenumber 6. For illustration, the top panels of Fig. 1 show surface θ' solutions using anelastic and compressible PDEs. Evident frontogenesis is similar to the observed weather systems at the planetary scales.[6]

---

[6] For a discussion of the differences in various soundproof and compressible solutions see [20] and references therein.

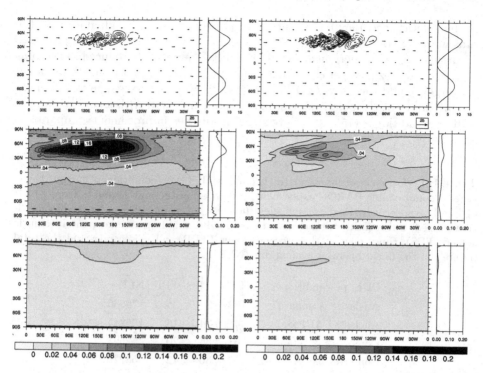

Fig. 1. Anelastic (left) and compressible (right) solutions for the baroclinic instability benchmark at day 8. Top panels display the isolines of surface potential temperature perturbation, overlaid with the flow vectors, on a regular longitude-latitude grid. Contour intervals are 4 K, dashed/solid lines correspond to negative/positive contour values, and zero contour lines are not shown. Zonally averaged profiles to the right of the contour plots depict mean kinetic energy over the full 10-day integration. Middle panels refer to the model run employing standard preconditioner (11) with 3 preconditioner iterations per each GCR iteration. The contour plots show the vertically averaged 10-day-mean residual errors, with their zonally averaged profiles appearing on the right. The bottom panels convey corresponding measures for the runs employing ADI preconditioner (14) with one preconditioner iteration per each GCR iteration. Contour values in the middle and bottom panels refer to the fraction of maximum allowed residual error $r_{max} = 10^{-7}$; cf. [9] and the accompanying discussion.

The reported calculations employ a coarse lon-lat "research" grid 128×64 (corresponding to 2.8^{o} horizontal resolution) and 23 Km deep atmosphere resolved with 48 equally spaced vertical levels. The implicit solver employs physically-based stopping criteria [16] and iterates until the L_∞ norm of the residual error is no larger than 10^{-7}; i.e., orders of magnitude smaller than maximal Courant number $\lesssim 1$. All calculations use 2880 timesteps with $\delta t = 300$ s. The resulting residual errors displayed in Fig. 1, for the runs using three iterations of the standard preconditioner (11) and the runs using one iteration of the two directional ADI preconditioner (14), document substantial improvement in the solution accuracy for the latter. In particular, mean residual error in the runs with the standard preconditioner evinces a signature of nonlinear processes occurring in the northern hemisphere, which is dramatically reduced in the runs with one ADI x-z pass per each GCR iteration. Similarly, effects due to meridian convergence and the refined Δx are practically removed, except in the immediate proximity of the poles. Interestingly, tests conducted with the fully implicit ADI preconditioner (15) essentially show no further improvement compared to ADI x-z [7].

Table 1. Performance of the standard (11) and ADI (14) preconditioners for the anelastic and compressible BAROC experiment, presented in the middle and bottom panels of Fig. 1, respectively. First column identifies the governing PDEs and the preconditioner used. Subsequent columns list: time-to-solution (TTS); time spent in elliptic solver (TES); time spent in the preconditioner (TPR), all in seconds; and the total number of the linear operator evaluations (NLE).

PDE; preconditioner	TTS	TES	TPR	NLE
Anelastic; standard	102	80	35	175386
Anelastic; ADI x-z	48	34	13	105245
Compressible; standard	61	36	20	52593
Compressible; ADI x-z	52	30	12	58225

Table 1 summarizes computational performance of the runs illustrated in Fig. 1. All experiments were performed on the Cray XC30 "Piz Daint" at CSCS, using Cray Fortran compiler with strong optimization options, "-O aggress, cache3, scalar3, fp3, ipa5, vector3, thread0". Standard preconditioner runs were distributed in a $16 \times 8 \times 1$ processor array, whereas ADI x-z runs used a $2 \times 16 \times 4$ array. The latter distribution was motivated by the fact that the tridiagonal solver is more expensive for periodic x direction than in the non-periodic z direction. The total cost of model integration is reflected in the second column, showing that the ADI x-z runs are altogether 50 % cheaper in the anelastic simulations, and 15 % cheaper in the compressible simulations. This discrepancy in the performance improvement results from the much lower overall number of the linear operator evaluations in the compressible case, regardless of the preconditioner used. Although the GCR performs 10 % more iterations in the compressible run with the ADI x-z than with the standard preconditioner, the overall performance is improved due to the lower cost of evaluating the preconditioner itself.

5 Remarks

It is often emphasized that regular lat-lon grids are disadvantageous for global modeling, because the meridians convergence near the poles necessitates small timesteps to assure computational stability and impairs conditioning of the linear operator in semi-implicit flow solvers (thus resulting in a poor performance of iterative elliptic solvers). The first aspect can be circumvented by resorting to semi-implicit semi-Lagrangian integrators. Here we addressed the second aspect and showed that it can be mitigated with ADI preconditioners employing implicit inversions in x and z directions. In particular, for highly anisotropic grids, this permits much larger preconditioner's pseudo-timestep τ (e.g., 40 and 10 larger with ADI x-z in the discussed anelastic and compressible runs, respectively), leading to significantly faster convergence of the elliptic solver and consequently to reduction of the time-to-solution. Benefits of ADI preconditioning are expected to be even more pronounced when MPI parallelization scheme of EULAG will be supplemented by the shared memory decomposition.

Acknowledgements. This work was supported by "Towards peta-scale numerical weather prediction for Europe" project realized within the "HOMING PLUS" programme of Foundation for Polish Science, co-financed from European Union, Regional Development Fund. Selected code optimizations were supported by the Polish National Science Center (NCN) under the Grant no.: 2011/03/B/ST6/03500. Piotr K. Smolarkiewicz is supported by funding received from the European Research Council under the European Union's Seventh Framework Programme (FP7/2012/ERC Grant agreement no. 320375). This work was supported by a grant from the Swiss National Supercomputing Centre (CSCS) under project ID d25 and by the Interdisciplinary Centre for Mathematical and Computational Modelling (ICM) University of Warsaw under grant no. G49-15.

References

1. Chorin, A.J.: Numerical solution of the Navier-Stokes equations. Math. Comp. **22**, 742–762 (1968)
2. Douglas, J.J.: On the numerical integration of $u_t = u_{xx} + u_{yy}$ by implicit methods. SIAM J. **3**(1), 42–65 (1955)
3. Dutton, J.A.: The Ceaseless Wind. McGraw-Hill, New York (1976)
4. Eisenstat, S.C., Elman, H.C., Schultz, M.H.: Variational iterative methods for nonsymmetric systems of linear equations. SIAM J. Numer. Anal. **20**(2), 345–357 (1983)
5. Jablonowski, C., Williamson, D.L.: A baroclinic instability test case for atmospheric model dynamical cores. Q. J. Roy. Meteorol. Soc. **132**(621C), 2943–2975 (2006)
6. Lipps, F.B., Hemler, R.S.: A scale analysis of deep moist convection and some related numerical calculations. J. Atmos. Sci. **39**(10), 2192–2210 (1982)
7. Matejczyk, B.: Preconditioning in mathematical weather forecasts. Master's thesis, University of Warsaw (2014)
8. Peaceman, D.W., Racheford, H.H.: The numerical solution of parabolic and elliptic numerical differential equations. SIAM J. **3**(1), 28–41 (1955)
9. Piotrowski, Z.P., Smolarkiewicz, P.K., Malinowski, S.P., Wyszogrodzki, A.A.: On numerical realizability of thermal convection. J. Comp. Phys. **228**(17), 6268–6290 (2009)

10. Piotrowski, Z.P., Wyszogrodzki, A.A., Smolarkiewicz, P.K.: Towards petascale simulation of atmospheric circulations with soundproof equations. Acta Geophys. **59**(6), 1294–1311 (2011)
11. Prusa, J.M., Smolarkiewicz, P.K.: An all-scale anelastic model for geophysical flows: dynamic grid deformation. J. Comput. Phys. **190**(2), 601–622 (2003)
12. Prusa, J.M., Smolarkiewicz, P.K., Wyszogrodzki, A.A.: EULAG, a computational model for multiscale flows. Comput. Fluids **37**(9), 1193–1207 (2008)
13. Prusa, J.M., Gutowski, W.J.: Multi-scale waves in sound-proof global simulations with EULAG. Acta Geophys. **59**(6), 1135–1157 (2011)
14. Saad, Y., Schultz, M.H.: GMRES: a generalized minimal residual algorithm for solving nonsymmetric linear systems. SIAM J. Sci. Stat. Comput. **7**(3), 856–869 (1986)
15. Skamarock, W.C., Smolarkiewicz, P.K., Klemp, J.B.: Preconditioned conjugate-residual solvers for Helmholtz equations in nonhydrostatic models. Mon. Weather Rev. **125**(4), 587–599 (1997)
16. Smolarkiewicz, P.K., Grubisic, V., Margolin, L.G.: On forward-in-time differencing for fluids: stopping criteria for iterative solutions of anelastic pressure equations. Mon. Weather Rev. **125**(4), 647–654 (1997)
17. Smolarkiewicz, P.K., Margolin, L.G., Wyszogrodzki, A.A.: A class of nonhydrostatic global models. J. Atmos. Sci. **58**(4), 349–364 (2001)
18. Smolarkiewicz, P.K., Temperton, C., Thomas, S.J., Wyszogrodzki, A.A.: Spectral preconditioners for nonhydrostatic atmospheric models: extreme applications. In: Proceedings of the ECMWF Seminar Series on Recent Developments in Numerical Methods for Atmospheric and Ocean Modelling, Reading, UK, pp. 203–220 (2004)
19. Smolarkiewicz, P.K., Charbonneau, P.: EULAG, a computational model for multiscale flows: an MHD extension. J. Comput. Phys. **236**, 608–623 (2013)
20. Smolarkiewicz, P.K., Kühnlein, C., Wedi, N.P.: A consistent framework for discrete integrations of soundproof and compressible PDEs of atmospheric dynamics. J. Comput. Phys. **263**, 185–205 (2014)
21. Smolarkiewicz, P., Margolin, L.: Variational solver for elliptic problems in atmospheric flows. Appl. Math. Comp. Sci **4**(4), 527–551 (1994)
22. Smolarkiewicz, P., Margolin, L.: Variational methods for elliptic problems in fluid models. In: Proceedings of ECMWF Workshop on Developments in Numerical Methods for Very High Resolution Global Models, pp. 137–159 (2000)
23. Thomas, S.J., Hacker, J.P., Smolarkiewicz, P.K., Stull, R.B.: Spectral preconditioners for nonhydrostatic atmospheric models. Mon. Weather Rev. **131**(10), 2464–2478 (2003)
24. Zhang, Y., Cohen, J., Owens, J.D.: Fast tridiagonal solvers on the GPU. ACM Sigplan Not. **45**(5), 127–136 (2010)

Author Index

Printed in the United States
By Bookmasters